FRACTALS IN NATURAL SCIENCES

FRACTALS IN NATURAL SCIENCES

Editors

T. VICSEK
Eötvös University, Budapest, Hungary

M. SHLESINGER
Office of Naval Research, Washington VA, USA

M. MATSUSHITA
Chuo University, Tokyo, Japan

World Scientific
Singapore • New Jersey • London • Hong Kong

Published by

World Scientific Publishing Co. Pte. Ltd.
5 Toh Tuck Link, Singapore 596224
USA office: 27 Warren Street, Suite 401-402, Hackensack, NJ 07601
UK office: 57 Shelton Street, Covent Garden, London WC2H 9HE

British Library Cataloguing-in-Publication Data
A catalogue record for this book is available from the British Library.

FRACTALS IN NATURAL SCIENCES

ISBN-13 978-981-02-1624-5
ISBN-10 981-02-1624-6

PREFACE

This book contains 64 papers selected from the contributions presented at the conference on Fractals in Natural Sciences, held in Budapest from 30 August to 2 September, 1993. This conference in Budapest was the first conference on fractals organized with the specific purpose of enabling simultaneous discussions to be held on results of discoveries in such diverse branches of sciences as astrophysics, biology, chemistry, geophysics and physics. The papers in this book are based on the materials of principal invited talks and some of the most interesting works presented as contributed talks or posters. It is envisaged that the present collection of papers will form a concise source of the "state-of-the-art" of the research on the fractal geometry in nature.

The main objective of the conference was that by bringing together the most recent developments and well known experts in the various fields of sciences, a fruitful interaction of the approaches and scientific views on the complex spatial and temporal behaviors would take place. Indeed, the high quality of the presentations given on topics including growth phenomena, self-organized criticality, turbulence, biological and geophysical pattern formation, iterative functions, colloidal aggregation and inhomogeneous materials resulted in a stimulating interaction among the participants during the meeting and this is expected to be helpful in finding the common principles emerging from the presence of fractals in a wide variety of natural phenomena.

In order to make the material more easily accessible, the papers discussing the above questions have been organized into five main sections: Biology, Chemistry, Earth Sciences, Methods and Physics. Each section begins with an Introduction written by a well known expert in the given field. These Introductions give a short account of the status and perspective on fractal research in the main areas of science and provide a guide for the readers when using this book or studying the specific area in corresponding literature.

The following list contains some of the major topics which are covered by the papers in this volume.

Fractal correlations in DNA sequences and biorythms
Complex patterns in bacterial colony formation and neuronal growth
Correlations of the galaxy distribution
Fractal tectonics, erosion and river networks
Geometry of large diffusion-limited aggregates
Wavelets and other methods of characterizing fractals beyond fractal dimension
Fixed-scale transformation
Anomalous diffusion in dynamical systems

- Transfer across fractal electrodes
- Fracture and statistics of crack branching
- Fractal structure of electrodeposits
- Vibrations of fractals
- Surface roughening in sedimentation experiments, molecular beam epitaxy and in the presence of pinning forces
- Theoretical and simulational results in the study of kinetic roughening
- Self-organized criticality
- Segregation and pattern formation in granular materials
- Nonlinear ocean waves

The conference on Fractals in Natural Sciences was generously sponsored by the following institutions and organizations: Accord-Phare Program; Eötvös University; Hungarian Academy of Sciences; Foundation for Industry; International Science Foundation; Office of Naval Research, USA; Office of Naval Research, Europe; US Geological Survey, and World Scientific Publishing Company.

Since this book is based on the material presented at the conference, we would like here to thank all who have helped to make the conference a success. The following individuals gave of their time and energy to provide the requisite conditions for a good scientific program and smooth running of the conference: members of the international and local organizing committees, the employees of SCOPE Meetings Co., several students from the Eötvös University involved in fractal research, and last, but not least, Dr. Mária Vicsek who was the conference secretary and who provided such great help in collecting the papers for this volume as well.

***by* Tamás Vicsek**
for the Editors of this book

Contents

Chemistry

Earth Sciences

Biology

INTRODUCTION

Any biological object is in general structurally very complex in every scale. In fact, despite the tremendous amount of studies on the structures of biological objects, their quantitative investigation had been very limited and unsuccessful until the introduction of the new concept of "fractal" proposed by B. B. Mandelbrot in the middle of the seventies. It is, therefore, understandable that it was some biologists and geologists who quickly accepted the new concept and analyzed the fractality of various complex objects such as trees, roots, bronchial tree of the mammalian lung, river networks and coastlines. They found fairly nice scaling and determined the fractal dimensions of these objects. This means that such complex objects can now be classified quantitatively in terms of their fractality.

However, many questions arose immediately: So what? What are the basic mechanisms to bring about these objects? What is the biological implication of these fractal dimensions? What is the reason why these biological objects take on fractal forms? Of course, the fractal dimension alone is not the answer to these questions. Its direct application only reveals how complex objects are organized. Initial enthusiasm faded out rather quickly.

Then came the physicists (mainly). Throughout the eighties they were busy proposing various simple models of randomly growing fractal patterns such as diffusion-limited aggregation (DLA) and cluster-cluster aggregation. They were carrying out large-scale computer simulations of these models and related experiments. In this regard, physicists and/or chemists tried to answer the question, "Which fractal is manifested and what is the basic mechanism?" by extracting the most essential process of any physical or chemical phenomenon; by making as simple models as possible to carry out simulations, carrying out related experiments in ideal circumstances; and by comparing numerical with experimental results. It seems that we now have "universality classes" of models that produce basic and widely seen random patterns such as DLA, Eden and dense-branching morphology. In particular, DLA was found to be relevant to many seemingly different phenomena such as electrochemical deposition, dielectric breakdown, dendritic solidification, viscous fingering and chemical dissolution.

Since the end of the eighties even biology-related phenomena have started to be studied and explained in terms of DLA, thanks to the good collaboration among physicists, biologists and medical scientists. These include the growth of bacterial colonies, retinal vasculature, neuronal outgrowth and viscous fingering behavior of gastric acid. Due partly to these successes, various phenomena in biology have once again attracted the attention of many scientists in various fields who are interested in fractals. These include the self-affine interface growth, through bacteria and fungi colonies, the fractal landscape of DNA sequences, the relation of some biorythms to self-organized criticality or chaos, the morphological change in various organisms, to the territorial development in ecology. Certainly we will find other interesting examples in the future. Also, owing to the surprisingly rapid development of computers, various biological organisms can now be finely mimicked through taking rather complex growth processes. Biology seems to be in the renaissance in the fractal world.

It should be noted, however, that an understanding of the complex structures and (possibly simple) growth processes of biological objects is not enough. In many cases we still do not know the biological implication of what are the functional and structural roles in larger-scale organisms or the environment, and why. In order to gain a deeper understanding of the trinity of theories, numerical computations and experiments, better and stronger collaboration is clearly needed among researchers in various fields. Also we should not restrict ourselves to fractals alone but should widen our research viewpoints.

***by* M. Matsushita**

Fractals, Vol. 2, No. 1 (1994) 15–44

COMMUNICATION, REGULATION AND CONTROL DURING COMPLEX PATTERNING OF BACTERIAL COLONIES

ESHEL BEN-JACOB, OFER SHOCHET, ADAM TENENBAUM and INON COHEN
School of Physics and Astronomy,
Raymond & Beverly Sackler Faculty of Exact Sciences,
Tel-Aviv University, Tel-Aviv 69978, Israel

ANDRAS CZIRÓK and TAMÁS VICSEK
Department of Atomic Physics, Eötvös University,
Budapest, Puskin u 5-7, 1088 Hungary

Received December 14, 1993; Accepted December 14, 1993

Abstract

We present a study of interfacial pattern formation during growth of bacterial colonies. Growth of bacterial colony bears similarities to but presents an inherent additional level of complexity compared to non-living systems. In the former case, the building blocks themselves are living systems each with its own autonomous self-interest and internal degrees of freedom. At the same time, efficient adaptation of the colony to adverse growth conditions requires self-organization on all levels — which can be achieved only via cooperative behavior of the bacteria. To do so, the bacteria have developed sophisticated communication channels on all levels. Here we present a non-local communicating walkers model to study the effect of local bacterium-bacterium interaction and communication via chemotaxis signaling. We demonstrate how communication enables the colony to develop complex patterns in response to adverse growth conditions. Efficient response of the colony requires self-organization on all levels, which can be achieved only via cooperative behavior of the bacteria. It can be viewed as the action of an interplay between the micro-level (the individual bacterium) and the macro-level (the colony) in the determination of the emerging pattern. Some qualitative features of the complex morphologies can be accounted for by invoking ideas from pattern formation in non-living systems together with a simplified model of chemotactic "feedback."

1. INTRODUCTION

Discovering how patterns emerge spontaneously from an orderless and homogeneous environment has been a challenge to researchers in the natural sciences throughout the ages.[1,2] Is the diversity of patterns found in nature a result of different causes and effects, or is there a unifying picture in which they all share the same underlying principles? Only during the last decade has a satisfying answer started to evolve.[3–9] These exciting new developments in the understanding of pattern determination in nature offer the promise that a unified theoretical framework is at hand, one which would also include processes in living systems.[10,11]

The most challenging evolving systems a scientist can investigate are those of living organisms. A biological system is in constant flux relative to the environment as it regulates its growth and survival. The energy balances at the cellular level involve a complicated interplay between the microscopic dynamics and the macroscopic environment, through which life at the intermediate mesoscopic scale[12] is maintained. The development of a multicellular structure requires nonequilibrium dynamics, as microscopic imbalances are translated into the macroscopic gradients which control collective action and growth.[13,14] Much effort is devoted to the search for basic principles of growth (communication, regulation and control) at the cellular and multicellular levels.[15–21] As in any new field, the first hurdle is to identify problems which are sufficiently simple, so that research can progress, but also well-motivated by the implications of their solutions.

For reasons to be clarified below, we have chosen to study complex patterning of bacterial colonies during growth under adverse conditions. Typically, bacterial colonies are grown on substrates with a high level of nutrient and intermediate agar concentrations. Under such "friendly" conditions, the colonies develop simple (almost structureless) compact patterns with a smooth envelope. This behavior fits well the contemporary view of the bacterial colonies as a collection of independent unicellular organisms (or non-interacting "particles" using terminology borrowed from physics). However, in nature, bacterial colonies regularly have to cope with hostile environmental conditions.[22] What will happen if we try to simulate such adverse growth conditions in a petri dish, say, by using very low level of nutrients or a hard surface (high concentration of agar), or both? The bacteria reproduction rate, which determines the growth rate of the colony, is limited by the level of nutrients concentration available for the bacteria. The latter is limited by the diffusion of nutrients towards the colony (for the appropriate adverse conditions as is explained below). Hence, the colony growth seems to be similar to diffusion limited growth in azoic systems, such as solidification from a supersaturated solution, growth in Hele-Shaw cell, electrochemical deposition, etc.[5] From the study of diffusive patterning in azoic systems, we understand that the diffusion field drives the system towards decorated (on many length scales) irregular fractal shapes. Moreover, we now understand the competition between the action of the diffusion field and that of the microscopic effects (surface tension and surface kinetics) in the determination of the evolved pattern.[3–9] The microscopic effects, when present, compete with and channel the diffusive instability towards regular and less decorated structures. Generally speaking, the morphology changes from fractal-like[23,24] when the diffusive instability dominates the growth, to compact patterns[25] when the diffusive instability is weaker than the action of the microscopic effects.

In view of the above, one would naively expect that whenever bacterial colonies are grown on very low-nutrient agar (diffusion limited) it will lead to a branching growth (instead of compact), so it will be easier for the colony to reach out for food. Indeed, bacterial colonies can develop patterns reminiscent of those observed during growth in azoic systems,[26–30,10,31,11,32] such as solidification from supersaturated solution, liquid-crystal solidification, electrochemical deposition, etc.[5] One wonders, whether complex patterning of bacterial colonies is simply an additional example (albeit more involved) of spontaneous emergence of patterns that may be explained according to the theory of patterning in azoic systems, or is this theory merely the key to a new understanding required for biological systems.

We will show that comparison of the behavior in the two worlds will enable us to identify both the common principles and the additional features of the bacterial colonies. As we demonstrate, growth of bacterial colony presents an inherent additional level of complexity compared to azoic systems. In the former case, the building blocks themselves are living systems, each having its own autonomous (at times "selfish") self-interest and internal degrees of freedom. At the same time, efficient adaptation of the colony to adverse growth conditions requires self-organization on all levels — which can be achieved only via cooperative behavior of the bacteria. It can be viewed as the action of a singular interplay[10,11,32] (singular perturbation or regulation and singular feedback or control) between the micro-level (the individual bacterium) and the macro-level (the colony) in the determination of the emerging pattern. In general, for more adverse growth conditions we observe a more complex global structure together with a higher micro-level organization.

To do so, the bacteria have developed sophisticated communication channels on all levels,[22,33–40] from direct (by contact) bacterium-bacterium physical interaction and chemical interaction, through indirect physical and chemical interactions via marks left on the agar surface and chemical (chemotactic) signaling, to genetic communication via exchange of genetic materials. The communication enables each bacterium to be both actor and spectator (using Bohr's expressions) during the complex patterning. In the azoic world, we are used to the particles-waves duality. The bacteria developed a particle-field duality: each of the bacterium is a localized (moving) particle which can produce a chemical and physical field around itself. For researchers in the pattern formation field, the above communication, regulation and control mechanism opens a new class of tantalizing complex models which exhibit a much richer spectrum of patterns than the models for azoic systems.

Here we present a communicating walkers model[41] to study the effect of local bacterium-bacterium interaction and communication via chemotaxis signaling. We will demonstrate how communication enables the colony to develop complex patterns in adaptation to adverse growth conditions. This model represents a generic modeling approach which is similar in spirit to the model proposed by Kessler and Levine[18] to describe aggregation of *Dictyostelium* and Kessler's approach[42] in the study of "microbial convection pattern".

An additional inherent level of complexity, in comparison with azoic systems, which we do not discuss here, is the potential of the individual bacterium to perform inheritable genetic metamorphosis in response to the growth conditions.[10,11,32] Indeed, as the bacteria are exposed to more adverse growth conditions, new modes (mutations) with more sophisticated communication strategies which enable more complex behavior were observed. Hence, pattern selection during complex patterning is directly related to the issue of genome cybernetics, as the colony organization (being the environment) can directly affect the genetic metamorphosis of the individual and vice-versa.[10,11,31]

2. COMPACT GROWTH WITH ROUGH INTERFACE

2.1 Experimental Observations

We have chosen to start our experimental endeavor with *Bacillus subtilis* 168 which is not a wild type strain. We have grown bacterial colonies under adverse growth conditions, using standard petri-dishes covered with a thin layer of substrate (12 ml). The growth conditions varied from extremely poor level of nutrient (0.1 gram pepton per liter) to a rich mixture of 10 g/l, and from soft substrate (about 1% agar concentration ≡ 1 gram per 100 ml) to a very hard substrate (4%). When the colonies were grown on rich substrate, simple compact structures evolved as expected. Next, we inoculated the colonies on poor agar, expecting to observe branching growth. Surprisingly, the experiments yielded different results. Compact patterns with rough interface were observed (Fig. 1). More adverse growth conditions led to a stronger roughening and more complex micro-level organization (Fig. 2). But the overall structure remained compact.

Seeking a clue to understand the above, and seeking the required strategy needed for branching behavior, we turn now to a study of simple models of the growth.

2.2 The "Diffusion-Growth" Model

The generic features we would like to test in the following simple model are the effects of diffusion limited growth (diffusion of nutrient) and the effect of the growth of non-communicating bacteria. The growth is restricted to being due to reproduction and no movement of the bacteria is allowed. For this, we define and solve numerically the following model inspired by the diffusion transition scheme[43,44] and our microscopic observations of the growth. The latter reveal that the bacteria are stationary and reproduction simply leads to occupation of a new area.

Let c be the local concentration of nutrients which is defined over a two-dimensional lattice.

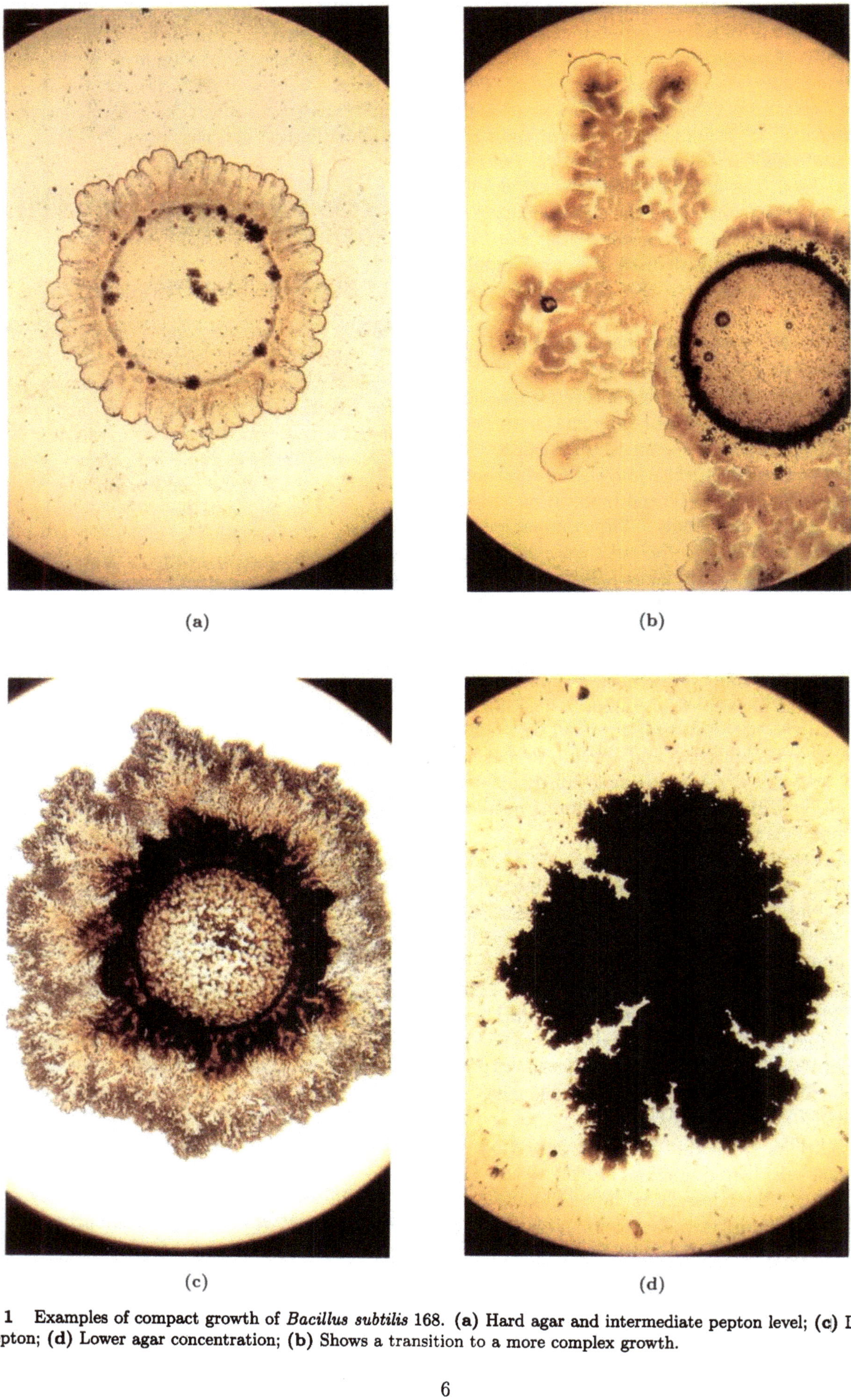

(a) (b) (c) (d)

Fig. 1 Examples of compact growth of *Bacillus subtilis* 168. **(a)** Hard agar and intermediate pepton level; **(c)** Lower level of pepton; **(d)** Lower agar concentration; **(b)** Shows a transition to a more complex growth.

(a)

(b)

Fig. 2 Microscopic observation of the interface of *Bacillus subtilis* 168 demonstrates the complexity of the interface; **(b)** shows larger arms formed at low pepton level.

Its dynamics is separated into two parts. It diffuses (and for numerical convenience we will assume linear diffusion) and it is being consumed by the bacteria in a manner which is described below.

The size of a single bacterium (projected on 2d) is about 1 μm^2, while the colony area may grow up to 10^9 μm^2 and the colony may consist of layers of bacteria (without accurate ordering). Hence, it is impractical to deal with modeling of a single bacterium and instead we will consider mesoscopic units (coarse graining of the colony) which we call "cells". Each of these "cells" consumes food at a certain rate. The food is being deducted from the c field wherever it is consumed. If there is not enough food in the location of a given "cell", it simply does not consume food. After the amount of food consumed by a specific "cell" over time reaches a specific threshold, the "cell" divides. The new "cell" will cover one (randomly chosen) of the available empty nearest neighbors sites of the cell. If all the

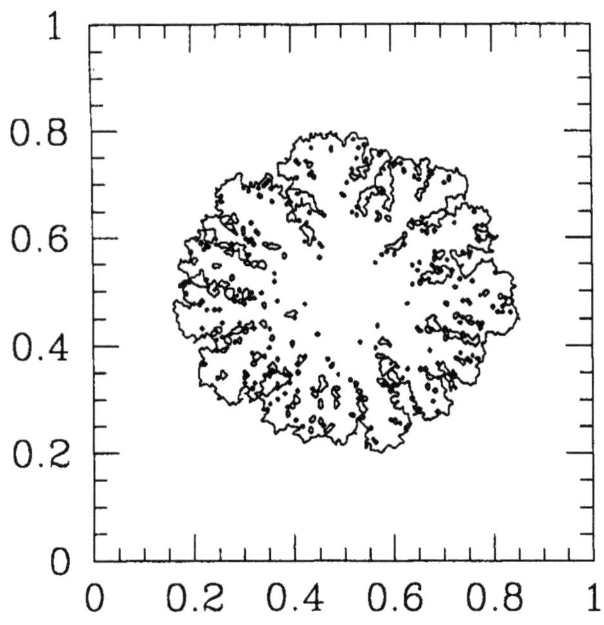

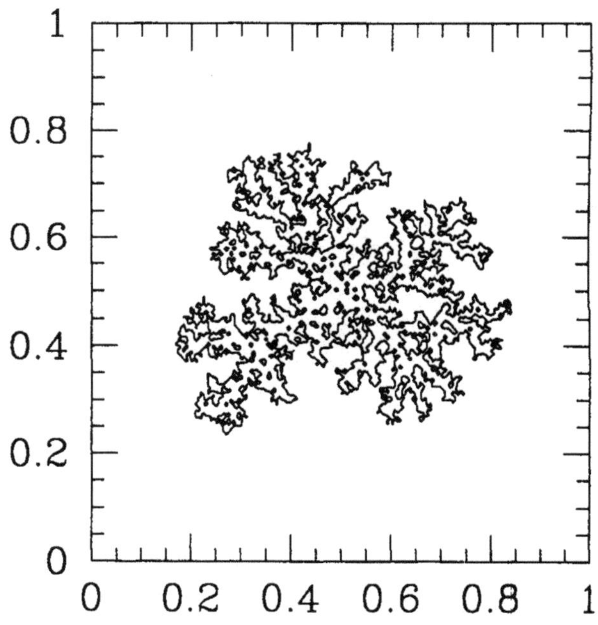

Fig. 3 Colonies grown using the "diffusion-growth" model. For all range of parameters, rough interface are observed as there is no surface tension like interaction between the cells. **(a)** For high food concentration and low rate of food consumption, compact rough pattern is observed ($c_\infty = 0.91$, $r = 10$, the system size is 200×200); **(b)** At low food concentration and high rate of food consumption kinetic tip-splitting with rough interface is observed ($c_\infty = 0.61$, $r = 100$, the system size is 200×200).

neighboring sites are occupied, the cell does not divide. For simplicity the cell sizes are the same as the numerical discretization size of the c field. The model has only two parameters (except for numerical discretization size): c_∞, the concentration of food at infinity, and r, the rate of food consumption, both measured in units of the threshold. This simple model does not require high computational power and can be simulated on a personal computer.

Numerical simulations of the model (Fig. 3) demonstrate that, if the only ingredients are diffusion of food and reproduction, the growth has indeed rough interface. However, compact rough growth is not always observed. For low food concentration and high rate of food consumption, rough branches are observed. Why don't the experimental patterns exhibit branches? The reason probably stems from the time scale of the growth process. The bacterium may divide at most every 20 minutes, leading to the advance of the colony's interface at a maximal velocity of a few μm per hour. Thus the diffusion field can be relaxed during growth (leading to an Eden-like process) and compact rough structure is observed.

2.3 Non-interacting Walkers in a Diffusion Field

In order to obtain additional insight, we now turn to study another model, which includes the additional element of bacteria mobility. It is similar to the previous model, but in addition each cell performs a random walk. The walkers move off lattice with a step size smaller than the 6-fold lattice size used for solution of the diffusion equation. However, movement "costs" energy, so in the absence of sufficient food for a period of time, the walker becomes stationary. Hence, there is indirect weak interaction between the walkers as they compete over the available food. Note that this model is also an extension (inclusion of the diffusion field) of the walkers model introduced and studied by Stanley et al.[45] Results of numerical simulations are shown in Fig. 4.

Clearly such tactics of non-interacting bacteria would impose severe limitation on the ability of the colonies to cope with adverse growth conditions. Can the bacteria learn to do better? Would they develop mobility? Would there be a change in the nature of the reproduction? Or, would communication be established leading to a cooperative behavior?

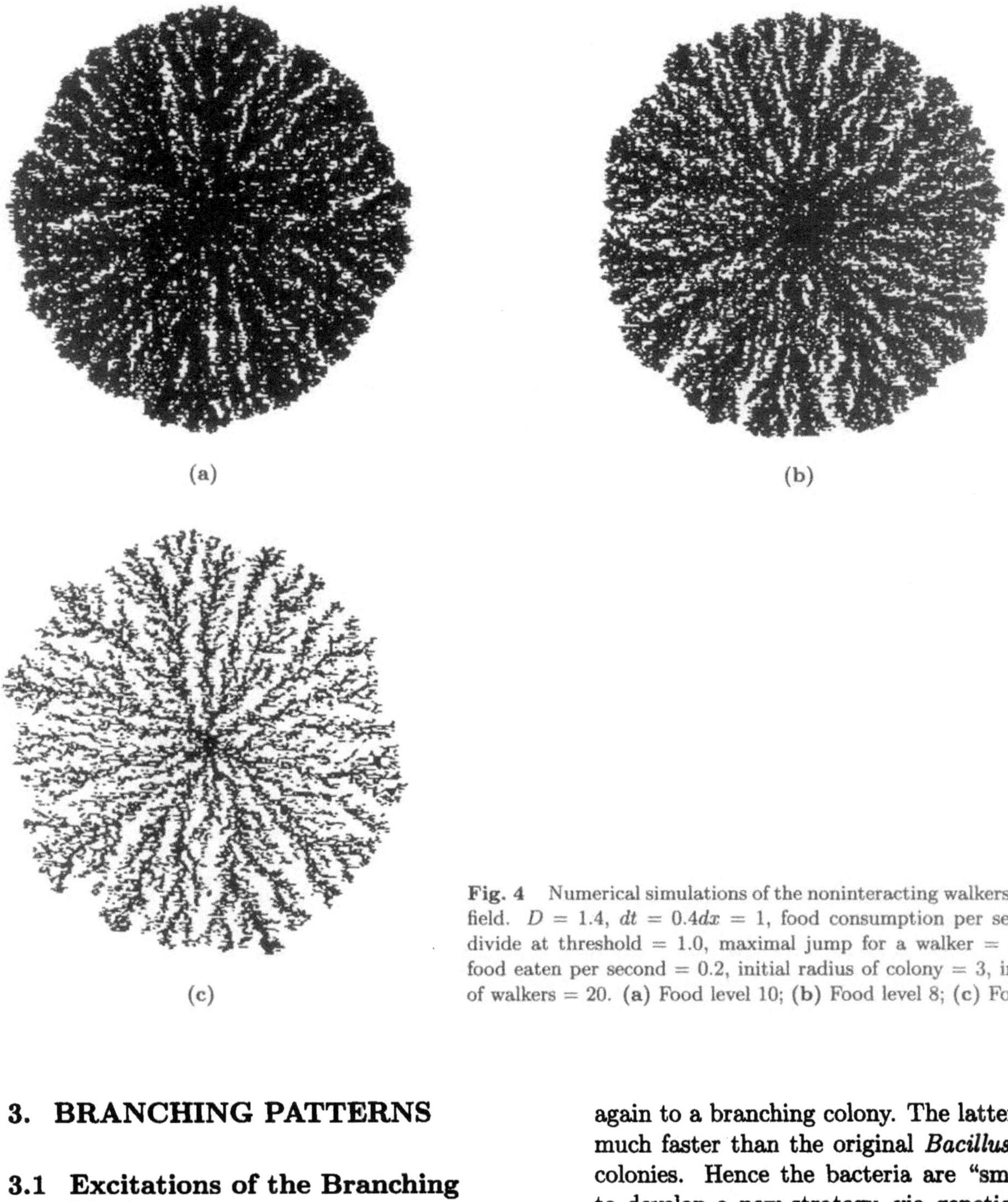

(a) (b) (c)

Fig. 4 Numerical simulations of the noninteracting walkers in a diffusion field. $D = 1.4$, $dt = 0.4dx = 1$, food consumption per second = 0.04, divide at threshold = 1.0, maximal jump for a walker = 0.2, maximal food eaten per second = 0.2, initial radius of colony = 3, initial number of walkers = 20. **(a)** Food level 10; **(b)** Food level 8; **(c)** Food level 4.

3. BRANCHING PATTERNS

3.1 Excitations of the Branching Modes

Seeking experimental answers to the above questions, Ben-Jacob et al.[30,10,31,11,32] have performed experiments in which numerous numbers of colonies of the *Bacillus subtilis* 168 were grown under a wide range of adverse conditions. Occasionally, bursts of spectacular new modes of growth exhibiting branching patterns were observed (Fig. 5). This new branching modes are inheritable, i.e., inoculation of bacteria (and even single bacterium after dilution in liquid) from the new bursting mode leads again to a branching colony. The latter propagates much faster than the original *Bacillus subtilis* 168 colonies. Hence the bacteria are "smart" enough to develop a new strategy, via genetic change, for better adaptation to the environment. Surprisingly, to do so they combine all three tactics (mentioned above): (1) Microscope studies reveal rapid random walk-like movement of the bacteria within a well defined envelope; (2) There is a change in the functional form of the growth rate, which we will not discuss here; (3) The similarity between the tip-splitting growth of the bacteria and that of azoic systems indicates bacteria-bacteria interaction (communication) which acts as an effective surface tension and surface kinetics (Sec. 4).

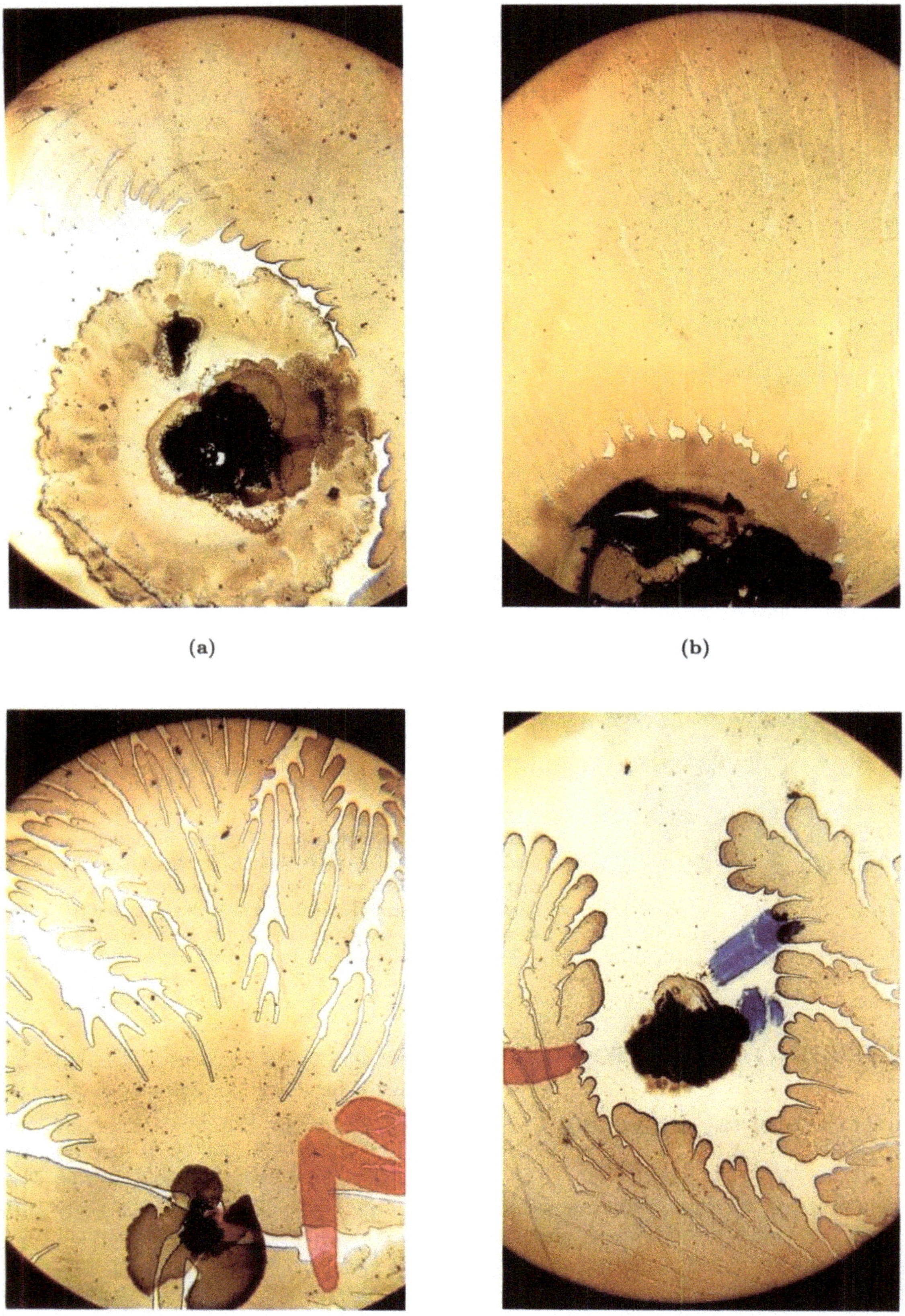

(a) (b) (c) (d)

Fig. 5 Excitation of the branching mode from *Bacillus subtilis* 168. **(a)** Localized transition; **(b)** Extended transition (all over the interface); **(c)** Reinoculation from the new mode leads to tip-splitting growth; **(d)** The growth in the center is inoculation from the *Bacillus subtilis* 168 and the surrounding tip-splitting growth is from inoculation from the new mode. It demonstrates the fact that the properties are inheritable and distinct.

3.2 The Observed Patterns and the Morphology Diagram

The new branching mode can adopt various shapes as the growth conditions are varied. In Figs. 6 and 7, we present some of the observed shapes. In general, the patterns are compact for high pepton levels and become more ramified (fractal like) at low levels. Surprisingly, at even lower pepton levels (below 0.25 g/l), the colonies again adopt a more organized (well defined circular envelope) structures. For the same pepton level, the patterns at higher agar concentration are more branched with more organized branches for higher agar concentration.

As we will discuss below, these phenomena probably indicate that the chemotaxis signaling is more dominant at this range of extreme conditions. The chemotaxis is also amplified in the limit of very hard agar (above about 2.5%). It is manifested in the observed ring structure (Fig. 8) reminiscent of the patterns observed by Shapiro and Trubatch,[15] and an overall twist always with the same handedness (Fig. 8).

Microscope studies reveal that the bacteria perform a random walk-like movement within a well defined envelope. The latter is formed by chemicals which are excreted by the bacteria and/or by fluid drawn by the bacteria from the agar. The envelope slowly propagates as the bacteria collide with it. Many collisions are required for the envelope to move. At very low pepton levels, the bacteria density is very low — the distance between the bacteria being up to several times the size of the individual bacterium. In this range, the bacteria are longer (about 5 μm in length) and the movement seems to be more organized. At high agar concentration, there is a boundary layer of high bacteria density at the leading tips of the growing branches. In this range the colonies also have a pronounced structure in the perpendicular direction.

The patterns observed in azoic systems can be organized using the concept of a morphology diagram, indicating the existence of a pattern selection principle.[5,46,47,8] Is this plausible for patterning of the bacterial colonies?[10,11] Does a morphology

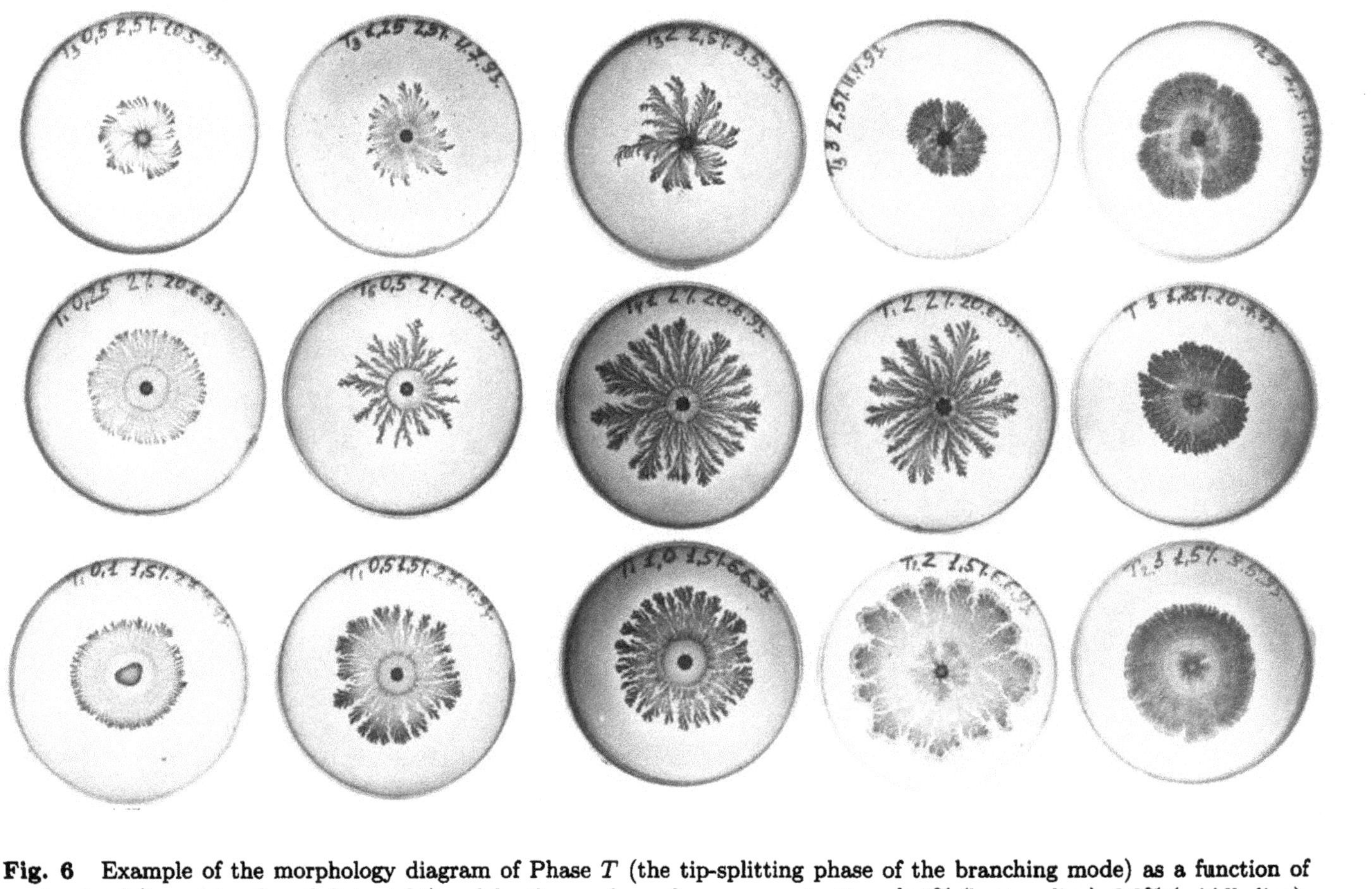

Fig. 6 Example of the morphology diagram of Phase T (the tip-splitting phase of the branching mode) as a function of pepton level (increasing from left to right) and for three values of agar concentration. [1.5% (bottom line); 2.0% (middle line); 2.5% (upper line)].

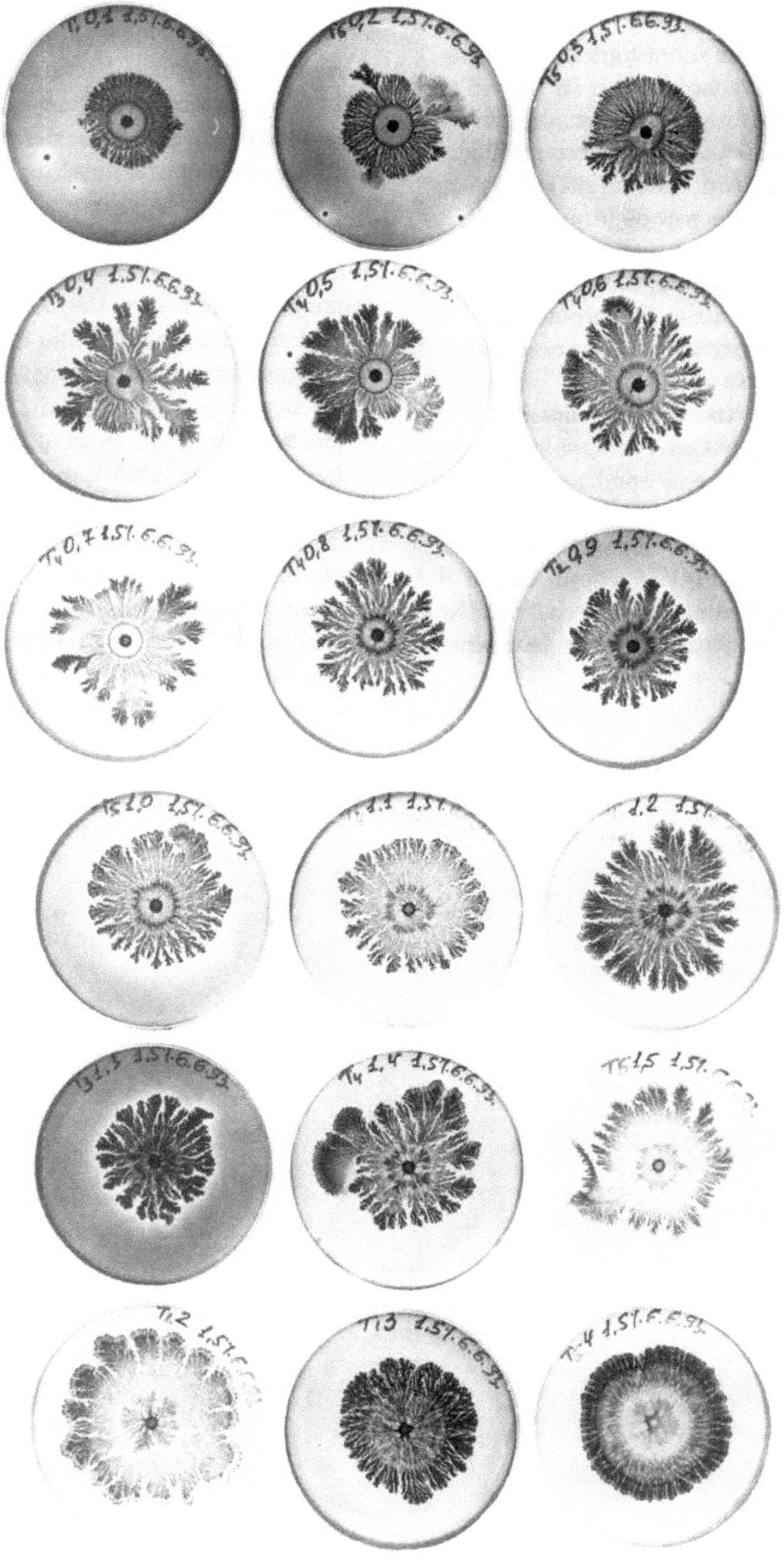

Fig. 7 Detailed sequence as a function of pepton level (steps of 0.1 g/l) for 1.5% agar concentration. The sequence changes from compact at 4 g/l to ramified as the pepton level is decreased. At very low levels (0.3 g/l, 0.2 g/l and 0.1 g/l) the growth is denser with thin radial branches and well defined circular envelope.

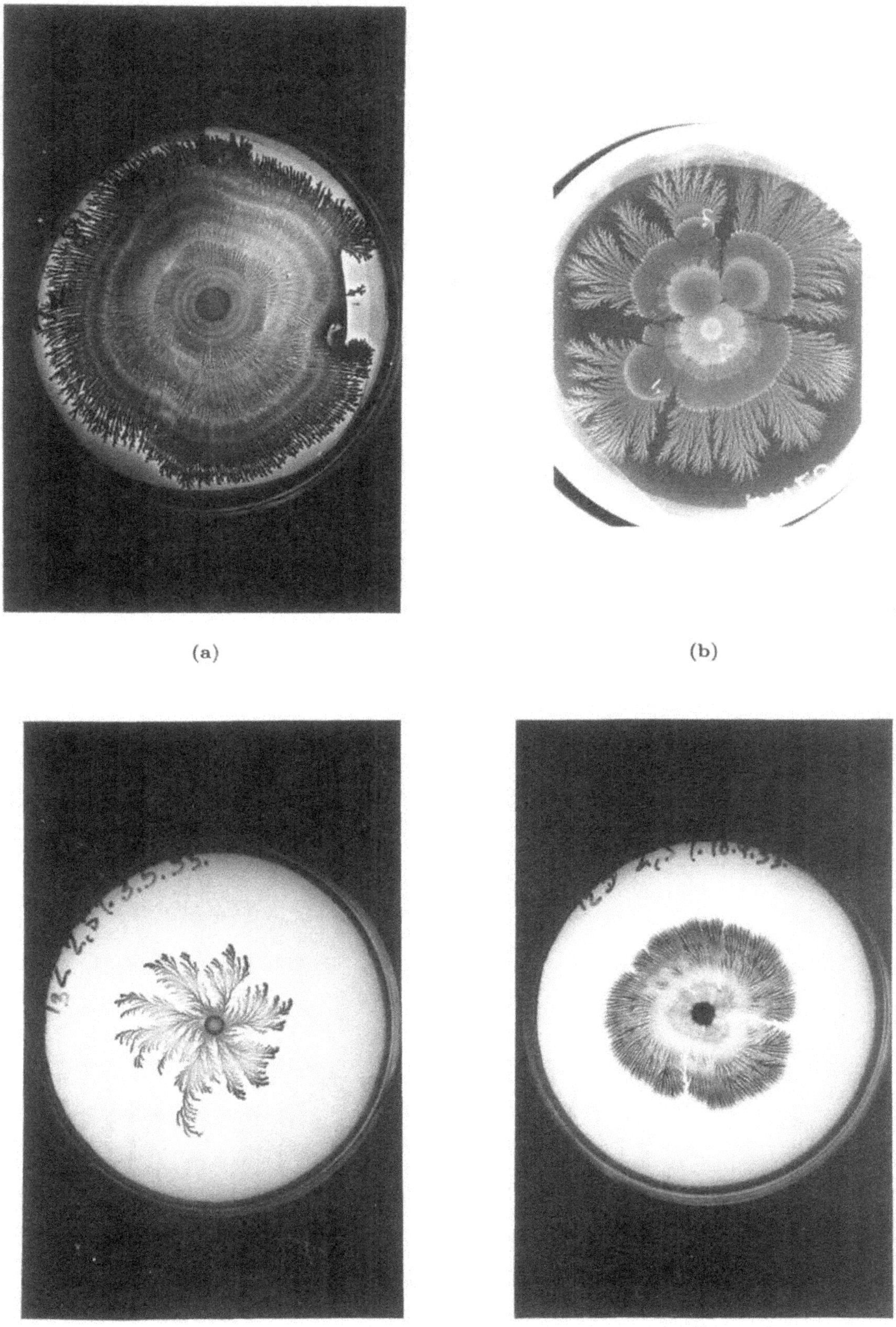

(a) (b) (c) (d)

Fig. 8 Examples of patterns observed at high agar concentration. There is a pronounced ring structure and overall twist of the colony (weak chirality).

selection principle exist in this case as well? Clearly, different shapes are observed for the branching mode as we vary the growth conditions. However, as is explained in Refs. 8, 11 and 48, observations of different shapes is a necessary but by no means a sufficient condition to imply the existence of a morphology diagram. For additional support, the growth velocities were studied in detail. Correlations between the patterns and the velocity, and sharp changes in the velocity with crossing of the boundaries between different regimes of different patterns were found[11] (Fig. 9). These results are promising but we have only just started with the more convincing observation of morphology coexistence.[48]

3.3 Bacterial "Snowflakes"

An essential part of the new understanding of interfacial pattern formation in azoic systems has to do with the role of anisotropy. Anisotropy is needed in order to have dendritic growth rather than a cascade of tip splitting. For example, in the case of solidification, the anisotropy is provided by the symmetry of the crystalline structure of the growing solid phase. In a Hele-Shaw experiment, anisotropy was imposed via grooves on the plexi-glass,[49] insertion of a string in the fluid,[50] addition of small bubbles,[51] etc. These experiments demonstrated that, irrespective of the form of the imposed anisotropy, it can lead to a similar dendritic growth.

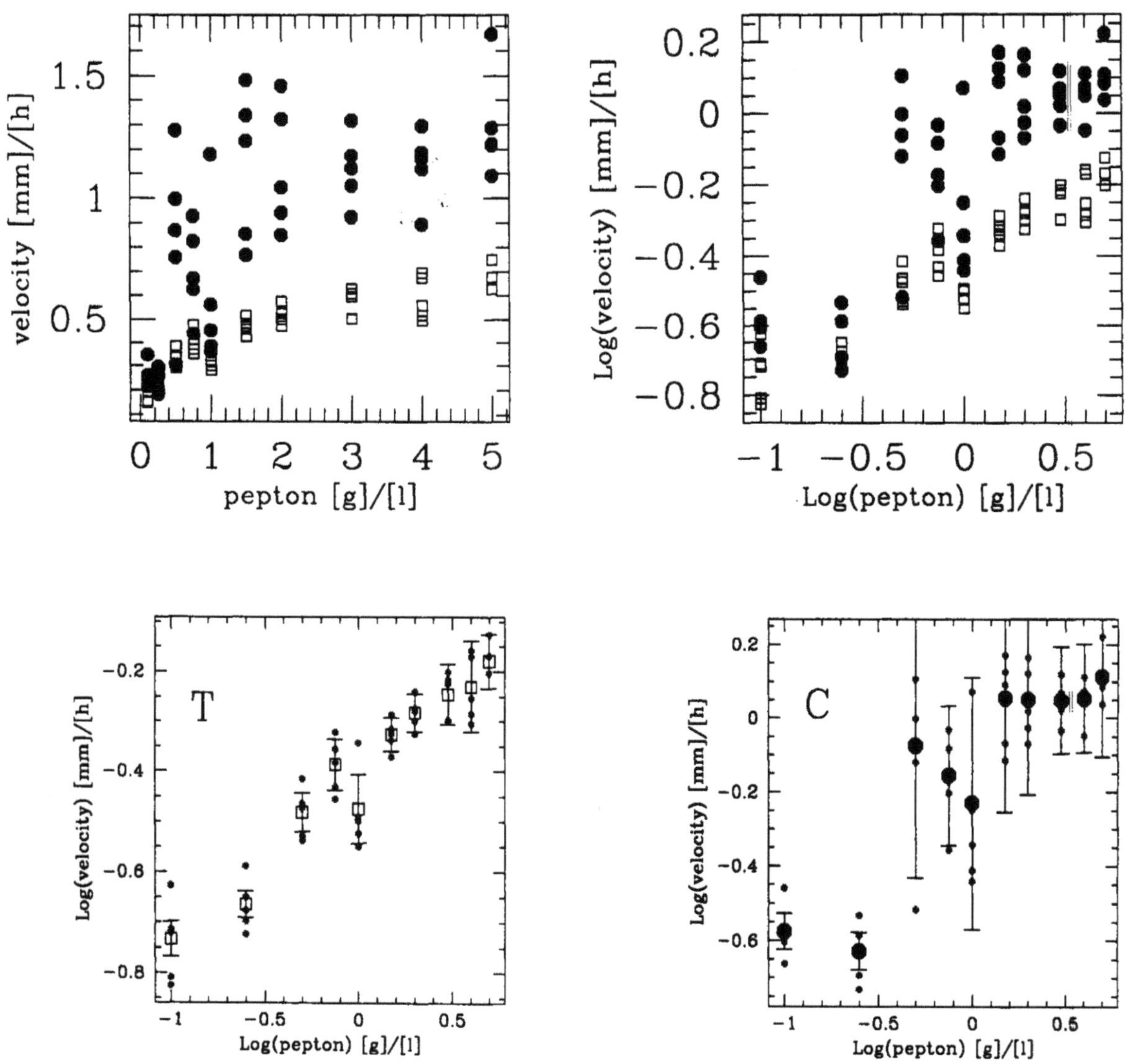

Fig. 9 Example of the velocity measurements of phase T (open squares) and of phase C (filled circles) for 1.5% agar. The bottom figures show also the averaged values of the velocities. For both phases, there are three distinct regimes of growth and sharp transitions between the regimes (between 0.25 g/l and 0.5 g/l pepton level and between 1 g/l and 1.5 g/l pepton level).

Hence, an additional important test of the similarity of bacterial growth to interfacial pattern formation would be to impose anisotropy on the growth of the bacteria, and to look for dendritic growth.

In Figs. 10 and 11, we show results of imposition of anisotropy via ripples on top of the agar. We have stamped the agar with grooved plexiglass to introduce the anisotropy. Figure 10 shows "bacterial snowflakes" patterns. Here a plexiglass with six grooved lines was used to stamp the agar. We used grooves of different widths. According to the grooves width and the substrate parameters, the leading tips are either dendrites or parity breaking dendrites[52,53] (see Ref. 54 for more details).

Fig. 10 Patterns observed when the agar was stamped with Plexiglass on which six lines were grooved.

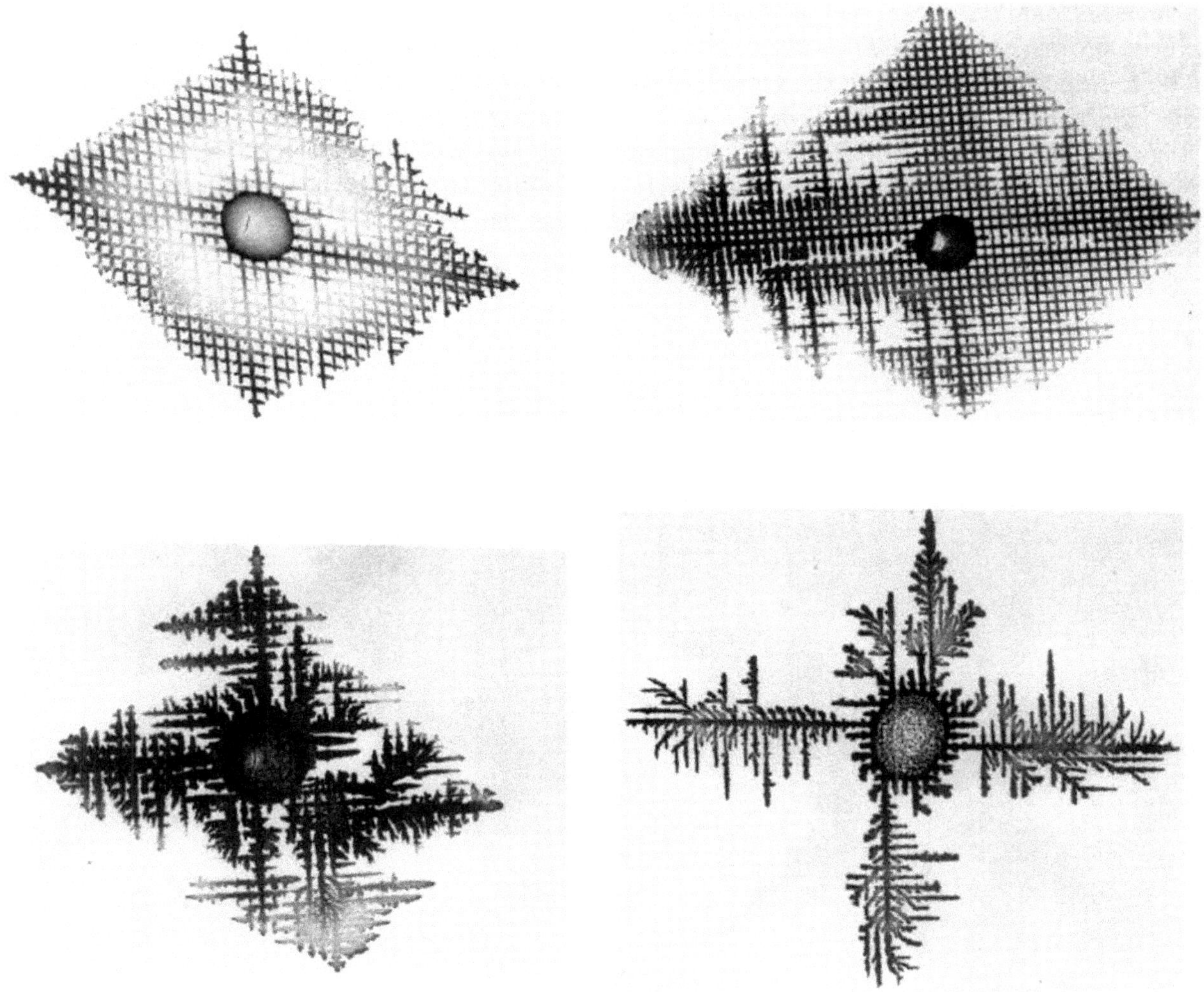

Fig. 11 Patterns observed when the agar was stamped with Plexiglass on which 4-fold lattice is grooved.

Figure 11 shows the results of stamping the agar with a 4-fold lattice. Here we show "crystal" growth of the colonies. At higher pepton levels the effect becomes weaker.[54]

These results provide additional support for the similarity between bacterial growth and interfacial pattern formation in non-living systems. They demonstrate that singular perturbation and singular feedback do play a crucial role during the growth of bacterial colonies. The ripples on the agar break the symmetry in the motion of only a few bacteria, in the immediate vicinity of the ripples. Yet, this minute asymmetry (or local anisotropy), being a singular perturbation, is sufficient for dendritic growth. At the same time, via the singular feedback, the macroscopic dendritic growth leads to a correlated motion of the bacteria on the micro-level in other parts of the colony.

4. THE "COMMUNICATING WALKERS" MODEL

4.1 Motivation

The models described in Sec. 2.2 are an oversimplification of the real bacterial colonies. Growth of the colony is initially exponential with time. After a few divisions, it becomes very complex, as a result of close proximity of the accumulated cells.[22] This proximity creates a situation that may be called "physiological crowding", in which the individual cells compete with each other for available nutrients and have indirect interaction via the accumulation of waste products and active release of chemotaxis materials. Since nutrients must diffuse to the cells from the surrounding medium, while waste products must diffuse away from the cells, the individual members of the colony are subjected to a variety of environmental conditions, depending on their location within the colony. Through communication, the bacteria may dynamically choose to grow in a strategy which is favorable to the colony but not necessarily the best tactics for each of the single bacterium. Hence, the pattern finally determined will result from singular interplay between the individual bacterium and the colony as a whole.

4.2 Description of the Model

In order to demonstrate the dramatic effect of the inclusion of local interaction (or communication), we have modified[41] the "walkers" model as follows: The walkers move within a well defined envelope (Fig. 12). Each segment of the envelope moves after it has been hit N_c times by the walkers. The above was motivated by our microscope observations which show random-like movement within a well defined envelope. The latter is probably formed by chemicals which are extracted by the bacteria and or by fluid drawn by the bacteria from the agar. Under the microscope it seems that the bacteria move in a thin layer of fluid which is pushed forward by their movement. Note that, to a first approximation, the level of N_c represents the agar concentration. More "collisions" are needed to move the envelope when the agar is harder, due to the roughness of the top layer of the gel. Here the presence of the joint envelope introduces local communication and effective surface tension and surface kinetics.

4.3 Results of Numerical Simulations

A typical morphology diagram produced by the communicating walkers model is shown in Fig. 13. As in the growth of bacterial colonies, the patterns are compact at high pepton levels, becoming more ramified with decreasing food level. For a given pepton level, the patterns are more ramified as the agar concentration increases. The distribution of the walker is shown in Fig. 14. In Fig. 15, we show the growth velocities, and the fractal dimensions as a function of pepton level, and in Fig. 16 — the branch width. Although the growth is limited by the diffusion of nutrients, the fractal dimension becomes lower than the DLA fractal dimension[23,55–57] at low nutrients. It results from the fact that the

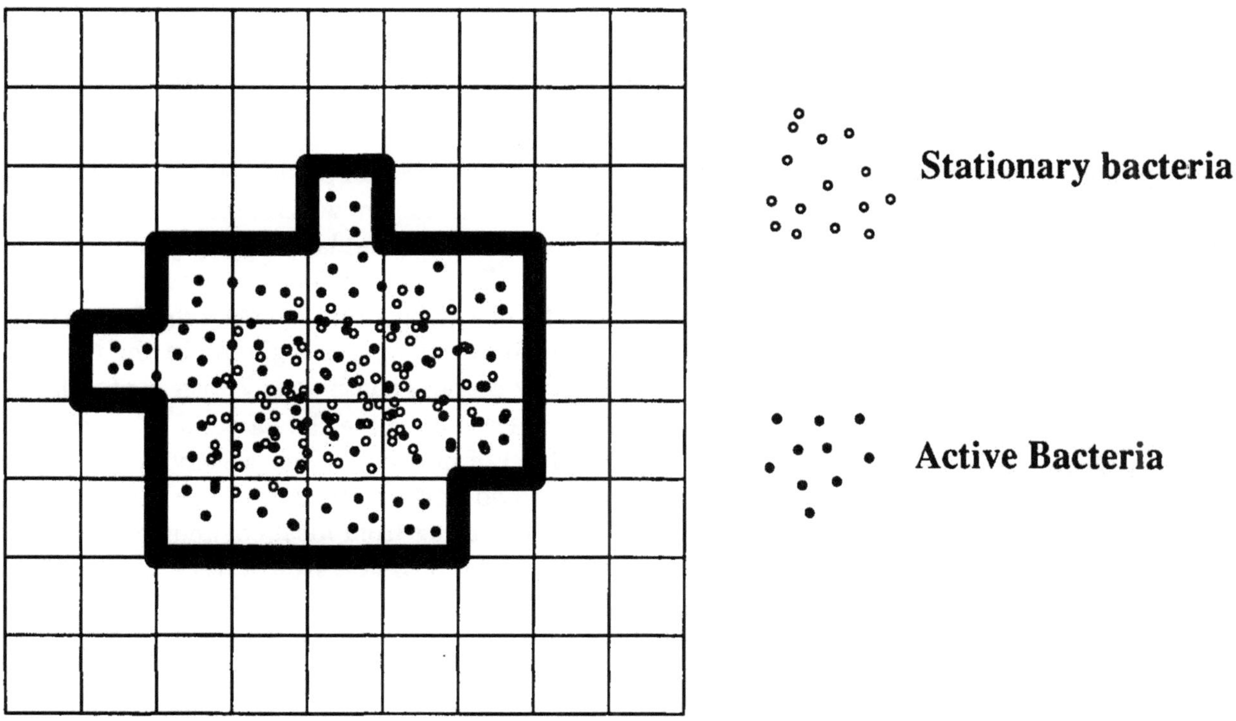

Fig. 12 Schematic illustration of the "communicating walkers" model. The active walkers are moving within a well defined envelope. The envelope moves after a walker try to cross into a nearest cell N_c times. The food concentration obeys a diffusion equation and is solved numerically on the same lattice.

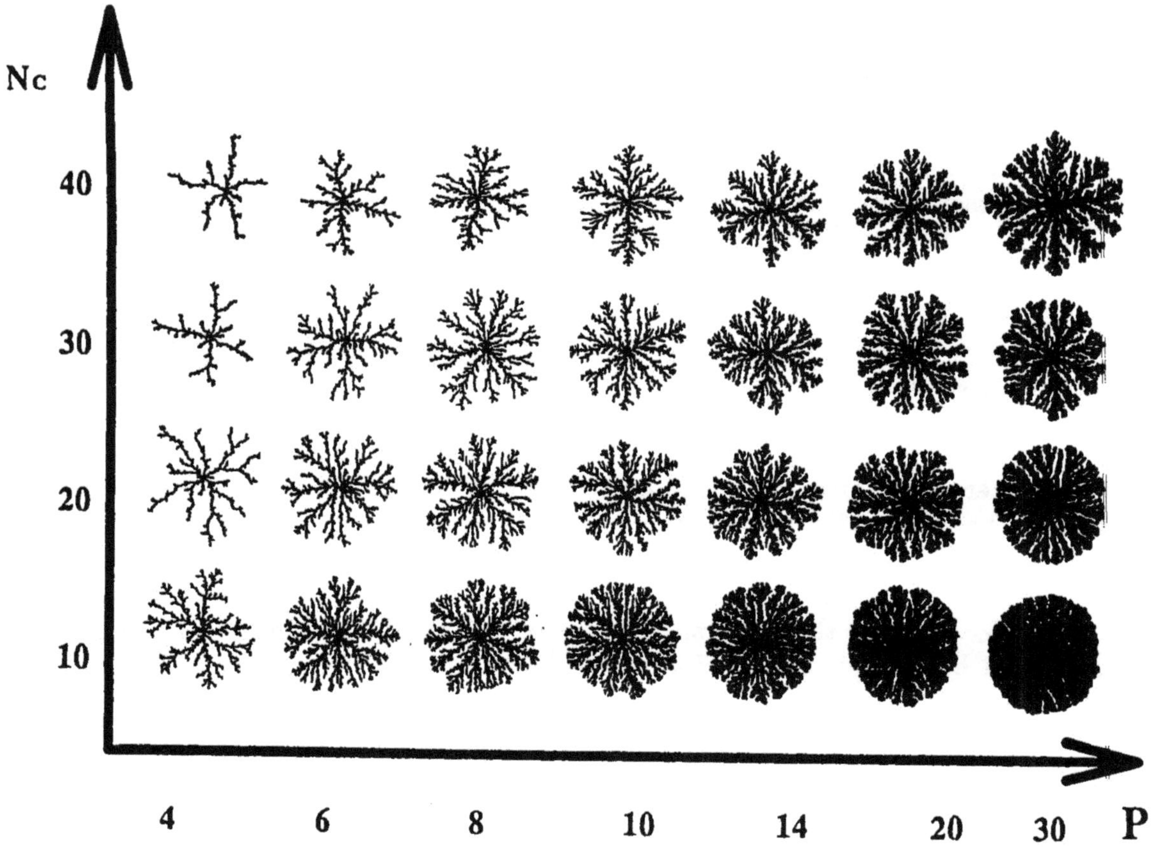

Fig. 13 Morphology diagram of the "communication walkers" model (without chemotaxis). $D = 1.4$, $dt = 0.4dx = 1$, food consumption per second = 0.04, divide at threshold = 1.0, maximal jump for a walker = 0.2, maximal food eaten per second = 0.2, initial radius of colony = 3, initial number of walkers = 20. In the late stage growth the number of walkers is about 10^5. P (represents the pepton level) is the initial concentration measured in units of the threshold.

local communication we have introduced acts both as surface tension and surface kinetics. It also introduces a nontrivial dependence of the interface motion on the gradient of the diffusion field. The latter is similar to the case of dielectric breakdown[58] and the Hele-Shaw cell with non-Newtonian fluids.[59]

5. INTRODUCTION OF CHEMOTAXIS SIGNALING

5.1 Motivation

Did we reach a statisfactory description of branching patterning during growth of bacterial colonies? Clearly the results shown in Fig. 13 are very encouraging and capture many features of the observed patterns. However, there are some crucial qualitative differences. First, at very low pepton (below about 0.5 g/l) the bacteria do adopt (again) a more organized structure — well defined circular envelope and a dense structure of radial thin branches, and not a fractal-like structure produced by the model. Second, at high agar concentration and high pepton levels the colonies adopt Hele-Shaw-like patterns of dense smooth fingers which differ from the Electrochemical deposition-like patterns produced by the model. Third, the velocity measurements during growth of the bacterial colony exhibit sharp transitions between various regimes. In the model, the velocity changes continuously with pepton level.

Suppose we expose the bacteria to an extremely low level of nutrients. Can we envision a more sophisticated tactic (strategy) than simple branching — resulting from local communication? Imagine a group of settlers on a very hostile terrain. First, it would be efficient to move forward in small groups without spending much time on exploring the sides in an attempt to find better conditions ahead. Every so often it will need to slow down, restore forces and then again move forward faster. In addition, the group has to develop a high level

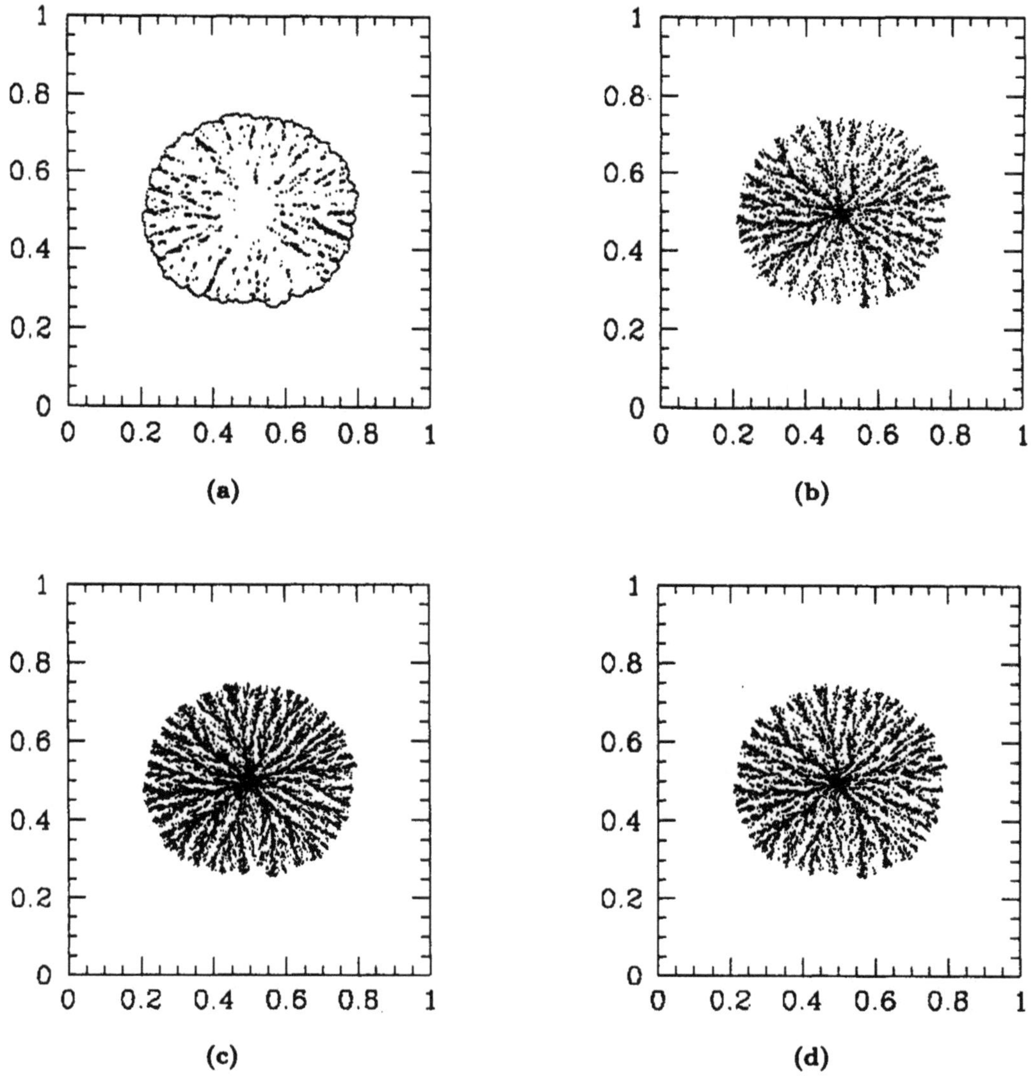

Fig. 14 Branching pattern is observed in the third dimension of the colony structure. **(a)** The colony at $N_c = 10$ and food level 30; **(b)** All cells in which there are more than 15 walkers; **(c)** All cells in which there are more than 10 walkers; **(d)** All cells in which there are more than 12 walkers.

of cooperative behavior including transfer of information (communication) between each of the individuals and the group as a whole (regulation). And the group has to develop means to bias the activity of the individuals (control) towards the group interests.

5.2 Chemotaxis Signaling

Bacterial colonies have developed the ability to behave exactly in such a manner, using chemotaxis signaling. Generally, chemotaxis means movement of the bacteria in response to a gradient of some chemicals field.[39] Ordinarily, the movement is along the gradient either in the opposite direction (surfing, or movement towards the regimes of low level of the chemical) or in the gradient direction (anti-surfing). The chemotaxis response can be to an external field or to a field produced by the bacteria. The latter case is chemotaxis signaling or chemotaxis communication.[39,18] Such non-local communication enables each bacteria to obtain information and response to the state of the colony as a whole.

Here we first include a simple version of chemotaxis communication to reveal the generic features. Each of the stationary bacteria produces a communication chemical at a fixed rate. (In reality, the rate should be a function of nutrients and of the concentration of the communication chemical.) The movement of the active bacteria changes from

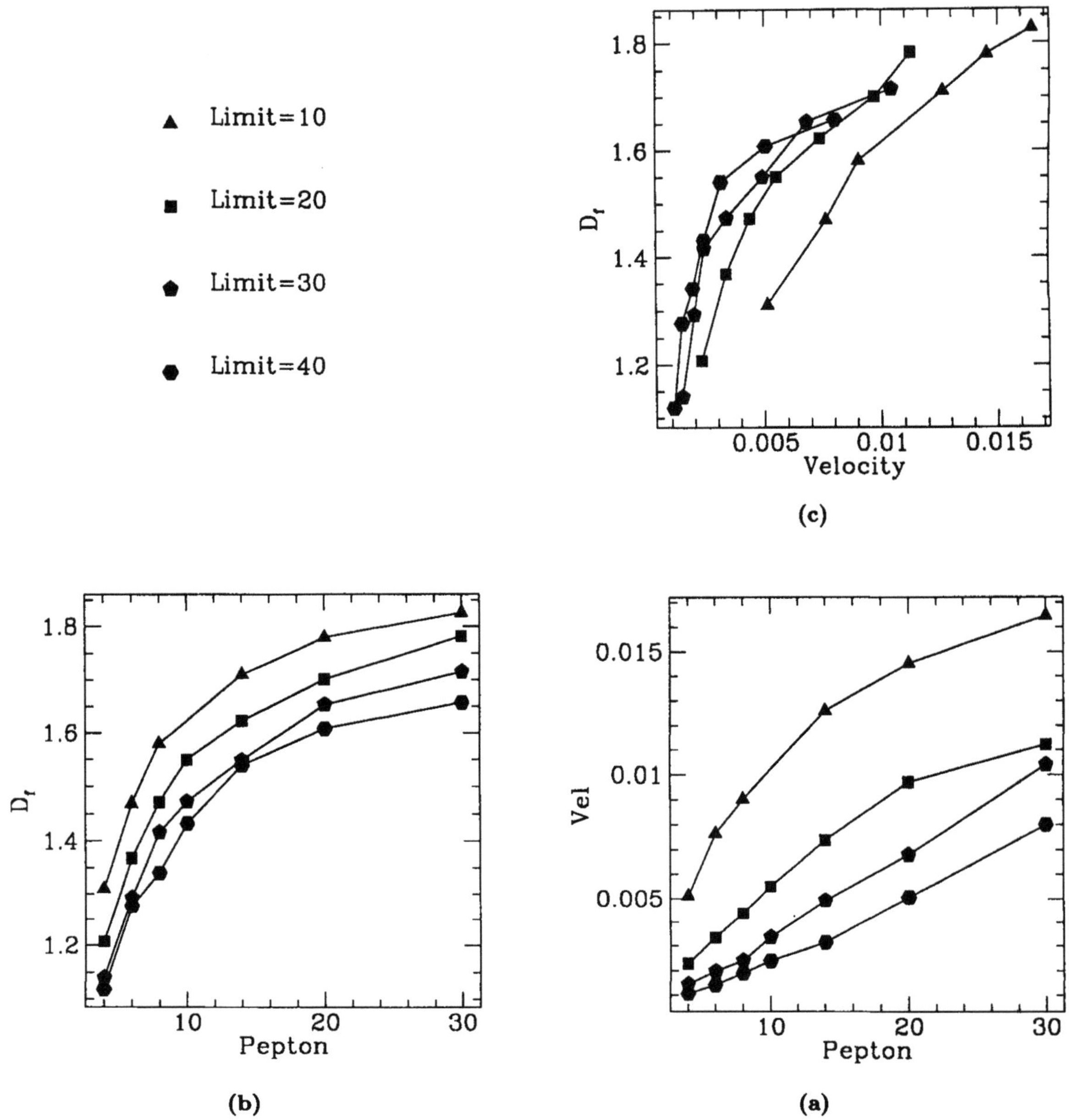

Fig. 15 **(a)** Growth velocity vs. food concentration for different N_c. The velocity increases as more food is available for the walkers. As N_c increases, it is harder for the colony to advance and the growth velocity decreases. In all cases the velocity appears to be a smooth function of the growth parameters. **(b)** The fractal dimension (of mass distribution) vs. food concentration for different N_c. The patterns are more compact for high pepton level. **(c)** The fractal dimension vs. growth velocity. A high correlation is observed between these two observables.

pure random walk to a random walk with a bias along the gradient. The probability that the walker will move in a direction θ is proportional to $\nabla_\theta c$, where ∇_θ is the derivative in the θ direction and c is the chemotaxis concentration. This is the simplest strategy, as the only element in the communication is sending out (broadcasting) a stress signal. This is the way in which the stationary bacteria "protect themselves". They send a chemical signal to affect the mobile bacteria to move away. Other bacteria receive the signal and respond accordingly. In the next stage, we add an element expressing that the active bacteria consumes the communication chemical. Again they do so from a "defensive" interest of eliminating a chemical that affects their free movement. As it turns out, these selfish signals serve the interest of the colony as a whole under conditions of a low level of nutrients. In another

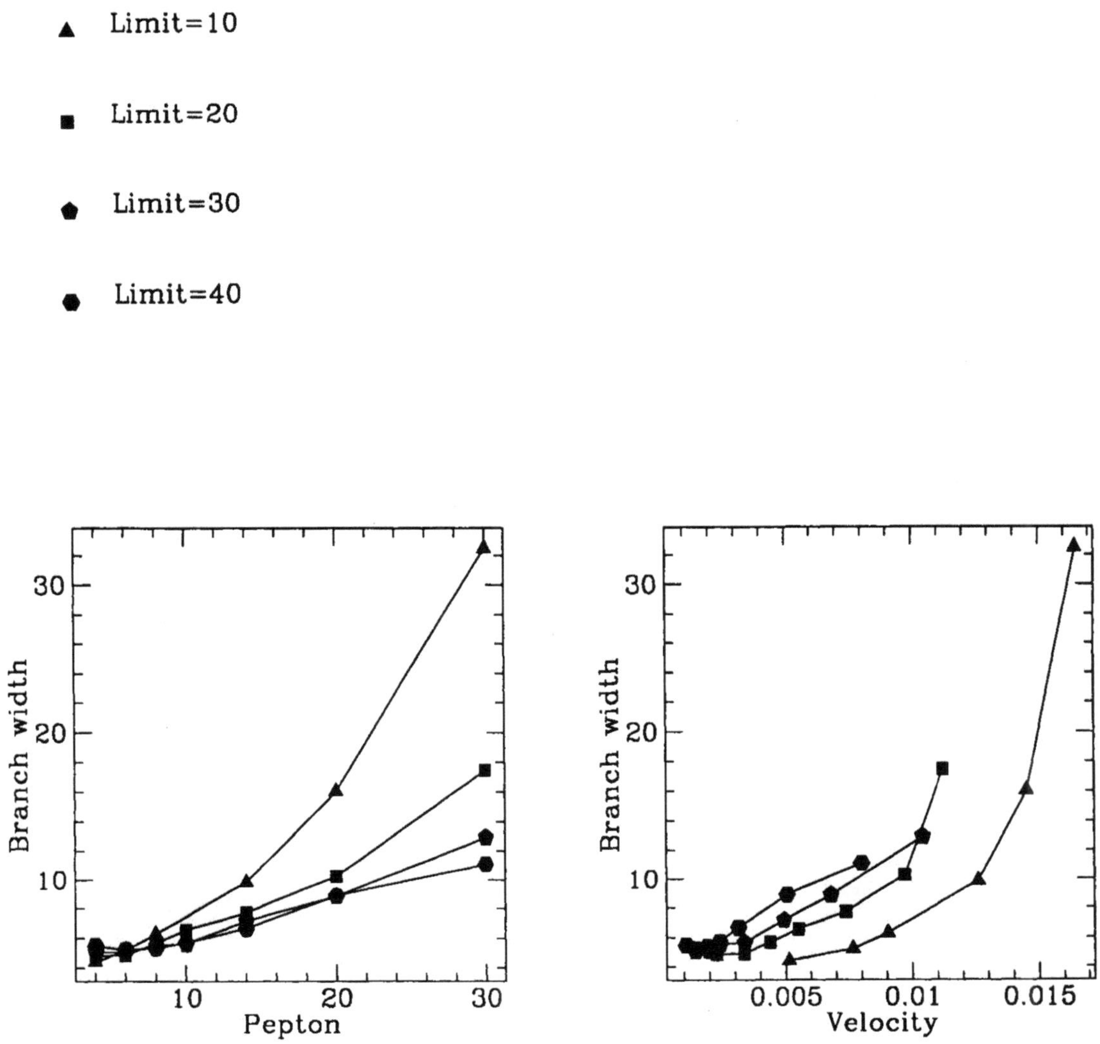

Fig. 16 Branch width as a function of pepton level and velocity.

publication we will show the effect of more complex tactics, in which each bacterium can both absorb and produce the signaling chemicals and does so according to the state condition (pepton level, chemotaxis materials and bacterial density).

5.3 Results of Numerical Simulations

In Fig. 17, we demonstrate the dramatic effect of the chemotactic signaling. The pattern becomes much denser, with more radially oriented thinner branches. Yet, comparison with the experimental observations reveals a qualitative difference. The bacterial colonies have a circular smooth envelope. The results shown in Fig. 17 are for the case in which the stationary bacteria emit the chemotaxis materials, and the active bacteria respond only via a change of the random walk. It is reasonable to assume that the active bacteria also respond by consuming the chemotaxis materials. The result of this tactics is shown in Fig. 18. Now the branches are even more oriented in the radial direction, and the pattern has a smooth circular envelope. Both are in excellent agreement with the experimental observations. In the presence of chemotaxis signaling, the colony propagates much faster with about the same number of walkers. Comparison of Figs. 17 and 18 with the experimental observations demonstrates the usefulness of the modeling. It enables us to determine the appropriate chemotactic strategy adopted by the bacteria. In Fig. 19, we show the morphology diagram as a function of pepton level and N_c. For all patterns, the same intensity of chemotaxis signaling was used. Again, comparison

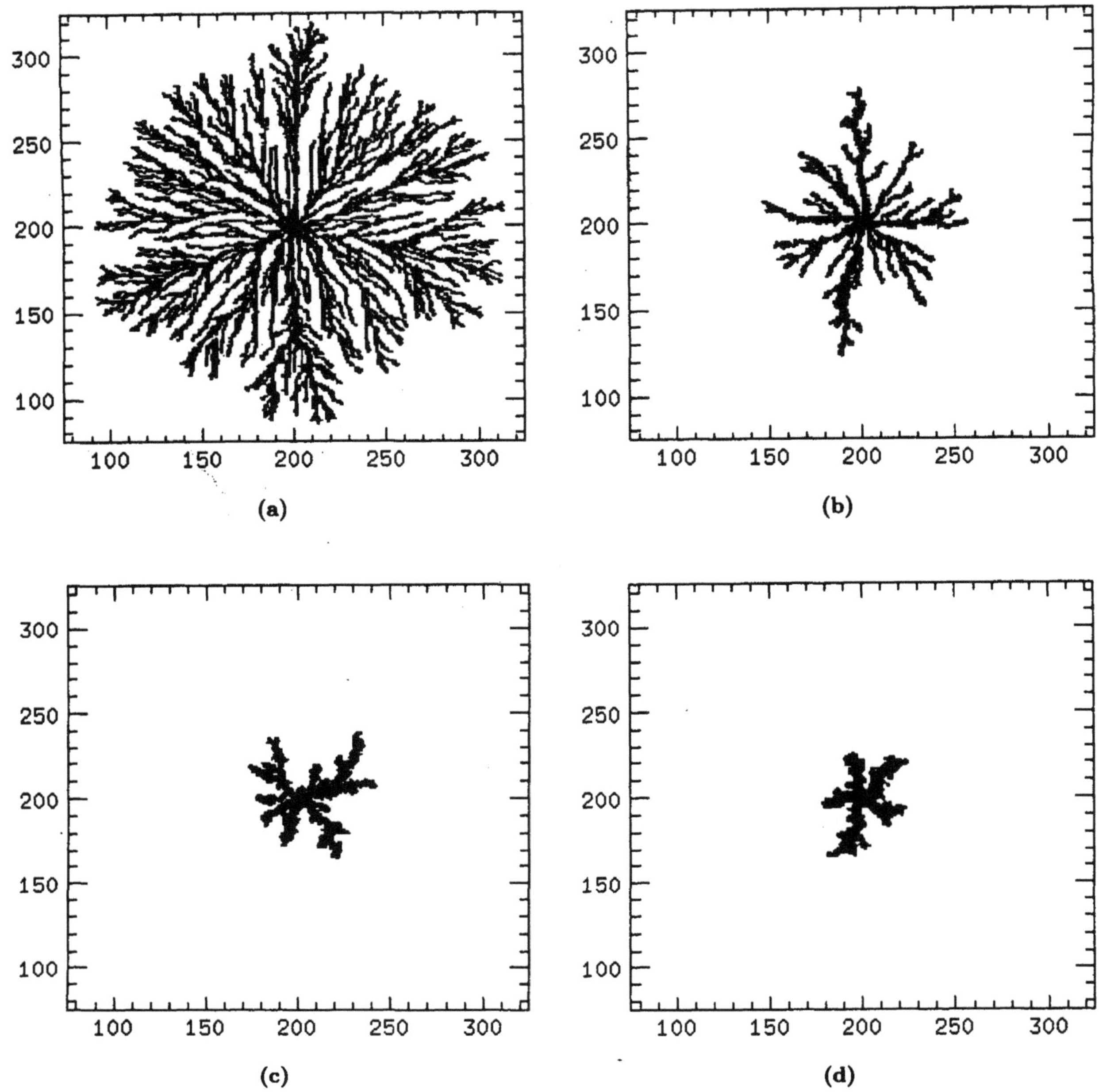

Fig. 17 Effect of chemotaxis signaling for $P = 2$ and $N_C = 10$ using the tactic in which only the stationary bacteria sends signals. The chemotaxis intensity is the highest for (a) and lowest for (d). Note that the envelope is not smooth.

with the experimental observations demonstrates that this is not the case. In reality, the intensity varies according to the growth conditions. Results of more realistic chemotactic signaling, including also the effect on the reproduction rates, will be presented elsewhere.

5.4 Comment about Modeling

How should we approach the modeling of the bacterial colonies patterning? One extreme approach would be to devise a set of computer rules which will mimic the observed patterns. Doing so does not reveal (directly) the biological functions and behavior. It does reflect though understanding of the geometrical nature of the patterns, which indirectly might help in revealing the biological functions. The results we show for chiral growth (next section) were obtained using such a model. Another approach would be to simply construct an algorithm which includes the known biological facts about the system. Such an approach sets a trajectory of ever

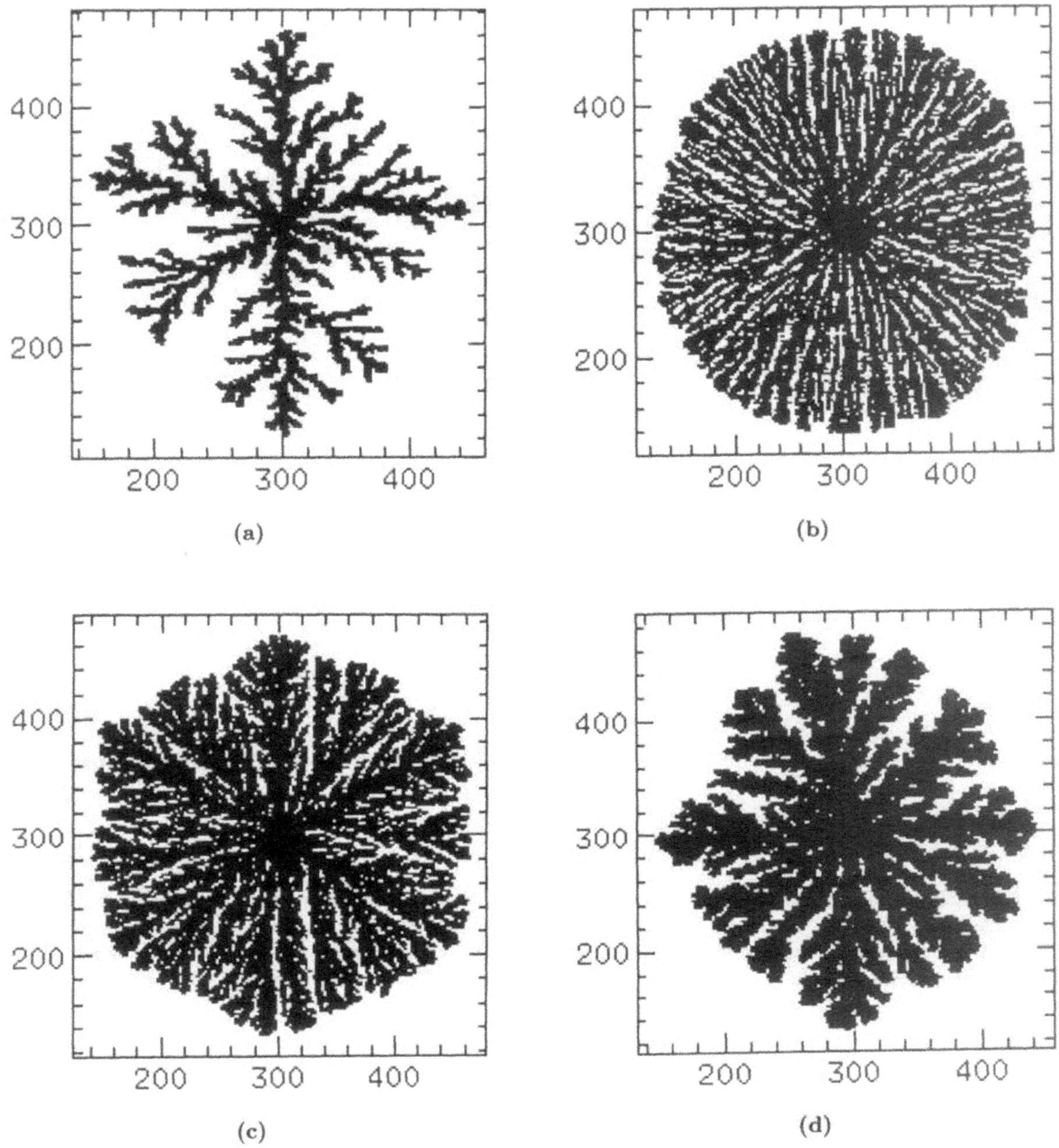

(c) **(d)**

Fig. 18 Effect of chemotaxis signaling for $P = 12$ and $N_C = 40$. **(a)** In the absence of signaling; **(b)** Using the tactic in which the active bacteria consumes the chemotaxis materials; **(c)** Using the tactic of Fig. 17(d) for $P = 30$, $N_C = 40$ and no chemotaxis.

including more and more details. The model keeps evolving to include so many parameters that it loses any predictability power.

Here we try to promote a different approach — one which is based on generic modeling.[18,60–62] In this approach we seek to reveal the generic features and the basic principles in order to explain the biological behavior. For example, here we studied the effect of local communication vs. the effect of nonlocal communication, in order to find out whether the nonlocal communication (such as chemotaxis) is essential for explaining the observed patterns irrespective of the exact details. We believe that generic modeling along with a close comparison of experimental observations will promote the development of a new understanding of biological systems. In other words, generic modeling is not using sophisticated (as it may) mathematical description

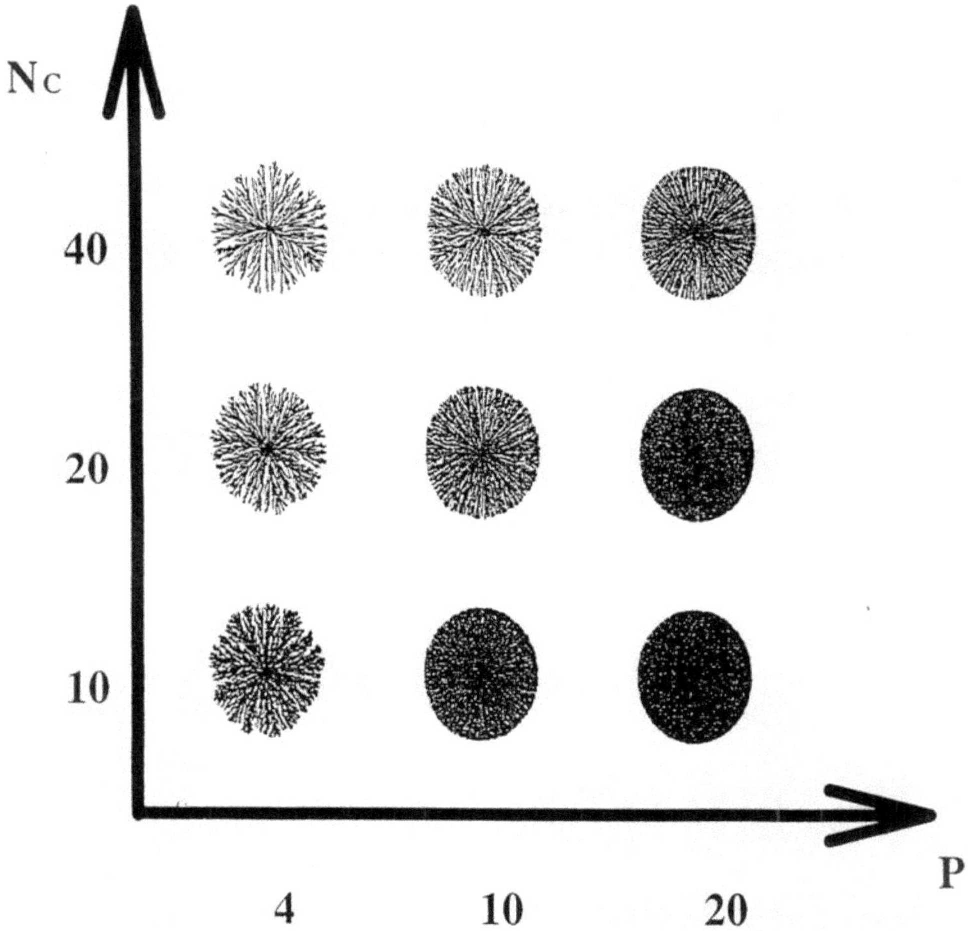

Fig. 19 Morphology diagram of the "communication walkers" model with chemotaxis. Chemotaxis at high pepton level leads to circular pattern which is almost compact unlike what is observed in experiments. $D = 1.4$, $dt = 0.4dx = 1$, food consumption per second = 0.04, divide at threshold = 1.0, maximal jump for a walker = 0.2, maximal food eaten per second = 0.2, initial radius of colony = 3, initial number of walkers = 20. 10^5. P (represents the pepton level) is the initial concentration measured in units of the threshold.

to dress pre-existing understanding of the biological behavior. Rather, it means a cooperative approach, using existing biological knowledge and understanding together with mathematical tools and synergetic point of view for evolving complex systems to reach a new understanding (which will be reflected in the model). Naturally, such an approach, in order to be efficient, calls for interdisciplinary collaboration of biologists, chemists, physicists and mathematicians.

6. EXOTIC PATTERNS

6.1 The Need for More Complex Tactic

Assume we expose the bacteria to a low level of nutrients on soft agar (easier mobility). Can we envision a more sophisticated tactics that will benefit from the easier mobility? Using the settlers metaphor, they are now facing soil that is poor but easy to move on. Hence, it is now possible to send forward small groups for fast exploration of further territories, with the hope of finding better conditions ahead. But, at the same time, the rest of the group left behind should use the resources available in the empty spaces not covered by the advancing groups. It turned out that the bacteria learned to do exactly so via the chiral growth described in the next section.

6.2 Chiral Patterns

During growth of the branching mode on soft agar (about 1%), bursts (inheritable) of new patterns which overgrow the branching ones are observed.[10,11] The new patterns consist of thinner branches all having the same handedness of twisting (Figs. 20 and 21). We name this type of growth chiral growth, and refer to the new inheritable phase as phase C. Chiral asymmetry exists in the whole range from subatomic particles to human beings, and seems to have played an important role in the evolution of living systems.[63,64] Hence, it is not

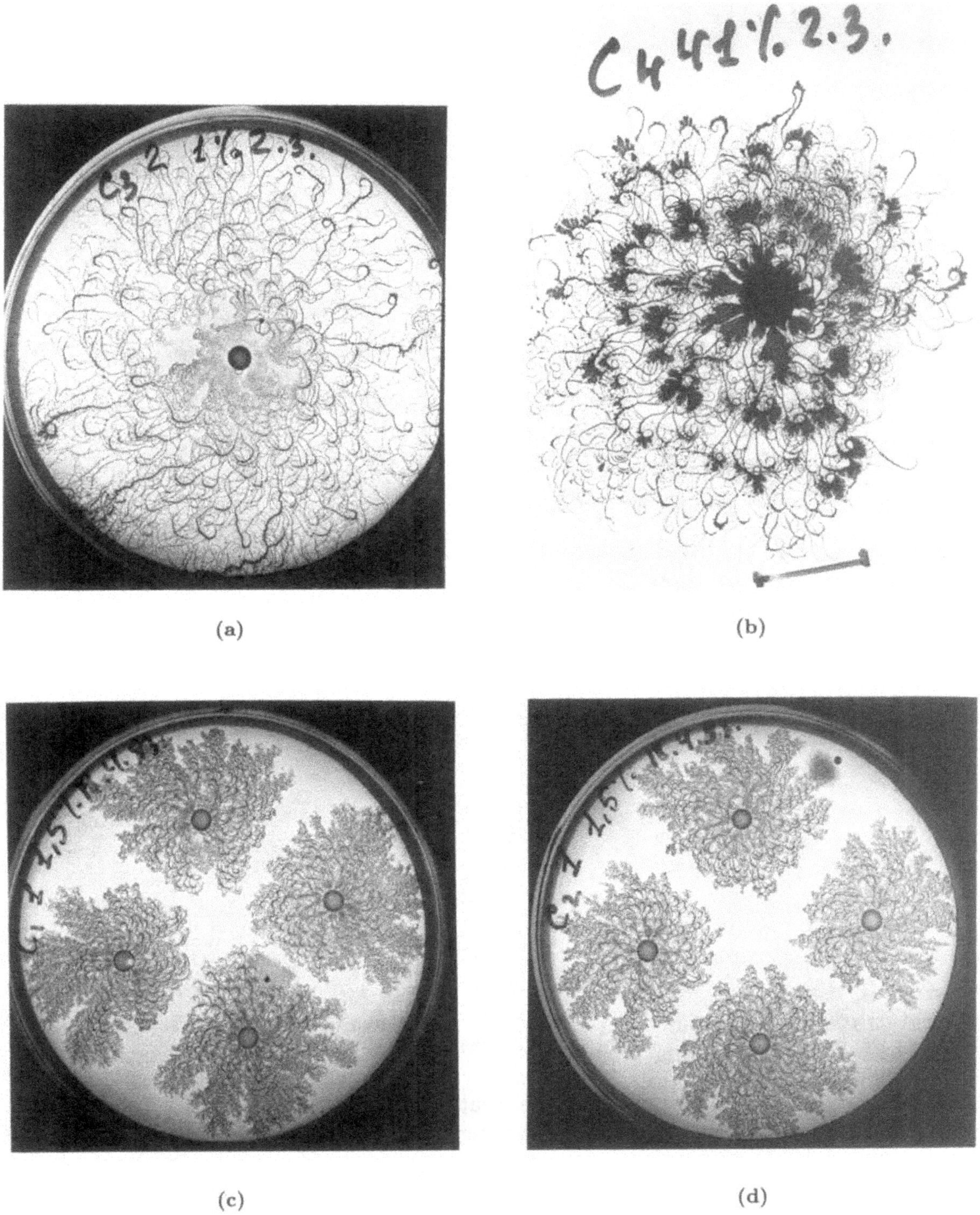

Fig. 20 Examples of patterns observed during growth of phase C (the chiral phase). (a), (b) and (c) are for different growth conditions. (c) and (d) are for the same growth conditions (different plates) to demonstrate the level of reproducibility of the observed patterns. As we described in Ref. 11, an extremely strict protocol has to be followed in order to reach such results.

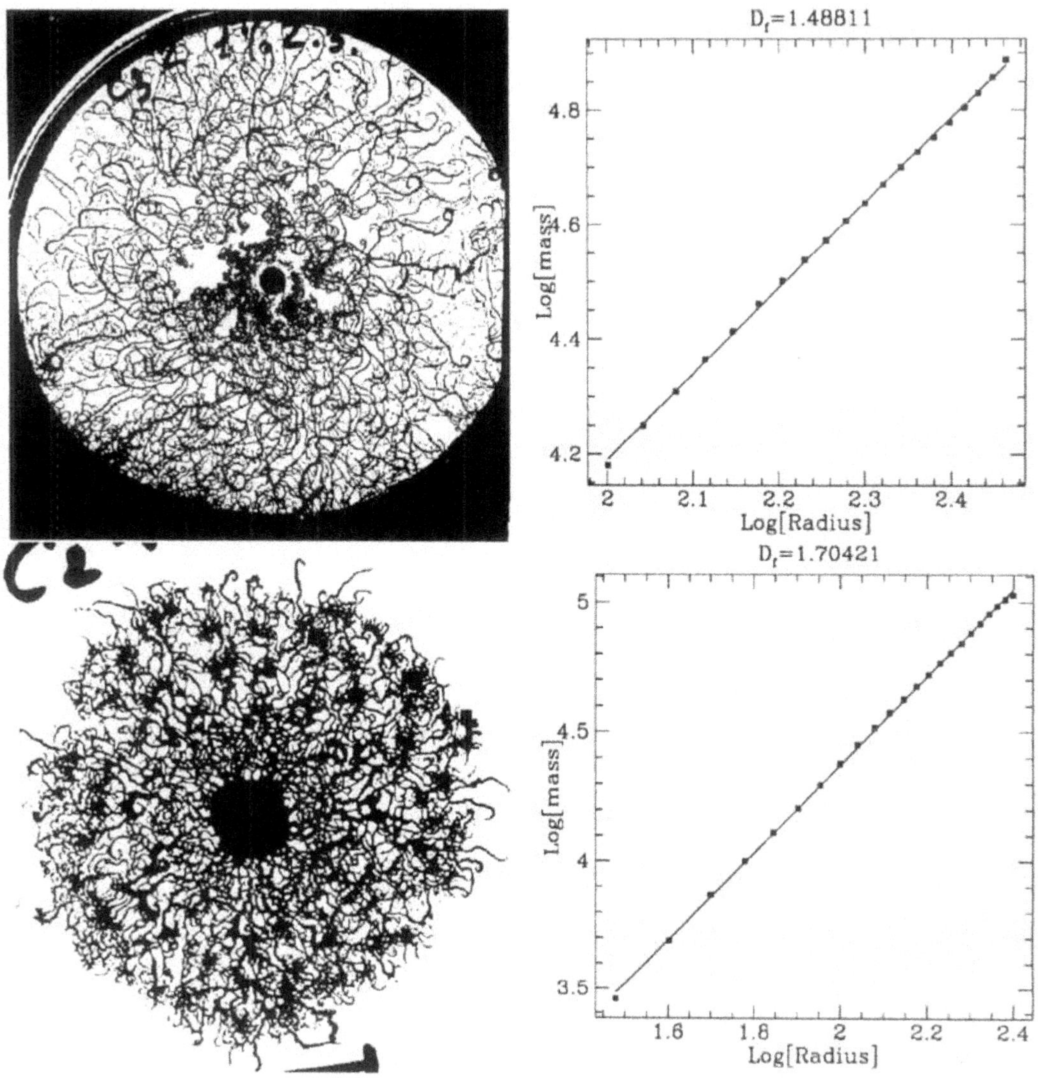

Fig. 21 Measurements of the fractal dimension of the chiral growth.

surprising that bacteria display chiral properties. Back in the seventies, Mendelson et al. showed[65–68] that long cells of ***Bacillus subtilis*** can grow in helices, in which the bacteria form long strings that are twisted around each other. These observations show that the bacteria possess chiral properties. In other observations, they have shown that the chiral characteristics affect the structure of the colony. Recently, we have found that chiral growth was actually reported long ago.[69,70,22]

Optical microscope observations indicate that during chiral growth the colony consists of moving bacteria. The individual bacteria are longer than those in tip-splitting growth (Fig. 22) and have a string-like shape, but the individual ones do not show (on this level) chiral structure (Fig. 23). The micro-level dynamics show a slow swarming of the long bacteria. Unlike in tip-splitting growth, the motion is coordinated. The bacteria seem to move along parallel trails following each other. For a defined handedness, the bacteria have to be able to distinguish between up and down. The growth in an upside-down petri dish shows the same chirality. Therefore, we think that the determination of up vs. down is done via the vertical gradient of the nutrient concentration or chemotaxis materials.

The chiral growth appears as a fairly sophisticated tactics. Do we need new features in order to explain this behavior? As we will show elsewhere, to develop the weak chirality of phase $\mathcal{T}$ on hard agar, we simply need to add to the random walk a component in a direction perpendicular to the direction of the gradient (Fig. 24).

In phase C the bacteria are much longer. Hence

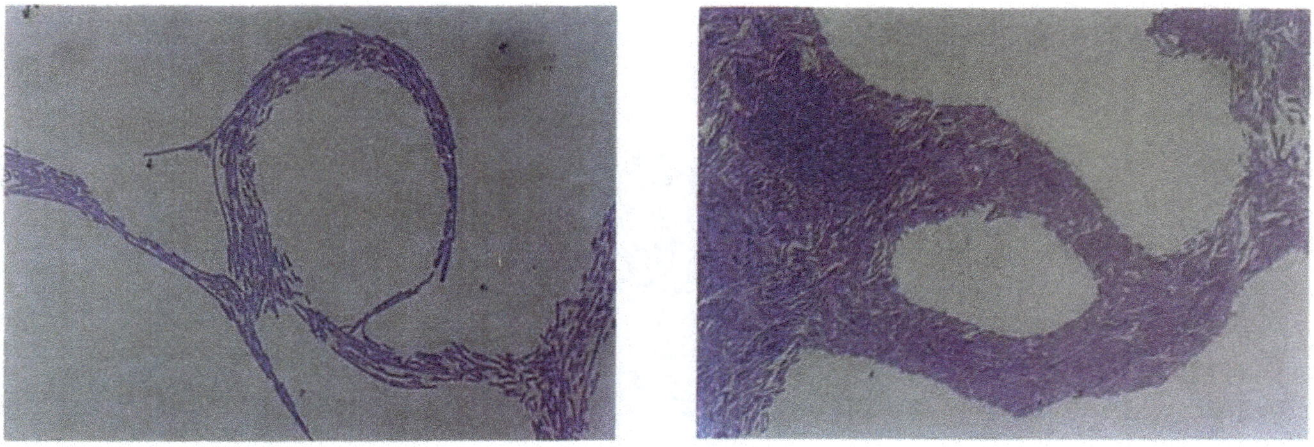

(a) (b)

Fig. 22 Optical microscope observations of the chiral growth demonstrating the high level of orientation correlations.

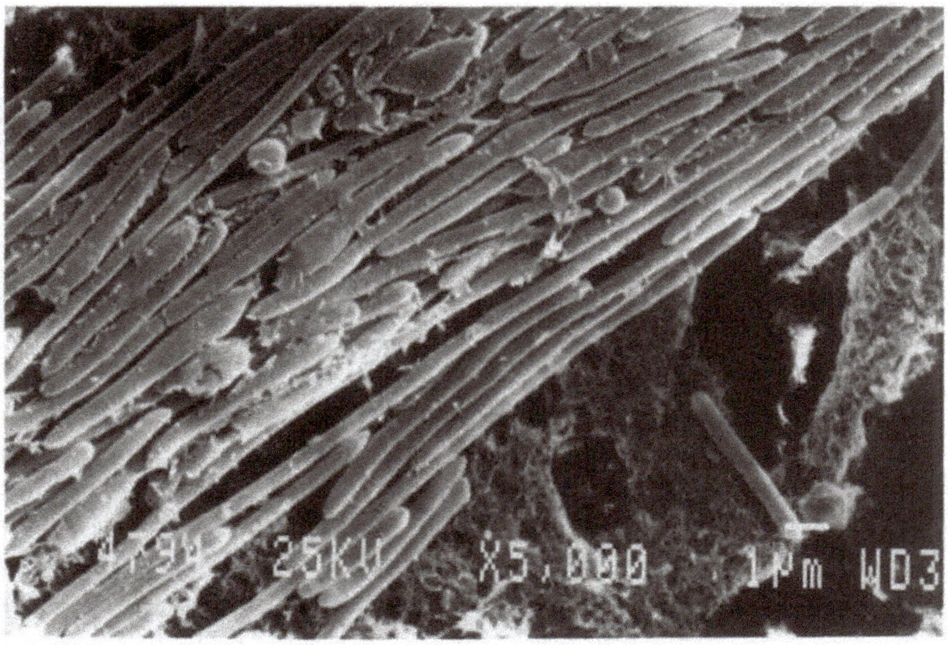

Fig. 23 Electron microscope observations of the chiral growth. No apparent chirality is observed.

each bacterium has a well defined orientation. The bacteria are no longer random walkers as there are no side movements; only forward and backward motion is observed with a weak twist. There is also a strong direct bacteria-bacteria interaction as the bacteria are attracted to move alongside each other. Actually, they should be referred to as gliders and not as walkers. In the introduction we have referred to the particle-field duality of the bacteria. Now each bacteria is not a point particle but an oriented particle. In Fig. 25, we show numerical simulations of a "diffusion-growth" like model which includes the above features (but not chemotactic signaling). We show the effect also if we include only the orientation and gliding features without the twist. In such a case, the colony will "shoot" out with few

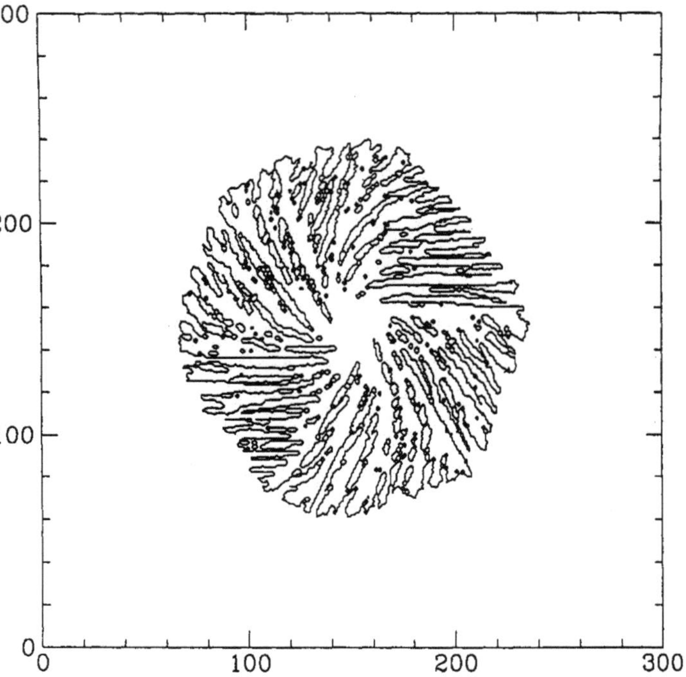

Fig. 24 Numerical simulations of the communicating walkers model including chemotaxis signaling and weak chirality.

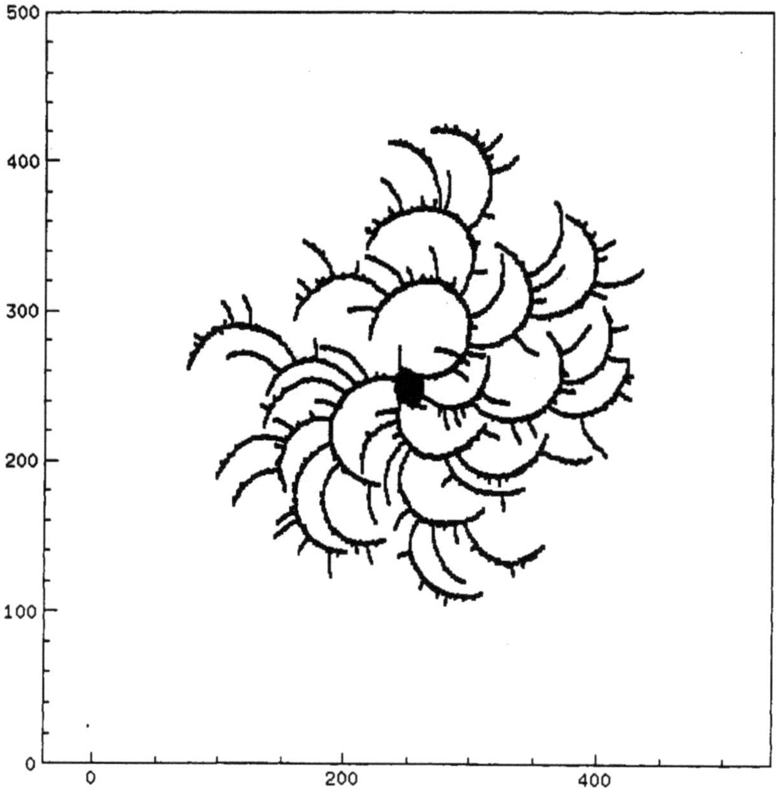

Fig. 25 Example of simulations of chiral growth of a "diffusion-growth"-like model with the additional features of bacteria orientation, gliding and twist (details are given in Ref. 54).

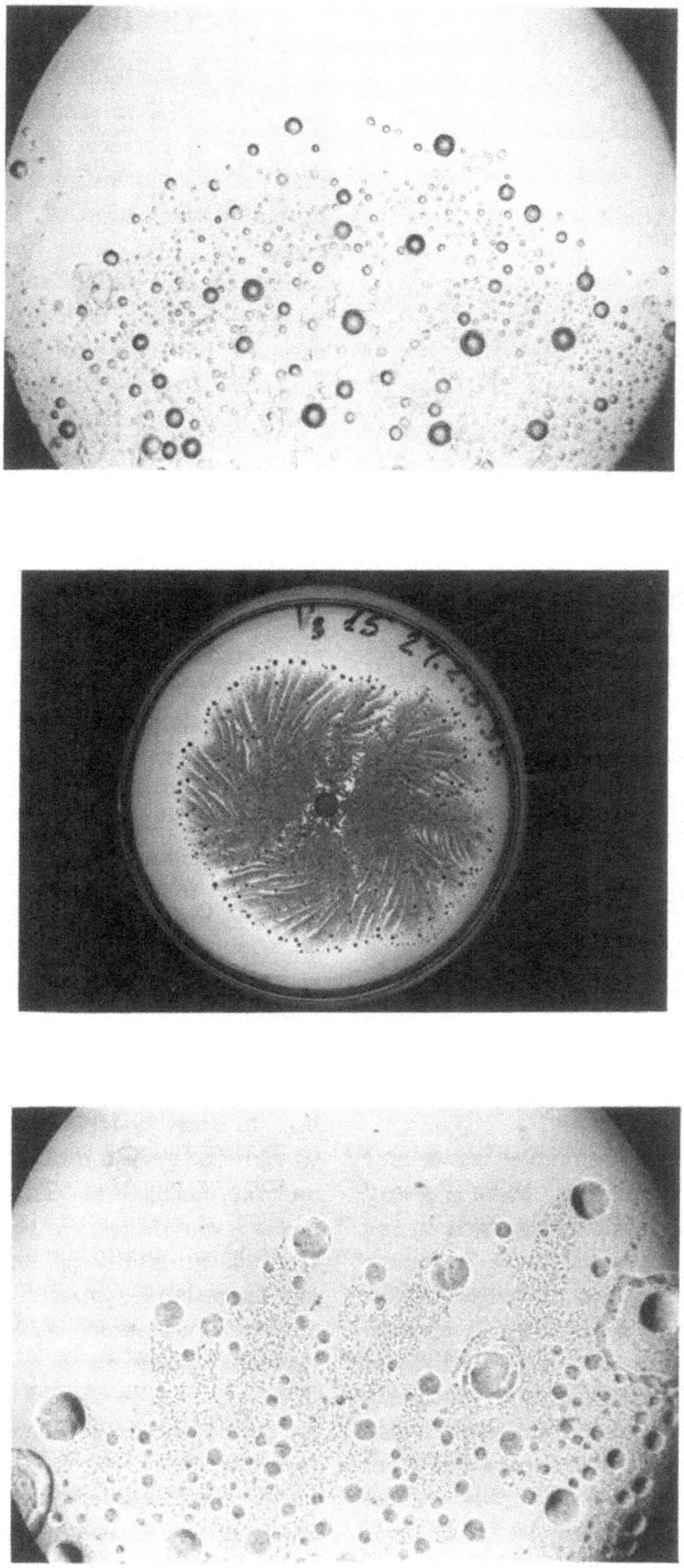

Fig. 26 Examples of vortex colonies (the preferred mode on hard agar) for different growth conditions. Each branch is formed by a spinning droplet of many bacteria all moving together in very high level of coordination.

"unprotected" branches, leaving behind much resources. In the model used to produce Fig. 25, the intensity of the twist was introduced as a fixed parameter. In reality, the chemotaxis materials and the other state conditions determine the twist.

6.3 The Vortex Mode

Growth of the bacteria on soft agar led to the complex chiral behavior. What would happen if we expose the bacteria to the other extreme — hard substrate? Under such conditions the "pioneering groups" have to be large enough to pave the way for the rest of the colony. Again it turned out that the bacteria are "smart" enough to develop the appropriate tactics, as is demonstrated by the vortex mode reported by Ben-Jacob et al.[30,10,31,11,32]

In Fig. 26 we show examples of the vortex colonies. These colonies spread via leading droplets (the darker dots) that leave behind a clear trail filled with bacteria on the agar. Each droplet consists of many bacteria that spin around a common center (hence named vortex) at typical velocities of 10 μm per second (the velocity is very sensitive to the growth conditions). Depending on the growth conditions, the number of bacteria in a vortex can vary from a mere few to millions. Within a given colony, both clockwise and anti-clockwise vortices are observed. The vortices also vary from being one layered to many layered. Usually, the pioneering droplets are the larger ones, and smaller droplets "explore" the empty areas left behind by the advancing front.

If we take the micro-level (individual bacterium) view point, we notice the following: there is a very high level of correlation, as all the bacteria in the vortex move together. At present, we do not know how to model this sophisticated collective behavior and many more studies are needed. It appears as if there is no relative motion of the individual bacterium and they are all drifted by a circulating self-generated flow field. The motion reminds one of the coordinated motion of many schools of other eucaryotes (e.g. schools of insects, squids, various types of fishes, etc.). Unfortunately, as the latter is also an open question, it cannot help in the modeling of the vortex motion. Actually, we expect that future understanding of the latter will help us in modeling school motion of eucaryotes.

7. CONCLUSIONS AND FURTHER DISCUSSIONS

It is now understood that bacteria paved the way and are crucial to maintaining life on earth. Yet, the view of bacteria as unicellular microbes has persisted for generations. Recently, Shapiro[16] has promoted the notion of "Bacteria as multicellular organisms". Following Shapiro, we would like to emphasize that understanding cybernetic processes (communication, regulation and control) during complex patterning of bacterial colonies can be a crucial step to understanding cybernetic processes in all organisms, both within a given organism and in groups (colonies, schools and societies) of organisms.

Shapiro concluded his paper saying: "Although bacteria are tiny, they display biochemical, structural and behavioral complexities that outstrip scientific description. In keeping with the current microelectronics revolution, it may make more sense to equate their small size with sophistication rather than with simplicity. There is little reason to doubt that insight gained from studying the interactions between billions of bacteria, living together in a volume of less than a few cubic millimeters, will enhance understanding of all forms of life."

At the same time, studies of complex patterning of bacterial colonies have a lot in common with studies of diffusive patterning in azoic systems. Hence, we believe that here in the definition of complexity[19,20] lies a bridge between synergetics and complexity of azoic systems and living systems. In particular we would like to emphasize the following: In azoic systems at equilibrium, we are used to view the system on two levels — the micro-level and the macro-level. The interplay between these levels is manifested via the introduction of the entropy as an additional variable (and a functional one for isolated systems) on the macro-level. The entropy is a measure of the number of possible microscopic states for a given state of the system. Hence, it can be viewed either as our lack of information about the micro-level (looking from the macro-level) or the freedom of the microdynamics for given imposed macroscopic conditions (looking from the micro-level). During the last decade we have learned that, as an azoic system is driven away from equilibrium, there is a singular interplay between the two levels which determines the self-organization of the entire system. The interplay can be viewed as balancing between the constraints that each level imposes on the other one. It is tempting to introduce the concept of complexity as a qualitative measure of the interplay. At present we lack a definition of complexity, and it is not clear that indeed such a new real variable does exist — real in the sense that a measurement of the variable can be defined. Ben-Jacob et al.[11] have argued that, to maintain singular interplay between the level of the individual bacterium and the colony, we need a regulating channel to a level below that of the individual bacterium. If correct, it implies that we need more than two levels to describe the interplay in the colony. We might use the lesson learned from the bacterial colonies as a hint for the development of the concept of complexity as a real variable for evolving azoic systems, in analogy to entropy for nonevolving processes (equilibrium).

Crossing the bridge will involve interdisciplinary activity, a level of risk and an ability to accept frustration in a way that is not faced in the regular activity within a discipline. However, we hope that the potential prospects will attract many to join us in the avenues of this new scholarly endeavor.

ACKNOWLEDGMENTS

We thank Inna Brainis, Dany Weiss, Hermann Leibovich and Orna Avidan for technical assistance. We have also benefited from conversations with Jim Shapiro, David Kessler, Herbert Levin, David Gutnik and Peter Garik. We thank D. Gutnik and A. Drukel for the joint work on the vortex. This research was supported in part by a grant from the G.I.F., the German-Israeli Foundation for Scientific Research and Development, by the Hungarian Research Foundation Grant No. T4439, and by the Program for Alternative Thinking at Tel-Aviv University.

REFERENCES

1. J. Kepler, *De Nive Sexangula Godfrey Tampach* (Frankfurt am Main, 1611).
2. D'Arcy W. Thompson, *On Growth and Form* (The University Press, Cambridge, 1964).
3. D. A. Kessler, J. Koplik and H. Levine, *Adv. Phys.* **37**, 255 (1988).
4. J. S. Langer, *Science* **243**, 1150 (1989).
5. E. Ben-Jacob and P. Garik, *Nature* **343**, 523 (1990).
6. H. Müller-Krumbhaar and W. Kurz, in *Phase Transformation in Materials*, ed. P. Haasen (VCH-Verlag, Weinheim, 1991).
7. E. Brener and V. I. Melnikov, *Adv. Phys.* **40**, 53 (1991).
8. E. Ben-Jacob, to be published in *Contemporary Physics.*
9. E. Ben-Jacob, O. Shochet and R. Kupferman, to be published in *Patterns in Nature: Fascination of their Origin and Simulation*, ed. A. Deutch (1993).
10. E. Ben-Jacob, H. Shmueli, O. Shochet and A. Tenenbaum, *Physica* **A187**, (1992).
11. E. Ben-Jacob, A. Tenenbaum, O. Shochet and O. Avidan, *Physica* **A202**, 1 (1994).
12. G. Albrecht-Buehler, *Int. Rev. Cyt.* **120**, 1919 (1990).
13. P. Pelce and A. Pocheau, *J. Theor. Biol.* **156**, 197 (1992).
14. P. Pelce and J. Sun, *Geometrical Models for the Growth of Unicellular Algae*, preprint 192.
15. J. A. Shapiro and D. Trubatch, *Physica* **D49**, 214 (1991).
16. J. A. Shapiro, *Scientific American*, June 1988.
17. E. Rosenberg, ed., *Myxobactera Development and Cell Interactions* (Springer-Verlag, New York, 1984).
18. D. Kessler and H. Levine, "Pattern Formation is Dictyostelium via Cooperative 'Bion' Dynamics," Preprint 1993.
19. H. Haken, *Information and Self-organization* (Springer-Verlag, Berlin, 1988).
20. G. Nicolis and I. Prigogine, *Exploring Complexity* (W. H. Freeman and Co., New York, 1989).
21. N. Wiener, *Cybernetics, or Control and Communication in the Animal and the Machine* (Wiley, New York, 1948).
22. R. Y. Stainer, M. Doudoroff and E. A. Adelberg, *The Microbial World* (Prentice-Hall, Inc., N.J., 1957).
23. T. Vicsek, *Fractal Growth Phenomena* (World Scientific, Singapore, 1989).
24. J. Feder, *Fractals* (Plenum, New York, 1988).
25. M. Eden, in *Proc. 4th Berkeley symp. on Mathematical Statistics and Probability*, Vol. 4, ed. F. Neyman (University of California Press, Berkeley, 1961) p. 223.
26. H. Fujikawa and M. Matsushita, *J. Phys. Soc. Jap.* **58**(11), 3875 (1989).
27. M. Matsushita and H. Fujikawa, *Physica* **A168**, 498 (1990).
28. M. Matsushita, J. Wakita and T. Matsuyama, *Proceedings of the NATO Advanced Research Workshop on Spatio-Temporal Patterns in Nonequilibrium Complex Systems* (Santa-Fe, April, 1993).
29. E. Ben-Jacob, P. Garik, S. Lipson, B. Orr and E. Raz, *Mat. Res. Soc.* **EA-25**, 11 (1990).
30. E. Ben-Jacob, H. Shmueli, O. Shochet and D. Weiss, in *Pattern Formation in Physical Systems and in*

Biology, eds. P. Meakin, L. Sander and P. Garcia (Plenum, 1993).
31. E. Ben-Jacob, O. Shochet, A. Tenenbaum and O. Avidan, *Proceedings of the NATO Advanced Research Workshop on Spatio-Temporal Patterns in Nonequilibrium Complex Systems* (Santa-Fe, April, 1993).
32. E. Ben-Jacob, O. Shochet and A. Tenenbaum, to be published in *Patterns in Nature: Fascination of their Origin and Simulation*, ed. A. Deutch (1993).
33. N. Sar, L. McCarter, M. Simon and M. Silverman, *J. Bacteriol.* **172**, 334 (1990).
34. J. Adler, *J. Gen. Microbiol.* **74**, 77 (1973).
35. M. Silverman and M. Simon, *Proc. Natl. Acad. Sci.* **74**, 3317 (1977).
36. W. W. Tso and J. Adler, *J. Bacteriol.* **118**, 560 (1974).
37. E. O. Budrene and H. C. Berg, *Nature* **349**, 630 (1991).
38. P. Devreotes, *Science* **245**, 1054 (1989).
39. J. M. Lackiie, ed., *Biology of the Chemotatic Response* (Cambridge Univ. Press, 1981).
40. P. N. Devreotes and S. H. Zigmond, *Ann. Rev. Cell Biol.* **4**, 649 (1988).
41. E. Ben-Jacob, O. Shochet, A. Tenenbaum, I. Cohen, A. Czirók and T. Vicsek, "Generic Modeling of Cooperative Growth Patterns in Bacterial Colonies," *Nature* (1994).
42. J. O. Kessler, *Contemp. Phys.* **26**, 147 (1985).
43. O. Shochet, K. Kassner, E. Ben-Jacob, S. G. Lipson and H. M. Müller-Krumbhaar, *Physica* **A181**, 136 (1992).
44. O. Shochet, K. Kassner, E. Ben-Jacob, S. G. Lipson and H. M. Müller-Krumbhaar, *Physica* **A197**, 87 (1992).
45. H. E. Stanley et al., *Bar-Ilan Conf. Proceedings* (1993).
46. E. Ben-Jacob, P. Garik, T. Muller and D. Grier, *Phys. Rev.* **A38**, 1370 (1988).
47. E. Brener, H. M. Müller-Krumbhaar and D. E. Temkin, *Europhys. Lett.* **17**, 535 (1992).
48. O. Shochet and E. Ben-Jacob, *Phys. Rev.* **E48**, R4168 (1993).
49. E. Ben-Jacob, R. Godbey, N. D. Goldenfeld, J. Koplik, H. Levine, T. Mueller and L. M. Sander, *Phys. Rev. Lett.* **55**, 1315 (1985).
50. G. Zocchi, B. E. Shaw, A. Libchaber and L. P. Kadanoff, *Phys. Rev.* **A86**, 1894 (1987).
51. Y. Couder, O. Cardoso, D. Dupuy, P. Tavernier and W. Thom, *Europhys. Lett.* **2**, 437 (1986).
52. T. Ihle and H. Müller-Krumbhaar, *Phys. Rev. Lett.* **70**, 3083 (1993).
53. R. Kupferman, O. Shochet, E. Ben-Jacob and Z. Schuss, "Numerical Study of Morphology Selection in Two-dimensional Phase-field model," submitted to *Phys. Rev. Lett.*
54. E. Ben-Jacob, O. Shochet, A. Tenenbaum, I. Cohen, A. Czirók and T. Vicsek (in preparation).
55. M. Sander, *Nature* **322**, 789 (1986).
56. T. A. Witten and L. M. Sander, *Phys. Rev. Lett.* **47**, 1400 (1981).
57. P. Meakin, *Phys. Rev.* **A27**, 1495 (1983).
58. J. Nittmann, G. Daccord and H. E. Stanley, *Nature* **314**, 141 (1985).
59. L. Niemeyer, L. Pietronero and H. J. Weismann, *Phys. Rev. Lett.* **52**, 1033 (1984).
60. M. Azbel, *Europhys. Lett.* **22**, 311 (1993).
61. H. Parnas and L. Segel, *J. Cell Sci.* **25**, 99 (1977); *J. Theor. Biol.* **71**, 1185 (1978).
62. S. A. Mackay, *J. Cell Sci.* **33**, 1 (1978).
63. R. A. Hegstrom and D. K. Kondepudi, *Sci. Am.* **108** (January, 1990).
64. V. A. Avetisov, V. I. Goldanskii and V. V. Kuzmin, *Phys. Today* **33** (July, 1991).
65. N. H. Mendelson, *Proc. Natl. Acad. Sci. USA* **75**(5), 2478 (1978).
66. N. H. Mendelson and S. L. Keener, *J. Bact.* **151**, 455 (1982).
67. N. H. Mendelson and J. J. Thwaites, *J. Bact.* **171**, 1055 (1989).
68. N. H. Mendelson, *Sci. Progress Oxford* **74**, 425 (1990).
69. N. R. Smith and F. E. Clark, *J. Bacteriol.* **35**, 59 (1938).
70. L. E. Shinn, *J. Bacteriol.* **39**, 419 (1938).

Fractals, Vol. 1, No. 3 (1993) 283–301

FRACTAL LANDSCAPES IN BIOLOGICAL SYSTEMS

H. E. STANLEY,* S. V. BULDYREV,* A. L. GOLDBERGER,† S. HAVLIN,*,‡
S. M. OSSADNIK,* C.-K. PENG† and M. SIMONS†
*Center for Polymer Studies and Department of Physics,
Boston University, Boston, MA 02215, USA
†Cardiovascular Division, Harvard Medical School,
Beth Israel Hospital, Boston, MA 02215, USA
‡Department of Physics, Bar Ilan University, Ramat Gan, Israel

Abstract

The purpose of this opening talk is to describe an example of recent progress in applying fractal concepts to biological systems. We first briefly review several biological systems, and then focus on the fractal features characterized by the long-range correlations found recently in DNA sequences containing *non-coding* material. We also discuss the evidence supporting the finding that for sequences containing only *coding* regions, there are no long-range correlations. Finally, we discuss the finding that the exponent α characterizing the long-range correlations increases with evolution.

1. INTRODUCTION

In the last decade it was realized that many biological systems have no characteristic length scale, thus having fractal or, more generally, self-affine properties.[1] In contrast to compact objects, fractal objects are almost entirely composed of "surface". Thus, fractals have a very large surface area. This observation explains why fractals are of great importance in biology, where surface phenomena are of crucial importance. For example, matter exchange takes place across membranes, and therefore requires large contact areas of the participating systems.

Lungs exemplify this feature. The surface area of a human lung is as large as a tennis court. A lung is made up of self-similar branches with many scale lengths, which is the defining attribute of a fractal surface. The efficiency of the lung is enhanced by this fractal property, since at each breath oxygen and carbon dioxide have to be exchanged at the lung surface. The structure of the bronchial tree has been quantitatively analyzed using fractal concepts.[1,2]

A second example is blood vessels. Blood must be carried to all the cells of the body. For this purpose blood vessels must have fractal properties. The diameter distribution of blood vessels ranging from capillaries to arteries follow a power-law distribution which is one of the main characteristics of fractals. Specific examples include the chick embryo circulatory system, and the human retinal circulatory system.[3]

Considerable attention in the biological community has arisen from the possibility that neuron shape can be quantified using fractal concepts. For example, Smith et al.[4] studied the fractal features of vertebrate central-nervous-system neurons in culture and found that the fractal dimension is increased as the neuron is more developed. Caserta et al.[5] showed that the shapes of quasi-two-dimensional retinal neurons can be characterized by a fractal dimension d_f. They found for fully developed neurons *in vivo*, $d_f = 1.68 \pm 0.15$, and suggest that the growth mechanism for neurite outgrowth bears a direct analogy with the growth model called diffusion-limited-aggregation (DLA).

One of the most amusing examples where fractal concepts come into play concerns the possibility[6] that HCl, when released under pressure by the secretory glands, indeed crosses the viscous lining of the stomach, using the principles of viscous fingering that govern the breakdown of any viscous liquid when a less viscous liquid is forced under pressure through it. The DLA-type model governing viscous fingering thereby serves to resolve the age-old paradox *"Why doesn't the stomach digest itself?"*

In recent years long-range power-law correlations have been discovered in a remarkably wide variety of systems. Such long-range power-law correlations are a physical fact that in turn gives rise to the increasingly appreciated "fractal geometry of nature".[7–12] So if fractals are indeed so widespread, it makes sense to anticipate that long-range power-law correlations may be similarly widespread. Indeed, recognizing the ubiquity of long-range power-law correlations can help us in our efforts to understand nature, since as soon as we find power-law correlations we can quantify these with a critical exponent. Quantification of different behavior allows us to recognize similarities between different systems, thereby eventually leading to recognizing underlying unifications that might otherwise have gone unnoticed.

Our intuition tells us that correlations should decay exponentially:

$$C_r = (C_1)^r = e^{-r/\xi} \tag{1a}$$

where the second equality in Eq. (1a) serves to define the correlation length $\xi \equiv -1/\ln C_1$. Our simple intuition, that correlations decay exponentially because of the fashion in which order is "propagated", seems to always work — except at the critical point,[13] where the exponential decay of Eq. (1a) gives over to a power law decay:

$$C_r \sim \left(\frac{1}{r}\right)^{d-2+\eta} \tag{1b}$$

The difference between Eqs. (1a) and (1b) is profound: Eq. (1a) states that there is a characteristic length ξ fixed by the strength of the nearest-neighbor correlation C_1, while Eq. (1b) states that there is no characteristic length at all.

Can we intuitively understand how it is possible to find a *non-exponential* decay of correlations? At first sight, it might appear that whenever we increase the distance between two molecules by one lattice constant, the correlation should decrease by roughly the same factor, but this intuition leads immediately to exponential decay. A possible resolution to this paradox stems from the fact that near a critical point, "information" propagates from a molecule at the origin to a molecule at position $\vec{r}$ *not via a single path* (as for $d = 1$), but *rather via an infinite number of paths*; some of these paths are explicitly enumerated in Fig. 9.4 of Ref. 13. Ornstein and Zernike[14,15] recognized this fact, but approximated the fashion in which "order is propagated" and so obtained predictions that today we call "classical" [Fig. 7.5 of Ref. 13]. Exact enumeration methods, such as high-temperature series expansions, take into account exactly such paths up to a certain length k_{max}, where k_{max} is typically 20. To obtain power-law correlations, the exact results for $k < 20$ are extrapolated to obtain an estimate of the behavior for all k. In some sense, although the correlation along each path *decreases* exponentially with the length of the path, the number of such path *increases* exponentially. Therefore, the net effect is that the dominant exponential decay is magically "canceled", leaving the sub-dominant but longer range power-law correlations — which are in fact observed.

At one time, it was imagined that the "scale-free" case of Eq. (1b) was relevant to only a fairly narrow slice of physical phenomena — only to systems that had been "tuned" by exceedingly painstaking experimental work to be exactly at a critical point.[13] Now we appreciate the ubiquity of systems displaying scale-invariant behavior. First of all, any system examined on length scales smaller than the correlation length is likely to display power-law behavior (because all paths between the origin and r are relevant up to the correlation length, and these cancel out the exponential decay for $r < \xi$). Moreover, the number and nature of systems displaying power-law correlations have increased dramatically, including systems that no one might ever have suspected as falling under the umbrella of "critical phenomena." The latter part of the century has witnessed a veritable explosion in the study, both experimental and theoretical, of such systems. The *1991 Nobel Prize* was awarded to P.-G. de Gennes in part for his recognition that polymer systems behave analogously to systems near their critical points. The *1993 Wolf Prize* was awarded to Benoit Mandelbrot for the recognition of the "fractal geometry of nature". Another very prestigious Israeli prize, the *1993 Israel Prize*, has been awarded this year to Shlomo Alexander, in large part for his discoveries that under appropriate conditions a wide range of systems obey scaling or scale invariance.

The list of systems in which power-law correlations appear has grown rapidly in recent years, including models of turbulence and even earthquakes.[16] What do we anticipate for biological systems? Generally speaking, when "entropy wins over energy" — i.e., randomness dominates the behavior — we find power laws and scale invariance. Biological systems sometimes are described in language that makes one think of a Swiss watch. Mechanistic or "Rube Goldberg" descriptions must in some sense be incomplete, since it is only some appropriately-chosen averages that appear to behave in a regular fashion. The trajectory of each individual biological molecule is of necessity random — albeit correlated. Thus one might hope that recent advances in understanding "correlated randomness"[17–20] could be relevant to biological phenomena. While there have been reports of scale invariant phenomena in isolated biological systems — ranging from the fractal shapes of neurons[4,5] to long-range correlations in heart beat intervals,[21] human writings,[22] and the stock market[23] — there has been no systematic study of a *biological* system that displays power-law correlations.

Here we will attempt to summarize the key findings of some recent work[24–43] suggesting that — under suitable conditions — the sequence of base pairs or "nucleotides" in DNA also displays power-law correlations. The underlying basis of such power-law correlations is not understood at present, but it is at least possible that this reason is of as fundamental importance as it is in other systems in nature that have been found to display power-law correlations.

2. DISCOVERY OF LONG-RANGE CORRELATIONS IN DNA

In order to study the scale-invariant long-range correlations of a DNA sequence, we first introduced a graphical representation of DNA sequences, which we term a *fractal landscape* or *DNA walk*.[24] For the conventional one-dimensional random walk model,[44] a walker moves either "up" $[u(i) = +1]$ or "down" $[u(i) = -1]$ one unit length for each step i of the walk. For the case of an uncorrelated walk, the direction of each step is independent of the previous steps. For the case of a correlated random walk, the direction of each step depends on the history ("memory") of the walker.[18–20]

One definition of the DNA walk is that the walker steps "up" $[u(i) = +1]$ if a pyrimidine (C or T) occurs at position a linear distance i along the DNA chain, while the walker steps "down" $[u(i) = -1]$ if a purine (A or G) occurs at position i. The question we asked was whether such a walk displays only short-range correlations (as in an n-step Markov chain) or long-range correlations (as in critical phenomena and other scale-free "fractal" phenomena).

The DNA walk provides a graphical representation for each gene and permits the degree of correlation in the base pair sequence to be directly visualized, as in Fig. 1. Figure 1 naturally motivates a quantification of this correlation by calculating the "net displacement" of the walker after ℓ steps, which is the sum of the unit steps $u(i)$ for each step i. Thus $y(\ell) \equiv \sum_{i=1}^{\ell} u(i)$.

An important statistical quantity characterizing any walk[44] is the root mean square fluctuation $F(\ell)$ about the average of the displacement; $F(\ell)$ is defined in terms of the difference between the average of the square and the square of the average,

$$F^2(\ell) \equiv \overline{[\Delta y(\ell) - \overline{\Delta y(\ell)}]^2} = \overline{[\Delta y(\ell)]^2} - \overline{\Delta y(\ell)}^2 , \quad (2)$$

of a quantity $\Delta y(\ell)$ defined by $\Delta y(\ell) \equiv y(\ell_0 + l) - y(\ell_0)$. Here the bars indicate an *average* over all positions ℓ_0 in the gene. Operationally, this is equivalent to: (a) taking a set of calipers set for a fixed distance ℓ; (b) moving the beginning point sequentially from $\ell_o = 1$ to $\ell_o = 2, \cdots$ and (c) calculating the quantity $\Delta y(\ell)$ (and its square) for each value of ℓ_o; and (d) averaging all of the calculated quantities to obtain $F^2(\ell)$.

The mean square fluctuation is related to the auto-correlation function

$$C(\ell) \equiv \overline{u(\ell_0)u(\ell_0 + l)} - \overline{u(\ell_0)}^2 \quad (3a)$$

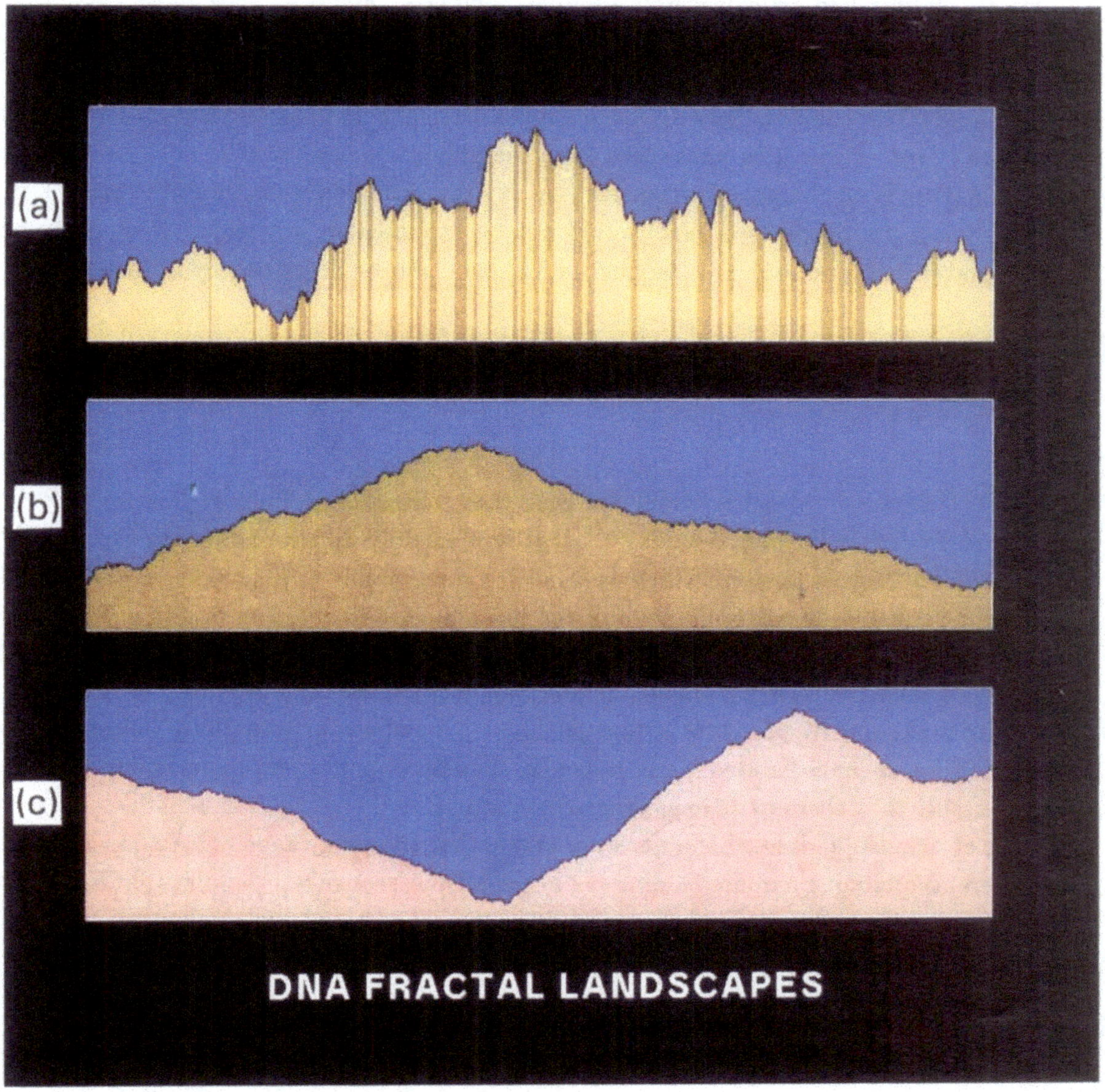

Fig. 1 The DNA walk representations of **(a)** human β-cardiac myosin heavy chain gene sequence, showing the coding regions as vertical golden bars, **(b)** the spliced together coding regions, and **(c)** the bacteriophage lambda DNA which contains only coding regions. Note the more complex fluctuations for (a) compared with the coding sequences (b) and (c). We found that for almost all coding sequences studied that there appear regions with one strand bias, followed by regions of a different strand bias. The fluctuation on either side of the overall strand bias we found to be random, a fact that is plausible by visual inspection of the DNA walk representations. We used different step heights for purine and pyrimidine in order to align the end point with the starting point. This procedure is for graphical display purposes only (to allow one to visualize the fluctuations more easily) and is not used in any analytic calculations. (After Ref. 24.)

through the relation:

$$F^2(\ell) = \sum_{i=1}^{\ell} \sum_{j=1}^{\ell} C(j-i)\,. \tag{3b}$$

The calculation of $F(\ell)$ can distinguish three possible types of behavior.

(i) If the base pair sequence were random, then $C(\ell)$ would be zero on average [except $C(0) = 1$], so $F(\ell) \sim \ell^{1/2}$ (as expected for a *normal* random walk).

(ii) If there were a local correlation extending up to a characteristic range R (such as in Markov chains), then $C(\ell) \sim \exp(-\ell/R)$; nonetheless *the asymptotic behavior* $F(\ell) \sim \ell^{1/2}$ *would be unchanged from the purely random case.*

(iii) If there is no characteristic length (i.e., if the correlation were "infinite-range"), then the scaling property of $C(\ell)$ would not be exponential, but would most likely be a power-law function, and the fluctuations will also be described by a power-law:

$$F(\ell) \sim \ell^{\alpha} \tag{4a}$$

with $\alpha \neq 1/2$.

Figure 1(a) shows a typical example of a gene that contains a significant fraction of base pairs that do *not* code for amino acids.[45,46] It is immediately apparent that the DNA walk has an extremely jagged contour which we shall see corresponds to long-range correlations. Figure 2 shows double logarithmic plots of the mean square fluctuation function $F(\ell)$ as a function of the linear distance ℓ along the DNA chain for a typical intron-containing gene.

The fact that the data are linear on this double logarithmic plot confirms that $F(\ell) \sim \ell^{\alpha}$. A least-squares fit produces a straight line with slope α substantially larger than the prediction for an uncorrelated walk, $\alpha = 1/2$, thus providing direct experimental evidence for the result that there exist long-range correlations.

Peng et al. also addressed the question of whether the long-range correlation properties are different for coding and non-coding regions of a DNA sequence,[24] a point that is currently

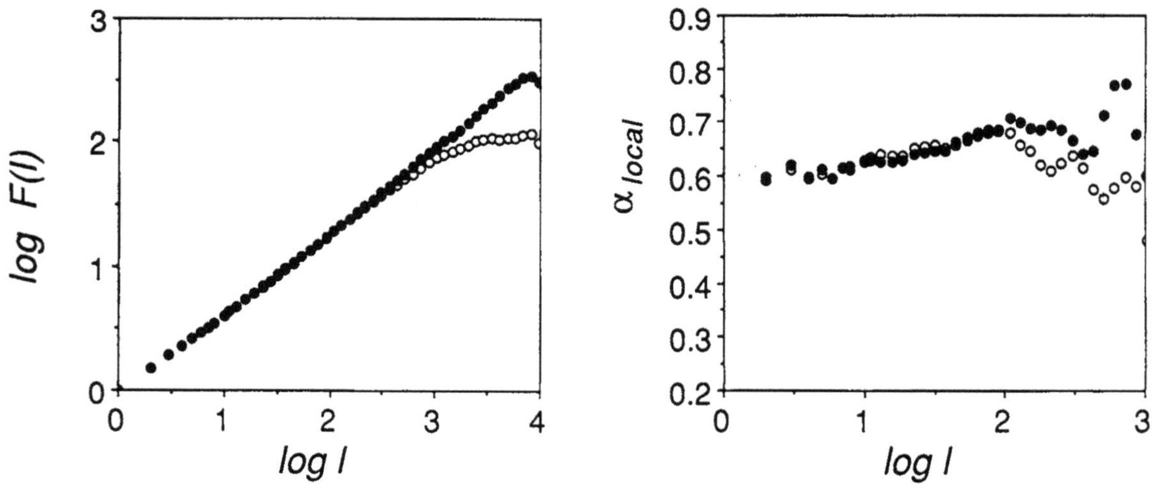

Fig. 2 (a) Double logarithmic plots of the mean square fluctuation function $F(\ell)$ as a function of the linear distance ℓ along the DNA chain for the rat embryonic skeletal myosin heavy chain gene (○) and its "intron-spliced sequence" (•); (b) The corresponding local slopes, α_{local}, based on pairs of successive data points of part (a). We see that the values of α are roughly constant. For this specific gene, the sequence with exons removed has an even broader scaling regime than the DNA sequence of the entire gene, indicated by the fact that part (a) is linear up to 10 000 nucleotides. (After Ref. 33.)

the subject of some continuing debate.[27,31,37] Figure 1(b) shows the DNA walk for a sequence formed by splicing together the coding regions of the DNA sequence of this same gene. Figure 1(c) displays the DNA walk for a typical sequence with only coding regions. In contrast to Fig. 1(a), these coding sequences have less jagged contours, suggesting a shorter range correlation.

3. OTHER METHODS OF MEASURING LONG-RANGE CORRELATIONS

One can also worry that the apparent long-range correlation is some artifact of the DNA walk method itself. To compare the fluctuations of α in our DNA walk method with those found in other methods, Peng et al. used two standard methods to study the correlation property of sequences, namely, the correlation function $C(\ell)$ and the power spectrum $S(f)$. The power spectrum density, $S(f)$, is obtained by: (a) Fourier transforming the sequence $\{u(i)\}$, and (b) taking the square of the Fourier component. For a stationary sequence, the power spectrum is the Fourier transform of the correlation function. If the correlation decays algebraically (not exponentially), i.e., there is no characteristic scale for the decay of the correlation, as we found in the non-coding DNA sequences, then we expect power-law behavior for both the power spectrum and the correlation function,

$$S(f) \sim \left(\frac{1}{f}\right)^{\beta}, \tag{4b}$$

and

$$C(\ell) \sim \left(\frac{1}{\ell}\right)^{\gamma}. \tag{4c}$$

The correlation exponents α, β and γ are not independent, since[18,19]

$$\alpha = \frac{1+\beta}{2} = \frac{2-\gamma}{2}. \tag{5}$$

For a typical DNA sequence of finite length, both the correlation function and power spectrum are fairly noisy, but the estimates of β and γ obtained are consistent with those calculated from the DNA walk method. The reason for the smaller fluctuations of α in the DNA walk method is due to the fact that $F^2(\ell)$ is a double summation of $C(\ell)$. Thus it would seem that the original DNA walk method is more useful due to reduced noise.

Apart from the reduced noise mentioned above, one additional advantage of the DNA walk method[24] is that to find the exponent characterizing the long-range correlation, one need not correct the data by subtracting the white noise, $S(\infty)$.[27] Since there is no unambiguous method of estimating $S(\infty)$, this need to correct the data introduces an uncontrollable source of uncertainty.

4. DIFFERENCE BETWEEN CORRELATION PROPERTIES OF CODING AND NON-CODING REGIONS

Our initial report[24] on long-range (scale-invariant) correlations in DNA sequences has generated contradicting responses. Some[25,27,29] support our initial finding, while others[26,31,35,37] disagree. Furthermore, the conclusions of Refs. 27 and Refs. 26, 31 and 35 are inconsistent *with one another* in that Refs. 26 and 37 doubt the existence of long-range correlations (even in non-coding sequences), while Refs. 26, 31 and 35 conclude that even coding regions display long-range correlations ($\alpha > 1/2$). Prabhu and Claverie[31] claim that their analysis of the putative *coding* regions of the yeast chromosome III[45] produce a *wide range of exponent values*, some larger than 0.5. The source of these contradicting claims may arise from the fact that, in addition to normal statistical fluctuations expected for analysis of rather short sequences, coding regions typically consist of only a few lengthy regions of alternating strand bias. Hence scaling analysis cannot be applied reliably to the entire sequence but only to sub-sequences.

Prabhu and Claverie[31] noted that nonlinear curves were obtained for *even some intron-containing sequences.* The DNA walks for intron-containing sequences do not show any apparent length scale of strand bias; as noted above, there seems to be a broad distribution of lengths of strand bias. In principle, for this type of DNA walk, the min–max procedure is not necessary since there is no characteristic length that needs to be taken into account. For most *intron-containing* sequences, our analysis shows that there is a broad range of scaling (with $\alpha > 0.5$) when we study the entire sequence as a whole. For certain intron-containing genes, however, we find that the min–max procedure can extend the scaling region, leading to a larger range of constant-α behavior.

One might also wonder if the selection of "max" and "min" segments gives a bias to the calculations. The "max" and "min" operation is a systematic method to treat all DNA sequences on equal footing without applying any *a priori* knowledge of the sequence itself. Notice that although this procedure eliminates the problem of large scale variation of concentration for the coding sequences, it will not alter the result of a truly self-similar sequence. To obviate the need to perform the min–max partitioning, Peng et al.[42] have recently applied the "bridge method" to DNA, and have also developed a new method specifically adapted to handle problems associated with non-stationary sequences which they term *detrended fluctuation analysis* (DFA).

To provide an "unbiased" test of the thesis that non-coding regions possess, but coding regions lack, long-range correlations, Ossadnik et al.[43] analyzed several uncorrelated and correlated control sequences of size 10^5 nucleotides using the GRAIL neural net algorithm.[46] The GRAIL algorithm identified about 60 putative exons in the uncorrelated sequences, but only about 5 putative exons in the correlated sequences. We also used the beachcomber method,[43] which shows pronounced dips in α in the region where genes are expected (Fig. 3).

Voss[27] has recently proposed that *coding* as well as non-coding DNA sequences display long-range power-law correlations in their base pair (bp) sequences. This finding disagreed with our earlier analysis,[24] claiming that coding DNA sequences do not display power-law correlations. However, the discrepancy between Refs. 27 and 24 could have arisen because the analysis in Ref. 24 was based on partitioning the entire coding sequence into a few large subsequences of constant overall compositional bias. It is important to resolve this discrepancy, since Voss based his scientific conclusion ("immunity to errors on all scales") on his claim of power-law correlations in *coding* sequences.[47]

Recently, Buldyrev et al. argued that the Voss proposal does not hold generally.[48] Specifically, they presented two counterexamples that clearly display *no* long-range correlations *when directly analyzed* (without partitioning into subsequences): (i) the complete genome of

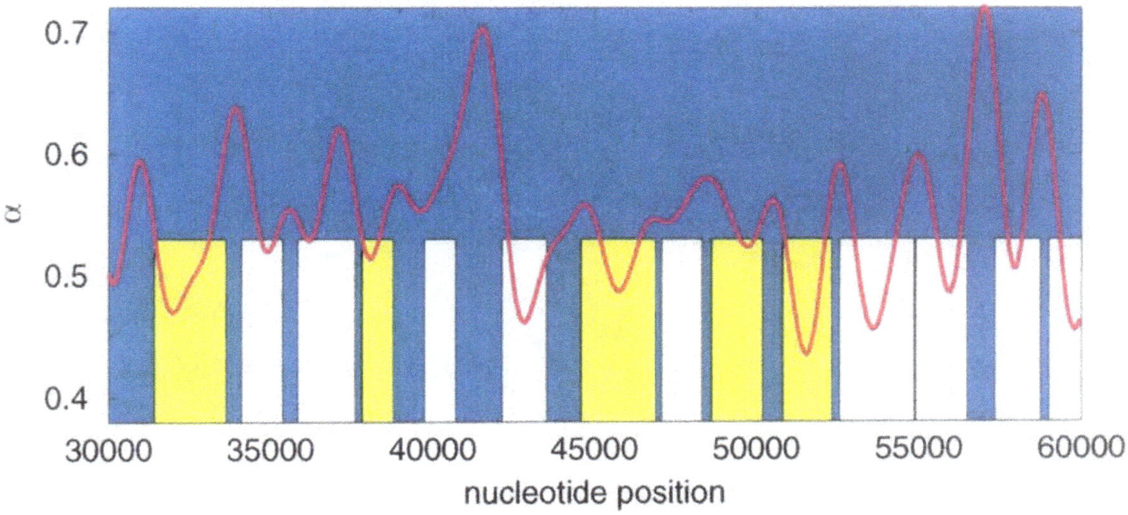

Fig. 3 Beachcomber plot for a typical section containing about 10% of the Yeast Chromosome III[45] — from base pair #30 000 to #60 000. The vertical yellow bars indicate the set of base pairs forming identified genes (GenBank Release #76), while the white bars indicate less certain "putative genes" determined from analysis of open reading frames.[45] The exponent α is calculated by the beachcomber method[43]: We form an observation box of length 800, place this box at the beginning of the chromosome, and calculate the long-range correlation exponent α for the 800 base pairs lying inside this box. Then we move the box 75 base pairs further along the chromosome, and again calculate α for the 800 base pairs lying inside this box. Iterating this procedure, we obtain $315\,000/75 = 4200$ successive values of α, each giving a "local" measurement of the degree of long-range correlation. The red curve is obtained using rule 1 — a "down" step for A or G (purines) and an "up" step for C or T (pyrimidine). We see that when the box is covering coding regions, the value of α is generally small, while in between coding regions, there is frequently a peak in α. If α were the same for coding and non-coding regions, we would expect the peaks and dips to occur with no evident correlation in the position of genes.

T7 bacteriophage (39 936 bp), which contains *only* coding regions; and (ii) the Ti plasmid fragment (24 595 bp), which is believed to consist almost entirely of coding regions.

Figure 4(a) shows the DNA walks for (i) and (ii). Figure 4(b) shows $F(\ell)$, the fluctuation in rms amplitude; the slopes of the log-log plots, *fit over 3 decades*, are 0.53 and 0.49, indicating the absence of long-range correlation for both cases. Figure 4(c) shows the power spectrum $S(f)$, which is almost perfectly flat (indicative of no correlation or "white noise"). The scaling behavior in Figs. 4(b) and 4(c) is markedly different than that found for genomic sequences containing substantial noncoding sub-regions, for which $F(\ell) \sim \ell^{\alpha}$ with $\alpha \approx 2/3$, and $S(f) \sim 1/f^{\beta}$ with $\beta \approx 1/3$.[24]

So why did Voss find $\beta = 1.02$ and 1.16, respectively, for phage and bacteria (which contain mostly coding regions)? A clue is apparent from comparing Voss' analysis for these cases (Fig. 3 of Ref. 27) with his fits for the non-coding segments. We see that all except a few small-f data points are well fit by a horizontal line, corresponding to $\beta = 0$ (no long-range correlation). The departure at small f likely corresponds to the fact that most coding sequences contain uncorrelated sub-segments — *with a characteristic length* — of alternating compositional bias. The DNA walks, therefore, resemble a spliced together string of *uncorrelated* but *biased* random walks. We confirmed this likely source of spurious low-f behavior by calculations on artificial "control" sequences.

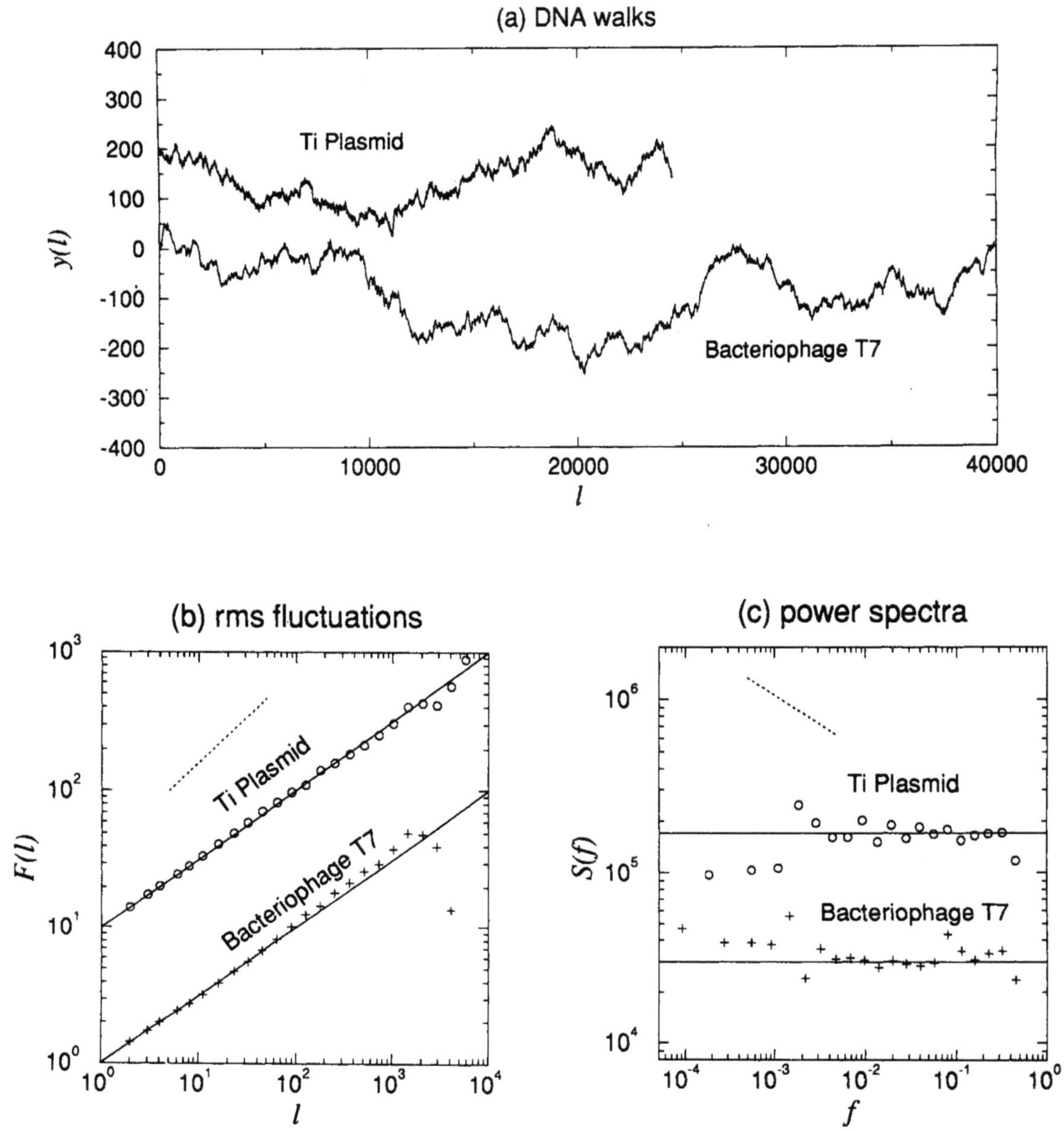

Fig. 4 (a) "DNA walks"[24] of nucleotide sequences; a random walker moves either up or down depending on whether the nucleotide at position ℓ is a pyrimidine or purine; (b) The rms fluctuation, $F(\ell)$, of the DNA walk displacement $y(\ell)$; (c) Power spectrum, $S(f)$, of the binary nucleotide sequences (pyrimidine = 1, purine = -1). Shown are two examples that display *no* long-range correlations: (i) the complete genome of T7 bacteriophage (GenBank name: PODOT7), which contains *only* coding regions, and (ii) the Ti plasmid fragment (ATACH5), which is believed to consist almost entirely of coding regions. The solid lines in (b) and (c) have slopes $\alpha = 1/2$ and $\beta = 0$ respectively; for comparison, dashed lines of slope $2/3$ and $-1/3$ are also shown, corresponding to the typical behavior found for sequences containing non-coding regions.[24] The data for Ti (o) shifted on the plots for visual comparison. (After Ref. 48.)

5. FRACTAL LANDSCAPES AND MOLECULAR EVOLUTION

Molecular evolutionary relationships are usually inferred from comparison of coding sequences, conservation of intron/exon structure of related sequences, analysis of nucleotide substitutions, and construction of phylogenetic trees.[49] The changes observed are

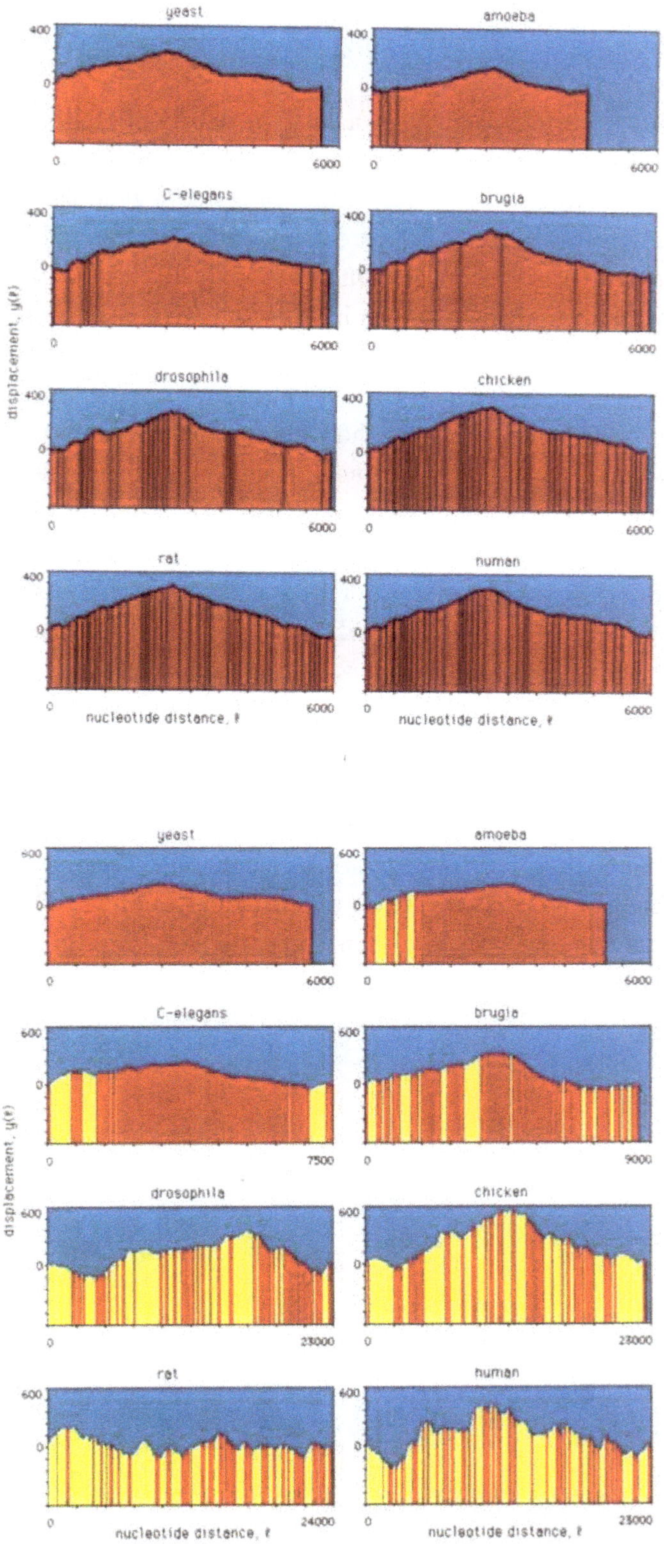

Fig. 5 The DNA walk representations of (a) 8 cDNA sequences from the MHC family; and (b) the corresponding genes. DNA landscapes are plotted so that the end points have the same vertical displacement as the starting points.[24] The graphs are for yeast, amoeba, *C. elegans*, brugia, drosophila, chicken, rat and human (from top to bottom, left to right). The shaded areas in (b) denote coding regions of the genes. The DNA walks for the genes show increasing "complexity" with evolution. In contrast, the cDNA walks all show remarkably similar crossover patterns due to sequential "up-hill" and "down-hill" slopes representing different purine/pyrimidine strand biases in the regions coding for the head and tail of the MHC molecule, respectively. (After Ref. 41.)

conventionally interpreted with respect to nucleotide sequence composition (mutations, deletions, substitutions, alternative splicing, transpositions, etc.) rather than overall genomic organization.

Very recently, Buldyrev et al.[41] sought to assess the utility of DNA correlation analysis as a complementary method of studying gene evolution. In particular, they studied the changes in "fractal complexity" of nucleotide organization of a single gene family with evolution. A recent study by Voss[27] reported that the correlation exponent derived from Fourier analysis was lowest for sequences from organelles, but paradoxically higher for invertebrates than vertebrates. However, this analysis must be interpreted with caution since it was based on pooled data from different gene families rather than from the quantitative examination of any single gene family.

Buldyrev et al.[41] tested the hypothesis that the fractal complexity of genes from higher animals is greater than that of lower animals, using single gene family analysis. They focused their analysis on the genome sequences from the conventional (Type II) myosin heavy chain (MHC) family. Such a choice limits potential bias that may arise secondary to non-uniform evolutionary pressures and differences in nucleotide content between unrelated genes. They also used this technique to study the MHC gene family because of the availability of completely sequenced genes from a phylogenetically diverse group of organisms, and the fact that their relatively long sequences are well-suited to statistical analysis.

The landscape produced by DNA walk analysis reveals that each MHC cDNA consists of two roughly equal parts with significant differences in nucleotide content (Fig. 5). The first part that codes for the heavy meromyosin or "head" of the protein molecule has a slight excess of purines (52% purines and 48% pyrimidines); the second part that codes for the light meromyosin or "tail" has about 63% purines and 37% pyrimidines. The *absolute nucleotide* contents are not shown in the graphical representation of Fig. 5(a) because we subtract the average slope from the landscape to make relative fluctuations around the average more visible. Indeed, one can easily see from Fig. 5(a) that the relative concentration of

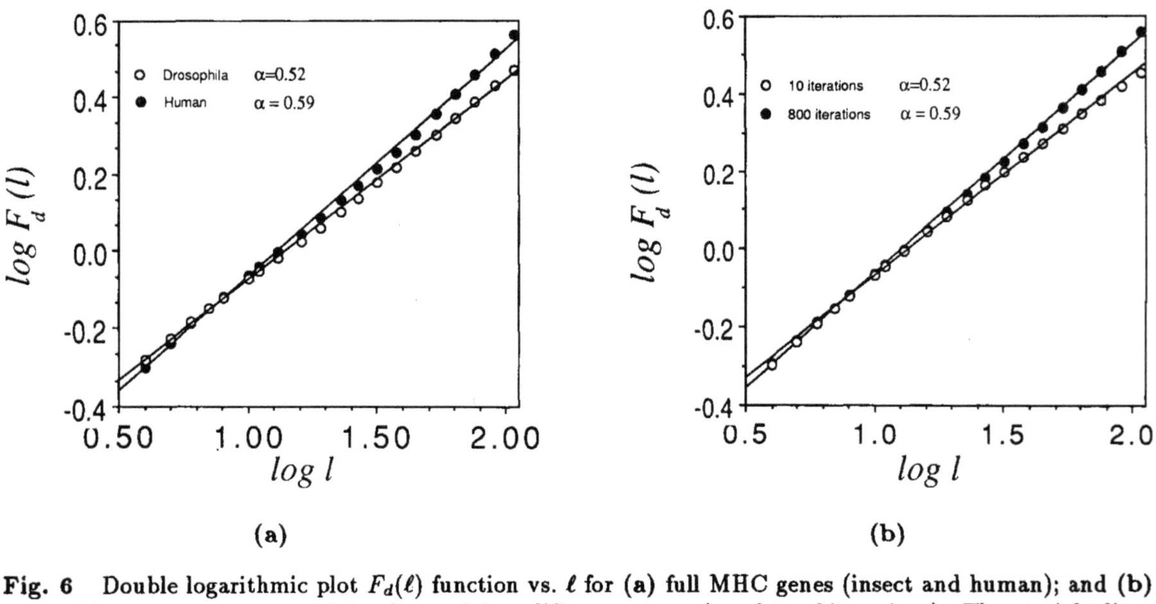

Fig. 6 Double logarithmic plot $F_d(\ell)$ function vs. ℓ for **(a)** full MHC genes (insect and human); and **(b)** nucleotide sequences generated by the model at different stages (number of iterations). The straight lines are linear regression fits from $\ell = 4$ to 108. (After Ref. 41.)

pyrimidines in the first part ("uphill" region) of the myosin cDNA is much higher than in the second ("downhill" region).

As previously reported,[24] we find that $\alpha \approx 1/2$ for all cDNAs (corresponding to no correlations or only short-range correlations), while all MHC genes containing introns have $\alpha > 1/2$, corresponding to long-range correlations. Figure 6 shows representative scaling plots of $F_d(\ell)$ vs. ℓ, where F_d denotes the fluctuation quantity of Eq. (3b) defined using the "detrended fluctuation analysis" developed in Ref. 42. Of note, the value of α is not strongly related to the presence of exons since "stitching together" intron sequences (by removing exons) produces a value similar to that of the full gene, for example, for the human MHC gene the value of α after the exons are removed is 0.593 versus 0.586 for the complete sequence, further supporting the view that the composition of non-coding elements is the principal source of long-range correlations in genomic sequences.

6. INSERTION-DELETION MODEL

To gain some insight into possible evolutionary mechanisms that could increase the complexity of the landscapes and generate long-range correlations of DNA sequences, Buldyrev et al.[41] introduce a simple model that simulates conversion of originally coding regions into noncoding introns. The model is based on the hypotheses that genetic information was originally encoded in an mRNA molecule which was subsequently converted into a DNA sequence, and that this sequence underwent modifications due to mutagenesis and insertion of non-coding genetic material (introns).[50]

(i) To simulate cDNA sequences, one starts with a biased random walk of length L with an overall excess of purines over pyrimidines corresponding to that observed in the cDNA sequences.

(ii) At each time step, one "mutates" the sequence by the following procedure:

  (a) Choose a random point in the sequence and cut a sub-sequence of length n starting from that point, where the length n is chosen from a power-law distribution $\phi(n) \sim n^{-\beta}$ with $\beta \approx 2$ (between $L_o \approx 20$ and $L/2$). The reason for this power-law distribution is that the cutting of a DNA segment most likely occurs when a loop is formed, and it is known that the distribution of loop sizes in a long polymer obeys a power law.[51] Choose another random point in the sequence at which we insert this length-n subsequence.

  (b) With probability 0.5, a *strand substitution* may occur in this sub-sequence (i.e., all purines are substituted by pyrimidines and vice versa, thereby inserting a complementary strand).

  (c) To simulate retroviral insertion occurring, with some small probability p_i, the subsequence to be inserted is substituted by a random sequence of equal length with the same percentage of purines and pyrimidines as in the initial cDNA sequence.

Representative purine-pyrimidine landscapes generated by this model are shown in Fig. 7. After the first few iterations [Fig. 7(a)], the landscape is remarkably similar to that of

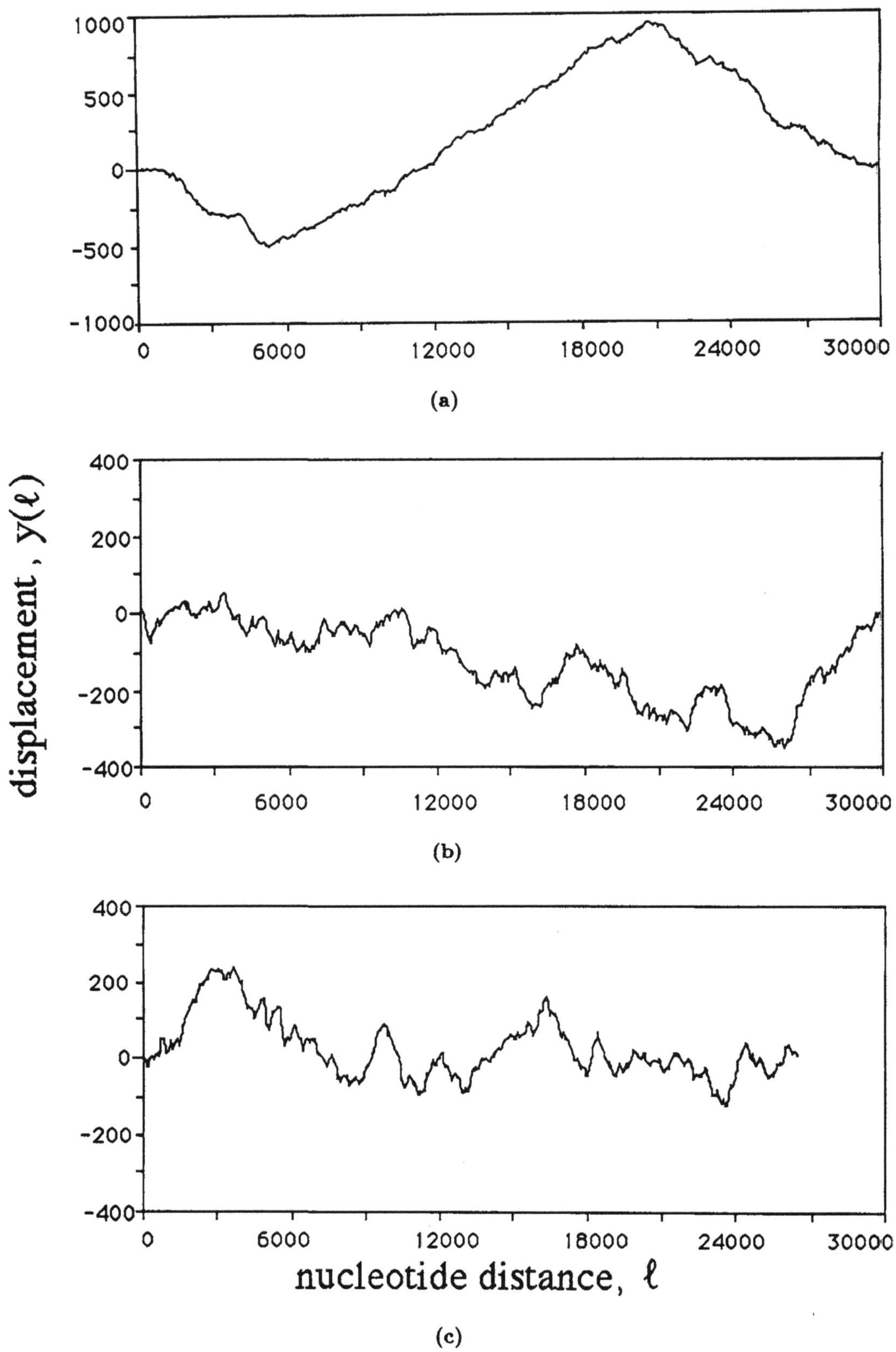

Fig. 7 DNA walk representations of artificial sequences generated by the stochastic model described in the text. The parameters used in this simulation are: $L = 30\,000$, $p_i = 0.2$, $L_o = 20$, and 62% purines in the initial sequence. (**a**) The early stage of "evolution" (after 100 iterations in the model simulation) shows a landscape with 2 to 3 large regions of different bias (the up-hill and down-hill regions); (**b**) After 1600 iterations, the landscape becomes visually more complex, resembling the actual DNA walk representation for the rat MHC sequence in (c). The values of p_i, L_o, and L given here are typical of those we used in our simulation. (After Ref. 41.)

primitive organisms such as phages (see Fig. 1 of Ref. 24) and *E. coli* which have only a small percentage of introns. The scaling behavior shows $F_d(\ell) \sim \ell^{1/2}$ as anticipated. After more iterations, the landscape seems visually more complex and the α measured for the first two decades increases, reminiscent of the increase noted in the MHC family from yeast to invertebrates to vertebrates. After roughly 1000 iterations, the value of α asymptotically approaches 0.60 as shown in Fig. 6(b). The landscapes for the model and for the rat MHC gene become quite similar [cf. Figs. 7(b) and 7(c)].

Of note, if the model is iterated without the insertion of random biased sequences as assumed in rule (iic), the value of α will return to 0.5, indicating a random sequence. Insertion of biased random regions (according to a power-law distribution) maintains the exponent $\alpha > 0.5$. The importance of rule (iic) of the model is consistent with the hypothesized role of retroviral insertions in the genomes of high animals.[52]

Furthermore, without strand substitution as implemented by rule (iib), no long-range correlation will appear. This mirror-image replacement mimics molecular evolution occurring by partial gene duplication or transposition[53] and the occurrence of "extinguished exons".[54] In order to test our assumption of strand substitution we also analyzed an alternative DNA landscape in which nucleotides cytosine (C) and guanine (G) result in an up step, while adenine (A) and thymine (T) correspond to a down step. Since such walks cannot be affected by strand substitution, our model would predict the absence of long-range correlations. Indeed, our analysis of the fluctuation $F_d(\ell)$ for this modified DNA landscape does not exhibit as robust a power-law correlation as for the original purine-pyrimidine rule. Another crucial assumption is the existence of an overall bias (either of purines or of pyrimidines) in the initial sequence; it is this bias that enables strand substitution to produce differences in nucleotide content. This assumption is consistent with our observation that most coding regions exhibit overall bias in their purine-pyrimidine concentration.

The mechanism of generating power-law correlations in this insertion-deletion model is related to the competition between two countervailing "forces". The deletion and insertion of segments in rule (iia) and (iib) tends to *randomize* the sequence, while the insertion of biased segment implemented by rule (iic) tends to *organize* the system. As the iteration proceeds, the newly inserted biased segment is then broken into smaller pieces of different bias (according to a power-law distribution). After a large (but finite) number of iterations (which depends on the parameters of the model), these two competing effects will tend to balance each other. At this point the system will exhibit power-law correlation.

The observed trend of α to increase with evolutionary status for the MHC family is also consistent with the predictions of the model: "higher" species that appeared more recently will tend to generate long-range correlations with a larger value of the parameter α. Thus, vertebrate myosin is likely to be more "complex" than invertebrate myosin because the former incorporated genetic material from the latter species. This view of molecular evolution is consistent with the theory of punctuated equilibrium[55] that postulates rather rapid periods of change (occurring during speciation) followed by periods of stasis.

The finding that α increases with evolution contradicts a recent study by Voss[27] which paradoxically reported that the strength of nucleotide correlations (quantified by the power spectral scaling exponent β, which is uniquely related to α) increases from organelle to invertebrates, but then *decreases* for primates. This apparent discrepancy is likely due to the fact that Voss[27] (i) did not analyze *single* gene families with evolution, (ii) did not distinguish intron-containing vs. intron-less sequences, and (iii) did not correct for large regions of "strand bias" (unequal numbers of purines and pyrimidines). We have found that if one does not take into account the *crossover* between two large (but uncorrelated) regions of strand bias as seen in all the MHC cDNAs [corresponding to the uphill and downhill regions in Fig. 1(a)], one can obtain a spuriously large value of α.

Nee[26] proposed that it is the alternation of introns and exons (regions containing different nucleotide content) which modulates the long-range correlations. This idea is somewhat similar to the proposed model, but its main conclusion — that the sequence from which all exons have been cut does not exhibit long-range correlations — appears to be incorrect. In fact, intron sequences show long-range correlations as robust as those of complete genes with approximately the same exponent α. In contrast, our model describes not only the intron-insertion process, but also the shuffling process within non-coding sequences (introns and intergenomic sequences). This shuffling process (not just the insertion of uncorrelated introns) leads to $\alpha > 0.5$ within single introns and intergenomic sequences, a fact that cannot be explained in the framework of the Nee hypothesis. Further, our model bears potential relevance to biological evolution, by providing a possible mechanism for transformation of primordial RNA molecules (currently considered to be the first to develop) into complex DNA sequences containing noncoding elements.

Finally, two major theories have been advanced to explain the origin and evolution of introns. One suggests that precursor genes consisted entirely of coding sequences and introns were inserted later in the course of evolution to help facilitate development of new structures in response to selective pressure, perhaps, by means of "exon shuffling".[56] The alternative theory suggests that precursor genes were highly segmented and subsequently organisms not requiring extensive adaptation or new development or, perhaps, facing the high energetic costs of replicating unnecessary sequences, lost their introns.[57,58] Support for these hypotheses has remained largely conjectural; no models have been brought forward to support either process. The landscape analysis of the MHC gene family and the stochastic model presented in this study are most consistent with the former view.

7. CONCLUDING REMARKS

What is the biological meaning of our finding? If two nucleotides whose positions differ by 10 000 were uncorrelated, then there might be no meaning. However, if they are correlated there must be a reason! The method we describe points out a new element of DNA structure and suggests a possible fundamental role for the non-coding regions (termed "introns"). Our work may also reveal an interesting feature of the coding regions (termed "exons").

Before concluding, we note that the long-range correlations in DNA sequences are of interest because they may be an indirect clue to its three-dimensional structure[36,40] or a reflection of certain scale-invariant properties of long polymer chains.[51] In any case, the statistically meaningful long-range "scale-invariant" (see Fig. 8) correlations in non-coding regions and their absence in coding regions will need to be accounted for by future explanations of global properties in gene organization and evolution.

It is awe-inspiring that remarkably complex objects in nature can be quantitatively characterized by a single number, α. It is equally awe-inspiring that such complex objects can be described by various models with extremely simple rules. Whether physicists can actually contribute to understanding of the principles on which biological fact is based, or whether we cannot, remains to be seen.

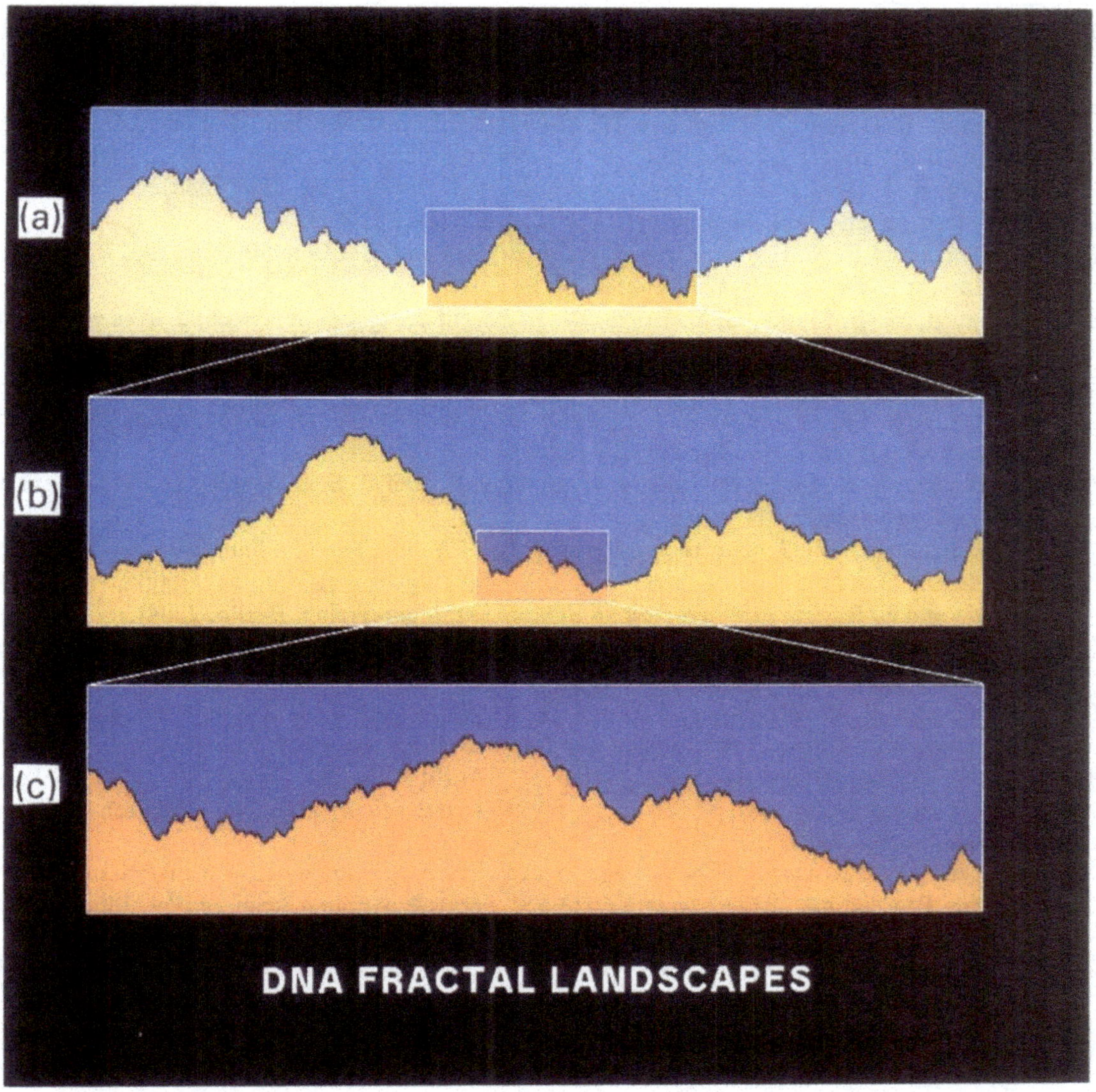

Fig. 8 The DNA walk representation for the rat embryonic skeletal myosin heavy chain gene ($\alpha = 0.63$). (a) The entire sequence; (b) The magnification of the solid box in (a); (c) The magnification of the solid box in (b). The statistical self-similarity of these plots is consistent with the existence of a scale-free or fractal phenomenon which we call a fractal landscape. Note that one must magnify the segment by different factors along the ℓ (horizontal) direction and the y (vertical) direction; since F has the same units (dimension) as y, these magnification factors M_ℓ and M_y (along ℓ and y directions respectively) are related to the scaling exponent α by the simple relation $\alpha = \log(M_y)/\log(M_\ell)$ [e.g. from (a) to (b), $\log(M_y)/\log(M_\ell) = \log(2.07)/\log(3.2) \cong 0.63$].

ACKNOWLEDGMENTS

We wish to thank F. Sciortino for important contributions in the initial stages of this project, and C. Cantor, C. DeLisi, J. M. Hausdorff, L. Liebovitch, R. D. Rosenberg, J. J. Schwartz, M. Schwartz, M. F. Shlesinger, and N. Shworak for valuable discussions. Partial support was provided to ALG by the G. Harold and Leila Y. Mathers Charitable Foundation, the National Heart, Lung and Blood Institute and the National Aeronautics and Space Administration, to C.-K.Peng by an NIH/NIMH Postdoctoral NRSA Fellowship, to M. Simons by the American Heart Association, and to HES by the National Science Foundation.

REFERENCES

1. B. J. West and W. Deering, *Phys. Reports* **00**, 000 (1994); B. J. West, *Fractal Physiology and Chaos in Medicine* (World Scientific, Singapore, 1990); A. L. Goldberger, D. R. Rigney and B. J. West, *Sci. Am.* **262**, 42 (1990); B. J. West and M. F. Shlesinger, *Am. Sci.* **78**, 40 (1990); B. J. West and A. L. Goldberger, *Am. Sci.* **75**, 354 (1987); L. S. Liebovitch, J. Freichbarg and J. P. Koniarek, *Math. Biosci.* **89**, 36 (1987); A. L. Goldberger and B. J. West, *Yale J. Biol. Med.* **60**, 421 (1987); B. J. West and A. L. Goldberger, *J. Appl. Physiol.*, **60**, 189 (1986).
2. M. F. Shlesinger and B. J. West, *Phys. Rev. Lett.* **67**, 2106 (1991).
3. A. A. Tsonis and P. A. Tsonis, *Perspectives in Biology and Medicine* **30**, 355 (1987); F. Family, B. R. Masters and D. E. Platt, *Physica* **D38**, 98 (1989).
4. T. G. Smith, W. B. Marks, G. D. Lange, W. H. Sheriff Jr. and E. A. Neale, *J. Neurosci. Meth.* **27**, 173–180 (1989).
5. F. Caserta, H. E. Stanley, W. D. Eldred, G. Daccord, R. E. Hausman and J. Nittmann, *Phys. Rev. Lett.* **64**, 95 (1990); F. Caserta, R. E. Hausman, W. D. Eldred, H. E. Stanley and C. Kimmel, *Neurosci. Lett.* **136**, 198–202 (1992).
6. K. R. Bhaskar, B. S. Turner P. Garik, J. D. Bradley, R. Bansil, H. E. Stanley and J. T. LaMont, *Nature* **360**, 458 (1992).
7. B. B. Mandelbrot, *The Fractal Geometry of Nature* (W. H. Freeman, San Francisco, 1982).
8. A. Bunde and S. Havlin, eds., *Fractals and Disordered Systems* (Springer-Verlag, Berlin, 1991); A. Bunde and S. Havlin, eds., *Fractals in Science* (Springer-Verlag, Berlin, 1994).
9. D. Stauffer and H. E. Stanley, *From Newton to Mandelbrot: A Primer in Theoretical Physics* (Springer-Verlag, Heidelberg and N.Y., 1990).
10. E. Guyon and H. E. Stanley, *Les Formes Fractales* (Palais de la Découverte, Paris, 1991); English translation: *Fractal Forms* (Elsevier North-Holland, Amsterdam, 1991).
11. T. Vicsek, *Fractal Growth Phenomena*, 2nd ed. (World Scientific, Singapore, 1992); J. Feder, *Fractals* (Plenum, NY, 1988); H. Takayasu, *Fractals in the Physical Sciences* (Manchester Univ. Press, Manchester, 1990).
12. H. E. Stanley and N. Ostrowsky, eds., *On Growth and Form: Fractal and Non-Fractal Patterns in Physics*, Proceedings 1985 Cargèse NATO ASI, Series E: Applied Sciences, Vol. 100 (Martinus Nijhoff, Dordrecht, 1985).
13. H. E. Stanley, *Introduction to Phase Transitions and Critical Phenomena* (Oxford University Press, London, 1971).
14. H. E. Stanley and N. Ostrowsky, eds., *Correlations and Connectivity: Geometric Aspects of Physics, Chemistry and Biology*, Proceedings 1990 Cargèse Nato ASI, Series E: Applied Sciences, Vol. 188 (Kluwer, Dordrecht, 1990).
15. F. Zernike, *Physica* **7**, 565 (1940).
16. P. Bak and M. Creutz, in *Fractals in Science*, eds. A. Bunde and S. Havlin (Springer-Verlag, Berlin, 1994).
17. H. E. Stanley and N. Ostrowsky, eds., *Random Fluctuations and Pattern Growth: Experiments and Models*, Proceedings 1988 NATO ASI, Cargèse (Kluwer Academic Publishers, Dordrecht, 1988).
18. S. Havlin, R. Selinger, M. Schwartz, H. E. Stanley and A. Bunde *Phys. Rev. Lett.* **61**, 1438 (1988); S. Havlin, M. Schwartz, R. B. Selinger, A. Bunde and H. E. Stanley, *Phys. Rev.* **A40**, 1717 (1989); R. B. Selinger, S. Havlin, F. Leyvraz, M. Schwartz and H. E. Stanley, *Phys. Rev.* **A40**, 6755 (1989).
19. C.-K. Peng, S. Havlin, M. Schwartz, H. E. Stanley and G. H. Weiss, *Physica* **A178**, 401 (1991); C.-K. Peng, S. Havlin, M. Schwartz and H. E. Stanley, *Phys. Rev.* **A44**, 2239 (1991).
20. M. Araujo, S. Havlin, G. H. Weiss and H. E. Stanley, *Phys. Rev.* **A43**, 5207 (1991); S. Havlin, S. V. Buldyrev, H. E. Stanley and G. H. Weiss, *J. Phys.* **A24**, L925 (1991); S. Prakash, S. Havlin, M. Schwartz and H. E. Stanley, *Phys. Rev.* **A46**, R1724 (1992).
21. C.-K. Peng, J. Mietus, J. Hausdorff, S. Havlin, H. E. Stanley and A. L. Goldberger, *Phys. Rev. Lett.* **70**, 1343 (1993); C. K. Peng, S. V. Buldyrev, J. M. Hausdorff, S. Havlin, J. E. Mietus, M. Simons, H. E. Stanley and A. L. Goldberger, in *Fractals in Biology and Medicine*, eds. G. A. Losa, T. F. Nonnenmacher and E. R. Weibel (Birkhauser Verlag, Boston, 1994).
22. A. Schenkel, J. Zhang and Y.-C. Zhang, *Fractals* **1**, 47 (1993).
23. R. N. Mantegna, *Physica* **A179**, 23 (1991).

24. C.-K. Peng, S. V. Buldyrev, A. L. Goldberger, S. Havlin, F. Sciortino, M. Simons and H. E. Stanley, *Nature* **356**, 168 (1992).
25. W. Li and K. Kaneko, *Europhys. Lett.* **17**, 655 (1992).
26. S. Nee, *Nature* **357**, 450 (1992).
27. R. Voss, *Phys. Rev. Lett.* **68**, 3805 (1992).
28. J. Maddox, *Nature* **358** 103 (1992).
29. P. J. Munson, R. C. Taylor and G. S. Michaels, *Nature* **360**, 636 (1992).
30. I. Amato, *Science* **257**, 747(1992).
31. V. V. Prabhu and J.-M. Claverie, *Nature* **357**, 782 (1992).
32. P. Yam, *Sci. Am.* **267**(3), 23 (Sept. 1992).
33. C.-K. Peng, S. V. Buldyrev, A. L. Goldberger, S. Havlin, F. Sciortino, M. Simons and H. E. Stanley, *Physica* **A191**, 25 (1992).
34. H. E. Stanley, S. V. Buldyrev, A. L. Goldberger, J. M. Hausdorff, S. Havlin, J. Mietus, C.-K. Peng, F. Sciortino and M. Simons, *Physica* **A191**, 1 (1992).
35. C. A. Chatzidimitriou-Dreismann and D. Larhammar, *Nature* **361**, 212 (1993).
36. A. Yu. Grosberg, Y. Rabin, S. Havlin and A. Neer, *Biofisika (Russia)* **26**, 1 (1993); *Europhys. Lett.* **23**, 373 (1993).
37. S. Karlin and V. Brendel, *Science* **259**, 677 (1993).
38. C.-K. Peng, S. V. Buldyrev, A. L. Goldberger, S. Havlin, M. Simons and H. E. Stanley, *Phys. Rev.* **E47**, 3730 (1993).
39. S. V. Buldyrev, A. L. Goldberger, S. Havlin, C.-K. Peng, M. Simons and H. E. Stanley, *Phys. Rev.* **E47**, 4514 (1993).
40. A. S. Borovik, A. Yu. Grosberg and M. D. Frank-Kamenetskii, *Nature* (in press).
41. S. V. Buldyrev, A. L. Goldberger, S. Havlin, C.-K. Peng, H. E. Stanley, M. H. R. Stanley and M. Simons, "Fractal Landscapes and Molecular Evolution: Modeling the Myosin Heavy Chain Gene Family," *Biophys. J.* **65**, 2675–2681 (1993).
42. C.-K. Peng, S. V. Buldyrev, S. Havlin, M. Simons, H. E. Stanley and A. L. Goldberger, *Phys. Rev.* **E49** 000 (1994).
43. S. M. Ossadnik, S. V. Buldyrev, A. L. Goldberger, S. Havlin, C.-K. Peng, M. Simons and H. E. Stanley (submitted).
44. E. W. Montroll and M. F. Shlesinger, in *Nonequilibrium Phenomena II. From Stochastics to Hydrodynamics*, eds. J. L. Lebowitz and E. W. Montroll (North-Holland, Amsterdam, 1984), pp. 1–121.
45. S. G. Oliver et al., *Nature* **357**, 38 (1992).
46. E. C. Uberbacher and R. J. Mural, *Proc. Natl. Acad. Sci. USA* **88**, 11261 (1991).
47. In the power spectrum analysis for those sequences containing non-coding regions, subtracting of the white noise, $S(\infty)$, as performed in Ref. 27, gives more weight to the non-coding segments (correlated) than the coding segments (uncorrelated). See also H. E. Stanley, S. V. Buldyrev, A. L. Goldberger, S. Havlin, C.-K. Peng and M. Simons, *Physica* **A200**, 4 (1993) and refs. therein.
48. S. V. Buldyrev, A. Goldberger, S. Havlin, C.-K. Peng, F. Sciortino, M. Simons and H. E. Stanley, *Phys. Rev. Lett.* **71**, 1776 (1993).
49. W.-H. Li and D. Graur, *Fundamentals of Molecular Evolution* (Sinauer Associates, Sunderland MA, 1991).
50. G. F. Joyce, *Nature* **338**, 217 (1989).
51. J. des Cloizeaux, *J. Physique (Paris).* **41**, 223 (1980); P. G. de Gennes, *Scaling Concepts in Polymer Physics* (Cornell Univ. Press, Ithaca NY, 1979).
52. R. H. Hwu, J. W. Roberts, E. H. Davidson and R. J. Britten, *Proc. Natl. Acad. Sci. USA.* **83**, 3875 (1986).
53. R. Schleif, *Science* **240**, 127 (1988).
54. C. J. Jaworski and J. Piatigorsky, *Nature* **337**, 752 (1989).
55. N. Eldredge and S. J. Gould, in *Models in paleobiology*, ed. T. J. M. Schopf, (Freeman and Cooper Inc., San Francisco, 1972), p. 82.
56. W. Gilbert, *Nature* **271**, 501 (1978).
57. P. Hagerman, *Ann. Rev. Biochem.* **59**, 755 (1990).
58. W. F. Doolitle, in *Intervening Sequences in Evolution and Development*, eds. E. Stone and R. Schwartz, (Oxford University Press, NY, 1990), pp. 42–62.

Fractals, Vol. 1, No. 3 (1993) 302–311

SELF-SIMILAR COLONY MORPHOGENESIS BY BACTERIA AS THE EXPERIMENTAL MODEL OF FRACTAL GROWTH BY A CELL POPULATION

TOHEY MATSUYAMA
Department of Bacteriology, Niigata University School of Medicine, Niigata 951, Japan

RASIKA M. HARSHEY
Department of Microbiology, The University of Texas at Austin, Austin, Texas 78712-1095, USA

MITSUGU MATSUSHITA
Department of Physics, Chuo University, Bunkyo-ku, Tokyo 112, Japan

Abstract

Bacteria point inoculated onto the solid medium exert two-dimensional spreading growth on the surface. The colony morphologies are rich in diversity and depend on various biological and physico-chemical factors. Since it is easy to pursue the growth processes macro- and microscopically and to carry out experiments by controlling specific factors, precise experimental approaches to morphogenesis by living cells have now become possible. Several critical factors for self-similar morphogenesis were identified.

1. INTRODUCTION

As established by R. Koch,[1] preparation of a bacterial pure culture and examination of its purity is based on the discrimination of different bacterial colonies on a solid medium. Thus, colony morphologies have been important in the technology of bacteriology.

Recently, however, the colony itself has been receiving scientific attention because of its pattern diversity[2–8] and the multicellular behavior of the constituent bacterial cells.[8–13] Now the bacterial colony is recognized as the simplest experimental system for the analysis of morphogenesis by living cells. In this paper, we review various experimental approaches valuable to elucidate how specific patterns (e.g. self-similar fractal) are being generated by interactions of physico-chemical and biological factors.

2. EXPERIMENTS

2.1 Macroscopic Examinations

Most fractal structures made by living cells such as bronchial tree are difficult to follow in their growth processes. On the other hand, bacteria point-inoculated onto the surface of an agar medium in a plastic dish develop two-dimensional spreading patterns which are observable anytime without disturbing the growth processes. Figures 1 and 2 show such growing fractal patterns recorded at different incubation times. Some growth characteristics (e.g. screening effect by extending protrusions) are evident by comparing Figs. 1(a) and 1(b).

It is also easy to prepare different cultural conditions and examine the effects of specific factors reproducibly. The developing patterns are quite different between Figs. 1 and 2, but they were made by the same strain of *Salmonella typhimurium*. Thus, we can specify important factors (concentration of agar and glucose, in this case), and analyze the pattern generation mechanisms by comparing the patterns.

Depending on the species of organisms, there are so many biological morphologies in nature. It is evident that genetic information codes various living morphologies. For the identification of such biological factors, comparative analyzes of isogenic mutants are quite useful. In this respect, bacteria are convenient organisms. It is easy to obtain mutants

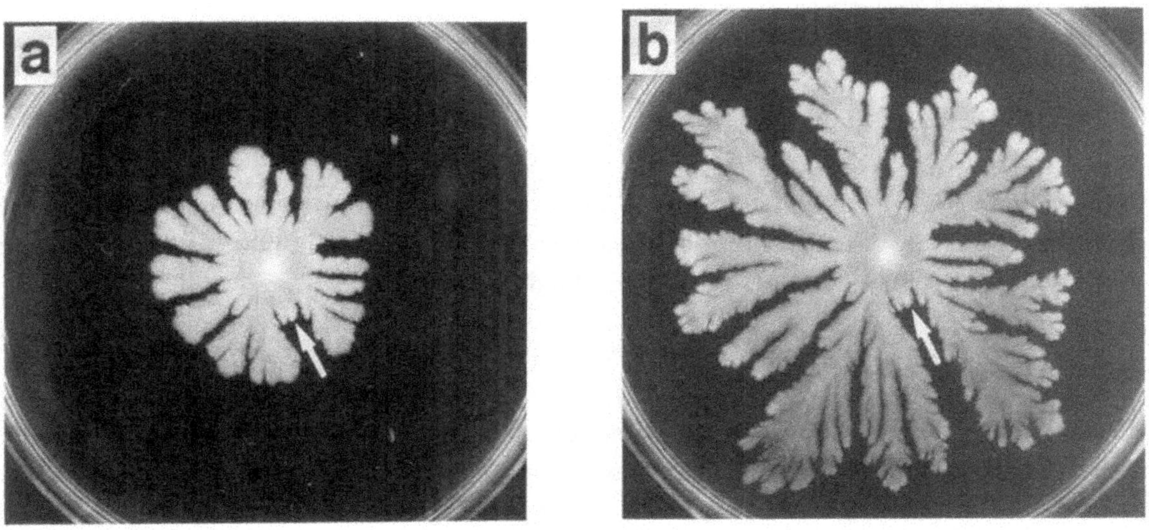

Fig. 1 A colony of *Salmonella typhimurium* 14028-2b growing on a Vogel-Bonner medium containing 1.4% agar and 0.4% glucose. Screening effects are occurring everywhere. A representative one is shown by an arrow. (a) A one-week culture at 37°C. (b) A two-week culture at 37°C.

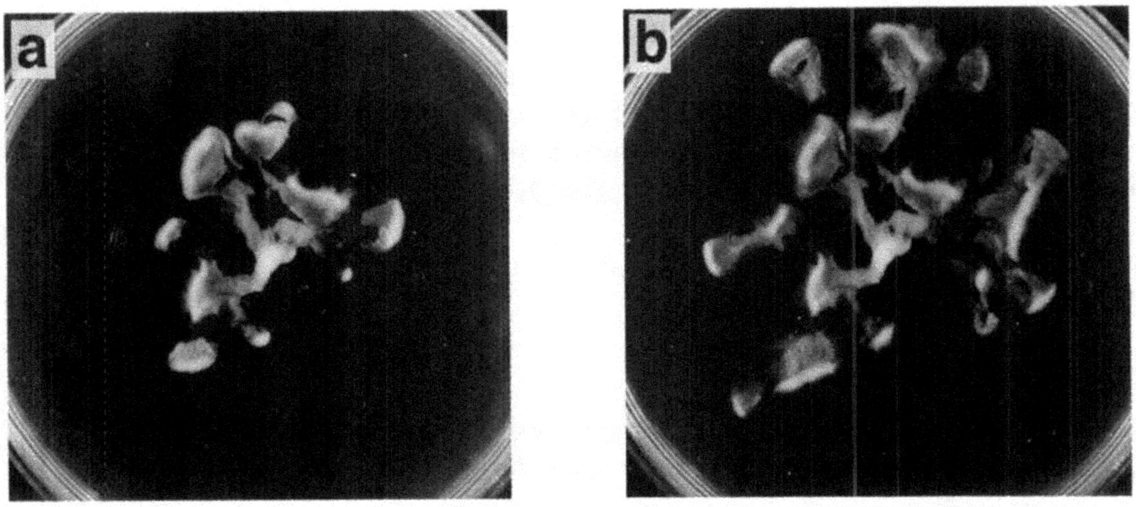

Fig. 2 A colony of *Salmonella typhimurium* 14028-2b growing on a Vogel-Bonner medium containing 1.0% agar and 0.8% glucose. (a) A 36-hour culture at 37°C. (b) A 48-hour culture at 37°C.

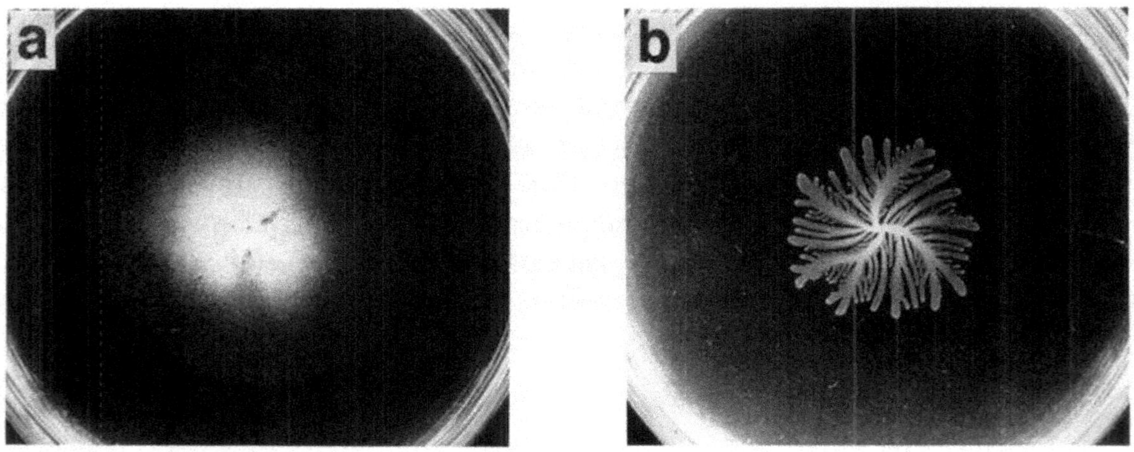

Fig. 3 A colony of ***Bacillus subtilis*** grown on a semi-solid nutrient agar after incubation for 14 hours at 37°C. (a) A wild type strain ATCC 21331. (b) A mutant 21331 b3 (defective in production of surfactin).

and examine their phenotypes. In Fig. 3, an example of such an outcome of ***Bacillus subtilis*** genotype is shown. A defect in production of some kind of exolipid induces drastic changes on the mutant colony (Fig. 3). Details will be described later.

2.2 Referring to a Model

Fractal patterns shown in Fig. 1, are common among many species of bacteria.[6] So, some general principles independent of species-specific factors seem to be operating. Experiments with ***B. subtilis*** on agar media with varying concentration of nutrients[5,7] demonstrated that low concentration of nutrients is critical for such self-similar morphogenesis. A two-dimensional field of nutrient diffusion was also shown to be necessary.[4–8] The growth rate of the colony was very slow and competitive screening effects among protrusions

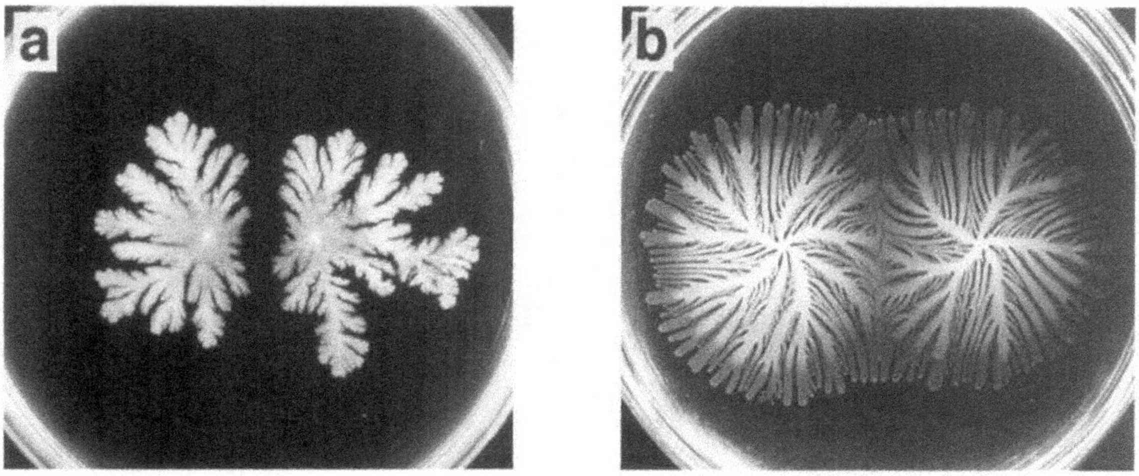

Fig. 4 Two neighboring colonies on a agar medium. (a) Two colonies of *S. typhimurium* ATCC 14028 on a Vogel-Bonner medium incubated at 37°C for 15 days. A repulsion effect is evident. (b) Two colonies of *B. subtilis* 21331-b3 on a semi-solid nutrient agar incubated at 37°C for 16 hours.

were recognized as shown in Fig. 1. Thus, diffusion-limited aggregation (DLA) model[14] was suggested as a compatible analogy for the basic processes. Fractal morphologies and dimensions (calculated by the box-counting method[6]) of such colonies were similar and close to those of the DLA model.[4–6] If the morphogenic growth of such colonies are progressing by the same principles of the DLA model, a repulsion between two neighboring colonies should occur as predicted by Meakin.[15] By parallel inoculations of *S. typhimurium* on a solid agar plate, the expected repulsion effect is clearly shown to occur [Fig. 4(a)]. In contrast, after parallel inoculation of motile bacteria on a semi-solid agar, such repulsive effect is absent and colonies are colliding with each other as shown in Fig. 4(b). So, distinct self-similar branching by a *B. subtilis* cell population on semi-solid agar seems to be progressing by mechanisms different from the DLA processes.

2.3 Microscopic Examinations

The smallest morphogenic unit of a bacterial colony is a single cell organism "bacterium". In contrast to invisible molecules in crystal growth, pleomorphic shapes and specific movements of the units are easily visible using a microscope connected to a video-recorder. In addition, the examination induces no destruction nor disturbance of a growing pattern.

In Fig. 5, a high magnification picture of an extending branch end of Fig. 3(a) is shown. Bacteria of wild type *B. subtilis* on a semi-solid medium are forming a very thin branch permitting successive new small shootings. The mutant shown in Fig. 3(b), on the other hand, is developing a thick branch (Fig. 6, a low magnification picture for overview of a whole branch end) with a rigid shell composed of immotile cells (clear figures at the margin of a branch); active swirlings of cells (blurred figures) are prominent at the inside of the branch end. The area just behind the tip is a special point crowded with actively moving cells. Densities of active and inactive cells gradually decrease toward the root of the branch. A schematic structure of such a branch is given in Fig. 7. By keeping such a structured

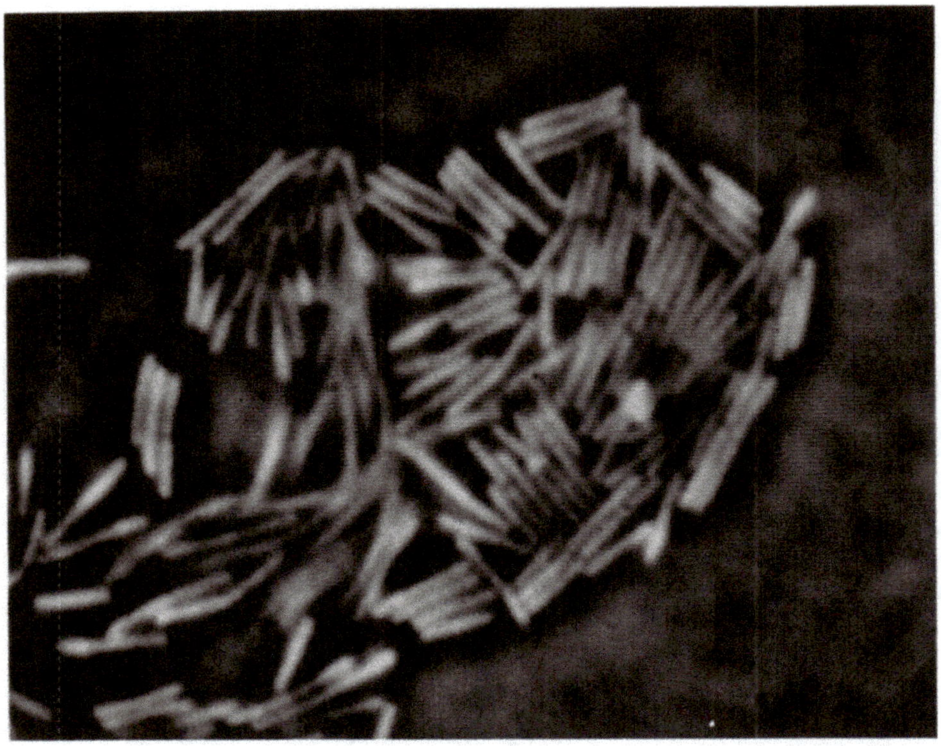

Fig. 5 A high magnification micrograph of the tip of a growing thin branch shown in Fig. 3(a).

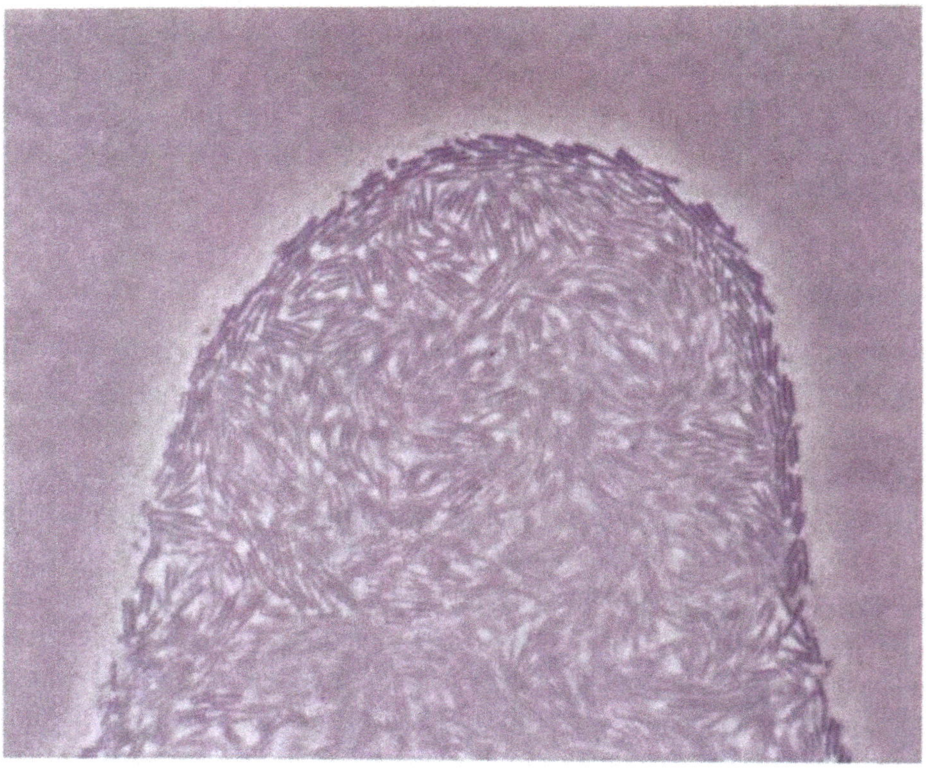

Fig. 6 A low magnification micrograph of the tip of an extending thick branch (turning right) shown in Fig. 3(b). Groups of bacteria are swirling clockwise and counterclockwise at random.

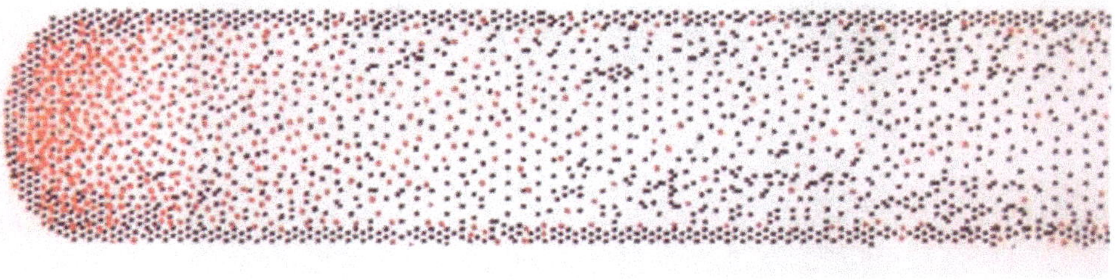

Fig. 7 A schematic figure of an extending branch. Red spots indicate actively moving bacteria. Black spots indicate inactive bacteria. Just behind the front end, there is a special area rich in active bacteria.

distribution of bacteria, the bacteria population branch is shown to extend outward and turning right almost exclusively.

2.4 Identification of Specific Bacterial Products Influencing Morphogenesis

In Fig. 8, pattern changes observed between *Serratia marcescens* wild type and its mutant strain are shown. The change is quite similar to that observed in the colonies of *B. subtilis* (Fig. 3). Since these mutants of *B. subtilis* and *S. marcescens* are both defective in the production of specific surface active exolipids (surfactin produced by *B. subtilis* and serrawettins produced by *S. marcescens*), these bacterial lipids seemed to be responsible for the drastic pattern change.[8,16] So, external effects of the purified ex-

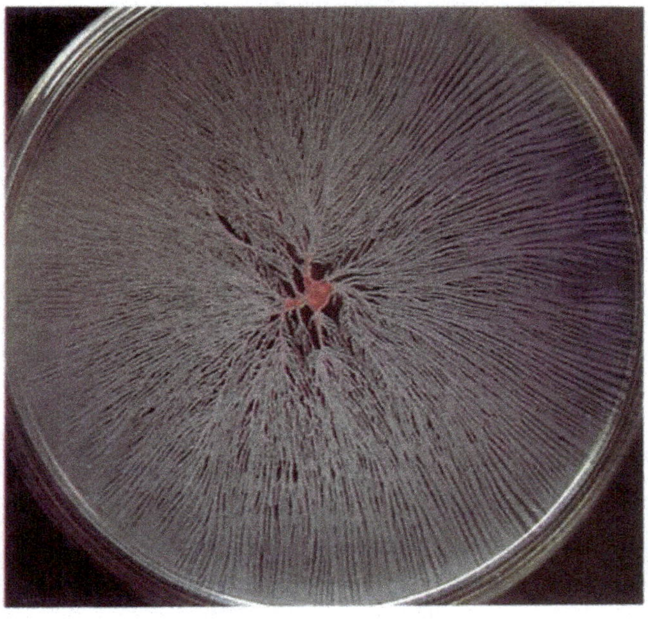

Fig. 8 A giant colony of *Serratia marcescens* grown on semi-solid nutrient agar after incubation at 30°C for 24 hours. (a) A wild type NS 38 strain.

Fig. 8(b) A serrawettin-less mutant NS 38-09.

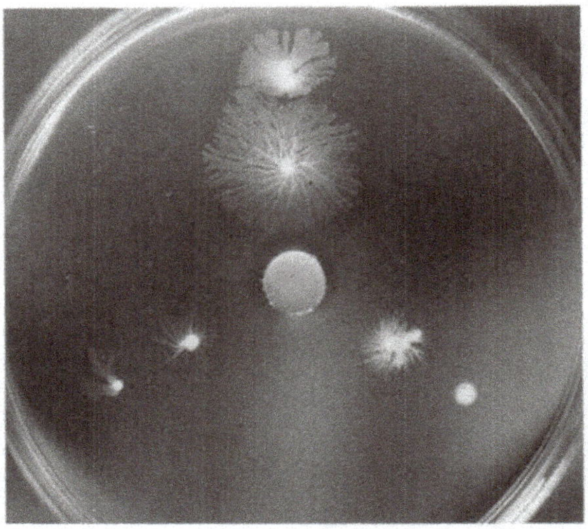

Fig. 9 Effect of serrawettin W2 on the spreading pattern of serrawettin-less mutants (Top, NS 38-09; right, NS 45-09; left, NS 25-04) growing on semi-solid nutrient agar at 30°C for 6 hours. A paper disk at the center contains 100 μg serrawettin W2. Colonies near to the disk are receiving the effect of serrawettin W2 and extending thin branches.

olipid "serrawettin W2" (D-3-hydroxydecanoyl-D-leucyl-L-seryl-L-threonyl-D-phenylalanyl-L-isoleucyl lactone)[16,17] were examined. The results (Fig. 9) clearly indicate a specific function for the exolipid in the morphogenesis by the bacteria on the agar surface. Surface active exolipid isolated from different species of bacteria (rhamnolipid from *Pseudomonas aeruginosa*) demonstrated the same effects (Fig. 10). Thus, the surface tension working on the bacterial population mass may have important influence on the morphogenic behavior of bacteria.

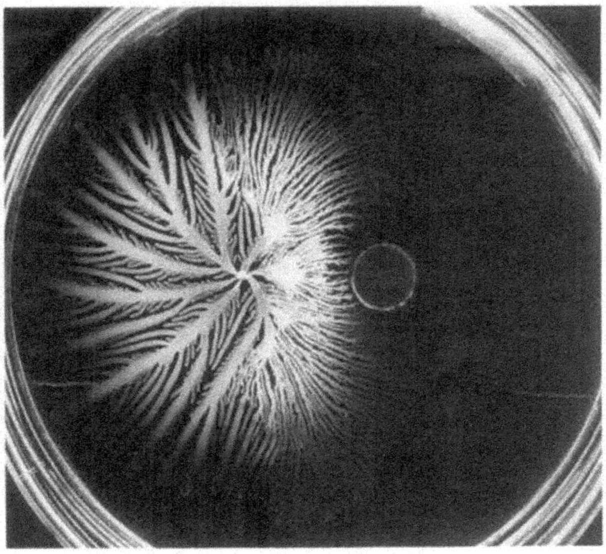

Fig. 10 A colony of *Bacillus subtilis* 21331 b3 under the influence of rhamnolipid (100 μg, in a right pap disk) on a semi-solid nutrient agar incubated at 37°C for 24 hours.

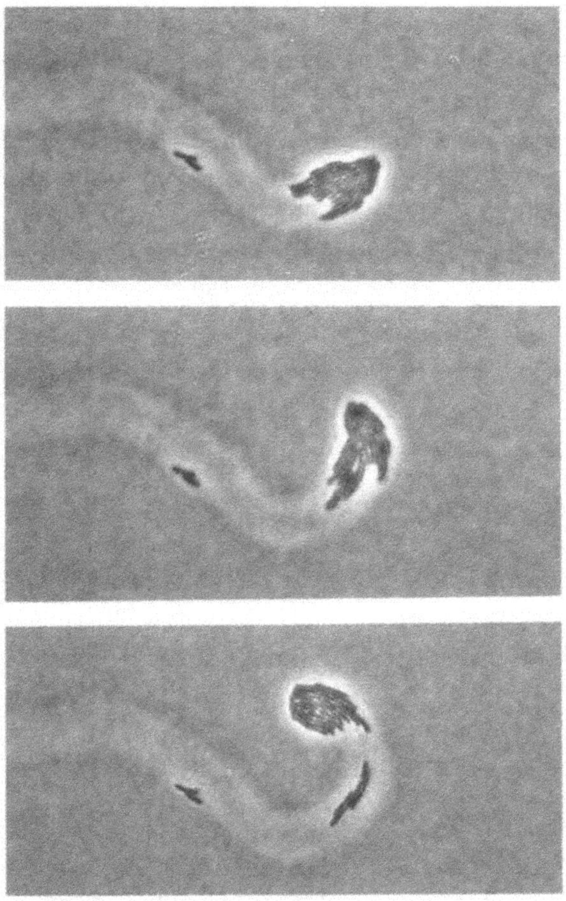

Fig. 11 Photomicrographs of a bacterial cell group translocating on a nutrient agar. Scenes are recorded at 1 min intervals. *Bacillus sp.* isolated from the mouth of a healthy adult was point inoculated and incubated at 37°C.

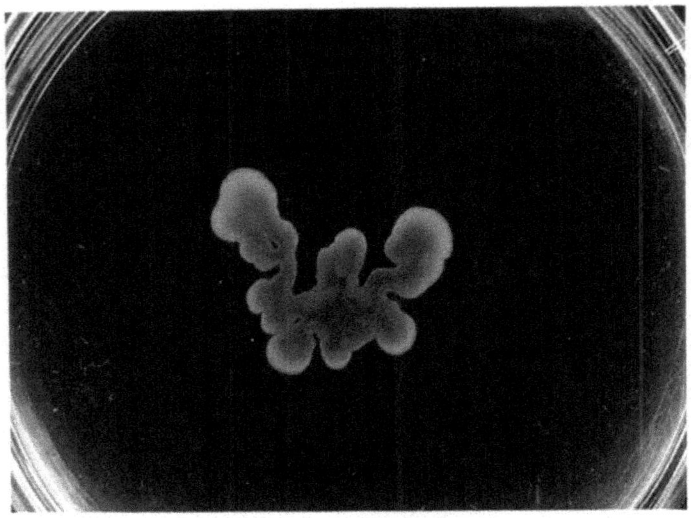

Fig. 12 A colony of *Serratia marcescens* Che^- mutant SMu 3262 grown on LB medium containing 0.75% agar after 12-hour incubation at 30°C.

3. PERSPECTIVES

Although a bacterium is a single celled organism, bacteria composing a colony are not homogenous in morphology and functional activity. Some multicellular behaviors of bacterial cells in a colony have been reported.[8–13,18,19] In Fig. 11, a coordinated translocation of bacterial cells is clearly observable. Communication and cooperation among cells are expected to be at work during growth and spreading of a colony population. Some characteristic colony morphologies may be the outcome of such biological activities. A colony in Fig. 12 is formed by Che^- mutant of *S. marcescens* which has functioning flagella but translocates erratically presumably due to defects in the system that senses a chemicals gradients and transmits signals.[20] In comparison with this colony morphology of Che^- mutant, giant colonies in Fig. 8 seem to indicate the presence of specialized processes to be examined from the viewpoint of biology and of physical sciences.

ACKNOWLEDGMENTS

We thank M. Ohgiwari and J. Wakita for their assistance in experiments.

This work was supported in part by Grants-in Aid for Scientific Research from the Ministry of Education, Science and Culture of Japan (No. 02808026, 02804019 and 04452053).

REFERENCES

1. T. D. Brock, *Robert Koch, A Life in Medicine and Bacteriology* (Science Tech Publishers and Springer-Verlag, New York, 1988).
2. A. L. Cooper, A. C. R. Dean and C. Hinshelwood, *Proc. Roy. Soc.* **B171**, 175 (1968).
3. T. Matsuyama, M. Sogawa and Y. Nakagawa, *FEMS Microbiol. Lett.* **61**, 243 (1989).
4. H. Fujikawa and M. Matsushita, *J. Phys. Soc. Jpn.* **58**, 3875 (1989).
5. M. Matsushita and H. Fujikawa, *Physica* **A168**, 498 (1990).
6. T. Matsuyama and M. Matsushita, *Appl. Environ. Microbiol.* **58**, 1227 (1992).

7. M. Ohgiwari, M. Matsushita and T. Matsuyama, *J. Phys. Soc. Jpn.* **61**, 816 (1992).
8. T. Matsuyama and M. Matsushita, *CRC Crit. Rev. Microbiol.* **19**, 117 (1993).
9. J. A. Shapiro and C. Hsu, *J. Bacteriol.* **171**, 5963 (1989).
10. J. A. Shapiro, *ASM News* **57**, 247 (1991).
11. J. A. Shapiro and D. Trubatch, *Physica* **D49**, 214 (1991).
12. B. J. Budrene and H. C. Berg, *Nature* **349**, 630 (1991).
13. J. A. Shapiro, *J. Bacteriol.* **174**, 7262 (1992).
14. T. A. Witten and L. M. Sander, *Phys. Rev. Lett.* **47**, 1400 (1981).
15. P. Meakin, in *Phase Transition and Critical Phenomena*, eds. C. Domb and J. L. Levowitz (Academic Press, New York, 1988).
16. T. Matsuyama, K. Kaneda, Y. Nakagawa, K. Isa, H. Hara-Hotta and I. Yano, *J. Bacteriol.* **174**, 1769 (1992).
17. Y. Nakagawa and T. Matsuyama, *FEMS Microbiol. Lett.* **108**, 99 (1993).
18. T. Muto, *Centr. Bakt. Orig.* **37**, 321 (1904).
19. R. B. Morrison and A. Scott, *Nature* **211**, 255 (1966).
20. J. O'Rear, L. Alberti and R. M. Harshey, *J. Bacteriol.* **174**, 6125 (1992).

Fractals, Vol. 1, No. 3 (1993) 312–320

DEMONSTRATION OF $1/f$ FLUCTUATIONS AND WHITE NOISE IN THE HUMAN HEART RATE BY THE VARIANCE–TIME–CURVE: IMPLICATIONS FOR SELF-SIMILARITY

MALTE MEESMANN, JAN BOESE, DANTE R. CHIALVO*,
PETER KOWALLIK, WOLFGANG R. BAUER, WERNER PETERS,
FERDINAND GRÜNEIS and KLAUS-DIETRICH KNIFFKI†
Medizinische Klinik und
†*Physiologisches Institut der Universität Würzburg, FRG*
**Santa Fe Institute, New Mexico, USA*

Abstract

Spectral analysis of heart rate variability is usually performed by Fast Fourier Transform. Here we demonstrate the self-affine properties of the human heart rate using a spectral analysis based on counting statistics. Each QRS complex is considered to be a point event and from the number of events $N(\Delta t)$ in consecutive time windows Δt the variance is calculated. From the finding that the variance of $N(\Delta t)$ follows a power law proportional to $(\Delta t)^{1+b}$ in case of $1/f^b$ noise, it is shown that the variance of the heart rate as determined for windows of length Δt, i.e., $N(\Delta t)/\Delta t$, is proportional to $(\Delta t)^{b-1}$. From a 12-day Holter recording, the scaling region could be determined to cover 0.16 to 0.000136 Hz.

Address for correspondence:
M. Meesmann, MD
Medizinische Universitätsklinik Würzburg
Josef-Schneider-Str. 2
D-97080 Würzburg, FRG
Phone: 49-931-201-2774
Fax: 49-931-201-3453
E-mail: medk234@rzbox.uni-wuerzburg.de

A function $X(t)$ is self-affine if $X(t)$ and $X(rt)/r^H$ have the same distribution functions. From the variance-time-curve, it can be shown that the exponent H is dependent on b with $b = 2H - 1$. In young healthy men, the parameter b fluctuates between 0.2 and 1.0 during 24 h and thus determines the self-affine scaling factor $H = (b-1)/2$ for the amplitude of heart rate, if the time axis is scaled by r. Thus, during periods of $1/f$ noise, the heart rate scales with $H = 0$, and for periods of almost white noise, with H close to $-\frac{1}{2}$.

1. INTRODUCTION

Traditionally, analysis of heart rate variability is performed by Fast Fourier Transform. This approach, however, requires substantial preprocessing of the data. This includes transformation of the point process (occurrence of QRS complexes) into a continuous signal, resampling to obtain equidistant samples and removal of extrasystoles.[1-3] Using this method, a few reports have indicated $1/f$ fluctuations over 10^{-1} to 10^{-4} Hz.[4-6] Saul et al. pointed to the "self-similarity" of heart rate data without giving detailed statistics.[5]

In this paper, we extend our observations with the variance-time-curve introduced recently for the spectral analysis of heart rate variability.[7] In particular, we analyze the self-similar and self-affine properties of time series obtained from the coupling intervals between heart beats from healthy male volunteers.

2. METHODS

2.1 Recording

Twenty-four hour Holter recordings were obtained from 10 healthy male volunteers (age 24 to 37 years). To exactly define the scaling region, especially its lower border, a 12-day Holter record was obtained from one of the authors (M. M.). The recordings were performed using the ELATEC system (ELA medical GmbH, München) which allows for recognition of QRS-complexes with an accuracy of ± 5 ms.

2.2 Data Processing

For further processing, the data were exported to ASCII-files. Due to selection of high quality recordings, there were no periods of artifacts longer than 10 s. These were substituted by coupling intervals with values similar to the ones preceding and following the artifacts by manual editing (linear splining). In the 12-day Holter recording, there were three periods of some 10 min during which no recording could be obtained due to repositioning of the electrodes and these periods were substituted with coupling intervals with mean and variance similar to neighboring values. To avoid possible distortion of low frequency fluctuations, no recording periods were discarded. There were less than 100 ventricular extrasystoles per 24 h and these were not removed, since the counting statistics is not influenced by infrequent extrasystoles.[7] All calculations were performing using programs written in BorlandC++ for OS/2 (Borland International).

2.3 Variance-Time-Curve

Each QRS complex is considered to be a point event. The number of events $N(\Delta t_i)$ is determined for consecutive windows i of length Δt. From these numbers, the mean

$$\bar{N}(\Delta t) = \frac{\sum_i N(\Delta t_i)}{n}$$

and the variance

$$\mathrm{Var}\{N(\Delta t)\} = \frac{\sum_i [N(\Delta t_i) - \bar{N}(\Delta t)]^2}{n-1}$$

are estimated. This procedure is repeated for increasing Δt. The minimum number of windows used for calculation of variance was 5, so the maximum Δt was one fifth of the total record length. The plot of $\mathrm{Var}\{N(\Delta t)\}$ vs. Δt is called the variance-time-curve (VTC). As can be seen in Fig. 1, there is a linear relation between $\log[\mathrm{Var}\{N(\Delta t)\}]$ vs. $\log[\Delta t]$ suggesting that:

$$\mathrm{Var}\{N(\Delta t)\} \propto (\Delta t)^{1+b} . \tag{1}$$

As has previously been demonstrated,[7,8] Eq. (1) corresponds to a spectrum:

$$S(f) \propto \frac{1}{f^b}, \tag{2}$$

if the scaling region comprises more than 3 decades and $b < 1$.

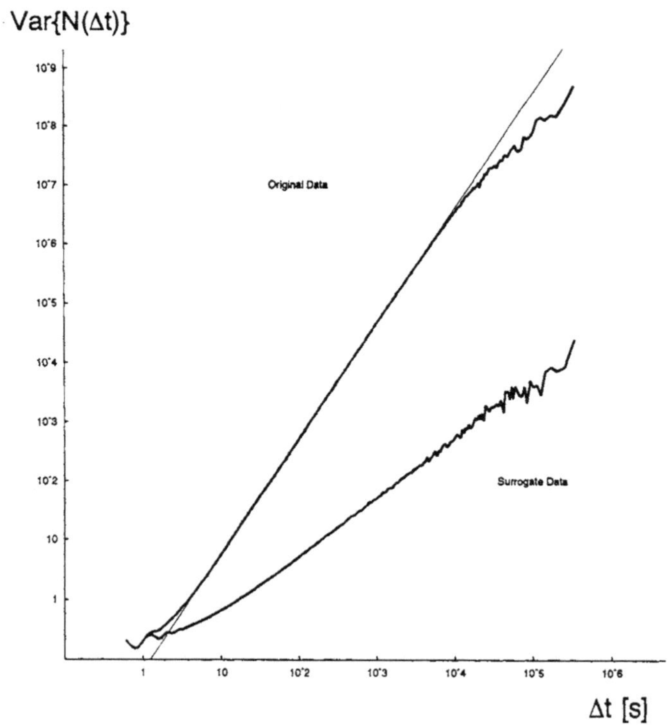

Fig. 1 Log-log plot of the variance–time–curve for the original and the surrogate data (random shuffling) of the 12-day Holter recording. The thin line indicates the regression line (slope 1.94) for the scaling region which extends from $\Delta t = 6$ s to $\Delta t = 7360$ s. (For further details, see text.)

To evaluate the dynamic spectral features of the time series, repeated measurements of the VTC were performed on segments of 150 min duration shifted in steps of 30 min as described previously.[7]

2.4 1/*f* Noise, Statistical Self-similarity and Self-affinity

The term "self-similarity" is frequently used in context with $1/f$ noise.[9] A continuous function $X(t)$ is statistically self-similar, if $X(t)$ and $X(rt)/r$ have the same statistical characteristics. If, however, a noise $X(t)$ is statistically identical to its transform by expansion in time followed by a different change in intensity, it is called self-affine.[9–12] This requires a scaling factor r and an exponent H so that $X(t)$ and $X(rt)/r^H$ have the same distribution functions. Unlike a statistical self-similar curve ($H = 1$), $X(t)$ requires scaling factors in two coordinates [r for t and r^{-H} for $X(t)$] with $H < 1$.

In the spectral domain, $X(t)$ gives rise to $1/f^b$ noise with $b = 2H - 1$. This relation has been derived for continuous functions $X(t)$.[12] A similar relation also holds for fractional point processes as shown in the following. In this case $X(t)$ represents a point process with self-affine properties. Denote $N(\Delta t)$ the cumulative number of events in Δt defined by:

$$N(\Delta t) = \int_t^{t+\Delta t} X(\tau)d\tau\,.$$

Since $X(t)$ is self-affine, $N(\Delta t)$ is also self-affine; this means that:

$$N(\Delta t) \quad \text{and} \quad r^{-H}N(r\Delta t)$$

have the same distribution functions giving rise to:

$$\mathrm{Var}\{N(\Delta t)\} = r^{-2H}\,\mathrm{Var}\{N(r\Delta t)\}\,. \tag{3}$$

According to Eq. (1):

$$\mathrm{Var}\{N(\Delta t)\} = C(\Delta t)^{1+b}\,, \tag{4}$$

C being a constant. Thus, scaling of the time axis by r results in:

$$\mathrm{Var}\{N(r\Delta t)\} = C(r\Delta t)^{1+b} = r^{1+b}\,\mathrm{Var}\{N(\Delta t)\}\,. \tag{5}$$

By comparison with Eq. (3), one finds $b = 2H - 1$.

In biology, and especially in the analysis of heart rate variability, the rate of a process is often investigated. This means, that $N(\Delta t)/\Delta t$ is observed rather than $N(\Delta t)$. In this case,

$$\mathrm{Var}\,\frac{N(r\Delta t)}{(r\Delta t)} = \frac{\mathrm{Var}\{N(r\Delta t)\}}{(r\Delta t)^2}$$

and the variance in Eq. (5) transforms to:

$$\mathrm{Var}\,\frac{N(r\Delta t)}{(r\Delta t)} = r^{b-1}\,\mathrm{Var}\,\frac{N(\Delta t)}{\Delta t} \tag{6}$$

implying:

$$\mathrm{Var}\,\frac{N(\Delta t)}{\Delta t} \propto (\Delta t)^{b-1}\,. \tag{7}$$

In case of a pure $1/f$ noise ($b = 1$), the scaling factor H is equal to zero and the variance of the heart rate remains unaffected under scaling. In case of white noise characteristics ($b = 0$), the variance decreases as $(\Delta t)^{-1}$ with the corresponding H equal to $-\frac{1}{2}$.

3. RESULTS

3.1 Tachogram

The interval tachogram, i.e., a plot of the coupling intervals as a function of interval number, of the 12-day Holter recording is shown in Fig. 2. To ease orientation, the end of each day rather than the number of beats is indicated on the abscissa. The recording of each day starts at 10 pm, the lengthening of coupling intervals during night time and the rather sudden increase during the morning hours can be seen (Fig. 2). In addition to the daily fluctuations in heart rate, there seems to be a low frequency oscillation with a general slowing down of the heart rate between days 4 and 8 in this particular case.

3.2 The Variance-Time-Curve

The VTC of the 12-day Holter record is depicted in Fig. 1. The log-log plot can be fitted by a straight line from $\Delta t = 6$ s to approximately $\Delta t = 7360$ s. These values define the scaling region within which the variance is proportional to $(\Delta t)^{1+b}$ with $b = 0.94$. The corresponding frequency range calculated by $1/\Delta t$ ranges from 0.16 to 0.000136 Hz. According to Eq. (2), the exponent of the spectrum b is 0.94 which is an almost pure $1/f$ noise. At Δt above 7360 s, the slope of the VTC is reduced and can be estimated to be 1.3 indicating a more random nature of the very low frequency fluctuations ($f < 0.000136$ Hz).

For comparison, the VTC of "surrogate" data is shown. For this purpose, the sequence of the original values was randomly shuffled to destroy any serial correlations but preserving

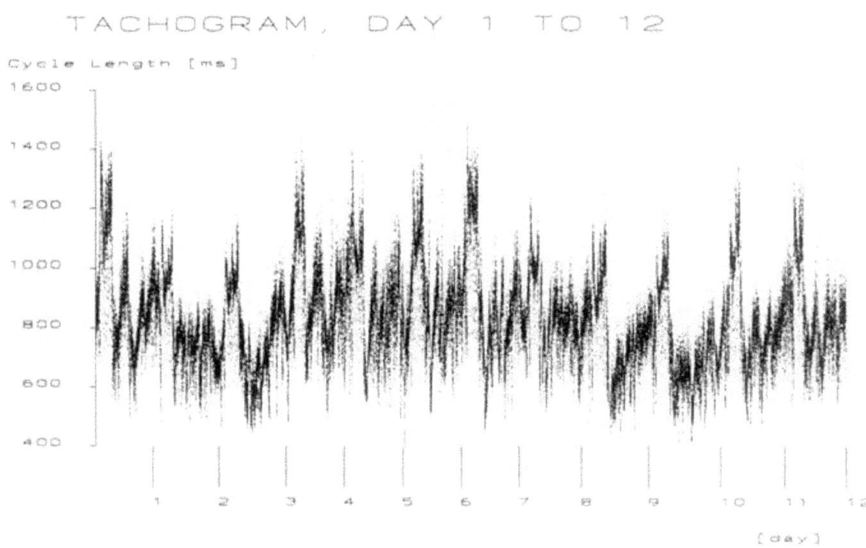

Fig. 2 Interval tachogram of 12-day Holter recording. For technical reasons only 1/8 of a total of 1 254 068 points is depicted. Recording starts at 10 pm. (For further details, see text.)

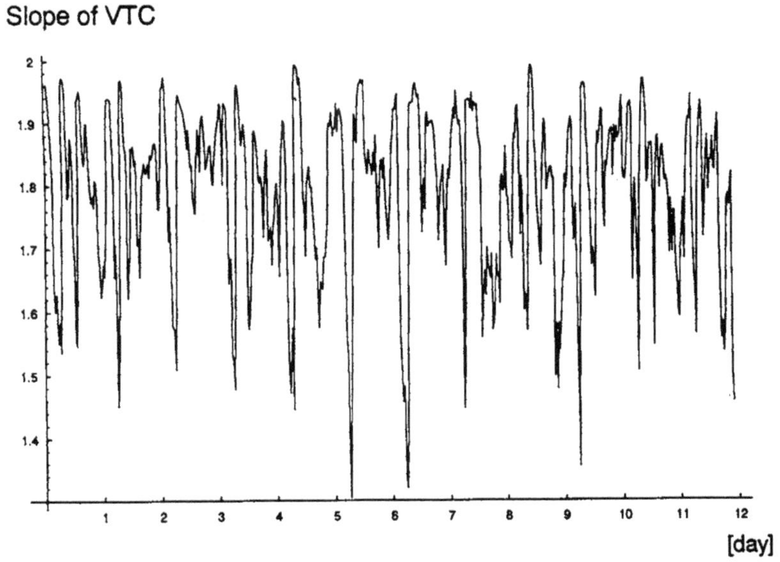

Fig. 3 Slope of the variance–time–curve for repeated (every 30 min) measurements of 150 min duration.

the mean and the standard deviation of the data. As expected from theory and simulation,[7] the slope of the VTC approaches 1 with $b = 0.003$.

The dynamic nature of the VTC for shorter time segments is demonstrated in Fig. 3. The slope of the VTC as measured from segments of 150 min is shown for every 30 min. Occasionally a slope of 1.99 is reached. Three to four times a day, the average value of 1.79 is separated by short reductions of the slope to local minima with a slope of 1.3 to 1.6. The scaling region for these short data segments extends from 0.13 to 0.0006 Hz. Comparison of the minima of Fig. 3 with the corresponding parts of Fig. 2 revealed rather low heart rates (long coupling intervals) during periods of a reduced slope of the VTC. If the slope of the VTC is less than 1.4, the heart rate is usually below 60 beats per minute.

For the ten healthy male volunteers, the mean slope of the VTC for the total 24 h Holter recording was 1.93, with a standard deviation of ± 0.02. Because of the shorter recording period, however, the lower limit of the scaling region could not be exactly defined but extended to frequencies near 0.0001 Hz. The dynamic changes in the slope of the VTC for 150 min segments, were similar to the changes observed in the 12-day Holter recording with the slope of the VTC ranging between 1.20 and 1.99.

3.3 Self-affinity

The self-affinity of the heart rate, i.e., $N(\Delta t)/\Delta t$, is demonstrated in Fig. 4. The heart rate of the original time series as observed for different Δt and expressed in beats per minute (bpm) is shown in panel A. Rescaling Δt by $r = 10$ had little influence on the variance of the heart rate as calculated for the depicted points and shown to the right of each tracing. However, there are many segments one could have chosen for the shorter segments and many of these show quite different values for the variance. Thus, the self-affinity in a formal way can only be demonstrated if all possible non-overlapping samples are used for calculation of variance as shown in Table 1. It can be seen that for $\Delta t = 6$ s to $\Delta t = 6000$ s, the variance is rather constant, whereas it decreases outside the scaling region ($\Delta t > 7360$ s). Panel B

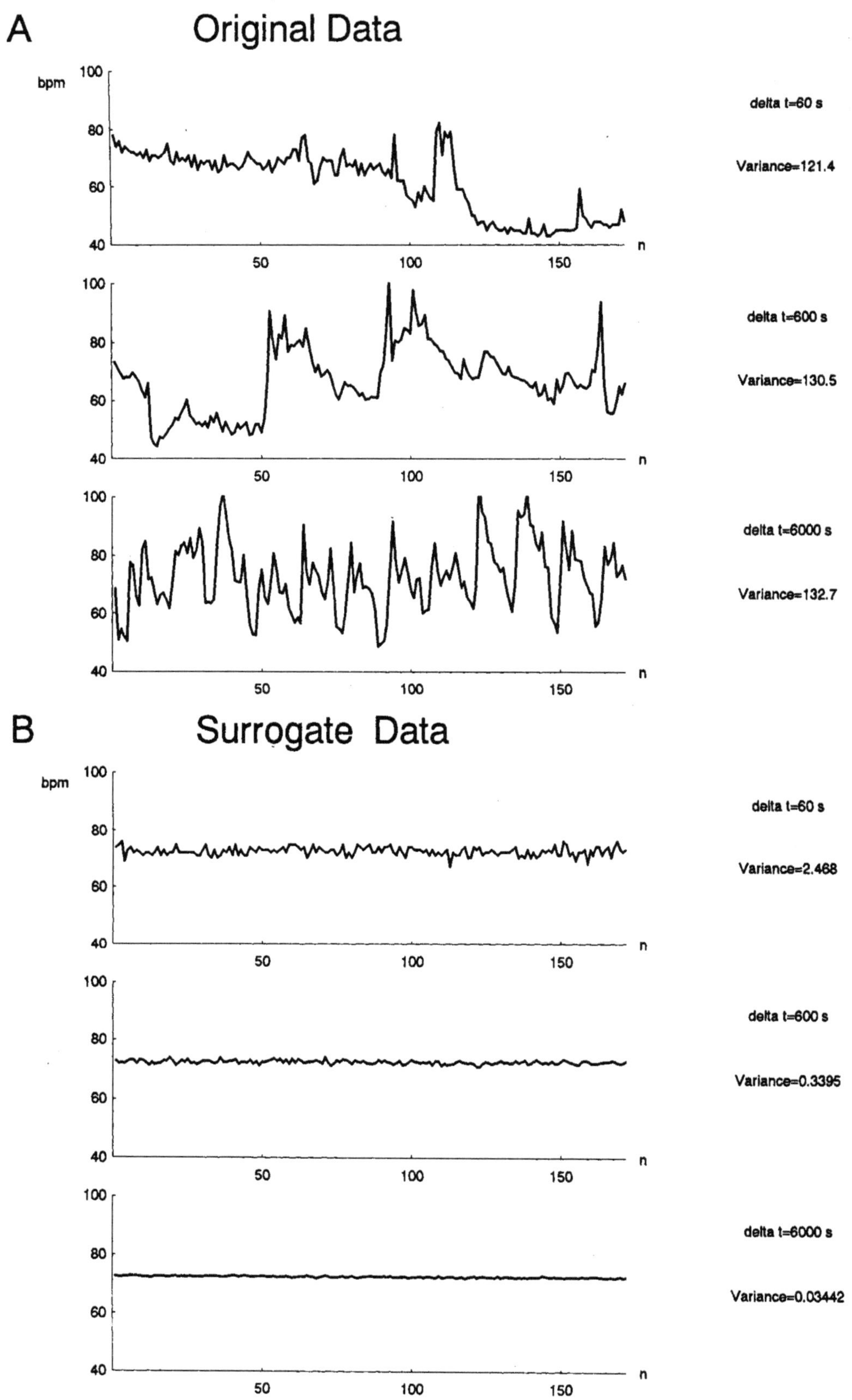

Fig. 4 Heart rate $[N(\Delta t)/\Delta t]$ for different Δt. The variance of the signals is indicated in the number to the right of each tracing. Panel A shows the original data, panel B the surrogate data. The variance as calculated from 170 values is indicated on the right.

Table 1 Variability of Heart Rate and Window Length

$\Delta t[s]$	VARIABILITY [(bpm)^2] Original Data	Surrogate Data	Number of Windows
6	217.50	46.385	172 800
60	185.24	3.091	17 280
600	164.83	0.293	1728
6000	132.68	0.034	172
60 000	38.83	0.004	17

The variance of the heart rate as measured in windows of different length Δt. The heart rate, $[N(\Delta t)/\Delta t]$, is normalized to beats per minute (bpm). Note the slight decrease in variance due to the fact that b is not 1.0 but rather 0.94, thus yielding a correction factor r^H for the heart rate equal to 1.072 for $r = 10$. For the surrogate data, the variance decreases by a factor of approximately 10 with each increase of Δt by a factor of 10 as expected from Eq. (7).

of Fig. 4 shows the corresponding traces for the surrogate data and it can be clearly seen that the variance decreases significantly with increasing Δt.

4. DISCUSSION

Applying counting statistics we have demonstrated empirically that the heart rate variability of healthy men follows a power law. The variance of the heart rate in a time interval Δt scales with $(\Delta t)^{b-1}$ [see Eq. (7)]. As with any power law observed in nature, there must be limits to the scaling region.[9] The lower limit of the scaling region as far as it can be defined from a 12-day record, was in the region of 7360 seconds and thus significantly shorter than the cycle of the day, i.e., 86 400 s. It should be noted, however, that even from this long record only a few windows of Δt larger than 10 000 s could be analyzed, thus reducing the accuracy of the estimate for the variance (see Table 1).

The minimum frequency that can be resolved in a 12-day recording is approximately 0.000 001 Hz. This reduces to 0.000 005 Hz if 5 averages of the periodogram — corresponding to 5 windows — are performed. The observed lower limit of the scaling region of 0.000 136 Hz does not seem to yield much more resolution than the 0.0001 Hz described by Saul et al.[5] which was obtained from Holter records of 24.27 h. However, the power spectra shown in their paper could be fitted by a straight line only down to frequencies of about 0.0004 Hz, whereas the power spectra were rather rough at lower frequencies.

The dynamic changes in the spectrum of heart rate variability are demonstrated in Fig. 3. These correspond to rather large short-term changes in the heart rate as can be seen in Fig. 2. Beat-to-beat changes in heart rate are most pronounced during periods of slow heart rate at night time, in which there are short phases of accelerations to day time values. This is one of the reasons why only an average of all shorter time segments shows self-similarity with the larger portion of the time series. The changes of the slope in the VTC (Fig. 3) reveal rhythmic fluctuations by themselves with 3 to 4 minima per day. It is conceivable that this reflects a biological rhythm with a period of about 6 to 8 hours.

The average slope of the VTC for repeated measurements is less than the slope of the VTC of the total record for the 12 days. This is consistent with the fact that long recording periods are required for the slope of the VTC to converge.[7] In case of $1/f$ fluctuations, with b between 0.95 and 1, the variance of the heart rate is almost independent of the time interval in which it is estimated. Such a behavior has been called "flicker floor" in a different context.[13] On the other hand, for white noise characteristics (b close to 0), the variance of the heart rate decreases approximately with $(\Delta t)^{-1}$ as expected from Eq. (7).

These phenomena point to complex patterns of control within the sinus node and by the autonomic nervous system reflecting long term correlations. With $1/f$ fluctuations, these control structures create effects on the heart rate, maintaining an almost constant variance over a wide range of time scales. The periods of white noise were most pronounced at heart rates markedly slower than the intrinsic heart rate,[14] i.e., the heart rate in the absence of autonomic drive, indicating increased vagal drive. It is conceivable that acetylcholine, the transmitter of the vagus nerve, may cause the irregular dynamics of the pacemaker cells within the sinus node as described in experiments and computer simulations by Michaels et al.[15] The pathophysiological and prognostic significance of our findings remains to be established.

In conclusion, the variance–time–curve is a valuable tool to delineate the dynamic changes in heart rate variability. From its measurement, the self-similar aspects of the human heart beat as a point process $[N(\Delta t)]$ and the self-affine aspects of the heart rate $[N(\Delta t)/\Delta t]$ can readily be derived.

ACKNOWLEDGMENTS

This study has been supported by the Wilhelm Sander-Stiftung, München.

REFERENCES

1. O. Rompelman, *The Study of Heart Rate Variability*, eds. R. I. Kitney and O. Rompelman (Clarendon Press, Oxford, 1980), p. 59.
2. P. Albrecht and R. J. Cohen, *Comp Cardiol IEEE* **15**, 311 (1989).
3. R. D. Berger, S. Akselrod, D. Gordon and R. J. Cohen, *IEEE Trans. Biomed. Eng. BME* **33**, 900 (1986).
4. M. Kobayoshi and T. Musha, *IEEE Trans. Biomed. Eng. BME* **29**, 456 (1982).
5. J. P. Saul, P. Albrecht, R. D. Berger and R. J. Cohen, *Comp Cardiol IEEE* **14**, 419 (1988).
6. H. E. Stanley, S. V. Buldyrev, A. L. Goldberger, J. M. Hausdorff, S. Havlin, J. Mietus, C. K. Peng, F. Sciortino and M. Simons, *Physica* **A191**, 1 (1992).
7. M. Meesmann, F. Grüneis, P. Flachenecker and K.-D. Kniffki, *Biol. Cybern.* **68**, 299 (1993).
8. F. Grüneis, M. Nakao, M. Yamamoto and T. Musha, *Noise in Physical Systems and 1/f Fluctuations*, eds. by T. Musha, S. Sato and M. Yamamoto, (Ohmsha, Tokyo, 1991), p. 583.
9. M. Schroeder, *Fractals, Chaos, Power Laws*, (Freeman, New York, 1991), p. 121.
10. J. Feder, *Fractals*, (Plenum Press, New York, 1988), p. 184.
11. R. F. Voss, *Fractal Images*, eds. H. O. Peitgen and D. Saupe (Springer-Verlag, New York, 1988), p. 21.
12. B. B. Mandelbrot, *The Fractal Geometry of Nature (Updated and Augmented)*, (Freeman and Company, New York, 1983), p. 349.
13. D. W. Allan, *Proc. IEEE* **54**(2), 221 (1966).
14. M. E. Josephson, *Clinical Cardiac Electrophysiology. Techniques and Interpretations*, 2nd ed. (Lea & Febiger, Philadelphia, 1993), p. 73.
15. D. C. Michaels, E. P. Matyas and J. Jalife, *Cardiac Electrophysiology. From Cell to Bedside*, eds. D. P. Zipes and J. Jalife (Saunders Company, Philadelphia, 1990), p. 182.

Fractals, Vol. 1, No. 3 (1993) 321–325

FRACTAL STOCHASTIC ELECTROMAGNETIC BACKGROUND AS A FACTOR STABILIZING PROCESSES OF VITAL ACTIVITY

VADIM M. URITSKY, NATALIA I. MUZALEVSKAYA
Nonequilibrium Systems Laboratory
Department of Science and Technology of SOFID Co.
196 140 *Russia, St. Petersburg,*
Pulkovskoe shosse 86-56

Abstract

The linkage between functional status of living systems and scale organization of environmental low-frequency fluctuations is studied and shown to be possible only for influences transferring biologically significant information. Fundamental relict form of such influences is background weak energy electromagnetic field. It is demonstrated experimentally that physiologically normal state is supported by scale-invariant (fractal) electromagnetic fluctuations with $1/f$ spectrum, which have pronounced stabilizing effect on homeostasis and can act as a powerful therapeutic agent. On the contrary, distortion of fractal structure of the background, i.e., during geomagnetic storms, is a destabilizing and pathogenic factor.

Primary response of living systems to fractal electromagnetic background appears as restoration and support of self-similarity of endogenic low-frequency fluctuations. Vitally important interaction of living systems with fractal fluctuations of non-living environment is called fractal stochastic coupling.

1. INTRODUCTION

Common structural organization of fractals in physical and biological worlds persistently reminds us that interdisciplinary boundaries in science are arbitrary, and algorithms of Nature universal. In this work we ask a question: Is there a direct causal relation between structures of random processes in living and nonliving Nature? Does specific contact between a biological system and its environment exist that results from self-similar properties of the latter, and what are the physiological role and principles of such contact?

A widely spread type of natural self-similar processes is low-frequency fluctuations with $1/f$ power spectrum. Time realizations of $1/f$ fluctuations are scale-invariant and belong to random fractals.[1] The facts suggest that these fluctuations fulfill an important physiological function,[2] maintaining stable homeostasis regulation of living systems.[3] Realization mechanism of this function appears to be universal for different organs and functional systems and can be understood in terms of interaction between an integral biosystem and its environment.

$1/f$ fluctuations have always been inherent to habitats of living systems. Can it be that self-similarity as an attribute of stable vital activity is an evolutionary imprint of self-similar properties of environment? If such an influence of environment did exist, its modern manifestation should be critical dependence of living systems on scale organization of external low-frequency fluctuations. Environmental factors that fluctuate according to $1/f$ law should be expected to stabilize processes of vital activity introducing into them standard of self-similarity. As for other factors, their influence should be destabilizing, giving additional load to homeostasis.

Theoretically, not all factors can transfer their fractal structure to the living system, since this structure is realized in principally very broad (several orders of magnitude) dynamic range. So fractal structure can be transferred only by stimuli that are used by the organism to obtain information: visible light (dynamic range of sensitivity of 6 orders of magnitude), audible sound (6 orders), as well as weak nonsensory stimuli, viz. background extremely low frequency electromagnetic fields (for example, sensitivity range for geomagnetic field is 5 orders of magnitude[4]).

Electromagnetic (EM) background has a number of unique features: its lower threshold of sensitivity (10^{-12} Wt/m^2) coincides with that for light and sound,[5] and its biological action results from its space-time nonuniformity[6]; however, it does not carry coded information and acts nonspecifically on all structures of the living system. Interaction with EM-background appears as the primary evolutionary form of informational (broad range) contact with the environment, and so it interests us most of all.

2. EXPERIMENTAL DATA AND DISCUSSION

2.1 Biological Action of Non-Fractal EM-Background: Dysfunctions and Pathologies

Natural extremely low frequency EM-background is determined to a large extent by contribution from geomagnetic fluctuations. One can ascertain that in geomagnetically quiescent periods, spectrum of these fluctuations approaches the $1/f$ type. On the contrary, during periods of disturbances (geomagnetic storms), it is markedly non-monotonous and unstable, showing distortion of self-similar structure of fluctuations.

Statistics shows that there is distinct correlation between geomagnetic storms and decrease in adaptation reserve in humans, aggravating various dysfunctions and pathologies. These periods are characterized by drastic increase in the number of myocardial infarctions.[7] Aggravated are also dysfunctions of the central nervous system and functions of attention, and rates of motor reactions in humans deteriorate, causing increase in traffic accidents.[8]

During the period of unstable EM-background, we registered synchronous deterioration of attention functions in two subjects operating a flight simulator.

Destabilizing influence of nonself-similar natural EM-background was confirmed in comprehensive investigations of biological response to artificial, extremely low frequency magnetic fields carried out by one of the authors (NM). Negative biological effects were observed in all the main systems of experimental animals. Circulation dynamics was deranged, blood clotting was either pathologically increased or decreased,[9] and destructive changes in myocardium were registered.[6] Ferments responsible for energy exchange in the brain were inactivated.[10] The increase in activity of neurosecretory system and adrenal cortex characteristic for response to strong stress stimuli was observed.[9] Functional reliability and vital capacity of T-cells belonging to the system of cell-bound immunity were decreased.

In model experiments of other authors with both magnetic and electrical fields of nonself-similar time structure, various effects of biological destabilization were also observed. Priorities of many discoveries in this field belong to scientists from Russia (see, for example, Ref. 11).

Paradoxically, electromagnetic shielding cannot be used for protection from geomagnetic storms. Moreover, long-term shielding may cause nonspecific structural and functional disturbances similar to those described above.[8,11] Indirectly, it shows that under natural conditions, nonstationary electromagnetic disturbances as a pathogenic factor act in combination with biologically stabilizing influence of EM-background, removal of which is harmful. Actually, the disturbances themselves are not pathogenic, but the distortion of some omnipresent life-supporting factor is. In the following, we consider the nature of this factor.

2.2 Fractally Organized EM-Fluctuations: Normalization of Physiological Functions of Living Systems

In model experiments we have shown for the first time that the stabilizing function of EM-background is realized by its self-similar $1/f$ fluctuations. EM-background was formed with a specially designed generator of $1/f$ fluctuations and with Helmholtz rings.

Effects of fractal EM-fluctuations appeared in various functional systems of experimental animals. However, in contrast to effects of geomagnetic storms and their simulations, they were biologically positive. Pronounced support of adaptation mechanism of homeostasis was obtained, as was testified by the increase in endurance of animals under stress conditions. Maximum duration of swimming of rats in cooled water was increased by 25%, duration of compulsory running by 30%. Fractal EM-background supported immunologic homeostasis. Resistance of immunocompetent cells to concentration hypoxia, a nonspecific indicator of the cell's functional reliability, was considerably strengthened. Namely, the density of nuclear material of T-cells increased and approached the value for cells before their exposure to hypoxia; survival of T-cells and other leukocytes under conditions of hypoxia was increased by 40–60%.

Direct influence of fractal EM-field on scale organization of endogenic biological processes was studied by analyzing ECG and EEG signals from rabbits. The primary stage of response of heart dynamics to fractal EM-background consisted of restoration of pathologically distorted self-similarity of ECG fluctuations (Fig. 1). Then functional normalization took place, manifested in stabilization of phase trajectories and decrease in dispersion of ECG parameters. Response of CNS revealed a shift to more pronounced scale-invariance of

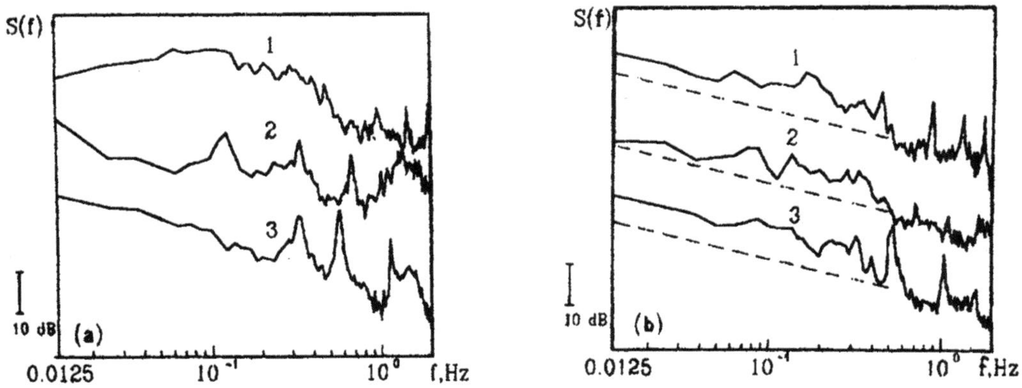

Fig. 1 Low-frequency power spectra $S(f)$ of ECG signals from three rabbits, before (a), and after (b), exposition to artificial fractal EM-background. Correction of self-similarity of fluctuations for rabbits 1, 2 is obvious (for convenience the spectra are shifted along the vertical axis; dependence of the $1/f$-type is shown by a dashed line).

attractors of EEG signals. The range of scale distances r in which strange attractors of the signals had strictly fractal structure, i.e., in which integral correlation function $C(r)$ varied as $C(r) \sim r^d$, was widened by 70–80%. According to modern psychophysiological models, such changes facilitate memorizing of images and forming of conditioned reflexes.

Life-supporting nature of fractal EM-background received its most dramatic proof in cases of radiation disease and cancer, pathologies that tax the organism's defences to the utmost. Survival of rats that received X-ray dose of 4.0–6.0 Gy was increased by 30–40%. Survival of mice with induced Ehrlich ascites cancer not curable by other methods reached 73% for exposures to artificial fractal EM-background at early stages of cancer, and 53% for exposures at late stages. In both cases, involution of ascites tumors were observed: 60% of animals in the first case and 40% in the second, showed no signs of cancer by the end of the test. Survival rate of controls was zero.

3. CONCLUSION

The above data show that living systems do depend critically on scale organization of EM-background. Stable self-similarity of background EM fluctuations proves to be necessary, and in some cases crucial, for stabilization of vital activity. If satisfied, this condition turns EM-background from being an aggressive agent of the environment to a powerful therapeutic factor.

The question asked in the beginning of this work can be answered positively: biological influence of the environment resulting from its specific self-similar properties does exist. The mechanism of this effect, appears as useful cooperation of living and nonliving random fractals. i.e., of endo- and exogenic scale-invariant fluctuations. We call this effect fractal stochastic coupling of a living system with its environment.

Fractal coupling differs principally from contacts of organisms with the environment traditionally studied in biology, since the leading agent of the fractal coupling is not energy nor substance, but a form of scale organization. Therefore, it is not accidental that fractal coupling can be realized only with fractals carrying biologically significant information, i.e., which are perceived through interpreted images. For extremely low frequency

EM-background, this process goes unconscious; in case of light and sound, consciousness takes part in it, and we admire classical music, masterpieces of architecture, landscapes, all of which are fractals.

Physical description of fractal coupling is bound to encounter difficulties. In the framework of Maxwell's electrodynamics, this problem is intractable because of non-differentiability of fractal realizations of field components. Relevant equations can be solved on limited and separated time scales, but the native self-similar structure of the field will be lost, as the essence of music is lost after spectral analysis. An alternative approach seems to be necessary that would organically take into account fractal properties of natural EM-background, an approach which could be called electrodynamics of life.

ACKNOWLEDGMENTS

We thank A. M. Reschikov and E. V. Korolyov for decisive help in carrying out a number of experiments, and G. P. Timoshinov for valuable discussions on this work.

REFERENCES

1. R. F. Voss, "Random Fractals: Self-affinity in Noise, Music, Mountains and Clouds," *Physica* **D38**, 362–371 (1989).
2. Y. Ichemaru and S. Katayama, "Central Nervous System and Heart Rate Variabilities," in *Proc. 11th Int. Conf. on Noise in Physical Systems and 1/f Fluctuations* (Ohmsha, Tokyo, 1991), pp. 691–694.
3. N. I. Muzalevskaya and V. M. Uritsky, "Stochastic Function of Homeostasis," in *Proc. of Djanelidze Res. Inst.* **1**, 77–86 (1993) (in Russian).
4. N. I. Muzalevskaya, "Classification of Stimuli and Study of Nonsensory Influences," in *Informational Interactions in Biology* (Izd. TGU, Tbilisi, 1987), pp. 70–77 (in Russian).
5. N. I. Muzalevskaya, "On Biological Activity of Disturbed Geomagnetic Field," in *Effects of Solar Activity on the Earth's Atmosphere and Biosphere* (Israel Program for Scientific Translations, Jerusalem, New York, 1977), pp. 171–177.
6. N. I. Muzalevskaya, "Nonuniformity of Low Frequency Electromagnetic Field Fluctuations and Its Ecological Significance," in *Adaptation System and Environment* (Izd. LSGMI, Leningrad 1974), pp. 117-121 (in Russian).
7. V. G. Sidyakin, N. A. Temuryants, V. B. Makeev and B. M. Vladimirsky, *Cosmic Ecology* (Naukova Dumka, Kiev, 1985) Chap. 2, pp. 59–64 (in Russian).
8. A. P. Dubrov, *The Geomagnetic Field and Life: Geomagnitobiology* (Plenum Press, N. Y. 1987).
9. N. I. Muzalevskaya, "Ultra-low Frequency Low Strength Magnetic Field and State of Adaptation Reserve in Experimental Animals", *Problems of Cosmic Biology* **43**, 82–98 (1982) (in Russian).
10. R. N. Pavlova, N. I. Muzalevskaya and V. V. Sokolovsky, "Some Biochemical Aspects of the Influence of Weak Low Frequency Magnetic Fields," in *Response of Living Systems to Magnetic Fields* (Nauka, Moscow, 1978), pp. 49–58 (in Russian).
11. N. A. Temuryants, B. M. Vladimirsky and O. G. Teshkin, *Extremely Low Frequency Electromagnetic Signals in Biological World* (Naukova Dumka, Kiev, 1992) (in Russian).

Fractals, Vol. 1, No. 3 (1993) 326–335

AN "ASYMPTOTIC FRACTAL" APPROACH TO THE MORPHOLOGY OF MALIGNANT CELL NUCLEI

GABRIEL LANDINI* and JOHN W. RIPPIN
Oral Pathology Unit, School of Dentistry, University of Birmingham, St. Chad's Queensway, Birmingham B4 6NN, UK

Abstract

To investigate quantitatively nuclear membrane irregularity, 672 nuclei from 10 cases of oral cancer (squamous cell carcinoma) and normal cells from oral mucosa were studied in transmission electron micrographs. The nuclei were photographed at ×1400 magnification and transferred to computer memory (1 pixel = 35 nm). The perimeter of the profiles was analysed using the "yardstick method" of fractal dimension estimation, and the log-log plot of ruler size vs. boundary length demonstrated that there exists a significant effect of resolution on length measurement. However, this effect seems to disappear at higher resolutions. As this observation is compatible with the concept of asymptotic fractal, we estimated the parameters c, L and B_m from the asymptotic fractal formula $B_r = B_m \{1 + (r/L)^c\}^{-1}$, where B_r is the boundary length measured with a ruler of size r, B_m is the maximum boundary for $r \to 0$, L is a constant, and c = asymptotic fractal dimension minus topological dimension $(D - Dt)$ for $r \to \infty$. Analyses of variance showed c to be significantly higher in the normal than malignant cases ($P < 0.001$), but $\log(L)$ and B_m to be significantly higher in the malignant cases ($P < 0.001$). A multivariate linear discrimination analysis on c, $\log(L)$ and B_m re-classified 76.6% of the cells correctly (84.8% of the normal and 67.5% of the tumor). Furthermore, this shows that asymptotic fractal analysis applied to nuclear profiles has great potential for shape quantification in diagnosis of oral cancer.

*E-mail: G.Landini@bham.ac.uk

1. INTRODUCTION

Malignant transformation of tissues is thought to involve a series of qualitative and quantitative changes in the cell genome. Qualitative changes include abnormal gene expression and loss of control of the cell dynamics, resulting in features such as increased number of mitoses, abnormal mitoses and disturbance in the tissue differentiation. The quantitative changes relate to an increase (or more rarely decrease) in the DNA contents, and are usually due to errors in DNA replication.

The abnormal DNA content of malignant nuclei is responsible for changes in the nuclear stainability and morphology. Classically, these changes are described as "hyperchromatism" (increase in nuclear stain uptake), "increased nuclear/cytoplasmic ratio" (increase in nuclear size) and "nuclear pleomorphism" (variation in nuclear size and shape). Nuclear pleomorphism in particular remains a subjective parameter, though there have been some attempts to quantify nuclei irregularity with such shape descriptors as form factor, shape factor, circularity departure, and PERBAS.[1–3] However, these descriptors provide a "single resolution" measure on objects which exhibit non-standard scaling properties, and hence they are incomplete as measures of irregularity. On the other hand, fractal geometry provides methods for quantifying irregular shapes that are "scale invariant", and so we have used this methodology to investigate the nuclear membrane irregularity of normal and cancerous (squamous cell carcinoma) oral epithelial cell nuclei in transmission electron micrographs.

2. MATERIALS, METHODS AND RESULTS

Freshly resected tissues from 10 cases of oral cancer (well differentiated squamous cell carcinoma), including normal surrounding tissue from the surgical margins, were fixed in 2.5% glutaraldehyde in 0.1 M cacodylate buffer at 4°C, and embedded in Epox 812 resin. Sections (60 nm in thickness) were stained with uranyl acetate-lead citrate, and nuclei of cancer and normal epithelial cells photographed at ×1400 magnification in a transmission electron microscope. A total of 672 sections of different nuclei were photographed and subsequently enlarged photographically 2.6 times; they were then hand traced, and transferred to computer memory as binary images (512×480 pixels) *via* an image scanner. The final resolution obtained was 1 pixel = 35 nm (Fig. 1).

The perimeter of each profile was analyzed using the "yardstick method" of fractal dimension estimation.[4] The nuclear perimeter was measured with yardsticks or rulers of different sizes (consecutive units from 1 to 10, steps of 5 from 10 to 40, and steps of 10 from 40 to 200 pixels). The fractal dimension can then be estimated as one minus the slope of the linear regression of $\log(r)$ on $\log(B_r)$:

$$B_r = Fr^{1-D} \tag{1}$$

where B_r is the boundary measured with a yardstick of size r, F is a pre-factor and D is the fractal dimension.

The log-log plots of boundary length of the sectioned nuclei demonstrated (B_r) on ruler size (r) show that there is a significant effect of resolution on length measurement (Richardson effect). However, this seems to disappear at higher resolutions,[5] as demonstrated by the curved plots (Figs. 2 and 3), and so the linear regression of $\log(B_r)$ on $\log(r)$ seems to be of

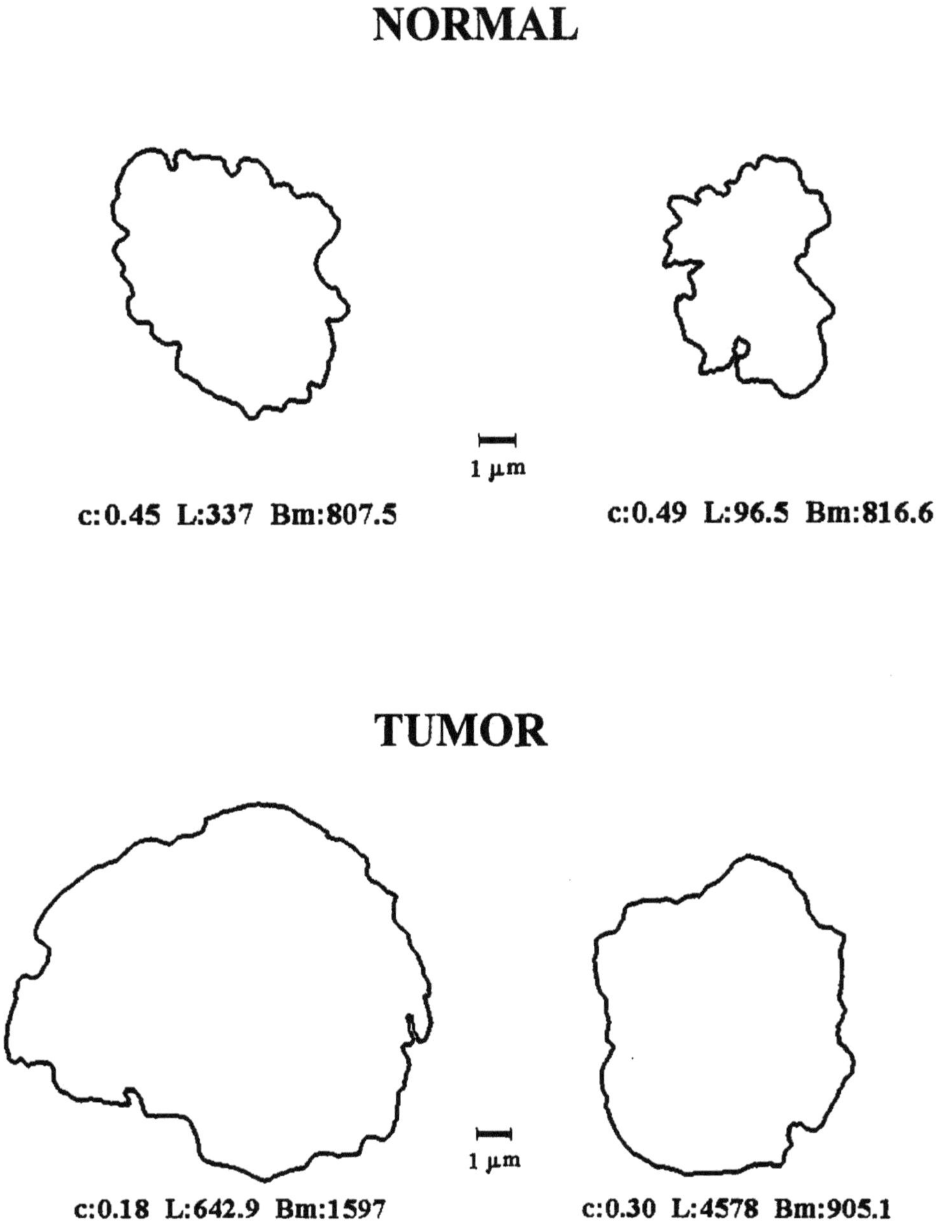

Fig. 1 Examples of normal and tumor cell nuclear profiles.

limited value. This is because, strictly speaking, although the cell outlines are irregular, the irregularity is not fully reproduced at all scales; hence, the outlines do not qualify as true fractal shapes. In other words, it is evident that at relatively low magnifications [×9600, Fig. 3(a)] the nuclear outlines look irregular, but at very high resolutions [×96 000, Fig. 3(b)] they appear smooth. Nevertheless, this observation is compatible with the concept of an asymptotic fractal[6–8] that assumes an asymptotic tendency towards a maximum boundary length at high resolution, and asymptotic fractal behavior at low resolution. Although the asymptotic fractal concept was derived empirically, it allows fitting of experimental data to which it is not possible to assign a true fractal dimension.

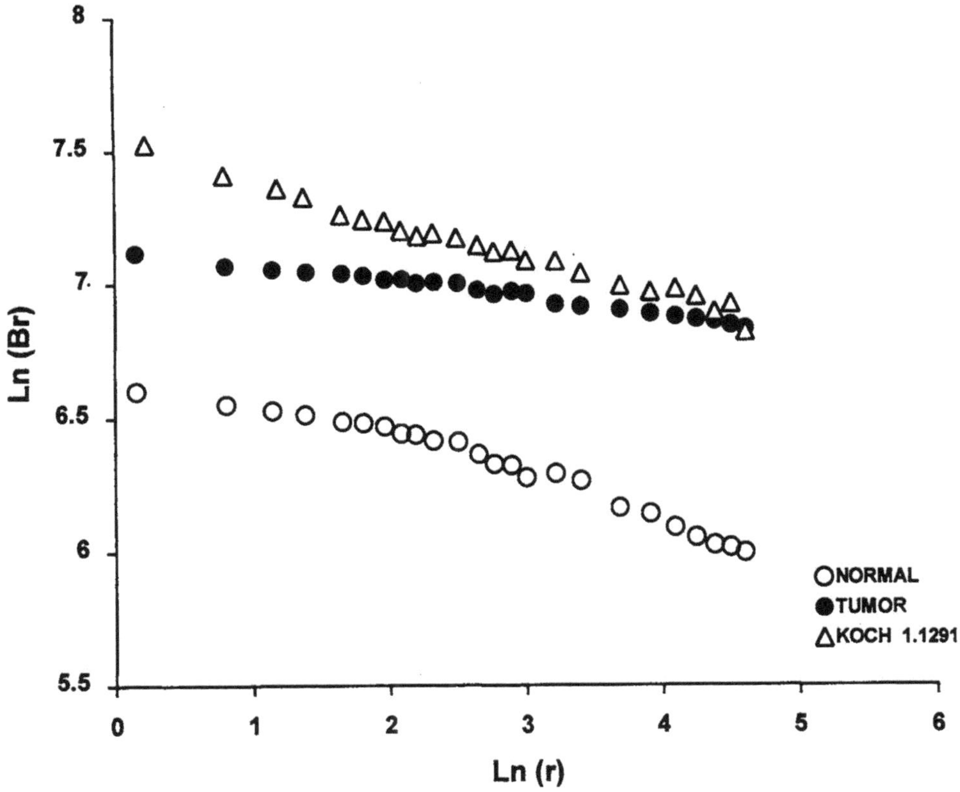

Fig. 2 Relation of the "yardstick" length (r) on the boundary length (B_r) of normal, ant tumor cell nuclear profiles, and of a Koch curve (fractal dimension = 1.1291). B_r and r in pixel units.

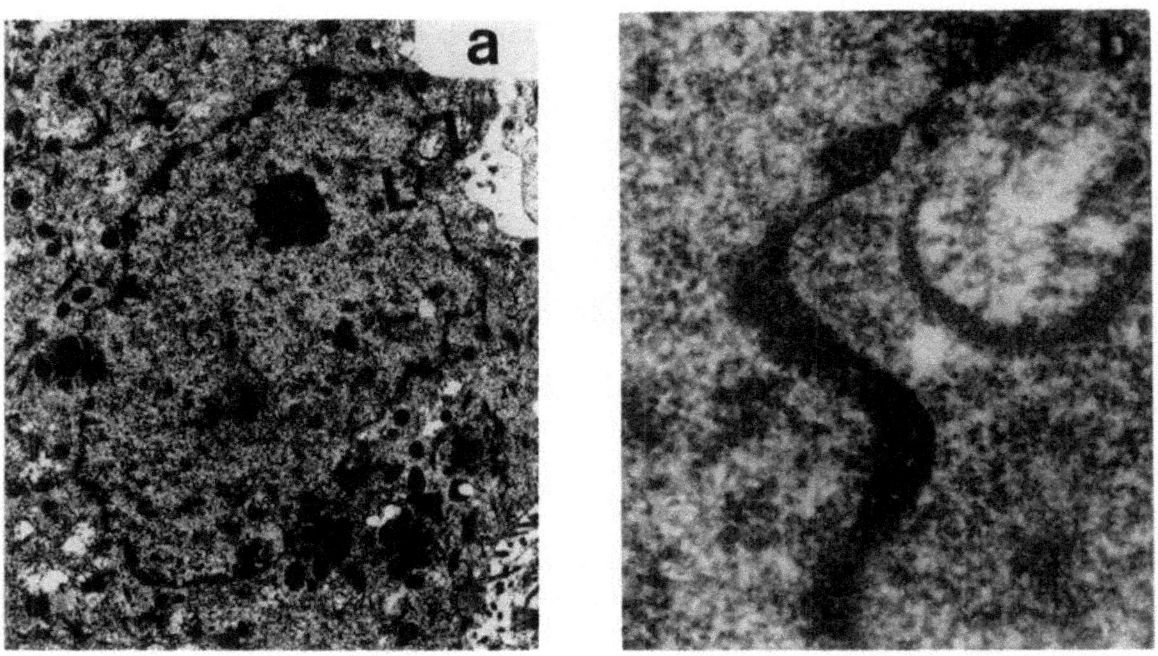

Fig. 3 (a) A ×9500 transmission electron micrograph of a malignant cell nucleus. (b) ×95 000 of the same cell. Note that the low scale irregularity of the nuclear membrane vanishes at this high resolution.

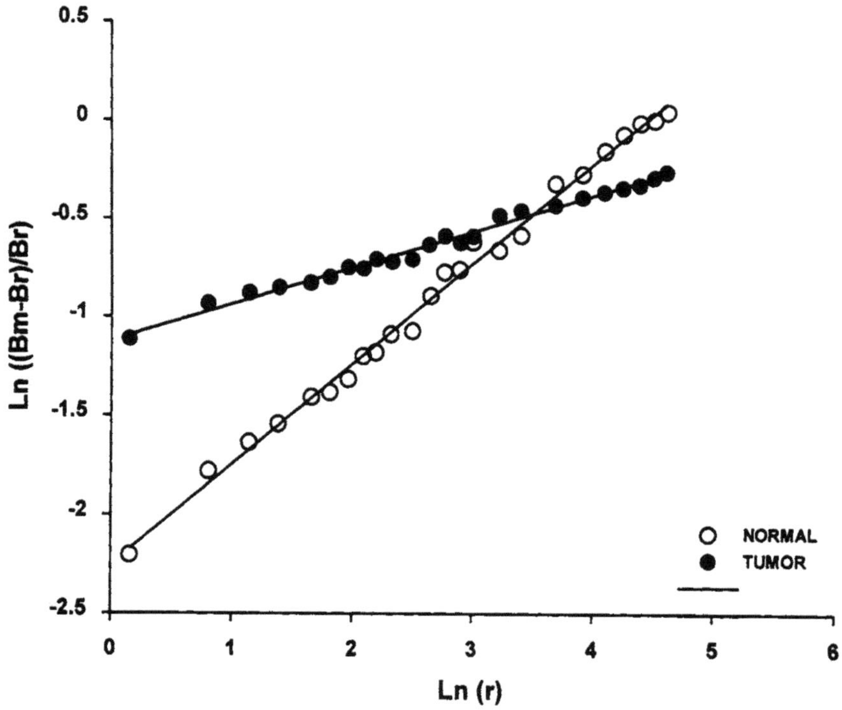

Fig. 4 Hill-Wyman plot of the normal and malignant nuclei data from Fig. 2.

The asymptotic fractal formula is defined by:

$$B_r = B_m \left\{ 1 + \left(\frac{r}{L} \right)^c \right\}^{-1} \tag{2}$$

where B_r is the boundary length measured with a ruler of size r, B_m is the asymptotic maximum boundary for $r \to 0$, L is a constant and c is the asymptotic fractal dimension minus the topological dimension, i.e., $(D - Dt)$ for $r \to \infty$.

Estimation of the values of B_m, L and c was carried out using an iterative approach according to the Hill-Wyman method of searching for the highest coefficient of determination (R^2) of the regression of $\ln[(B_m - B_r)/B_r]$ on $\ln(r)$ (Fig. 4). The iteration is on B_m, starting with $B_m = B_{r=1} + 1$ [to avoid $\ln(0)$], using an optimisation searching subroutine rather than monotone increase of B_m. When the highest R^2 is found, the linear regression line with formula $y = ax + b$ is used to estimate the parameters c and L:

$$c = a \tag{3}$$

and

$$L = e^{(-b/a)} \tag{4}$$

Interestingly, when the object is fractal no solution is possible, since $B_m \to \infty$ for $r \to 0$ and the R^2 increases very slowly. In such a case, an arbitrary value of B_m (for example $B_m > B_1 * 10$) can be used to terminate the procedure and then the object is considered as fractal rather than asymptotic fractal. An extensive chapter on this subject can be found in Ref. 8.

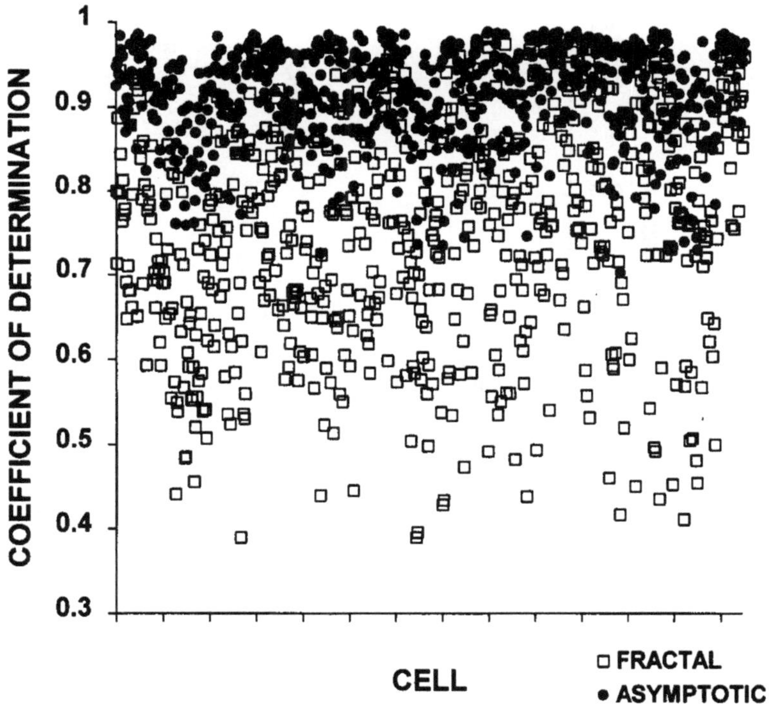

Fig. 5 Coefficient of determination of the fractal (logarithmic) and asymptotic fractal (Hill-Wyman) plots.

From examinations of the R^2 of the linear regression of the logarithmic (fractal model) and Hill-Wyman plots (asymptotic fractal model), it is evident that the asymptotic fractal model fits the experimental data better than the fractal (Fig. 5).

Analyses of variance showed c to be significantly higher in the normal than malignant cases ($P < 0.001$), ranging from (mean) 0.3251 to 0.5738, but $\log(L)$ to be significantly higher in the malignant cases ($P < 0.001$), mean ranging from (mean) 2.409 to 3.745. B_m was also significantly higher in the malignant cases. For these variables taken individually, there is overlap between the two populations of cells (normal and tumor), and hence there is no clear cut distinction between them (Fig. 6). However, a multivariate linear discriminant analysis on c, $\log(L)$ and B_m re-classified 76.6% of the cells correctly (84.8% of the normal and 67.5% of the tumor) (Table 1).

Table 1 **Linear Discriminant Analysis based on c, log(L) and B_m**

	True Groups	
Group	Normal	Tumor
Normal	301	103
Tumor	54	214
Total N	355	317
N Correct	301	214
Proportion	0.848	0.675

$N = 672$ N Correct = 515 Proportion Correct = 0.766

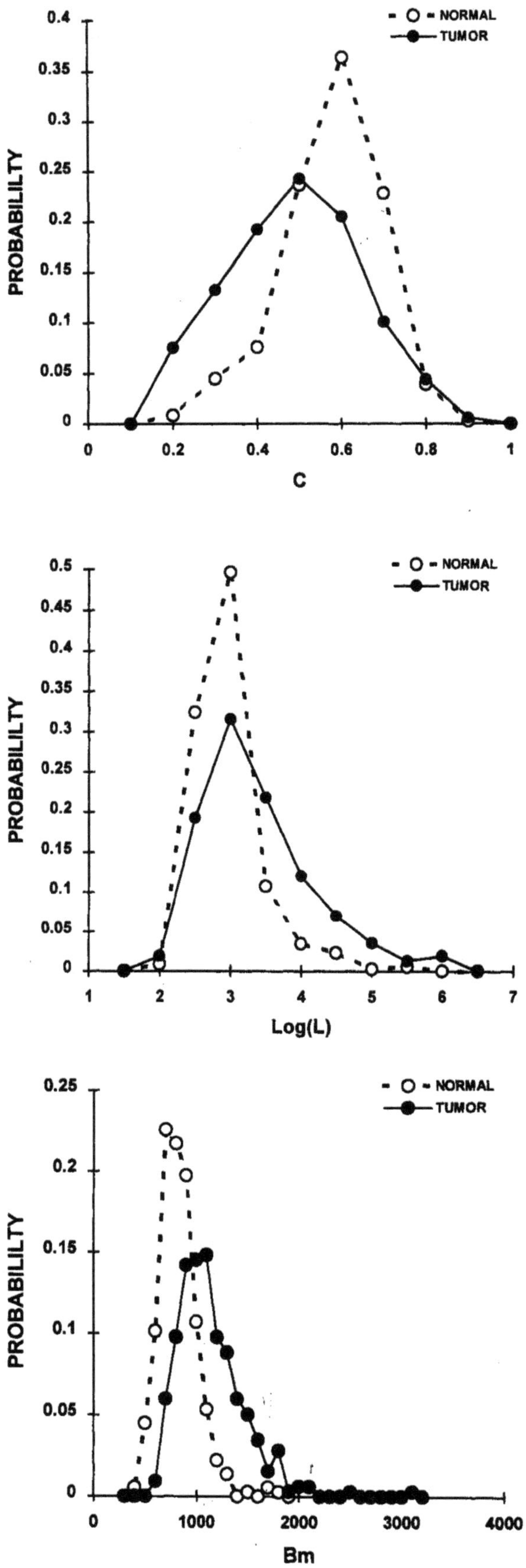

Fig. 6 Distribution of the values of c, $\log(L)$ and B_m in the normal and tumor groups. B_m and L in pixel units.

3. DISCUSSION

Oral cancer is thought to be preceded by first a "premalignant" stage and then a preinvasive one which is termed "intraepithelial neoplasia"; either of them may last for years. The importance of the identification of these stages is that it may allow early surgical removal of the tissues which carries a very high probability of complete cure, or, when this is not possible, the use of effective chemopreventive agents to promote regression (for example, the use of beta-carotene in the treatment of dysplastic oral leukoplakia).[9]

In order to understand the process of tumor development and to improve diagnosis, major efforts are aimed at the study of biological and morphological markers of the disease. These may allow analytical and quantitative methods of diagnosis which are, of course, objective and reproducible; eventually the methods may become semi- or fully automatic laboratory routines.

Computer technology allows the screening of large groups of data sets and the application of multivariate statistics in order to find which variables are the most important in the discrimination of normal from diseased tissue.[10–13] However, problems remain; for example the assessment of shape irregularity is largely a subjective judgement that has been shown to be poorly reproducible, with high intra- and inter-observer variation, and therefore no weight can be assigned to such variables during the diagnostic procedure. But fractal geometry provides a new set of tools for quantifying complex patterns which applied to some unsolved problems in medical imaging and histopathology have shown very promising results.[14–17]

Attempts to assess the irregularity of the outlines of cell nuclei have been made by assessing the projected outlines in routine paraffin sections (usually 5 μm thick) using light microscopy. Some parameters, such as departure from circularity, form factor, shape factor and PERBAS[1–3,18] have been found to be statistically different in normal and malignant cells. However, these parameters do not quantify irregularity itself for two reasons.

Firstly, they apply resolution bounded measurements to structures which exhibit the "Richardson effect": when the magnification is changed, these parameters change in a non-linear mode. For example "form factor" and "shape" descriptors are calculated as:

$$\text{form\_factor} = \frac{\text{area}^* 4\pi}{\text{perimeter}^2} \tag{5}$$

$$\text{shape} = \frac{\text{perimeter}^2}{\text{area}} \tag{6}$$

involving area (which remains finite for area measuring units $r \to 0$) and perimeter (which increases for yardsticks $r \to 0$ as we showed previously). For true fractals, however, the *area-perimeter relation*[4] accounts for the fractal dimension of the perimeter.

The PERBAS (perimeter ratio before and after smoothing) calculates the ratio between the perimenter and the length of the smoothed perimeter (the convex hull).[3] Again this parameter is dependent on the resolution used (some reports even do not state at which magnification the analyses were done!)

Secondly, previous studies have assessed the "projected outline" of the nuclei, a technique which masks the small detail due to the non-negligible thickness of the histological section. This can be avoided by analysing the "zeroset": one dimension is reduced by the intersection of the shape with a plane[4,19] of the nuclei using very thin sections such as the ones used here (60–70 nm).

In conclusion, we have shown the important effect of resolution on the estimation of the boundary length of the nuclear perimeter. This finding questions the value of classical "shape factors" as objective measures of irregularity.

The asymptotic fractal formula permits the empirical fitting of the data. Multivariate statistics (linear discriminant function) on the asymptotic fractal parameters c, $\log(L)$ and B_m allows a correct reclassification of 76.6% of the nuclear profiles into the "normal" and "tumour" groups, using the perimeter data alone. This shows that asymptotic fractal analysis of nuclear profiles has great potential for diagnosis and prognosis of malignancy. It is objective and reproducible, and it can be automated once the object has been isolated.

ACKNOWLEDGMENT

This study was supported by The Leverhulme Trust grant No. F.94AO.

REFERENCES

1. M. Abdel-Salam, B. H. Mayall, K. Chew, S. Silverman and J. S. Greenspan, "Prediction of Malignant Transformation in Oral Epithelial Lesions by Image Cytometry," *Cancer* **62**, 1981–1987 (1988).
2. M. Abdel-Salam, B. H. Mayall, K. Chew, S. Silverman and J. S. Greenspan, "Which Oral White Lesions will become Malignant? An Image Cytometric Study," *Oral Surg. Oral Med. Oral Path.* **69**, 345–350 (1990).
3. C. M. Payne, C. G. Bjore, D. W. Cromey and F. Roland, "A Comparative Mathematical Evaluation of Contour Irregularity Using Form Factor and PERBAS, a New Analytical Shape Factor," *Anal. Quant. Cytol. Histol.* **11**(5), 341–352 (1989).
4. B. B. Mandelbrot, *The Fractal Geometry of Nature* (San Francisco, Freeman, 1982).
5. D. Paumgartner, G. Losa and E. R. Weibel, "Resolution Effect on the Stereological Estimation of Surface and Volume and Its Interpretation in Terms of Fractal Dimensions," *J. Microsc.* **121**(1), 51–63 (1981).
6. J. P. Rigaut, "An Empirical Formulation Relating Boundary Lengths to Resolution in Specimens Showing 'Non-ideally Fractal' Dimensions," *J. Microsc.* **133** Pt 1, 41–54 (1984).
7. J. P. Rigaut, "Fractals, Semi-fractals et Biomètrie," in *Fractals, Dimensions Non Entières et Applications*, ed. G. Cherbit (Paris, Masson, 1987).
8. J. P. Rigaut, "Fractals in Biological Image Analysis and Vision," in *Gli Oggetti Frattali in Astrofisica, Biologia, Fisica e Matematica*, Vol. 3, eds. G. Losa, D. Merlini and R. Moresi (CERFIM Locarno, 1991).
9. C. W. Boone, G. J. Kelloff and V. E. Steele, "The Natural History of Intraepithelial Neoplasia: Relevance to the Search for Intermediate Endpoint Biomarkers," *J. Cell. Biochem. Suppl.* **G16**, 23–26 (1992).
10. H. E. Dycht, M. Bibbo, P. H. Bartels, J. H. Puls and G. L. Wied, "A PC-based System for the Objective Analysis of Histologic Specimens through Quantitative Contextual Karyometry," *Applied Optics* **26**(16), 3270–3279 (1987).
11. P. H. Bartels, "The Diagnostic Pattern in Histopathology," *Am. J. Clin. Pathol.* **91**(1), S7–S13 (1989).
12. G. L. Wied, P. H. Bartels, M. Bibbo and H. E. Dytch, "Image Analysis in Quantitative Cytopathology and Histopathology," *Human Pathol.* **20**(6), 549–571 (1989).
13. I. R. H. Kramer, N. El-Labban and S. Sonkodi, "Further Studies on Lesions of the Oral Mucosa Using Computer-aided Analyses of Histological Features," *Br. J. Cancer* **29**, 223–231 (1974).
14. G. Landini and J. W. Rippin, "The Fractal Dimensions of the Epithelial-Connective Tissue Interfaces in Premalignant and Malignant Epithelial Lesions of the Floor of the Mouth," *Anal. Quant. Cytol. Histol.* **15**(1), 1993 (in press).

15. G. Landini, G. Misson and P. I. Murray, "Fractal Properties of Herpes Simplex Dendritic Keratitis," *Cornea* **11**(6), 510–514 (1992).
16. G. Landini, G. Misson and P. I. Murray, "Fractal Analysis of the Normal Human Retinal Fluorescein Angiogram," *Curr. Eye Res.* **12**(1), 23–27 (1993).
17. G. A. Losa, G. Baumann and T. F. Nonnenmacher, "The Fractal Dimension of Pericellular Membrane from Lymphocytes and Lymphoblastic Leukemic Cells," *Acta. Stereol.* **11**(1), 335–341 (1992).
18. J. W. Gamel, I. W. McLean, R. A. Greenberg, L. E. Zimmermann and S. J. Lichtenstein, "Computerized Histologic Assessment of Malignant Potential: A Method for Determining the Prognosis of Uveal Melanomas," *Human Pathol.* **13**(10), 893–897 (1982).
19. R. F. Voss, "Characterization and Measurement of Random Fractals," *Physica Scripta* **T13**, 27–32 (1986).

Fractals, Vol. 1, No. 3 (1993) 336–345

FORMATION OF RAMIFIED COLONY OF FUNGUS *ASPERGILLUS ORYZAE* ON AGAR MEDIA

SHU MATSUURA and SASUKE MIYAZIMA*
School of High-Technology for Human Welfare, Tokai University, Nishino 317, Numazu, Shizuoka 410-03, Japan
**Department of Engineering Physics, Chubu University, Kasugai, Aichi 387, Japan*

Abstract

Ramified colonies of fungus *Aspergillus oryzae* have been found to grow at a low growth rate on "liquid-like" agar media with low concentrations of agar and glucose. Box-counting fractal dimensions of the individual colony branches have been found to decrease with the time of incubation. Addition of glucose solution in the interior of branched colonies has brought about the production of the hyphal filaments almost only at the apical region of the colony branches. Active growth of the ramified colonies is localized in the peripheral zone, and this growth manner implies that the fungus is exhibiting a positive exploitation.

1. INTRODUCTION

The growth of fungi is sensitive to environmental conditions, such as the nutrient content, pH, water, temperature, and gaseous conditions. Several physical conditions such as the stiffness of substrate, osmotic pressure on the cell wall, etc. also influence the regulation of colony formation. Further, fungal colonies might change their shape depending on the components of the substrate and the cultivation conditions.

Generally, the fungal mycelia exhibit thick circular colonies on artificial media. Mycelium is known to be one giant cell consisting of the basal hypha, the aerial hypha, and other organs for reproduction such as conidia. Nutrients absorbed in one part of the thallus are transported to other parts. Also, as the colony expands, partitioning of the tasks occurs. Such internal communications are supposed to contribute to the pattern formation of colonies.

In our previous study, we have found that *A. oryzae* forms compact colonies on solid agar media or on glucose-rich media with various agar concentrations (i.e., various stiffness), and that the roughness of the growth front of the colonies is characterized by self-affine fractals.[1,2] Further, we have found fairly ramified colonies generated on media with low agar concentrations when the glucose concentration is lowered from about 0.1 to 0.01 wt%. Particularly, the shapes of the branched colonies grown at 24°C from line seeds on the "liquid-like" medium of 0.15 wt% agar, with 0.01 wt% glucose, resemble diffusion-limited aggregations (DLA).

Hypha secretes various enzymes to decompose and absorb nutrient materials. Diffusion of nutrient toward the mycelium or the individual hypha is supposed to be one of the important factors which contribute to the mycelial pattern formation. In the above system, the diffusion of glucose is thought to be easier in the medium of low agar concentrations.

In this study, we focus our attention to the shape and growth of the ramified colonies, measuring the box-counting fractal dimensions to characterize their shape and growth. Box-counting fractal dimension D_{box} is determined by the following method. Digitized images of colonies are meshed with the square lattices of size ε. Then, we count the number of lattices $N(\varepsilon)$ occupied by the colony image. When the colony shape is fractal, a scaling form hold as,

$$N(\varepsilon) \sim \varepsilon^{-D_{\text{box}}} . \tag{1}$$

Further, to check the effect of nutrient diffusion, we add the glucose solution in the interior of the ramified colony and observe the response of colony growth. We also discuss the formation of ramified colonies as an active strategy to survive an unfavorable environment.

2. EXPERIMENTAL METHODS

The strain used is *Aspergillus oryzae* (supplied by the Institute for Fermentation, Osaka), cultivated on the Czapek synthetic agar medium containing $NaNO_3$ (0.3 wt%), K_2HPO_4 (0.1 wt%), $MgSO_4 \cdot 7H_2O$ (0.05 wt%), KCl (0.05 wt%), $FeSO_4 \cdot 7H_2O$ (0.001 wt%), glucose (0.01 wt%), and Difco Bact-Agar (0.15 wt%). Stiffness of the medium is controlled by the amount of agar. The medium of the above agar content is viscous fluid, and let us call it "liquid-like" medium. 25 and 75 ml of sterile medium are poured into petri plates of 9 and 15 cm in diameter, respectively. For inoculation, a sterile string is inserted into spore suspension, and an inoculation line is drawn manually on the surface of medium. The plates are incubated at 24°C. No tape seal is applied to the plates, in order to sustain the ventilation for long period cultivations. For glucose addition experiments, the solution of 2 wt% glucose is prepared and dropped on the medium using a sterile pipette after 15 and 20 days incubation.

For the box-counting fractal analysis, macro photographs of colonies are taken with a 35 mm camera. Colonies are illuminated from the bottom side by a halogen lamp. The black-and-white prints of colony images are digitized using an image-scanner. The resolution of images is approximately 600 × 400 pixels at the maximum. The images of individual branch colonies of different sizes are enlarged and digitized with the above resolution.

3. RESULTS AND DISCUSSION

3.1 Shape and Growth of Ramified Colony

Examples of *A. oryzae* colonies formed on the liquid-like media with 0.01 wt% glucose are shown in Fig. 1. As seen in the colony from line inoculum [(Fig. 1(a)], branched sub-colonies extend from the inoculation line perpendicularly to the uncolonized medium space. There seems to exist a competition among the sub-colonies. Many sub-colonies cease growing in

(a)

(b)

Fig. 1 *Aspergillus oryzae* colonies cultivated on the Czapek agar media with 0.01 wt% glucose and 0.15 wt% agar at 24°C. (a) 46 days after line inoculation, and (b) 50 days after point inoculation. The diameter of plates is 9 cm.

the earlier stages, and a few of them continue to grow sharing the medium space with each other. In the individual sub-colonies, although small branching occurs rather frequently, only a few strong branches keep extending.

It is known that different compartments of a hypha have each their own age.[3] For example, along the hypha, the nuclear synthesis decreases in the older portions. Also, each compartment of the hypha has its physiological age, corresponding to its biosynthetic activity. Physiological age is not necessarily identical to the absolute age of the compartment, because the biosynthetic activity can be recovered in older compartments when the branching occurs.

In our ramified colonies, the hyphae in the apical zone of strong branch continue to extend and branch out. In other positions of the mycelium, hyphae cease extending and form the chains of conidia at their apical portions. Occurrence of these phenomena implies that the task of the hypha has changed from vegetative growth to reproduction through physiological ageing. Under an unfavorable condition, it may be advantageous for the mycelium to localize the active growth area for exploring the substrate, rather than to sustain all hyphae active, delivering nutrients absorbed to the hyphae.

Figure 1(b) shows a colony grown from a point inoculum. Sub-colonies in this culture branch out rather widely and the hyphae spread more homogeneously in comparison with the line inoculation culture. In addition, the distance between neighboring branch colonies is apparently smaller in the point inoculation cases. There might be a long range inhibitory interaction between neighboring branch colonies extending from the line inoculum. These branch colonies from the line inoculum consist of denser hyphae and have clearer outlines than the effusing colonies found in the circular colony.

The mean heights of the colonies are plotted against the cultivation time in Fig. 2, and the growth of ramified colonies on liquid-like medium is compared with that of homogeneous

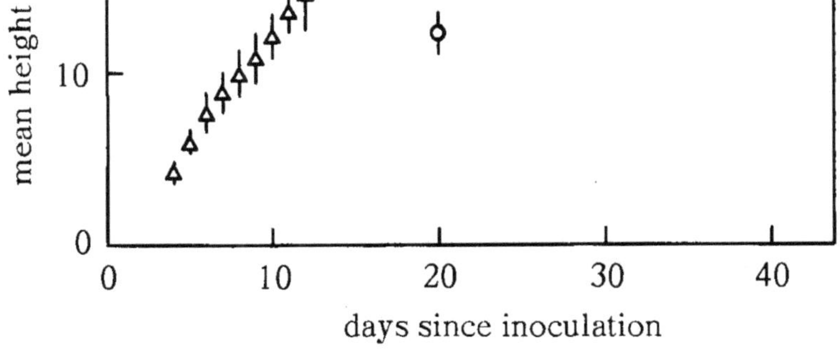

Fig. 2 The mean height of *A. oryzae* colonies vs. days after inoculation, for colonies cultivated on media with 0.01 wt% glucose at 24°C. ○: 0.15 wt% agar, and △: 1.5 wt% agar. The symbols indicate mean height averaged over 26 or 25 colonies for ○, and 10 colonies for △, respectively. The error bars indicate standard deviations.

colonies on solid agar medium with 1.5 wt% agar. The mean height of colony is defined by the mean value of the distance from the inoculation line to the front point over the width of inoculation line. For ramified colonies, the colony height is measured for growing branch colonies. The growth rate of colony height on liquid-like medium is 0.58 ± 0.01 (mm/day), whereas that on solid medium is 1.26 ± 0.04 (mm/day). Thus, it can be stated that the liquid-like medium is unfavorable for fungal growth, despite the fact that diffusion of the nutrient must be faster.

Stiffness of the liquid-like medium may not be substantial enough to support the thallus. As a result, the basal hypha tends to extend into the substrate, and the production of the aerial hyphae, which contributes mainly to the respiration of the mycelium, is not found to be frequent.

Examples of ln-ln plot of N and ε obtained from the individual branch colony are shown in Fig. 3. The scaling relation of Eq. (1) is well satisfied, so that the colony shapes are statistically self-similar. Further, the averaged box-counting fractal dimension D_{box} of the individual branch colonies is plotted against the time of cultivation in Fig. 4. Although the value of D_{box} is near to that of the DLA, D_{box} obtained between 20 and 40 days is found to decrease.

To check whether or not the decrease in D_{box} results from the shortage of the medium volume (i.e., the amount of nutrient), we measured D_{box} of the colonies grown on a petri plate of 15 cm in diameter, and plotted the results in Fig. 4. As seen in the figure, D_{box} of colonies on 15 cm plate does not significantly differ from that on 9 cm plate, and is found to decrease with the incubation time. Thus, the decrease in D_{box} for branched colonies from line seed is not attributable to the boundary effect of the medium plate.

When the hyphal production is diminished, the amount of nutrient absorbed by the mycelium will become insufficient for the production of dense hyphae. Rather, the hyphae in the interior of the colony will lose their growth ability through physiological ageing, and only the hyphae in the apical region, where the nutrient is not yet consumed, will continue to extend.

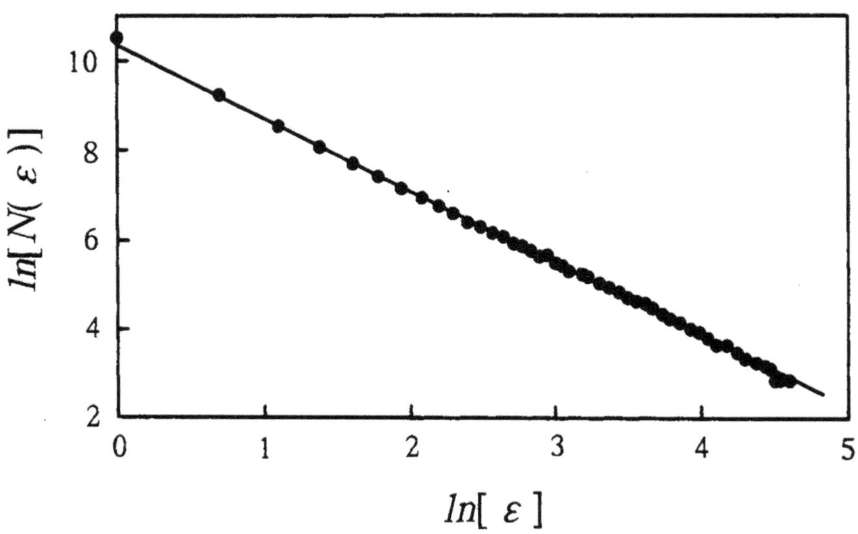

Fig. 3 An example of ln-ln plot of the number of lattice cells N vs. cell size ε for a single colony branch of *A. oryzae* grown for 40 days. The range of ε is from 1 to 100 pixels. The line shows the slope of −1.61.

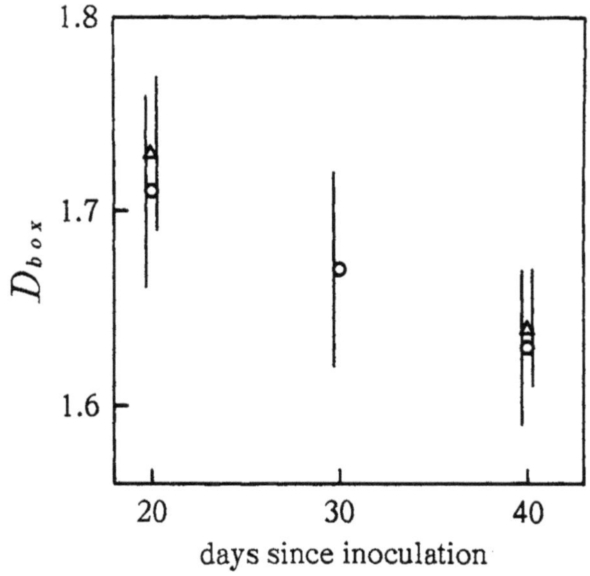

Fig. 4 Box-counting fractal dimension D_{box} of *A. oryzae* colony branches cultivated on the liquid-like media plotted against the days after inoculation. ○: for colonies formed on 9 cm petri plates, and △: 15 cm plates, respectively. The points indicate the averaged values over 26 or 25 colony branches for ○, and 7 branches for △. The error bars indicating standard deviations for ○ is drawn at the left side of the symbol, and at the right for △.

Figure 5 shows an example of the growth of a single colony branch. Here, parts of a branch are photographed through the microscope, and they are reconstructed into an image of the branch in a mosaic way. Leading hyphae extend from apical hyphal system toward the outer space, and secondary hyphae fill the apical growth zone. Particularly, as clearly seen in the photographs of 25 to 28 days, leading hyphae branch out to form a small thallus opening toward the outer space. Then, the strong branches extend out of the local thallus, and they form the next open-shaped thallus. Colony pattern formation might be regarded as a repetitive generation of this formative unit of open-shaped thallus.

3.2 Response of Colony Shape to the Addition of Glucose

From the view point of nutrient diffusion, the peripheral portions of branches will absorb the nutrient from the outer space, and they will shield the nutrient diffusion into the interior of the colony. However, there might exist some peculiar control of mycelial growth. To check if such a control exists, we observe the response of colony growth to the addition of nutrient at a position of the colony.

Figure 6 shows the growth of a single branch after addition of 0.1 ml of 2 wt% glucose solution near the top of the apex. In the first 2 days, hyphae are produced densely at the apical portion of the branch (i.e., the formative unit becomes thick). Then, a few leading hyphae are found to extend. These leading hyphae do not have dense secondary hyphae. These facts suggest that the branching frequency firstly increases and absorb nutrient, and then, strong hyphae are produced using materials which are produced and transported from the dense secondary hyphae.

Finally, let us observe how the colony grows when 0.2 ml of 2 wt% glucose solution is added in the interior (i.e., older portions) of the branched colony. Figure 7 shows the mosaic

Fig. 5 Growth of a single colony branch of *A. oryzae*. Days of cultivation is 19, 20, 21, 22, 23, 25, 26, and 28 days, respectively, from left to right.

Fig. 6 Growth of a single branch colony of *A. oryzae* after addition of 0.1 ml of 2 wt% glucose solution near the top of the colony (upside of the photographs). Upper left is the colony before glucose addition cultivated for 15 days. Time since glucose addition is 28, 40, and 50 hours from upper left to right, and 64, 90, 97, 119 hours from bottom left to right, respectively. The arrows indicate the leading hyphae.

photographs of two neighboring colony branches before and after the addition of glucose, i.e., glucose solution was added at the vacant space between two branches. After the addition, hyphal production occurs at the apical portions in one day, and then, leading branches extend from dense apical area toward the outer space in 3 days. Meanwhile, there is little change in the older portions. These facts are quite equivalent to those found in Fig. 6.

As the new hyphae are not produced in the interior of the colony, the inner hypha is thought to be inactive as the vegetative organs. Therefore, the growth ability of any positions of the colony is not equivalent to each other, and the inhomogeneity of the nutrient concentration seems not to be a main factor which causes the ramification. Thus, the growth of ramified colony is not necessarily stated as diffusion-limited.

Besides, the strong leading hyphae produced at the apical portion are directed toward the outer fresh medium space, rather than toward the position where glucose was added. In the hyphal growth, the cytoplasmic flow toward the top of the hypha, turgor pressure, chemical autotropism, accumulation of the toxic materials, etc. is thought to contribute to the determination of growth directions. The active portions are sited in the peripheral zone of the whole colony, so that the whole colony always expands and the peripheral area always exploits outside.

(a)

(b)

Fig. 7 Growth of two single branch colonies of *A. oryzae* before and after addition of 0.2 ml of 2 wt% glucose solution at a position between and near the bottom of branches (the position indicated by the letters "1 day after addition" in (b)). **(a)**: colonies before glucose addition cultivated for 20 days, **(b)**: 1 day after addition, and **(c)**: 2 days, respectively. The arrows indicate the leading hyphae.

(c)

Fig. 7 (*Continued*)

4. CONCLUDING REMARKS

The shape of the ramified colony is similar to the DLA. However, the change of shape due to the additional nutrient exhibits a positive exploitation of whole colony. The low stiffness of the substrate is an unfavorable condition for growth. The hyphae in the interior of the colony are made inactive and differentiate into the reproducible organs. Absorbed nutrient will be transported to the active peripheral region to sustain the growth of strong leading hyphae.

Formation of open-shaped thallus of apical hyphae and the production of leading hyphae within the thallus may play an important role in forming the branched structure. Also, the shape and size of the local thallus might be related to the amount of nutrient actually utilized.

Many fungi are known to produce peculiar root-like aggregation of hyphae, called rhizomorph, in the soil. The strand hyphae within the rhizomorph are thought to take part mainly in the transport of nutrients, etc. Although it is not certified that fungi have a potentiality to construct the structures of rhizomorph, the ramified colony we found might be a direct example of the prototype of such structures.

REFERENCES

1. S. Matsuura and S. Miyazima, *Fractals* **1**, 11 (1993).
2. S. Matsuura and S. Miyazima, to appear in the proceedings of *1st Int. Symp. on Fractals in Biology and Medicine*, (Birkhauser-Verlag, Basel).
3. G. T. Cole and B. Kendrick (editors), *Biology of Conidial Fungi*, Vols. 1 and 2, (Academic Press, New York, 1981).

Fractals, Vol. 1, No. 3 (1993) 375–379

SIMULATION OF ENVIRONMENTALLY INDUCED GROWTH FORMS IN MARINE SESSILE ORGANISMS

JAAP A. KAANDORP
Department of Computer Science, University of Amsterdam,
Kruislaan 403, 1098 SJ Amsterdam, The Netherlands

Abstract

In marine sessile organisms (e.g. sponges and stony-corals), a distinction can be made in organisms that use filter-feeding, light or a combination of both, as a source of energy. In many of these organisms, growth takes place by the addition of layers of new material on top of the preceding growth stages that remain unchanged. The growth process of an organism that uses exclusively light and one that uses only filter-feeding for energy intake, are simulated using an iterative geometric construction in combination with, respectively, a light intensity function and a diffusion equation.

1. THE MODEL

Marine sessile organisms, as for example sponges and stony-corals, use light, filter-feeding by trapping suspended material from the environment, or a combination of both as an energy source. In the case where light is exclusively being used for photosynthesis, only light and inorganic substances are required for the growth process, and the organism is indicated as phototrophic. In the case where filter-feeding is the only energy input, the growth process is influenced by the local concentration of (organic) suspended material and the organism is called heterotrophic. In many of these organisms (e.g. many stony-corals) a combination of heterotrophic and the phototrophic energy source is used.

The growth process of these organisms is mimicked with a three-dimensional morphological model.[1] In this model, the addition of new layers of skeleton material to the organism as well as the influence of the physical environment on the growth process is simulated. The simulation is done in an iterative geometric construction, where in each iteration step, a layer of triangles is constructed on top of the layers of the preceding iteration steps. The basic construction is shown in Fig. 1. The construction starts with a triangulated sphere. In the simulations, the objects are basically represented as layers of triangles in continuous space and for computational reasons also a discrete version of the geometric model is used. The discrete version is obtained by mapping the triangles[2] situated at the outer surface of the model on a 3D lattice. The edges of the triangles, as well as the connections between two succeeding layers, represent skeleton elements (for example, spicules in sponges or edges of corallites in stony-corals) of the actual organisms.[3,1] The connecting edge between two succeeding layers is indicated as the longitudinal element. The length l of this element is, in the simulations of the exclusively heterotrophic organisms, determined by:

$$l = \begin{cases} s \cdot k(c) \cdot \bar{r} & \text{for } k(c) \cdot \bar{r} > 0.02s \\ 0.0 & \text{for } k(c) \cdot \bar{r} \leq 0.02s \end{cases} \tag{1}$$

In the simulations of the exclusively phototrophic organisms, l is determined by the function:

$$l = \begin{cases} s \cdot I(\theta) \cdot \bar{r} & \text{for } I(\theta) \cdot \bar{r} > 0.02s \\ 0.0 & \text{for } I(\theta) \cdot \bar{r} \leq 0.02s \end{cases} \tag{2}$$

The length l in both Eqs. (1) and (2) is expressed in a constant s, representing the size of the basic building elements, the two functions $k(c)$, $I(\theta)$ and the variable $\bar{r}$ being all within the range 0.0–1.0. In both equations, the threshold $0.02s$ is used for l. In the actual objects, the longitudinal elements are constructed from discrete elements which cannot become arbitrarily short. Below this threshold no material is added; vertices in the geometric construction of Fig. 1 with a longitudinal element with l above this threshold are indicated as active. The direction of the longitudinal element (V_1, V_2) is determined in the basic

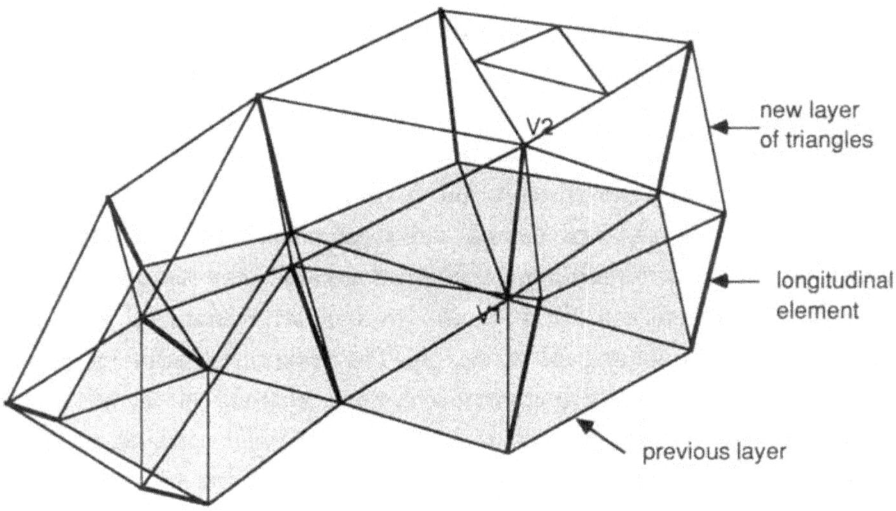

Fig. 1 Construction of a new layer of triangles on top of the preceding layer. The triangles of the preceding layer are displayed in gray.

construction of Fig. 1, by the mean value of the normal vectors of the triangles that surround the vertex V_1 of the preceding layer.

The function $k(c)$ in Eq. (1) represents the influence of the local concentration gradient of suspended material on the growth process. The local concentration gradient are determined by using the lattice version of the simulated object. The object is first mapped on a lattice; for this purpose in the simulations of the heterotrophic organisms, a lattice of 100^3 sites is used, where each lattice site contains a concentration value in the range 0.0–1.0. After this, a solution Laplace equation is approximated in a similar procedure as discussed in a 2D version of the model[4] and the concentration for each lattice site and the gradients are estimated. In the approximation, it is assumed that the concentration on the object itself is zero. The lattice sites situated on the surface and in the inside of the object, as well as the lattice sites at the bottom plane of the lattice are set to the value zero. In the approximation, a linear source is assumed by setting the lattice sites of the top plane to the value 1.0. With this approximation, the concentration distribution of suspended material around a filter-feeding organism under sheltered conditions is mimicked.

With the function $I(\theta)$, the influence of the local light intensity on the growth process of the phototrophic organisms is modeled. In the determination of the local light intensity on the surface of the object, a lattice version of 400^3 sites of the geometric model is used. The local light intensity is calculated in each triangle using the cosine of the angle of incidence θ between the normal vector of the triangle and the light source. In the simulations, a parallel light source is assumed and the light direction corresponds to the vertical. The lattice version is used to estimate the part of each triangle that is shaded by parts of the object situated above the considered triangle. After the local light intensities are corrected for this shading effect, $I(\theta)$ is used to determine l.

The growth velocities that develop in a newly added layer in both the phototrophic[5] and heterotrophic[3] organisms, depend on the capabilities of the system with which nutrients are transported through the tissue of the organism. The access to the suspended material in the environment in the heterotropic organisms and the access to inorganic substances necessary for building the skeleton in the phototrophic ones can be described as a quotient of the surface through which the substance can pass and the volume (with a certain distance from the surface) being supplied with nutrient. The amount of contact with the environment is estimated by measuring the local radius of curvature in a vertex in several tangential directions on the surface of the object. These local radii of curvature are measured by selecting a set of pairs of vertices surrounding the considered vertex at a certain distance. First, all radii of curvature are normalized, the normalization function being the same as used in the 2D models and described in Ref. 3. The amount of contact with the environment is expressed as the average normalized radius of curvature $\bar{r}$. In the actual organism, the amount of contact with environment will decrease when the local radius of curvature increases, while $\bar{r}$ decreases. For $\bar{r} = 0.0$, the contact is minimal and for the value 1.0, maximal. When a certain critical value, i.e., the maximum radius of curvature, which is related to the capabilities of the transport system, no nutrient at all will be transported and the growth velocity will become zero. The consequence is that as soon as the object flattens too much, locally the thickness of the new layer decreases and the object splits into branches. This process of branch formation is depicted in Fig. 2.

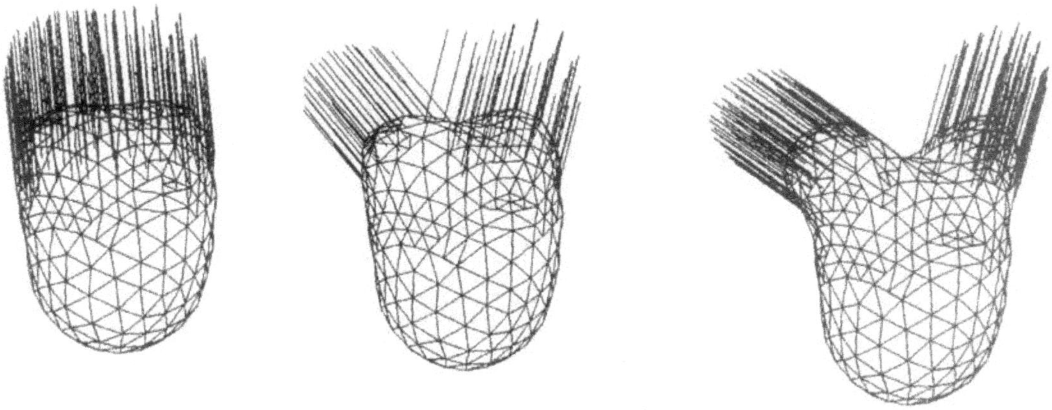

Fig. 2 Objects subsequently emerging in the iteration process. The active vertices are displayed with an edge. After a few iteration steps, the original patch of active vertices is separated into two new ones.

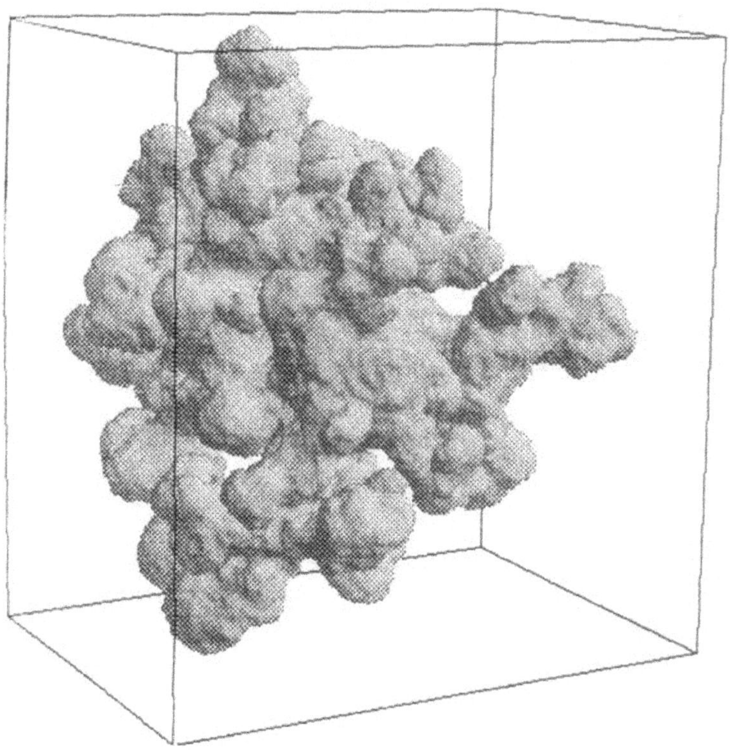

Fig. 3 Simulation of a heterotrophic organism using Eq. (1).

In Fig. 3, a simulated growth form of a heterotrophic organism is shown; in the geometric construction Eq. (2) is applied. The box dimension[6] of this object on the scale range $1.0s$–$8.5s$ is 2.3. Figure 4 which shows a simulated form of a phototrophic organism, was obtained by applying Eq. (2). The box dimension of this object on the scale range $1.0s$–$8.5s$ is 2.1.

Fig. 4 Simulation of a phototrophic organism using Eq. (2).

ACKNOWLEDGMENTS

This research was done partly under a Government of Canada Award at the University of Calgary in Canada. The author would like to thank Przemyslaw Prusinkiewicz (University of Calgary) for his helpful comments and Marcel Wijkstra (University of Amsterdam) for his help with the visualization.

REFERENCES

1. J. A. Kaandorp, in *Applications of Fractals and Chaos*, eds. A. J. Crilly, R. A. Earnshaw and H. Jones, (Springer-Verlag, New York, 1993), pp. 41–61.
2. A. Kaufman, *Comput. and Graphics*, **12**, 213–219 (1988).
3. J. A. Kaandorp, *Mar. Biol.* **110**, 203–215 (1991).
4. J. A. Kaandorp, in *Growth Patterns in Physical Sciences and Biology*, eds. E. Louis, L. M. Sander, P. Meakin and J. M. Garcia-Ruiz (Plenum Press, New York, 1993), pp. 331–340.
5. B. Rinkevich and Y. Loya, *Coral Reefs*, **1**, 243–247 (1983).
6. J. Feder, *Fractals* (Plenum Press, New York, 1988).

Fractals, Vol. 1, No. 3 (1993) 380–387

TEMPORAL FLUCTUATIONS IN BIORHYTHMS: EXPRESSION OF SELF-ORGANIZED CRITICALITY?

KLAUS-D. KNIFFKI, WOLFGANG MANDEL* and PHUOC TRAN-GIA*
Physiologisches Institut and Institut für Informatik,*
Röntgenring 9 and Am Hubland, Universität Würzburg,
D-97070 Würzburg, Germany

Abstract

Recently, a general organizing principle has been reported connecting $1/f$-noise with the self-similar scale-invariant 'fractal' properties in space, hence reflecting two sides of a coin, the so-called *self-organized critical state.* The basic idea is that dynamical systems with many degrees of freedom operate persistently far from equilibrium at or near a threshold of stability at the border of chaos. Temporal fluctuations which cannot be explained as consequences of statistically independent random events are found in a variety of physical and biological phenomena. The fluctuations of these systems can be characterized by a power spectrum density $S(f)$ decaying as f^{-b} at low frequencies with an exponent $b < 1.5$. We present a new approach to describe the individual biorhythm of humans using data from a colleague who has kept daily records for two years of his state of well-being applying a fifty-point magnitude category scale. This time series was described as a point process by introducing two discriminating rating levels R for the occurrence of $R \geq 40$ and $R \leq 10$.

For $b < 1$ a new method to estimate the low frequency part of $S(f)$ was applied using counting statistics without applying Fast Fourier Transform. The method applied reliably discriminates these types of fluctuations from a random point process, with $b = 0.0$. It is very tempting to speculate that the neural mechanisms at various levels of the nervous system underlying the perception of different values of the subjective state of well-being, are expressions of a self-organized critical state.

1. INTRODUCTION

A variety of phenomena in nature exhibit temporal fluctuations in the absence of intentional stimulation which cannot be explained as a consequence of statistically independent random events. It has been shown that temporal fluctuations found in phenomena as different as membrane currents, earthquakes, sunspot activity, light emitted from quasars, sand falling through an hour glass, traffic flow, heart beat or breathing activity, can be characterized by their power spectrum density $S(f)$ decaying as f^{-b} at low frequencies with $0.5 \leq b \leq 1.5$. This behavior of the temporal fluctuations of a system described by its $S(f)$ is called $1/f$-noise.

Recently, it has been suggested[1] that the large fluctuations in time characterized as $1/f$-fluctuations and the self-similarity in space might both be manifestations of a self-organized critical state. Self-organized criticality (SOC) describes the tendency of some open dissipative many-body systems to drive themselves spontaneously to a critical state with no characteristic time or length scales without any fine-tuning by external fields: hence the criticality is self-organized. This is in contrast to the criticality of equilibrium systems undergoing phase transition only at a critical external field, such as temperature, pressure, electrical or magnetic field. The idea provides a unifying concept for large scale behavior in systems with many degrees of freedom operating persistently far from equilibrium at or near a threshold of instability, so to speak on the "border to chaos".[2]

The SOC phenomenon is expected to be quite universal and we assume that it is the underlying principle of some biological many-body systems. We present a new approach to describe the individual biorhythm in humans using data from a colleague who has kept standardized daily records for two years of his state of general well-being applying a fifty-point magnitude category scale and analyze the temporal fluctuations by estimating the power spectrum density in its low frequency range to characterize the self-similar temporal rating sequences.

2. METHODS

The subjective intensity of well-being was measured with a linear category scaling procedure (category partitioning)[3] with five categories each subdivided in ten steps: 1–10: very strong "down", 11–20: strong "down", 21–30: moderate "down"/moderate "up", 31–40: strong "up", 41–50: very strong "up". Thus, the subject could, after choosing a major category, fine-tune the rating of well-being by choosing a number within that main category. In general, the daily ratings were performed at 6.00 a.m. and stored for subsequent analysis. Occasionally, fluctuations within a day of the subjective well-being were observed, but were not monitored and therefore neglected in this analysis.

As shown in Fig. 1, the time series of the daily ratings R can be described as a point process by introducing discriminating rating levels for the occurrence of $R \geq r$, e.g. for the occurrence of "ups" (cf. Fig. 2) and $R \leq s$, e.g. for the occurrence of "downs".

Usually $S(f)$ is obtained by Fast Fourier Transform (FFT). To avoid the well-known problems in using FFT for the obtained point process, we used a new simple method based on counting statistics[4] to analyse the low frequency part of $S(f)$ of the recorded ratings of human general well-being.

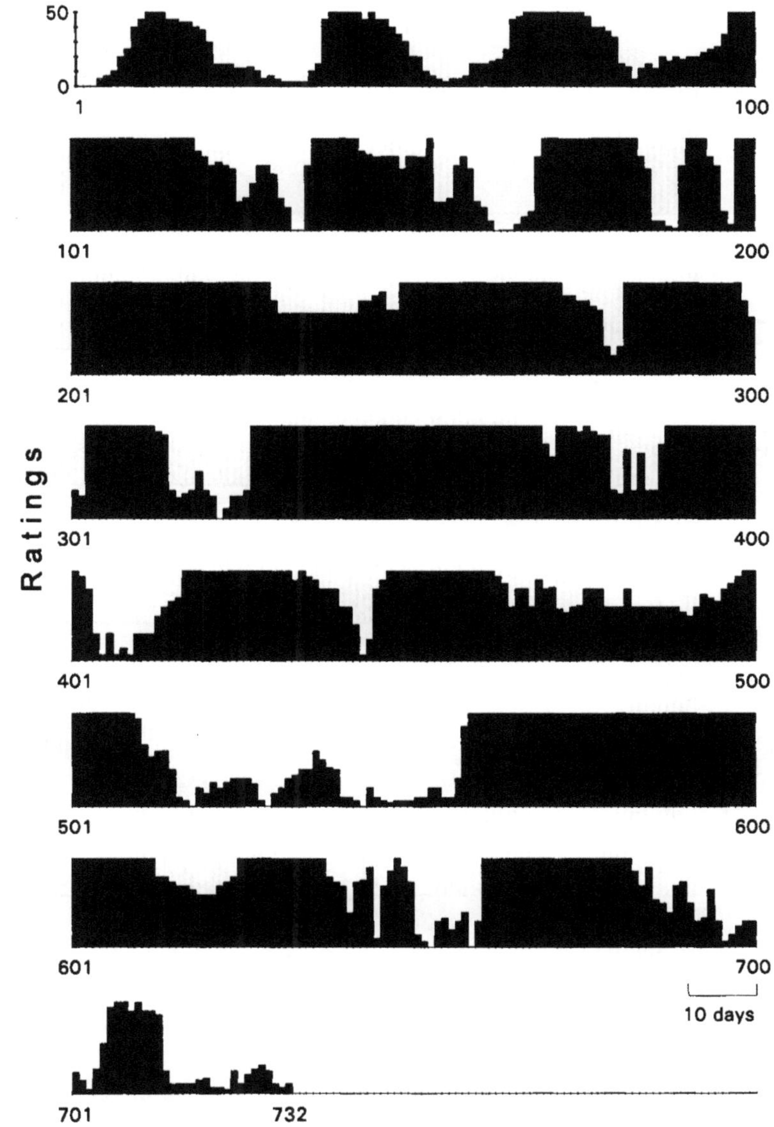

Fig. 1 Daily ratings of the subjective well-being (biorhythm) for two years using a linear category scaling procedure.

The series of recorded ratings after introducing a discriminating rating level is considered to be a point process described as

$$y(t) = \sum_{i=1}^{n} \delta(t - t_i) \tag{1}$$

in which $\delta(t - t_i)$ represents Dirac's delta function and t_i is the time of occurrence of a particular $R \geq r$ or $R \leq s$ within the train of n events. In the absence of severe intentional stimulation $y(t)$ is assumed to be statistically stationary. Another statistical variable derived from Eq. (1) is the actual number of events $N(\Delta t)$ occurring in a time interval Δt ranging from t_1 to t_2. Thus, $N(\Delta t)$ can be expressed as

$$N(\Delta t) = \int_{t_1}^{t_2} \sum_{j}^{\nu} \delta(t - t_j) dt \,. \tag{2}$$

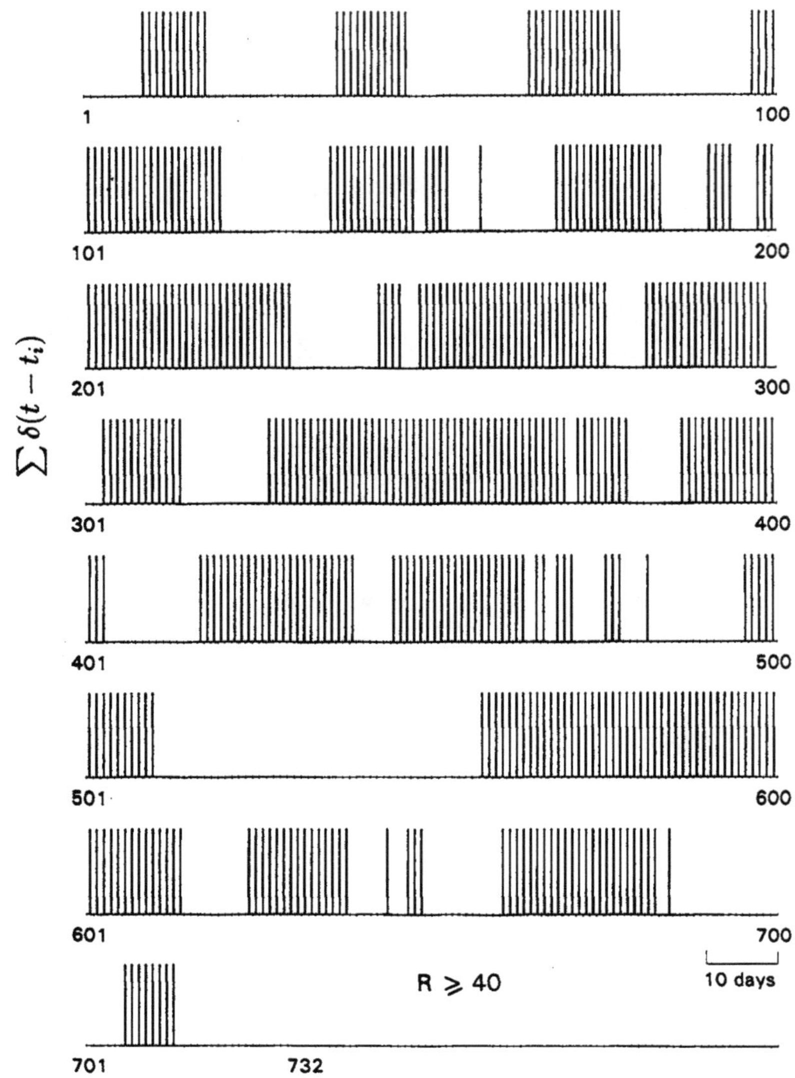

Fig. 2 Occurrence of ratings of the subjective well-being with $R \geq 40$ of the entire data set shown in Fig. 1. The corresponding days are marked by Dirac's delta functions $\delta(t - t_i)$.

The second time derivative of the variance of counts $Var[N(\Delta t)]$, the so-called variance-time curve, is related to the auto-covariance function of y, $C_y(\Delta t)$ by[5]

$$C_y(\Delta t) = \frac{1}{2}(Var[N(\Delta t)])'' \tag{3}$$

and therefore the key to determine the low frequency part of the spectrum $S_y(f)$ is to experimentally obtain $Var[N(\Delta t)]$.[4] If the variance-time curve follows within certain limits a power law

$$Var[N(\Delta t)] \sim (\Delta t)^{1+b} \quad \text{with } b < 1\,, \tag{4}$$

then it can be shown using the Wiener-Chinchin theorem that the spectrum $S_y(f)$ scales as

$$S_y(f) \sim f^{-b} \tag{5}$$

within $f_{\min} < f < f_{\max}$.[4,6]

The variance-time curve is defined by the variance of counts for time intervals of length Δt as

$$Var[N(\Delta t)] = \langle N^2(\Delta t)\rangle - \langle N(\Delta t)\rangle^2 \tag{6}$$

with $\langle\cdots\rangle$ denoting expectation values. For estimating $Var[N(\Delta t)]$, the entire observation time T is divided into k counting windows of duration Δt with $T = k\Delta t$ and the variance of counts is determined for this particular window Δt. This is repeated for different values of Δt. The results were plotted as $Var[N(\Delta t)]$ vs Δt on a log-log scale and fitted by linear regression using the least square method.

3. RESULTS

In Fig. 1 the whole data set is shown, i.e., the daily ratings for two years are displayed. It is obvious from the data, that the state of subjective well-being is not constant but fluctuates in general from day to day. By no means these fluctuations taken as a whole are simple oscillations describable by a sine function as it is often assumed by performing the so-called biorhythm analysis.[7] In a rough approximation the data look as if the basic underlying mechanism responsible for the subjective well-being is a two-state ("up"–"down") system with a certain endogenic dynamics.

By introducing discriminating rating levels for the occurrence of $R \geq r$ to reveal the fluctuations in the "up", the data shown in Fig. 1 were transformed into a point process. To obtain Fig. 2 the discriminating rating level was set to $R \geq 40$, i.e. the point process shows the fluctuations of the strong "ups" of the subjective well-being irrespective of their actual rating. Similar point processes showing the clusters of events described as $\delta(t-t_i)$ for other discriminating rating levels were obtained and analyzed. In particular, for determining the fluctuations in the occurrence of the "downs" $R \leq 10$ were chosen.

After introducing a certain $R \geq r / \leq s$, for the resulting point process the low frequency part of the spectrum $S(f)$ was determined by using counting statistics as described in Methods. Fig. 3 shows the result of the point process shown in Fig. 2, i.e. the $Var[N(\Delta t)]$ is plotted on a log-log scale versus the counting windows Δt. From the straight lines fitted to the data points it is demonstrated that the variance-time curve follows the power law

$$Var[N(\Delta t)] \sim (\Delta t)^{1+0.65} \tag{7}$$

for the scaling region $1d \leq \Delta t \leq 15d$ and thus the low frequency part of the spectrum scales as

$$S(f) \sim f^{-0.65}\,. \tag{8}$$

For $\Delta t > 15d$ a second scaling region was observed showing an almost random behavior ($b = 0.17$).

Similar results, i.e. similar scaling behavior for the variance-time curve and for the spectrum was obtained for other discriminating levels, in particular for $R \leq 10$, i.e. for the "downs".

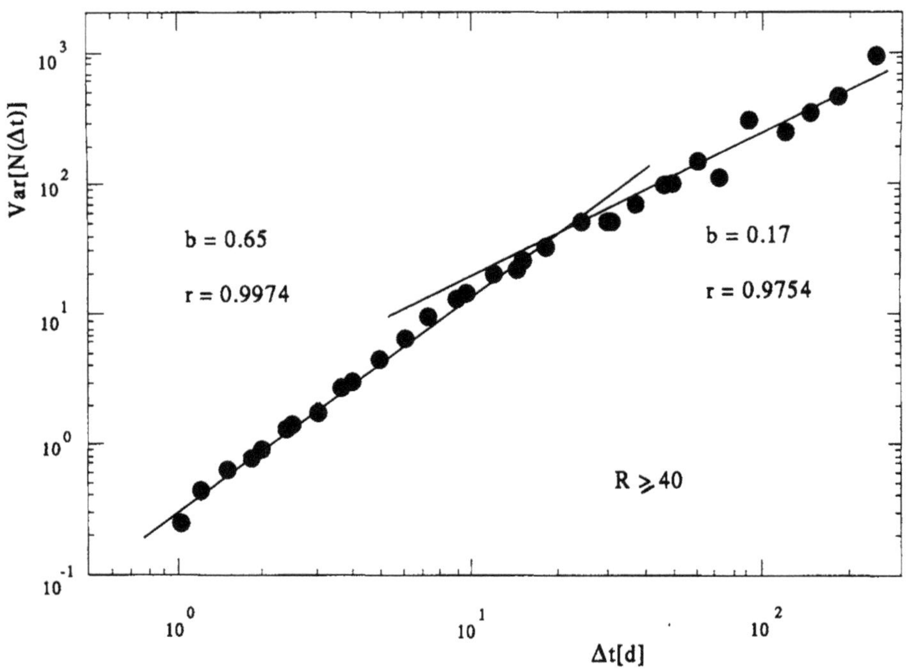

Fig. 3 The variance-time curve $Var[N(\Delta t)]$ for the point process shown in Fig. 2 plotted on a log-log scale versus the counting windows Δt. The variance-time curve scales as $(\Delta t)^{1+b}$ for $1d \leq \Delta t \leq 15d$ with $b = 0.65$ obtained by linear regression using the least square method with the indicated correlation coefficient r. For the second scaling region, i.e. for $\Delta t \geq 15d$ the exponent $b = 0.17$ indicates more random fluctuations of the ratings $R \geq 40$.

4. DISCUSSION

Recently, also due to the introduction of the concept of self-organized criticality,[1] attention has been drawn to the characterization of temporal fluctuations in a number of physical and biological systems. In the following the discussion will be focused on the fluctuations of endogenous biological rhythms.

The human heart rate, even in the healthy resting subject, displays a considerable amount of fluctuations, which have been characterized as $1/f$-fluctuations.[8–11] Furthermore, it was demonstrated that the heart rate variability of healthy men shows periods of $1/f$-fluctuations with interpolated periods of white noise within 24 hours.[4,13]

In animal experiments it has been demonstrated that the fluctuations in respiratory intervals also exhibited $1/f$-fluctuations, but these characteristic types of fluctuation disappeared into white noise fluctuations when the end-tidal p_{CO_2} was raised to 50 or 60 mmHg.[14]

The fluctuating insulin requirements of an unstable diabetic over an eight-year period have been subjected to spectral analysis and it was demonstrated that the low frequency part of the spectrum did exhibit $1/f$ characteristics.[15]

Recently, the spectral analysis of the discharge of neurones located in the mesencephalic reticular formation during paradoxical sleep of the cat has revealed that in this state of the animal $1/f$-fluctuations of the neuronal discharge do exist. However, the low frequency spectral profile became flat, i.e. white noise was found during slow-wave sleep.[16,17] So far, also the thalamic neuronal discharge exhibited $1/f$-fluctuations in the absence of intentional stimulation, but we have not seen the transition into white noise fluctuations.[18,19] Earlier, even for the discharges in primary afferent auditory fibres $1/f$ characteristics have been reported.[20]

It is tempting to speculate that the basic mechanisms underlying the neuronal and humoral activity in the central nervous system responsible for the subjective state of well-being in the absence of intentional stimulation are expressions of a self-organized critical state as introduced for physical systems.[1] Self-organized criticality (SOC) describes the tendency of dissipative systems with many degrees of freedom to drive themselves to a critical state with a wide range of length and time scales without any fine-tuning of external fields. The idea complements the concept of chaos, wherein simple systems with a small number of degrees of freedom can display quite complex behavior.[21]

Currently, it is hard to give a rigorous definition for SOC; however, usually this name is given to those systems which do not need fine-tuning of external fields to give power-law characteristics for the parameters describing the system. The canonical example of SOC is the cellular automaton model called "sand-pile model".[1] The critical state is characterized by "avalanches" (activity) with power-law spatial and temporal distribution functions limited only by the size of the system. We assume that the subjective well-being dynamics can be described as a self-organized critical process and characterize the temporal fluctuations by its low frequency part of the power spectrum. The method applied reliably discriminates f^{-b} fluctuations with $b = 0.65$ in our case in the first scaling region (cf. Fig. 3) from a random point process, which would result in $b = 0.0$.

The exponent b will converge to the theoretical value only for infinite long sequences of $R\delta(t - t_i)$.[4] Since in praxi the ratings are confined to limited time periods $T_{\max}$, the exponent b will be a function of $T_{\max}$, i.e. $b = b(T_{\max})$. In our case $T_{\max} = 2$ years, which could result in an underestimation of the exponent b (cf. Ref. 4). In any case, the time sequence of events with $R \geq 40$ (Fig. 2) is not the representation of a Poisson process, i.e., of white noise. Therefore, without external stimulation based on one's own monitored biorhythm for a given time period it should be possible to predict future episodes with a certain probability by applying modified feed-forward neural networks learning with the backpropagation algorithm.

If the neuronal/humoral system responsible for the subjective well-being is indeed operating at a critical state, an external perturbation can create either a small effect or a large one. There is in principle no limit on how long the effect may last. The degree of unpredictability is actually less severe than for chaotic systems; SOC systems are operating at the "border of chaos".[2] In SOC systems due to an external perturbation the maximum predictability decays as a power law, t^{-a}, where a is some constant.[22] Fluctuations due to external stimulation are much stronger in SOC systems than those being realized in an equilibrium system and cannot be prevented. In case of the described biorhythm this would mean that a transition from the "up" state to the "down" state due to a severe external perturbation is inevitable for the individual.

As a very rough approximation the biorhythm displayed in Fig. 1 may be described by a two-state, i.e., an "up"–"down" system with an intrinsic dynamics. Currently, with the limited amount of data it is impossible to decide whether the neuronal/humoral system responsible for the biorhythm is a representation of a general process which has been studied under the name *stochastic resonance*[23,24] or is the realization of an *alternating fractal renewal process.*[25]

REFERENCES

1. P. Bak, C. Tang and K. Wiesenfeld, "Self-organized Criticality: An Explanation of $1/f$ Noise," *Phys. Rev. Lett.* **59**, 381–384 (1987).
2. P. Bak, "Is the World at the Border of Chaos?" *Ann. N. Y. Acad. Sci.* **581**, 110–118 (1990a).
3. O. Heller, "Hörfeldaudiometrie mit dem Verfahren der Kategorienunterteilung (KU)," *Psychol. Beitr.* **27**, 478–493 (1985).
4. M. Meesmann, F. Grüneis, P. Flachenecker and K.-D. Kniffki, "A New Method of Analysis of Heart Rate Variability: Counting Statistics of $1/f$ Fluctuations," *Biol. Cybern.* **68**, 299–306 (1993a).
5. D. R. Cox and V. Isham, *Point Processes* (Chapman and Hall, London 1980).
6. M. Meesmann and K.-D. Kniffki, in preparation.
7. W. A. Appel, *Biorhythmik*, Falken-Verlag, Niedernhausen, 1982.
8. M. Kobayashi and T. Musha, "$1/f$-Fluctuations of Heartbeat Period," *IEEE Trans. Biomed. Eng. BME* **29**, 456–457 (1982).
9. J. P. Saul, P. Albrecht, R. D. Berger and R. J. Cohen, "Analysis of Long Term Heart Rate Variability: Methods, $1/f$-Scaling and Implications," *Comput. Cardiol.* **14**, 419–422 (1987).
10. J. P. Zbilut, G. Mayer-Kress, P. A. Sobotka, M. O'Toole and J. X. Thomas Jr., "Bifurcations and Intrinsic Chaotic and $1/f$ Dynamics in an Isolated Perfused Rat Heart," *Biol. Cybern.* **61**, 371–378 (1989).
11. R. E. Kleiger, J. T. Bigger Jr., M. S. Bosner, M. K. Chung, J. R. Cook, L. M. Rolnitzky, R. Steinman and J. L. Fleis, "Stability Over Time of Variables Measuring Heart Rate Variability in Normal Subjects," *Am. J. Cardiol.* **68**, 626–630 (1991).
12. J. T. Bigger, J. L. Fleiss, R. C. Steinman, L. M. Rolnitzky, R. E. Kleiger and J. N. Rottman, "Frequency Domain Measures of Heart Period Variability and Mortality after Myocardial Infarction," *Circulation* **85**, 164–171 (1992).
13. M. Meesmann, J. Boese, D. Chialvo, P. Kowallik, W. R. Bauer, W. Peters, F. Grüneis and K.-D. Kniffki, "Demonstration of $1/f$ Fluctuations and White Noise in the Human Heart Rate by the Variance-Time-Curve: Implications for Self-Similarity," *Fractals* **1**, 000–000 (1993b) in press.
14. K. Kawahara, Y. Yamauchi, Y. Nakazono and Y. Miyamoto, "Spectral Analysis on Low Frequency Fluctuation in Respiratory Rhythm in the Decerebrate Cat," *Biol. Cybern.* **61**, 265–270 (1989).
15. M. J. Campbell and B. W. Jones, "Cyclic Changes in Insulin Needs of an Unstable Diabetic," *Science* **177**, 889–891 (1972).
16. M. Yamamoto, H. Nakahama, K. Shima, K. Aya, T. Kodama, H. Mushiake and M. Inase, "Neuronal Activities During Paradoxical Sleep," *Adv. Neurolog. Sci.* **30**, 1010–1022 (1986).
17. F. Grüneis, M. Nakao and M. Yamamoto, "Counting Statistics of $1/f$ Fluctuations in Neuronal Spike Trains," *Biol. Cybern.* **62**, 407–413 (1990).
18. K.-D. Kniffki, M. K. C. Mengel and C. Vahle-Hinz, "A New Tool to Characterize Ongoing Neuronal Discharge: Counting Statistics of f^{-b}-Fluctuations," *Soc. Neurosci. Abstr.* **18**, 1393 (1992).
19. M. K. C. Mengel, C. Vahle-Hinz and K.-D. Kniffki, "Is the Activity of Neurones in the Somatosensory Thalamus an Expression of Self-organized Criticality?" *Europ. J. Neurosci. Suppl.* **5**, 138 (1992).
20. M. C. Teich, "Fractal Character of the Auditory Neural Spike Train," *IEEE Trans. Biomed. Eng.* **36**, 150–160 (1989).
21. K. Christensen, Z. Olami and K. Bak, "Deterministic $1/f$ Noise in Nonconservative Models of Self-organized Criticality," *Phys. Rev. Lett.* **68**, 2417–2420 (1992).
22. P. Bak, "Simulation of Self-organized Criticality," *Physica Scripta.* **T33**, 9–10 (1990).
23. F. Moss, "Stochastic Resonance," *Ber. Bunsenges. Phys. Chem.* **95**, 303–311 (1991).
24. A. Longtin, A. Bulsara and F. Moss, "Time-interval Sequences in Bistable Systems and the Noise-induced Transmission of Information by Sensory Neurons," *Phys. Rev. Lett.* **67**, 656–659 (1991).
25. S. B. Lowen and M. C. Teich, "Fractal Renewal Processes Generate $1/f$ Noise," *Phys. Rev.* **E47**, 992–1001 (1993).
26. B. B. Mandelbrot, *The Fractal Geometry of Nature* Freeman, New York (1983).

Chemistry

INTRODUCTION

While chemistry is traditionally thought of in terms of sharing, or transferring valence electrons, aspects of geometry can enter as well. The possibilities for fractal geometry are plentiful. As a first example, consider the bonding of amino acids into polypeptides. The multitude of polypeptide sequences, of course, generates the many proteins whose functions allowed the evolution of complex organisms. It is perhaps not well known that the set of alpha-carbons in the chain backbone fractally fills space. The dimension for myoglobin is 1.91. The vibrational spectra of iron containing proteins exhibits a different kind of fractal property deduced from the temperature dependence of a paramagnetic relaxation time. This is interpreted in terms of a fracton dimension which governs the low frequency vibrations of the protein. The list of other areas for the investigation of fractals in biology is growing. For example, the chemically driven opening and closing of neural membrane channels has been argued to have fractal statistics. As we see fractals enter the chemistry of life in unexpected ways, and point to an exciting area for future work.

Proteins are short unbranched polymers of only a few hundred units. Polymers themselves, in general, are another area replete with fractal scaling, the simplest case being the linear self-avoiding random walk polymer fractal structure. A more sophisticated example is that the structural relaxation of polymeric glasses can be a fractal time process.

Another example where fractal geometry mixes with chemistry is chemical reactions at surfaces/interfaces, such as a porous electrode or a catalytic surface. If the interface is fractal the active surface area is increased, in a scaling fashion, and the nature of the reaction can be quite different from standard first or second order reaction kinetics. Several fractal surfaces have been identified, including carbon black and coal. The question of the growth of fractal surfaces and interfaces has intrinsic interest in and of itself. Aggregation phenomena can lead to fractal structures such as diffusion-limited aggregates and cluster-cluster aggregates. Motion of particles on fractal structures or reacting with such structures, is again an area which exhibits interesting fractal scaling laws. Aggregation of particles into clusters can play a serious role in environmental concerns from determining how long particles will remain in the atmosphere following a volcanic eruption, to effecting the carbon balance in the ocean.

The best known fractal aggregate is the diffusion-limited aggregate (DLA). While the random walk model for growing DLA is simple to construct and its understanding in terms of an underlying Laplacian is appreciated, DLA remains an active area of investigation. Although DLA has a hierarchical (branched) structure, a straightforward renormalization group approach of its detailed properties fails because the hierarchy is not simply ordered. This is because if two branch ends of the DLA grow close together they dramatically lower the probability for a new particle to enter the fjord formed by the area between the branches. This probability may in fact be lower than the probability for a particle to reach a more interior point. This complication keeps the field open and active. The multifractal nature of DLA also points to its complexity. The concept of a distribution of fractal dimensions (its multifractality) has expanded upon the concept of a discrete number of critical exponents, which is the staple of critical phenomena. The opposite of aggregation is dissolution. Corrosion is an example of a dissolution process which can create fractal structures with as much appeal and importance as aggregates.

One of the most important applications of fractals in chemistry is to give credibility to the concept of a time dependent reaction rate. Electron scavenging reactions in frozen chemical solutions is a prime example of a system with a time dependent reaction rate. The transport of electrons or holes in xerographic films is another. Much of science has focused on deriving a definite answer to a problem, rather than a distribution of possible answers (much less a fractal distribution of answers). The Arrhenius reaction rate is a concept deeply held by many in chemistry, and a fractal distribution of rate constants takes some getting used to, but it is well worth the effort. An understanding of fractal rates has been used to explain the empirical Kohlrausch stretched exponential law from the mid-1850s which governs mechanical relaxation in amorphous material. The basis of this law was never understood until the fractal time concept was introduced. Now its applicability to volumetric, NMR, dielectric, and other types of relaxation in glassy structures is well studied.

Fractal aspects can also appear in a variety of situations from mixing in diffusion fronts to the time dependent population size of segregated reactants in reaction-diffusion systems. Fractal properties have been reported from protein structures up to macroscopic sized clusters visible to the eye. As the following papers illustrate, the role of fractals in chemistry is an exciting area with many avenues of adventure to explore.

by **M. Shlesinger**

Fractals, Vol. 1, No. 3 (1993) 389–404

FRACTAL DESCRIPTION OF ANOMALOUS DIFFUSION IN DYNAMICAL SYSTEMS

J. KLAFTER
School of Chemistry, Tel-Aviv University,
Tel-Aviv, 69978 Israel

G. ZUMOFEN
Laboratorium für Physikalische Chemie, ETH-Zentrum,
CH-8092 Zürich, Switzerland

M. F. SHLESINGER
Physics Division, Office of Naval Research,
800 North Quincy Street, Arlington,
Virginia 22217-5660, USA

Abstract

Anomalous diffusion properties are common in a broad spectrum of systems including dynamical systems. In this paper we review an approach based on Lévy scale-invariant distributions to describe transport in dynamical systems. We introduce the basic ingredients that make the approach useful in describing anomalous diffusion and demonstrate the applicability in the cases of one-dimensional iterated maps and of the standard map.

1. INTRODUCTION

It is by now appreciated that complex systems can exhibit a wide range of behaviors most readily described by the popular terms of chaos and fractals. Under these circumstances one tries to determine distribution functions which govern the systems, and especially, as precisely as possible, to determine the exponents which characterize the asymptotic behavior of the distributions. In many instances these distributions represent probabilities and reflect a scaling behavior. The well known Cauchy distribution

$$p(x) = \frac{1}{\pi}\frac{1}{1+x^2} \tag{1}$$

is an early example of a probability distribution whose first moment is zero, but whose second and higher moments are infinite. Furthermore, if N identically distributed random variables, each with a Cauchy distribution, are added the sum $x_1 + x_2 + \cdots + x_N$ also has a Cauchy distribution, $P_N(x)$ with

$$P_N(x) = \frac{1}{\pi N}\frac{1}{1+(x/N)^2}, \tag{2}$$

where $P_N(x)$ and $p(x)$ are related by a change of scale x to x/N. This is the hallmark of fractals, where the whole $P_N(x)$ (the distribution of the sum) looks like $p(x)$ (the distribution of a single variable). In viewing the sum of random variables as adding random walk steps, this scaling is equivalent to finding the distribution for the position of a random walker after one step and after N steps, i.e.,

$$p(x) = N^{\alpha} P_N(xN^{\beta}), \tag{3}$$

with $\alpha = 1$ and $\beta = 1$. P. Lévy in the 1920s attacked more generally the problem of when a random walk distribution is independent (up to a scale factor) of the number of steps taken. For an unbiased one-dimensional random walk he found that:

$$P_N(k) = \int_{-\infty}^{\infty} P_N(x)e^{jkx}dx = e^{-N|k|^{\alpha}}, \quad \alpha \leq 2. \tag{4}$$

Here and in what follows the argument indicates in which space the function is meant to hold, e.g. $P_N(k)$ is the Fourier ($x \to k$) transform of $P_N(x)$. Expression (4) implies the scaling equation:

$$p(x) = N^{1/\alpha} P_N(xN^{1/\alpha}). \tag{5}$$

When $\alpha < 2$, the second moment of the probability diverges so no proper characteristic scale exists, and so both $p(x)$ and $P_N(x)$ behave asymptotically as a power law $|x|^{-\alpha-1}$. Again this is the signature of fractal behavior. The random-walk process which follows Eqs. (4) and (5) is known as the Lévy flight.[2]

Nonlinear dynamical systems under appropriate conditions can exhibit complex scale-invariant behavior in their transport properties.[3–6] Features on one scale reveal finer features under higher resolution and so on and so on. This may result in anomalous diffusion, dispersive or enhanced.[4–9] Here we will be interested only in the enhanced case. The obvious question to ask is whether Lévy's scale-invariant distributions are applicable to describing the behavior of dynamical systems in, or near, chaotic regimes.[6] We answer that question in the affirmative for our initial investigations in a tangent bifurcation map and for a Hamiltonian map, the standard map. These applications of Lévy distributions are proposed to be a new useful tool for characterizing complex dynamical systems.

2. LÉVY WALKS IN THE CONTINUOUS TIME-RANDOM WALK FRAMEWORK

In the Lévy flights mentioned above, jumps occur between successively visited sites, however distant, obeying a power-law distribution of distances, which leads to a divergence of the mean-square displacement. The flights cannot be readily used to study diffusional behavior. One is interested therefore in modifying the Lévy-flights so that the motion still follows Eq. (4), namely a power-law distribution of distances, but with a finite mean-squared displacement. We shall explore a stochastic process which we call Lévy walk and which visits the same sites as in the Lévy flight, but with a time "cost" which depends on the distance. Taking into account the time to complete the move in the Lévy-walk case removes the divergence of the mean-squared displacement.[9-11] We formulate the Lévy-walk process in the framework of the continuous-time random walk (CTRW) theory. For this purpose, we introduce a probability distribution $\psi(r, t)$ to move a distance r in time t in a single motion event such that:

$$\psi(r, t) = \phi(r|t)p(r) \tag{6a}$$

$$\phi(r|t) = \delta(|r| - t) . \tag{6b}$$

where $p(r)$ is the distribution of distances and where $\phi(r|t)$ is the conditional probability which accounts for the time t it takes to move a distance r. Equations (6a) and (6b), within the CTRW approach, define the Lévy walks which result in finite mean-squared displacements at time t and which account for the possibility of enhancement. Alternatively, one can write Eq. (6) as[9]:

$$\psi(r, t) = \frac{1}{2}\delta(|r| - t)\psi(t) , \tag{7}$$

where now $\psi(t)$ is the waiting time distribution which is the distribution of times spent for moving over distances r. $\psi(t)$ is considered to follow a power-law. Other forms of conditional probability densities were discussed in Ref. 10. Lévy walks, as we see later, are characterized by rich dynamical behavior depending on the power-law exponents of $p(r)$ or $\psi(t)$. They seem to arise naturally in many dynamical systems as an interesting and new mode of transport.[6]

We now briefly outline the CTRW approach on which we base our study of anomalous diffusion in dynamical system.[9,13] In the CTRW framework, the random walk process is entirely specified by the probability density $\psi(r, t)$ for a single motion event. $\psi(r, t)$ can either be decoupled $\psi(r, t) = \psi(t)p(r)$, as commonly encountered in theories and applications of CTRW,[12-15] or coupled through some space-time ansatz.[9-11] In the decoupled scheme, usually assumed in the jump model, $\psi(t)$ determines whether the motion is asymptotically regular or dispersive. The only requirement on $p(r)$ is that its second moment is finite. In order to derive an enhanced diffusion behavior within the CTRW formulation, one has to introduce coupled memories[9-11] proposed in Eq. (6) or (7). This space-time coupling scheme can be introduced into different models: jump, velocity and two-state.[9] While in the jump model no information is given on how the particle spends its time between jumps; the velocity picture with the coupled memory $\psi(r, t)$ of Eqs. (6) and (7) provides a clear physical origin for the time spent between turning points.

In the velocity model, the walker moves at a constant velocity for a given time (or displacement), then stops and chooses a new direction and a new sojourn time at random from given distributions. Here we are interested in calculating the Lévy-walk propagators, $P(r, t)$, in the velocity picture. In order to obtain the propagator we define the probability $\Psi(r, t)$ to move a distance r in time t in a single motion event but not necessarily stop at r[9,16,17]:

$$\Psi(r, t) = \phi(r|t) \int_t^\infty dt' \int_{|r|}^\infty dr' \psi(r', t') . \tag{8}$$

By inserting Eq. (6) or (7) into Eq. (8) we find for the coupled memory:

$$\Psi(r, t) = \delta(|r| - t) \int_{|r|}^\infty dr' p(r') = \frac{1}{2}\delta(|r| - t) \int_t^\infty dt' \psi(t') . \tag{9}$$

In order to derive recursive expressions for $P(r, t)$ we consider $Q(r, t)$, the probability to arrive at r exactly at time t and to stop before randomly choosing a new direction:

$$Q(r, t) = \int_{-\infty}^\infty \int_0^t Q(r - r', t - t')\psi(r', t')dr'dt' + \delta(r)\delta(t) . \tag{10}$$

In the velocity model, we assume that $P(r, t)$ is the probability to stop or pass at location r at time t and thus we have:

$$P(r, t) = \int_{-\infty}^\infty dr' \int_0^t dt' Q(r - r', t - t')\Psi(r', t') . \tag{11}$$

In the Fourier ($r \to k$) and Laplace ($t \to u$) spaces, Eq. (11) simplifies to:

$$P(k, u) = \Psi(k, u)/[1 - \psi(k, u)] . \tag{12}$$

Equation (12) is the central result in this section and will be used to calculate the propagator in some simple nonlinear maps.[9,16,17]

3. LÉVY WALK DESCRIPTION FOR ONE-DIMENSIONS MAPS

We now concentrate on the application of Lévy walks to a family of iterated maps. We show that Lévy walks, introduced above, provide an appropriate probabilistic description of deterministic motion resulting from such maps. We study iterated maps that have been shown to generate enhanced diffusion by intermittent chaotic motion.[8,9,16,18] We denote the map by $g(x)$ and write for the nth iteration:

$$x_{n+1} = g(x_n) . \tag{13}$$

Assuming periodic and inversion symmetries we have:

$$g(x+N) = g(x) + N\,, \quad g(-x) = -g(x)\,, \tag{14}$$

where N is integer and denotes the box number.

For the mean-squared displacement after t iterations we have:

$$\langle r^2(t)\rangle = \langle (x_{n+t} - x_n)^2\rangle \tag{15}$$

with the initial coordinate x_0 chosen arbitrarily. The problem of how to take the average was discussed elsewhere.[9,16,17] The propagator $P(r, t)$ is calculated accordingly:

$$P(r, t) = \langle \delta(r - x_{n+t} + x_n)\rangle\,. \tag{16}$$

We follow a map studied by Geisel, Nierwetberg, and Zacherl[8] which generates enhanced diffusion. For a reduced range the definition is:

$$g(x) = (1+\varepsilon)x + ax^z - 1\,, \quad 0 \le x \le 1/2\,, \tag{17}$$

where a is set to $a = 2^z(1 - \varepsilon/2)$. The quantity ε in Eq. (17) is supposed to be small and serves as a cutoff to enable the normalization of the invariant measure for the reduced map for $z > 2$.

This map which has been extensively discussed[8,9,16,17] gives rise to intermittent behavior which can be interpreted as a motion at a constant velocity where the laminar phases are interrupted by short chaotic motions. In order to model this behavior we consider the following assumptions: $v(t)$ denotes a stochastic time dependent function. The motion takes place at a constant velocity for some time (number of iterations) after which the direction and the length of the next motion event is chosen randomly. The motion events are uncorrelated and the duration time of the motion events follow independently a probability distribution $\psi(t)$. This idealized picture of the motion defines the probabilistic description of the motion which was introduced for the velocity model in the CTRW approach discussed above.[9,16] We choose Eq. (7) to represent $\psi(r, t)$. In Eq. (7), $\psi(t)$ can be interpreted as the probability distribution of being trapped on all scales in laminar phases under the assumption that $\psi(t)$ follows a power-law. Figure 1 shows the distribution $\psi(t)$ of waiting times and the corresponding $p(r)$ obtained directly from the map [Eq. (17)] for two values of z. The power-law behavior of both is evident. This power-law behavior combined with the space-time coupled velocity picture suggests that the Lévy walk description is appropriate.

In our analysis, we use for the waiting time distribution:

$$\psi(t) = \gamma b^\gamma / [b + t]^{\gamma+1}\,, \tag{18}$$

where we set $\gamma = 1/(z-1)$ and $b = 2^{z-1}/[a(z-1)]$. We consider γ as the relevant exponent for the description of the motion events. For more details, see Refs. 9, 16 and 17.

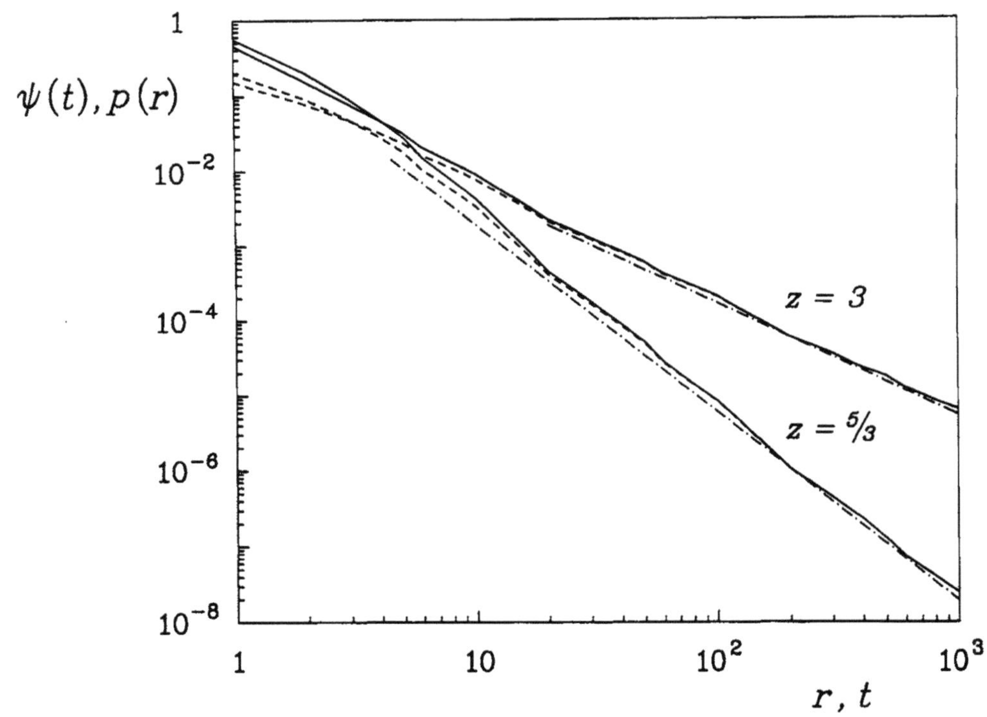

Fig. 1 Waiting-time distributions obtained from the iterated map, [Eq. (17)] for $z = 3$ and $z = 5/3$. The full lines give the waiting-time distribution $\psi(t)$ and the dashed lines denote the corresponding distribution of distances $2p(r)$, where the factor 2 is introduced to account for normalization. The dash-dotted lines give the theoretical slopes of 1.5 and 2.5 for $z = 3$ and $z = 5/3$, respectively.

Depending on the exponent γ, it has been found that the motion shows a range of behaviors, namely regular diffusion for $\gamma > 2$, intermediate enhanced diffusion for $1 < \gamma < 2$ and ballistic type of motion for $0 < \gamma < 1$.[8,9,16,17] The ballistic type regime is characterized by a diverging mean sojourn time $\bar{t}$, where the mean sojourn time is defined as:

$$\bar{t} = \int_0^\infty t\psi(t)dt\,. \tag{19}$$

A diverging $\bar{t}$ is associated with an invariant density for the reduced map that is not normalizable and consequently a cutoff has to be introduced.[8,17]

Depending on the way of averaging, two slightly different propagators can be determined. We have derived expressions for both cases.[17] Here we apply the expression for $P(k, u)$ [Eq. (12)] derived in the previous section. For the asymptotic long-time behavior of the propagator in these regimes, we obtain the following scaling representation[9]:

$$P(r, t) = f(\xi)/\tau\,, \tag{20}$$

where ξ is the scaling variable, $\xi = |r|/\tau$, and where

$$\tau = \begin{cases} t\,, & 0 < \gamma < 1 \\ t^{1/\gamma}\,, & 1 < \gamma < 2 \\ t^{1/2}\,, & 2 < \gamma\,. \end{cases} \tag{21}$$

$f(\xi)$ in Eq. (20) denotes the scaling function which we studied analytically and numerically. Here we present only the main results. For the ballistic-type regime, $0 < \gamma < 1$, $f(\xi)$ is constant for small ξ and diverges when ξ approaches the value 1. This becomes obvious for the particular case of $\gamma = 1/2$ for which we obtained the asymptotic expression:

$$f(\xi) = \pi^{-1}(1-\xi^2)^{-1/2}, \quad \xi \leq 1, \quad \gamma = 1/2. \tag{22}$$

The divergence at $\xi = 1$ is expected for a ballistic motion but we do not obtain a "pure" δ-function due to the random walk nature of the process.

For the intermediate enhanced regime, $1 < \gamma < 2$, the expression of $P(r, t)$ follows closely a Lévy-stable distribution, which is characterized by a Gaussian behavior for small ξ and by a power-law wing:

$$f(\xi) \sim \begin{cases} \exp(-c\xi^2), & \xi < 1 \\ \xi^{-(\gamma+1)}, & 1 < \xi \text{ and } r < t \\ 0, & r > t. \end{cases} \tag{23}$$

For $\gamma > 2$, the propagator follows a Gaussian in its main part; scaling is not obeyed for $\xi \gg 1$ and the decay follows a power law, namely $P(r, t) \sim t/r^{\gamma+1}$.

In Figs. 2–5, we compare the propagators obtained from the iterated maps [Eqs. (16) and (17)], with those obtained from a numerical calculation of $P(r, t)$, and with the asymptotic

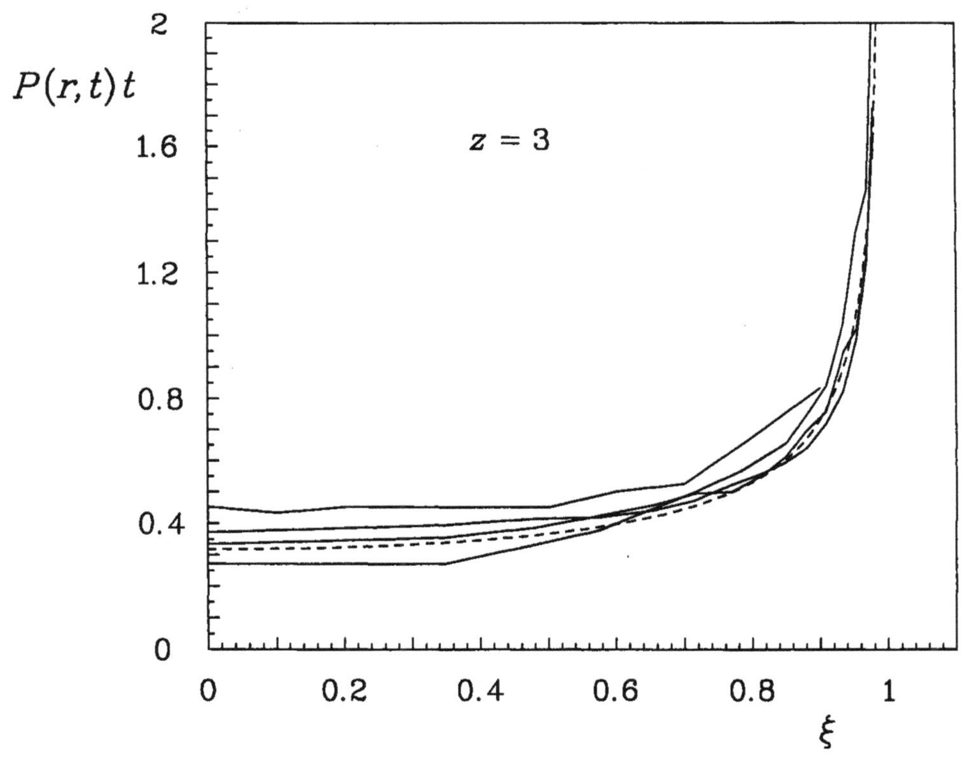

Fig. 2 The propagator in the ballistic-type regime. Results obtained from iterated map [Eq. (17)] for $z = 3$ are plotted as $tP(r, t)$ vs. ξ for times $t = 10, 10^2, 10^3, 10^4$, and 10^5. The scaling variable is $\xi = r/t$. The dashed line is the analytic result of Eq. (22).

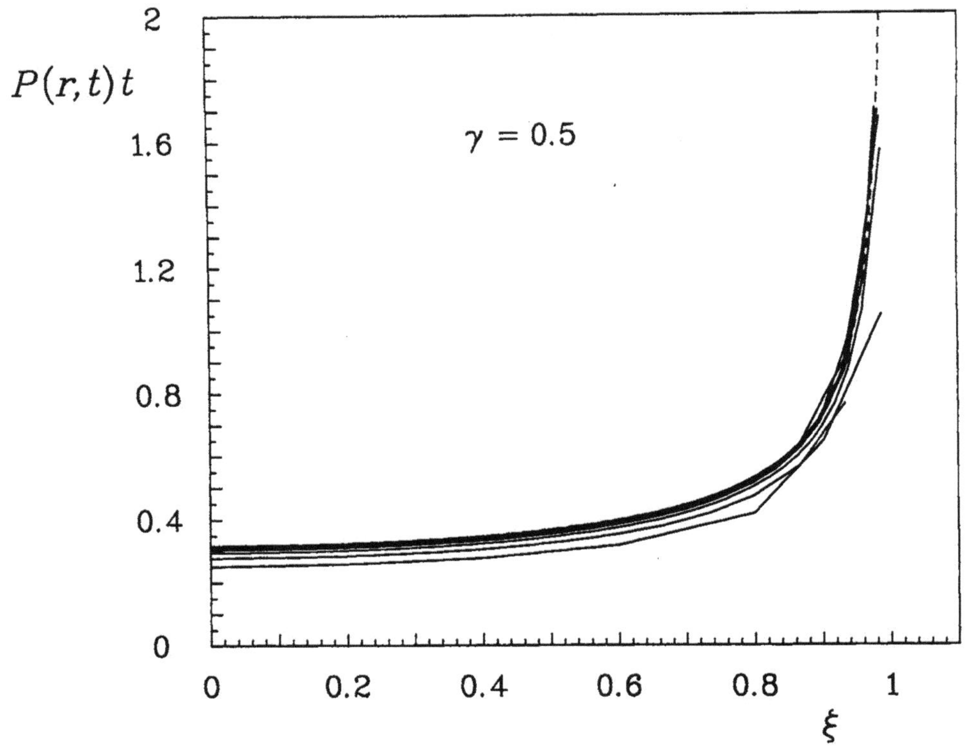

Fig. 3 Same as in Fig. 2, but with results obtained from the numerical inversion of Eq. (12) for $\gamma = 1/2$.

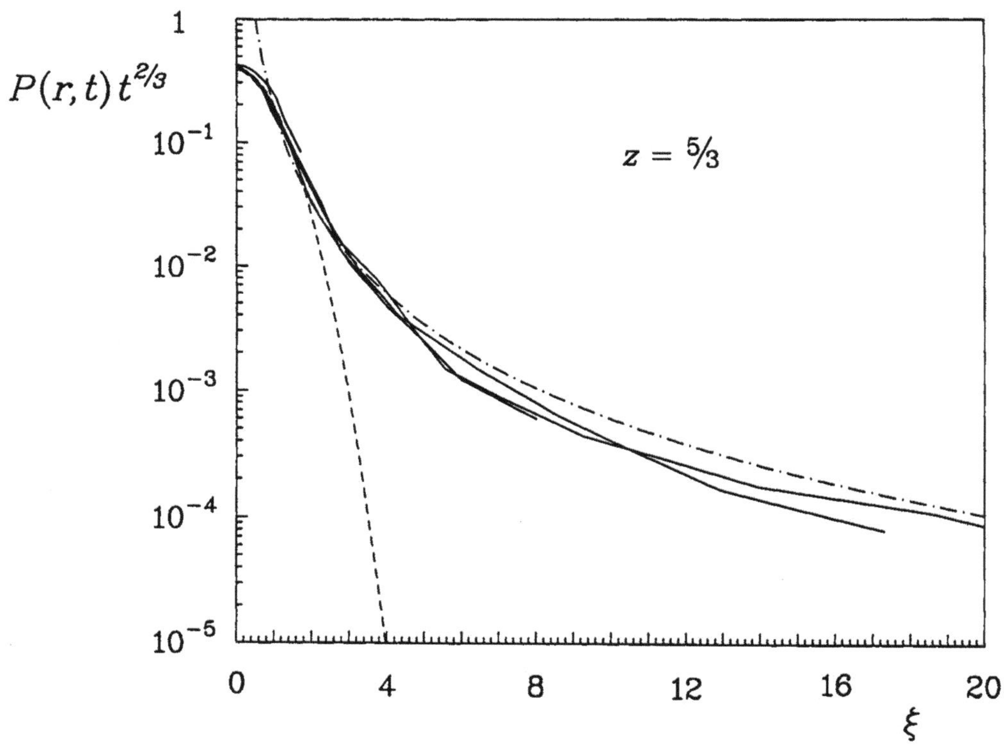

Fig. 4 The propagator in the enhanced diffusion regime. The results obtained from iterate maps for $z = 5/3$ are plotted as $t^{2/3}$ $P(r, t)$ vs. ξ for times $t = 10$, 10^2, 10^3, 10^4, and 10^5. The dashed line gives the Gaussian short ξ behavior and the dashed-dotted line is large ξ power-law [Eq. (23)]; coefficients are taken from Ref. 16. The scaling variable is $\xi = r/t^{2/3}$.

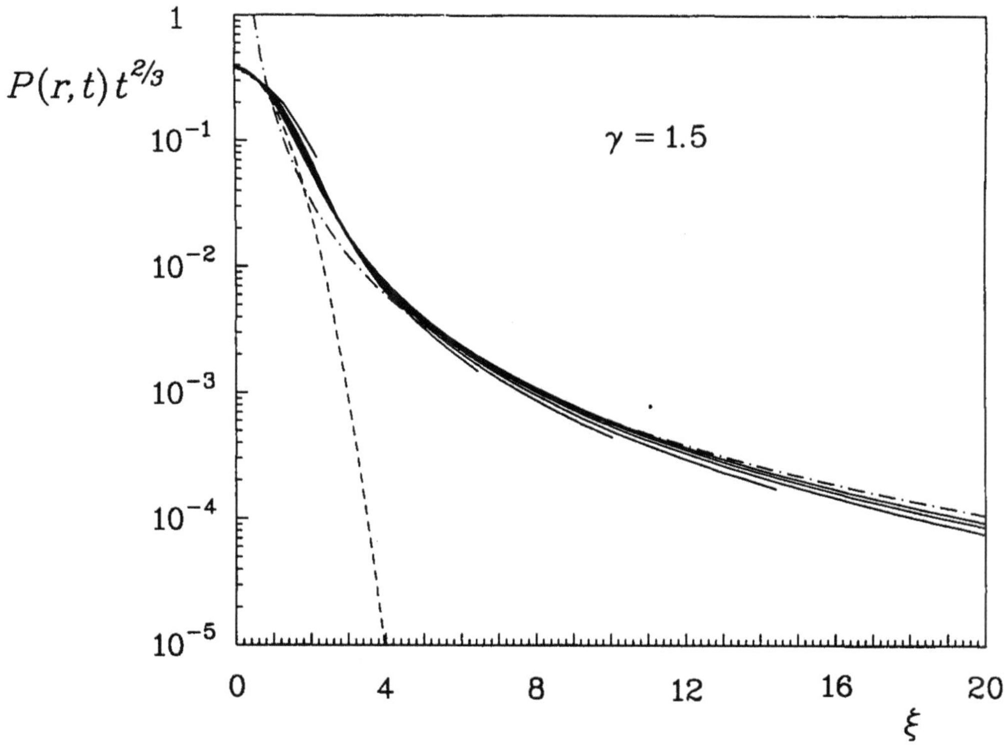

Fig. 5 Same as in Fig. 4, but with results obtained from the numerical solution of Eq. (12) for $\gamma = 3/2$.

scaling forms [Eqs. (20)–(23)]. For the maps 10^{10} iterations were performed. In Figs. 2 and 3, the numerical results for the ballistic type of behavior are compared with the analytic form [Eq. (22)]. In Figs. 4 and 5, the propagator in the intermediate enhanced regime is shown; the numerical results are compared with asymptotic behaviors for short and large ξ values [Eq. (23)]. In all cases presented, the data collapse for various times onto single curves indicates that scaling is obeyed.

The asymptotic forms for the mean-squared displacement corresponding to the above regimes are[9]:

$$\langle r^2(t)\rangle \sim \begin{cases} t^2\,, & 0<\gamma<1 \\ t^2\ln(t/b)\,, & \gamma=1 \\ t^{3-\gamma}\,, & 1<\gamma<2 \\ t\ln(t)\,, & \gamma=2 \\ t\,, & 2<\gamma\,. \end{cases} \tag{24}$$

4. LÉVY WALKS IN THE STANDARD MAP

While the one-dimensional map discussed in the previous section belongs to a family of dissipative systems, here we apply the Lévy statistics to conservative systems. We will analyze the transport obtained from the standard map in terms of the same Lévy-walk framework developed above for motion dominated by power-law distributions.

The standard map, which can be regarded as the discretized Hamiltonian of the kicked-rotor, is a two-dimensional map of the form[3,19–21]:

$$\begin{aligned} x_{n+1} &= x_n + K \sin(2\pi\theta_n) \\ \theta_{n+1} &= \theta_n + x_{n+1} \,. \end{aligned} \tag{25}$$

This map has been shown to have ranges of the stochasticity parameter K for which the motion along the x-direction is enhanced.[21] This is due to the accelerating modes characterized by the fixed points[22]:

$$\begin{aligned} |x_p^* - x_0^*| &= l \\ |\theta_p^* - \theta_0^*| &= 0 \,, \end{aligned} \tag{26}$$

with l being an integer. When $l = p$, the modes are called fundamental accelerating modes of period p. In the neighborhood of these fixed points, it is known that there is a complicated mixture of stable islands and of chaotic regions (termed chaotic sea). These islands are encircled by cantori, which represent barriers of slow penetration for the orbits. Islands and cantori show some hierarchical self-similar structure.[4,5,6,20] Orbits approaching these islands tend to stick to the critical tori and become quasi stable for some time such that laminar phases result[3] (similar to those of the one-dimensional map when an orbit is injected into the neighborhood of the marginal stable fixed point). Hence due to repeated visits to the self-similar cantori, an intermittent behavior results characterized by broad distributions of sticking times. These broad distributions cause the motion to be anomalous and to be a promising candidate for a description in terms of Lévy walks.[6,20,21,23]

Depending on the stochasticity parameter K, there are ranges where the motion is dominated by the fundamental accelerating modes. In these cases, the sticking to the cantori leads to motion at a constant velocity, since a particle moves along x by l units in p iterations; namely, the velocity is approximately l/p. Therefore, sticking to the cantori means being trapped in a mode of constant velocity. Thus the motion is characterized by the laminar phases of constant velocity interrupted by short chaotic motion in the chaotic sea, which makes the space-time coupled waiting time distribution apply. Furthermore, since the sticking (trapping) times are broad, the motion is anomalous, and considering power-law distributed sticking times the Lévy walk picture is appropriate.

In our numerical investigations, we considered the $p = 3$ and $p = 5$ (period three and period five fundamental) accelerating modes with K values 1.1 and 1.03, respectively.[21] In Fig. 6 is shown the orbit in phase space (x, θ) for the period-three mode. We notice the chaotic behavior of the orbit and the embedded accelerating mode islands. Figure 7 shows a particular realization of a laminar phase motion of length 10^5 number of iterations. With increasing magnification, the structure of the orbits is resolved on a smaller scale and we observe finer and finer structures of islands around islands. In particular, the hierarchical arrangement of islands becomes visible. As mentioned above, this hierarchy is thought to be the origin of the broadness in the sticking time distribution.

Our analysis of the waiting time distribution revealed that the sticking time distributions follow approximately power laws and that the asymptotic behavior starts to hold for times $t > 10^2$ as displayed in Fig. 8, both for $p = 3$ and $p = 5$. Figures 9 and 10 demonstrate

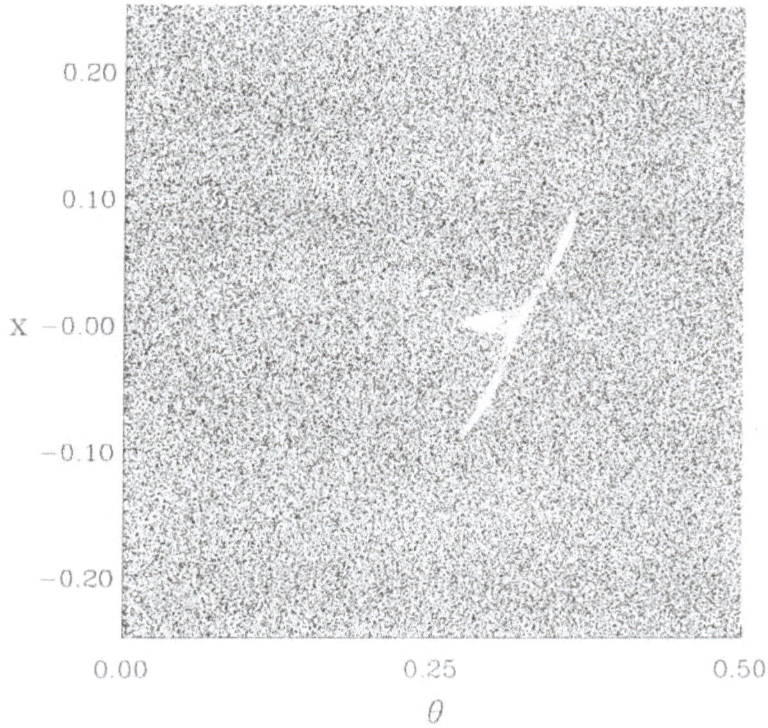

Fig. 6 Chaotic motion for the period-three mode. Plotted is an orbit of 10^6 number of iterations with initial coordinates chosen close to the origin.

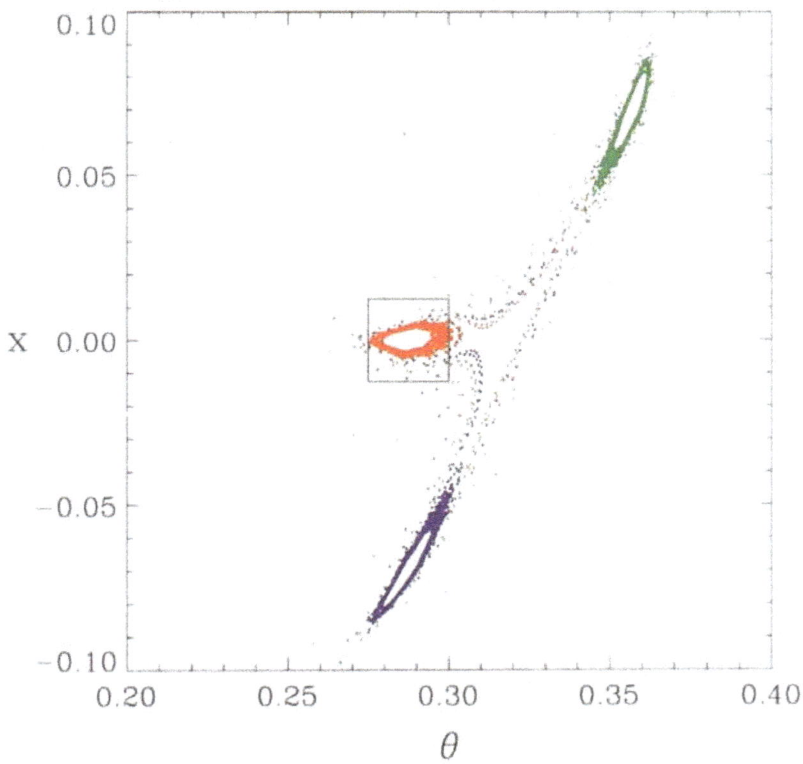

(a)

Fig. 7 The orbit of a laminar phase of 10^5 number of iterations for the period-three mode showing islands of three generations (a)–(c).

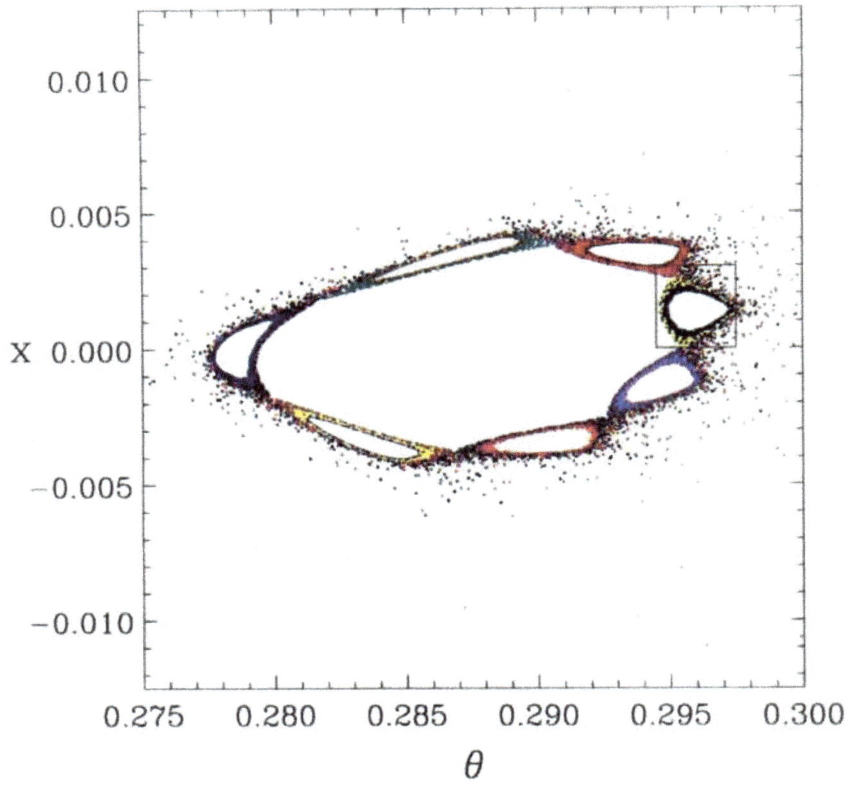

(b)

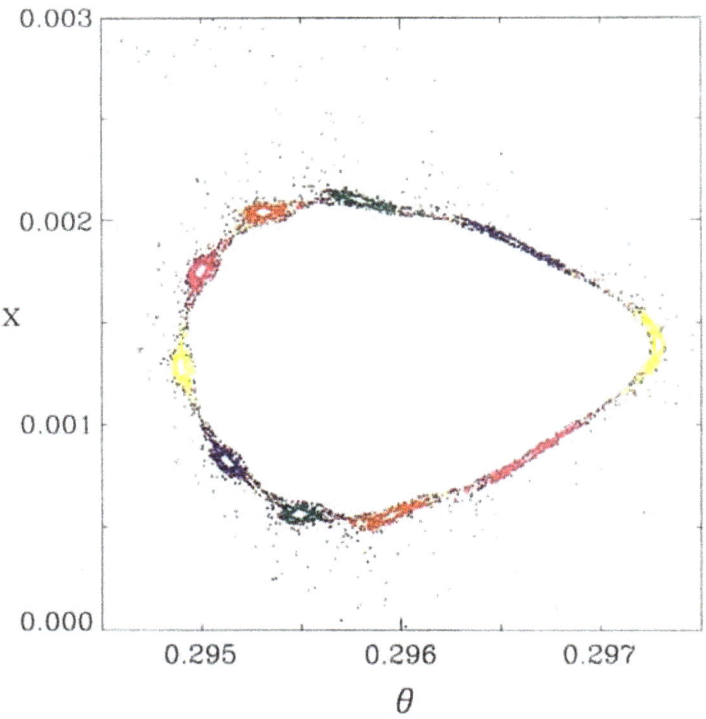

(c)

Fig. 7 (*Continued*)

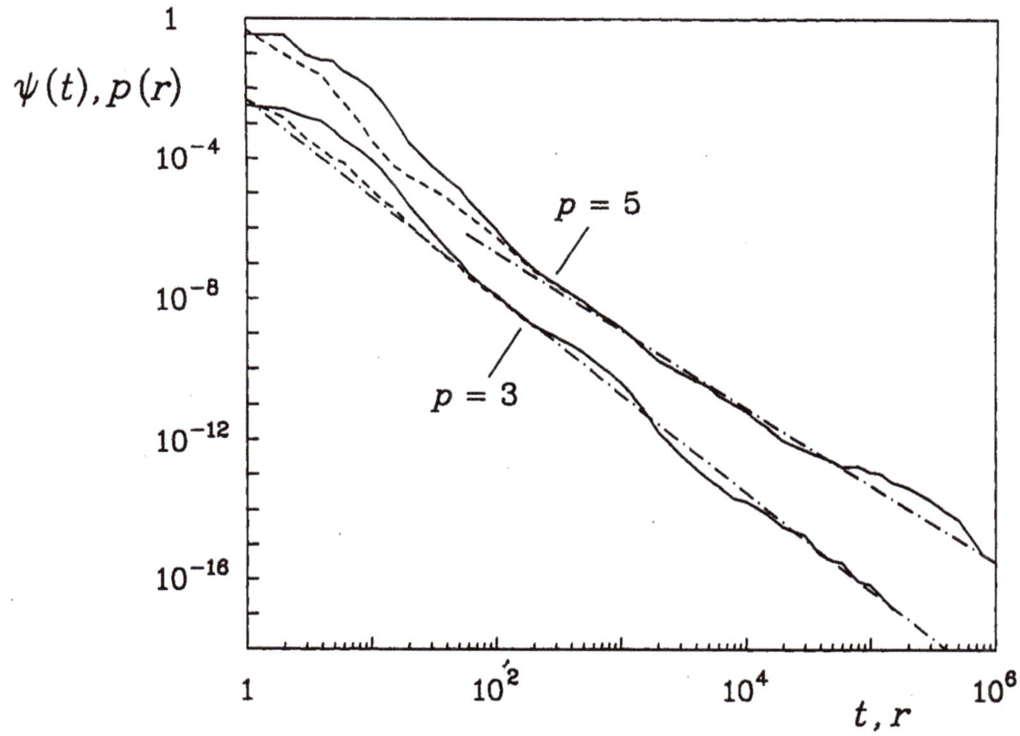

Fig. 8 Waiting time distributions for the $p = 3$ and $p = 5$ accelerating modes. Full lines give the $\psi(t)$ and dashed lines give $2p(r)$, where the factor 2 is introduced to account for normalization. The dashed-dotted lines indicate fitted slopes of 1.2 and 1.8, respectively.

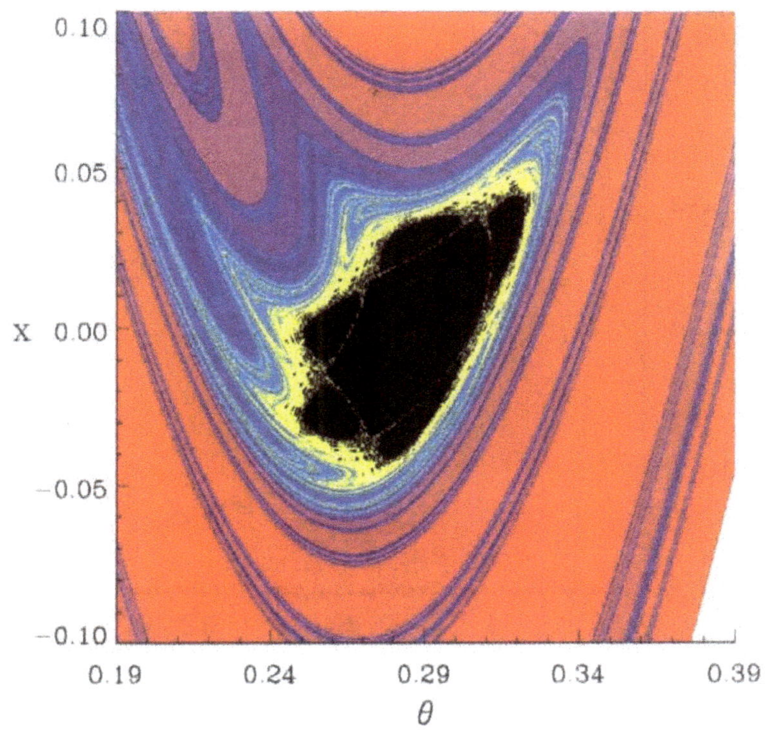

Fig. 9 Exit times for initial coordinates chosen in a window of the phase space (x, θ). The black areas mark the islands of stability and the other colors demonstrate the decrease of the exit times away from the boundary of the islands.

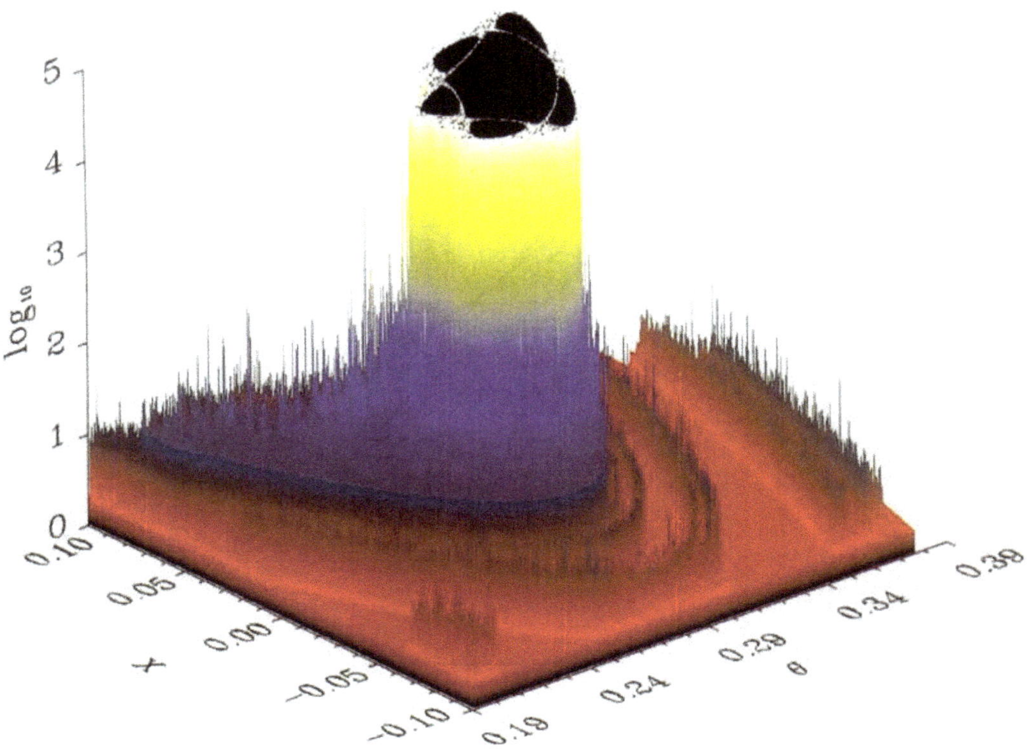

Fig. 10 Exit times for initial coordinates chosen in the phase space (x, θ) presented in 3-D plot. The black areas mark the islands of stability as in Fig. 9, and the heights give the exit times on a logarithmic scale.

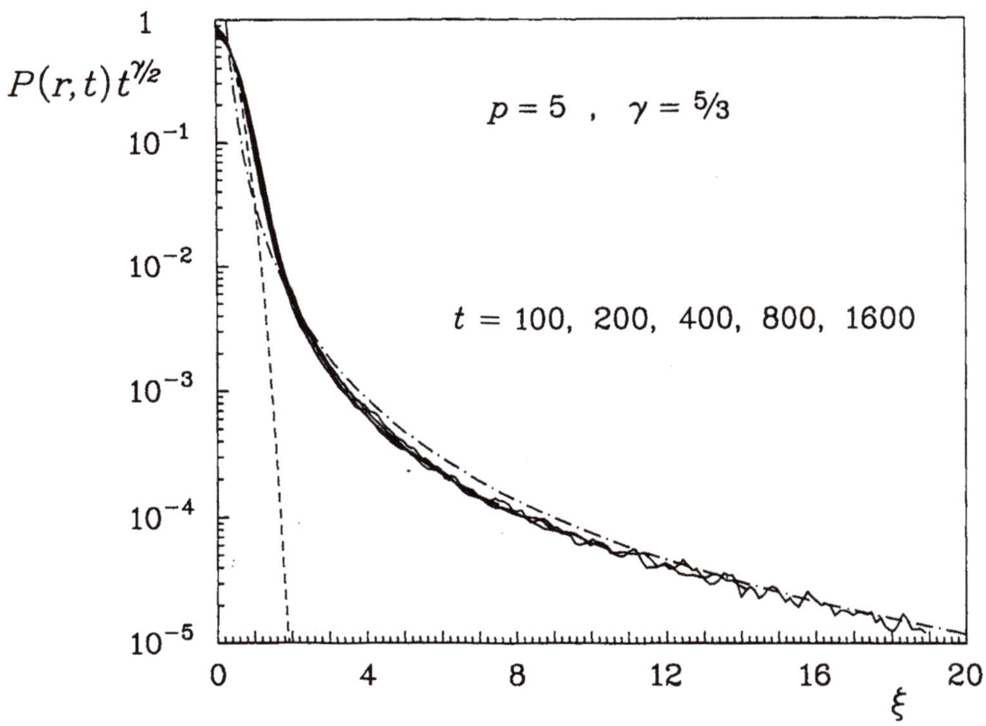

Fig. 11 The propagator for the five-period mode in the standard map. The results are plotted analogously to those in Fig. 4, but for times as indicated. The relevant exponent is $\gamma = 5/3$. The dashed line gives the Gaussian short ξ behavior and the dashed-dotted line is the large ξ power-law [Eq. (23)]; coefficients are taken from Ref. 19 but with the time rescaled by a constant which was determined independently. The scaling variable is $\xi = r/t^{1/\gamma}$.

the broad distribution of exit times for $p = 5$, which implies a broad distribution of waiting times. In the figure, the colors indicate different exit times; the black areas correspond to the islands of stability, i.e., of infinite exit times and of zero Lyapunov exponents. The decrease in the exit times away from the zone of stability is given by a variation of the colors. Similarly, the exit times are shown in Fig. 10 as a 3-D plot. Again the black zones are the islands of stability and the height gives the exit time on a logarithmic scale and the coloring is chosen to improve the impression.

The analysis of the mean-squared displacement follows power-laws which demonstrate enhancement and for which the asymptotic behavior sets in for times $t > 10^2$. For the propagator, we chose times in the range $10^2 < t < 2 \times 10^3$. The relevant exponents are thus intermediate and not that of the asymptotic behavior. In Fig. 11, we show the propagator $P(r, t)$ for the five-period mode in the scaling representation on log vs. linear scales. The dashed and dashed-dotted lines give the short and large ξ forms of Eq. (23) for the exponent $\gamma = 5/3$, which is equal to that chosen in Ref. 21. In order to avoid complications, we did not consider the problem of stationarity which, depending on how averages are taken, concerns the sharp peaks at $|r| \simeq t$.[16] We observe a remarkable data collapse on a single curve indicating that scaling holds. Furthermore, the power-law wing shows that the propagator follows the Lévy distribution. The standard map provides yet another manifestation of the usefulness of the Lévy-walk framework in describing the deterministically generated diffusion.

5. CONCLUSIONS

We have studied a probabilistic approach to the enhanced diffusion generated by 1-D and 2-D iterated maps using Lévy walks within the CTRW velocity model. The quantitative and qualitative analyses of the time dependent mean-squared displacements and of the propagators have demonstrated that iterated maps serve as useful examples for the study of enhanced diffusion obtained in dynamical systems and that the CTRW Lévy walk approach appears as a consistent probabilistic description of anomalous diffusion as obtained from these complex systems. Lévy walks are expected to provide a statistical description of transport in more complicated and even realistic problems and may also describe, on some level, the dynamical aspects of turbulent diffusion and mixing.

ACKNOWLEDGMENTS

We thank Prof. K. Dressler for helpful discussions and F. Weber for technical assistance. J. K. thanks the ETH for the hospitality during the time this work was carried out. A grant from the Rechenzentrum der ETH-Zürich is gratefully acknowledged.

REFERENCES

1. P. Lévy, *Théorie de l'addition des variables aléatoires* (Gauthier Villars, Paris, 1937).
2. B. B. Mandelbrot, *The Fractal Geometry of Nature* (Freeman, San Francisco, 1982).
3. B. V. Chirikov, *Phys. Rep.* **52**, 264 (1979).
4. C. F. F. Karney, *Physica* **D8**, 360 (1983).
5. T. Geisel, A. Zacherl and G. Radons, *Z. Phys.* **B71**, 117 (1988).
6. M. F. Shlesinger, G. M. Zaslavsky and J. Klafter, *Nature* **263**, 31 (1993).
7. T. Geisel and S. Thomae, *Phys. Rev. Lett.* **52**, 1936 (1984).

8. T. Geisel, J. Nierwetberg and A. Zacherl, *Phys. Rev. Lett.* **54**, 616 (1985).
9. G. Zumofen and J. Klafter, *Phys. Rev.* **E47**, 851 (1993); G. Zumofen, J. Klafter and A. Blumen, *Phys. Rev.* **E47**, 2183 (1993).
10. M. F. Shlesinger, J. Klafter and Y. M. Wong, *J. Stat. Phys.* **27**, 499 (1982).
11. M. F. Shlesinger and J. Klafter, *Phys. Rev. Lett.* **54**, 2551 (1985).
12. G. H. Weiss and R. J. Rubin, *Adv. Chem. Phys.* **52**, 363 (1983).
13. J. W. Haus and K. W. Kehr, *Phys. Rep.* **150**, 263 (1987).
14. A. Blumen, J. Klafter and G. Zumofen, in *Optical Spectroscopy of Glasses*, ed. I. Zschokke (Reidel, Dordrecht, 1986), p. 199.
15. J.-P. Bouchaud and A. Georges, *Phys. Rep.* **195**, 127 (1990).
16. J. Klafter and G. Zumofen, *Physica* **A196**, 102 (1993).
17. G. Zumofen and J. Klafter, *Physica D* **xxx** (1993).
18. X.-J. Wang, *Phys. Rev.* **A45**, 8407 (1992).
19. A. J. Lichtenberg and M. A. Lieberman, *Regular and Stochastic Motion* (Springer, New York, 1982).
20. B. V. Chirikov, *Chaos, Solitons, and Fractals*, **1**, 79 (1991).
21. R. Ishizaki, H. Hata, T. Horita and H. Mori, *Prog. Theor. Phys.* **84**, 179 (1990); R. Ishizaki, T. Horita, T. Kobayashi and H. Mori, *Prog. Theor. Phys.* **85**, 1013 (1991).
22. Y. H. Ichikawa, T. Kamimura and T. Hatori, *Physica* **D29**, 247 (1987).
23. G. M. Zasklavsky, *Fluid Dynamics Res.* **8**, 127, (1991); G. M. Zaslavsky and M. K. Tippett, *Phys. Rev. Lett.* **67**, 3251 (1991).
24. G. Zumofen and J. Klafter (to be published).

Fractals, Vol. 1, No. 3 (1993) 405–415

REACTION-FRONT DYNAMICS IN $A+B\rightarrow C$ WITH INITIALLY-SEPARATED REACTANTS

S. HAVLIN*†, M. ARAUJO*, H. LARRALDE*, A. SHEHTER† and H. E. STANLEY*
*Center for Polymer Studies and Department of Physics, Boston University, Boston, MA 02215, USA
†Department of Physics, Bar-Ilan University Ramat-Gan, Israel

Abstract

We review recent developments in the study of the diffusion reaction system of the type $A+B\rightarrow C$ in which the reactants are initially separated. We consider the case where the A and B particles are initially placed uniformly in Euclidean space at $x>0$ and $x<0$ respectively. We find that whereas for $d\geq 2$ a single scaling exponent characterizes the width of the reaction zone, a multiscaling approach is needed to describe the one-dimensional system. We also present analytical and numerical results for the reaction rate on fractals and percolation systems.

1. INTRODUCTION

The dynamics of diffusion controlled reactions of the type $A+B\rightarrow C$ has been studied extensively since the pioneering work of Smoluchowski.[1,2] Most studies have focused on homogeneous systems, i.e., when both reactants are initially uniformly mixed in a d-dimensional space, and interesting theoretical results have been obtained. When the concentrations of the A and B reactants are initially equal, i.e., $c_A(0)=c_B(0)=c(0)$, the concentration of both species is found to decay with time as, $c(t)\sim t^{-d/4}$ for Euclidean $d\leq 4$-dimensional systems[3–10] and as $c(t)\sim t^{-d_s/4}$ for fractals[5,6] with fracton dimension $d_s\leq 2$. Also, self-segregated regions of A and B in low dimensions $(d\leq 3)$[4] and in fractals[9] have been found. Quantities such as the distributions of domain sizes of segregated regions and interparticle distances between species of the same type and different types have been calculated.[11–13] These systems were also studied theoretically and numerically under steady state conditions and interesting predictions have been obtained.[14–17] However, the above numerical and theoretical predictions have not been observed in experiments, in part because of difficulties to implement the initially uniformly-mixed distributions of reactants.

In recent years, it was realized that diffusion reaction systems in which the reactants are initially separated,[18] can be studied experimentally[19,20] and that the dynamics of such a system have many surprising features.[20–27] These systems are characterized by the presence of a dynamical interface or a front that separates the reactants. Such a reaction front appears in many biological, chemical and physical processes.[28–34]

Gàlfi and Ràcz[18] studied diffusion-controlled reactions with initially separated reactants. They studied the kinetics of the reaction diffusion process by a set of mean-field (MF) type equations,

$$\frac{\partial c_A}{\partial t} = D_A \frac{\partial^2 c_A}{\partial x^2} - k c_A c_B \tag{1a}$$

$$\frac{\partial c_B}{\partial t} = D_B \frac{\partial^2 c_B}{\partial x^2} - k c_A c_B \,, \tag{1b}$$

where $c_A \equiv c_A(x, t)$ and $c_B \equiv c_B(x, t)$ are the concentrations of A and B particles at position x at time t respectively, $D_{A,B}$ are the diffusion constants and k is the reaction constant. The rate of production of the C-particles at site x and time t, which we call the reaction-front profile, is given by $R(x, t) \equiv k c_A c_B$. The initial conditions are that the A species are uniformly distributed on the right-hand side of $x = 0$ and the B species are uniformly distributed on the left-hand side.

Using scaling arguments Gàlfi and Ràcz[18] find that the width w of the reaction front $R(x, t)$ scales with time as, $w \sim t^{\alpha}$ with $\alpha = 1/6$ and the reaction rate at the center of the front, which is called the reaction height, scales as $h \sim t^{-\beta}$ with $\beta = 2/3$.

Experiments[19] and simulations[19,21–24] for $d \geq 2$ systems in which both reactants diffuse, support the above predicted values for α and β. Indeed, Cornell et al.[23] argue that the upper critical dimension is $d = 2$ and the MF approach should therefore be valid for $d \geq 2$. Moreover, numerical simulations of 1D systems show that the width exponent appears to be $\alpha \simeq 0.3$ and the height exponent $\beta \simeq 0.8$.[23,24] Recently,[25] it was argued that α varies between 1/4 and 3/8, depending on the moment at which the width is calculated. For a more detailed discussion, see Sec. 3. The origin of the difference between the exponents of 1D systems and those of higher dimensional systems is due to fluctuations in the location of the front which are important in low dimensions and are neglected in the MF approach.

Taitelbaum et al.[20,22] studied analytically Eqs. (1) and presented experiments for the limit of small reaction constant or short time. The main results are that several measurable quantities undergo interesting crossovers. For example, the global reaction rate changes from $t^{1/2}$ in the short time limit to $t^{-1/2}$ at the asymtotic time regime. The center of the front can change its direction of motion as found in experiments.[20] Ben-Naim and Redner[22] studied the solution of Eqs. (1) under steady-state conditions.

2. THE FORM OF THE REACTION FRONT, $R(x, t)$, IN THE MEAN-FIELD APPROACH

In a recent work,[27] we consider the symmetric case in which both diffusion constants and concentrations are equal, i.e., $D_A = D_B \equiv D$ and $c_A(x, 0) = c_B(x, 0) = c_0$. If we define $F(x, t) \equiv c_A(x, t) - c_B(x, t)$, then from Eqs. (1) it follows,

$$\frac{\partial F}{\partial t} = D\frac{\partial^2 F}{\partial x^2}, \tag{2}$$

subject to the conditions that initially the A particles are uniformly distributed to the right of the origin while the B particles are uniformly distributed to the left of the origin. Equation (2) has the solution $F(x, t) = c_0 \operatorname{erf}(x/\sqrt{4Dt})$.

We rewrite the concentrations of A and B particles as (see Fig. 1):

$$c_A(x, t) = G_1(x, t) + \delta c_1(x, t), \quad c_B(x, t) = G_2(x, t) + \delta c_2(x, t) \tag{3}$$

where

$$G_1(x, t) = \begin{cases} F(x, t) & [x > 0] \\ 0 & [x < 0], \end{cases} \tag{4}$$

and $G_2(x, t) = G_1(-x, t)$. It is easy to see that under the above conditions, $\delta c_1(x, t) = \delta c_2(x, t) \equiv \delta c(x, t)$. Substituting Eqs. (3) into Eqs. (1) yields:

$$\frac{\partial(\delta c)}{\partial t} = D\frac{\partial^2(\delta c)}{\partial x^2} - k\left(c_0 \operatorname{erf}\left(\frac{x}{\sqrt{4Dt}}\right) + \delta c\right)\delta c \tag{5}$$

The asymptotic solution for this equation that vanishes as $x \to \infty$ is:

$$\delta c(x, t) \sim t^{-1/3}\left(\frac{x}{t^{1/6}}\right)^{-1/4} \exp\left[-\frac{2}{3}\left(\frac{\lambda x}{t^{1/6}}\right)^{3/2}\right], \quad t^{1/6} \ll x \ll (4Dt)^{1/2}, \tag{6}$$

where $\lambda = (ka/D)^{1/3}$, $a \equiv c_0/(\pi D)^{1/2}$. As may be confirmed by direct substitution, this expression is a solution of Eq. (5) up to terms of order $(\delta c)/t$, which can be neglected for large t.

Using Eq. (6) we can write an expression for the reaction-front profile $R(x, t)$ defined in Eqs. (1) as:

$$R(x, t) \simeq \frac{kax}{t^{1/2}}(\delta c) \sim t^{-2/3}\left(\frac{x}{t^{1/6}}\right)^{3/4} \exp\left[-\frac{2}{3}\left(\frac{\lambda x}{t^{1/6}}\right)^{3/2}\right]. \tag{7}$$

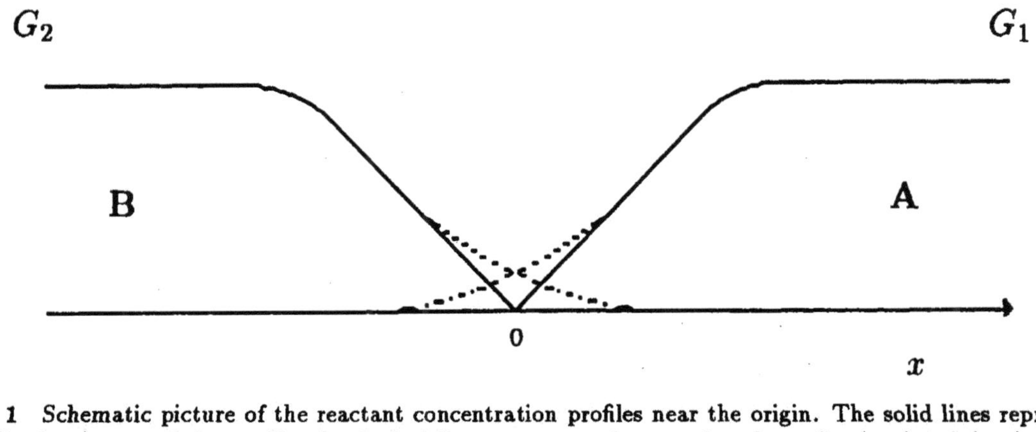

Fig. 1 Schematic picture of the reactant concentration profiles near the origin. The solid lines represent the $G_{1,2}(x, t)$ part of the profile, the dashed lines represent the complete form $G_{1,2}(x, t) \pm \delta c(x, t)$. Note that the profile of species A is given solely by $\delta c(x, t)$ on the left of the origin. (From Ref. 27.)

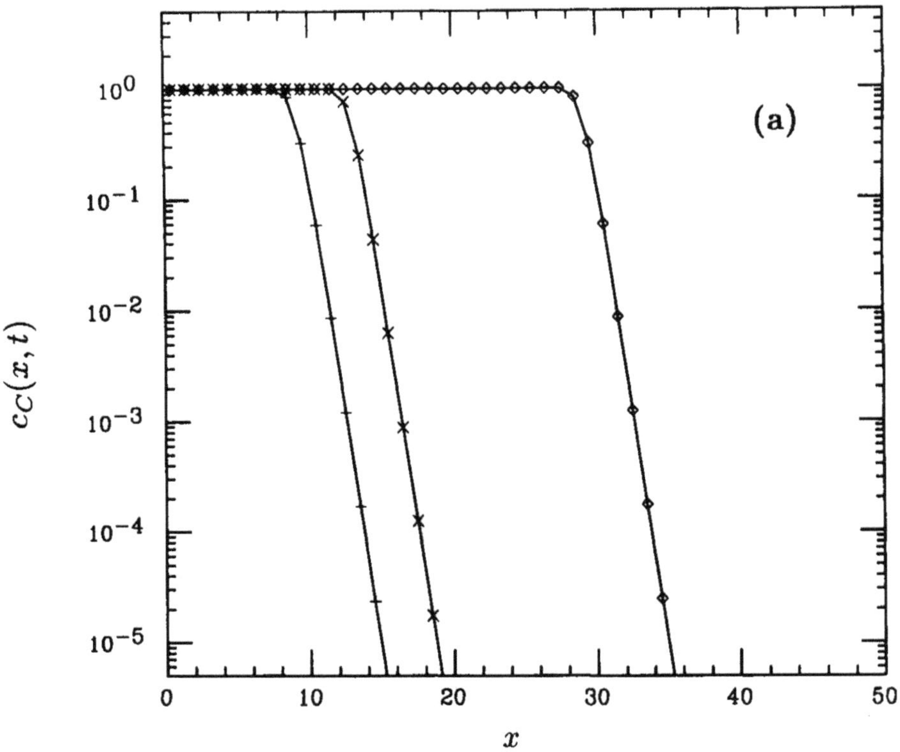

(a)

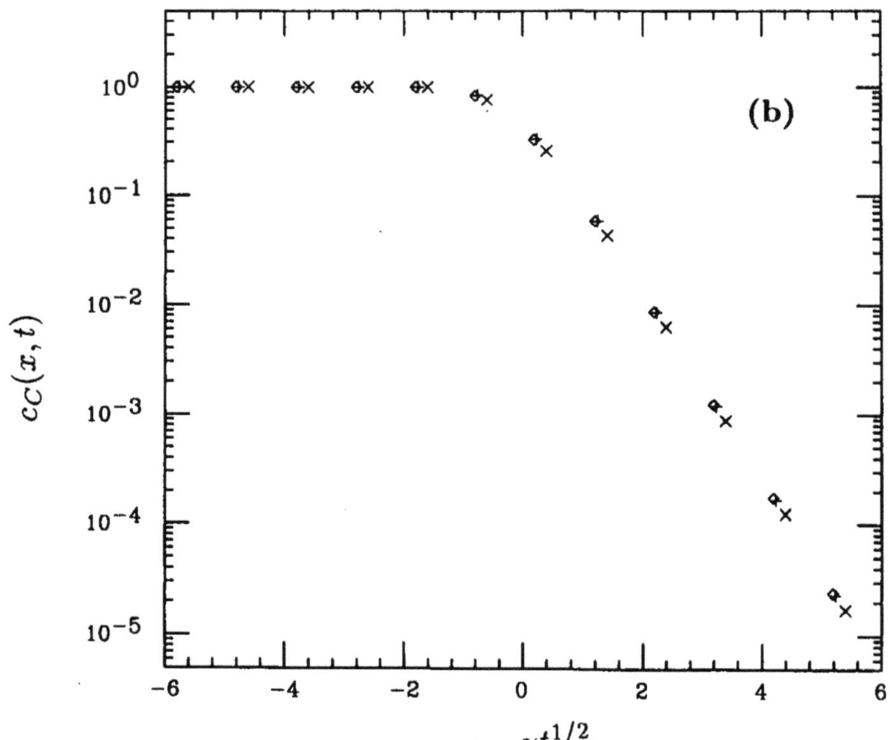

(b)

Fig. 2 Numerical solution of Eq. (1) for the case $D_B = 0$, $D_A \neq 0$: (a) plot of $c(x, t)$ as a function of x for $t = 500$, 1000 and 5000; (b) the good scaling plot of $c(x, t)$ as a function of $(x - \langle x \rangle)$ indicates that $\alpha = 0$ and $\beta = 1/2$.

It is seen that the width of the reaction front grows as $t^{1/6}$, whereas the height can be identified with the prefactor $t^{-2/3}$ in Eq. (7), consistent with the exponents found by Gàlfi and Ràcz.[18] Equation (7) provides a more quantitative solution to Eqs. (1) than the previous scaling arguments,[18] as well as information on the dependence of the form of the reaction front on the parameters c_0, k, and D, for the symmetric case.

For the case in which one reactant is static no analytical solution (of Eqs. 1) exists for the form of the reaction front. However, numerical solutions of Eqs. (1) with $D_B = 0$ shown in Fig. 2, suggest that $R(x,t) \sim t^{-\beta} g(x/t^{\alpha}) \exp(-|x|/t^{\alpha})$. The excellent scaling in Fig. 2(b) suggests that the width does not increase with time, i.e., $w \sim t^{\alpha}$ with $\alpha = 0$ and $h \sim t^{-\beta}$ with $\beta = 1/2$.

3. THE FRONT, $R(x, t)$, in $d=1$

The spatial distribution of the C particles in $d = 1$ systems, $R(x,t)$, where both reactants are diffusing with the same diffusion constant, $D_A = D_B \neq 0$, has been calculated numerically.[24] The data shown in Fig. 3 suggests that:

$$c_C(x,t) \equiv \int_0^t R(x,t')dt' \sim \exp(-a|x|/t^{\delta}), \tag{8}$$

with $\delta = 0.37 \pm 0.05$. In a recent work[25] it was found that the reaction front in one dimension is characterized by several characteristic length scales. Let $m(t)$ be the midpoint between the rightmost A particle (RMA) and the leftmost B particle (LMB). We find that $\xi(t) \equiv \langle m^q(t)\rangle^{1/q} \sim t^{\delta}$ with $\delta \approx 3/8$ independent of q. Another length scale that characterizes the reaction front is the distance $l_{AB}(t)$ between the RMA and LMB. At time t, a reaction will occur at position $m(t)$ if and only if $l_{AB}(t) = 0$. We find that $\langle l^q_{AB}(t)\rangle^{1/q}$ scales as t^{σ} with $\sigma \approx 1/4$ for all q. This result is similar to that found by Weiss et al.[35] for a similar quantity in the one-dimensional trapping problem.

Next we study the spatial moments of the reaction front[25]:

$$\langle |x|^q \rangle \equiv \frac{\int_0^{\infty} x^q R(x,t)dx}{\int_0^{\infty} R(x,t)dx} \sim t^{1/2}\frac{d}{dt}\int_0^{\infty} x^q c_C(x,t)dx \equiv t^{q\alpha_q}. \tag{9}$$

We find that while both the distributions of $m(t)$ and $l_{AB}(t)$ can be described in terms of a single length scale, the moments of the reaction front do not follow a simple scaling relation. Indeed, for small moments, $q \to 0$, $\alpha(q)$ approaches the value $\alpha_0 = 1/4$, whereas for large moments $\alpha(q)$ approach the value $\alpha_{\infty} = 3/8$ (Fig. 4). In particular the exponent of the second moment coincides with the value ($\simeq 0.3$) obtained in previous studies.[23,24]

To explain the behavior of α_q, we studied numerically the form of $R(x,t)$. Our data suggest that:

$$\bar{R}(x,t) \equiv \frac{R(x,t)}{\int_0^{\infty} R(x,t)dx} \sim u^{-2} e^{-|u|/t^{1/8}}, \tag{10}$$

where $u = x/t^{1/4}$. Substituting this expression into Eq. (9) we obtain:

$$\alpha_q = \begin{cases} 1/4 & q < 1 \\ (3/8) - (1/8q) & q \geq 1 \end{cases} \tag{11}$$

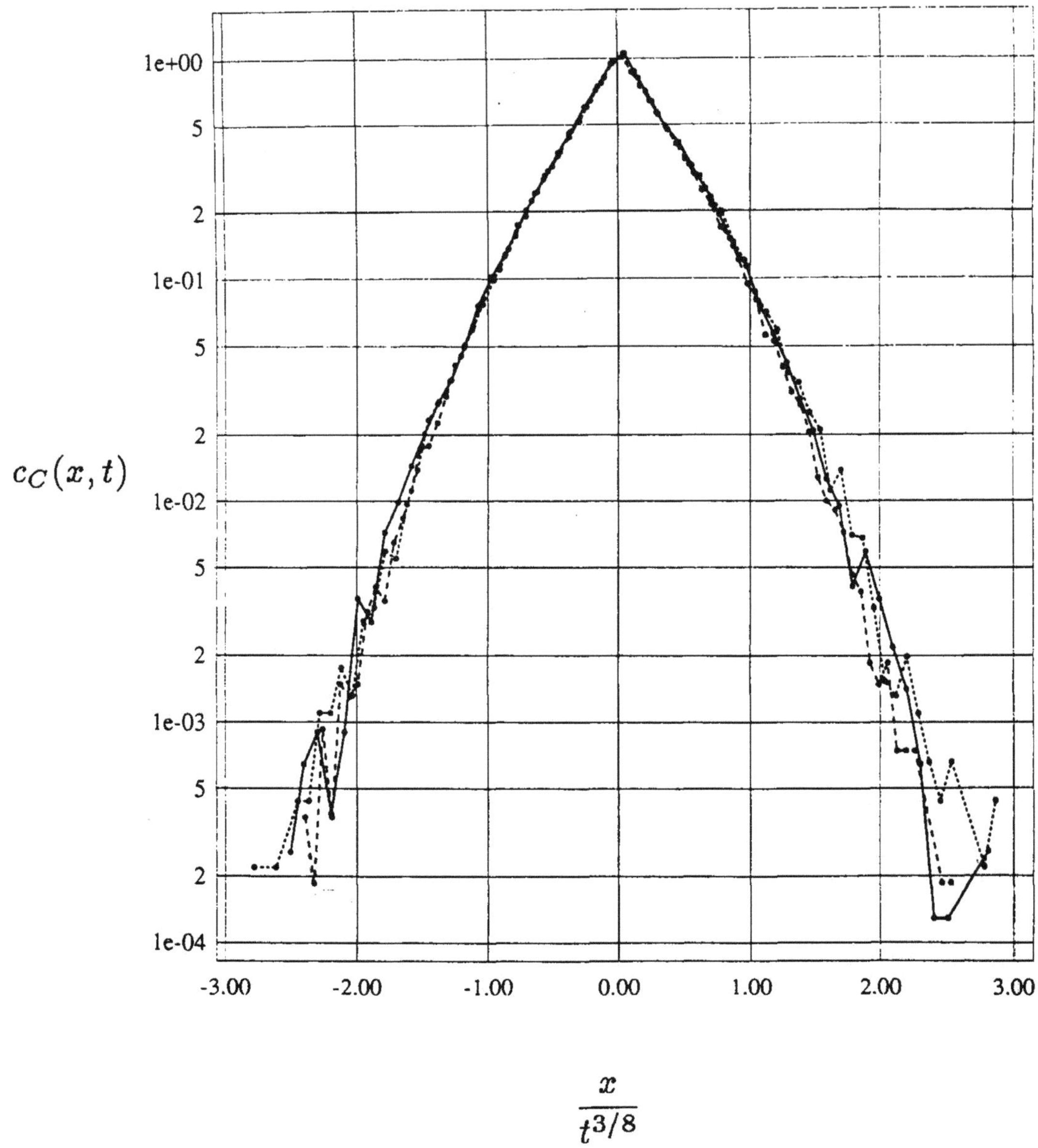

Fig. 3 Plot of $c_C(x, t)$ defined in Eq. (8) for $d = 1$ system with system size $L = 10\,000$ and $t = 1000$, 5000 and 10 000.

where for $q = 1$

$$\langle x \rangle \sim t^{1/4} \ln t \,. \tag{12}$$

Our results for $q \geq 1$ are in very good agreement with these theoretical predictions (see Fig. 4). However for $q < 1$, our results are slightly larger than 1/4, which is probably due to not reaching the asymptotic limit.[36]

For the case $D_A \neq 0$, $D_B = 0$, analytical and numerical studies[37] yield for the reaction front:

$$R(x, t) = \frac{1}{4t^{3/4}} \left(\frac{2\gamma^2}{\mu\pi} \right)^{1/2} \exp\left[-\frac{(x - \gamma t^{1/2})^2}{2\mu t^{1/2}} \right] \left[1 + \frac{x - \gamma t^{1/2}}{2\gamma t^{1/2}} \right] , \tag{13}$$

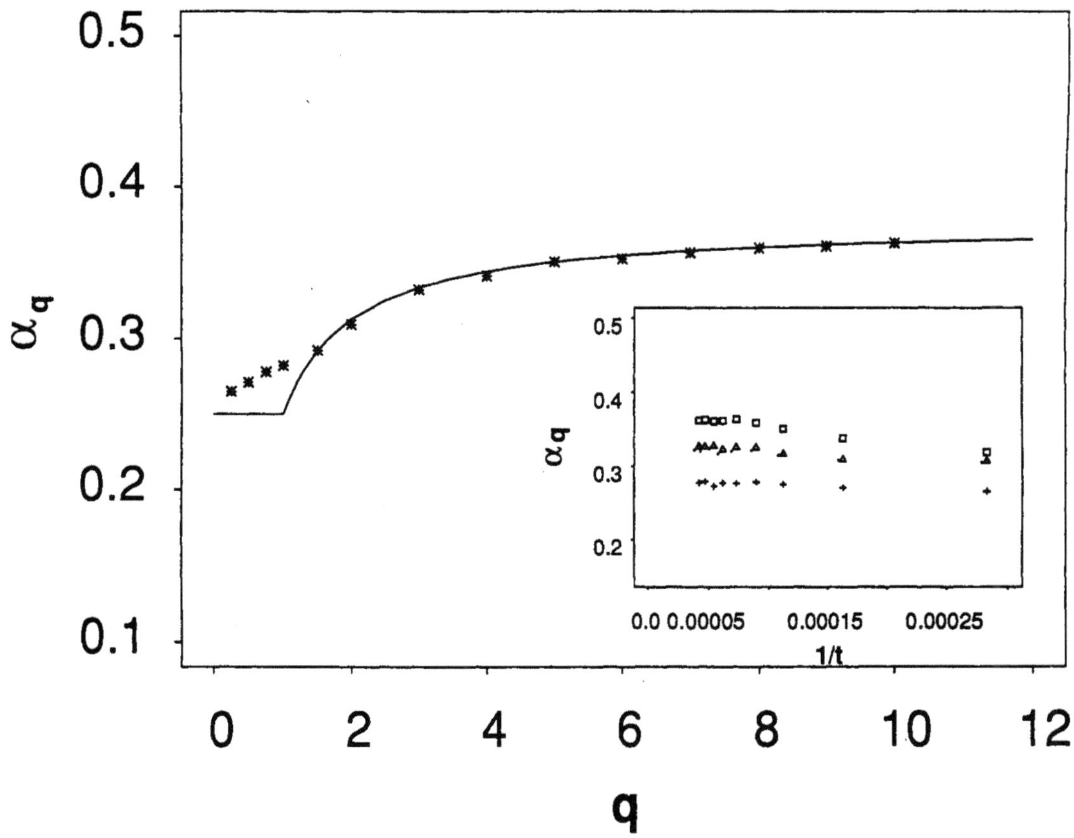

Fig. 4 Plot of α_q as a function of q obtained from extrapolation to infinite times for values of q ranging from 1/4 to 10. The solid line represents the theoretical prediction of Eq. (11). Inset shows successive values of the exponents α_q which characterize the moments $\langle x^q \rangle^{1/q}$ of the reaction as a function of $1/t$ for $q = 1/2(+)$, $2(\times)$, $8(\square)$.

where γ and μ are constants. From Eq. (13), it follows that $\alpha = 1/4$ and $\beta = 3/4$. It is interesting to note that the time integral of $R(x, t)$, which is the total production of the C particles at x up to time t, is given by:

$$c_C(x, t) = \int_0^t R(x, \tau) d\tau = \frac{1}{2} \text{erfc} \left(\frac{x - \gamma t^{1/2}}{\sqrt{2\mu t^{1/2}}} \right) . \tag{14}$$

We performed analogous calculations of $\langle m^q(t) \rangle^{1/q}$, for $\langle l_{AB}^q \rangle^{1/q}$, for the case where one of the reactants is static. We find numerically and theoretically that in this case the two exponents coincide: $\delta = \sigma = 1/4$.

To summarize, we find four sets of exponents (see Table 1).

Table 1

	$d = 1$	MF ($d > 2$)
[both moving]	$1/4 \le \alpha_q \le 3/8$	$\alpha = 1/6$
	$\delta = 3/8$, $\sigma = 1/4$	$\beta = 2/3$
[one static]	$\alpha = 1/4$	$\alpha = 0$
	$\beta = 3/4$	$\beta = 1/2$
	$\delta = \sigma = 1/4$	

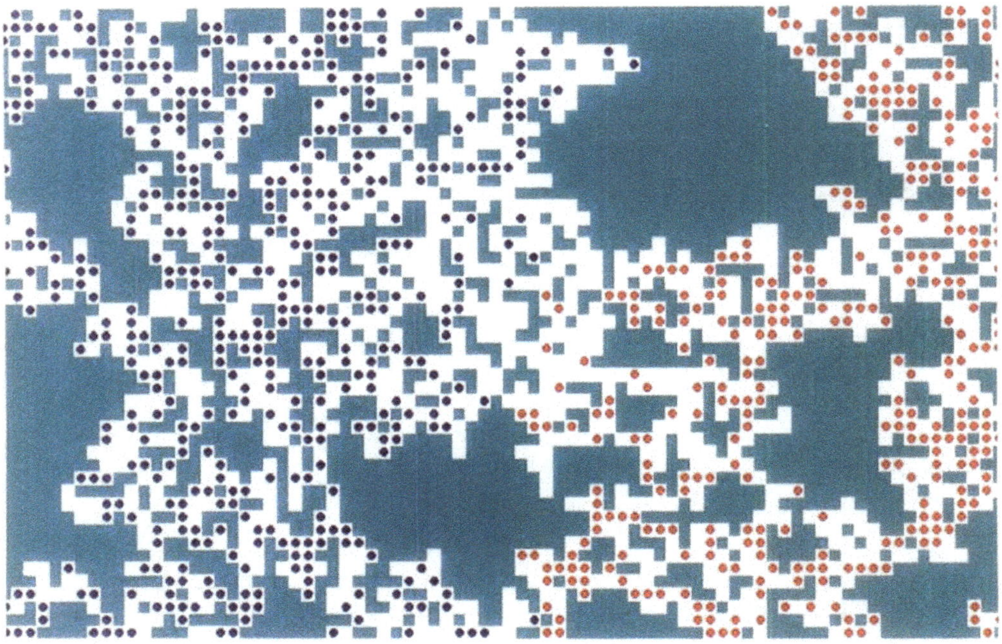

Fig. 5 Reaction-diffusion on the 2-dimensional infinite percolation cluster at criticality for $t = 2000$. White squares represent sites of the infinite cluster. Blue and red circles represent the A and B particles. Initially all sites of cluster in the right half plane and left half plane were occupied by A and B particles, respectively.

4. THE REACTION RATE IN $d = 2$ PERCOLATION

The case of $A + B \to C$ with initially-separated reactants on fractal systems was studied on the $d = 2$ infinite percolation cluster at criticality[22]; for a demonstration of the system see Fig. 5. It is expected that the total number of reactants up to time t,

$$\int_{-\infty}^{\infty} c_C(x, t)dx\,,$$

scales as the mean displacement of a random walker on a fractal, i.e., $\langle r^2 \rangle^{1/2} \sim t^{1/d_w}$, where d_w is the anomalous diffusion exponent.[38] From this, it follows that the reaction rate:

$$R(t) \equiv \int_{-\infty}^{\infty} R(x, t)dx \sim t^{-\gamma}\,, \quad \gamma = 1 - 1/d_w\,. \tag{15}$$

One has to distinguish between reaction-diffusion on the infinite cluster and in the percolation system, containing also small clusters.[39] The reaction rate on the infinite cluster is smaller and decreases slower compared with the system containing clusters of all sizes. This can be understood as follows. At any finite time we can divide all clusters into two groups according to their sizes: *active* — clusters of mass $s > s^*$, in which at time $t < t^* \sim (s^*)^{d_w/d_f}$ particles are not aware of the finiteness of their cluster (this group contains the infinite cluster), and *inactive* clusters of mass $s < s^*$ on which at least one of the reactants has vanished and the reaction rate is zero. According to this picture, in the full percolation system at any time there are active clusters of finite size, that can contribute to the reaction rate. Therefore, the rate of reaction in the percolation system is always higher than on the infinite cluster. Also, at any time there are some finite clusters that become inactive causing additional (comparing with the infinite cluster network) decrease of the rate of reaction in the percolation system. Since the system is self-similar, one expect a change in the reaction rate exponent.

To quantify the above considerations, we can look on each cluster of mass s and linear size $r \sim s^{1/d_f}$ as a reservoir of particles divided by the *front line* into two A- and B-parts. We define the *active front* of a cluster as the sites belonging both to the cluster and to the front line. The length l_s of the *active front* of a single cluster of size s is expected to be:

$$l_s \sim r^{d_f-1} \sim s^{\frac{d_f-1}{d_f}} \tag{16}$$

Next we assume that the rate of reaction on a cluster of mass s per unit length of active front is:

$$R_0(t) \sim \begin{cases} t^{-\gamma} & t < t^* \\ 0 & t > t^*, \end{cases} \tag{17}$$

where $t^* = s^{d_w/d_f}$. Therefore, the total contribution of active clusters of size s to the reaction rate is:

$$R_s(t) \sim \varphi_s s^{\frac{d_f-1}{d_f}} t^{-\gamma}, \tag{18}$$

where φ_s is number of clusters of size s that intersect the front line. One can estimate φ_s as follows. In a percolation system of size $L \times L$, there are n_s clusters of mass s. Only a small part of them intersects the front line, those in a strip of width $w \sim s^{1/d_f}$ around the front line. Their fraction is w/L. Therefore, $\varphi_s \sim \frac{s^{1/d_f}}{L} n_s \sim s^{1/d_f} n_s$. Substituting this in Eq. (18) we get:

$$R_s(t) \sim s^{1/d_f} n_s s^{\frac{d_f-1}{d_f}} t^{-\gamma} = t^{-\gamma} s n_s\,. \tag{18a}$$

Thus, the reaction rate in the percolation system is:

$$R(t) = \sum_{s=s^*}^{\infty} R_s(t) = t^{-\gamma}(s^*)^{2-\tau} = t^{-\gamma}t^{-\delta} = t^{-\gamma'}$$

$$\delta = \gamma' - \gamma = \frac{d_f}{d_w}(\tau - 2)\,. \tag{19}$$

These results are in good agreement with our simulations (see Fig. 6).

We also study the finite size effects on the reaction-diffusion system. For a percolation system of size $L \times L$, we expect that for the infinite cluster:

$$R(t) = L^{d_f-1}t^{-\gamma}, \tag{21a}$$

while for the percolation system:

$$R(t) = Lt^{-\gamma'}\,. \tag{21b}$$

We expect that at time $t^* \sim L^{d_w}$ these two rates become equal, since no "small" active cluster exist in the system above t^*. Indeed, equating the last two expressions (21) reproduce Eq. (19). The prefactor L in Eq. (21b) assures that the reaction rate for percolation system is larger for $t < t^*$ than the reaction rate on the infinite cluster. The ratio of the reaction rates at $t = 1$ is L^{2-d_f}.

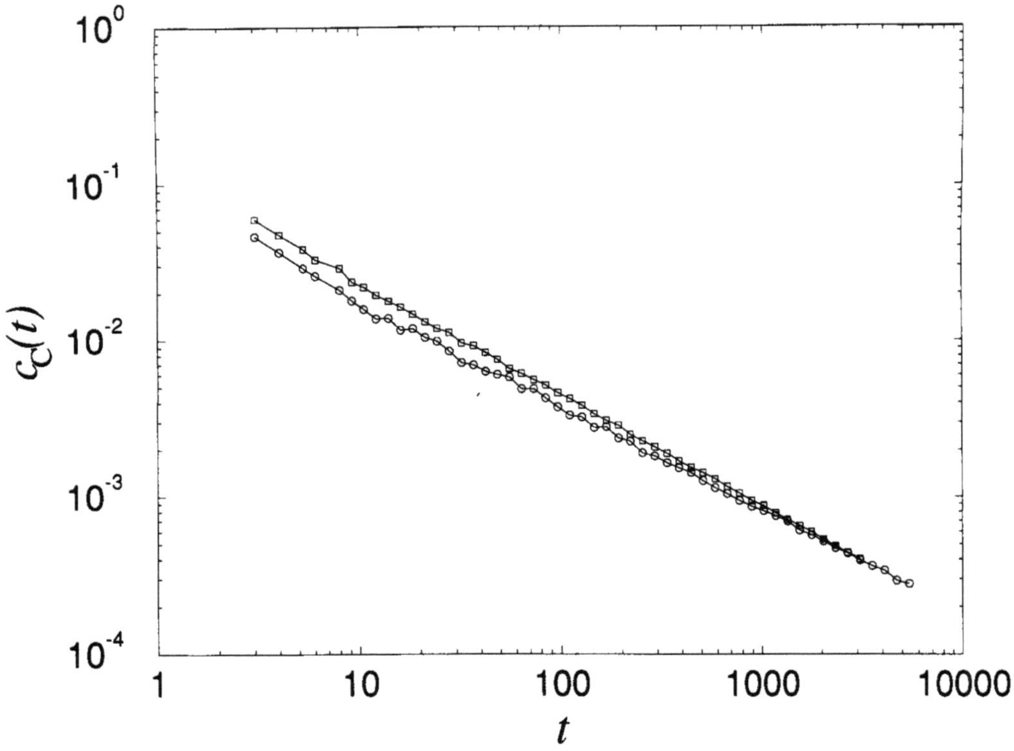

(a)

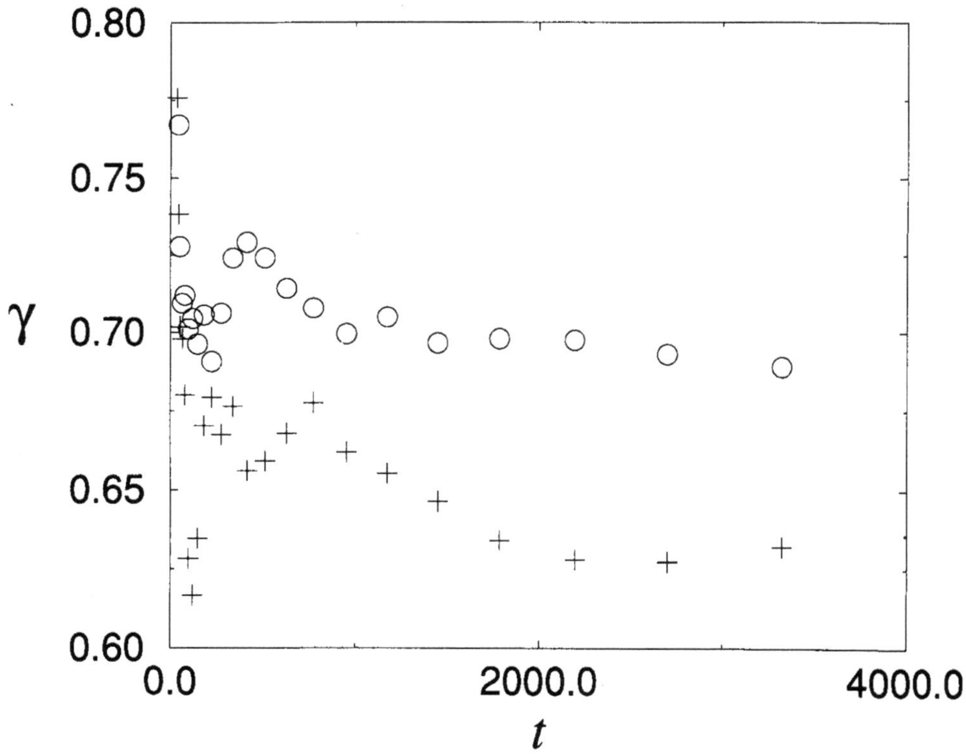

(b)

Fig. 6 **(a)** Plot of the rate $c_C(t)$ for the percolation system (□) and for the infinite percolation cluster (o). **(b)** Plot of successive slopes of the data in (a). The exponent γ of reaction rate on the infinite cluster (+) and on the percolation system (o).

ACKNOWLEDGMENTS

We wish to thank F. Leyvraz, S. Redner, H. Taitelbaum and G. H. Weiss for useful discussions, and NSF for financial support.

REFERENCES

1. M. V. Smoluchowski, *Z. Phys. Chem.* **92**, 129 (1917).
2. See, e.g. S. A. Rice, *Diffusion-Limited Reactions* (Elsevier, Amsterdam, 1985).
3. A. A. Ovchinnikov and Y. B. Zeldovich, *Chem. Phys.* **28**, 215 (1978).
4. D. Toussiant and F. Wilzek, *J. Chem. Phys.* **78**, 2642 (1983).
5. P. Meakin and H. E. Stanley, *J. Phys.* **A17**, L173 (1984).
6. K. Kang and S. Redner, *Phys. Rev. Lett.* **52**, 955 (1984); *Phys. Rev.* **32**, 435 (1985).
7. K. Lee and E. J. Weinberg, *Nucl. Phys.* **B246**, 354 (1984).
8. G. Zumofen, A. Blumen and J. Klafter, *J. Chem. Phys.* **82**, 3198 (1985).
9. R. Kopelman, *Science*, **241**, 1620 (1988).
10. M. Bramson and J. L. Lebowitz, *Phys. Rev. Lett.* **61**, 2397 (1988); *J. Stat. Phys.* **65**, 941 (1991).
11. P. Argyrakis and R. Kopelman, *Phys. Rev.* **A41**, 2121 (1990).
12. G. H. Weiss, R. Kopelman and S. Havlin, *Phys. Rev.* **A33**, 466 (1989).
13. F. Leyvraz and S. Redner, *Phys. Rev. Lett.* **66**, 2168 (1991); S. Redner and F. Leyvraz, *J. Stat. Phys.* **65**, 1043 (1991).
14. L. W. Anacker and R. Kopelman, *Phys. Rev. Lett.* **58**, 289 (1987); *J. Chem. Phys.* **91**, 5555 (1987).
15. K. Lindenberg, B. J. West and R. Kopelman, *Phys. Rev. Lett.* **60**, 1777 (1988).
16. D. Ben-Avraham and C. R. Doering, *Phys. Rev.* **A37**, 5007 (1988).
17. E. Clément, L. M. Sander and R. Kopelman, *Phys. Rev.* **A39**, 6455 (1989).
18. L. Gálfi and Z. Rácz, *Phys. Rev.* **A38**, 3151 (1988).
19. Y. E. Koo, L. Li and R. Kopelman, *Mol. Cryst. Liq. Cryst.* **183**, 187 (1990); Y. E. Koo and R. Kopelman, *J. Stat. Phys.* **65**, 893 (1991).
20. H. Taitelbaum, Y. E. Koo, S. Havlin, R. Kopelman and G. H. Weiss, *Phys. Rev.* **A46**, 2151 (1992).
21. Z. Jiang and C. Ebner, *Phys. Rev.* **A42**, 7483 (1990).
22. H. Taitelbaum, S. Havlin, J. Kiefer, B. L. Trus and G. H. Weiss, *J. Stat. Phys.* **65**, 873 (1991).
23. S. Cornell, M. Droz and B. Chopard, *Phys. Rev.* **A44**, 4826 (1991).
24. M. Araujo, S. Havlin, H. Larralde and H. E. Stanley, *Phys. Rev. Lett.* **68**, 1791 (1992).
25. M. Araujo, H. Larralde, S. Havlin and H. E. Stanley, *Phys. Rev. Lett.* **71**, 3592 (1993).
26. E. Ben-Naim and S. Redner, *J. Phys.* **A25**, L575 (1992).
27. H. Larralde, M. Araujo, S. Havlin and H. E. Stanley, *Phys. Rev.* **A46**, 855 (1992).
28. D. Avnir and M. Kagan, *Nature*, **307**, 717 (1984).
29. G. T. Dee, *Phys. Rev. Lett.* **57**, 275 (1986).
30. B. Heidel, C. M. Knobler, R. Hilfer and R. Bruinsma, *Phys. Rev. Lett.* **60**, 2492 (1986).
31. R. E. Liesegang, Naturwiss, *Wochensch*, **11**, 353 (1896).
32. T. Witten and L. M. Sander, *Phys. Rev. Lett.* **47**, 1400 (1981).
33. R. A. Ball, *Ausr. Gemmol.* **12**, 89 (1984).
34. K. F. Mueller, *Science*, **255**, 1021 (1984).
35. G. H. Weiss, R. Kopelman and S. Havlin, *Phys. Rev.* **A39**, 466 (1989); S. Redner and F. Leyvraz, *J. Stat. Phys.* **65**, 1043 (1991).
36. Note that a similar multiscaling behavior was found for the moments of $\langle l_{AA}^q \rangle$ in the initially homogeneous reaction, $A + B \to C$, by F. Leyvraz and S. Redner, *Phys. Rev.* **A46**, 3132 (1992).
37. H. Larralde, M. Araujo, S. Havlin and H. E. Stanley, *Phys. Rev.* **A46**, R6121 (1992).
38. A. Bunde and S. Havlin, eds., *Fractals and Disordered Systems* (Springer, Heidelberg 1991).
39. S. Havlin and D. Ben-Avraham, *Adv. in Phys.* **36**, 695 (1987).

Fractals, Vol. 1, No. 3 (1993) 416–424

CORROSION PATTERN FORMATION IN ALUMINIUM THIN LAYERS

L. BALÁZS, L. NYIKOS, I. SZABÓ* and R. SCHILLER
Central Research Institute for Physics,
Institute for Atomic Energy Research
**Research Institute for Material Sciences,*
P.O.B. 49, H-1525 Budapest, Hungary

Abstract

The morphology of pitting corrosion, brought about by Cl^- and Fe^{3+} containing acidic solutions on aluminium thin layers evaporated onto glass substrate was investigated by image analysis. The influence of electrolyte composition on the morphology was studied. In a wide range of liquid composition, patterns with uniform fractal properties were grown. The dimension of patterns was found to be between 1.9–1.97, the lower value referring to the least aggressive case. Results indicate similarities with spreading percolation processes.

1. INTRODUCTION

The characterization of any heterogeneous chemical process usually requires both a description of the chemical reactions and an evaluation of mass transport of the reactants and reaction products at the interface. Mathematical treatment of transport becomes rather difficult if reactions occur at an irregular interface or the morphology of the interface changes with time.[1] Only arbitrary interpretation of experimentally determined quantities averaged over the whole system (e.g. weight loss of a corroding piece of metal) is possible if there is no detailed information on the interface geometry. Temporal variation and final shape of the interface, however, can be regarded as a footprint of the chemical transformation. The aim of this paper is to present certain quantitative results on chemically effected variations of interface morphology in a well-known heterogeneous chemical process.

Pitting corrosion,[2] a most dangerous form of local corrosion, is quite common with metals such as aluminium or stainless steel, whose corrosion resistance is due to a spontaneously formed, thin, compact and adherent oxide layer. In this process, the surface of a macroscopically homogeneous piece of metal exposed to a solution containing aggressive anions (e.g. Cl^- ions) is not attacked uniformly; after some induction period, the passive oxide film is removed at certain sites of the surface due to multistep chemical reactions. Since these metals themselves readily dissolve in an aqueous medium, once the oxide film loses its protective ability at a certain site, a pit starts to grow on the spot of attack.

A number of mechanisms have been suggested for pit initiation and propagation.[2] They usually rest on the formation of a soluble metal complex at a dislocation site of the oxide layer, a process followed by metal dissolution sustained by the reduction on the surrounding, still passive metal surface. Pit propagation is strongly influenced by local concentration changes at the metal/electrolyte interface. Stable pit growth requires the maintenance of the balance between the dissolution rate of the metal and the diffusion of ions in the liquid phase.

In order to simplify processes and evaluation, a two dimensional phenomenon, the corrosion of an aluminium layer evaporated onto glass substrate, was investigated. The layer was sufficiently thin to ensure that total in-depth dissolution of the metal is much faster than any other processes, in particular much faster than lateral growth of the corrosion pattern. This system provides an excellent opportunity for investigating pit growth since the whole pattern is visible throughout the corrosion process and the pit morphology can be determined directly by means of image analysis.[3,4]

Spontaneous pit formation and propagation can be observed on aluminium layers exposed to solutions containing chloride ions and some kind of oxidizing agent, e.g. Fe^{3+} ions. A preliminary series of experiments has shown that acidity, chloride and Fe^{3+} content, and the total ion concentration all influence the final geometry of the pit. Layer thickness, stirring of the liquid, and temperature also play a role here. We examine the influence of the electrolyte composition on the morphology of corrosion patterns.

2. EXPERIMENTAL

Aluminium layers of thickness of 45 nm were evaporated onto a thoroughly cleaned glass substrate from 99.99% pure Al filament. The evaporation rate was about 4 nm/s and the evaporator base pressure was less than 10^{-3} Pa. Electrolytes were made from reagent grade chemicals and triply distilled water. All the experiments were carried out under "open circuit" conditions, i.e., no contacts were made to the films so that the potential of the metal layer was determined solely by electrochemical reactions taking place at the metal/electrolyte interface. The electrolyte was deaerated before the experiment with N_2, but gas bubbling was stopped after the immersion of the samples and a N_2 blanket was maintained above the solution. Pits were grown to the size of a few millimeters in stagnant liquid at ambient temperature. The corrosion pattern covered only a small part of the total sample area. The time of the pit growth ranged from 1 to 10 minutes but this was not considered in this investigation. After each experiment, the samples were rinsed with distilled water and dried in a hot air stream. Images of corrosion patterns were taken by a video camera attached to a microscope. Several images for each sample at magnifications $\times150$, $\times300$, $\times600$ and $\times1500$, respectively, were recorded. The video signal was digitized by a frame grabber card into a 512×512 pixel image with multilevel gray scale. The picture was binarized by cutting the gray scale histogram at the local minimum between the two peaks corresponding to bright and dark pixels and so dividing the image into white (intact Al) and black (glass substrate) areas.

Three different methods[5] were used for the morphological characterization of corrosion patterns which were regarded as contiguous clusters consisting of black pixels. Besides box-counting for the dissolved area (black pixels), and the area distribution of undissolved Al islands (white pixels), i.e., the number of islands, $N(A)$ in area interval A and $A + dA$, we determined the variation of cluster mass, $M(r)$, i.e., the area covered by black pixels within a square of side $2r$. Images taken in the proximity of the edge of the pattern were omitted from the analysis.

3. RESULTS AND DISCUSSION

Useful information can be obtained by mere microscopic inspection of the corroding sample. Pit growth always starts from one single point and spreads laterally in the metal layer. Particularly if corrosion proceeds under mild conditions, distinct growth regions on the pit perimeter can be distinguished. If the front ceases to grow at a certain spot, the growth never starts again.

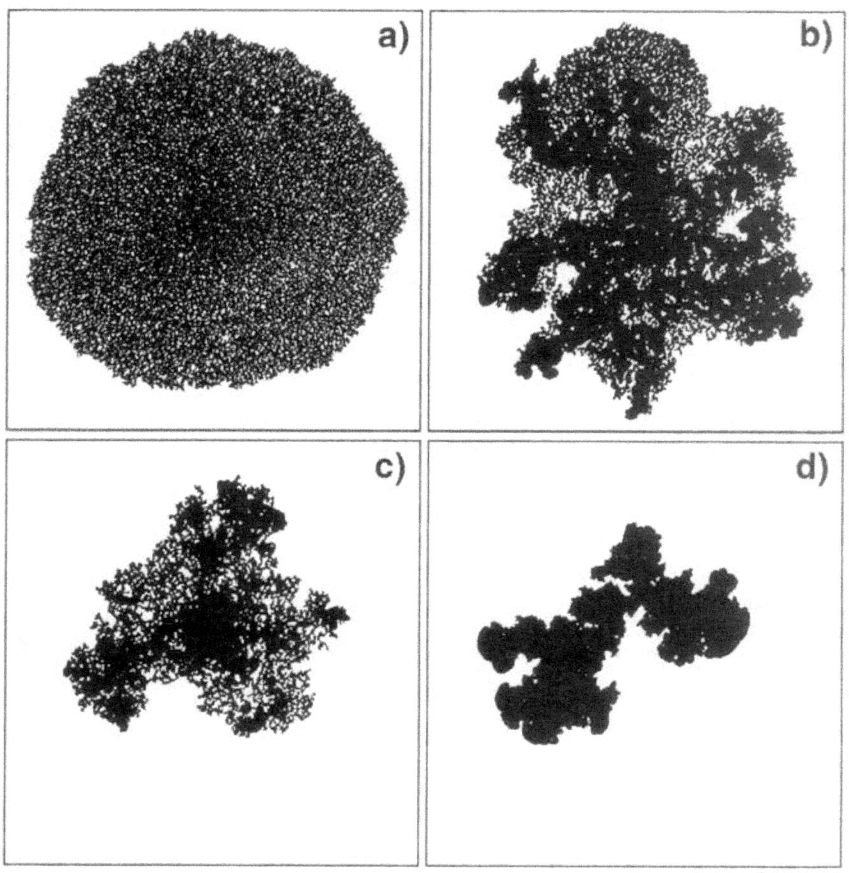

Fig. 1 Digitized images (512 × 512 pixels) of corrosion patterns formed in solutions containing 5 mM $Fe_2(SO_4)_3$ and (a) 0.001 M, (b) 0.01 M, (c) 0.1 M, and (d) 1 M HCl. White: undissolved aluminium; Black: glass substrate. Full width on an image: 1.126 mm.

Another observation worth mentioning is that as dissolution of the aluminium film proceeds, undissolved islands are formed. Once an island becomes detached from the bulk of the surrounding metal layer, it is no longer exposed to any further attack, i.e., pattern morphology behind the growth front remains intact throughout the experiment. Hence not only the shape of the pit perimeter, but the entire final pit morphology can be used to characterize the process.

The digitized images in Fig. 1 represent pits created in HCl solutions of concentration between 1 mM and 1 M. All of the solutions contained 5 mM of $Fe_2(SO_4)_3$. In the most dilute, hence the least aggressive, of the solutions a round hole filled with Al islands is left behind, the size of the islands being much smaller than the diameter of pattern 1(a), whereas at the highest concentration, pattern 1(d) is observed which contains but a few islands.

Regarding the set of black pixels as a connected cluster pattern, 1(d) is obviously an object with mass dimension close to 2. $M(r)$ function determined for pattern 1(a) is given in Fig. 2. $M(r)$ is seen to vary at small distances as $r^{1.81\pm0.02}$ which goes over to the quadratic dependence at higher distances. In dilute acid, the pattern formed seems to be a cluster with a correlation length of about 5 micrometers, below which it has fractal-like properties, whereas above this value, the cluster is homogeneous. Both visual inspection and numerical evaluation show that the two samples exposed to hydrochloric acid of medium concentrations [Figs. 1(b) and 1(c)] exhibit a certain mixture of morphologies of patterns 1(a) and 1(d). The perimeter of the pattern changes its shape as dissolution proceeds: Pit growth starts in a way similar to case 1(d), but at a later stage a rough, fuzzy front will be formed at certain points of the pit perimeter, leaving behind practically the same structure as that of pattern 1(a).

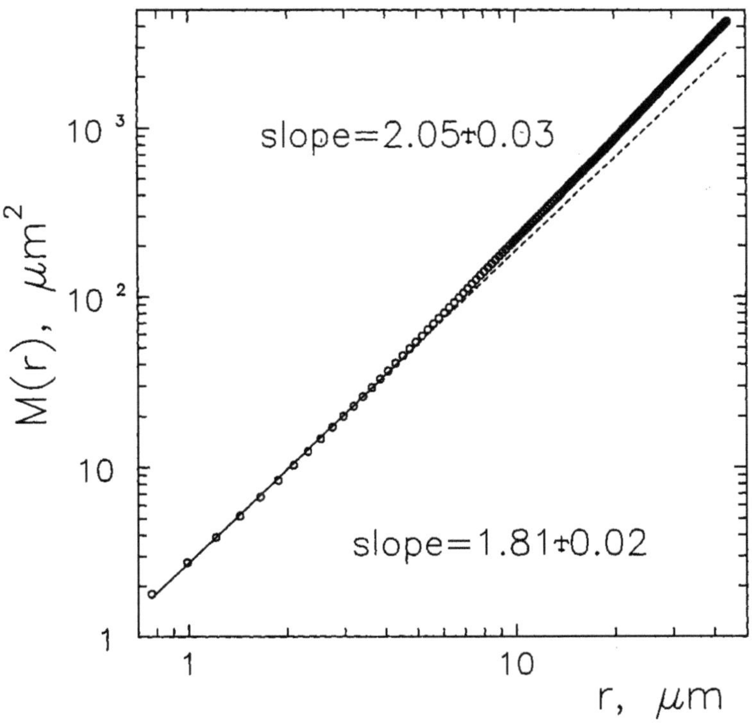

Fig. 2 Variation of the area covered by black pixels in Fig. 1(a), as a function of size. Note the crossover at about 5 μm from fractal to homogeneous behavior.

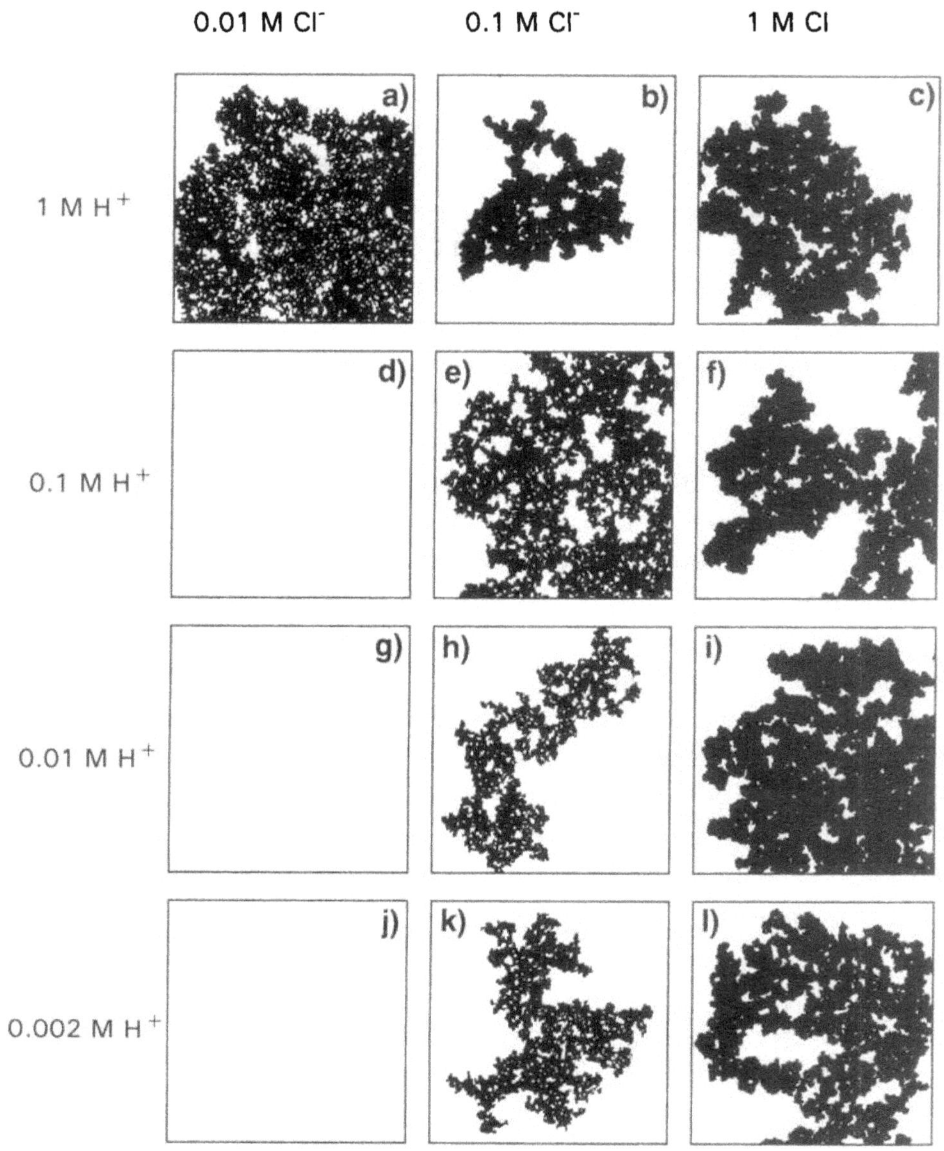

Fig. 3 Typical corrosion patterns formed in solutions containing 0.1 mM $Fe_2(SO_4)_3$ as well as H_2SO_4, NaCl and Na_2SO_4 in various concentrations, as indicated in the frame. The total amount of ions was held constant. White: unattacked aluminium layer; Black: glass substrate. Full width of an image: 1.126 mm. Three different concentration ranges yielding essentially three distinct morphologies (cf. Table 1) can be recognized: range I: (e), (h) and (k). Range II: (b), (c), (f), (i), (l). Range III: (a).

In another series of experiments, H^+ and Cl^- ion concentrations were varied independently with Na_2SO_4 added in order to maintain the total ion concentration, 1 M, constant in all the solutions. The concentration of oxidizing agent, $Fe_2(SO_4)_3$, was 0.1 mM. Pictures of etched surfaces are given in Fig. 3. The most dilute solutions do not attack the metal [Figs. 3(d), 3(g) and 3(j)], whereas with more aggressive solutions, three different concentration regions, yielding three distinct pit morphologies, can be recognized. The fraction of black pixels in patterns formed under not too aggressive conditions, concentration range I [Figs. 3(e), 3(h) and 3(k)] is significantly lower than in the most concentrated solutions, range II [Figs. 3(b), 3(c), 3(f), 3(i) and 3(l)]. A mixture of the two extreme morphologies can be recognized in Fig. 3(a), denoted as range III. Detailed analysis of pit morphologies

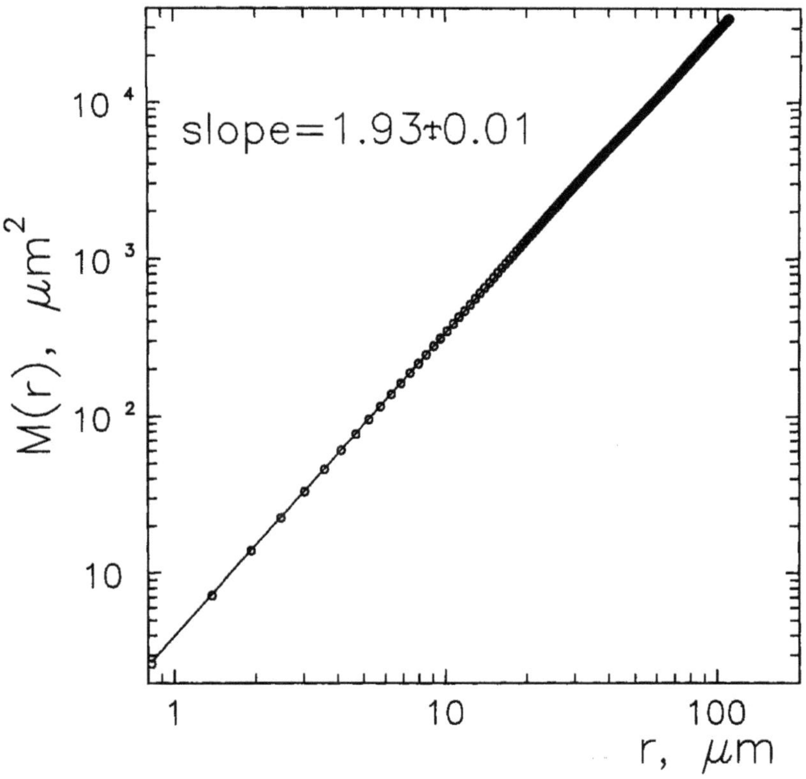

Fig. 4 Variation of the area covered by black pixels in Fig. 3(h) as a function of size.

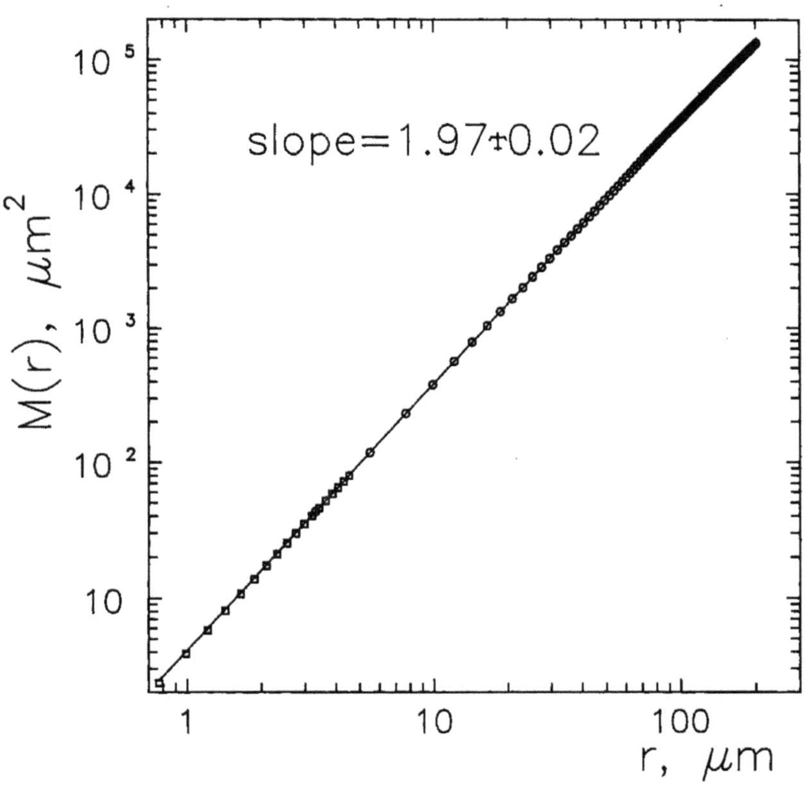

Fig. 5 Variation of the area covered by black pixels in Fig. 3(c) as a function of size.

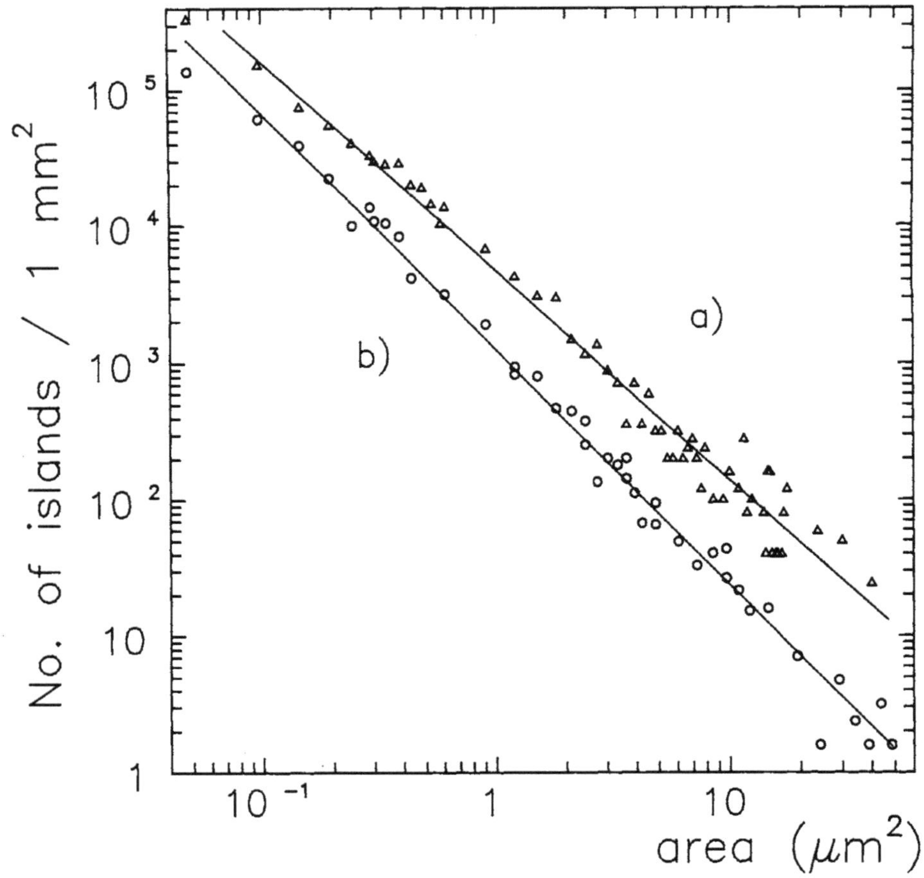

Fig. 6 Area distribution of undissolved Al islands for patterns shown in (a) Fig. 3(h), and (b) Fig. 3(c). Slope of the straight lines fitted to the data, (a) -1.5 ± 0.2, and (b) -1.7 ± 0.3.

indicates that all the patterns except 3(a) exhibit uniform fractal properties. As representative examples, mass distribution function for patterns 3(h) and 3(c) are plotted in Figs. 4 and 5, respectively. Area distributions of undissolved Al islands are compared in Fig. 6. Results of morphological analysis for all of the patterns shown in Fig. 3 are summarized in Table 1.

As is shown in Fig. 4, $M(r)$ in range I can well be approximated by a power function over a length scale of two orders of magnitude, from the minimal size determined by optical resolution up to the size of the cluster. The exponent determined by means of least squares parameter fitting was found to be 1.93 ± 0.01. The estimated error represents the error of curve fitting only.

For the patterns in range II, the scaling interval is shifted to somewhat larger distances and the mass exponent is closer to but still significantly less than 2, as is shown for pattern 3(c) in Fig. 5. The mass exponent alone does not reflect the obvious difference between patterns in concentration ranges I and II. Figure 6 reveals the striking difference between the two distinct morphologies. The area distribution of the undissolved islands can be approximated by a power function in both cases, but in the case of patterns in range I, the exponent is smaller and the number of islands significantly larger than in patterns grown in the most aggressive solutions (range II).

Table 1 Estimated characteristics of corrosion patterns shown in Fig. 3. Arrangement of the data is the same as for the images in Fig. 3. Numbers in a box (from top to bottom) stand for mass dimension, box-dimension, fraction of black pixels in the cluster, and the estimated exponent of $N(A)$, respectively

	0.01 M Cl^-	0.1 M Cl^-	1 M Cl^-
1 M H^+	1.95 ± 0.01 1.88 ± 0.02 0.78 ± 0.03 -1.7 ± 0.2	1.97 ± 0.03 1.95 ± 0.02 0.95 ± 0.02 -1.6 ± 0.3	1.97 ± 0.02 1.95 ± 0.02 0.95 ± 0.04 -1.7 ± 0.3
0.1 M H^+		1.93 ± 0.01 1.88 ± 0.01 0.78 ± 0.02 -1.5 ± 0.2	1.98 ± 0.01 1.95 ± 0.01 0.93 ± 0.05 -1.4 ± 0.3
0.01 M H^+		1.93 ± 0.01 1.87 ± 0.01 0.75 ± 0.06 -1.5 ± 0.2	1.97 ± 0.01 1.96 ± 0.01 0.94 ± 0.05 -1.6 ± 0.2
0.002 M H^+		1.90 ± 0.02 1.88 ± 0.01 0.80 ± 0.03 -1.5 ± 0.3	1.96 ± 0.02 1.94 ± 0.01 0.93 ± 0.3 -1.5 ± 0.3

4. CONCLUSIONS

Pit growth in aluminium thin layers exhibits several remarkable properties:

(i) There is a threshold in composition below which no pitting occurs;
(ii) Above this threshold, a contiguous system of channels is etched which, under extremely aggressive conditions, goes over to a round hole; and
(iii) Aluminium islands left behind the growth front are safe from subsequent corrosion.

These features cause the process to resemble spreading percolation phenomena [5,6], e.g. propagation of forest fire, epidemics, etc.

In certain concentration ranges, pits with uniform morphology are formed in the sense that $M(r)$ and $N(A)$ functions are statistically the same all over the pattern. This is an intriguing result because the parameters which influence the growth process are left uncontrolled after dissolution has started. This indicates the effective role of some self-regulating mechanism.

ACKNOWLEDGMENTS

The authors are grateful to Dr. T. Pajkossy and Prof. T. Vicsek for stimulating and encouraging discussions. This work was partly supported by the Hungarian Research Foundation OTKA under Contract No. 2981. L. B. is indebted to the MHB-Hungarian Scientific Foundation for financial support.

REFERENCES

1. D. Avnir, ed., *The Fractal Approach to Heterogeneous Chemistry* (Wiley, Chichester, 1989).
2. F. Mansfeld, ed., *Corrosion Mechanisms* (Marcel Dekker, New York, 1987).
3. G. S. Frankel, *Corros. Sci.* **30**, 1203 (1990).
4. G. S. Frankel, J. O. Dukovic, V. Brusic, B. M. Rush and C. V. Jahnes, *J. Electroanal. Chem.* **139**, 2196 (1992).
5. T. Vicsek, *Fractal Growth Phenomena* (World Scientific, Singapore, 1989).
6. A. Bunde and S. Havlin, eds., *Fractals and Disordered Systems* (Springer-Verlag, Berlin, 1991).

Fractals, Vol. 1, No. 3 (1993) 439–450

FRACTAL ELECTRODEPOSITS: MORPHOLOGY, GROWTH DYNAMICS AND DIFFUSION-LIMITED RESPONSE FUNCTIONS

PEDRO PABLO TRIGUEROS, JORDI MACH, JOSEP CLARET, FRANCESC MAS and FRANCESC SAGUÉS[1]

Departament de Química Física, Facultat de Química, Universitat de Barcelona (UB), C/Martí i Franquès, 1.E-08028 BARCELONA (Catalonia, Spain)

Abstract

Fractal characterization of zinc electrodeposits obtained in a quasi two-dimensional film cell under suitable applied potential and zinc sulphate concentrations is reported. Our results show that these deposits are formed according to a Laplacian growth mode, and moreover, their self-similar fractal nature allows the generalization of the Cottrell law for diffusion-limited electrochemical processes occurring on these surfaces.

1. INTRODUCTION

Quasi-two dimensional electrochemical deposition (ECD) is a typical example of pattern formation under non-equilibrium conditions.[1] Typically in these experiments, a metallic deposit is formed either in a thin layer of an aqueous electrolyte solution containing the metal cation (film cell),[2] or in the interface between this solution and an organic solvent (interface cell).[3] It has been clearly shown that different electrodeposit morphologies

PAC's numbers: 68.70 + w; 61.50 Cz; 82.45 + z

[1]To whom correspondence should be addressed.
E-mail: UBAQFT09@EBCESCA1; sagues@daphne.qf.ub.es

are obtained when the control parameters, including applied potential or current, cation concentration, cell dimensions, etc.,[2,4–6] are adequately modified. The influence of other parameters, mainly external convection[7,8] and electrolyte solution composition, i.e., the effect of an inert electrolyte,[9] are presently studied in our group.

Among the different electrodeposit morphologies, open fractal structures are formed in zinc electrodeposition experiments using a film cell at small applied potential values and high zinc sulphate concentration. In this case, fractal concepts are necessary, not only to morphologically characterize these aggregates, but most interestingly to capture the subtle details of their dynamical evolution and general behavior.

With this perspective, this paper aims at presenting recent work of our group on fractal zinc electrodeposits addressing specifically three different questions: (1) what are the appropriate exponents for their morphological characterization? (2) what is the intrinsic scaling of their locally distributed growth probability measure? and (3) how do they behave against simple heterogeneous processes simulating standard diffusion-controlled electrochemical reactions? These subjects are discussed in Sec. 4. In Sec. 2, the experimental conditions to obtain open fractal zinc electrodeposits are briefly reported. Section 3 is devoted to a short introduction of the basic concepts of fractal geometry and multifractal characterization. Finally, Sec. 5 summarizes some conclusions.

2. EXPERIMENTAL SECTION

Open fractal zinc electrodeposits are grown in a parallel film cell, where a thin layer of an aqueous zinc sulphate solution is sandwiched between two glass plates, under a constant potential difference applied between the electrodes. The cathode is a copper (70 μm) or platinum (25 μm) wire, and the anode is a zinc bar placed external to the thin layer of electrolyte solution and parallel to the cathode.[6] Experimental conditions, i.e., zinc sulphate concentration and applied potential values have been selected from the corresponding region of the diagram of morphologies for zinc electrodeposition, extensively studied in our previous work.[2,6]

3. FRACTAL AND MULTIFRACTAL ANALYSIS OF A DISTRIBUTION MEASURE

An extended analysis of the multifractal characteristics of an aggregation pattern is formulated in terms of the generalized fractal dimensions D_q.[10] Several numerical routines have been developed to compute the spectrum of dimensions D_q. The one that will be used below, previously used for the mass distribution measure[2,11,12] of electrodeposition aggregates, is based on a box-counting algorithm especially appropriate to the evaluation of D_q with $q \geq 0$. One then introduces the hierarchy of exponents τ_q according to:

$$Z_q(\varepsilon) \equiv \sum_{i=1}^{N(\varepsilon)} p_i^q \underset{\varepsilon\to 0}{\propto} \varepsilon^{\tau_q} \tag{1}$$

The set of generalized fractal dimensions D_q are related to τ_q through:

$$D_q \equiv \frac{\tau_q}{(q-1)} \tag{2}$$

where p_i is the relative measure of the cluster contained in the ith cell

$$\sum_{i=1}^{N(\varepsilon)} p_i = 1 \tag{3}$$

and $N(\varepsilon)$ is the total number of boxes necessary to cover the perimeter of the aggregate.

The generalized dimensions are then obtained from the relation:

$$D_q \equiv \lim_{\varepsilon \to 0} (q-1)^{-1} \frac{\ln \left\{ \sum_{i=1}^{N(\varepsilon)} [p_i(\varepsilon)]^q \right\}}{\ln(\varepsilon)} \tag{4}$$

For the particular value $q = 1$, the previous definition must be rewritten as:

$$D_1 \equiv \lim_{\varepsilon \to 0} \frac{\left(\sum_{i=1}^{N(\varepsilon)} p_i(\varepsilon) \ln p_i(\varepsilon) \right)}{\ln(\varepsilon)} \tag{5}$$

When all $D_q = D$, independently of the q-value, the distribution measure is a self-similar measure.

The scaling properties of any distribution measure are analyzed in relation to its moments Z_q, defined in Eq. (1). The next step in this standard strategy consists in resolving the exponents τ_q into a density of singularities f_q with singularity strengths α_q. In doing so, we avoid the common use of a Legendre transformation, and its numerical complications, to resort to a more direct procedure[13] based on the following equations:

$$\alpha_q \equiv \frac{d}{dq}\tau_q = \lim_{\varepsilon \to 0} \left(\frac{\sum_{i=1}^{N(\varepsilon)} \hat{p}_i(q, \varepsilon) \ln p_i(\varepsilon)}{\ln \varepsilon} \right) \tag{6}$$

$$f_q \equiv q\alpha_q - \tau_q = \lim_{\varepsilon \to 0} \left(\frac{\sum_{i=1}^{N(\varepsilon)} \hat{p}_i(q, \varepsilon) \ln \hat{p}_i(q, \varepsilon)}{\ln \varepsilon} \right) \tag{7}$$

which follows from Eqs. (1)–(3) and the definition: $\hat{p}_i(q, \varepsilon) \equiv \frac{p_i^q(\varepsilon)}{\sum_{j=1}^{N(\varepsilon)} p_j^q(\varepsilon)}$

4. RESULTS AND DISCUSSION

4.1 Self-Similar Characterization of Fractal-Like Zinc Electrodeposits

In this section, a multifractal characterization of the mass distribution measure of open fractal zinc aggregates is reported in terms of the generalized fractal dimensions, D_q ($q =$ 0, 1, 2, 6). Our previous results[2] show that in spite of small variations of the averaged D_q values under different experimental conditions, the aggregates corresponding to the open-fractal region of the diagram of morphologies of zinc electrodeposition display a self-similar behavior, within 40 and 600 μm, with an averaged final value for $\langle D_q \rangle = 1.61 \pm 0.02^2$. This is in good agreement with previous results[11,12] and DLA simulations.[11]

However, our aim is now to extend this conclusion to a wider range of length scales of the deposit. For this purpose, this study has been repeated at several magnifications of the zinc deposit obtained at 0.4 M zinc sulphate solution and 5 V, and the cell dimension being 4 cm, 3 cm and 25 μm for electrode length, electrode separation and cell thickness, respectively

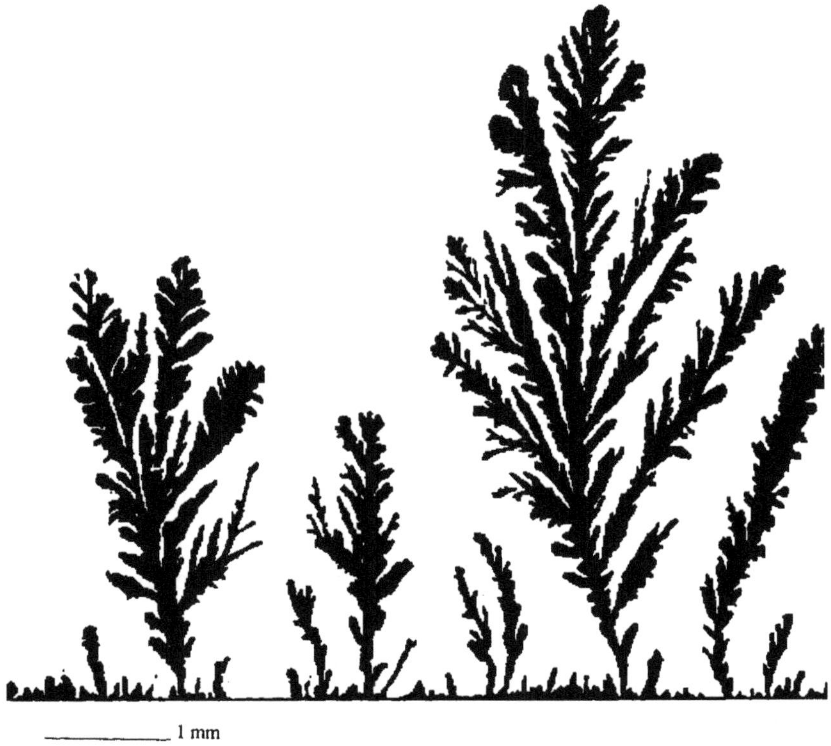

Fig. 1 Quasi-twodimensionally open-fractal electrodeposit in a parallel cell (25 μm thickness), digitized at 15 × (512 pixels ≡ 4.76 mm), and grown at $c \equiv [ZnSO_4(aq.)] = 0.4$ M and $\Delta V = 5$ V.

(Fig. 1). For the same magnification (15×) used in Ref. 2 (Fig. 1), a mean value for the generalized fractal dimension $\langle D_q \rangle = 1.646 \pm 0.003$ is obtained; for a half magnification (7.5×), $\langle D_q \rangle = 1.638 \pm 0.001$; and for twice the magnification (30×), $\langle D_q \rangle = 1.622 \pm 0.004$. These values show that the open fractal zinc electrodeposits behave as a true self-similar fractal structure with respect to its mass distribution measure, with an average value for the fractal dimension of $D_f = \langle D_q \rangle = 1.64 \pm 0.02$[14] which extends the self-similar behavior to length scales from 25 μm until 12 mm (2–3 orders of magnitude).

4.2 Non-Uniform Scaling of the Growth Probability Distribution

In order to gain a better understanding of the growth dynamics governing fractal electrodeposition, we compare the multifractal scaling of the experimentally computed growth probability distribution (GPD), with that corresponding to a Laplacian growth mode. In this case, open-fractal electrodeposits are grown at 6 V from a 0.4 M zinc sulphate solution in a cell whose dimensions are 9 cm, 3 cm and 70 μm for electrode length, electrode separation and cell thickness, respectively [(Fig. 2(a)].

The analysis here presented is original within the context of ECD, the closest similarities being those reported earlier on the dynamics of viscous fingering[15] and crystal growth.[16] Here the totally experimental procedure to compute the GPD is performed as follows. Successive images of the growing pattern are taken at intervals Δt [(Fig. 2(b)]. In this picture, some details of the interface profile are lost due to a smoothing procedure necessary to connect the irregular boundary of the electrodeposit in order to calculate the GPD as

(a)

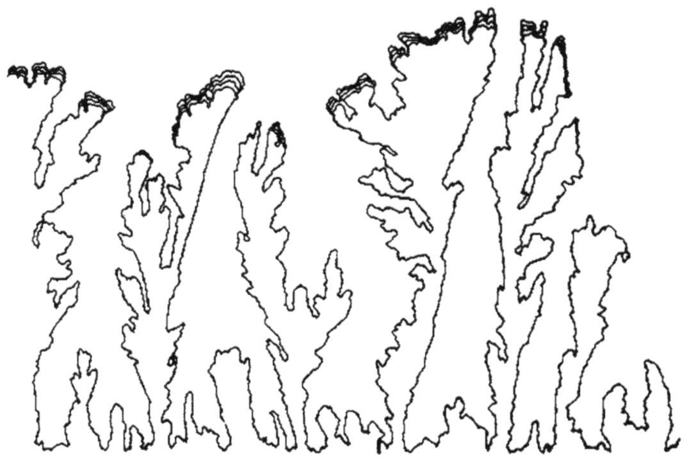

(b)

Fig. 2 (a) Quasi-twodimensionally open-fractal electrodeposit in a parallel cell (70 μm thickness) after 2,7 min growth, digitized at 15 × (512 pixels ≡ 4.76 mm), and grown at $c \equiv [ZnSO_4(aq.)] = 0.4$ M and $\Delta V = 6$ V. (b) Contour regularization of several patterns grown at different times $\Delta t \approx 5 \sim 6$ s.

commented below. The length of a segment drawn between two successive perimeters and taken on a line perpendicular to the surface contour at each point of the metallic boundary $\mathbf{r}_s$, gives us a measure of the local growth velocity $\mathbf{v}_n(\mathbf{r}_s)$, readily normalizable to a growth probability distribution $P_g(\mathbf{r}_s)$ (for more technical details, see Ref. 17). For the sake of comparison, the same experimental pattern is also tested with respect to the harmonic measure, $P_g(\mathbf{r}_s) \propto |\nabla_n \Phi(\mathbf{r}_s)|$, here computed by iteratively solving the discretized Laplace equation, following the same procedure as in Ref. 18.

Plots of τ_q and D_q are shown in Figs. 3(a) and 3(b), respectively, both for the experimentally computed GPD and the Laplacian growth measure. The different growth properties corresponding to the more visited ($q \to \infty$) and more ramified ($q \to -\infty$) regions of the electrodeposit are clearly exhibited. The expected non-trivial scaling of the GPD, as evidenced in previous related studies,[19] is here apparent in the departure from the linear dependence

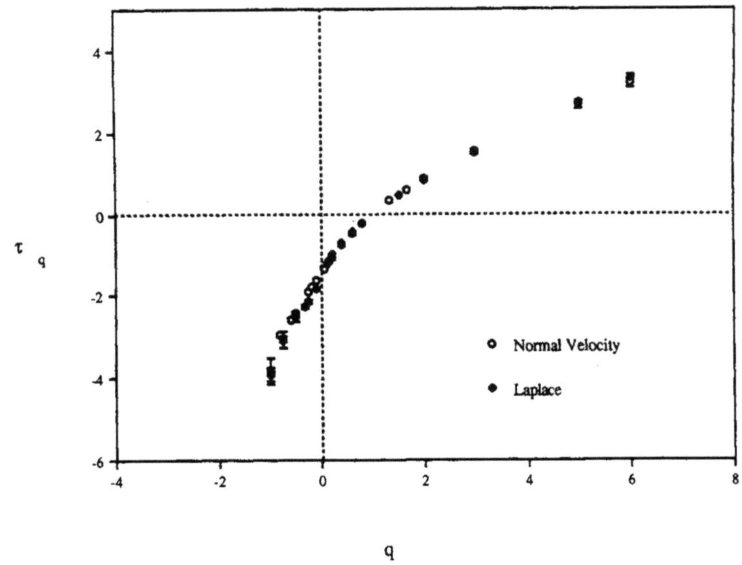

(a)

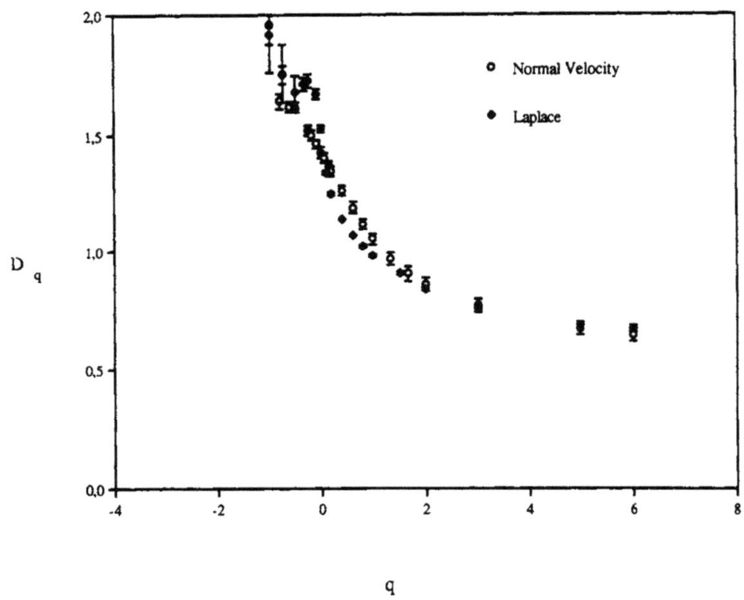

(b)

Fig. 3 **Multifractal scaling analysis of the GPD for the open-fractal morphology: (a) $\tau(q)$, and (b) D_q representations. Normal velocities distribution (o) and harmonic (Laplacian) measure (•) (60 μm $\leq l \leq$ 1100 μm).**

of τ_q on q when crossing around $q = 0$ from the region of positive q. In Figs. 4(a) and 4(b), both the $\alpha(q)$, $f(\alpha(q))$ and the common $f(\alpha)$ curve are shown for the experimentally computed GPD and the Laplacian growth measure. Deliberately we have not attempted to extend this analogy to negative q-values. The present unavoidable limitations in the acquisition and digitization techniques employed to image the growth of the electrodeposits

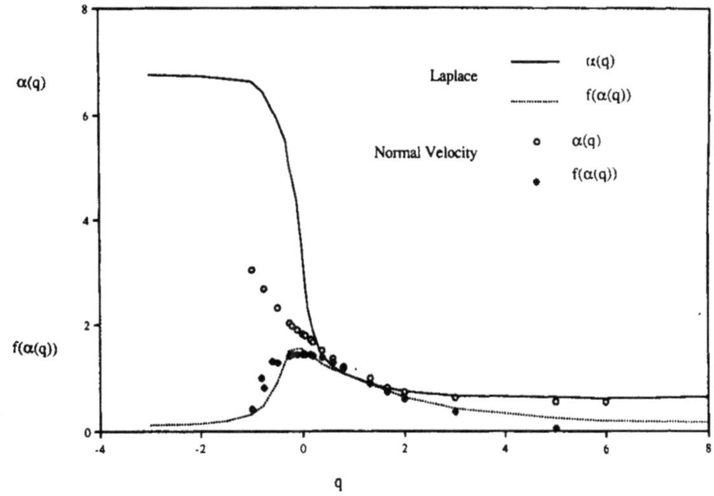

(a)

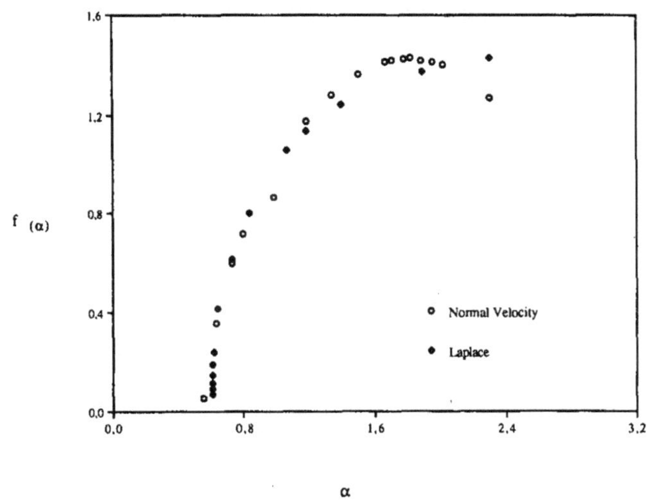

(b)

Fig. 4 **Multifractal scaling analysis of the GPD for the open-fractal morphology: (a) $\alpha(q)$ and $f(\alpha(q))$ representations (60 μm $\leq l \leq$ 1100 μm). Harmonic (Laplacian) measure are drawn in continuous line for $\alpha(q)$ curve and dotted line for $f(\alpha(q))$ curve and normal velocities distribution are drawn in markets: (○) for $\alpha(q)$ curve and (•) for $f(\alpha(q))$ curve. (b) $f(\alpha)$ curve. Normal velocities distribution (○) and harmonic (Laplacian) measure (•).**

do not allow us to capture a large enough subset of active points with minimum growth probabilities to ensure the statistical confidence needed in our multifractal analysis.

The most representative values of the above multifractal analysis applied to the fractal electrodeposition growth mode, together with related ones existing in the literature are summarized in Table 1. The computed D_0 value for the harmonic measure $D_0 = 1.53 \pm 0.01$ falls, for example, somewhat lower than expected ($D_0 = 1.64 \pm 0.01$ for this particular growth conditions was computed in Ref. 2). This is not surprising in view of the limitations imposed by the contour regularization[17,18] needed to assure the convergence of the Laplacian

Table 1 Comparison of Scaling Indices for Experimental and Simulated Fractal Growth Probability Distributions

q	Singularity Exponents	ECD[a] Normal Velocity	ECD[a] Laplace	CG[b] Normal Velocity	CG[b] Laplace	VF[c] Laplace	DLA[d] Green	DLA[e] Laplace	DLA[f] Laplace
$-\infty \equiv \inf$	$q_{\inf}$	-1	-3	—	—	-2	-5	-5	-1
	$D_{\inf}$	1.96 ± 0.08	1.9 ± 0.2	—	—	~ 3.5	~ 7.0	~ 7.0	6.8 ± 0.5
	$D_{-\infty} = \alpha_{-\infty}$	3.02 ± 0.07	6.8 ± 0.2	—	9.4 ± 0.2	~ 5.3	~ 9.0	~ 9.0	13.9 ± 1.2
	$f(\alpha - \infty)$	0.40 ± 0.07	0.01 ± 0.01	—	~ -0.1	~ 0.0	~ -2.0	~ 0.0	-0.1 ± 0.2
0	$f(\alpha_0) = D_0$	1.43 ± 0.02[g]	1.53 ± 0.01	—	1.63 ± 0.01	~ 1.1	~ 1.5	1.64 ± 0.01	1.75 ± 0.04
	α_0	1.82 ± 0.02	3.41 ± 0.09	—	2.97 ± 0.04	~ 1.6	~ 2.4	~ 4.0	4.49 ± 0.10
1	$f(\alpha_1) = \alpha_1 = D_1 = 1$[h]								
	D_1	1.06 ± 0.02	0.98 ± 0.00	1.02 ± 0.02	1.13 ± 0.02	—	—	1.04 ± 0.01	0.97 ± 0.04
3	$\tau_3 = 2D_3$	1.53 ± 0.06	1.51 ± 0.02	—	1.63 ± 0.02	—	—	—	1.54 ± 0.08
$\infty \equiv \sup$	$q_{\sup}$	15	15	8	8	3	7	10	5
	$D_{\sup}$	0.64 ± 0.02	0.63 ± 0.01	~ 0.68	~ 0.72	~ 1.0	~ 0.8	~ 0.65	0.70 ± 0.02
	$D_{\infty} = \alpha_{\infty}$	0.53 ± 0.03	0.61 ± 0.01	0.59 ± 0.03	0.60 ± 0.04	~ 1.0	~ 0.7	$0.64 - 0.70$	0.60 ± 0.04
	$f(\alpha\infty)$	0.03 ± 0.01	0.07 ± 0.01	~ 0.0	~ 0.0	~ 0.8	~ 0.2	~ 0.0	0.22 ± 0.06

[a] Presents results for Electrochemical Deposition.
[b] Crystal growth of NH_4Cl (Ohta and Honjo in Ref. 16)
[c] Viscous fingering (Nittmann, Stanley, Touboul and Daccord in Ref. 15)
[d] DLA obtained using Green's functions (Amitrano, Coniglio and di Liberto in Ref. 20)
[e] DLA with numerical solution of Laplace equation and Legendre transformation (Hayakawa, Sato and Matsushita in Ref. 21)
[f] DLA with numerical solution of Laplace equation and without numerical Legendre transformation (Ball and Spivack in Ref. 22)
[g] Fractal dimension of the experimental growing sites. For all perimeter sites $D_0 = 1.53 \pm 0.01$.
[h] Theoretical result (Makarov in Ref. 23)

solution, and the use itself of a fixed-size box-counting algorithm, whose suitability for $q \leq 0$ is somewhat dubious. In any case, and restricting to $q \geq 0$, particularly well-known identities established for relevant exponent of the harmonic measure, are also fulfilled with a maximum few percent inaccuracy when applied to the corresponding experimental values: $D_1 = 1.06 \pm 0.02$ ($D_1 = 1$, Ref. 23); $\tau_3 = 1.53 \pm 0.06 \approx 1 + \alpha_{sup} = 1.53 \pm 0.03$ ($\tau_3 = 1 + \alpha_{sup}$, Refs. 24 and 25). Both the specific tests just quoted and the general behavior shown in Figs. 3 and 4 are, to our understanding striking and robust evidences that support the Laplacian nature of the growth dynamics for open-fractal ECD, at least taking the GPD measure as indicator and given the spatial resolution, and corresponding length-scale range, here analyzed.

4.3 Diffusion-Limited Response Functions

Heterogeneous reactions taking place at solid-gas or solid-liquid interfaces are influenced by the surface structure of the catalyst. This has been clearly demonstrated for surfaces with different structures at the atomic level, i.e., different crystallographic orientations, and for surfaces with different structures at larger length scales, i.e., porous and rough surfaces. In this case, fractal terms are useful for a geometrical characterization of rough, porous or partially active surfaces,[26] which exhibit a non-classical length scale.

In this context, our aim is now to study the behavior of these surfaces against a simple diffusion-controlled electrode process $R + ne^- \Leftrightarrow O$, and in particular, the effect of the fractal dimension of the surface on the law governing the current, i.e., the well-known Cottrell law.[27] This can be extended to fractal electrodes, taking into account that it holds locally for every box of length ε, as if it were a flat microelectrode, and the total current is obtained in an associative way from the local current of each box of length ε. Then, the current will depend on the length scale, ε,

$$I(\varepsilon;\, d) = nFA(\varepsilon;\, d)c_R^*\sqrt{\frac{D_R}{\pi t}} \tag{8}$$

where $A(\varepsilon; d)$ is the generalized area of the electrode in a d-dimensional space where the diffusion process is embedded and c_R^* and D_R are respectively the bulk concentration and diffusion coefficient of electroactive especies R.

For diffusion-controlled processes, it is a common procedure to choose as a length scale, ε, the diffusion layer length.[26,28,29] However, there are two temporal cutoffs to this power law.[26] One at the microscopic level, when the electrode profile shows all its closest details and its area adopting its maximum value, and the other cutoff is found at the macroscopic level, for which the electrode area has its minimum value, and it becomes planar and perpendicular to the relevant spacial coordinate.

Thus, Eq. (8) can be written in the usual way,[14,26] as a power-law, between the two temporal cutoffs mentioned above:

$$I(t) \propto t^{-\alpha}, \quad \alpha \equiv \frac{D_f - (d-2)}{2} \tag{9}$$

where $\alpha = D_f/2$ for $d = 2$.

Monte Carlo (MC) simulation methods (among which we chose MDLA (Multiparticle Diffusion-Limited Aggregation)[30,31]) have been applied to electrode processes in order to analyze the diffusion towards a rough surface.[14,32] The default simulation conditions are as follows:

(i) The square 2d-lattice ($L \times Y$), with periodic lateral boundaries, has been divided into two regions: the electrode and the solution. The size of the solution particle could be equal than or lower than that of the electrode. In this case, since the minimum size of one electrode particle has necessarily to be one pixel (with dimensions that come from the experimental conditions), the linear size of one solution particle has been taken equal to 1, $\frac{1}{2}$, $\frac{1}{3}$ or $\frac{1}{4}$ of one pixel (see below). The concentration of solution particles (ρ^*) does not significantly influence the exponent of the scaling law [Eq. (9)].[14]

(ii) Each particle is moved to any one of its four neighbours sites, if this is unoccupied, and the unit time corresponds to a complete movement of each particle of the lattice.

(iii) Solution particles, that in each iteration come into contact with the electrode surface, are removed from the lattice. Therefore, current values at time t will be given by this number of solution particles.

(iv) Each simulation has been carried out several times in order to obtain the average of the results and an estimation of the statistical errors.

The values of the exponent α, which generalize the classical Cottrell law, are summarized in Table 2(a). We also show the experimental fractal dimension D_{exp}, obtained from the exponent α, according to Eq. (9) with $d = 2$. Its value is compared with the fractal dimension computed from the mass distribution measure $\langle D_q \rangle$. The values of the exponent α

Table 2

(a) Values of Exponents of Scaling Law (Eq. 9) for Open-fractal Pattern

Magnification	α	D_{exp}	$D_f = \langle D_q \rangle$
7.5×	0.62 ± 0.02	1.24 ± 0.04	1.638 ± 0.001
15×	0.63 ± 0.02	1.26 ± 0.04	1.646 ± 0.003
30×	0.69 ± 0.03	1.38 ± 0.06	1.622 ± 0.004

$L = 512$, $Y = 768$, $L/l = 512$ and $\rho^* = 0.2$

(b) Values of Exponents of Scaling Law (Eq. 9) for Open-fractal Pattern at Different Ratios of L/l

Magnification	L/l	α	D_{exp}	$D_f = \langle D_q \rangle$
7.5×	512	0.62 ± 0.02		
	1024	0.69 ± 0.01		
	1536	0.69 ± 0.01		
	2048	0.69 ± 0.03	1.38 ± 0.06	1.41 ± 0.03
15×	512	0.63 ± 0.02		
	1024	0.71 ± 0.01		
30×	512	0.69 ± 0.03		
	1024	0.73 ± 0.01		
	2048	0.71 ± 0.03	1.42 ± 0.06	1.42 ± 0.03

$Y = 768$ and $\rho^* = 0.2$

shown in Table 2(a) support the self-similar nature of the open-fractal morphology although the α and D_{exp} values are slightly different, and always smaller, than the fractal dimension of the mass distribution measure $\langle D_q \rangle$. These low D_{exp} values could be explained, considering that this simulation has been made with the same size as the diffusing and the aggregate particles belonging to the rough electrode (the digitized pixel). This assumption implies that many parts of the electrode surface are not accessible to the diffusing particles. This yields an error of finite length scale, because the effective electrode surface decreases and leads to a lower value of the fractal dimension, D_{exp}.[33]

These limitations of our simulation procedure can be overcome by considering two kinds of particles of different size: the diffusing and the aggregate particles. Their size relationship can be expressed by means of the ratio, L/l, between the total electrode length L and the lattice length for the diffusion particle l. We can expect that by taking this L/l ratio as high as possible, i.e., considering the diffusing particle lower than the aggregate particle, the screening effects will be minimized. In order to increase this ratio L/l in the simulations, we expand the digitized pattern by a length factor of $n = 2$, 3 and 4. This means that, as commented above, we are performing simulations with a length size of the diffusing particles of $\frac{1}{2}$, $\frac{1}{3}$ and $\frac{1}{4}$ of the length size of the aggregate particle, respectively. We have to remark that this expansion does not follow the fractal law of the mass distribution measure of the rough electrode, but it is made by a dilatation of the aggregate by the length factor n mentioned above.

Table 2(b) summarizes the results obtained at several magnifications and ratios L/l. When the size of the diffusing particles with respect to the electrode particles is decreased, the electrode surface accessible to solution particles increases and thus, higher α and D_{exp} values are reached. The values of α and D_{exp} obtained with this procedure tend to a limit, which is very similar at each magnification so that, an averaged limit of $\alpha = 0.70 \pm 0.04$ ($D_{exp} = 1.40 \pm 0.08$) can be calculated. This is indeed a consequence of the self-similar behavior of the open morphology.

However, this average limit for D_{exp} continues being lower than the fractal dimension of the mass distribution measure $D_f = \langle D_q \rangle = 1.64 \pm 0.02$. This result could be interpreted taking into account the procedure used to increase the ratio L/l. We have to think that the digitized experimental electrodeposit is simply expanded by a fixed length factor and therefore, the fractal dimension of its mass distribution measure is modified. In this way, the number of planar regions of the electrode surface is increased (decreasing its roughness), and its fractal dimension $D_f = \langle D_q \rangle$ must decrease. We have computed the generalized fractal dimensions D_q for two digitized experimental open patterns, expanded by a factor $n = 4$ at two different magnifications (7.5× and 30×). The results also shown in Table 2(b) agree with our assumptions, and an average value $D_f = \langle D_q \rangle = 1.41 \pm 0.05$ for both magnifications is obtained, which coincide with the limit value for $D_{exp} = 1.40 \pm 0.08$ found from the α exponent of the generalized Cottrell Eq. (9) computed by MC simulation.

5. CONCLUDING REMARKS

A multiple investigation of the fractal characteristics of the Zn electrodeposition morphology at low applied potential and intermediate concentrations is reported.

First, all the mass distribution measure of this morphology displays a self-similar fractal nature within a range of length scales which extends from tens of microns to a few millimeters.

Secondly, the scaling structure of the growth probability distribution for this morphology has been investigated. Results for the more active growth regimes show a robust multifractal behavior which, within the length scale range examined, is compatible with a Laplacian growth mode.

Finally, the generalization to fractal electrodes of the classical Cottrell law, valid for a diffusion-limited electrode process, has been tested and verified by means of MC simulation, taking as the electrode surface a quasi two-dimensional open-fractal zinc electrodeposit.

ACKNOWLEDGMENTS

Two of us P. P. Trigueros and J. Mach benefitted from a F.P.I. grant from Ministerio de Educación y Ciencia (Spain). Financial support from Dirección General de Investigación Científica y Tecnológica (DGICYT, Spain) under project PB90-0455 is acknowledged.

We are grateful to the "Centre de Supercomputació de Catalunya" (CESCA, Catalonia, Spain) for providing us with computation facilities.

REFERENCES

1. L. M. Sander, in *Physics of Structure Formation*, eds. W. Guttinger and G. Dangelmayr (Springer Verlag, Berlín, 1987), p. 257.
2. P. P. Trigueros, J. Claret, F. Mas and F. Sagués, *J. Electroanal. Chem.* **312**, 219 (1991) and references quoted therein.
3. R. Tamamushi and H. Kaneko, *Electrochim. Acta* **25**, 391 (1980); M. Matsushita, M. Sano, Y. Hayakawa, H. Honjo and Y. Sawada, *Phys. Rev. Lett.* **53**, 286 (1984).
4. Y. Sawada, A. Dougherty and J. P. Gollub, *Phys. Rev. Lett.* **56**, 1260 (1986).
5. D. G. Grier, E. Ben-Jacob, R. Clarke and L. M. Sander, *Phys. Rev. Lett.* **56**, 1264 (1986).
6. P. P. Trigueros, J. Claret, F. Mas and F. Sagués, *J. Electroanal. Chem.* **328**, 165 (1992).
7. L. López-Tomàs, J. Claret, F. Mas and F. Sagués, *Phys. Rev.* **B46**, 11495 (1992).
8. L. López-Tomàs, J. Claret and F. Sagués, submitted.
9. P. P. Trigueros, F. Sagués and J. Claret, submitted.
10. P. Grassberger, *Phys. Lett.* **A97**, 227 (1983); H. G. E. Hentschel and I. Procaccia, *Physica* **D8**, 435 (1983); P. Grassberger and I. Procaccia, *Physica* **D13**, 34 (1984).
11. F. Argoul, A. Arneodo, G. Grasseau and H. L. Swinney, *Phys. Rev. Lett.* **61**, 2558 (1988).
12. F. Sagués, F. Mas, M. Vilarrasa and J. M. Costa, *J. Electroanal. Chem.* **278**, 351 (1990).
13. A. Chhabra and R. V. Jensen, *Phys. Rev. Lett.* **62**, 1327 (1989).
14. P. P. Trigueros, F. Mas, J. Claret, F. Sagués, J. Galceran and J. Puy, *J. Electroanal. Chem.* **348**, 221 (1993).
15. J. Nittmann, H. E. Stanley, E. Touboul and G. Daccord, *Phys. Rev. Lett.* **58**, 619 (1987).
16. S. Ohta and H. Honjo, *Phys. Rev. Lett.* **60**, 611 (1988).
17. J. Mach, F. Mas and F. Sagués, submitted.
18. F. Mas and F. Sagués, *Europhys. Lett.* **17**, 541 (1992).
19. H. E. Stanley and P. Meakin, *Nature* **335**, 405 (1988).
20. C. Amitrano, A. Coniglio and F. di Liberto, *Phys. Rev. Lett.* **57**, 1016 (1986).
21. Y. Hayakawa, S. Sato and M. Matsushita, *Phys. Rev.* **A36**, 1963 (1987).
22. R. C. Ball and D. R. Spivack, *J. Phys.* **A23**, 5295 (1990).
23. N. G. Makarov, *Proc. London Math. Soc.* **51**, 369 (1985).
24. L. A. Turkevich and H. Scher, *Phys. Rev. Lett.* **55**, 1026 (1985).
25. T. C. Halsey, *Phys. Rev. Lett.* **59**, 2067 (1987).
26. T. Pajkossy, *J. Electroanal. Chem.* **300**, 1 (1991).
27. A. J. Bard and L. R. Faulkner, *Electrochemical Fundamentals and Applications* (Wiley, New York, 1980).
28. P. G. de Gennes, *C. R. Acad. Sci.* **295**, 1061 (1982).
29. P. Pfeifer, D. Avnir and D. Farin, *J. Statis. Phys.* **36**, 699 (1984).
30. R. F. Voss and M. Tomkiewicz, *J. Electrochem. Soc.* **132**, 371 (1985).
31. F. Sagués and J. M. Costa, *J. Chem. Educ.* **66**, 502 (1989).
32. S. Zalis, L. Pospisil and N. Fanelli, *J. Electroanal. Chem.* **349**, 443 (1993).
33. S. Havlin, in *The Fractal Approach to Heterogeneous Chemistry*, ed. D. Avnir (J. Wiley, Chichester, 1989), Chap. 4.1.

Fractals, Vol. 1, No. 3 (1993) 451–459

INFLUENCE OF CHEMICAL PERTURBATIONS ON THE SURFACE ROUGHNESS OF THIN LAYER ELECTRODEPOSITS

A. KUHN and F. ARGOUL
Centre de Recherche Paul Pascal, Avenue Schweitzer, F-33600 Pessac

Abstract

We have studied the origin of macroscopic and microscopic changes of the morphology of metal electrodeposits in thin layer cells. We concentrate on zinc electrodeposition to show that a part of the variety of growth patterns reported in the literature (dense radial, stringy, dendritic and DLA-like fractal morphologies) is probably due to pollution of the electrolyte. One common pollution in these experiments is molecular oxygen, which is dissolved in the electrolyte during its preparation and conservation in contact with air. Its cathodic reduction can seriously perturb the metallic host lattice by formation and intercalation of metal oxide. Other perturbations with equal dramatic effects on the morphology and the roughness of the deposit are alkali cations. By well-controlled addition of one of these ions or molecules, we are able to generate a wide variety of morphologies, from highly ordered dendritic growth to completely disordered growth. Other metals such as copper or tin are shown to be also highly sensitive towards impurities. Based on this sensitivity upon chemical perturbation, we finally propose a new explanation for the morphological changes, called Hecker effect, which can occur during the growth process.

1. INTRODUCTION

During the past ten years, nonlinear growth phenomena have been the subject of a great number of theoretical and experimental investigations,[1] because they can generate highly ramified structures (fractals) which occur in many natural systems. One of the most commonly studied experimental systems of fractal pattern formation is the electrochemical deposition of metals in quasi two dimensions.[2,6] By varying the kind and the concentration of the metal salt, the applied current (or voltage),[2,3,6] and the geometry of the cell,[7] one can explore a wide variety of morphologies like dense radial, stringy, dendritic and DLA (Diffusion Limited Aggregation)-like patterns. However, the reproducibility of the experiments seems somewhat disappointing and the so-called phase-diagrams, published by several authors[2,3,6] show inexplicable discrepancies. This probably confirms the well-admitted high sensitivity of electrodeposition processes to impurities, but, to our knowledge, in none of the cited experiments a purification of dissolved gas such as oxygen was undertaken. We report in the following the experimental evidence for the role played by molecular oxygen, dissolved in the electrolyte, on the morphology of deposits obtained in thin layer electrodeposition experiments. Furthermore, the influence of a second kind of impurity, alkali cations, which are also likely to be found in electrolytes, is studied.

1.1 Experimental Set-Up

All the experiments described herein are performed under galvanostatic conditions in a two-electrode configuration. The rectangular cell is obtained by confining two metal wires (Zn 99.999% purchased from Aldrich) with a diameter of 0.25 mm between two glass plates. The electrode distance is 2 cm and the width of the cell 5 cm. The volume is filled with an aqueous solution of zinc sulphate (99% $ZnSO_4 7H_2O$, ACS reagent, Aldrich) which has been treated, prior to use, by bubbling nitrogen or oxygen through it. Well-defined amounts of hydrogen peroxide or alkali metal salts are added when necessary. The pictures of the growing crystals are recorded with a video camera, and an image analyzer is used for digitization on a computer.

2. INFLUENCE OF MOLECULAR OXYGEN

The generic structure of the zinc system, observed in highly purified solutions, is dendritic [(Fig. 1(a)] over a wide range of concentration and current density as we have shown in a very recent publication.[8] However, the aging of the solution when it stays for hours or days in contact with atmosphere, can induce dramatic changes in the morphology as in [Fig. 1(b)]. Other metals like copper and tin show the same behavior [Figs. 1(c)–1(f)]. This means that the absorption of a gas, coming from air must perturb the growth. We have checked that for a clean solution of zinc sulphate saturated with carbon dioxide, the morphology remains dendritic; therefore, it seems that oxygen is responsible for this kind of changes. Starting with a solution that we have saturated by bubbling oxygen through it for one full day, and diluting this solution with different amounts of a oxygen free solution (saturated with nitrogen), we can get solutions with different concentrations of dissolved oxygen. We observe that, when the dissolved proportion of oxygen is increased, the regular dendritic growth transforms continuously towards a very disordered, dense morphology when saturation is reached.

A finer tuning of the oxygen content is possible by replacing molecular oxygen by hydrogen peroxide, which seems to be a reasonable substitute because it can easily disproportionate into O_2 and H_2O and the cathodic reduction of oxygen involves the intermediate production

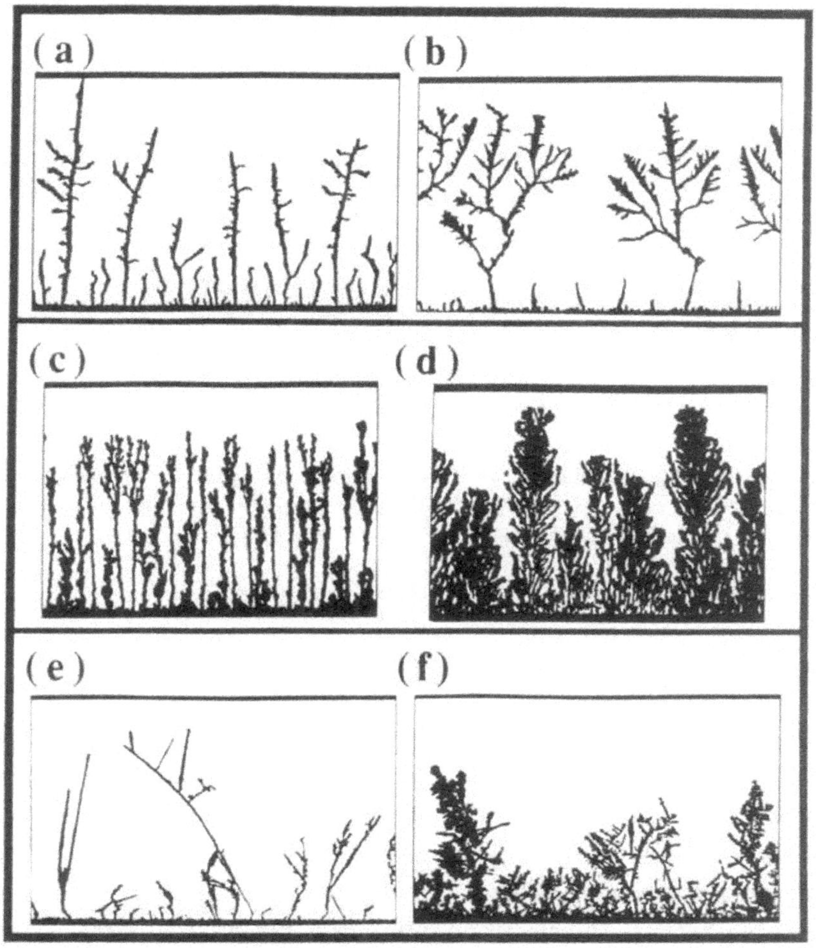

Fig. 1 Influence of aging in contact with atmosphere on different electrolytes: (a) Zinc system, j = 20 mA cm^{-2}, [$ZnSO_4$] = 0.03 mol l^{-1}, nitrogen has been bubbled through the solution for one hour; (b) Same as (a) but the electrolyte was left in contact with air for one day; (c) Copper system, j = 120 mA cm^{-2}, [$CuSO_4$] = 0.1 mol l^{-1}, purified by nitrogen bubbling for one hour; (d) Same as (c) but the electrolyte has been in contact with air for one day; (e) Tin system, j = 30 mA cm^{-1}, [$SnCl_2$] = 0.1 mol l^{-1}, pH has been adjusted to 2 by adding HCl and nitrogen has been bubbled through the solution for one hour; (f) Same as (e) but the electrolyte has been in contact with air for one day.

of hydrogen peroxide. Figure 2 describes the evolution from sharp stalks to highly dense pattern as $r = [H_2O_2]/[Zn^{2+}]$ is increased. One can clearly see the important role of oxygen, because it can inhibit locally the crystalline lattice axis of the metal by its adsorption, reduction and partial transformation into metal oxide. A more quantitative description of these changes can be found elsewhere.[8] When there is as much hydrogen peroxide as zinc ions, no more zinc deposition occurs but an opaque band of amorphous zinc oxide forms at the cathode. This band, represented in black in Fig. 2(h), remains close to the cathode and passivates irreversibly the growth. The ratio of oxygen to metal salt concentration determines the structure of the deposit; the smaller the concentration of the electrolyte, the more sensitive the morphology to its pollution. These observations have important consequences upon morphological diagrams in the parameter space of concentration and current density (or voltage) which have been constructed in the past without taking into account this effect.

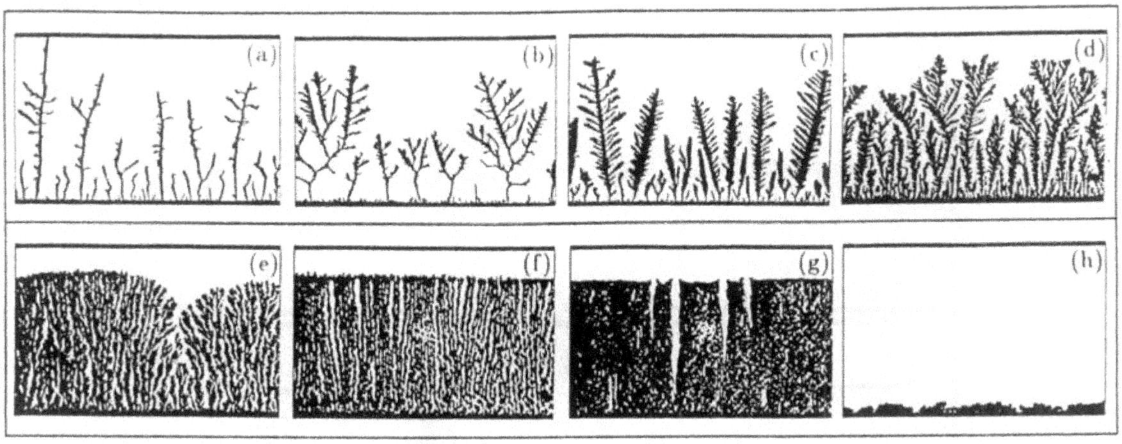

Fig. 2 Zinc electrodeposition morphology vs. the proportion of hydrogen peroxide, $j = 20$ mA cm^{-2}, $[ZnSO_4] = 0.03$ mol l^{-1}, the ratio $r = [H_2O_2]/[ZnSO_4]$ is changed by adjusting $[H_2O_2]$: $a : r = 0$, $b : r = 0.015$, $c : r = 0.021$, $d : r = 0.03$, $e : r = 0.06$, $f : r = 0.12$, $g : r = 0.3$, $h : r = 1$.

3. INFLUENCE OF ALKALI CATIONS

In thin layer electrodeposition, the reduction of the metal ions is not the only interfacial process which can happen at the cathode. Besides the reduction of oxygen, treated above, the parasitic production of hydrogen by the reduction of protons is one of the most important side reactions.[9] In three-dimensional stirred cells, the evolution of hydrogen has been shown to occur in concurrence with electrodeposition of metal.[10] In quasi two-dimensional cells, its role is still enhanced because the electrode interface is depleted by metal ions, which favors the reduction of other chemical species, like protons. This reaction can be catalyzed by adding alkali cations,[11] which increases the occupation of active reduction sites by hydrogen on a growing dendrite. Instead of very few dendrites emerging from the cathode as shown in Fig. 3(a), a continuously increasing number of stalks is observed when the alkali ion content becomes higher. In the limit of too high alkali salt concentrations, the production of hydrogen is favored in such a way that large bubbles appear and metal deposition is quite impossible.

At a microscopic scale, one can also observe an amplification of the ramification rate (finer and denser dendritic side branches). This leads, in a slightly different way than in the case of oxygen pollution, to a drastic increase of the surface roughness. In a preliminary study we have computed the fractal dimension of the external boundary of the deposit, to quantify the complexity of the surface accessible to particles coming from the anode. From the values in Fig. 4, obtained by the box-counting method,[12,13] it is clearly visible that for the sodium ions, as well as for the oxygen pollution, the ramification of the interface goes through a maximum for an intermediate concentration of impurities. Furthermore, the growth morphology is more sensitive to the cationic impurities than to oxygen. This might be due to the positive charge of the alkali ions which increases the efficiency of the perturbation because they are attracted by the growing interface. We plan to apply more sophisticated analytical techniques like the determination of the generalized fractal dimensions and the singularity spectrum $f(\alpha)$,[14–16] in order to understand the relation between the complexity of these structures and their catalytic activity. We conclude that alkali ions perturb strongly the growth even in small concentrations and therefore at least

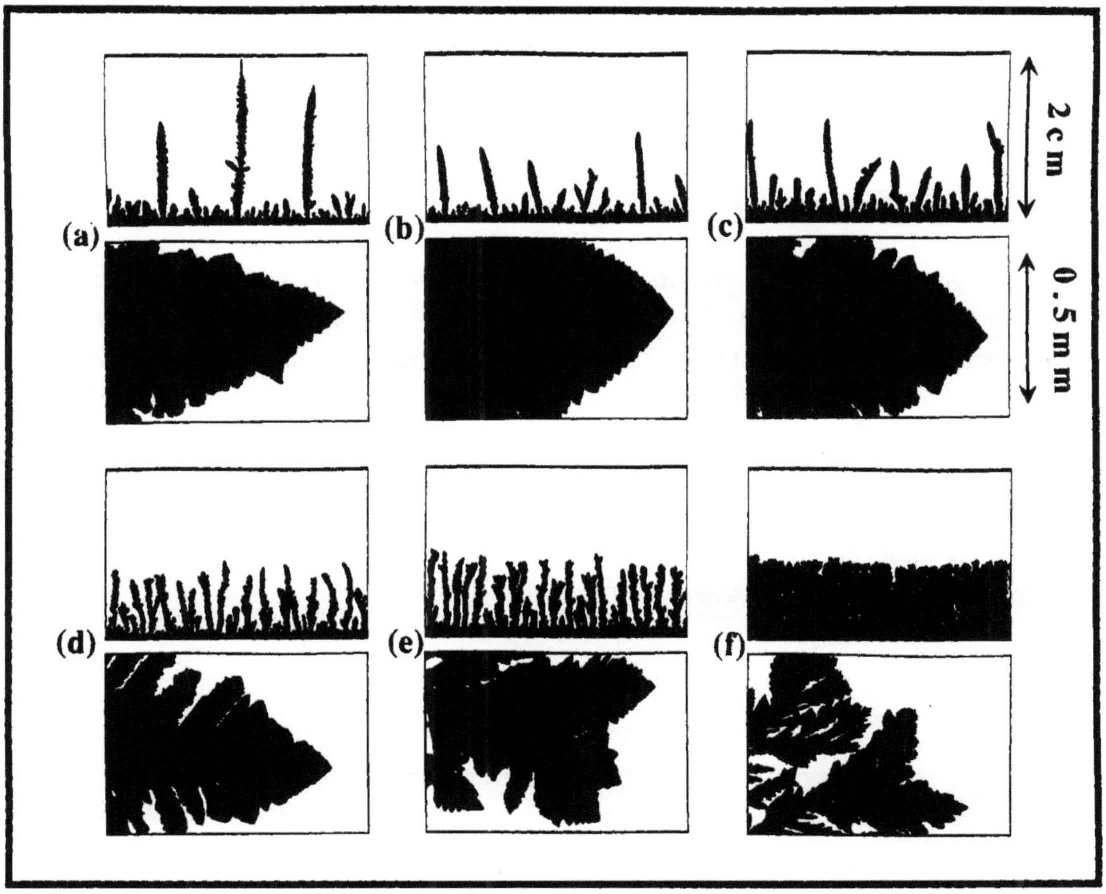

Fig. 3 Evolution of the microscopic and macroscopic morphology with the sodium content of the electrolyte, $j = 60$ mA cm^{-1}, [$ZnSO_4$] = 0.1 mol l^{-1}. The lower picture of each couple is taken with a magnification of 20, $r = [Na_2SO_4]/[ZnSO_4]$: $a : r = 0$, $b : r = 2.5 \times 10^{-4}$, $c : r = 8.5 \times 10^{-4}$, $d : r = 1.3 \times 10^{-3}$, $e : r = 1.7 \times 10^{-3}$, $f : r = 0.025$.

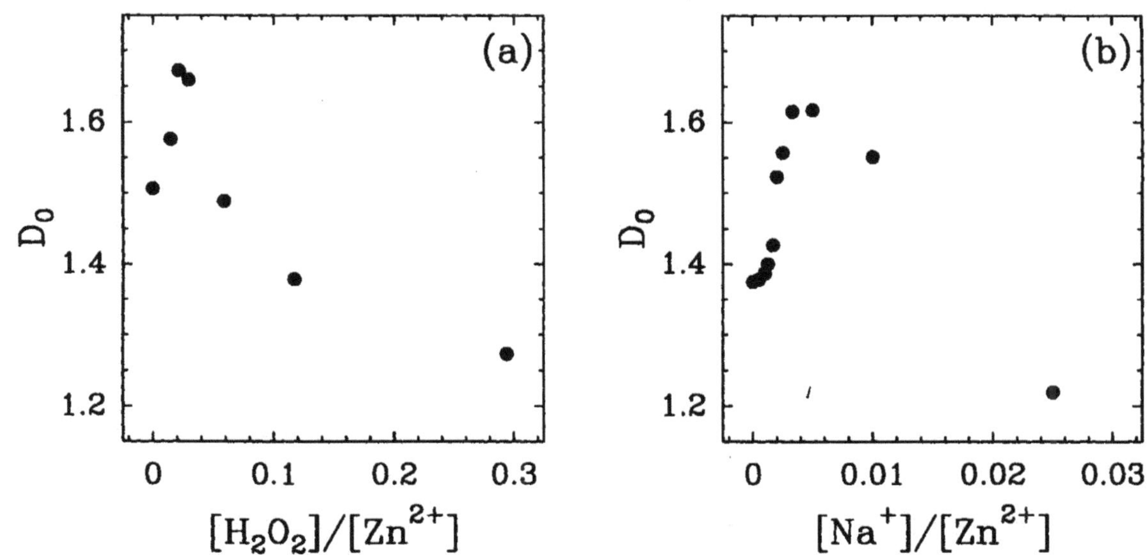

Fig. 4 Fractal dimension of the outer boundary of the deposit for different percentages of impurities. (**a**) Series of experiments of Fig. 2; (**b**) Series of experiments of Fig. 3.

for experiments where pH indicators, which nearly always contain alkali cations, have been added to the solution,[17] the structure of the deposit is mostly determined by the amount of indicator and much less by other parameters like the current density.

4. SPATIO-TEMPORAL TRANSITIONS

During the growth of copper and zinc electrodeposits, some striking morphological changes have been observed in the past by several authors,[6,17-21] but up to now there has been only little agreement concerning the number, the position and the origin of these transitions. Better known as the Hecker effect, they have been described as a sudden change of the color, the roughness and the number of branches of the deposit at a certain distance between the cathode and the anode, which scales with the total size of the cell. In the following, we briefly report on two transitions which are related to the chemical perturbations treated above. A detailed description and explanation of these transitions, including two other ones, has been published elsewhere.[22]

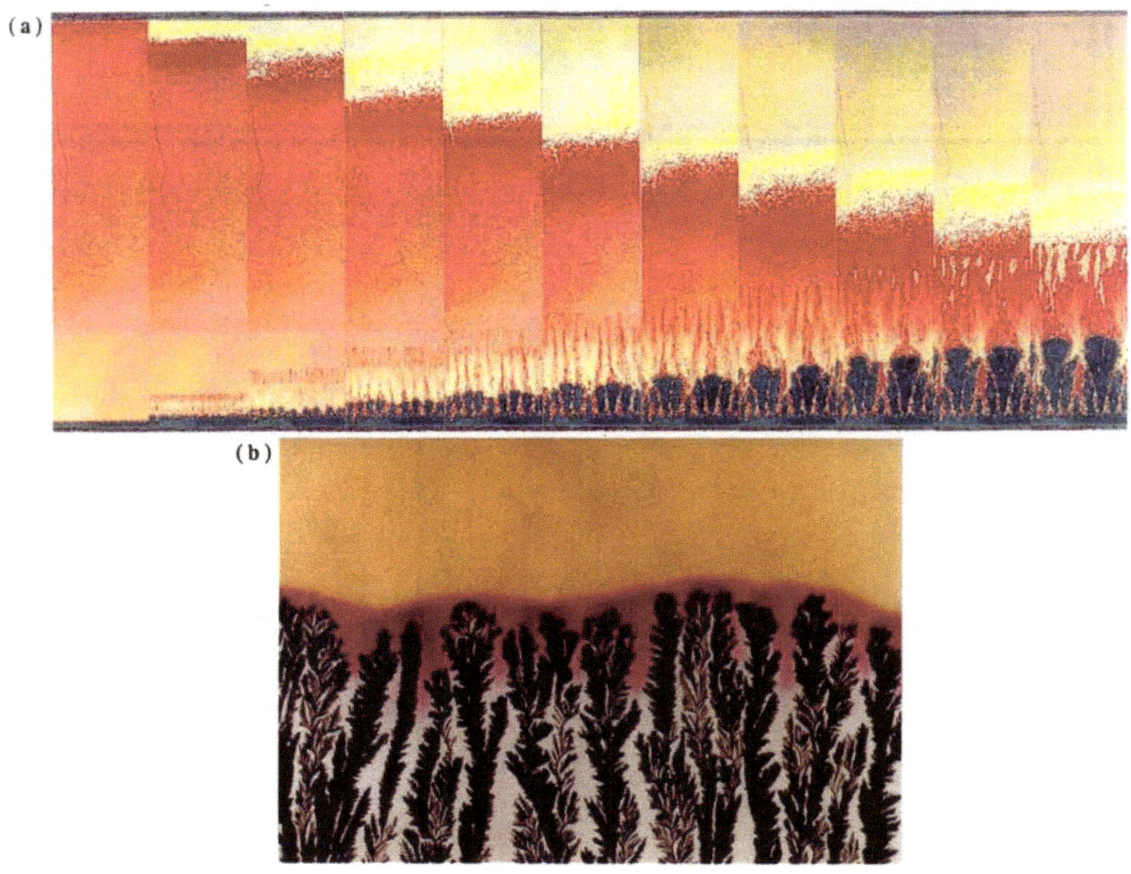

Fig. 5 Change of the pH during the electrodeposition of zinc: From the left to the right: (a) yellow front of more basic pH is migrating from the anode towards the cathode and meets at 20% the growing deposit. The colors do not represent the real colors of the indicator (Bromocresolgreen) because the images have been treated by a computer in order to make the front better visible; (b) After the first transition, the neighborhood of the growth becomes more basic, represented by the real color of the basic form of bromocresolpurple.

4.1 The pH Transition

As has already been discussed in Sec. 3, protons play a fundamental role in thin layer electrodeposition and we have shown in a previous publication[9] that during the electrodeposition experiment the pH of the solution changes. Strictly speaking, in the case of zinc deposition a front of more basic pH meets the growth at about 20% because the protons, consumed by hydrogen production at the cathode are not replaced at the anodic side of the cell. In Fig. 5(a), this front has been visualized by adding a pH indicator to the solution and the measured velocity coincides with the proton migration.[22] In this case, the transition at 20% is not visible because the added indicator perturbs the growth via its sodium cations. Normally at 20%, the bubble formation in the deposit stops, because the solution surrounding the interface becomes too basic for hydrogen evolution, as can be seen from Fig. 5(b). In the series of pictures in Fig. 5(a), one can wonder about the stripes appearing in front of the growth. They are due to some strong convection near the cathode, which will be treated in a forthcoming publication.[23]

If the same experiment is done without indicator but with traces of oxygen in the solution, a well pronounced transition can be observed [Fig. 6(a)]. This is due to the fact that the

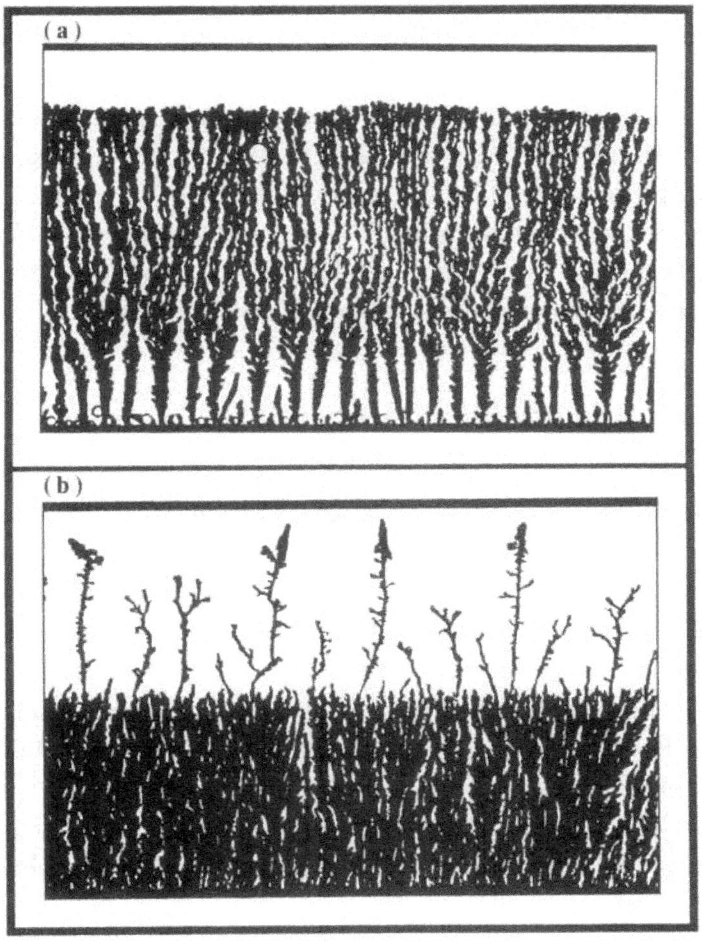

Fig. 6 (a) The pH transition in an experiment without indicator, $j = 20$ mA cm^{-1}, $[ZnSO_4] = 0.03$ mol l^{-1}, $r = [H_2O_2]/[ZnSO_4] = 0.1$, the pH of the solution has been adjusted to 2.7 by addition of sulphuric acid; (b) The sodium transition at about 60%, $j = 20$ mA cm^{-1}, $[ZnSO_4] = 0.05$ mol l^{-1}, $[Na_2SO_4] = 2.5 \times 10^{-4}$ mol l^{-1}.

oxygen ions, formed by the cathodic reduction of molecular oxygen, can combine before the transition with protons to give simply water.

$$2H^+ + O^{2-} \rightarrow H_2O \tag{1}$$

After the transition, when the surrounding of the growth is more basic, the oxide ions produced form a zinc oxide rather than water because of the missing protons.

$$O^{2-} + Zn^{2+} \rightarrow ZnO \tag{2}$$

This zinc oxide perturbs the dendritic growth, as mentioned above, by blocking active reduction sites.

With the help of the different ionic mobilities, straightforward calculations[20] show that in a parallel geometry, such as in our experiments, one should actually expect this transition at about 20%, whereas in a circular geometry, often used by other authors, this transition should happen at about 45%. We must emphasize here that this morphological change is much more visible when there is some oxygen in the solution, which has been the case for most of the quasi two-dimensional electrodeposition experiments reported in the literature.

4.2 The Alkali Ion Transition

Alkali ions migrate like protons from the anode to the cathode and they cannot be renewed during the experiment by anodic dissolution. As a consequence, the growth meets at some percentage of its development $t_c = \mu_{an}/(\mu_{an} + \mu_{al})$[20] (where μ_{an} is the anionic mobility and μ_{al} is the mobility of the alkali cation), a zone containing no more alkali ions and a transition from the former disordered structure to a more dendritic one occurs. For a sodium impurity, like our indicator salt, and a zinc sulphate solution, one can predict from standard table values a transition at about 60% in parallel geometry, which corresponds quite well to the experimental observations as shown in [Fig. 6(b)]. Other alkali ions also provoke morphological transitions whose location is directly related to their mobility.[24]

5. CONCLUSION

We have shown that the growth morphology and the surface roughness in thin layer electrodeposition of metals is very sensitive to chemical perturbations. Small amounts of oxygen or minute quantities of alkali ions are sufficient to change completely the microscopic and macroscopic appearance of the growth. This high sensitivity is essentially due to the quasi two-dimensional, very ramified structure of the growth. In three-dimensional growth experiments, the impurities have less influence on the interfacial process, for instance, industrial electrodeposition is often done by addition of a supporting electrolyte in high concentrations, whereas in thin layer experiments, this inhibits completely the metal growth because of a strong hydrogen evolution.

Controlled addition of these chemical species could be used to produce very thin metal catalysts, containing a well-defined amount of metal oxide, with a tunable specific surface.

ACKNOWLEDGMENTS

We are grateful to A. Arneodo, J. Elezgaray, J. F. Muzy and J. Huth for useful discussions. This work has been supported by the Centre National des Etudes Spatiales under contract

92/CNES/0225, NATO under grant RG 900685 and the DRET under grant 92/097. The work of A. Kuhn was supported by CEC under contract S/SC1*/915114.

REFERENCES

1. T. Vicsek, ed., *Fractal Growth Phenomena* (World Scientific, Singapore, 1989).
2. D. Grier, E. Ben-Jacob, R. Clarke and L. M. Sander, *Phys. Rev. Lett.* **56**, 1264 (1986).
3. Y. Sawada, A. Dougherty and J. P. Gollub, *Phys. Rev. Lett.* **56**, 1260 (1986).
4. M. Matsushita, Y. Hayakawa and Y. Sawada, *Phys. Rev.* **A32**, 3814 (1985).
5. F. Argoul, A. Arneodo, G. Grasseau and H. L. Swinney, *Phys. Rev. Lett.* **61**, 2558 (1988).
6. P. P. Trigueros, J. Claret, F. Mas and F. Sagues, *J. Electroanal. Chem.* **328**, 165 (1992).
7. W. Y. Tam and J. J. Chae, *Phys. Rev.* **A43**, 4528 (1991).
8. A. Kuhn and F. Argoul, "A New Mechanism for Morphological Changes in Thin Layer Electrodeposition," *Electrochim. Acta* (1993) submitted.
9. F. Argoul and A. Kuhn, "Experimental Demonstration of the Origin of Interfacial Rhythmicity in Electrodeposition of Zinc Dendrites," *J. Electroanal. Chem.* (1993) in press.
10. J. St-Pierre and D. L. Piron, *J. Electrochem. Soc.* **137**, 2491 (1990).
11. L. Antropov, *Electrochimie Théorique* (Mir, Moscow, 1975).
12. J. D. Farmer, E. Ott and J. A. Yorke, *Physica* **D7**, 153 (1983).
13. P. Grassberger, R. Badii and A. Politi, *J. Stat. Phys.* **51**, 135 (1988).
14. A. Renyi, *Probability Theory* (North-Holland, Amsterdam, 1970).
15. P. Grassberger, *Phys. Lett.* **A97**, 227 (1983).
16. P. Grassberger and I. Procaccia, *Physica* **D13**, 34 (1984).
17. J. R. Melrose, D. B. Hibbert and R. C. Ball, *Phys. Rev. Lett.* **65**, 3009 (1990).
18. N. Hecker, D. G. Grier and L. M. Sander, in *Fractal Aspects of Materials*, eds. R. B. Laibowitz, B. B. Mandelbrot and D. E. Passoja (Materials Research Society, University Park, PA, 1985).
19. P. Garik, D. Barkey, E. Ben-Jacob, E. Bochner, N. Broxholm, B. Miller, B. Orr and R. Zamir, *Phys. Rev. Lett.* **62**, 2703 (1989).
20. V. Fleury, M. Rosso and J. N. Chazalviel, *Phys. Rev.* **A43**, 6908 (1991).
21. E. Ben-Jacob, G. Deutscher, P. Garik, N. D. Goldenfeld and Y. Lareah, *Phys. Rev. Lett.* **57**, 1903 (1986).
22. A. Kuhn and F. Argoul, "Spatio-temporal Morphological Transitions in Thin Layer Electrodeposition: The Hecker Effect Revisited," *Phys. Rev. E.* (1993) submitted.
23. J. Huth, H. L. Swinney, A. Kuhn, F. Argoul and A. Arneodo, "Transport Processes and Reaction Mechanism in Thin Layer Electrodeposition: A Subtle Competition," to be submitted to *J. Chem. Phys.*
24. A. Kuhn and F. Argoul, "Determination of Ionic Mobilities by Thin Layer Electrodeposition," *J. Chem. Ed.* (1993) submitted.

Fractals, Vol. 1, No. 3 (1993) 460–469

INTERFACIAL AGGREGATION OF FLOATING MICROPARTICLES UNDER THE CONTROL OF SHORT-RANGE COLLOID AND VERY LONG-RANGE CAPILLARY FORCES

Z. HÓRVÖLGYI and M. ZRINYI
Department of Physical Chemistry, Technical University of Budapest, H-1111 Budapest, Egry József utca 20-22., Hungary

Abstract

Interfacial aggregation of surface modified glass beads (62–74 μm diameter) at water/air interface was carried out by using two differently hydrophobic samples, respectively. The effect of aggregation time on the self-similar structure of forming aggregates was studied comparing the actual results to those obtained previously.[1]

The time dependence of restructuring from the point of view of fractal geometry has been proved but the results call attention to another time dependent process — orientation of growing clusters during their collisions due to anisotropy of cluster-cluster interactions.

1. INTRODUCTION

Formation of colloid aggregates usually takes place in a complex way. Primary growth can be accompanied by secondary process (restructuring) in a number of cases showing the aspiration of forming objects for equilibrium situation.

During the past decade, it has been realized that non-equilibrium colloid aggregates could well be described in terms of fractals.[2–4] Although several computer simulations as well as experiments were carried out to show the connection of restructuring with the fractal geometry of growing aggregates,[5–18] and both equilibrium and non-equilibrium aspects of structure formation for diffusing as well as non-diffusing particles were studied experimentally in two-dimension,[19,20] the role of this secondary process in the formation of aggregates has not been clear enough.

Earlier, two-dimensional aggregation of hydrophobed, floating microparticles (diameter: 62–74 μm) was studied experimentally at different liquid-fluid interfaces, and the structure of forming aggregates was analyzed from the fractal geometry point of view.[1,21,22] The unique advantage of this model system is that the short-range (max. some hundred nanometers) colloid and extremely long-range (some particle diameters) capillary interactions[23,24] can *control together* the aggregation. In order to calculate the fractal dimension, interfacial number density of microparticles of a selected aggregate vs. aggregate size was determined and plotted in log-log way for the series of different sized growing aggregates. These "growth (density) functions" only revealed self-similar fractal structures for small aggregates. Crossover from fractal to non-fractal growth was attributed to restructuring caused by attractive capillary forces resulting in quasiequilibrium structures for medium sized as well as large aggregates. In this model system, each collision of particles and clusters leads to sticking, so the mechanism of aggregation can be considered diffusion limited from this point of view and characterized by clustering of cluster process. However, after the sticking, rotation of (single- or double-row) particle segments (for higher and medium hydrophobic samples) or migration of particles (for medium and lower hydrophobic samples) can occur, controlled by short-range colloid (structural[25] and electric double layer) interactions, causing interaction dependent fractal structure formation.[1,21,22] Recently, this process was analyzed from the aspect of colloid interactions, and universality of primary growth could be proved by screening the double layer interparticle repulsion in the systems investigated.[22]

In the present work, the effect of restructuring on the aggregate geometry is studied from its time-dependent aspect. The surface concentration of aggregating particles was increased which resulted in faster growth, so restructuring was hindered in kinetic way and the influence of this on the fractal property of forming aggregates could be investigated. In order to obtain some information about the effect of colloid interactions under the actual circumstances, aggregation experiments were carried out by using two kinds of particle samples having a lower and a higher degree of hydrophobicity, respectively. The lower hydrophobic particles are supposed to come to secondary potential energy minimum as a result of aggregation, while the higher hydrophobic ones can reach primary potential energy minimum due to strong hydrophobic[25] and weaker double layer interactions between them,[1] leading to significant difference in the structural strengths of their aggregates.[26] (The greater the structural strength of aggregate, the weaker its ability to restructure.)

2. MATERIALS AND METHODS

Materials. Distilled water. Spherical Pyrex glass particles (62–74 μm diameter) which were chemically hydrophobed and characterized previously. Water contact angle was found to be 99° and 41° for the higher and the lower hydrophobic particles, respectively[1,22] (according to another measuring methods, the advancing water contact angle may be close to 90° for the higher hydrophobic system[27]).

Methods. *Aggregation experiments* were carried out in Petri dishes of diameter 12.0 cm. The particles (20 or 40 mg which means surface concentration of 800 and 1600 pieces/cm^2, respectively) were scattered uniformly on the water surface over a period of some seconds, yielding individual beads at the beginning of the aggregation. They began to move as a result of temperature inhomogeneity in the boundary layer and to aggregate owing to the capillary and colloid (adhesion) forces. For analysis, the different sized, forming aggregates were photographed during the time of growth. *In order to determine the fractal dimension,* the number of primary particles, characteristic aggregate size (R) and aggregate area were calculated for a selected aggregate. R was approximated by the geometric mean method[28] and the area was accepted as R^2. The ratio of particles' number and aggregate area gives the interfacial particle number density (ρ). Fractal dimensions were determined from the $\ln \rho$ vs. $\ln R$ functions obtained for the series of different sized aggregates.

3. EXPERIMENTAL RESULTS

As a result of aggregation, usually one large aggregate formed taking about 5–10 minutes at every surface concentration of particles. (It should be noted that, formerly, in the case of much lower surface concentration of particles, the whole process took place over 1–2 hours[1,21,22]) Instead of simple segment rotation found previously,[1,21,22] a crack-like reorganization of the higher hydrophobic particle segments was observed in several cases. During this segment motion, the particles remained together. Sometimes the large aggregates of both samples became holed suddenly. For demonstration, different stages of growth for the higher and the lower hydrophobic systems are given in Figs. 1 and 2, respectively. Experimentally obtained "growth functions" are depicted in Figs. 3(a)–3(c). A comparative evaluation of fractal dimensions, aggregate size ranges of fractal growth considering the previous results[1] are given in Table 1.

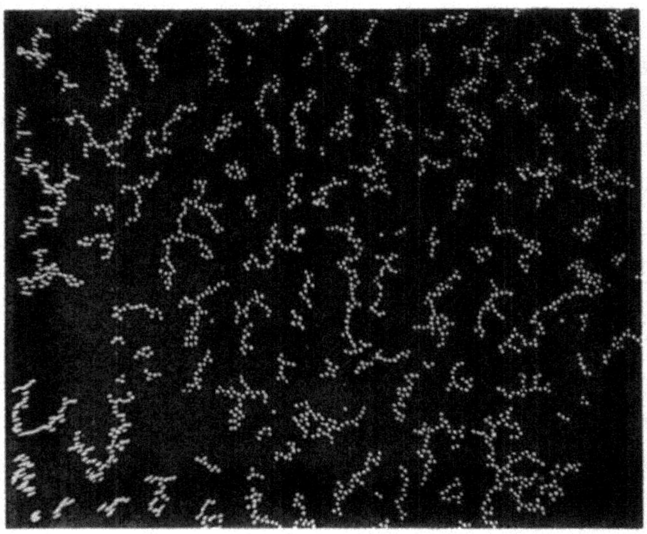

(a)

Fig. 1 Different stages of interfacial aggregation of the higher hydrophobic particles, (a)–(d).

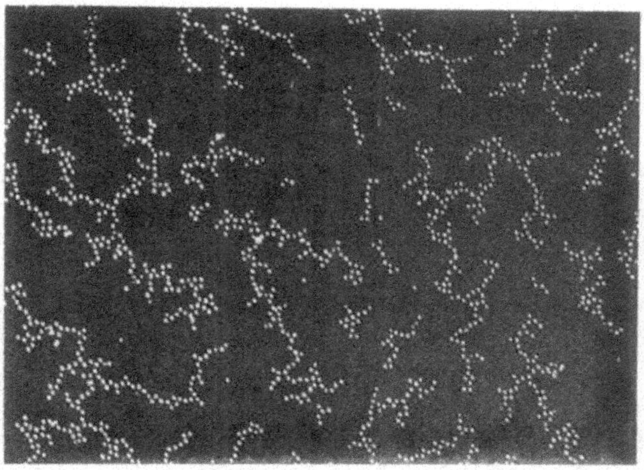

(b)

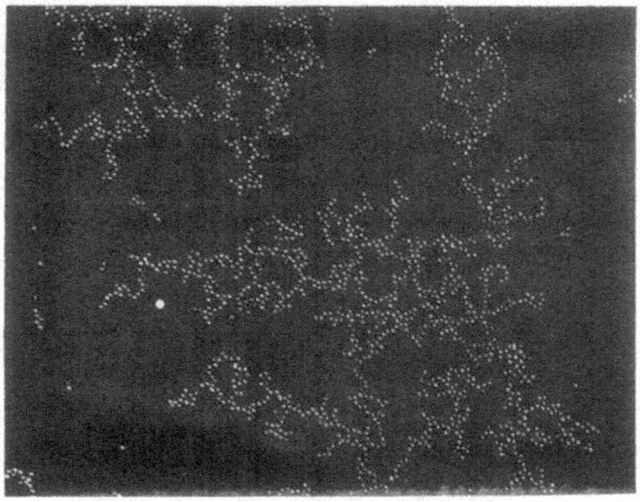

(c)

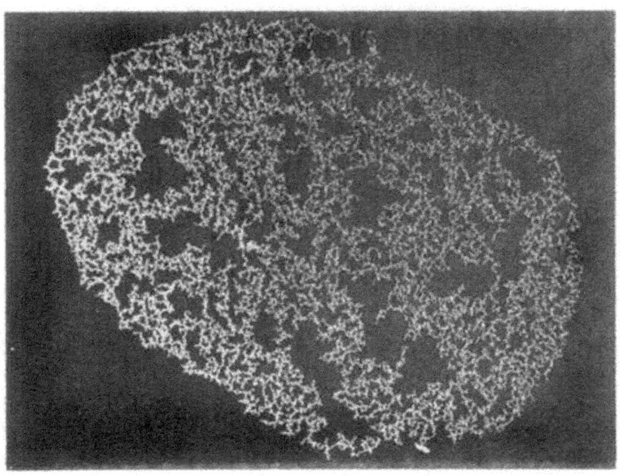

(d)

Fig. 1 (*Continued*)

(a)

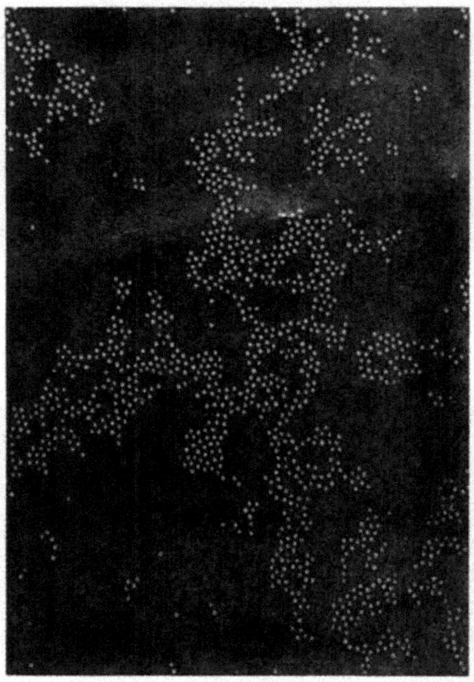

(b)

Fig. 2 Different stages of interfacial aggregation of the lower hydrophobic particles, (a)–(d).

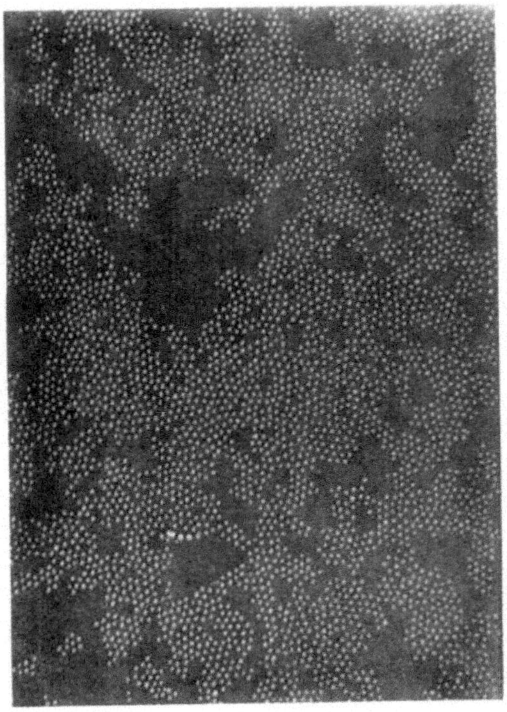

(c)

(d)

Fig. 2 (*Continued*)

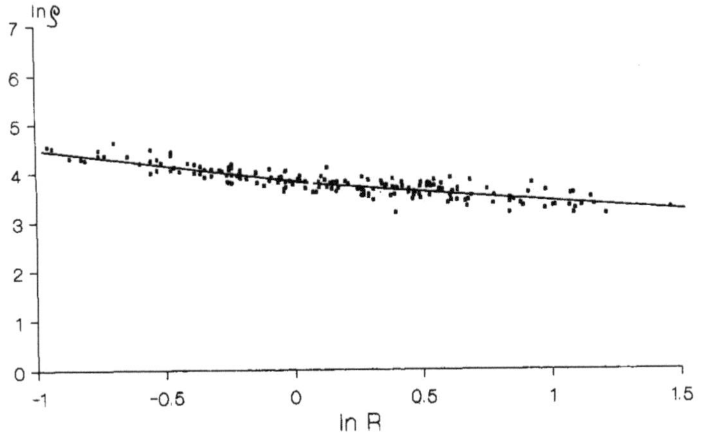

(a)

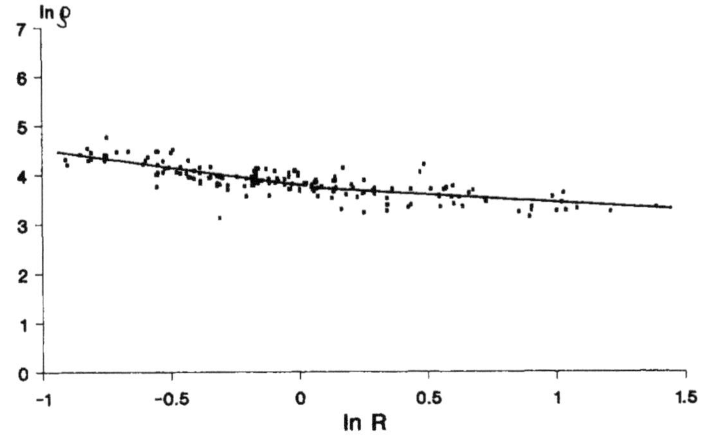

(b)

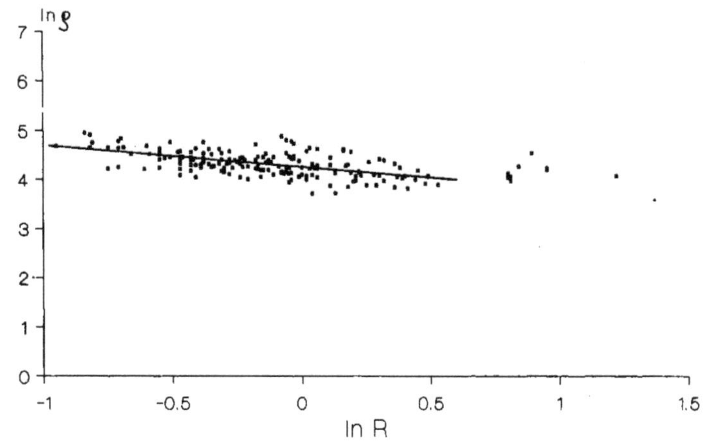

(c)

Fig. 3. **(a)** Experimentally obtained "growth function" for the higher hydrophobic sample (surface concentration of particles is 1600 pieces/cm^2); **(b)** experimentally obtained "growth function" for the higher hydrophobic sample (surface concentration of particles is 800 pieces/cm^2); **(c)** experimentally obtained "growth function" for the lower hydrophobic sample (surface concentration of particles is 1600 pieces/cm^2).

Table 1 Parameters determined from "growth (density) functions" obtained experimentally (surface coverage calculated by a mean particle diameter of 68 μm is also shown)

	Surface concentration of particles (surface coverage)		
	$1600/cm^2$(5.6%)	$800/cm^2$(2.8%)	$30–40/cm^2$(≈0.12%)[1]
Sample $\theta = 99°$			
	$D = 1.20(-0.79)$	1.22(−0.77)	1.12(−0.94)
	$R° = 0–1.0$ mm	0–1.0 mm	0–1.5 mm
	$N = 50$	55	90
	$D = 1.62(-0.64)$	1.65(−0.59)	non-fractal
	$R° = 1.0–4.5$ mm	1.0–4.5 mm	> 1.5 mm
	$N = 479$	541	> 90
Sample $\theta = 41°$			
	$D = 1.56(-0.57)$	—	1.58(−0.80)
	$R° = 0–1.6$ mm	—	0–1.0 mm
	$N = 131$	—	95
	The structures were non-fractal for $R > 1.6$ mm		The structures were non-fractal for $R > 1.0$ mm

Standard deviations of fractal dimensions were found to be ±0.1.
D: fractal dimensions (correlation coefficients are given in the brackets); $R°$: size range of aggregates in which fractal growth characterized by above structure parameter occurs; N: maximum number of particles which form an aggregate in the above range.

4. DISCUSSION

The visual observations indicated that relatively high surface concentration of particles led to non-equilibrium structure formation. Nevertheless, significant difference can be seen between the aggregates of different hydrophobic samples indicating the role of short range colloid interactions (Figs. 1–2). In accord with former results,[1] the lower hydrophobic particles can form much denser aggregates, i.e., in spite of the relatively fast growth, the weak particle-particle bonds permit their structures to reorganize. Although some short particle segments can form at the beginning, the structures are rather compact [(Figs. 2(a)–2(d)].

"Growth (density) functions" similarly to former results[1,21,22] also indicate crossover phenomenon for all systems investigated [(Figs. 3(a)–3(c)]. The fractal dimensions obtained for the first part of these functions are equal within experimental errors to those obtained previously at lower surface concentration of particles[1] (Table 1). The effect of aggregation time can only be observed on the aggregate size range of fractal growth, although the results are contradictory. In the case of faster aggregation, decrease of aggregate size range of fractal growth ($D = 1.20$ and 1.22) is observed for the higher hydrophobic sample, while for the other one, the extension of that range ($D = 1.58$) can be found (Table 1). The reason for this contradiction may be that, beside restructuring, there is another time dependent phenomenon, orientation of colliding clusters due to some kind of anisotropy originated in e.g. colloid repulsion,[29,30] which can also be responsible for the relatively low values of fractal dimensions obtained for the higher hydrophobic systems. In this case, there is not enough time for the clusters to reorientate during colliding, i.e., crossover occurs at lower aggregate sizes which is the case of the higher hydrophobic system. However, for the lower hydrophobic system, where restructuring is the decisive phenomenon because of the relatively weak particle-particle bonds, fast aggregation which hinders restructuring slightly, can cause the extension of aggregate size range in which fractal growth occurs. In this case, the restructuring can overcome the effect of anisotropy even at small aggregate sizes, leading to higher fractal dimension than that obtained for the "tip-to-tip" model developed for anisotropic cluster-cluster aggregation.[31,32]

The effect of faster aggregation appears more convincingly in the case of the second part of "growth functions" for the higher hydrophobic systems (Figs. 3(a) and 3(b) and Table 1). Instead of non-fractal growth found previously at lower concentration of primary particles,[1] fractal growth characterized by fractal dimensions of 1.62–1.65 was revealed. It should be noted that the surface concentration of primary particles cannot influence significantly the fractal geometry of forming objects in the present concentration range (Table 1). The experimentally obtained fractal dimensions are between the simulated result of two-dimensional diffusion limited cluster-cluster aggregation including cluster rotation ($D = 1.485 \pm 0.015$),[5] and that obtained for two-dimensional simulation of aggregation at high surface concentration of particles ($D = 1.75 \pm 0.07$),[33] indicating a transition from low to high cluster concentration aggregation in which restructuring can also play a role.

5. CONCLUSION

Time dependency of restructuring in the light of fractal growth of interfacial aggregates can be shown. Nevertheless, the results call attention to the fact that, beside restructuring, there may be another time dependent phenomenon, orientation of growing clusters due to anisotropy of cluster-cluster interaction which can also influence the fractal growth through the aggregation rate, and in certain cases, the results can only be interpreted by taking into account both opposite effects.

ACKNOWLEDGMENTS

This work was supported by the National Foundation for Scientific Research (OTKA F4216). Technical assistance of Mr. Árpád Dobszai is gratefully acknowledged.

REFERENCES

1. Z. Hórvölgyi, G. Medveczky and M. Zrinyi, *Colloids Surfaces* **60**, 79 (1991).
2. D. A. Weitz and M. Oliveria, *Phys. Rev. Lett.* **52**, 1433 (1984).
3. P. Meakin, *Adv. Colloid Interface Sci.* **28**, 249 (1988).
4. T. Vicsek, *Fractal Growth Phenomena* (World Scientific, Singapore, 1989).
5. P. Meakin and R. Jullien, *J. Phys. (Paris)* **46**, 1543 (1985).
6. R. Botet and R. Jullien, *Phys. Rev. Lett.* **55**, 1943 (1985).
7. P. Meakin, *J. Chem. Phys.* **83**, 3645 (1985).
8. P. Meakin and J. Deutch, *J. Chem. Phys.* **83**, 4086 (1985).
9. P. Meakin, *J. Colloid Interface Sci.* **112**, 187 (1986).

10. M. Kolb, *J. Phys. A: Math. Nucl. Gen.* **19**, L263 (1986).
11. M. Kolb, *Phys. Rev. Lett.* **56**, 2769 (1986).
12. R. Jullien and P. Meakin, *J. Colloid Interface Sci.* **127**, 265 (1989).
13. P. Dimon and S. K. Sinha, D. A. Weitz, C. R. Safinya, G. S. Smith, W. A. Varady and H. M. Lindsay, *Phys. Rev. Lett.* **57**, 595 (1986).
14. C. Aubert and D. S. Cannell, *Phys. Rev. Lett.* **56**, 738 (1986).
15. G. Medveczky, M. Sc. Thesis, Loránd Eötvös University, Budapest (1990)..
16. Y. Adachi and S. Ooi, *J. Colloid Interface Sci.* **135**, 374 (1990).
17. J.-F. Roussel, C. Camoin and R. Blanc, *J. Phys. France* **50**, 3259 (1989).
18. J.-F. Roussel, C. Camoin and R. Blanc, *J. Phys. France* **50**, 3269 (1989).
19. A. T. Skjeltorp, *Phys. Rev. Lett.* **58**, 1444 (1987).
20. A. T. Skjeltorp, G. Helgesen, "Random Fluctuations and Pattern Growth: Experiments and Models," in *Proc. of the NATO ASI*, eds. H. E. Stanley and N. Ostrowsky, (Kluwer Academic Publishers, Dordrecht, Netherlands, 1988), p. 56.
21. Z. Hórvölgyi, G. Medveczky and M. Zrinyi, *Colloid Polym. Sci.* **271**, 396 (1993).
22. Z. Hórvölgyi, M. Máté and M. Zrinyi, *Colloids Surfaces* (in press).
23. D. Y. C. Chan, J. D. Henry, Jr. and L. R. White, *J. Colloid Interface Sci.* **79**, 410 (1981).
24. P. A. Kralchevsky, V. N. Paunov, I. B. Ivanov and K. Nagayama, *J. Colloid Interface Sci.* **151**, 79 (1992).
25. B. V. Derjaguin and N. V. Churaev, *Colloids Surfaces* **41**, 223 (1989).
26. Z. Hórvölgyi, S. Németh and J. H. Fendler, *Colloids Surfaces A: Physiochem. Eng. Asp.* **71**, 327 (1993).
27. Z. Hórvölgyi and R. Csalavári, *XXIII. Int. Symp. on Colouristics* (Balatonszéplak (Hungary), 1991) Abstr. 9.
28. R. J. Samson, G. W. Mulholland and J. W. Gentry, *Langmuir* **3**, 272 (1987).
29. I. L. Thomas, K. H. McCorkle, *J. Colloid Interface Sci.* **36**, 110 (1971).
30. A. J. Hurd and D. W. Schaefer, *Phys. Rev. Lett.* **54**, 1043 (1985).
31. R. Jullien, *Phys. Rev. Lett.* **55**, 1967 (1985).
32. R. Jullien, *J. Phys. A: Math. Nucl. Gen.* **19**, 2129 (1986).
33. M. Kolb and H. Herrmann, *J. Phys. A: Math. Nucl. Gen.* **18**, L435 (1985).

Fractals, Vol. 1, No. 3 (1993) 470–474

KINETIC ROUGHENING AND INTERACTING REACTING SPECIES: NONLINEAR ASPECTS

I. M. SOKOLOV*†, P. ARGYRAKIS‡ and A. BLUMEN*
*Theoretische Polymerphysik, Universität Freiburg, Rheinstraße 12, 79104 Freiburg i.Br., Germany
‡Department of Physics, University of Thessaloniki, 54006 Thessaloniki, Greece
†P. N. Lebedev Physical Institute of the Academy of Sciences of Russia, Leninsky prosp. 53, Moscow 117924, Russia

Abstract

We consider the $A + B \to 0$ reaction, in which particles interact through short-range forces. The analysis leads to expressions akin in form to those which describe kinetic roughening. In a situation in which particles are generated with a constant rate j_0, their concentration $n(t)$ grows as $j_0^{\beta} t^{\gamma}$ in $d = 1$. Here the theoretical analysis predicts $\gamma = 1/5$ and $\beta = 2/5$, in very good agreement with direct Monte-Carlo simulations of the reaction-diffusion process.

1. INTRODUCTION

Pioneering investigations of diffusion-controlled chemical reactions are due to Smoluchowski (1917),[1] but only recent years have led to new insights into the problem.[2–10] By now, the $A + B \to 0$ reaction between *uncharged* particles is well-understood; less clear, however, is the situation where the A and B particles are subject to electrostatic forces. As noted in Ref. 11, reaction-diffusion schemes have much in common with interface-growth processes. As stressed in Ref. 12, fluctuation effects in reaction kinetics have their counterpart in the kinetic roughening of growing interfaces.

The present article studies in a 1d system the $A + B \to 0$ reaction under steady particle generation; furthermore, the particles exert forces on each other. An analytical model (nonlinear diffusion equation) for such a situation was put forward in Ref. 12; here we support the analytical findings through Monte-Carlo simulations of the reaction process. During their motion, particles interact through exponential forces; these are of opposite sign so that particles of the same kind repel and particles of different kinds attract each other. The reaction is assumed to take place on contact.

We start from the analytical description of the $A + B \to 0$ reaction.[12] In a continuum approximation, the equations for the local concentrations of particles, $n_A(\mathbf{x}, t)$ and $n_B(\mathbf{x}, t)$, read:

$$\frac{\partial n_A}{\partial t} = \nabla(D\nabla n_A - \alpha \mathbf{f}_A\{n_A, n_B\}n_A) - R\{n_A, n_B\} + j_A(\mathbf{x}, t) \tag{1a}$$

and

$$\frac{\partial n_B}{\partial t} = \nabla(D\nabla n_B - \alpha \mathbf{f}_B\{n_A, n_B\}n_B) - R\{n_A, n_B\} + j_B(\mathbf{x}, t) \tag{1b}$$

Here D is the diffusion coefficient (equal for both reacting species), α is the particles' mobility in the force field $\mathbf{f}_A\{n_A, n_B\}$ or $\mathbf{f}_B\{n_A, n_B\}$, respectively, $R\{n_A, n_B\}$ is the local reaction term and j_A and j_B represent local particle generation rates. We take j_A and j_B to be Gaussian and δ-correlated with $\langle j_A \rangle = \langle j_B \rangle = j_0$.

In the analytic framework, the next steps are as follows: For extremely fast reactions (say $R \to \infty$) in low-dimensional spaces, there exist only nonpenetrating A- and B-clusters, so that reactions occur only at the clusters' boundaries. Setting $q(\mathbf{x}, t) = n_A(\mathbf{x}, t) - n_B(\mathbf{x}, t)$ for the local concentration difference, we obtain:

$$n_A = q\theta(q)\,; \quad n_B = -q\theta(-q) \tag{2}$$

In fact the form Eq. (2) may hold even for finite local reaction rates R when the times are long enough.[8] Evidently, the forces vanish for a distribution of A- and B-particles which is homogeneous. Furthermore, for short-range interactions at longer times when clusters are formed, the forces are to a very good approximation proportional to ∇q:

$$\mathbf{f}_A(\mathbf{x}) = k\nabla q(\mathbf{x}) \tag{3a}$$

and

$$\mathbf{f}_B(\mathbf{x}) = -k\nabla q(\mathbf{x}) \tag{3b}$$

Now we obtain by inserting Eqs. (3) into Eqs. (1), by subtracting Eq. (1b) from Eq. (1a) and by taking $n_A + n_B = |q|$:

$$\frac{\partial q}{\partial t} = \nabla(D + \mu|q|)\nabla q + j\,, \tag{4}$$

where $j(\mathbf{x}, t) = j_A(\mathbf{x}, t) - j_B(\mathbf{x}, t)$ and $\mu = k\alpha > 0$. Evidently, from the properties of j_A and j_B it follows that $\langle j(\mathbf{x}, t)\rangle = 0$ and $\langle j(\mathbf{x}, t) j(\mathbf{x}', t')\rangle = 2j_0\delta(\mathbf{x} - \mathbf{x}')\delta(t - t')$. As noted in Ref. 12 we can rewrite Eq. (4) as:

$$\frac{\partial q}{\partial t} = (D + \mu|q|)\Delta q + \mu\,\mathrm{sign}(q)(\nabla q)^2 + j \tag{5}$$

a form akin to the KPZ-equation,[13] which depicts the growth of an interface of local height h. The dynamics of the processes described by Eq. (5) and by the KPZ-equation is, however, dissimilar, since, distinct from KPZ, Eq. (5) has an additive nonlinear term, namely, $\mu|q|\Delta q$. On the other hand, as in the KPZ-case, neglecting all nonlinear terms reduces Eq. (5) to the Edwards-Wilkinson (EW) form.[14] In the following, we will focus on the mean particle concentration, $n(t) = \langle |q(x, t)| \rangle$ and on its dependence on t and j_0.

An understanding of the reaction can be achieved by using scaling arguments as in Ref. 4. We focus on a characteristic correlation length $\mathcal{L}(t)$ (e.g. typical cluster size), and assume that inside a box of sidelength $\mathcal{L}(t)$, all particles can react up to time t with their counterparts (if any). At t the total number of surviving particles in such a box is given by the difference (excess) in the number of particles of the two kinds. We calculate the number of excess particles N_{ex} from the total number of particles generated up to time t; the latter is $j_0 t\mathcal{L}^d(t)$ in d-dimensions. Following Poisson statistics, one has $N_{ex} \sim (j_0 t\mathcal{L}^d(t))^{1/2}$. The concentration $n(t)$ of the surviving particles is therefore proportional to $N_{ex}/\mathcal{L}^d(t)$, i.e.,

$$n(t) \sim (j_0 t)^{1/2}\mathcal{L}(t)^{-d/2} \tag{6}$$

In the absence of nonlinear terms, one has $\mathcal{L}(t) \sim D^{1/2}t^{1/2}$, so that in a one-dimensional situation it follows that $n(t) \sim j_0^{1/2}t^{1/4}$. In the nonlinear case, we approximate $D+\mu|q(x, t)|$ by its spatial average $D+\mu n(t) \equiv \bar{D}(t)$, so that in a strongly nonlinear situation, $\mu n(t) \gg D$, one obtains:

$$\mathcal{L}^2(t) \simeq \int_0^t \bar{D}(t)dt \simeq \mu \int_0^t n(t)dt \tag{7}$$

As before, we assume that $n(t)$ depends algebraically on t, $n(t) \sim t^\gamma$. From Eqs. (6) and (7) it follows, self-consistently, that:

$$\gamma = \frac{2-d}{4+d} \tag{8}$$

This means $\gamma = 1/5$ in 1d, a result established in Ref. 12 by solving Eq. (4) numerically. Furthermore, from Eqs. (5) and (6) the dependence of $n(t)$ on j_0 also follows; we find then $n(t) \sim j_0^\beta$ with $\beta = \frac{2}{d+4}$, i.e., $\beta = 2/5$ in 1d.

Now we present explicit Monte-Carlo simulation of the process. We consider a one-dimensional chain of length $L = 10^5$. On this lattice A- and B-particles are generated pairwise, but spatially uncorrelated. The motion of each particle is modeled through hops to one of its nearest-neighbor sites. The probabilities that the i-th particle hops in the next step to the right (+) or to the left (−) are:

$$p_\pm^i = \frac{1}{2} \pm f^i \tag{9}$$

where f^i stands for the force acting on the i-th particle. Now f^i is a sum over forces of the form $f_{ij} = \pm f_0 \exp(-\lambda|j-i|)$. For $j > i$, the negative (positive) sign holds for j of the same (different) kind as i; for $j < i$, the signs are reversed. For the parameters used one has rarely $|f^i| > 1/2$; in such occurrences we set $p_\pm$ in Eq. (9) to be equal to 0 or 1, respectively. Furthermore, Eq. (9) treats diffusion only approximately, but we use it for simplicity. Now a time step of the procedure is accomplished when each particle has moved once. In our simulations, cyclic boundary conditions are used. We do not implement excluded volume conditions so that several particles of the same type can occupy the same lattice site. To assess the role of forces restricted to nearest-neighbors only, we also simulated such situations and found that this kind of restriction does not lead to qualitative changes.

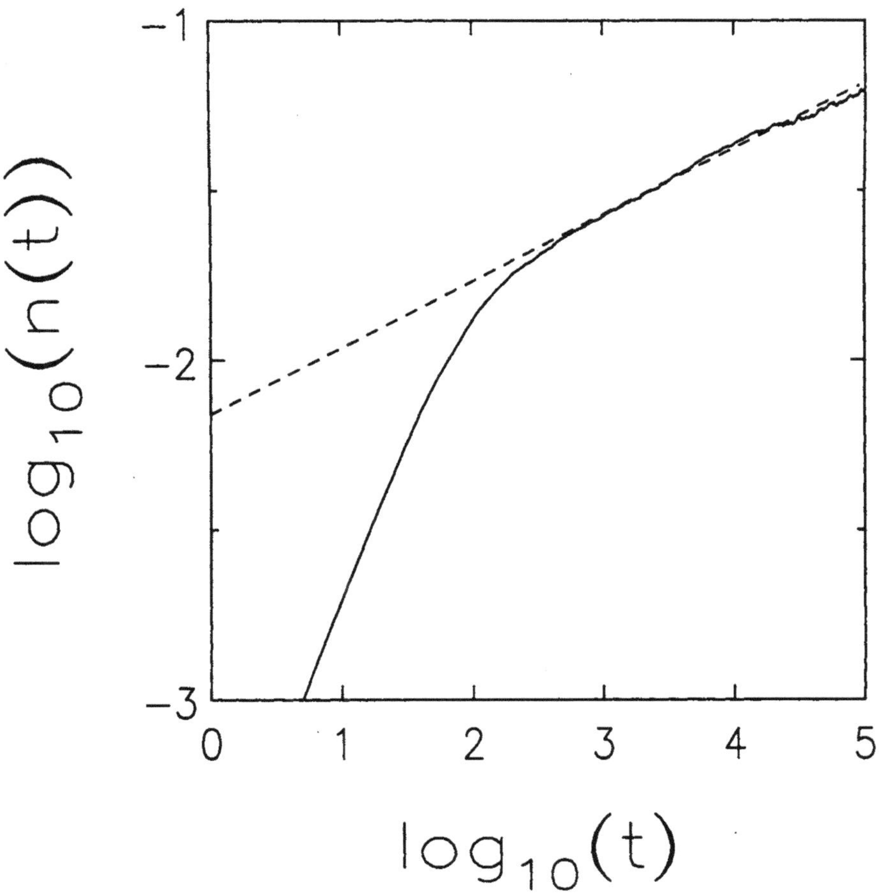

Fig. 1 The particles' concentration $n(t)$ as a function of time in double logarithmic scales. See text for details. The dashed line has the slope 0.196.

We now present the results of our simulations. In the absence of forces between particles ($f_0 = 0$), we recover the expected behavior, $n(t) \sim t^{1/4}$. Turning now to exponentially decaying forces, we plot in Fig. 1 in double-logarithmic scales $n(t)$ as obtained from 8 independent realizations of the process with $f_0 = 0.5$, $\lambda = 0.2$ and $j_0 = 10^{-4}$ as parameters. As may be seen from Fig. 1, for $t > 300$ the relation $n(t) \sim t^\gamma$ is well-obeyed. We establish for γ through a least-squares fit in the range $300 < t < 10^5$ the value $\gamma = 0.196 \pm 0.01$; the error is estimated by evaluating γ for each realization separately. Hence γ is in good agreement with the theoretical analysis, leading to $\gamma = 0.2$.

To obtain the exponent β, we confine ourselves to models with nearest-neighbor forces only, a restriction which allows much faster simulations. In Fig. 2 we plot in double-logarithmic scales n vs. j_0 for $t = 10^4$; here the values of j_0 extend from 10^{-5} to 10^{-3}. The straight line is the least-squares-fit to the data and has a slope of 0.40, in very good agreement with the analytical scaling arguments presented above.

In summary, reaction-diffusion systems in 1d are indeed well-described by the nonlinear equation proposed in Ref. 12. The kinetics of accumulation of particles follows the power-law $n(t) \sim j_0^\beta t^\gamma$, where the values of β and γ are consistent with $\beta = 2/5$ and $\gamma = 1/5$, as predicted by simple scaling arguments.

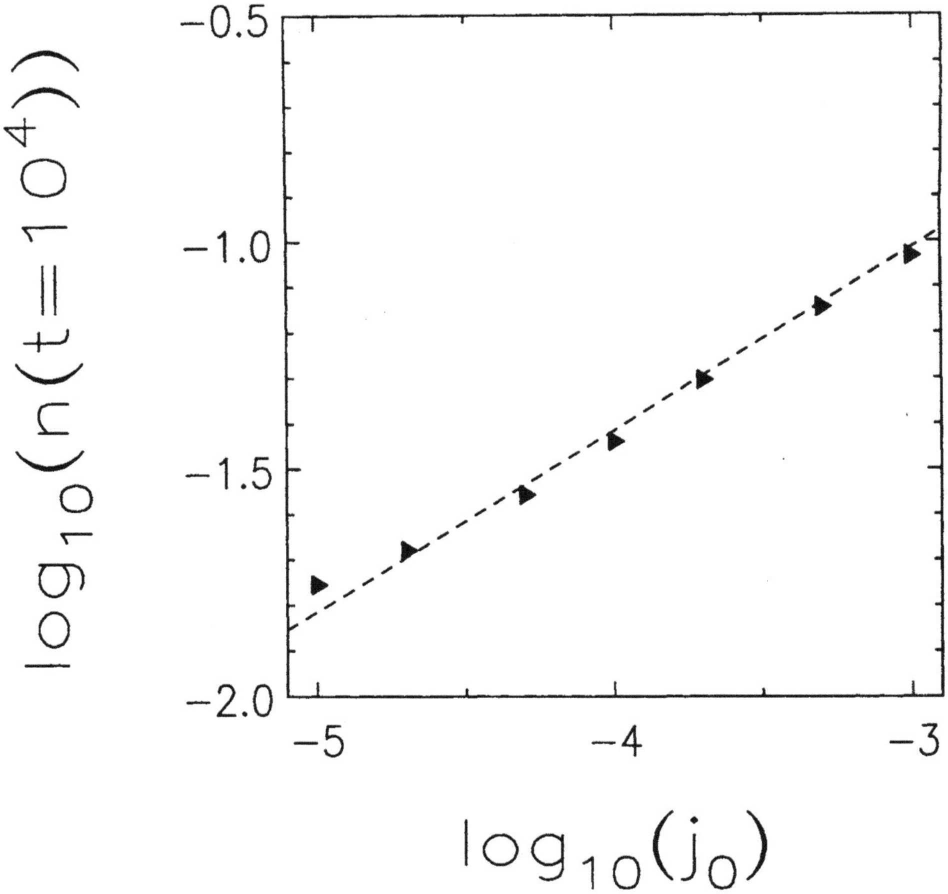

Fig. 2 $n(t)$ plotted in double-logarithmic scales as a function of the generation rate j_0 for $t = 10^4$. Here the forces are between nearest-neighbors only; see text for details. The dashed line has a slope 0.400.

ACKNOWLEDGMENTS

We acknowledge many discussions with Profs. J. Klafter and G. Zumofen, and support by the German-Greek Collaboration Grant (KFA Jülich 6I3A6D01), by the SFB 60 of the DFG and by the Fonds der Chemischen Industrie.

REFERENCES

1. M. von Smoluchowski, *Zeit. Phys. Chem.* **92**, 129 (1917).
2. A. A. Ovchinnikov and Ya. B. Zeldovich, *Chem. Phys.* **28**, 215 (1978).
3. D. Toussaint and F. Wilczek, *J. Chem. Phys.* **78**, 2642 (1983).
4. K. Kang and S. Redner, *Phys. Rev. Lett.* **52**, 955 (1984).
5. I. M. Sokolov, *Zh. Eksp. Teor. Fiz.* **94**, 199 (1988) (*Sov. Phys. JETP*, **67**, 1846 (1988)).
6. H. Schnörer, V. Kuzovkov and A. Blumen, *Phys. Rev. Lett.* **63**, 805 (1989).
7. I. M. Sokolov, H. Schnörer and A. Blumen, *Phys. Rev.* **A44**, 2388 (1991).
8. G. Zumofen, J. Klafter and A. Blumen, *J. Stat. Phys.* **65**, 1015 (1991); *Phys. Rev.* **A44**, 8390 (1991).
9. P. Argyrakis and R. Kopelman, *Phys. Rev.* **A41**. 2121 (1990).
10. P. Argyrakis and R. Kopelman, *Phys. Rev.* **A45**, 5814 (1992).
11. Y.-C. Zhang, *Phys. Rev. Lett.*, **59**, 1726 (1987).
12. I. M. Sokolov and A. Blumen, *Europhys. Lett.* **21**, 855 (1993).
13. M. Kardar, G. Parisi and Y.-C. Zhang, *Phys. Rev. Lett.* **56**, 889 (1986).
14. S. F. Edwards and D. R. Wilkinson, *Proc. R. Soc. Lond.* **A381**, 17 (1982).

Fractals, Vol. 1, No. 3 (1993) 480–490

UNIVERSALITY OF 1-DIMENSIONAL REACTION MODELS

N. INUI, A. Y. TRETYAKOV and H. TAKAYASU
School of Information Sciences,
Tohoku University, Aramaki, Aoba-ku, Sendai, 980, Japan

Abstract

We investigate the general case of a nearest neighbor interacting particle system in 1-dimension by using a real space-time renormalization approach. Discrete time branching annihilation random walk and basic contact process are included as special cases. We present an evidence that all nearest neighbor interacting particle systems involving 2nd order extinction-survival dynamical phase transition converge to contact process in the large scale limit, which is consistent with numerical results for contact process and branching annihilating random walk.

1. INTRODUCTION

Simple reaction-diffusion models, representing a number of processes in nature, like heterogeneous catalysis, spread of disease and similar phenomena, including creation and demolition of particles, describing spatial configuration of a system, are now firmly established as a subject of research, involving both theoretical studies and computer simulations.[1–11] Here we are going to discuss a general case of an interacting particle system with nearest-neighbor interactions on a one-dimensional lattice, involving only one sort of particles. The most widely known members of this class are contact process (in short CP)[6–11] and branching annihilating random walk (in short BARW),[3–5] both of which have been shown numerically to belong to the same directed percolation (in short DP) universality class, on the basis of behavior in the vicinity of an extinction-survival second order dynamical phase transition.[5,9] Using a renormalization-group approach, we are going to bring some additional evidence that all interacting particle systems with nearest-neighbor interactions in 1-dimension belong to DP universality class.

2. NUMERICAL EVIDENCE OF UNIVERSALITY

2.1 Critical Exponents of CP and BARW

CP is defined as follows: a randomly chosen particle either creates a particle on a random neighbor site, if it is empty, or it is annihilated, leaving an empty site. Depending on the ratio between creation and annihilation rates, particles eventually get extinct or some non-zero concentration of particles is maintained indefinitely. There is a strong theoretical evidence that CP in d-dimension belongs to the same universality class as directed percolation in $d+1$ dimension,[9] which is consistent with simulation results in 1-dimension.[7]

In BARW, a randomly chosen particle jumps to a nearest neighbor site with probability p or creates a new particle on a random nearest neighbor site with probability $1-p$. If two particles try to share a site, they annihilate leaving an empty site. Unlike CP, in BARW process at small concentration of particles, if long-range correlations are not taken into account, the rate of extinction goes to zero, so that simple mean field treatment does not predict the existence of extinction region. This discrepancy with simulation results[4] reflects the fact that in BARW, particles are being removed as a result of interaction, not spontaneously, as in CP.

Similarly in CP and other particle systems involving 2nd order dynamical phase transitions, if p is close to the threshold point p_c, the particle density ρ follows:

$$\rho \sim |p-p_c|^{\beta}, \tag{1}$$

where β is the concentration critical exponent.

The first estimate of β for 1-dimension[4] was a little larger than that of CP, $\beta = 0.2769(2)$[10]; however, recent simulations by Jensen[5] show that the values of dynamical exponents Δ, ν and z, related to time behavior at the critical point, are very close to DP values, which indicates that BARW and CP should be in the same universality class.

Our own investigation of BARW using finite size scaling approach (averaging stationary concentration over surviving realizations on lattices of size N changing from 32 to 1024, taking $32768/N$ runs for each N, with 16^*N steps per lattice site for each run, and collecting data over the last quarter of a run as long as there are particles in the system), confirmed the value of p_c, obtained by Jensen,[5] and gave for the ratio β/ν (where ν is the spatial correlation radius exponent), $\beta/\nu = 0.257$ (see Fig. 1), which is very close to $\beta/\nu = 0.2524(5)$ for directed percolation.[10,11] This result encouraged us to recalculate β averaging particle density over surviving realizations for a number of values of p close to the threshold (on 1024 site lattice, in conditions similar to the ones described for the finite-size scaling simulation), and fitting data by Eq. (1). This improved calculation gave $\beta = 0.28$ on the assumption that $p_c = 0.107$,[5] much closer to the DP value than the less computer-time consuming simulations of Ref. 4.

2.2 Fractal Dimensions of CP and BARW

One more approach to universality is to compare fractal dimensions of particle distribution at the threshold. The estimations of correlation dimension D_c and box counting dimension D_B [see Figs. 2(a) and 2(b)] are (the lattice size is 2048, simulation under conditions similar to the ones described in Sec. 2.1):

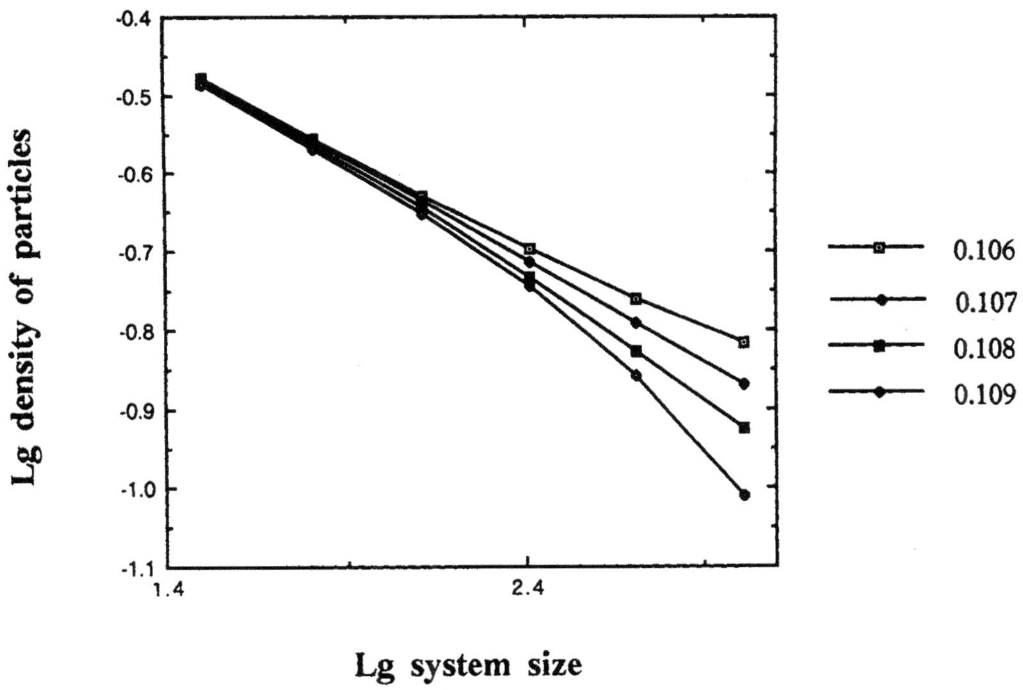

Fig. 1 Logarithm of particle density averaged over surviving realizations at different values of p. The best line fit at $p = 0.107$, indicates that this is the transition point; β/ν ratio is given by the slope.

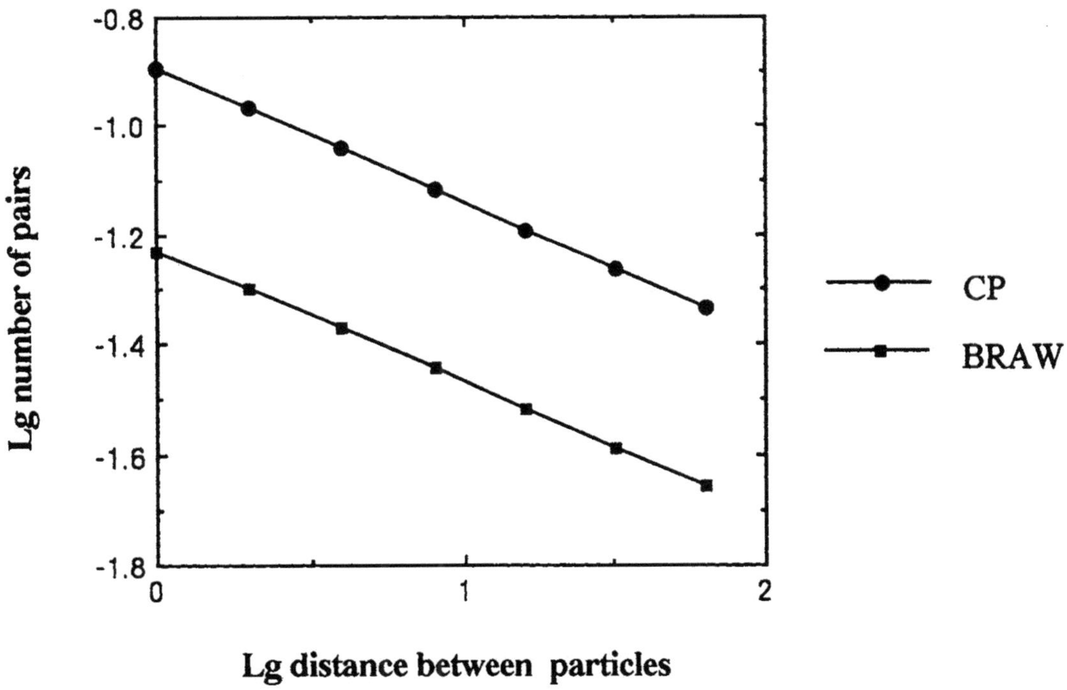

Fig. 2(a) Logarithm of the number (per lattice site) of pairs of particles located at a fixed distance from each other, plotted vs. logarithm of the distance. BARW and CP clearly give the same slope, indicating that the correlation dimension is the same.

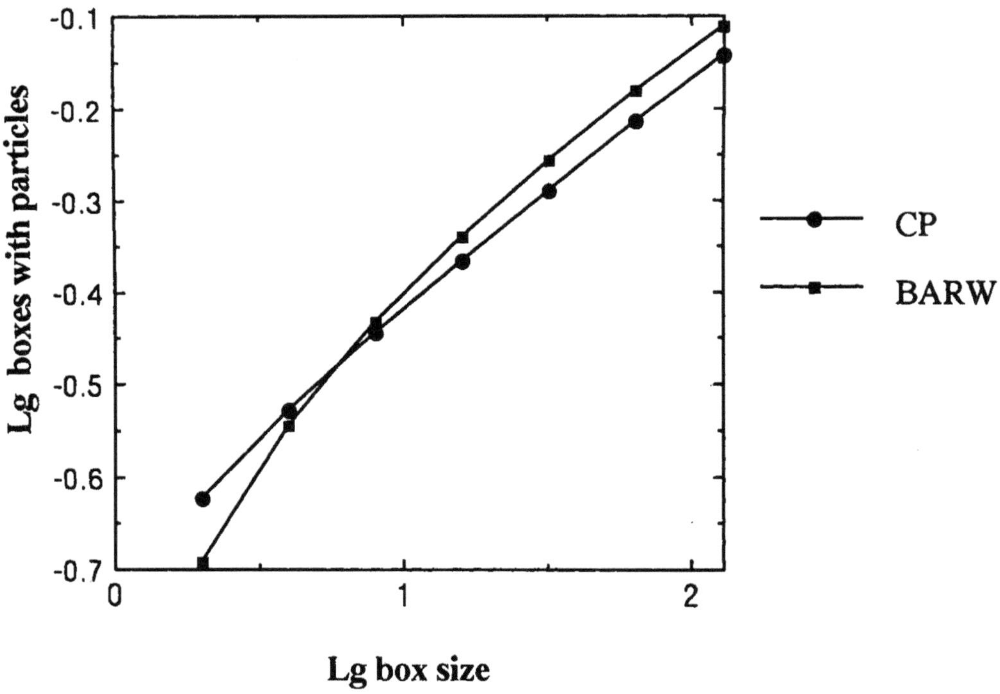

Fig. 2(b) Logarithm of the fraction of boxes, containing particles, vs. logarithm of the box size. At small box sizes the line fit is poor; fractal box dimension has been calculated on the basis of a slope, given by two biggest points.

$$D_c = 0.76 \pm 0.01 \text{ [BARW]}, \qquad D_c = 0.75 \pm 0.01 \text{ [CP]},$$
$$D_B = 0.76 \pm 0.02 \text{ [BARW]}, \qquad D_B = 0.75 \pm 0.02 \text{ [CP]}.$$

Thus, we see that the exponents, related to spatial distribution of particles at criticality, also seem to have the same values for CP and BARW.

3. RENORMALIZATION PROCEDURE

We define nearest-neighbor interacting particle systems as discrete time Markov processes in $\{0, 1\}^{z^1}$; at each site $x \in z^1$ (here z^1 is the set of all integer numbers) random variable $\eta(x)$ takes value 0 or 1, representing a vacant site and a particle on a site.

By assuming that an adsorption point is the extinction state, the dynamics of a system which is composed of only k sites is represented as:

$$\mathbf{P}_k(t+1) = \mathbf{P}_k M_k \tag{2}$$

$$\mathbf{P}_k(t) = \{P(j, t);\ 0 \le j \le 2^k - 1\} \tag{3}$$

$$P(j, t) = \text{Prob.}\left\{ \sum_{x=0}^{k-1} 2^x \eta(x) = j,\ t \right\} \tag{4}$$

Matrix element $M_k(i, j)$, representing the probability of transition from state i to state j is defined as:

$$M_k(i, j) = \text{Transition prob.}\left\{\sum_{x=0}^{k-1} 2^x \eta(x) = i\,, \quad \sum_{x=0}^{k-1} 2^x \eta(x) = j\right\} \tag{5}$$

By assuming left-right symmetry, the transition matrix $M2$ can be written as:

$$M_2 = \begin{pmatrix} 1 & 0 & 0 & 0 \\ a & 1-a-b-c & b & c \\ a & b & 1-a-b-c & c \\ d & e & e & 1-2e-d \end{pmatrix}, \tag{6}$$

where the parameters have the following physical meaning:

$a:$	spontaneous annihilation,	$01 \to 00$,	or $10 \to 00$
$b:$	random walk,	$01 \to 10$,	or $10 \to 01$
$c:$	branching,	$01 \to 11$,	or $10 \to 11$
$d:$	pair annihilation,	$11 \to 00$	
$e:$	single annihilation,	$11 \to 10$,	or $11 \to 01$.

In general, transition matrix M_k for a system of size k, is a function of a, b, c, d and e.

CP and BARW are characterized by one-parameter as:

$$\text{CP;} \quad b = 0\,, \quad c = 1-a\,, \quad d = 0 \quad \text{and} \quad e = \frac{a}{2}\,, \tag{7a}$$

$$\text{BARW;} \quad a = 0\,, \quad c = 1-b\,, \quad d = b \quad \text{and} \quad e = \frac{1-b}{2}\,. \tag{7b}$$

It should be noted that in this notation for BARW, b is equal to the jump probability p, introduced in Sec. 2.1.

Now we introduce the idea of renormalization group method for this general transition matrix. By coarsening space, we regard every 2 sites of a lattice as one super-site of a coarse-grained lattice. A super-site is occupied if one of the two corresponding sites of the original lattice is occupied; it is empty if (and only if) both the corresponding original sites are empty. Thus, the coarse random variable is defined as:

$$\bar{\eta}(x) = \begin{cases} 0 & \eta(2x) + \eta(2x+1) = 0 \\ 1 & \text{otherwise} \end{cases}\,. \tag{8}$$

As for the time steps, we regard 4 steps in the original system as 1 step in the coarse-grained dynamics (see Fig. 3). The elements of renormalized transition matrix are defined as:

$$\bar{M}_2(i, j) = \text{Transition Prob.}\{2\bar{\eta}(1) + \bar{\eta}(0) = i\,, \; 2\bar{\eta}(1) + \bar{\eta}(0) = j\}\,. \tag{9}$$

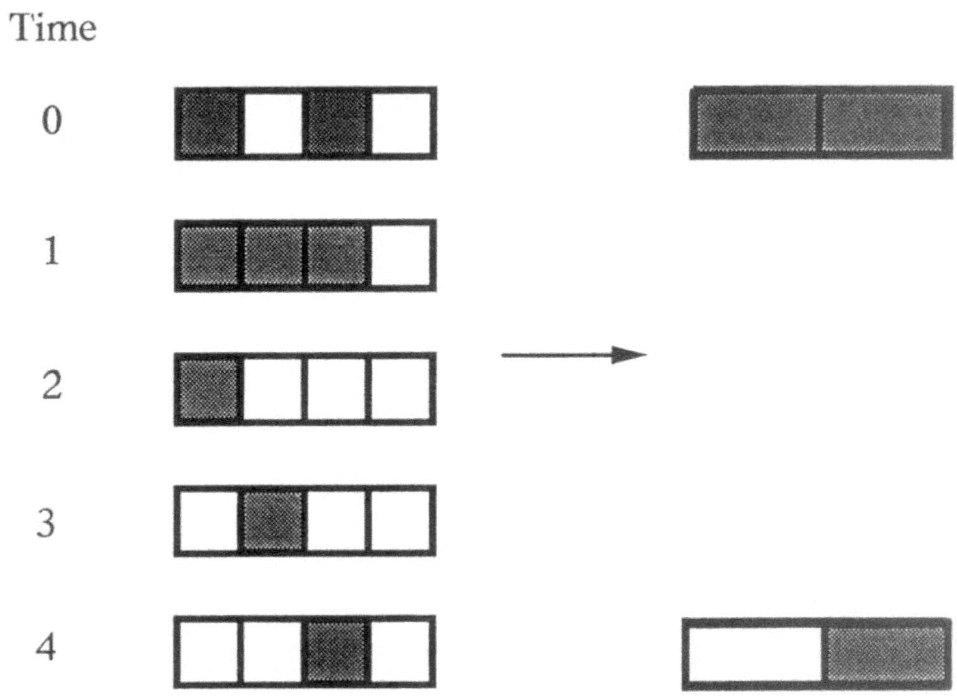

Fig. 3 Coarse-graining of space-time.

In order to determine the coarse-grained transition matrix, we count up all possible events in an original lattice of 4 sites over 4 times steps. A renormalized site of the super-lattice can, in general, be realized by several micro states (for example, for a coarse-grained pair 01, we have 3 configurations, 0011, 0001 and 0010), which need to be summed up with weights. This way, renormalized matrix (1,0) element a, corresponding to transition from 01 to 00 is defined as:

$$a = M_2 \equiv \omega_4(0, 1)M_4(1, 0) + \omega_4(0, 2)M_4(2, 0) + \omega_4(0, 3)M_4(3, 0)\,, \tag{10a}$$

$$\omega_4(0, i) = \frac{p(i, \infty)}{\sum_{k=1}^{3} p(k, \infty)}\,, \tag{10b}$$

where $\omega_4(0, i)$ are the weights of corresponding micro states, expressed via probabilities $p(i, \infty)$ to encounter them in the stationary state.

We approximate weights by using the stationary solution of master equation on 4-site lattice, $P_4(t+1) = M_4 P_4(t)$, with the following boundary conditions (mean field method):

$$p(\eta(0), \eta(1), \ldots, \eta(4), t) = \frac{p(\eta(0), \eta(1), \eta(2), \eta(3), t)p(\eta(1), \eta(2), \eta(3), \eta(4), t)}{p(\eta(1), \eta(2), \eta(3), 0, t) + p(\eta(1), \eta(2), \eta(3), 1, t)}\,. \tag{11}$$

Now, general renormalized matrix elements are represented as follows:

$$\bar{M}_2(i,j) = \sum_{\eta_2(x)\in 2\bar{\eta}(1)+\bar{\eta}(0)=j} \; \sum_{\eta_1(x)\in 2\bar{\eta}(1)+\bar{\eta}(0)=i} \times \left\{ \omega_4\left(\sum_{x=0}^{3} 2^x \eta_1(x), i\right) M_4\left(\sum_{x=0}^{3} 2^x \eta_1(x), \sum_{x=0}^{3} 2^x \eta_2(x)\right)\right\}, \qquad (12)$$

$$\omega_4\left(\sum_{x=0}^{3} 2^x \eta(x), i\right) = \frac{p_4\left(\sum_{x=0}^{3} 2^x \eta(x), \infty\right)}{\sum_{\eta(x)\in 2\bar{\eta}(1)+\bar{\eta}(0)=i} p_4\left(\sum_{x=0}^{3} 2^x \eta(x), \infty\right)}. \qquad (13)$$

4. RESULTS

Phase transition points can be estimated by watching the behavior of renormalized particle concentration. Figure 4 shows the stationary particle density in the renormalized system. We can clearly see the existence of a phase transition around $b = 0.1048$. This value is in reasonable agreement with the known simulation results. Figure 5(a) shows the values of renormalized parameters starting from BARW parameter values with $b = 0.103$ as functions of the number of renormalizations n. The increase of b and the decrease of a, c, and d indicate

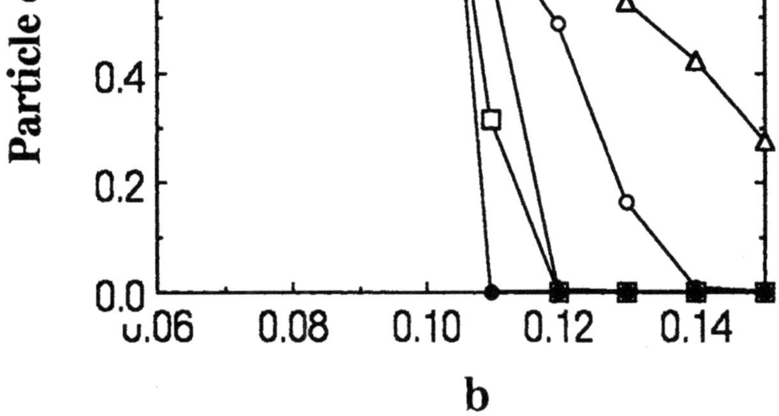

Fig. 4 Stationary particle density as a function of b. The phase transition becomes evident as the number of renormalization increases. Open triangles ($n = 3$), open circles ($n = 4$), solid triangles ($n = 5$), open squares ($n = 6$) and solid circles ($n = 7$).

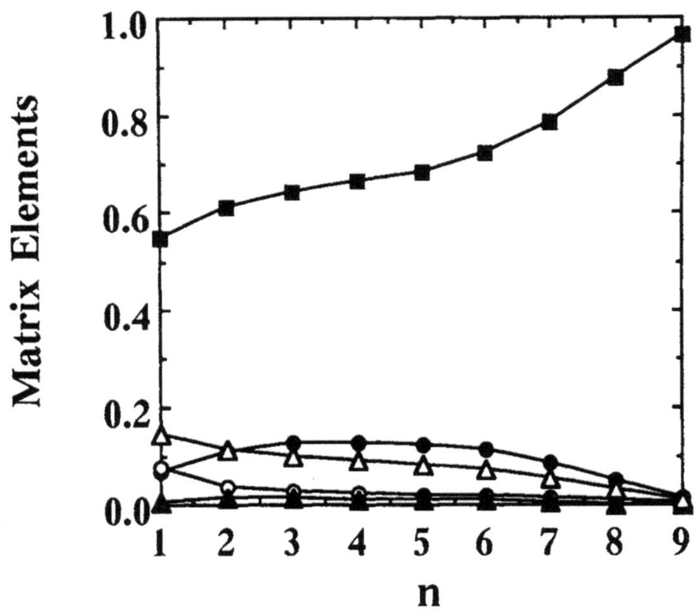

(a)

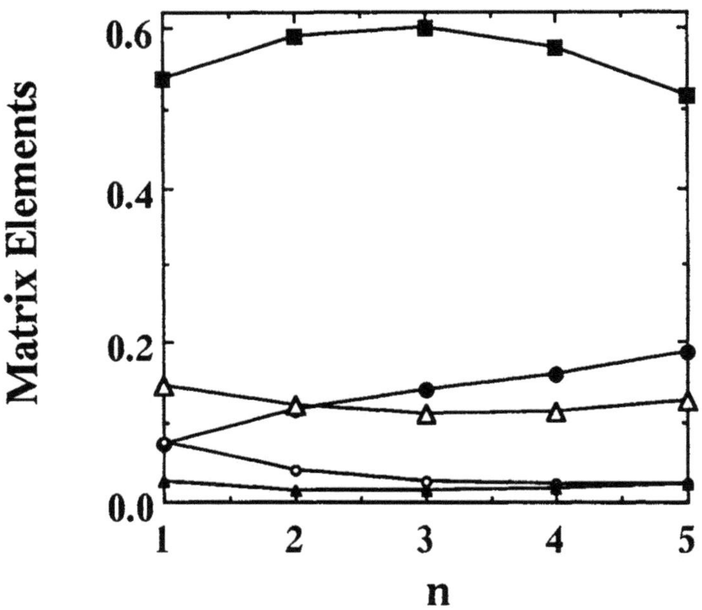

(b)

Fig. 5 (a) Change of transition probability matrix elements in a survival state, where n denotes the number of renormalization. a: solid circles, b: open circles, c: solid squares, d: solid triangles, and e: open triangles; (b) Change of transition probability matrix elements in an extinction state. Notations are the same as in Fig. 5(a).

that we are in survival state. Figure 5(b) shows the same plot starting with a little larger value $b = 0.105$, illustrating the parameters change in the extinction state.

We find that in the macroscopic limit, the values of b and d always become much smaller than the other parameters a, c and e. Quick decrease of parameters b and d is observed in Monte Carlo simulation, too. Figure 6 shows an example of super-particle dynamics renormalized by the same rules. We can clearly see that the number of pair annihilation and diffusion events decrease with n. This result is intuitively recognized as follows. Assume that we renormalize L sites as a super-site and L time steps as one super-step. A random walker needs L^2 time steps to move across L sites, so for large L the contribution of random walk is less effective. By adjusting the time scale at each step, which does not affect the stationary state behavior, it is possible to remove one more parameter, so that the number of remaining parameters is two.

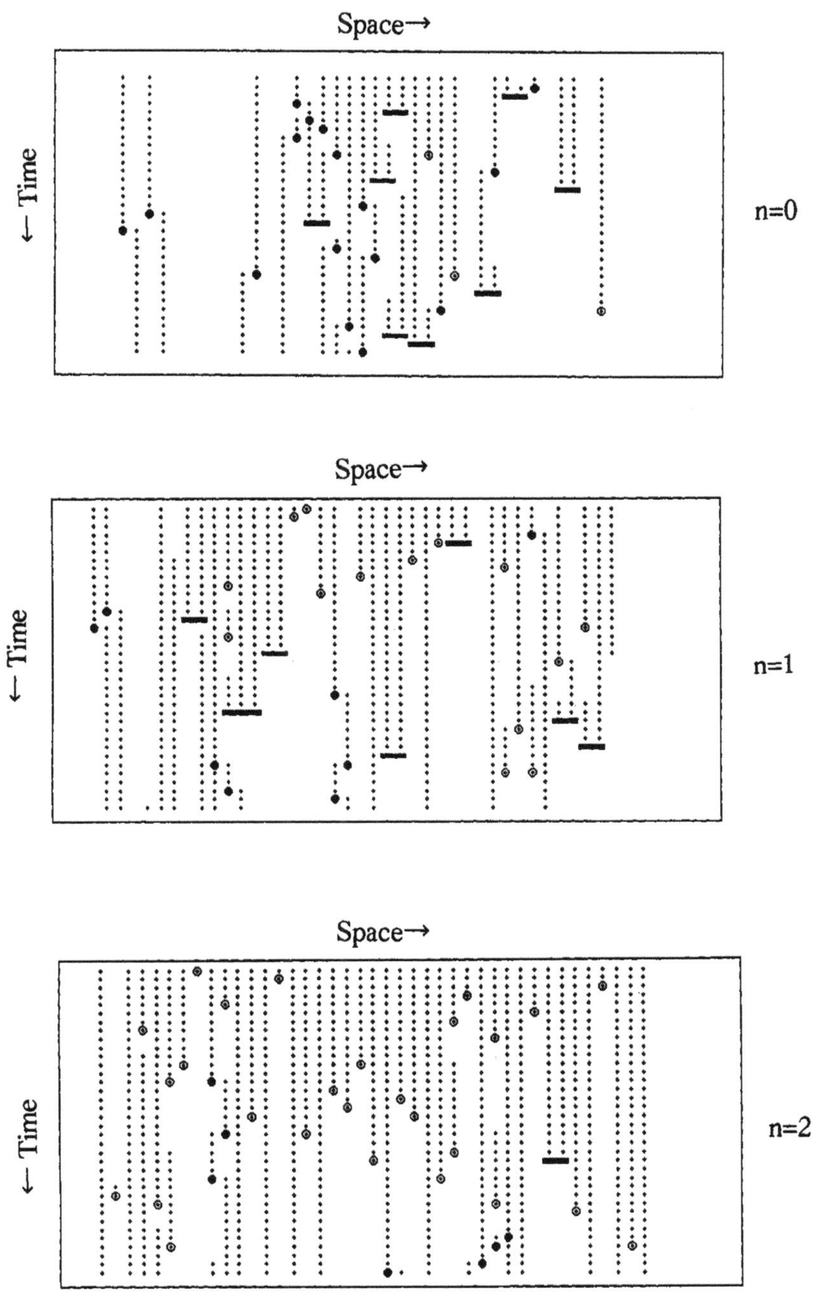

Fig. 6 Demonstration of renormalization procedure by Monte Carlo simulation starting from BARW parameter, $b = 0.8$. We show 40 sites, part of a 5000 sites lattice. Solid circle (diffusion event), rectangle (pair annihilation) and double circle (single annihilation).

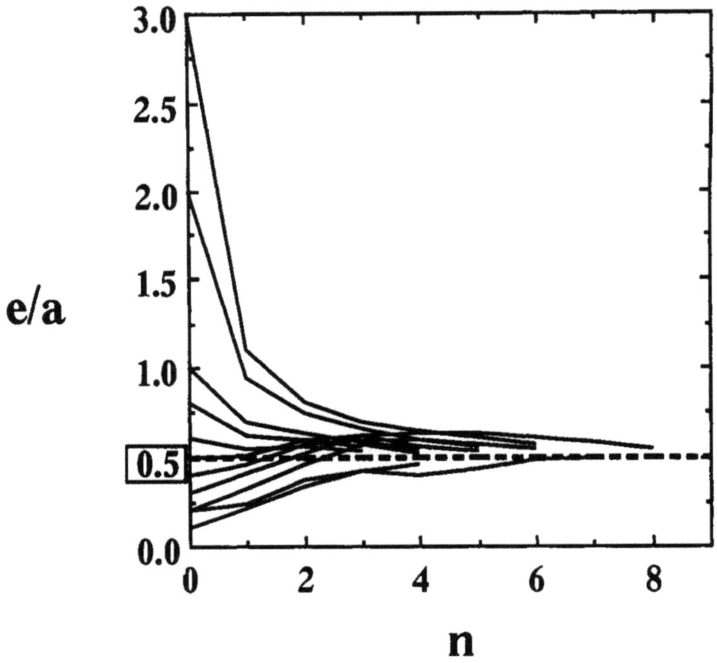

Fig. 7 Change of the ratio e/a by renormalization for various initial combinations of a, c and e, with $b = d = 0$. For different curves a is in the interval from 0.1 to 0.2, while e takes values between 0.02 and 0.3.

Suggesting that, in a sense, all processes should be equivalent to CP, which is defined by one parameter, we expect that, starting from a general transition matrix containing 5 parameters, in the infinite number of renormalizations limit the number of essential parameters should be one, and, if we choose a as the main parameter, e should be given by $e = 2a$, as for CP [Eq. (7a)]. Figure 7 shows that the ratios e/a, starting from several different combinations of a, c and e, gather around 1/2 after repeating the renormalization procedure for long enough, indicating that the systems covered by Eq. (2) and Eq. (6) tend to converge to CP in the macroscopic limit.

5. CONCLUSION

We introduce a general representation of nearest neighbor interacting particle systems in one dimension.

By directly renormalizing the space-time, we find that any transition matrix converges to that of CP in the macroscopic limit. This is consistent with numerical evidence related to CP and BARW, including the finite size scaling and fractal dimension data, presented in this paper, and supports the hypothesis that all nearest neighbor interacting particle systems, involving 2nd order extinction-survival phase transition, belong to the same universality class.

REFERENCES

1. R. Ziff, E. Gulari and Y. Barshad, *Phys. Rev. Lett.* **56**, 2553 (1986); P. Meakin and D. J. Scalpino, *J. Chem. Phys.* **87**, 731 (1987); D. Considine, H. Takayasu and S. Redner, *J. Phys.* **A23**, L1181, (1990).
2. R. Dickman, *Phys. Rev.* **B40**, 7005 (1989).

3. M. Bramson and L. Gray, *Z. Wahrsch. verw. Gebiete* **68**, 447 (1985).
4. H. Takayasu and A. Yu. Tretyakov, *Phys. Rev. Lett.* **68**, 3060 (1992).
5. I. Jensen, *Phys. Rev.* **E47**, R1 (1993).
6. T. E. Harris, *Ann. Prob.* **2**, 969 (1974).
7. P. Grassberger and A. de la Torre, *Ann. Phys.* **122**, 373 (1979).
8. N. Konno and M. Katori, *J. Phys. Soc. Jpn.* **59**, 1581 (1990).
9. J. L. Cardy and R. L. Sugar, *J. Phys.* **A13**, L423 (1980).
10. R. Dickmann and I. Jensen, *Phys. Rev. Lett.* **67**, 2391 (1991).
11. J. W. Essam, K. De'Bell, J. Adler and F. M. Bhatti, *Phys. Rev.* **B33**, 1982 (1986); J. W. Essam, A. J. Gutmann and K. De'Bell, *J. Phys.* **A21**, 3815 (1988).

Earth Sciences

INTRODUCTION

"How long is the coast of Britain?" This is the title of Mandelbrot's paper in which the concept of fractional dimension is applied to the real world for the first time. The graphics of computer-created landscapes in his book, *The Fractal Geometry of Nature*, attracted the interest of millions of people and evoked the fractal fever in almost all fields of sciences. These examples symbolize the importance of earth sciences in the progress of fractals related research.

Compared with other fields of natural sciences, earth sciences can be characterized by gigantic scales: A mountain of 10^3 m high is about 10^{13} times larger than a typical atomic scale; one million years, 3×10^{13} sec, can be an "instant" in geology; a huge volcanic activity is equivalent to 10^{14} candle flames in energy; and the earth itself is composed of nearly 10^{50} atoms.

These gigantic scales have been making the earth sciences difficult to approach. Theoretical physicists who made a big success in microscopic phenomena may try to solve the equations of atoms. However, the number of atoms involved for almost all objects in earth sciences is so huge that computer simulations based on atomic scale are impossible at present and seem to be unrealizable forever. The computer capacity has been growing at an incredible rate, 10^3 times per 10 years since the birth of computers, but this does not change the situation drastically. For example, even a rock of 1 m^3 contains about 10^{30} atoms while the largest computer at present has a memory size of about 10^{12} bytes, which implies that we have to wait for nearly 60 years before we can get a large enough memory size, even if the high growth rate continues.

The scale gaps are a headache for experimentalists, too. Experiments in a laboratory are generally too small in size to mimic the whole real geophysical phenomena, so the targets are inevitably limited to very small elementary processes.

The only steady scientific approach in earth sciences is observations. Observation data can be accumulated which are nearly proportional to our efforts; however, we also meet with a major difficulty, that is, large fluctuations. In the real world there are no two events which occur with identical initial and boundary conditions, so the separation of noises from the data is generally not easy. According to the theory of chaos, a small initial perturbation can change the whole system, and it is very likely that many of the macroscopic real phenomena such as atmospheric turbulence, are under chaotic regimes. Therefore, the separation of noises is especially hard in the earth sciences and we have to tackle the large fluctuations.

The concept of fractals, or scale-invariance, has changed the situations. In the case where we check fractality for objects in the real world, the accuracy achieved is better for objects with wider observation scale ranges. So, the gigantic scales have become attractive characteristics from the point of view of fractals.

The success of fractals in natural sciences is due to the robustness of analysis methods. For example, the box-counting method is so powerful that no process of noise separation is generally required when the fractal dimension has been obtained. This strong applicability is especially favorable for data in earth sciences with large fluctuations as typically seen in the case of coastlines.

Fractals are also encouraging computer simulations in earth sciences. As known from the well-established simulation of diffusion-limited aggregation, we can simulate with a reasonably constructed computer model that is defined on a macroscopic scale. We can totally forget the atomic dynamics; moreover, we do not have to use differential equations. Roughly speaking, the scale-invariance guarantees that we can model fractal objects at any scale length, so the gigantic scales are now not serious troubles and the present computers can produce a lot of meaningful results.

For fractal researchers, the field of earth sciences is very fertile. There are many empirical laws showing fractal behaviors which are not established firmly. For example, for earthquakes it is known that the space distribution of epicenters is fractal, the frequency of aftershocks decay following a power law in time, and the energy distribution also follows a power law, namely, earthquakes are fractal in space-time-energy. These laws have been confirmed by observation quite widely; however, computer models of earthquakes are still in bud.

In this section several topics are selected from earth sciences: Earthquakes, atmospheric turbulence, ocean waves, rivers and landscapes, rock fractures, and mineral deposits. Of course, this limited selection cannot cover all of the research areas. This is the result of a sampling from the set of current works which belong to the intersection of fractals and earth sciences.

A geologist once told me after my lecture on fractals: "I understand that fractals are so rich in nature. But, I think we have been treating them as background noises, and non-fractals have been our main research subjects." The current intensive researches on fractals will soon reverse the ground, from the background to the foreground!

***by* H. Takayasu**

Fractals, Vol. 1, No. 3 (1993) 491–512

FRACTAL TECTONICS AND EROSION

DONALD L. TURCOTTE
Department of Geological Sciences,
Cornell University, Ithaca, NY 14853, USA

Abstract

Tectonic processes build landforms that are subsequently destroyed by erosional processes. Landforms exhibit fractal statistics in a variety of ways; examples include (1) lengths of coast lines; (2) number-size statistics of lakes and islands; (3) spectral behavior of topography and bathymetry both globally and locally; and (4) branching statistics of drainage networks. Erosional processes are dominant in the development of many landforms on this planet, but similar fractal statistics are also applicable to the surface of Venus where minimal erosion has occurred. A number of dynamical systems models for landforms have been proposed, including (1) cellular automata; (2) diffusion limited aggregation; (3) self-avoiding percolation; and (4) advective-diffusion equations. The fractal statistics and validity of these models will be discussed. Earthquakes also exhibit fractal statistics. The frequency-magnitude statistics of earthquakes satisfy the fractal Gutenberg-Richter relation both globally and locally. Earthquakes are believed to be a classic example of self-organized criticality. One model for earthquakes utilizes interacting slider-blocks. These slider block models have been shown to behave chaotically and to exhibit self-organized criticality. The applicability of these models will be discussed and alternative approaches will be presented. Fragmentation has been demonstrated to produce fractal statistics in many cases. Comminution is one model for fragmentation that yields fractal statistics. It has been proposed that comminution is also responsible for much of the deformation in the earth's crust. The brittle disruption of the crust and the resulting earthquakes present an integrated problem with many fractal aspects.

1. INTRODUCTION

In this paper a variety of applications of fractals to tectonics and erosion will be discussed. Both self-similar and self-affine fractals will be considered. There are several definitions of self-similar fractals but the simple number-length scaling can form the basis to almost all natural applications; this can be written:

$$N = Cr^{-D} \tag{1}$$

where D is the fractal dimension, and N is the number objects with a linear dimension r for a discrete distribution and the number with a linear dimension greater than r for a continuous distribution.

The concept of fractals was introduced in terms of the length of the west coast of Great Britain by Mandelbrot[1] and his result is given in Fig. 1(a). The length P of the coast line is given as a function of the length r of the measuring rod. Similar results are obtained for the length of topographic contours on topographic maps, three examples being given in Fig. 1 for diverse geological settings. This application of fractals illustrates several features of most applications to natural phenomena. First, the applicable range for fractal statistics has both upper and lower bounds. Second, the fractal dimension is not diagnostic and the fractal dimensions of topographic contours are not sensitive to the geological setting or age. The fractal dimension of relatively smooth old mountains (i.e., the Adirondacks) is not significantly different from that of jagged young mountains (i.e., the Transverse Ranges).

Another striking application of self-similar fractals is to the frequency-size distribution of earthquakes. Seismicity is clearly the result of complex interactions between faults over a wide range of scales. Despite the complexity, there is order — in particular, the frequency-magnitude statistics of earthquakes are given by the Gutenberg-Richter relation:

$$\ln \dot{N} = -bm + a\,, \tag{2}$$

where $\dot{N}$ is the number of earthquakes per unit time in a specified region with magnitudes greater than m. The b-value is generally in the range $b = 0.9 \pm 0.1$. Equation (2) is equivalent to a fractal (power-law) relation between frequency and rupture area A[2]:

$$\dot{N} = CA^{-D/2} \tag{3}$$

which is essentially identical to the fractal relation.[1] We have $D = 2b$ so that the fractal dimension of distributed seismicity is $D = 1.8 \pm 0.2$. The fractal earthquake relation Eq. (2) is recognized to be applicable under a wide variety of tectonic settings from mid-ocean ridges to collison zones such as Tibet. Again, the fractal dimension (b or D) is not diagnostic since it has such a small variation.

Coastlines and topographic contours are examples of self-similar fractals. Topography is also an example of self-affine fractals. If a spectral analysis results in a power-law relation between the spectral power density S and the wave number k[3]:

$$S = Ck^{-\beta} \tag{4}$$

a self-affine fractal result with constraints on the value of β and $D = \frac{1}{2}(5 - \beta)$. Spectral expansions of the global topography of the earth[4] and Venus have been carried out, the results being given in Fig. 2. In both cases, a good correlation with $\beta = 2$ ($D = 1.5$) is found, i.e., Brown noise. This is somwhat surprising since erosion is dominant in the

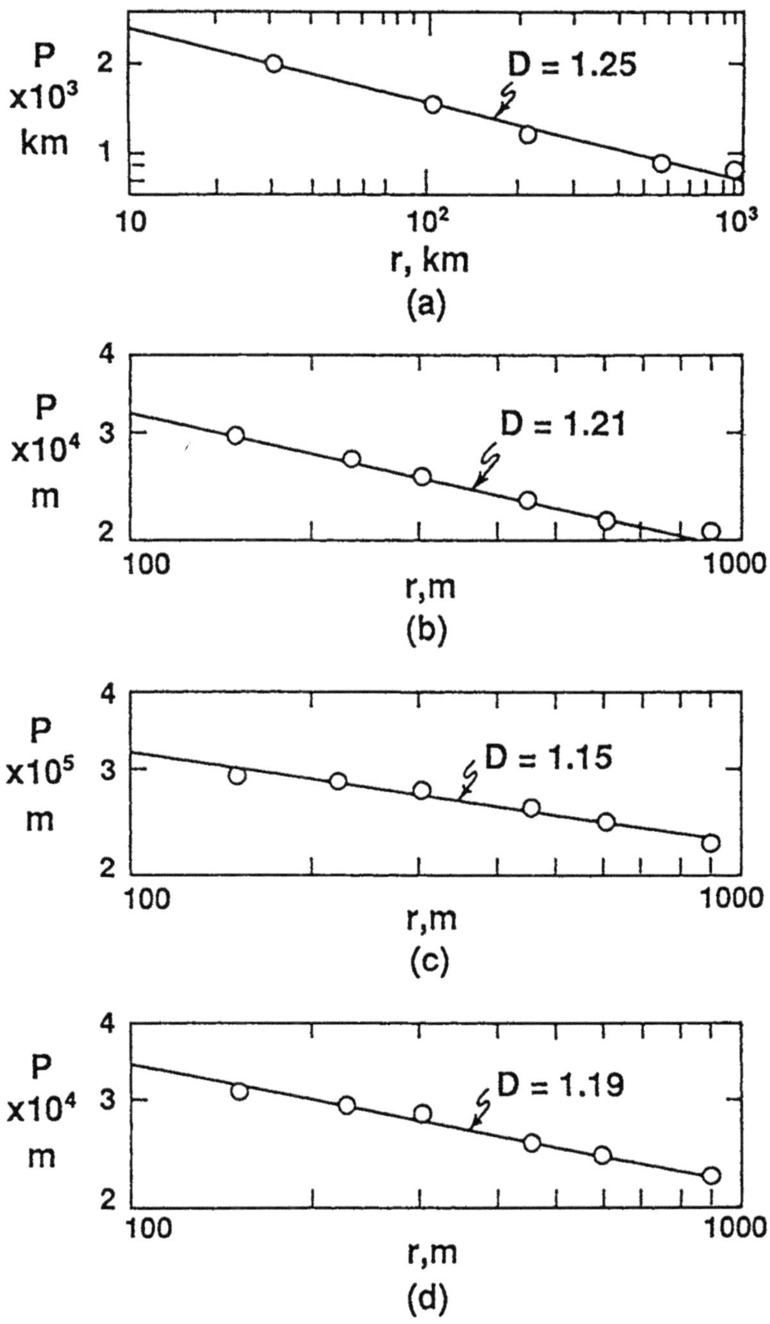

Fig. 1 (a) Length P of the west coast of Great Britain as a function of the length r of the measuring rod [1]; (b) to (d) lengths P of specified topographic contours in several mountain belts as a function of the length r of the measuring rod; (b) 3000 ft. contour of the Cobblestone Mountain quadrangle, Transverse Ranges, CA; (c) 10 000 ft. contour of the Byers Peak quadrangle, Rocky Mountains, Colorado; (d) 1000 ft. contour of the Silver Bay quadrangle, Adirondack Mountains, NY. Correlations are with the fractal relation [Eq. (1)].

evolution of many landforms on the earth, while erosion is essentially absent on Venus so that tectonic processes are dominant. This suggests that the tectonic processes that build topography and the erosional processes that destroy topography both give Brown noise statistics. Once again the fractal dimension of topography does not appear to be diagnostic.

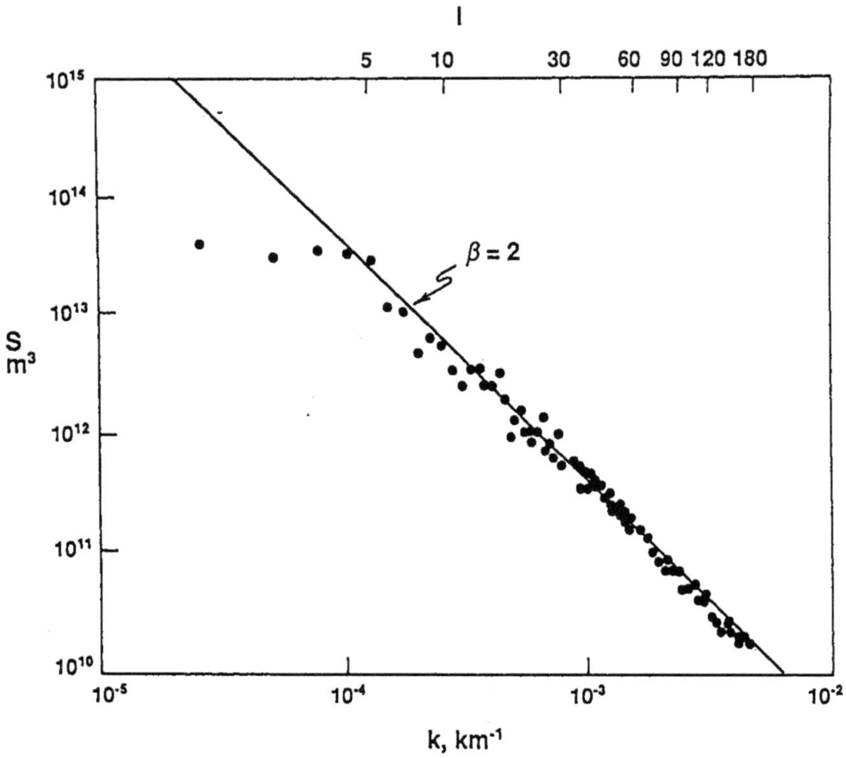

(a)

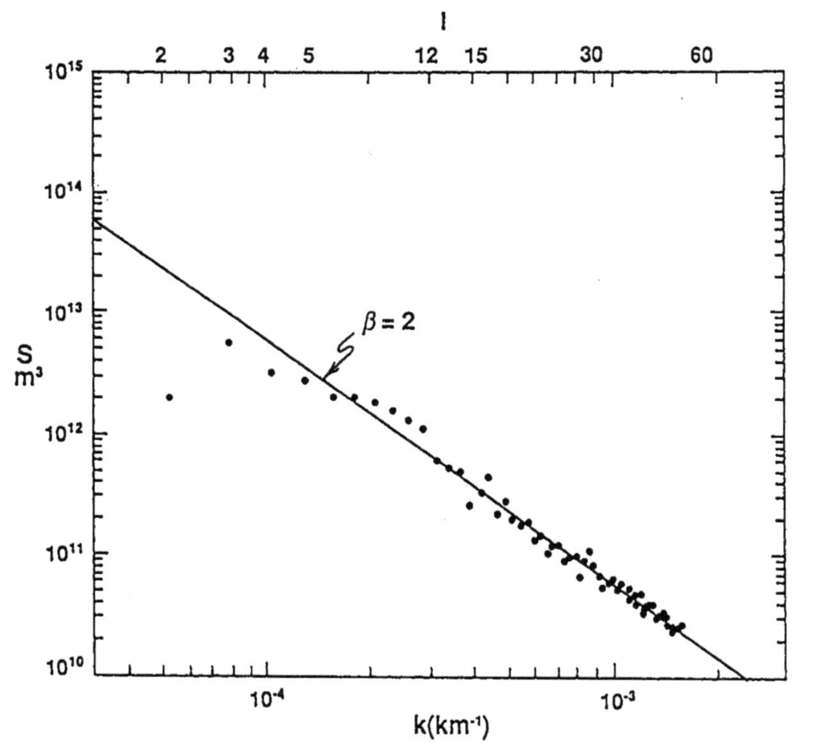

(b)

Fig. 2 Power spectral density, S, as a function of wave number, k, for spherical harmonic expansions of topography (degree ℓ) for (a) the earth and (b) Venus. Correlations are with [Eq. (4)] taking $\beta = 2$.

2. DRAINAGE NETWORKS

Essential features of most landforms on the earth are drainage networks. Drainage networks are classic examples of fractal trees. It is standard practice to use the Strahler[5] ordering system. When two like-order streams meet they form a stream with one higher order than the original. Thus two first-order streams combine to form a second-order stream, two second-order streams combine to form a third-order stream, and so forth. The bifurcation ratio R_b is defined by:

$$R_b = \frac{N_n}{N_{n+1}} \tag{5}$$

where N_n is the number of streams of order n. The length-order ratio R_r is defined by:

$$R_r = \frac{r_{n+1}}{r_n} \tag{6}$$

where r_n is the mean length of streams of order n. Both R_b and R_r are found to be nearly constant for a range of stream orders in a drainage basin.[6] From Eq. (1), the fractal dimension of a drainage network is:

$$D = \frac{\ln R_b}{\ln R_r} \tag{7}$$

An example of a drainage network is given in Fig. 3(a); this is the drainage network in the Volfe and Bell Canyons, San Gabriel Mountains near Glendora, California.[7] The number-

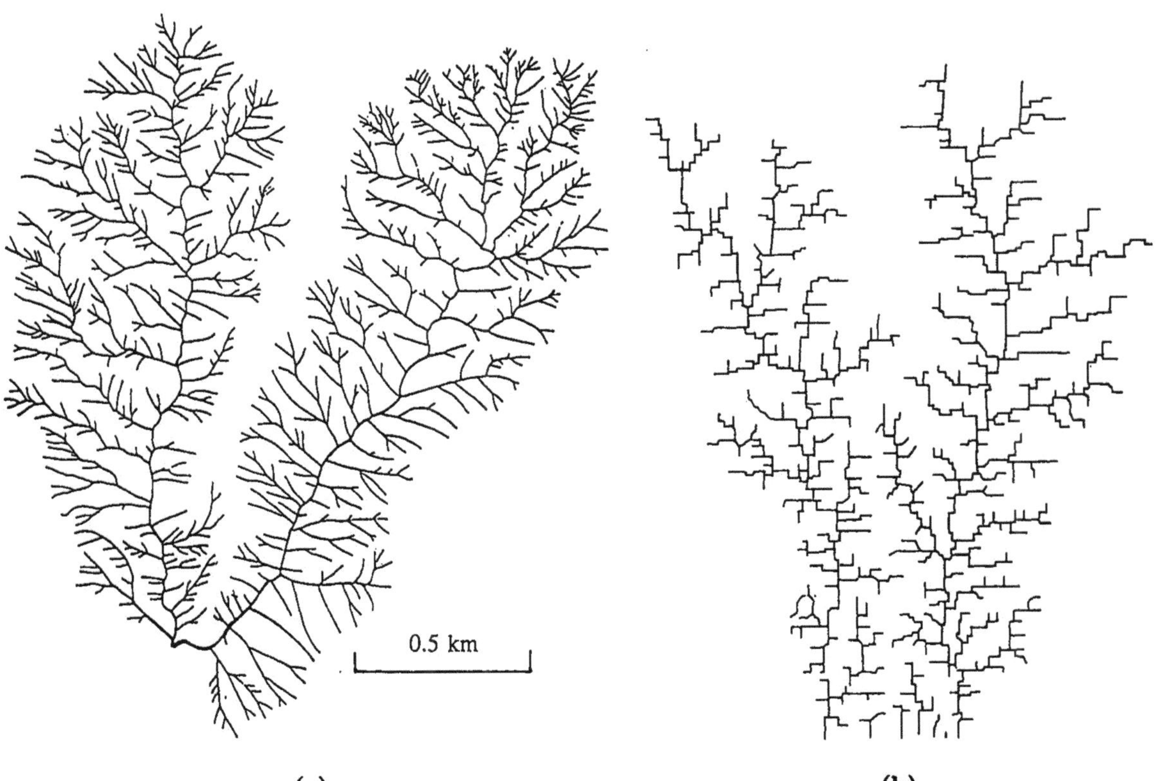

(a)

(b)

Fig. 3 **(a)** The drainage network in the Volfe and Bell Canyons, San Gabriel Mountains, near Glendora, California obtained from field mapping. **(b)** Illustration of a DLA model for a drainage network.

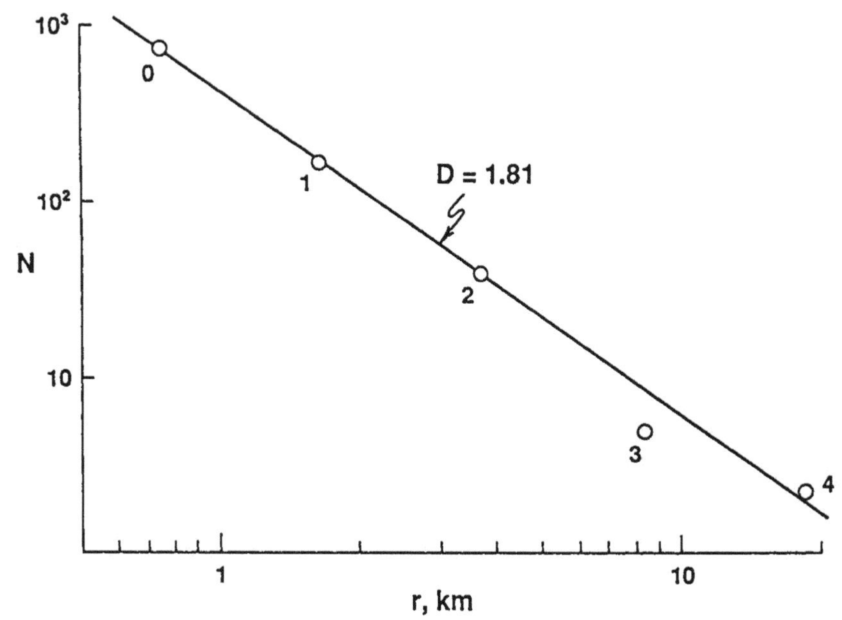

(a)

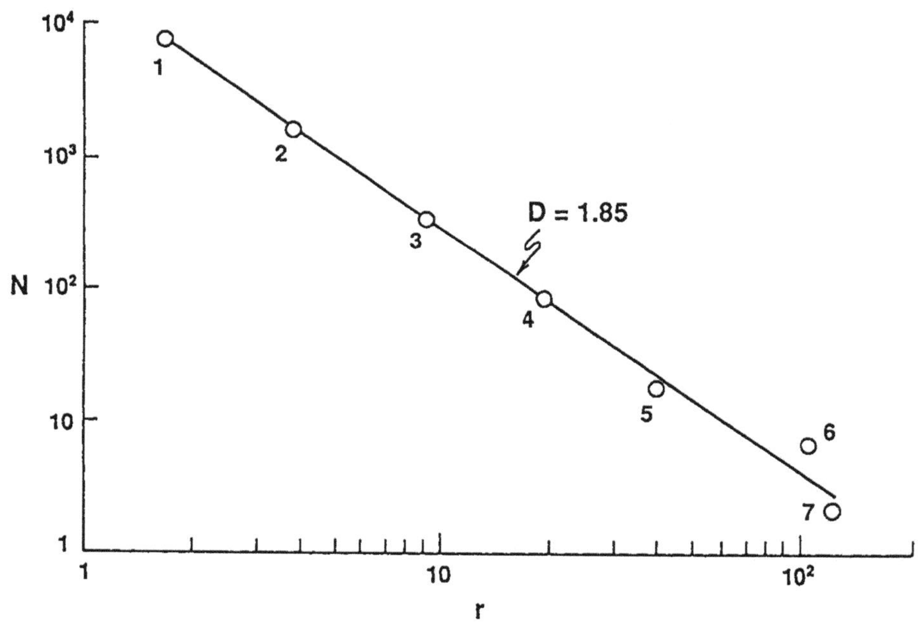

(b)

Fig. 4 Dependence of the number of streams of various orders on their mean length for (a) the example illustrated in Fig. 3(a) and (b) the model illustrated in Fig. 3(b).

length statistics for this network are given in Fig. 4(a); a good correlation with Eq. (7) is obtained taking $D = 1.81$. Leopold et al.[84] have obtained similar results for the entire United States; a good correlation with Eq. (7) was also found with $D = 1.83$. Drainage networks are in general fractal with little variation in the fractal dimension. Once again the fractal dimension of the drainage network is not diagnostic of its geological setting.

A variety of models have been proposed to simulate drainage networks. Leopold and Langbein[8] and Schenck[9] hypothesized that streams follow random walks, but the results bore little resemblance to actual drainage networks. Other probabilistic models for the generation of channel networks were introduced by Shreve,[10,11] Scheidegger,[12] Smart and Surkan,[13] Seginer[14] and Howard.[15] Smart[16] reviewed much of this work.

In the past five years there has been a rebirth of interest in the problem. Kondoh and Matsushita,[17] Meakin et al.[18] and Masek and Turcotte[19] have introduced diffusion-limited aggregation (DLA) models. Stark[20] developed a model based on self-avoiding percolation clusters. Chase[21] and Inaoka and Takayasu[22] introduced models based on self-organized criticality. Willgoose et al.[23] and Kramer and Marder[24] developed advection-diffusion models based on equations that yield a network structure.

In their DLA model, Masek and Turcotte[19] considered a square grid of points and introduced seed cells on one boundary of the grid. Cells are allowed to accrete to the evolving network if one (and only one) of the four nearest neighbor cells is part of the network. A cell is introduced randomly on the grid and follows a random walk on the grid until it either (i) accretes to the evolving grid, (ii) is lost from the grid, or (iii) occupies a forbidden site. A simulation carried out on a 256×256 grid is illustrated in Fig. 3(b); 20 000 random walkers have been introduced. The simulated and actual drainage networks are reasonably similar. The number-length statistics for the simulated network are given in Fig. 4(b); a good correlation with Eq. (7) is obtained taking $D = 1.85$. Again good agreement is obtained between the simulated and real networks.

The random walkers can be interpreted as floods that flow over a relatively flat surface until they find a gully. When the flood enters the gully, it further erodes the gully and extends the network, headwards. This type of headward gully evolution has often been proposed for actual drainage networks.[25]

3. FLOOD STATISTICS

An important question in geomorphology concerns which floods dominate erosion. Is erosion dominated by the 10 year, the 100 year, or the very largest floods? The answer to this question depends upon whether extreme flood probilities have an exponential or power-law dependence on time. Flood-frequency statistics also have a variety of other implications. Land-use regulations and flood-control projects are based on extrapolations for future floods. A wide variety of statistical techniques have been applied to the evaluation of the flood hazard, but the data base for establishing the validity of the statistics is quite limited. A speculative hypothesis is that floods also obey fractal statistics.[26]

The peak discharge during a flood $\dot{V}$ is a measure of its intensity. If floods are fractal then:

$$\dot{V} = CT^H \tag{8}$$

where $\dot{V}$ is the peak discharge in a time interval T (under some circumstances, H can be referred to as the Hausdorff measure). This scale invariant distribution can also be expressed in terms of the flood intensity factor F, the ratio of the peak discharge over a 10-year interval to the peak discharge over a one-year interval. With self-similarity the parameter F is also the ratio of the 100-year peak discharge to the 10-year peak discharge. The parameters H and F can be related by:

$$F = 10^H \tag{9}$$

Care must be taken in defining individual floods and avoiding annual variability.

As two specific examples, we consider station 1-1805 on the Middle Banch of the Westfield River in Goss Heights, MA for the period 1911–1960, and station 11-0980 in the Arroyo Seco near Pasadena, CA for the period 1914–1965. The reason that these stations were chosen is that they were used as benchmarks by Benson[27] who applied a variety of geostatistical distributions to flood-frequency forecasting. Floods are considered independent only if the peak flows are separated by more than one month. For a 50-year record, the 50 largest values of $\dot{V}_m$ are ordered, the largest $\dot{V}_m$ is assigned a period $T = 50$ years; the second largest flood a period $T = 50/2 = 25$ years; the third largest flood a period $T = 50/3 = 16.7$ years, and so forth. The results for the two stations are given in Fig. 5.

For station 1-1805, the fractal fit (F) using Eqs. (8) and (9) gives $H = 0.51$ and $F = 3.3$. For station 11-0980, the fractal fit gives $H = 0.87$ and $F = 7.4$. Also included in these figures are six statistical correlations given by Benson[27]: the two parameter gamma (Ga), Gumbel (Gu), log Gumbel (LGu), log normal (LN), Hazen (H), and log Pearson type III (LP). For large floods, the fractal predictions (F) correlate best with the log Gumbel (LGu), while the other statistical techniques predict longer recurrence time for very serious floods. The log Gumbel is approximately a power-law correlation, whereas the others are approximately exponential. The log Pearson type III (LP) is the federally approved distribution for evaluating the flood hazard in the United States. As can be seen in Fig. 5; if the fractal relation is valid, then the federally approved log Pearson type III distribution substantially underestimates the severity of the 100-year flood.

The values of the Hausdorff measure H and flood intensity factor$F$ for the ten benchmark stations considered by Benson[27] are given in Table 1. These results show that there are clear regional trends in the values of F. The values in the southwest are systematically high; this can be attributed to the arid conditions and the rare tropical storm that causes severe flooding. The values in the Pacific northwest are low; this can be attributed to the maritime climate. Since F is equivalent to a fractal dimension, $D = 2 - \log F$, this may be a case in which the fractal dimension of floods is diagnostic of climate.

Table 1 Values of the Hausdorff Measure *H* and Flood Intensity Factor *F* for the 10 Benchmark Stations considered by Benson (1968)

Station	Site	H	F
1–1805	Westfield River, Goss Heights, MA	0.513	3.31
2–2185	Oconee River, Greenboro, GA	0.540	3.47
5–3310	Mississippi River, St. Paul, MN	0.470	2.95
6–3440	Little Missouri River, Alzado, WY	0.520	3.31
6–8005	Elkhorn River, Waterloo, NE	0.540	3.47
7–2165	Mora River, Golondinas, NM	0.630	4.27
8–1500	Llano River, Junction, TX	0.719	5.24
10–3275	Humboldt River, Comus, NV	0.616	4.13
11–0980	Arroyo Seco, Pasadena, CA	0.875	7.4
12–1570	Wenatchee River, Plain, WA	0.310	2.04

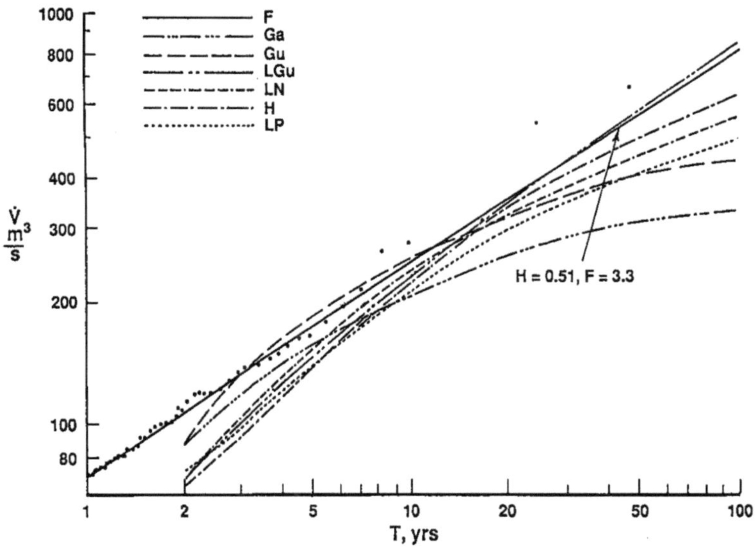

(a)

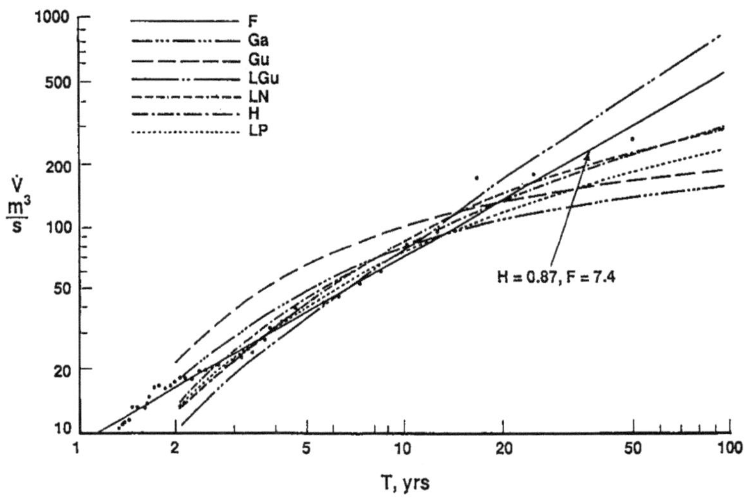

(b)

Fig. 5 Peak discharge $\dot{V}$ during a flood associated with the period T. **(a)** Middle Branch of the Westfield River at station 1-1805 in Goss Heights, MA; **(b)** Arroyo Seco at station 11-0980 near Pasadena, CA. Correlation lines are fractal (F), two parameter gamma (Ga), Gumbel (Gu), log Gumbel (LGu), log normal (LN), Hazen (H) and log Pearson type III (LP).

4. SEDIMENTARY STATISTICS

The statistics of floods are directly related to the statistics of the sedimentary record. An important question is whether erosion and sedimentation are dominated by the very largest floods. With power-law flood statistics, the very largest floods are likely to be dominant. With exponential flood statistics, intermediate sized floods will dominate.

Sedimentary rocks are generally composed of a sequence of layers, each layer representing a distinct sedimentation event. Each layer is often composed of an upward gradation from coarse-grained sediments to fine-grained sediments; individual layers are generally separated by well defined bedding planes. Depending on the sedimentary environment, each sedimentary event may be associated with different causes. In some cases sediments may be directly deposited in floods. Sediments may be redeposited by storms or slumps. Some sedimentary sequences have a fractal distribution of layer thicknesses, others do not.

Several recent studies of the thickness statistics of turbidite deposits have been carried out. Rothman et al.[28] carried out direct measurements on an outcrop of the Kingston Peak Formation near the southern end of Death Valley, CA. These results are given in Fig. 6(a); an excellent correlation with Eq. (1) is obtained taking $D = 1.39$. Hiscott et al.[29] have studied a volcaniclastic turbidity deposit in the Izu-Bonin forearc basin off shore of Japan. Layer thicknesses were obtained from formation-microscanner images in the middle to upper Oligocente part of the section. Results for two DSDP holes located 75 km apart are given in Fig. 6(b); a good correlation with Eq. (1) is obtained taking $D = 1.12$.

Turbidite deposits are associated with slumps off the continental margin. These avalanche-like events can be directly associated with the sandslide analogy to self-organized criticality.[28] Thus the fractal distribution of turbidite layers can be directly associated with a self-organized critical mechanism.

Fractal correlations of sedimentary sequences are not restricted to turbidite deposits. Stolum[30] has obtained fractal statistics for the Middle Jurassic Tilje Formation in the North Sea Haltenbanken basin; these sediments were deposited in a marine shelf environment. He found values of $D = 0.71, 0.80$. Malamud and Turcotte[31] obtained fractal statistics for sandy-bed thicknesses in the shallow marine environment of the Late Devonian Ithaca Formation, New York, with $D = 1.41$.

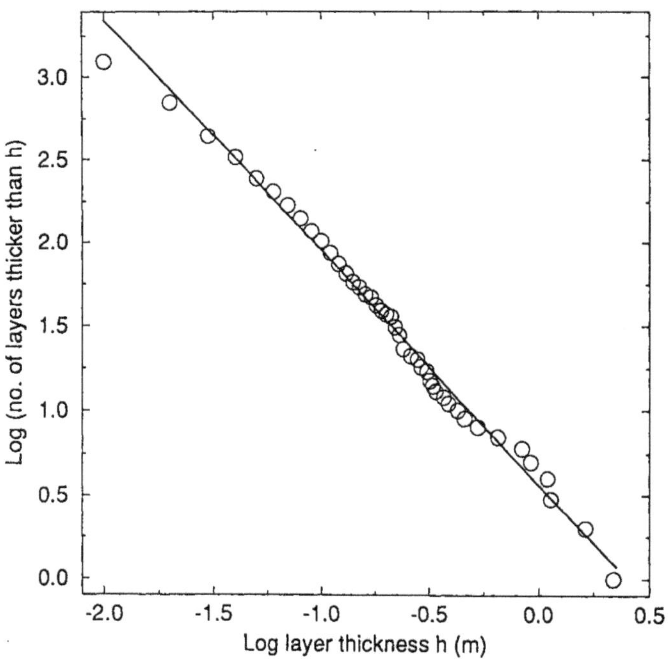

Fig. 6 Frequency-thickness statistics for turbidite sequences of sedimentary layers. (a) Kingston Peak Formation near the southern end of Death Valley, CA.

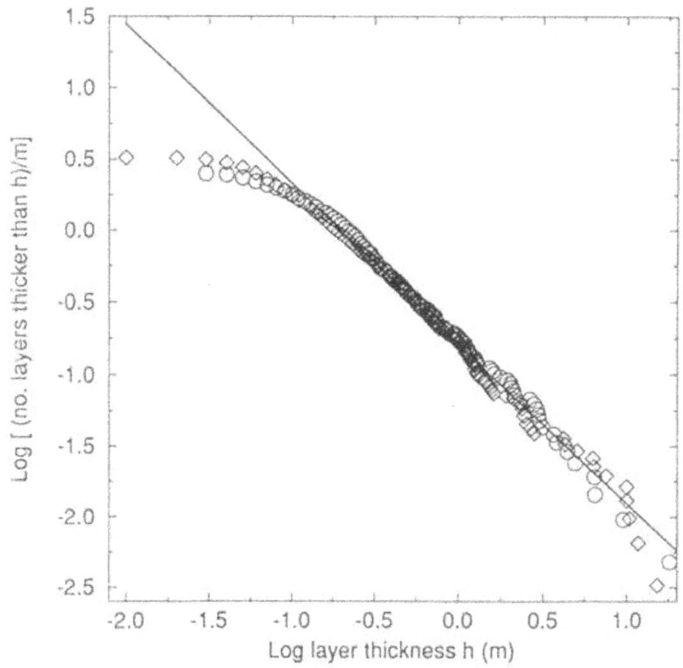

Fig. 6(b) Izu-Bonin forearc basin off shore of Japan.

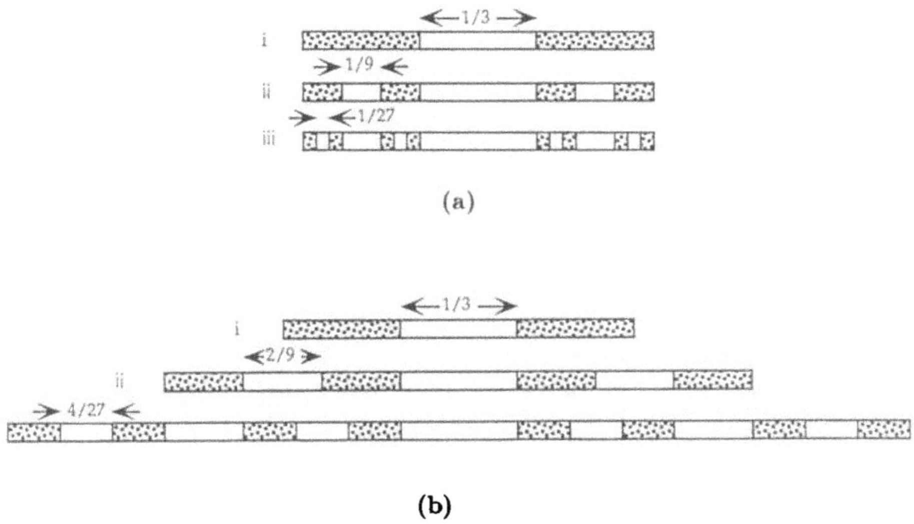

Fig. 7 (a) Cantor set $N_1 = 1$, $r_1 = 1/3$; $N_2 = 2$, $r_2 = 1/9$; $N_3 = 4$, $r_3 = 1/27$; $D = \ln 2/\ln 3 = 0.6309$. (b) Stretched Cantor set $N_1 = 1$, $r_1 = 1/3$; $N_2 = 2$, $r_2 = 2/9$; $N_3 = 4$, $r_3 = 4/27$; $D = \ln 2/\ln(3/2) = 1.710$.

In several cases, fractal dimensions for thickness statistics are greater than one. The constructions illustrated in Fig. 7 show that D can in fact be greater than one. The standard Cantor set is illustrated in Fig. 7(a); one layer, $N_1 = 1$, with thickness $r_1 = 1/3$, two layers, $N_2 = 2$, with thickness $r_2 = 1/9$, four layers, $N_3 = 4$, with thickness $r_3 = 1/27$. From Eq. (1), $D = (\ln 2)/(\ln 3) = 0.6309$. In Fig. 7(b), a stretched cantor set is illustrated. At each step the remaining segments are stretched by a factor of two before being further subdivided. This gives $N_1 = 1$ with $r_1 = 1/3$, $N_2 = 2$ with $r_2 = 2/9$, $N_3 = 4$ with $r_3 = 4/27$.

From Eq. (1), $D = (\ln 2)/[\ln(3/2)] = 1.710$. The length L of the set is unbounded; $L \to \infty$ as $r \to 0$. For real linear data sets with both upper and lower bounds on r, this construction illustrates that values of D greater than one are acceptable.

The fractal behavior of stratigraphic sequences has also been demonstrated using spectral techniques. Hewett[32] gave results for a density-porosity log in a well through a late Miocene-early Pleistocene sandstone formation deposited in a deep submarine fan. He showed that the spectra of the well log correlated with Eq. (2) taking $\beta = 0.71$. Similar results have been reported by Todoeschuck and Jensen[33] and by Todoeschuck et al.[34]

Hewett[32] also developed a fractal based interpolation technique for determining the porosity distribution in reservoirs. The vertical spectra was taken from analyses of well logs, the horizontal spectra from the typical Brown noise spectra of topography. The phases in the three-dimensional spectral expansion were constrained from well-log data; accurate and realistic sedimentary structures were found. These results provide further evidence that sedimentary structures obey fractal statistics.

An interesting possibility is that the statistics of sedimentary layers can be directly related to the statistics of floods. If this can be done it will provide an improved basis for flood hazard assessments. It may also be possible to study paleoclimates in this way.

5. TECTONICS AND SELF-ORGANIZED CRITICALITY

Despite limits and possible variations, the near universal applicability of the fractal relation [Eq. (2)] to regional seismicity is a remarkable feature. A simple explanation is that seismicity does not have a characteristic length scale and thus must be self-similar. The only statistical distribution that is self-similar is a power law. But this would not explain why the fractal dimension is nearly constant under wide variations in tectonic regime.

It is generally accepted as a working hypothesis that each fault is associated with a characteristic earthquake. Thus a fractal distribution of earthquakes implies a fractal distribution of faults. It does not follow, however, that the fractal dimension for the frequency-size distribution of faults is the same as that for earthquakes since this would imply that the interval of time between earthquakes is independent of scale.

Sammis et al.[35] have proposed that a comminution model for fragmentation is applicable to tectonic deformation. This hypothesis states that the direct contact between two blocks of near equal size during tectonic deformation will result in the breakup of one of the blocks. It is unlikely that small blocks will break large blocks or that large blocks will break small blocks. A discrete model for comminution is given in Fig. 8. Two diagonally opposed blocks are retained at each scale so that no two blocks of equal size are in contact with each other. We have two blocks with $r = 1/2$ and 12 blocks with $r = 1/4$ so that $D = \ln 6/\ln 2 = 2.585$. Turcotte[36] has shown that many experimental studies of fragmentation yield fractal frequency-size distributions with $D \approx 2.5$. Considering each side of a block in Fig. 8 to be a fault, we have a fractal distribution of fault sizes with $D = \ln 3/\ln 2 = 1.585$. Barton and Hsieh[37] considered the distribution of exposed joints and fractures near Yucca Mountain; Nevada and found good agreement with a fractal distribution of block exposure taking $D = 1.6$.

The concept of self-organized criticality was introduced by Bak et al.[38] and is defined to be a natural system in a marginally stable state, which when perturbed from that state it will evolve naturally back to the state of marginal stability. Energy input to the system is

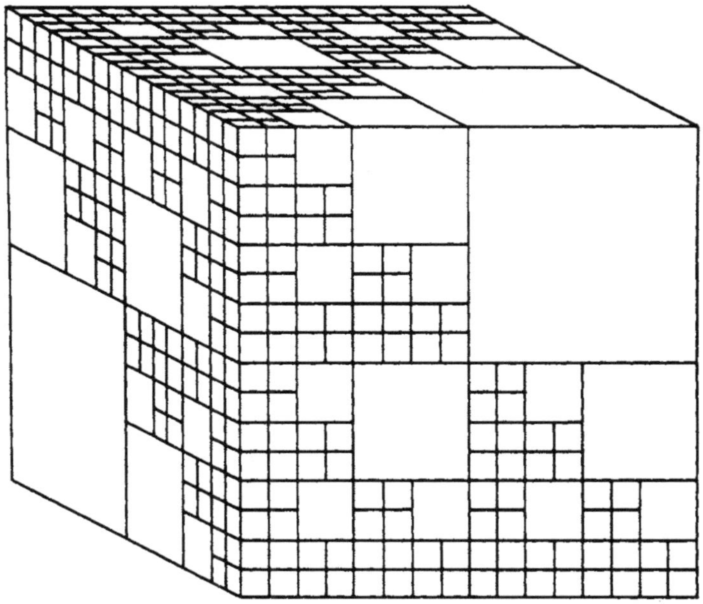

Fig. 8 Illustration of a discrete model for comminution. Diagonally opposite blocks are retained at each scale. The blocks satisfy a fractal relationship with $D = \ln 6/\ln 2 = 2.585$. The block surfaces satisfy a fractal relationship with $D = \ln 3/\ln 2 = 1.585$.

continuous but the energy loss is in a discrete set of events that satisfy fractal frequency-size statistics. Distributed seismicity is often taken as the classic example of a natural system that exhibits self-organized criticality. There is a continuous input of energy (strain) through the relative motion of tectonic plates. This energy is dissipated in a fractal distribution of earthquakes. Scholz[39] has argued that the earth's crust is in a state of self-organized criticality. He makes the point that wherever a large dam is built, induced seismicity results from the filling of the reservoir. Thus the crust is everywhere on the brink of failure. Similar arguments have been given by Sornette et al.[40]

In the original cellular-automata model for self-organized criticality, a two-dimensional grid of boxes was considered. Particles were added randomly to the boxes until there were four particles in a box. The box was then considered unstable and the four particles were redistributed to the four adjacent boxes. If any of those boxes had four or more particles, a further redistribution of elements was carried out. Particles were lost only from the sides of the grid.

In the steady state, the number of particles in the grid of boxes fluctuates about a mean value that is considerably less than the maximum allowed value $3N^2$. The behavior of the model is characterized by the statistical frequency-size distribution of events. The size of a multiple redistribution event can be quantified in several ways. One measure is the number of boxes N_b that become unstable in a multiple event. Another measure is the number of particles N_p lost from the grid in a multiple event. Bak et al.[38] and Kadanoff et al.[41] have carried out extensive studies of the behavior of this model as a function of grid size. They found that the number of events N of a particular size, either N_b or N_p, satisfies the fractal relation [Eq. (2)] with $D \approx 2$.

Bak et al.[38] made an analogy between their cellular-automata model and a sand pile on a table. Sand grains are dropped on the pile continuously and are lost from the table in

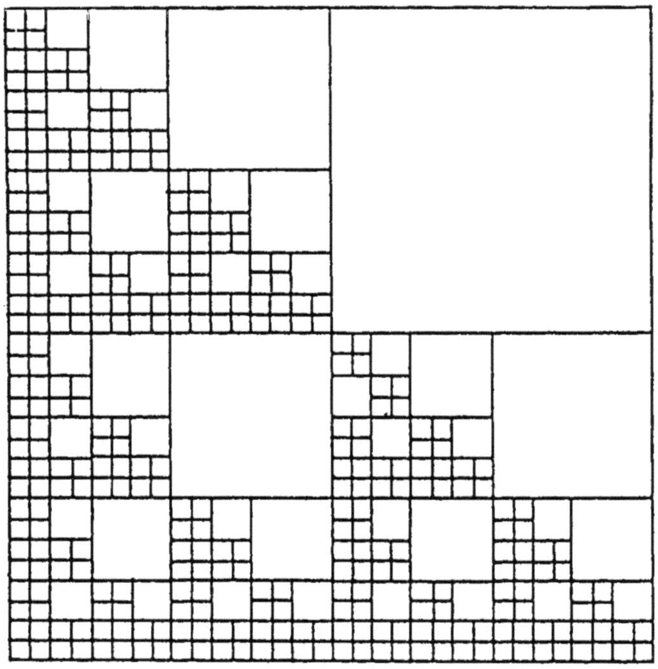

Fig. 9 Illustration of the fractal cellular-automata model corresponding to the discrete model for comminution illustrated in Fig. 8 carried to fifth order, $D = \ln 3/\ln 2 = 1.585$.

discrete sand slides; some sand slides are large and some are small. A number of groups have studied actual sandpiles and in some cases fractal distributions of sandslides have been found.[42,43]

In order to model crustal seismicity, Barriere and Turcotte[44] introduced a cellular-automata model in which the boxes have a scale-invariant distribution of sizes. The objective was to model a scale-invariant distribution of fault sizes. When a redistribution from a box occurs, it is equivalent to a characteristic earthquake on the fault. A redistribution from a small box (a foreshock) may trigger an instability in a large box (the main shock). A redistribution from a large box always triggers many instabilities in the smaller boxes (aftershocks).

A model of this type is illustrated in Fig. 9. A square box is divided into four equal sized boxes at first order. At second order one box is retained and the other three are further divided into four boxes; the model has been extended to 5th order with a 32×32 grid. This model corresponds to the discrete model for comminution illustrated in Fig. 8. For this example we have $N_1 = 108$ for $r_1 = 1$; $N_2 = 27$ for $r_2 = 2$; $N_3 = 9$ for $r_3 = 4$; $N_4 = 3$ for $r_4 = 8$; and $N_5 = 1$ for $r_5 = 16$. Except for N_1 the N_r are related to the r_i by Eq. (2) with $D = \ln 3/\ln 2 = 1.5850$, in agreement with our previous discussion of Fig. 8.

We apply the standard cellular automata rules to this model:

1. Particles are added one at a time to randomly selected boxes. The probability that a particle is added to a box is proportional to the area A of the box.
2. A box becomes unstable when it contains $4A$ particles.
3. Particles are redistributed to immediately adjacent boxes or are lost from the grid. The number of particles redistributed to an adjacent box is proportional to the linear dimension of the box.

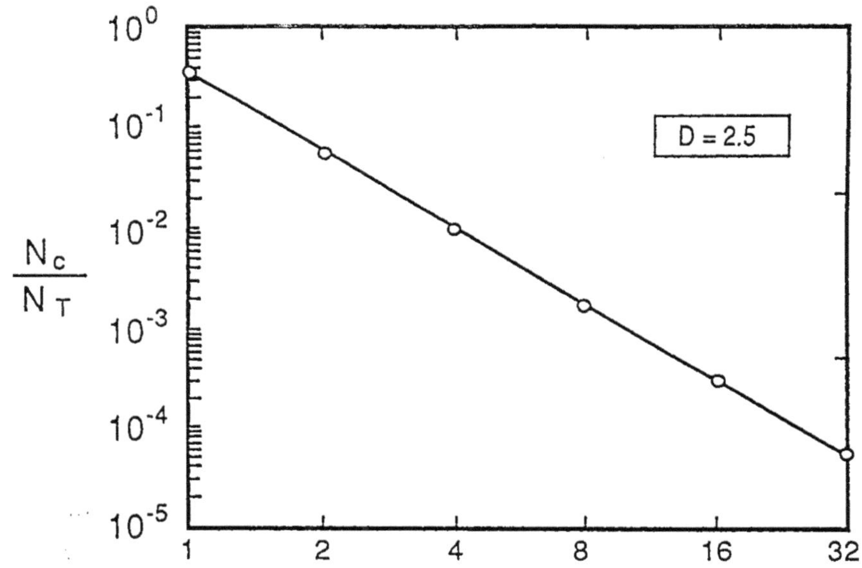

Fig. 10 Cumulative frequency-magnitude statistics for unstable events. The number of events N_c in boxes equal to or smaller than r is divided by the total number of events N_T and given as a function of r. The correlation is with Eq. (1).

4. If, after a redistribution of particles from a box, any of the adjacent boxes are unstable, one or more further redistributions are carried out. In any redistribution, the critical number of particles is redistributed. Redistributions are continued until all boxes are stable.

The cumulative frequency-magnitude statistics for main shocks of a seventh order (128×128) version of the model are given in Fig. 10. We find an excellent correlation with the fractal relation [Eq. (1)] taking $D = 2.50$ ($b = 1.25$). This is significantly higher than the observed values for distributed seismicity. Evernden[45] has obtained b-values for regional seismicity and concludes that $b = 0.85 \pm 0.20$. It was also found that 31.5% of the largest events had foreshocks. This is in reasonable agreement with studies of actual earthquakes; von Seggern et al.[46] found that 21% of the earthquakes studied had foreshocks and Jones and Molnar[47] found that 44% of larger shallow earthquakes that could be recorded teleseismically had foreshocks. The aftershocks also correlate well with Eq. (1) taking $D = 2.02$ ($b = 1.01$).

6. SLIDER-BLOCK MODELS

Although the simple cellular-automata models have a number of quite remarkable similarities to distributed seismicity, they lack much of the basic physics such as stick-slip behavior and elastic rebound. A model that includes both was introduced by Burridge and Knopoff.[48] They considered a series of slider blocks which were pulled over a surface by driver springs and connected to each other by connector springs. The simplest form of this model consists of two sliding blocks coupled to each other and to a constant velocity driver by springs. If the friction law is velocity weakening, the slider-blocks exhibit stick-slip behavior. Huang and Turcotte[49–51] and Narkounskaia and Turcotte[52] have shown that any asymmetry in this model results in classical chaotic behavior; the Feigenbaum period-doubling route to chaos is observed.

The low-order system can certainly be criticized as being too simple a model for the earth's crust. However, Huang and Turcotte[50] have shown that two observed sequences of earthquakes can be simulated in a realistic manner:

1. It is observed that major fault segments (examples are the Nankai and Ecuador subduction zones and the southern locked region of the San Andreas fault) sometimes break as a single unit and sometimes experience alternating earthquakes on subsections. This behavior was duplicated using a pair of slider-blocks with nearly equal properties.
2. It is observed that the Parkfield section of the San Andreas fault experiences a near periodic sequence of events between great southern California earthquakes. This behavior was simulated using two slider-blocks with two quite different properties. Two alternative scenarios for the next great southern California earthquake were observed:

   (i) A Parkfield event triggers simultaneously the great earthquake (the generally accepted scenario); and
   (ii) the Parkfield event transfers stress to the locked section but not sufficient stress to trigger the great earthquake. Subsequent regional strain triggers the great earthquake which also results in displacements on the Parkfield section, thus a great earthquake can occur anytime after a Parkfield event.

The probabilities of the two scenarios are about equal.

A model that combines the analog features of slider blocks and the high-order aspects of cellular-automata models involves the use of many slider blocks. Carlson and Langer[53] considered linear arrays of slider blocks with each block connected by springs to the two neighboring blocks and to a constant velocity driver. They used a velocity-weakening friction law and considered up to 400 blocks. Slip events involving large numbers of blocks were observed, the motion of all blocks involved in a slip event were coupled and the applicable equations of motion had to be solved simultaneously. Although the system is completely deterministic, the behavior was apparently chaotic. Frequency-size statistics were obtained for slip events and the events fell into two groups: smaller events obeyed a power-law (fractal) relationship and an anomalously large number of large events that included all the slider blocks. This model was considered to be a model for the behavior of a single fault, not a model for distributed seismicity. The large events were associated with characteristic earthquakes on the fault and smaller events with background seismicity on the fault between characteristic earthquakes.

Nakanishi[54,55] proposed a model that combined features of the cellular automata model and the slider-block model. A linear array of slider-blocks was considered but only one block was allowed to move in a slip event. The slip of one block could lead to the instability of either or both of the adjacent blocks, which would then be allowed to slip in a subsequent step or steps, until all blocks were again stable. Brown et al.[56] proposed a modification of this model involving a two-dimensional array of blocks. Other models of this type have been considered by Takayasu and Matsuzaki,[57] Ito and Matsuzaki,[58] Langer and Tang,[59] Carlson,[60,61] Carlson et al.,[62,63] Matsuzaki and Takayasu,[64] Rundle and Brown,[65] and Shaw et al.[66]

We have also considered two-dimensional arrays of slider-blocks (Fig. 11) in some detail.[67] We assumed that during the sliding of one block, all other blocks are stationary. This

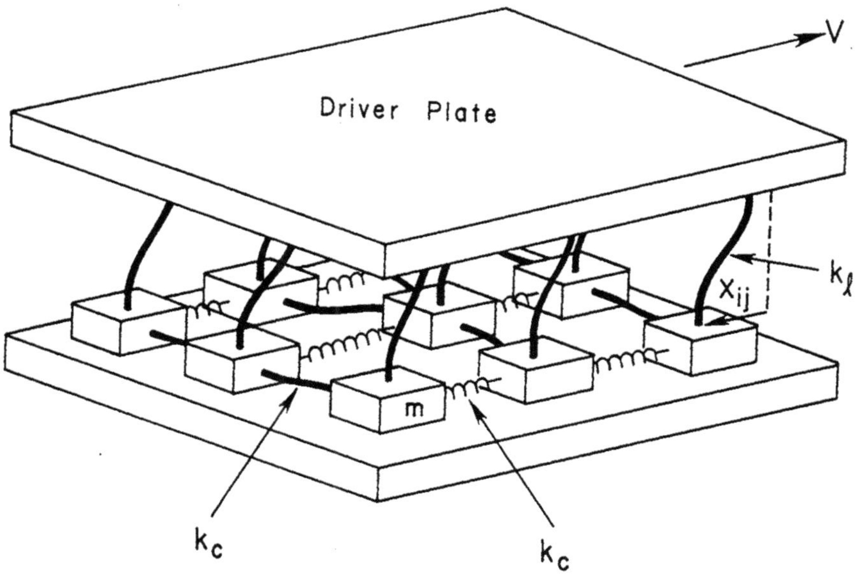

Fig. 11 Illustration of the two-dimensional slider-block model. An array of blocks each with mass m is pulled across a surface by a driver plate at a constant velocity V. Each block is coupled to the adjacent blocks with either leaf or coil spring constant k_c, and to the driver plate with a leaf spring of spring constant k_l. The extension of the (i, j) pulling spring is x_{ij}.

requirement limits the system to nearest neighbor interactions, which is a characteristic of cellular-automata systems. To minimize the complexity we considered a discontinuous static-dynamic friction law. After non-dimensionalization of the governing equations, the governing parameters are $\alpha = k_c/k_l$ (k_c is the spring constant of the connector springs, k_l is the spring constant of the puller springs), α is a measure of the stiffness of the system, $\phi = F_s/F_d$ (the ratio of the static F_s to dynamic F_d friction), and N the number of blocks considered. In this model the parameter ϕ can be eliminated by rescaling. Thus for large systems (N very large) the only scaling parameter is the stiffness α. Frequency-size statistics for a 50×50 ($N = 2500$) array are given in Fig. 12 for several values of the stiffness parameter α. A good correlation is obtained with the fractal relation [Eq. (3)] with $D = 2.72$. The frequency-size relation shows a roll-off from the power-law near the larger end of the scaling region. This deviation is reduced as the parameter α increases. Frequency-size statistics for several different size arrays are given in Fig. 13. When the parameter $\alpha/N^{1/2}$ is greater than one, we observe an excess number of catastrophic events that include the failure of all blocks. The failure statistics of these multiple-block systems clearly indicate a self-organized critical behavior and are remarkably similar to distributed seismicity.

The fundamental aspects of self-organized criticality are also directly relevant to earthquake prediction and hazard assessment strategies. The thermodynamic and statistical mechanics aspects of critical phenomena appear to be directly relevant to distributed seismicity.[68–70] The trajectories of a spring-block model in phase space are abelian, i.e., all allowed points in phase space are occupied with equal probability. This is also the case for thermodynamic systems. This strongly suggests that the earth's crust resembles a thermodynamic system near a critical point. The fluctuations are earthquakes.

In order to better understand earthquakes, it is essential to determine whether concepts of self-organized criticality are applicable. If these concepts are applicable, then all

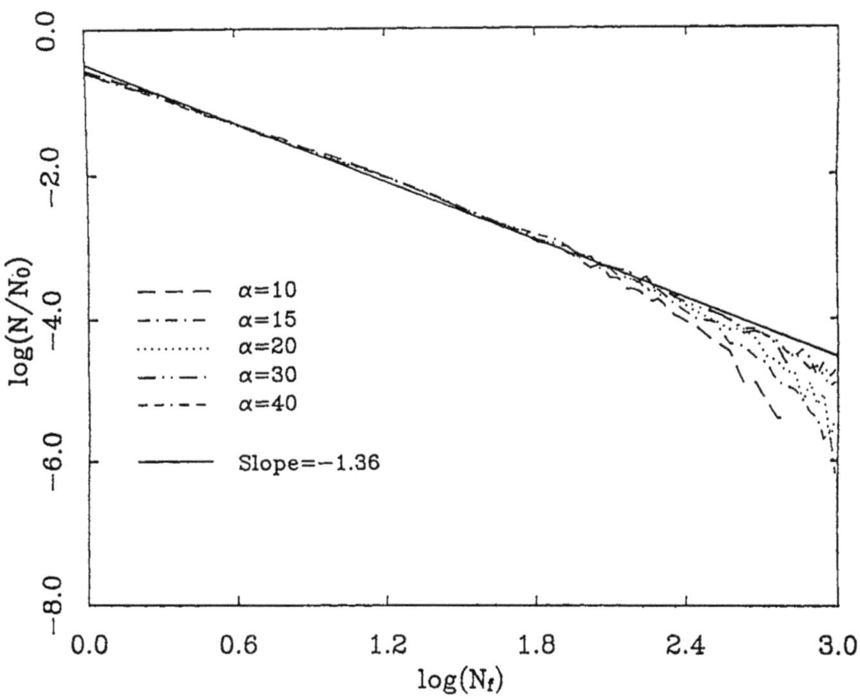

Fig. 12 The ratio of the number of events N with size N_f to the total number of events N_0 is plotted against N_f (N_f is the number of blocks that participate in an event and is a measure of the area of an event). Results are given for $\phi = 1.5$ and $\alpha = 10$, 15, 20, 30, and 40. The solid line is the correlation with the power-law (fractal) relation, Eq. (1).

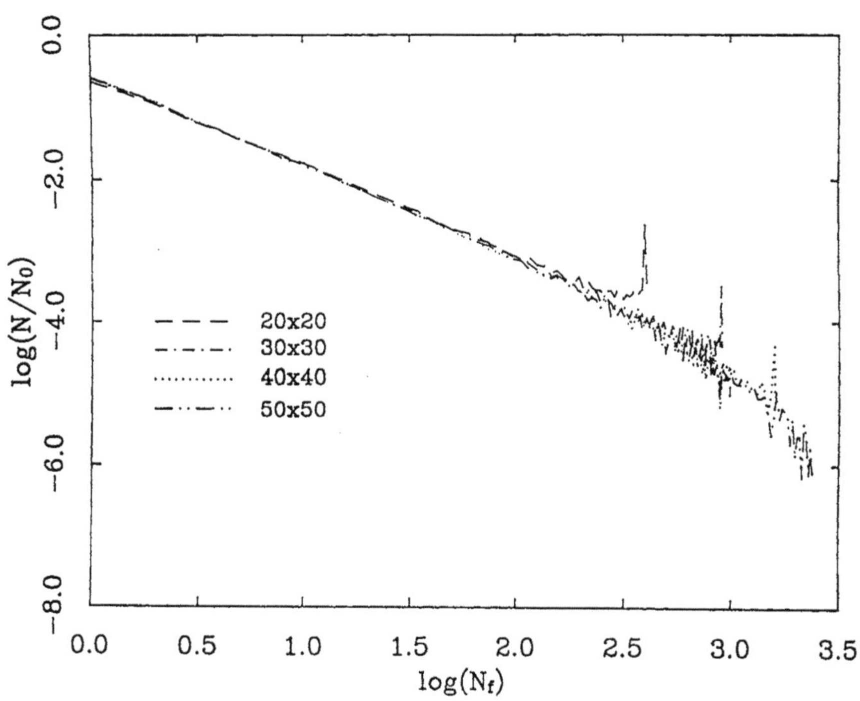

Fig. 13 The ratio of the number of events N with size N_f to the total number of events N_0 is plotted against N_f. Results are given for systems of size 20×20, 30×30, 40×40, and 50×50 with parameters $\phi = 1.5$ and $\alpha = 50$. The peaks at $\log N_f = 2.60$, 2.95, and 3.20 correspond to catastrophic events involving the entire system.

earthquakes, including the largest, represent noise on a background level of stress. Self-organized criticality may result in a near-constant level of regional seismicity between major earthquakes. The temporal variation in seismicity prior to a major earthquake is a subject of considerable controversy. Some authors accept precursory quiescence as a widely applicable phenomena.[71–73] But others support a regional buildup in seismicity as a precursor to a major earthquake. The buildup of activity in the San Francisco Bay area prior to the 1906 earthquake has been discussed by Sykes and Jaume.[74]

Precursory seismicity is the basis for intermediate-term prediction (1–5 years) pattern recognition methods developed at the International Institute for the Theory of Earthquake Prediction and Theoretical Geophysics in Moscow under the direction of Academician V.I. Keilis-Borok.[75–80] The pattern recognition includes increases in regional seismicity, increases in the clustering of earthquakes, and changes in aftershock statistics. Premonitory seismicity patterns were found for strong earthquakes in California and Nevada (algorithm CN) and for earthquakes with $M \geq 8$ worldwide (algorithm $M8$). Assuming self-similarity of the properties of seismicity, both algorithms were tested in seismically active regions. During the last 5 years, at least 3 strong earthquakes (Spitak, Loma-Prieta, Costa-Rica) were predicted in advance.

An important question is the association of large earthquakes with small earthquakes. Do large earthquakes only occur where small earthquakes occur? If so, the small earthquakes can be used to assess the regional seismic hazard. Johnson and Nava[81] have used this approach to assess the hazard in the New Madrid area. Turcotte[82,83] has discussed the approach in more general terms. A general association of large earthquakes with small earthquakes provides a rational basis for probabilistic earthquake hazard assessment. Such an approach would follow naturally from the applicability of self-organized criticality.

ACKNOWLEDGMENTS

This research was partially supported by the US Geological Survey (USGS), Department of the Interior, under USGS award number 1434-92-G-2165. The views and conclusions contained in the document are those of the authors and should not be interpreted as necessarily representing the official policies, either expressed or implied of the US Government. This is contribution #862 of the Department of Geological Sciences, Cornell University.

REFERENCES

1. B. Mandelbrot, "How Long is the Coast of Britain? Statistical Self-similarity and Fractional Dimension," *Science* **156**, 636–638 (1967).
2. K. Aki, "A Probabilistic Systhesis of Precursory Phenomena," in *Earthquake Prediction*, eds. D. W. Simpson and P. G. Richards (American Geophysical Union, Washington, D.C., 1981), pp. 566–574.
3. D. L. Turcotte, *Fractals and Chaos in Geology and Geophysics* (Cambridge University Press, Cambridge, 1992), p. 221.
4. R. H. Rapp, "The Decay of the Spectrum of the Gravitational Potential and the Topography of the Earth," *Geophys. J. Int.* **99**, 449–455 (1989).
5. A. N. Strahler, "Quantitative Analysis of Watershed Geomorphology," *Transactions, Am. Geophys. Un.* **38**, 913–920 (1957).
6. R. E. Horton, "Erosional Development of Streams and their Drainage Basins; Hydrophysical Approach to Quantitative Morphology," *Bull. Geol. Soc. Am.* **56**, 275–370 (1945).
7. J. C. Maxwell, "Quantitative Geomorphology of the San Dimas Experimental Forest, California, Project NR 389-042, Technical Report 19, Columbia University, Dept. of Geology," (1960).

8. L. B. Leopold and W. B. Langbein, "The Concept of Entropy in Landscape Evolution," *U.S. Geol. Survey Prof. Paper* **A500**, 20 (1962).
9. H. Schenck, "Simulation of the Evolution of Drainage-basin Networks with a Digitial Computer," *J. Geophys. Res.* **68**, 5739–5745 (1963).
10. R. L. Shreve, "Statistical Law of Stream Numbers," *J. Geol.* **74**, 17–37 (1966).
11. R. L. Shreve, "Infinite Topologically Random Channel Networks," *J. Geol.* **75**, 178–186 (1967).
12. A. E. Scheidegger, "A Stochastic Pattern for Drainage Patterns into an Intermontane Trench," *Int. Assoc. Sci. Hydrol. Bull.* **12**, 15–20 (1967).
13. J. S. Smart and A. J. Surkan, "The Relation Between Mainstream Length and Area in Drainage Basins," *Water Resour. Res.* **3**, 963–974 (1967).
14. I. Seginer, "Random Walk and Random Roughness Models of Drainage," *Water Resour. Res.* **5**, 591–607 (1969).
15. A. D. Howard, "Simulation Model of Stream Capture," *Geol. Soc. Am. Bull.* **82**, 1355–1376 (1971).
16. J. S. Smart, "Channel Networks," *Adv. Hydrosci* **8**, 305–346 (1972).
17. H. Kondoh and M. Matsushita, "Diffusion-limited Aggregations with Anisotropic Sticking Probability: A Tentative Model for River Networks," *J. Phys. Soc. Jpn.* **55**, 3289–3292 (1986).
18. P. Meakin, J. Feder and T. Jossang, "Simple Statistical Models for River Networks," *Physica* **A176**, 409–429 (1991).
19. J. G. Masek and D. L. Turcotte, "A Diffusion-limited Aggregation Model for the Evolution of Drainage Networks," *Earth Planet Sci. Let.* **119**, 379–386 (1993).
20. C. Stark, "An Invasion Percolation Model of Drainage Network Evolution," *Nature* **352**, 423–425 (1991).
21. C. G. Chase, "Fluvial Landsculpting and the Fractal Dimension of Topography," *Geomorphology* **5**, 39–57 (1992).
22. H. Inaoka and H. Takayasu, "Water Erosion as a Fractal Growth Process," *Phys. Rev.* **E47**, 899–910 (1993).
23. G. Willgoose, R. L. Bras and I. Rodriguez-Iturbe, "A Coupled Channel Network Growth and Hillsope Evolution Model 1: Theory," *Water Resour. Res.* **27**, 1671–1684 (1991).
24. S. Kramer and M. Marder, "Evolution of River Networks," *Phys. Rev. Let.* **68**, 205–208 (1992).
25. S. A. Schumm, M. P. Mosley and W. E. Weaver, *Experimental Fluvial Geomorphology* (John Wiley, New York, 1987), p. 413.
26. D. L. Turcotte and L. Greene, "A Scale-invariant Approach to Flood-frequency Analysis," *Stochastic Hydrol. Hydraul.* **7**, 33–40 (1993).
27. M. A. Benson, "Uniform Flood-frequency Estimating Methods for Federal Agencies," *Water Resour. Res.* **4**, 891–908 (1968).
28. D. H. Rothman, J. Grotzinger and P. Flemings, "Scaling in Turbidite Deposition," *J. Sed. Petrol* (1993) (in press).
29. R. N. Hiscott, A. Colella, P. Pezard, M. A. Lovell and A. Malinverno, "Sedimentology of Deep-water Volcaniclastics, Oligocene Izu-bonin Forearc Basin, Based on Formation Microscanner Images," *Proc. Ocean Dril. Prog. Sci. Results* **126**, 75–96 (1992).
30. H. H. Stolum, "Fractal Heterogeneity of Clastic Reservoirs," in *Reservoir Characterization II*, eds. L. W. Lake, H. B. Carroll and T. C. Wesson (Academic Press, San Diego, 1991), pp. 579–612.
31. B. D. Malamud and D. L. Turcotte, "Frequency-size Distributions of Sandy-bed Thicknesses in the Shallow Marine Environment of the Late Devonian Ithaca Formation, New York State, Abstracts with Programs, 1992 Annual Meeting," *Geol. Soc. Am.* p. A357 (1992).
32. T. A. Hewett, "Fractal Distributions of Reservoir Heterogeneity and their Influence on Fluid Transport," *Soc. Petrol. Eng. Paper* **15386**, 16 (1986).
33. J. P. Todoeschuck and O. G. Jensen, "Joseph Geology and Seismic Deconvolution," *Geophys.* **53**, 1410–1414 (1988).
34. J. P. Todoeschuck, O. G. Jensen and S. Labonte, "Gaussian Scaling Noise Model of Seismic Reflection Sequences: Evidence from Well Logs," *Geophys.* **55**, 480–484 (1990).
35. C. G. Sammis, R. H. Osborne, J. L. Anderson, M. Banerdt and P. White, "Self-similar Cataclasis in the Formation of Fault Gauge," *Pageoph.* **124**, 53–78 (1986).
36. D. L. Turcotte, "Fractals and Fragmentation," *J. Geophys. Res.* **91**, 1921–1926 (1986).
37. C. C. Barton and P. A. Hsieh, "Physical and Hydrological-flow Properties of Fractures," *28th International Geological Congress Field Trip Guidebook T385* (American Geophysical Union, Washington, D.C., 1989), p. 36.

38. P. Bak, C. Tang and K. Wiesenfeld, "Self-organized Criticality," *Phys. Rev.* **A38**, 364–374 (1988).
39. C. H. Scholz, "Earthquakes and Faulting: Self-organized Critical Phenomena with a Characteristic Dimension," in *Spontaneous Formation of Space-Time Structures and Criticality*, eds. T. Riste and D. Sherrington, (Kluwer Academic Publishers, Netherlands, 1991), pp. 41–56.
40. D. Sornett, P. Davy and A. Sornette, "Structuration of the Lithosphere in Plate Tectonics as a Self-organized Critical Phenomenon," *J. Geophys. Res.* **95**, 17,353–17,361 (1990).
41. L. P. Kadanoff, S. R. Nagel, L. Wu and S. M. Zhou, "Scaling and Universality in Avalanches," *Phys. Rev.* **A39**, 6524–6533 (1989).
42. S. R. Nagel, "Instabilities in a Sandpile," *Rev. Mod. Phys.* **64**, 321–325 (1992).
43. J. Rosendahl, M. Vekie and J. Kelley, "Persistent Self-organization of Sandpiles," *Phys. Rev.* **E47**, 1401–1404 (1993).
44. B. Barriere and D. L. Turcotte, "A Scale-invariant Cellular-automata Model for Distributed Seismicity," *Geophys. Res. Let.* **18**, 2011–2014 (1991).
45. J. F. Evernden, "Study of Regional Seismicity and Associated Problems," *Seis. Soc. Am. Bull.* **60**, 393–446 (1970).
46. D. von Seggern, S. S. Alexander and C. E. Baag, "Seismicity Parameters Preceding Moderate to Major Earthquakes," *J. Geophys. Res.* **86**, 9325–9351 (1981).
47. L. M. Jones and P. Molnar, "Some Characteristics of Foreshocks and their Possible Relationship to Earthquake Prediction and Premonitory Slip of Faults," *J. Geophys. Res.* **84**, 3596–3608 (1979).
48. R. Burridge and L. Knopoff, "Model and Theoretical Seismicity," *Seis. Soc. Am. Bull.* **57**, 341–371 (1967).
49. J. Huang and D. L. Turcotte, "Are Earthquakes An Example of Deterministic Chaos?" *Geophys. Res. Let.* **17**, 223–226 (1990).
50. J. Huang and D. L. Turcotte, "Evidence for Chaotic Fault Interactions in the Seismicity of the San Andreas Fault and Nankai Trough," *Nature* **348**, 234–236 (1990).
51. J. Huang and D. L. Turcotte, "Chaotic Seismic Faulting with a Mass-spring Model and Velocity-weakening Friction," *Pure. Ap. Geophys.* **138**, 569–589 (1992).
52. G. Narkounskaia and D. L. Turcotte, "A Cellular-automata, Slider-block Model for Earthquakes 1. Demonstration of Chaotic Behavior for a Low Order System," *Geophys. J. Int.* **111**, 250–258 (1992).
53. J. M. Carlson and J. S. Langer, "Mechanical Model of Earthquakes Generated by Fault Dynamics," *Phys. Rev.* **A40**, 6470–6484 (1989).
54. H. Nakanishi, "Cellular Automation Model of Earthquakes with Deterministic Dynamics," *Phys. Rev.* **A41**, 7086–7089 (1990).
55. H. Nakanishi, "Statistical Properties of the Cellular Automata Model for Earthquakes," *Phys. Rev.* **A43**, 6613–6621 (1991).
56. S. R. Brown, C. H. Scholz and J. B. Rundle, "A Simplified Spring-block Model of Earthquakes," *Geophys. Let.* **18**, 215–218 (1991).
57. H. Takayasu and M. Matsuzaki, "Dynamical Transition in Threshold Elements," *Phys. Let.* **A131**, 224–247 (1988).
58. K. M. Ito and M. Matsuzaki, "Earthquakes as Self-organized Critical Phenomena," *J. Geophys. Res.* **95**, 6853–6860 (1990).
59. J. S. Langer and C. Tang, "Rupture Propagation in a Model of An Earthquake Fault," *Phys. Rev. Let.* **67**, 1043–1046 (1991).
60. J. M. Carlson, "Time Intervals between Characteristic Earthquakes and Correlations with Smaller Events: An Analysis Based on a Mechanical Model of a Fault," *J. Geophys. Res.* **96**, 4255–4267 (1991).
61. J. M. Carlson, "Two-dimensional Model of a Fault," *Phys. Rev.* **A44**, 6226–6232 (1991).
62. J. M. Carlson, J. S. Langer, B. E. Shaw and C. Tang, "Intrinsic Properties of a Burridge-Knopoff Model of an Earthquake Fault," *Phys. Rev.* **A44**, 884–897 (1991).
63. J. M. Carlson, E. R. Granna, G. H. Swindle, "Self-organizing Systems at Finite Driving Rates," *Phys. Rev.* **E47**, 93–105 (1993).
64. M. Matsuzaki and H. Takayasu, "Fractal Features of the Earthquake Phenomenon and a Simple Mechanical Model," *J. Geophys. Res.* **96**, 19,925–19,931 (1991).

65. J. B. Rundle and S. R. Brown, "Origin of Rate Dependence in Frictional Sliding," *J. Stat. Phys.* **65**, 403–412 (1991).
66. B. E. Shaw, J. M. Carlson and J. S. Langer, "Patterns of Seismic Activity Preceding Large Earthquakes," *J. Geophys. Res.* **97**, 479–488 (1992).
67. J. Huang, G. Narkounskaia and D. L. Turcotte, "A Cellular Automaton, Slider-block Model for Earthquakes 2. Demonstration of Self-organized Criticality for a Two Dimensional System," *Geophys. J. Int.* **111**, 259–269 (1992).
68. J. B. Rundle, "A Physical Model for Earthquakes. 1. Fluctuation and Interaction," *J. Geophys. Res.* **93**, 6237–6254 (1988).
69. J. B. Rundle, "A Physical Model for Earthquakes. 2. Applications to Southern California," *J. Geophys. Res.* **93**, 6255–6274 (1988).
70. J. B. Rundle, "A Physical Model for Earthquakes. 3. Thermo-dynamical Approach and Its Relation to Nonclassical Theories of Nucleation," *J. Geophys. Res.* **94**, 2839–2855 (1988).
71. C. H. Scholz, "Mechanisms of Seismic Quiescences," *Pure Ap. Geophys.* **126**, 701–718 (1988).
72. M. Wyss and R. E. Haberman, "Precursory Seismic Quiescence," *Pure Ap. Geophys.* **126**, 319–332 (1988).
73. S. Yu Schreider, "Formal Definition of Premonitory Seismic Quiescence," *Phys. Earth Planet Int.* **61**, 113–127 (1990).
74. L. R. Sykes and S. C. Jaume, "Seismic Activity on Neighboring Faults as a Long-term Precursor to Large Earthquakes in the San Francisco Bay Area," *Nature* **348**, 595–599 (1990).
75. V. I. Keilis-Borok, L. Knopoff, I. M. Rotwain and C. R. Allen, "Intermediate-term Prediction of Occurrence Times of Strong Earthquakes," *Nature* **335**, 690–694 (1988).
76. V. I. Keilis-Borok, "The Lithosphere of the Earth as a Nonlinear System with Implications for Earthquake Prediction," *Rev. Geophys.* **38**, 19–34 (1990).
77. V. I. Keilis-Borok and I. M. Rotwain, "Diagnosis of Time of Increased Probability of Strong Earthquakes in Different Regions of the World: Algorithm CN," *Phys. Earth Planet Int.* **61**, 57–72 (1990).
78. V. I. Keilis-Borok and V. G. Kossobokov, "Premonitory Activation of Earthquake Flow: Algorithm M8," *Phys. Earth Planet* **61**, 73–83 (1990).
79. V. I. Keilis-Borok and V. G. Kossobokov, "Times of Increased Probability of Strong Earthquakes ($M \geq 7.5$) Diagnosed by Algorithm M8 in Japan and Adjacent Territories," *J. Geophys. Res.* **95**, 12,413–12,422 (1990).
80. G. M. Molchan, O. E. Dmitrieve, I. M. Rotwain and J. Dewey, "Statistical Analysis of the Results of Earthquake Prediction, Based on Bursts of Aftershocks," *Phys. Earth Planet Int.* **61**, 128–139 (1990).
81. A. C. Johnston and S. J. Nava, "Recurrence Rates and Probability Estimates for the New Madrid Seismic Zone," *J. Geophys. Res.* **90**, 6737–6753 (1985).
82. D. L. Turcotte, "A Fractal Approach to Probabilistic Seismic Hazard Assessment," *Tectonophys.* **167**, 171–177 (1989).
83. D. L. Turcotte, "Earthquake Prediction," *An. Rev. Earth Planet. Sci.* **19**, 263–281 (1991).
84 L. B. Leopold, M. G. Wolman and J. P. Miller, *Fluvial Processes in Geomorphology* (W. H. Freeman, San Francisco, 1964), p. 522.

Fractals, Vol. 1, No. 3 (1993) 513–520

A CASCADE MODEL OF WAVE TURBULENCE WITH APPLICATIONS TO SURFACE GRAVITY AND CAPILLARY WAVES

ROMAN E. GLAZMAN
Jet Propulsion Laboratory, California Institute of Technology
4800 Oak Grove Drive, Pasadena, CA 91109-8099, USA
E-mail: reg@pacific.jpl.nasa.gov

Abstract

A heuristic approach to the derivation of power spectra of wave motion is described and applied to capillary waves. The case of gravity waves studied earlier is briefly reviewed. In contrast to the previous studies, the nonlinearity of the wave motion is not required to be small, and the mean number of resonantly interacting wave harmonics is not limited to a smallest possible number (which is 4 for gravity waves on a deep fluid and 3 for capillary waves). The main external parameter of the problem is the input flux Q of the wave energy related to the mean wind velocity as $Q \propto U^3$. Depending on its value, wave spectra $S(\omega) \propto \omega^{-q}$ and $F(k) \propto k^{-p}$ take various forms — from that corresponding to the weak-turbulence limit and to that corresponding to the saturated (Phillips') wave spectra. Fractal dimensions of the sea surface are related to external factors, specifically to the mean wind speed. The theoretically predicted spectra are in qualitative agreement with the observations on capillary waves conducted in a large wave tank.

1. INTRODUCTION

The subject of this paper is turbulence of surface gravity and capillary waves, although the formalism presented here can be applied to other oceanic and atmospheric wave motions, such as internal waves, Rossby waves, etc. The nonlinearity of these wave systems tends to be weak; hence, the corresponding problems are amenable to a treatment by small perturbation techniques.[1] Derivation of turbulence spectra involves an additional assumption regarding statistical quasi-normality of a wave field. As a result, possible events of strongly nonlinear behavior (e.g. intermittently occurring steep and breaking waves, or jets, fronts, etc. — in a case of Rossby waves) would violate the basic assumptions of the theory. Apparently, while the strongly nonlinear events are rare, they may have an appreciable effect on the overall energy transfer, hence on the spectra of wave turbulence. In the present work, we employ a heuristic approach which goes beyond the limit of weak nonlinearity and attempts to account for such events. The well-known results of the weak turbulence theory (WTT) are derived as a special case.

The most important accomplishment of WTT is its prediction of power-law spectra for fields' oscillations in space and time. For example, the wavenumber spectrum for spatial variation of a fluid's surface, $\eta(\mathbf{x}; t)$, on scales of gravity waves is given by[2]:

$$F_\eta(k) = \alpha_g g^{-1/2} Q^{1/3} k^{-7/2} \tag{1.1}$$

In the range of capillary waves, the spectrum is[3] :

$$F_\eta(k) = \alpha_c \sigma^{-3/4} Q^{1/2} k^{-19/4} \tag{1.2}$$

where α_g and α_c are nondimensional universal ("Kolmogorov") constants, Q is the rate of energy transfer down the spectrum (divided by the fluid density, ρ), g is the acceleration of gravity, and σ is the coefficient of surface tension, T, divided by ρ. The angular, Θ, dependence of these two-dimensional spectra is omitted for simplicity.

The spectral energy transfer can be envisaged as a cascade process which manifests itself in the surface topography[4] or in the surface temporal oscillations: wavelets of a continuously decreasing size ride on top of a large-scale wave form. As a result, power spectra of surface height variations tend to be rather broad and, at frequencies well above the spectral peak frequency, they exhibit a power-law behavior. For a wavenumber spectrum of form $F(k) \propto k^{-p}$, the exponent, p, is related to the Hausdorff (fractal) dimension, D_H, of the surface elevation field: $D_H = (8-p)/2$, which affords an instructive interpretation of the quantities involved. In particular, wave spectra can be conveniently presented as[4,5]:

$$F(k) \propto k^{-4+2\mu} \tag{1.4}$$

The value $\mu = 1/4$ corresponds to Eq. (1.1), while $\mu = 0$ yields Phillips' saturated spectra discussed below. The meaning of μ follows from the relationship:

$$D_H = 2 + \mu \tag{1.5}$$

Hence, μ represents the "fractal index". This quantity invariably emerges in analytical expressions for various geometrical properties of a random surface, such as its mean Gaussian curvature, root-mean-square slope, etc. (e.g. Ref. 5).

Apparently, the surface geometry characterized by Eq. (1.2) is not fractal because the spectrum rolls off too rapidly giving $\mu < 0$. However, as shown in Sec. 2, the surface's temporal oscillations due to capillary waves, recorded at a fixed location, represent a fractal process. In Sec. 2 the range of fractal dimensions for $\eta(t)$ with an arbitrary degree of nonlinearity is estimated and D_H is related to external parameters of the problem.

At a sufficiently high external input of the wave energy, both the capillary and the gravity wave turbulence becomes stronger, and WTT is not applicable. The limiting spectra of strongly nonlinear wave fields are known as "saturated" spectra.[6] They are derived on the assumption that the energy transfer down the spectrum occurs primarily due to intermittent events of wave breaking rather than due to a Kolmogorov-type inertial cascade. The latter is too slow to support a high rate of energy input, while the breaking of steep waves allows the energy to be transferred to small scales by sporadic "jumps" from much larger scales. Respectively, the energy flux Q must not emerge in expressions for the wave spectra, and the limiting form is:

$$F(k) = \beta k^{-4}, \tag{1.6}$$

— as based on the analysis of dimensions.[7] This is valid for both capillary and gravity waves,[6] and β is known as the Phillips constant.

Experimental observations consistently show that the actual spectra of surface waves, while exhibiting a power-law behavior [Eq. (1.4)], are not limited to the forms in Eqs. (1.1), (1.2) and (1.6): the exponent (hence, μ in Eq. (1.4)) spans the entire range — from the weak turbulence limit [Eqs. (1.1) and (1.2)] to the saturated spectrum limit [Eq. (1.6)], — depending on the external conditions (the energy input and its partition between the component going down the spectrum and that consumed by other mechanisms of the total energy budget (see, e.g. Ref. 8). In what follows we suggest a simple heuristic theory explaining this behavior, and apply it to the case of capillary waves.

2. MULTIWAVE-INTERACTION THEORY FOR SURFACE GRAVITY AND CAPILLARY WAVES

The potential energy of the wave motion includes two components:

$$U = \frac{1}{2}\rho g \int \eta^2 d\mathbf{x} + \sigma\rho \int (\sqrt{1 + |\nabla\eta|^2} - 1) d\mathbf{x} \tag{2.1}$$

The first one is due to the gravity force and the second due to the surface tension. Here, $\eta = \eta(\boldsymbol{x}, t)$ is the elevation of the fluid surface above the zero-mean level. An approximate equipartition of energy between the kinetic and the potential parts allows one to approximate the surface density of the total wave energy, E, by:

$$E_g \approx \rho g \eta^2 \tag{2.2a}$$

$$E_\sigma \approx \sigma\rho(\nabla\eta)^2 \tag{2.2b}$$

where component E_g is due to gravity waves and E_σ is due to capillary waves. These quantities are related to the spectral density of the wave energy by:

$$E = \int S(\omega) d\omega = \int\int F(k, \theta) k d\theta dk = \int G(k) dk, \tag{2.3}$$

where the integration is carried out over all wavenumbers/frequencies. Here, $S(\omega)$ is the frequency spectrum, $F(k, \theta)$ is the two-dimensional wavenumber spectrum and $G(k)$ is the wavenumber modulus spectrum of the wave energy. Obviously, for gravity waves, the relationship between the energy spectrum, F of Eq. (2.3), and the surface height spectrum, F_η of Eq. (1.1), is: $F = \rho g F_\eta$. For capillary waves, this relationship is: $F = \sigma \rho k^2 F_\eta$.

Inertial flux of wave energy can be envisaged as a cascade process in the frequency space: at each nth step of the cascade, the amount of energy, e_n, transferred from the previous step is estimated as:

$$e_n = \int_{\omega_{n-1}}^{\omega_n} S(\omega)d\omega = \int_{k_{n-1}}^{k_n} G(k)dk \tag{2.4}$$

where (ω_{n-1}, ω_n) is the width of a cascade step (which is much smaller than the width of the inertial range). Suppose, the characteristic time (the "turnover time") of nonlinear wave-wave interaction is t_n. Then, the rate Q of energy transfer through the spectrum is given by:

$$\rho Q = e_n / t_n \tag{2.5}$$

For both gravity and capillary waves, we assume that the energy flux down the spectrum remains constant, hence Q is independent of k and ω. (A more complicated case of a non-conservative cascade is treated in Ref. 9). Provided e_n and t_n can be expressed in terms of k, ω and wave amplitude a, Eq. (2.5) allows one to derive the spectrum by means of elementary algebra (e.g. Ref. 10). Hence, we need to express these parameters in terms of the relevant quantities.

From Eq. (2.2a) it is obvious that e_n for gravity waves can be written as:

$$e_n \approx \rho g a_n^2 \tag{2.6a}$$

For capillary waves, Eq. (2.2b) leads to the following expression for e_n:

$$e_n \approx \sigma \rho (a_n k_n)^2 \tag{2.6b}$$

Here, a_n is the Fourier amplitude of surface oscillation at the frequency/wavenumber scales ω_n and k_n, corresponding to the nth step in the spectral cascade.

The derivation of the turnover time is formally based on the scaling of the collision integral in the kinetic equation[11–13] for the wave action spectral transfer. For the subsequent development it is useful to introduce the turnover time in a more general, although less formal, fashion. To this end we notice that the nonlinearity of wave processes is measured by the ratio, ε, of the fluid particle velocity, u, to the wave phase velocity, ω/k.[14] Since fluid particles in a surface wave execute an approximately orbital motion in the vertical plane with the radius equal to the wave amplitude and the period $2\pi/\omega$, the value of u at a given scale is found as $a_n\omega_n$. Respectively, the ratio $u/(c/\omega)$ is:

$$\varepsilon_n \equiv \frac{u_n}{\omega_n/k_n} \approx \frac{a_n\omega_n}{\omega_n/k_n} = a_n k_n \tag{2.7}$$

This quantity represents the small parameter in deterministic perturbation theories. However, since the kinetic equation describing wave action, $N(\mathbf{k}) = F(\mathbf{k})/\omega$, (or wave energy,

$F(\mathbf{k})$) spectral transfer is derived for second statistical moments of the fields, the perturbation equations of statistical theory are developed in powers of ε^2. In such equations, terms (i.e., collision integrals) of order ε^2 correspond to three-wave interactions, terms ε^4 correspond to four-wave interactions,[1] etc. In general, each additional Fourier component accounted for in the interaction integral adds new terms which are ε^2 times as great as a preceding term. The νth term is of order $\varepsilon^{2(\nu-2)}$. Respectively, the characteristic time of nonlinear wave-wave interactions increases as the number of interacting harmonics grows. For 3-wave interactions, this time is estimated as $t^{-1} \approx \omega\varepsilon^2$, and for an arbitrary number, ν, we have:

$$t^{-1} \approx \omega\varepsilon^{2(\nu-2)} \tag{2.8}$$

Formidable mathematical difficulties limit the kinetic equation to accounting only for the lowest-order resonant interactions. However, one can formally present it as:

$$\partial N/\partial t + \nabla_k \bullet T(\mathbf{k}) = p(\mathbf{k}), \tag{2.9}$$

where $p(\mathbf{k})$ is the spectral density of the input flux of wave action (from wind), and $\nabla_k \bullet T(\mathbf{k})$ denotes the spectral density of the action flux due to all wave-wave interactions to order ν:

$$\nabla_k \bullet T(\mathbf{k}) = I_3 + I_4 + \cdots + I_\nu \tag{2.10}$$

I_ν are collision integrals accounting for interactions among ν waves satisfying resonance conditions:

$$\begin{aligned} \omega_o \pm \omega_1 \pm \cdots \pm \omega_\nu &= 0 \\ \mathbf{k}_0 \pm \mathbf{k}_1 \pm \cdots \pm \mathbf{k}_\nu &= 0 \end{aligned} \tag{2.11}$$

(non-resonant terms can be eliminated by appropriate canonical transformations.[1]) For gravity waves, the minimum number of resonantly interacting components is 4, while for capillary waves it is 3.

It has been argued earlier[9] that intermittently occurring, rare events of steep and breaking waves (characterized by a locally high nonlinearity, hence a large, or even infinite, number of interacting Fourier components forming individual cusp-like wavelets), result in an increased mean (over a large time interval and large surface area) value of ν. While this ν may be substantially greater than the minimum resonant number appearing in WTT, the energy and action transfer may still be dominated by the weakly nonlinear inertial cascade. Thus, the "effective" ν is introduced as an unknown function of the problem, the assumption of locality of wave-wave interactions in the wavenumber space remaining in force. Let us notice that the turnover time given by Eq. (2.8) for the highest-order term in Eq. (2.10) is the slowest of all the times for "partial" fluxes associated with individual terms in Eq. (2.10). Therefore, although the total flux of the wave action, $\nabla_k \bullet T(\mathbf{k})$, is comprised of many partial fluxes I_3, I_4, etc., the appropriate characteristic time for the integral transfer is given by Eq. (2.8).

We consider the case when the external input is concentrated at wavenumbers below certain k_0 marking the high-wavenumber boundary of the "generation range". Therefore, at $k > k_0 : p(\mathbf{k}) = 0$, and a spectral flux is purely inertial. Obviously, the energy flux down the spectrum is given by:

$$\rho Q = \int_0^{k_0} \omega(k) k dk \int_{-\pi}^{\pi} p(k, \theta) d\theta \tag{2.12}$$

Respectively, Eq. (2.9) for the inertial range corresponds to:

$$e_n \omega_n (a_n k_n)^{2(\nu-2)} \approx \rho Q (= const) \tag{2.13}$$

where $n \geq 1$.

Using the dispersion relation for capillary waves, $\omega^2 = \sigma k^3$, and Eq. (2.6b), Eq. (2.13) yields $e_n \approx \rho Q^{1/(\nu-1)} \sigma^{(\nu-2)/(\nu-1)} \omega^{-1/(\nu-1)}$. The frequency spectrum is obtained based on the fact that Eq. (2.4) yields a scaling relationship: $S(\omega) \propto e_n/\omega$. Introducing a proportionality constant, α_c, playing the role of the Kolmogorov constant, we ultimately arrive at the energy spectrum for surface capillary waves:

$$S(\omega) = \alpha_c \rho Q^{1/(\nu-1)} \sigma^{(\nu-2)/(\nu-1)} \omega^{-\nu/(\nu-1)} \tag{2.14}$$

This is related to the frequency spectrum of the wave slope by $S_{\nabla\eta}(\omega) = (\sigma\rho)^{-1} S(\omega)$, whereas the surface elevation spectrum is found as $S_\eta(\omega) = S_{\nabla\eta}(\omega) k^{-2}$, i.e.:

$$S_\eta(\omega) = \alpha_c Q^{1/(\nu-1)} \sigma^{(2\nu-5)/[3(\nu-1)]} \omega^{-(7\nu-4)/[3(\nu-1)]} \tag{2.15}$$

The one-dimensional process described by Eq. (2.15) is fractal for all $\nu \geq 3$: the Hausdorff dimension of $\eta(t)$ whose spectrum falls off as ω^{-q} is given by (see, e.g. Ref. 4) $D_H = (5-q)/2$ which translates into

$$D_H = \frac{8\nu - 11}{6(\nu - 1)} \tag{2.16}$$

Therefore, $D_H > 1$ for all $\nu \geq 3$. The "Phillips spectrum" ($\nu \to \infty$) takes the form

$$S_\eta(\omega) = \beta \sigma^{2/3} \omega^{-7/3} \tag{2.17}$$

which yields $D_H = 4/3$, while the Zakharov-Filonenko spectrum in the frequency domain, $S(\omega) \propto \omega^{-17/6}$, yields $D_H = 13/12$.

The surface elevation spectrum in the wavenumber domain (omitting the angular spread factor) is found as:

$$F_\eta(k) = k^{-1} \left[S_\eta(\omega) \frac{d\omega}{dk} \right]_{\omega=\omega(k)} = \alpha_c' Q^{1/(\nu-1)} \sigma^{-9/[6(\nu-1)]} k^{-(8\nu-5)/[2(\nu-1)]} \tag{2.18}$$

(Factor 3/2 is included into the Kolmogorov constant α'_c). This coincides with the Zakharov-Filonenko spectrum [Eq. (1.2)] in a special case of $\nu = 3$, and it produces the Phillips spectrum [Eq. (1.6)] at $\nu \to \infty$. In the latter case, the surface becomes marginally fractal, that is: $D_H = 2$.

It is also easy to show that the use of the gravity wave dispersion relation, $\omega^2 = kg$, and of Eq. (2.6a) in Eq. (2.13) yields spectra of surface gravity waves. In particular the frequency spectrum of wave energy is[9]:

$$S(\omega) = \alpha_g Q^{1/(\nu-1)}(\rho g^3)^{(\nu-2)/(\nu-1)}\omega^{-(5\nu-8)/(\nu-1)} \tag{2.19}$$

This reduces to the Zakharov-Filonenko [Eq. (1.1)] and Phillips [Eq. (1.6)] spectra at $\nu = 4$ and $\nu \to \infty$, respectively.

3. DETERMINATION OF ν FOR THE INERTIAL RANGE

Provided ν is an increasing function of wind, Eqs. (2.15) and (2.18) would qualitatively agree with the laboratory observations[15] which show a conspicuous monotonic increase of spectral density levels as the wind increases. Therefore, we now must prove that ν increases as the wind grows. In order to avoid considering the entire problem of wave generation by wind, let us assume both the energy flux, Q, arriving from the lower-frequency range and the wave spectrum in that range to be known. Our task is to match that spectrum with Eqs. (2.15) and (2.18). Let ω_0 designate the lowest-frequency boundary of the capillary range at which $S_\eta(\omega_0)$ and Q are specified based on the matching of gravity and capillary spectra at a certain frequency ω_0 . Taking Eq. (2.15) at $\omega = \omega_0$ and solving it for ν yields:

$$\nu = a_0 + a_1 V \tag{3.1}$$

where

$$a_0 = \frac{3(Z_0 - A) - 5\Gamma + 4\Omega_0}{3(Z_0 - A) - 2\Gamma + 7\Omega_0}, \quad a_1 = \frac{3}{3(Z_0 - A) - 2\Gamma + 7\Omega_0} \tag{3.2}$$

and

$$Z_0 = \ln S_\eta(\omega_0), \quad A = \ln \alpha_c, \quad \Gamma = \ln \sigma, \quad \Omega_0 = \ln \omega_0 \text{ and } V = \ln Q .$$

Assuming that the lower-frequency wave spectrum obeys the Phillips law [Eq. (1.6)], both a_0 and a_1 become independent of wind. Therefore, since Q is proportional to the cube of wind velocity, Eq. (3.1) confirms a gradual (actually, logarithmic) increase of ν with an increasing wind.

ACKNOWLEDGMENTS

This work was performed at the Jet Propulsion Laboratory, California Institute of Technology, under contract with the National Aeronautics and Space Administration. Financial support was provided by the Office of Naval Research, Grant No. N0001492J1343.

REFERENCES

1. V. E. Zakharov, V. S. L'vov and G. Falkovich, *Kolmogorov Spectra of Turbulence I: Wave Turbulence* (Springer Verlag, Berlin, 1992), p. 264.
2. V. E. Zakharov and N. N. Filonenko, "The Energy Spectrum for Stochastic Oscillation of a Fluid's Surface," *Doklady Academii Nauk S.S.S.R.* **170**(6), 1292-1295, (1966). (In Russian)
3. V. E. Zakharov and N. N. Filonenko, "Weak Turbulence of Capillary Waves," *J. Appl. Math. Tech. Phys.* **4**, 506-515 (1967).
4. R. E. Glazman and P. Weichman, "Statistical Geometry of a Small Surface Patch in a Developed Sea," *J. Geophys. Res.* **94**(C4), 4998–5010 (1989).
5. R. E. Glazman, "Near-nadir Radar Backscatter from a Well-developed Sea," *Radio Science* **25**(6), 1211–1219 (1990).
6. O. M. Phillips, *The Dynamics of the Upper Ocean* (Cambridge Univ. Press, New York, 1977), p. 336.
7. O. M. Phillips, "The Equilibrium Range in the Spectrum of Wind-generated Waves," *J. Fluid Mech.* **4**, 426–434 (1958).
8. R. E. Glazman, Wave spectra of developed seas in. *Nonlinear Waves and Weak Turbulence*, eds. N. Fitzmaurice, D. Gurarie, F. McCaughan and W.A. Woyczynski. The Birkhauser Series on Nonlinear PDE's, ed. H. Brezis. (Birkhauser-Boston, 1993.)
9. R. E. Glazman, "Multiwave Interaction Theory for Wind-Generated Surface Gravity Waves," *J. Fluid Mech.* **243**, 623–635 (1992).
10. U. Frisch, P.-L. Sulem and M. Nelkin, "A Simple Dynamical Model of Intermittent Fully Developed Turbulence," *J. Fluid Mech.* **87**, 719–736 (1978).
11. V. E. Zakharov and V. S. L'vov, "Statistical Description of Nonlinear Wave Fields," *Radiophys. Quant. Electron.* **18**, 1084-1097 (1975).
12. S. A. Kitaigorodskii, "On the Theory of the Equilibrium Range in the Spectrum of Wind-generated Gravity Waves," *J. Phys. Oceanogr.* **13**, 816–827 (1983).
13. A. Larraza, S. L. Garrett and S. Putterman, "Dispersion Relations for Gravity Waves in a Deep Fluid: Second Sound in a Stormy Sea," *Phys. Rev.* **A41**(6), 3144–3155 (1990).
14. G. B. Whitham, *Linear And Nonlinear Waves* (John Wiley & Sons, New York, 1974.)
15. B. Jähne and K. S. Riemer, "Two-dimensional Wave Number Spectra of Small-Scale Water Surface Waves," *J. Geophys. Res.* **95**(C7), 11, 531–11, 546 (1990).

Fractals, Vol. 1, No. 3 (1993) 521–528

POWER-LAW DISTRIBUTION OF RIVER BASIN SIZES

HIDEKI TAKAYASU
Graduate School of Information Sciences, Tohoku University,
Aramaki, Aoba-ku, Sendai 980, Japan

Abstract

River models are reviewed with emphasis on the power-law nature of basin size distributions. From a general point of view, the whole river pattern on a surface can be regarded as a kind of tiling by random self-affine branches. Applying the idea of stable distributions, we show that the self-affinity and tiling condition naturally derive the power-law basin size distribution.

1. INTRODUCTION

Since the birth of the notion of fractals, rivers have been recognized as one of the most typical fractals in nature. They are characterized by several power-law exponents, such as Hack's law, the fractal dimensions, self-affine exponents, Horton's exponents for branching, and the exponent for the drainage basin size distributions. Here, we focus our attention on the size distribution of drainage basins, and discuss why it tends to follow a power law.

Simplified numerical models of rivers support these power-law behaviors. For example, it has been proved for Scheidegger's river model that the basin size distribution rigorously follows a power law.[1] A numerical simulation of water erosion[2] clearly shows that a nearly flat landscape automatically evolves into a fractal state where landscapes and river networks show power-law behaviors. An interesting point is that the river network patterns are frozen after the surface is completely covered by river patterns. The stable river patterns are confirmed to show a self-affinity and their basin size distribution follows a power law. It has also been discussed by considering a mean-field version that the space-filling condition in the river models is equivalent to the critical conditions.[3] Namely, the tiling by rivers can be recognized as a kind of self-organized criticality.[4]

From a mathematical point of view, power-law distributions are very deeply related to the stable distributions.[5] The stable distributions are defined by a kind of invariance under summation that any sum of independent random variables follows the same distribution as that of a summand. The Gaussian distribution is the most well-known example of the stable distributions, but there is a continuous class of other stable distributions. Among the stable distributions, the Gaussian is a very special exception and all others are known to have power-law tails. The central limit theorem for the Gaussian distribution has been generalized and we know that if the distribution of the sum of random variables with divergent variance converges, the limit distribution is a stable distribution with power-law tails. Namely, power-law distributions can be obtained by simple summation like the case of Gaussian distribution.

In this paper, we first review numerical models of rivers in the next section. We introduce the stable distributions in Sec. 3, and apply them to the river basin size distributions in Sec. 4. The final section is devoted to concluding remarks.

2. NUMERICAL RIVER MODELS

2.1 Scheidegger's Model

One of the simplest river models may be Scheidegger's lattice model which describes the situation where rain falls uniformly on a slope. At each point on the slope, the direction of water flow is fixed randomly either right down or left down with probability 1/2. We have randomly branched river patterns as shown in Fig. 1.

As typically demonstrated by this figure, river patterns are generally composed of branches without any loop. A complete river pattern fully covers the whole surface to drain all rain water to the boundary.

Fig. 1 River pattern by Scheidegger's model.

The shape of each drainage basin is characterized by the following self-affine relation,

$$h \sim w^{\nu}, \quad \nu = 2, \tag{1}$$

where w and h denote the width and the height of a basin, respectively. The exact value, $\nu = 2$, is easily derived because the boundaries of the basin follow the usual Brownian motions.

Since the basin size, s is estimated by the product of width and height, we have the following relation,

$$s^{n} \sim h, \quad n = \nu/(1+\nu). \tag{2}$$

Here, the exponent, n, is equivalent to Hack's exponent because h is proportional to the length of main stream. The asymptotic behavior of the distribution function of s has been estimated rigorously as[1]

$$P(>s) \sim s^{-\alpha}, \quad \alpha = 1/3, \tag{3}$$

where $P(>s)$ denotes the probability that a randomly chosen basin size is larger than s.

2.2 The Erosion Model

Although Scheidegger's model shows exact power laws, the model is too simple and neglecting many important physical effects. For example, the model is static and no time evolution is considered. Intuitively, river patterns are considered to be automatically created by water erosion; therefore, we should at least take into account the effect of water erosion.

Inaoka and the author have been working on numerical modeling of water erosion and found that the following simple rules are sufficient for the creation of random river patterns showing fractal behaviors.[2]

1. The direction of water flow is fixed at each time step by finding the lowest site among its nearest neighbors. If the elevation of destination is lower, then the water drains into the neighboring site; otherwise, water accumulates until the water level exceeds the destination height.
2. By the effect of erosion, the surface height is decreased by the running water. The effect of accumulation of eroded soil is neglected.

Starting with a nearly flat landscape with small randomness, we have a fully developed river pattern as shown in Fig. 2. This fully developed river pattern is very stable and the pattern does not change after it covers the whole space, although the landscape keeps changing due to erosion.

The size distribution of drainage basins follow the power law as:

$$P(>s) \sim s^{-\alpha}, \quad \alpha = 0.41 \pm 0.01, \tag{4}$$

and the shape of each drainage basin is characterized by Hack's law as:

$$s^{n} \sim h, \quad n = 0.62 \pm 0.02, \tag{5}$$

The fractal dimension of the main stream is confirmed to be very close to 1.0, which indicates that the drainage basins are self-affine with the exponent $\nu = n/(1-n) = 1.6 \pm 0.2$ in Eq. (1).

Fig. 2 River pattern by the erosion model.

The symmetry breakdown of isotropy in our model is due to the boundary condition, because the evolution rule and the initial condition are completely isotropic. In an open system like this, the boundary condition is very important and can control the whole system.

2.3 The Mean-Field Model

A mean-field version of the river model can be constructed by neglecting the spatial correlation. We assume that the drainage size distribution density, $p(s)$, satisfies a self-consistent condition. Let us consider the case that k streams with flow intensities $s_1, s_2, \ldots, s_k$ join at a site making an outflow of intensity $s = s_1 + s_2 + \cdots + s_k + 1$. The mean-field self-consistent condition becomes:

$$p(s) = \sum_k q_k \sum p(s_1)p(s_2)\ldots p(s_k)\,, \tag{6}$$

where q_k denotes the probability of finding a junction of k branches and the second summation is taken over all possible combinations of $\{s_1, s_2, \ldots, s_k\}$. By introducing the characteristic function which is a Fourier or Laplace transform of $p(s)$, Eq. (6) can be solved analytically. For general $\{q_k\}$, the probability density $p(s)$ decays exponentially at large s; however, there is a critical case showing a power-law decay, $p(s) \sim s^{-3/2}$, i.e., $\alpha = 1/2$ in Eq. (3). The critical condition is given as:

$$\sum_k k\, q_k = 1\,. \tag{7}$$

This equality has a clear physical meaning, i.e., the expected number of inflowing streams per site is equal to 1. Since every site has one outflow stream in the fully-developed steady-state, this condition simply means that the number of inflowings and outflowings are the same, which is equivalent to the condition that there remains no lake. Thus, the steady state is a kind of self-organized critical state.

3. STABLE DISTRIBUTIONS

The notion of stable distribution was invented in the early 1930s. It is known that there are some distributions which reproduce themselves up to a linear change of variable under the operation of summation. A mathematical definition is given as follows[5,6]:

"Let x, x_1, ..., x_n be independent random variables with a common distribution R. The distribution R is stable if and only if for $Y_n \equiv x_1 + x_2 + \cdots + x_n$ there exist constants c_n and γ_n such that:

$$y_n \stackrel{d}{=} c_n x + \gamma_n \,, \tag{8}$$

where $\stackrel{d}{=}$ indicates that the random variable of both sides have the same distribution."

The total family of stable distributions is characterized essentially by two real parameters. The probability density $p(x;\, a, \theta)$ is given as:

$$p(x;\, \alpha, \theta) = \frac{1}{\pi} Re \int_0^\infty dz \exp\{-ixz - z^\alpha e^{i\pi\theta/2}\} \,, \tag{9}$$

where Re denotes taking the real part and the parameters are in the region:

$$|\alpha - 1| + |\theta| \leq 1 \,. \tag{10}$$

Gaussian distribution is included as a special case of $\alpha = 2$, $\theta = 0$. The parameter α is called the characteristic exponent and it is known that all stable distributions other than Gaussian have power-law tails as:

$$p(x;\, \alpha, \theta) \sim |x|^{-\alpha-1} \,. \tag{11}$$

A lot of theorems are known about the stable distributions; the following two are especially important.

The generalized central limit theorem: "For random variables x_1, x_2, ..., x_n, let Y_n be the sum of these variables. If the distribution of Y_n, with an appropriate normalization, converges to some distribution R in the limit $n \to \infty$, then the distribution R is stable. In particular, when its variance is finite the limit distribution exists and it is Gaussian." This theorem tells us that the power-law tails can be as universal as Gaussian tails.

The other theorem is the following. "Let x_a be a random variable with a real parameter a. If x_a satisfies the next two relations, then x_a follows a stable distribution with the characteristic exponent α;

$$x_a \stackrel{d}{=} a^{1/\alpha} x_1 \,, \tag{12}$$

$$x_a + x_b \stackrel{d}{=} x_{a+b} \,, \tag{13}$$

where x_a and x_b are assumed to be independent." We are going to apply this theorem to the basin size problem. Equation (12) is a kind of scaling relation and Eq. (13) is a simple summation. This theorem has been applied to several power-law behaviors, for example, fracture size distribution.[7]

4. RANDOM SELF-AFFINE TILING

As we have seen in Sec. 2, the river patterns show both the self-affinity and power law size distribution. In the fully developed state, the plain is completely covered by rivers, so the river patterns can be viewed as a random tiling by self-affine shapes.

Here, we introduce a generalized drainage size, s_w, which is defined by the area draining water into a line segment of width w (see Fig. 3). In the discrete model s_w is equivalent to the sum of basin sizes of w rivers in sequence.

By assuming that the horizontal and vertical directions are scaled as $h \sim w^{\nu}$, we have the following scaling:

$$s_w \stackrel{d}{=} w^{1+\nu} s_1 . \tag{14}$$

Because the plain is completely covered by basins, a sum of two adjacent basins s_a and s_b makes one basin s_{a+b} as shown in Fig. 4. So, we can naturally expect that the following relation holds:

$$s_a + s_b \stackrel{d}{=} s_{a+b} . \tag{15}$$

Here, s_a and s_b are not independent; however, it is known rigorously for Scheidegger's model that the correlation of s_a and s_b are very weak, and as a first order approximation, we can treat them as independent.[8] Therefore, we apply the second theorem in Sec. 3 to the general situation as a first order approximation. According to the theorem, we directly find from Eqs. (14) and (15) that the distribution of s follows a stable distribution with the characteristic exponent $\alpha = 1/(1+\nu)$; consequently, we have the power-law tail as:

$$P(> s_w) \sim s_w^{-\alpha}, \quad \alpha = 1/(1+\nu) . \tag{16}$$

Fig. 3 The generalized river basin size, s_w.

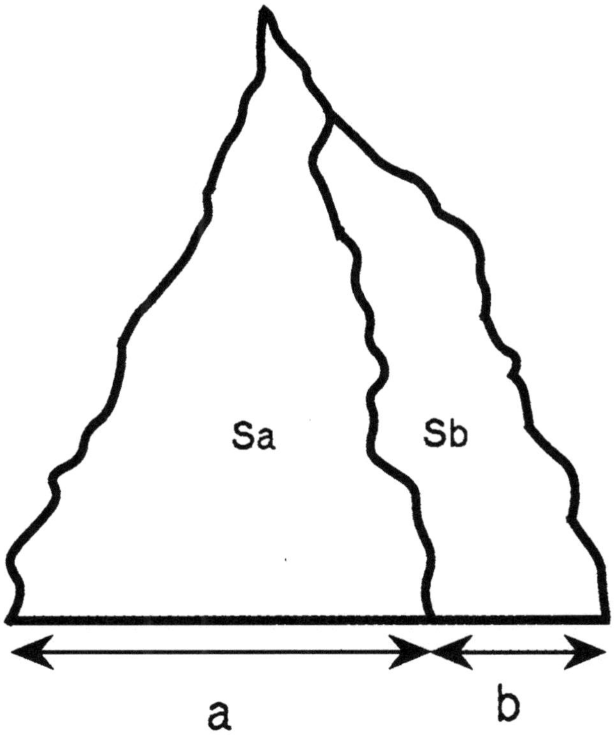

Fig. 4 A sum of two basins.

For Scheidegger's model, this result gives the exact relation ($\nu = 2$, $\alpha = 1/3$). Using the Hack's exponent, n, we have a much simpler relation:

$$\alpha = 1 - n\,. \tag{17}$$

This relation is confirmed in the erosion model within error range. An equivalent relation has been proposed by Nagatani for a generalized Scheidegger model, which is derived by assuming the distribution form as a power law.[9]

5. DISCUSSION AND SUMMARY

In river patterns, the symmetry of isotropy is broken by the boundary condition; consequently, the drainage basins show the self-affine scaling. As we have seen in Secs. 3.3 and 4, a complete covering of plain by rivers, i.e., the tiling, automatically makes the basin size distribution follow a power law. The power-law exponent is directly expressed by the self-affine exponent or by the Hack's exponent, namely, these exponents are not independent.

ACKNOWLEDGMENTS

The author is grateful for helpful discussions with Dr. Misako Takayasu and Mr. Hajime Inaoka.

REFERENCES

1. G. Huber, *Physica* **A170**, 463 (1991): H. Takayasu, M. Takayasu, A. Provata and G. Huber, *J. Stat. Phys.* **65**, 725 (1991).
2. H. Takayasu and H. Inaoka, *Phys. Rev. Lett.* **68**, 966 (1992): H. Inaoka and H. Takayasu, *Phys. Rev.* **E47**, 899 (1993).
3. H. Takayasu, *Fractals in the Physical Sciences* (Manchester University Press, Manchester, 1990), Sec. 4.5.
4. P. Bak, C. Tang, and K. Wiesenfeld, *Phys. Rev. Lett.* **59**, 381 (1987).
5. W. Feller, *An Introduction to Probability Theory and Its Applications* (Wiley, New York, 1966), Vol. 2.
6. Ref. 3, Section 5.2.
7. H. Takayasu, *J. Phys. Soc. Jpn* **57**, 2585 (1988).
8. M. Takayasu and H. Takayasu, *Phys. Rev.* **A39**, 4345 (1989).
9. T. Nagatani, private communication.

Fractals, Vol. 1, No. 3 (1993) 529–538

STRUCTURES GENERATED BY MODEL NONLINEAR HAMILTONIANS BASED ON OCEAN WAVES

JORGE F. WILLEMSEN
University of Miami, Rosenstiel School of Marine and Atmospheric Science, 4600 Rickenbacker Causeway, Miami, Fl., 33149, USA

Abstract

When surface gravity waves in deep water are driven to a balance between energy input and dissipation, it has been observed that the spectrum of wave-height correlations exhibits power-law behavior consistent with the interpretation that over a range of length scales the surface is a self-affine fractal. Zakharov and L'vov long ago invoked a Kolmogorov cascade hypothesis to describe the nonlinear modal transfer of energy which might explain this observation, even in the absence of driving and dissipation. If their hypothesis is correct, a purely autonomous Hamiltonian system would be capable of driving a surface to fractal roughness, in analogy to dynamical system chaos. In this report I summarize a body of recent work aimed at testing the Z-L hypothesis through numerical integration of Zakharov's Hamiltonian equations. I investigate both scaling and the existence of postulated emergent dynamical times which scale with modenumber. The results are mixed, in ways not easily abstracted. They lead to hard questions regarding the roles of driving and dissipation, and the process of equilibration of the waveheight statistical distribution, which cannot be answered at our present state of knowledge. But I hope that the form of the questions will prove stimulating.

1. INTRODUCTION: SURFACE WAVES ON WATER

Waves at the surface of a body of water are seemingly infinite in their variety, from the rings emanating about the spot where a pebble is dropped, to the glimmering patches as a breeze picks up, to the heaving swell in deep waters, to the crashing rollers so beloved of surfers at the beach. Or: the crashing rollers so potentially damaging to the coastline and to other activities relating to the coasts. The first "modern" studies of wave dynamics evidently stem directly from the desire to save lives during the Allied invasion of North Africa during World War II through improved seastate forecasting.

Clearly "surface waves on water" encompasses a wide range of phenomena. In this paper I discuss a specific theme, gravity waves in deep water. These waves have been the subject of intense study by physical oceanographers, both theoretically[1-7] and experimentally.[8-11] In spite of this research, many interesting features of their dynamics are only partially understood. The most important include the fundamental mechanisms by which the wind exchanges energy and momentum with the sea, and through which the sea relaxes. In addition, there exist diverse theoretical foci. Reference 12, e.g. contains an up-to-date review of approaches which focus on wavefields with narrow spectral bandwidths. These interesting studies will not be discussed here for reasons which will become apparent shortly.

2. SCALING OF THE WAVEHEIGHT SPECTRUM

An especially interesting dynamical regime occurs after persistent driving by a windfield. Evidently an "equilibrium" between driving and dissipation through mechanisms such as wave breaking is established. In this regime, the angle averaged spectrum of waveheight correlations scales as $F(k) \sim k^{-7/2}$ independent of windspeed U for $k \gg g/U^2$. (Here g is the gravitational acceleration, and U^2/g is the only length that can be formed from the parameters of the problem.) As is often true in geophysical contexts, the scaling law applies over an order of magnitude at best. Other general features of the spectrum include the existence of a spectral peak at $k \sim g/U^2$ at large "fetch" (see below), and $F \to 0$ as $k \to 0$ (there are no waves of infinite wavelength).

Most recent research[11] involves issues such as the angular dependence of the spectrum, its dependence on the spatial regime over which a well-characterized windfield has been acting (fetch), and on the degree to which the above scaling law actually holds; these issues are in addition to the fundamental ones mentioned above. Yet the theoretical work cited in Refs. 1–7 focused on understanding of the dynamical mechanisms which produce scaling in the first place, and identifying which specifics determine the scaling exponent. I take the position that this remains a legitimate avenue of inquiry.

Viewed appropriately, the problem has much in common with more recent research regarding surface roughening,[13] although there are differences. The most important difference is a bit paradoxical. From preceding remarks, driving and dissipation play a crucial role in establishing the "equilibrium" regime, hence one expects the mechanisms should be crucial in determining scaling. Indeed, Phillips[7] establishes the scaling exponent precisely through matching the nonlinear spectral transfer to a wind driving function obtained from measurements.[10]

In a much earlier paper,[14] however, Zakharov and L'vov (ZL) advanced the seemingly remarkable hypothesis that the nonlinear Hamiltonian dynamics of such waves might lead to the experimentally observed asymptotic scaling exponent *in the absence of driving and dissipation.* Thus if the ZL hypothesis is correct, roughening leading to a fractal surface can be produced by an autonomous nonlinear Hamiltonian system. We see in this the analogue to the once surprising statement that autonomous nonlinear Hamiltonian systems support chaotic solutions. In the next section I introduce the Hamiltonian which describes gravity waves in deep water, and then the ZL argument.

3. FULL AND APPROXIMATE DYNAMICS

3.1 The Waterwave Hamiltonian

Assume that the liquid in the waves is described by potential flow, so that the energy contained in the waves in the limit of an infinitely deep ocean is:

$$H = \frac{1}{2}\int d\mathbf{x} \int_{-\infty}^{h} dz (\nabla\phi)^2 + \frac{g}{2}\int d\mathbf{x}\, h^2 \tag{1}$$

As before g is the gravitational acceleration, and we choose units $g = 1$; $\mathbf{x}$ are two dimensional Cartesian surface coordinates; $h(\mathbf{x}, t)$ is the wave amplitude; and $\phi(\mathbf{x}, z, t)$ is the velocity potential throughout the fluid. Denote $\Psi(\mathbf{x}, t) = \phi(\mathbf{x}, z, t)|_{z=h}$, the potential evaluated right at the surface. Then[4] Ψ and h are canonically conjugate variables, and Hamilton's equations read $\partial h/\partial t = \delta H/\delta\Psi$, $\partial\Psi/\partial t = -\delta H/\delta h$.

Next, introduce conjugate variables in wavenumber space. If h_k and Ψ_k refer to the conventional Fourier components of $h(\mathbf{x}, t)$ and $\Psi(\mathbf{x}, t)$, then the variables $a_{\mathbf{k}}(t)$ and $a^*_{\mathbf{k}}(t)$ defined by:

$$\begin{aligned} h_k &= \frac{k^{1/4}}{\sqrt{2}}(a_{\mathbf{k}}(t) + a^*_{-\mathbf{k}}(t)) \\ \Psi_k &= \frac{-i}{\sqrt{2}k^{1/4}}(a_{\mathbf{k}}(t) - a^*_{-\mathbf{k}}(t)) \end{aligned} \tag{2}$$

are canonically conjugate and satisfy $i\,\partial a_{\mathbf{k}}(t)/\partial t = \delta H/\delta a^*_{\mathbf{k}}(t)$, together with the complex conjugate relation.

Expansion of Eq. (1) in powers of $a_{\mathbf{k}}$ and $a^*_{\mathbf{k}}$ leads to $H = H_2 + H_3 + H_4 + \cdots$. The term H_3 can be eliminated[4] by a further canonical change of variables $a_{\mathbf{k}} \rightarrow a'_{\mathbf{k}}$ because the linearized dispersion relation $\omega_{\mathbf{k}} = \sqrt{gk}$, $k = |\mathbf{k}|$ does not permit a three-wave resonance. Through a well-documented procedure[12] one finds after this change:

$$H_2 = \int d\mathbf{k}\, \omega_{\mathbf{k}} a^*_{\mathbf{k}}(t) a_{\mathbf{k}}(t) \tag{3}$$

$$H_4 = \frac{1}{2}\int\cdots\int d\mathbf{k}_1, \ldots, d\mathbf{k}_4 T(\mathbf{k}_1, \mathbf{k}_2; \mathbf{k}_3, \mathbf{k}_4) a^*_{\mathbf{k}1}(t) a^*_{\mathbf{k}2}(t) a_{\mathbf{k}3}(t) a_{\mathbf{k}4}(t) \delta(\mathbf{k}_1 + \mathbf{k}_2 - \mathbf{k}_3 - \mathbf{k}_4) \tag{4}$$

The coupling term describing the resonant nonlinear interactions of four gravity waves was first derived by Hasselmann[1] within the context of transport theory. It can be expressed[4] as T^2. I will nevertheless refer throughout to T as the Hasselmann interaction.

Term T is too lengthy to reproduce here. However, three of its features are indispensable for what follows. First, T is symmetric under interchange of arguments ($\mathbf{k}_1 \leftrightarrow \mathbf{k}_2$) and/or ($\mathbf{k}_3 \leftrightarrow \mathbf{k}_4$), and it is symmetric under interchange of the first pair of arguments with the second pair. These symmetries guarantee that the interaction is real in the form of Eq. (4). In addition, T is a homogeneous function of degree three with respect to its arguments, $T(\lambda\mathbf{k}_1, \lambda\mathbf{k}_2; \lambda\mathbf{k}_3, \lambda\mathbf{k}_4) = \lambda^3 T(\mathbf{k}_1, \mathbf{k}_2; \mathbf{k}_3, \mathbf{k}_4)$. Such homogeneity is possible because there are no intrinsic length or time scales in the free theory. Finally, T is pronouncedly largest when all the arguments are approximately equal in magnitude and direction.[15] This property is referred to as "quasi-locality" of the interaction.

3.2 The Zakharov-Lvov Argument

The full ZL argument is based upon machinery developed in Ref. 2, and draws upon then contemporaneous conjectures from statistical physics also invoked in Ref. 13. What follows is a heuristic re-write along lines they themselves devised.

From Eq. (1) it follows that:

$$i\,\partial a_{\mathbf{k}}/\partial t = \omega_{\mathbf{k}} a_{\mathbf{k}} + \int \cdots \int d\mathbf{k}_1 d\mathbf{k}_2 d\mathbf{k}_3 T(\mathbf{k}, \mathbf{k}_1; \mathbf{k}_2, \mathbf{k}_3) a^*_{\mathbf{k}1} a_{\mathbf{k}2} a_{\mathbf{k}3} \delta(\mathbf{k} + \mathbf{k}_1 - \mathbf{k}_2 - \mathbf{k}_3) \tag{5}$$

Heuristically, rewrite this as:

$$i\,\partial a_{\mathbf{k}}/\partial t \cong S(\mathbf{k}) \langle a^*_{\mathbf{k}} a_{\mathbf{k}} \rangle a_{\mathbf{k}}\,, \tag{6}$$

where $S(\mathbf{k})$ is a quasi-local approximation to T to be described below. It has a degree of homogeneity equal to 5. Now, quasi-locality implies that only nearby modes are coupled, and these have roughly equal $a_{\mathbf{k}}$ values, so we can drop the subscripts. In the same spirit, the linear term is dropped because the modes are roughly on resonance ($\Sigma\omega_i \cong 0$). So if $\langle a^*_{\mathbf{k}} a_{\mathbf{k}} \rangle = (\text{const.}) k^{-2\alpha}$ in the sense of a "mean field approximation", the time constant over which dynamic changes occur is roughly:

$$\tau(k) \sim \frac{k^{2\alpha}}{S(\mathbf{k})} \sim k^{2\alpha - 5}\,. \tag{7}$$

We need one further crucial assumption: quasi-locality of the interaction suggests a Kolmogorov cascade for the rate of change of the energy flux. Denote the dominant part of the flux $(H_4 k^d/\tau_k)$. If this rate is a constant, we arrive at the consistency condition $\alpha = 1 + (d/2)$ in d spatial dimensions. If $d = 2$, $\alpha = 2$, so taking note of Eq. (2), the true spectrum will scale according to the (7/2) law. Application of similar considerations to the cascade of the quantity $H_4/\omega_{\mathbf{k}}$ (instead of H_4) leads to a (10/3) law also predicted in Refs. 4–5.

Two observations follow immediately. First, the ZL argument makes use of the structural properties of the Hasselmann interaction T, but does not invoke specific details. Thus it is natural to conjecture that scaling would follow within *any model* for T which incorporates its "universal" properties. Secondly, the argument implements the Kolmogorov cascade through the clever introduction of dynamical times $\tau(k)$, which are totally absent from the autonomous dynamics and must "emerge" through the solution itself. The interesting question arises as to whether the ZL hypotheses, involving both scaling and the emergence of dynamical time scales, are indeed realized in the dynamics using "universal" models for the nonlinear interaction T.

4. RESULTS FROM A SPECIFIC MODEL HAMILTONIAN

In order to address the question just raised, in Ref. 16 (NLI) I introduced models of the Hasselmann interaction and solved Eq. (5) numerically. In general terms, quasi-locality is imposed by writing T in the (shorthand) form:

$$T_{k123} \equiv S_{k123}\{[\delta(\mathbf{k}-\mathbf{k}_1-\mathbf{p})+\delta(\mathbf{k}-\mathbf{k}_1+\mathbf{p})][\delta(\mathbf{k}_2-\mathbf{k}_3-\mathbf{q})+\delta(\mathbf{k}_2-\mathbf{k}_3+\mathbf{q})]+[\mathbf{p}\leftrightarrow\mathbf{q}]\} \tag{8}$$

where $\mathbf{p}$ and $\mathbf{q}$ are treated as infinitesimal quantities, and S is a suitably symmetric algebraic function. In the simulations reported in NLI, I chose $d = 1$, $p = 2$, $q = 0$, and function:

$$S_{k123} = (\mathbf{k} \cdot \mathbf{k}_1 + \mathbf{k}_2 \cdot \mathbf{k}_3)^2 \sqrt{|\mathbf{k} \cdot \mathbf{k}_1| + |\mathbf{k}_2 \cdot \mathbf{k}_3|} \tag{9}$$

with $\mathbf{k}_3$ (e.g.) constrained by a delta function. I briefly discuss general choices later.

4.1 Numerical Solutions: Scaling Behavior

Initial values were assigned for integer k in the range 1–10 with equal amplitudes and random phases. The $k = 0$ mode grows at the very first time step because it can couple to modes $k = 1$ and $k = 2$. Similarly mode $k = -1$ couples at the next time step, and so forth. But this particular choice of S creates a bottleneck right at small negative k values, these modes stay small, and for most of what follows they are neglected. With different initial conditions/different S the cascade would propagate into the $k < 0$ regime as well. The choice does not affect $k > 0$, where new modes appear at each time step without bottlenecks.

The results reported in NLI indicated that after about 2.2×10^6 computer time steps, which is appr. $2\pi/\omega(k = 1)$, the quantity $|a_{\mathbf{k}}|^2$ scaled approximately as $k^{-3.87}$ for k in the range 5–50. This is reasonably close to the ZL predicted value (4.0) for this quantity in $d = 1$. It was also noted in that reference that the implied averaging $\langle \cdots \rangle$ required to define the autocorrelation function could sensibly be construed as a running time average (RTA).

Since the time of that publication further simulations have been carried out, explicitly using the RTA. Figure 1 shows the results of reinitializing the entire calculation, running out to 5×10^6 steps, then running a further 10^5 steps and forming 5 separate equally binned RTAs. The good news vis-a-vis NLI is that many more modes have successfully been grown,

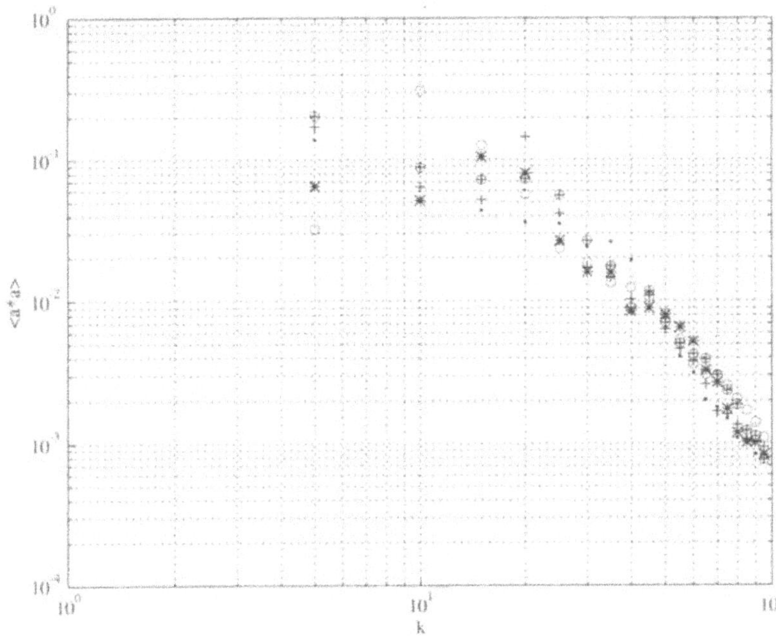

Fig. 1 $\langle a_{\mathbf{k}}^*(t) a_{\mathbf{k}}(t) \rangle_T$ vs. k as generated by the model Hamiltonian dynamics. See Sec. 4.1 for details of the calculation. This is a "new run", i.e., different data from that reported in Ref. 16.

and that RTA indeed reduces the scatter in the data considerably. There is more good news, but first we must note the bad news: the best fit over modes 20–96 is to $k^{-2.8}$! This must be addressed, but later.

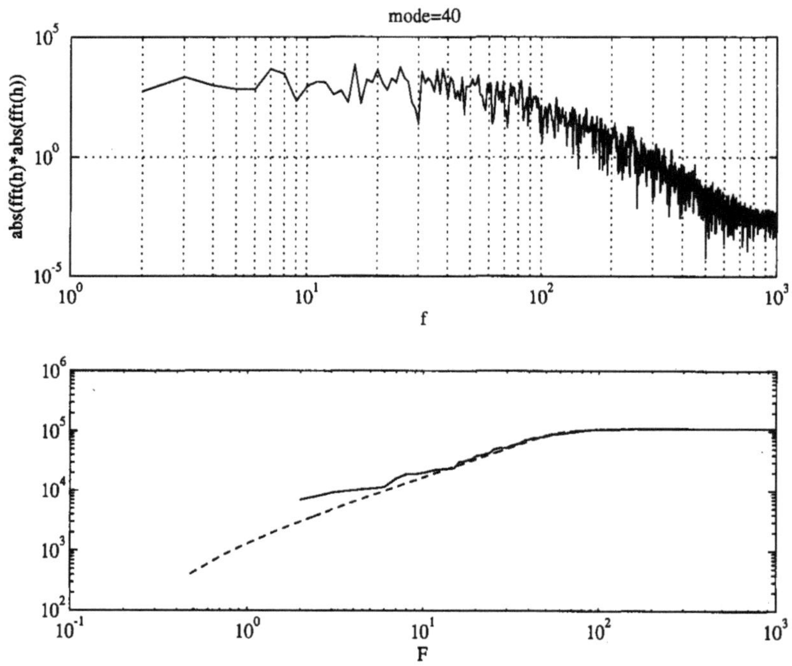

(a)

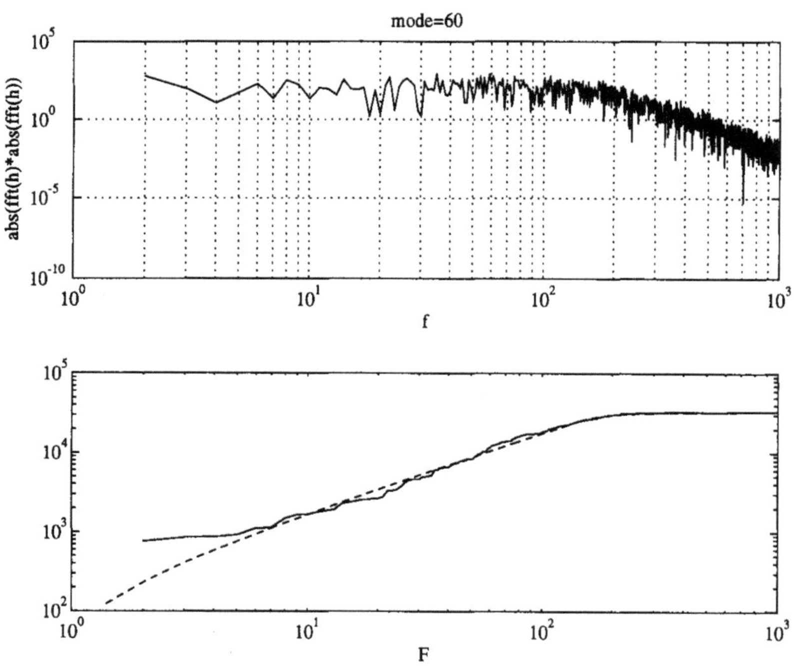

(b)

Fig. 2 The absolute value of the FFT with respect to time of the real part of the complex amplitude (here indicated as h) for Mode 40 (a), and Mode 60 (b). For details see Sec. 4.2.

4.2 Numerical Solutions: The Dynamic Interaction Times

A first attempt to isolate the Zakharov-L'vov dynamic interaction times was published recently.[17] In this work, the last previously generated a_k data from NLI were used to reinitialize the calculation, then selected modes were recorded at 2^{13} time intervals each containing 16 steps. Figure 2 displays the Fourier spectra (with respect to time) for two typical modes. The most striking feature of these spectra is the presence of a knee which separates a region of power-law decay from a nearly "white" region. Thus it was suggested that the position of the knee is a natural candidate for the desired $\tau(k)$, since it corresponds to the longest time over which the temporal autocorrelation function remains power-law correlated.

The methodology then consisted of identifying the position of the knee accurately, and investigating its dependence on wavenumber. I found it useful to perform a simple integration of the power spectra, followed by visual comparison with synthetic data. The method is fully described in Ref. 17 but too lengthy to repeat here.

The result of applying the procedure to modes 5–40 of the NLI data set is displayed in Fig. 3. The modes clearly split into high and low k groupings. For modes 5–20 a least squares fit gives $\tau(k) \sim k^{-2.31}$. Modes 20–40 can be fit with a linear dependence, but the coefficient (.02) is so small that $\tau(k) \sim$ constant seems equally reasonable. Recall from Eq. (6) that the ZL argument predicts $\tau(k) \sim k^{5-2\alpha}$ for $|a_k|^2 \sim k^{-2\alpha}$. Thus, returning to the NLI spectral exponent reported, we would predict that $5-2\alpha = 1.13$, which is not close to the numbers cited above.

To resolve this discrepancy, I argued that *the time series might provide information, otherwise not available, regarding the degree of development of the waveheight autocorrelation*

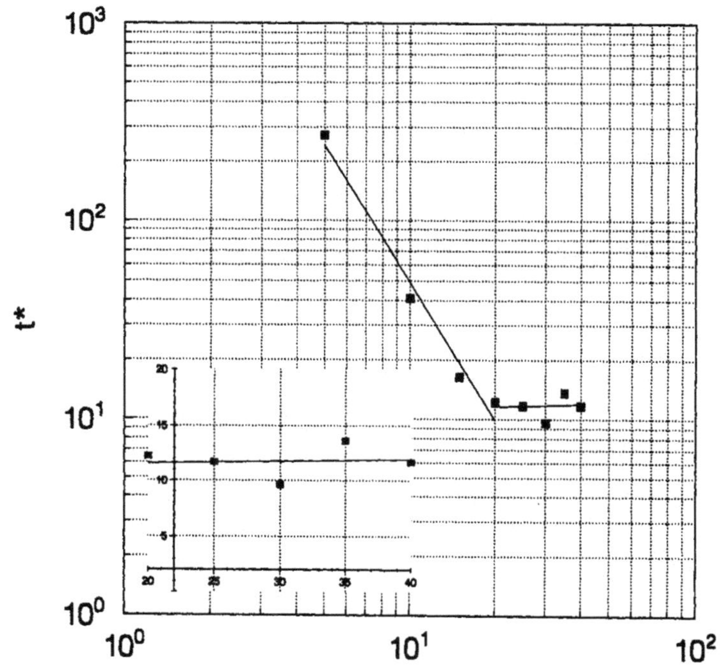

Fig. 3 The times corresponding to the knees in the power spectra in the "old" data sets published in Refs. 16, 17 and 19. The insert is a blow-up on a linear scale of modes 20–40.

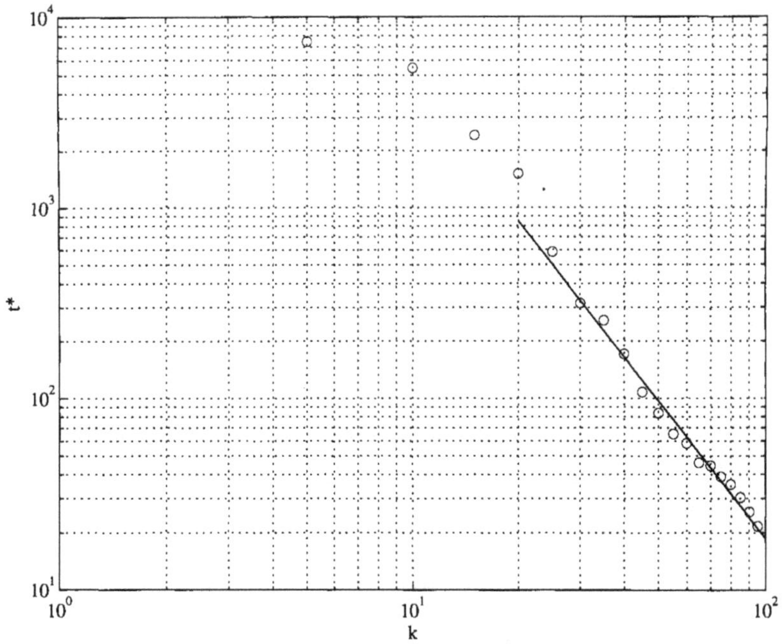

Fig. 4 The times corresponding to the knees in the power spectra in the "new" data sets.

function. Thus it might be legitimate to revert to the original data and re-interpret it in light of this new information. New fits to the data were attempted over the ranges 5–20 and 20–40 respectively. The first range was best fit by an exponent $2\alpha = 2.38$, while for the latter $2\alpha = 4.8$. These are clearly in much better agreement with the behavior of the observed transition times (knees), with (using now α as input) $5 - 2\alpha = 2.62$ (cf. 2.31) and 0.2 (cf. 0), respectively.

The upshot of this discussion is that the time series are consistent with "local" scaling exponents in accord with the ZL hypothesis. Naturally, the very concept of "local" scaling regions is anethema to a purist — but for a *growing* system it is perhaps not unreasonable to admit that at finite times modes present in the initial conditions may not yet have relaxed sufficiently, while higher modes have not yet developed up to the expected asymptotic behavior.

One way to settle this issue is through further simulation. Figure 4 illustrates the application of the time series analysis to the new data described in Sec. 4.1. To parallel the discussion there, the good news is that a single fit to the broad range of modes 25–95 is achieved, with exponent (−2.37). (Actually the fit from modes 5–95, −2.26, is not too different.) This number is in good agreement with the spectral slope (2.80) noted earlier. The bad news is the same as cited earlier: 2.37 is not a ZL exponent.

The super-good news, to an optimist, is that even though the spectral exponent is not the one predicted, there is consistency between the spectral exponent and the dynamic time. One interpretation is that the scaling hypothesis works, as does the mean field argument of Eq. (6). The Kolmogorov cascade hypothesis, on the other hand, which closes the exponent relations and so leads to unique exponents, fails. It could also be, however, that the "universality" hypothesis and/or the analysis in $d = 1$ are to blame. I briefly remark on analytic studies of these possibilities.

5. ANALYTIC INVESTIGATIONS IN $d = 1$

Equation (8) represents a general hypothesis regarding the implementation of quasi-locality of the Hasselmann interaction, whereas Eq. (9) specifies a particular model. In view of the observations above, that distinct numerical integrations of Eq. (5) subject to Eq. (9) produce significantly different spectral exponents, it is interesting to inquire whether analytical computations might be formulated which can clarify the situation. I have room only for an extremely abbreviated discussion of results described in detail in Ref. 18.

The central idea is to push Eq. (9) to its logical limit, so that under the impact of the Dirac delta functions selecting k_i values infinitesimally close to the central k, the function S (not restricted to the model of Eq. (9)) can legitimately be Taylor expanded about k. The complex fields a_{ki} are also to be Taylor expanded. This is shown to be legitimate for those values of central k well away from the edges of the cascade, where of course even the first derivative diverges. Upon reflection it is evident that in dimensionless units this requires that $t \gg k$. After a simple rescaling, the end result of this procedure, consistent with second order in parameters p and q, is a pair of real equations for the amplitude and phase of each field a_k:

$$\rho_t = -2^{9/2} k^3 F(\rho;\, \{\sigma\}) \sin\left\{\phi_{kk} + \frac{t}{4k^{3/2}}\right\} ; \tag{10}$$

$$\phi_t = -2^{9/2} k^5 \rho^2 \cos\left\{\phi_{kk} + \frac{t}{4k^{3/2}}\right\} ; \tag{11}$$

$$F(\rho;\, \{\sigma\}) = [(\sigma_3 + \sigma_5 + 2\sigma_4)\rho^3 + 3\sigma_1 \rho k^2 \rho_k^2 + \sigma_1 \rho^2 k^2 \rho_{kk} + 8\sigma_2 \rho^2 k \rho_k] . \tag{12}$$

The parameters σ appearing here relate to derivatives up to second order of an arbitrary appropriately symmetric function S.

In Ref. 18, I report that the Eqs. (10)–(12) support two classes of solutions (all for $d = 1$). One is $\rho(k, t) = (2k)^{-9/4}$ with a corresponding phase solution, and this is found to be linearly stable to a small perturbation. The second class is of the form $\rho = k^{-\beta}$ with the exponent determined through the equation $4\beta^2 - (B - 1)\beta + 4A = 0$; the parameters A and B may be expressed in terms of the first and second derivatives of the model for S. Viewed differently, two distinct real roots corresponding to the energy and action exponents discussed in Sec. 3 may emerge for suitable values of A and B, i.e., given specific constraints on the approximation to the nonlinear interaction T. Thus these exponents are not "universal". In addition, these solutions are linearly unstable in $d = 1$.

The numerical simulations reported earlier are not inconsistent with these findings. A numerical solution might well start off in one or another unstable regime, then bounce into a different regime due to computational roundoff error, for example. In this scenario, the postulated logical limit of quasi-local modeling reveals a possibility not heretofore considered: is the Kolmogorov state stable? Said differently, and perhaps within a context which transcends linear stability, one might ask if intrinsic dynamical fluctuations render the "mean field" cascade fundamentally erroneous, at least in $d = 1$. (The situation in $d = 2$ is more complicated.)

6. WHAT HAPPENED TO DRIVING AND DISSIPATION?

I have focused on free *Hamiltonian* dynamics, yet most theoretical work leading to scaling[1–7] invokes a transport formalism in which it is assumed *a priori* that the waveheight statistics are Gaussian. Thus, since the (7/2) law is obtained within this formalism once wind driving is taken into account,[7] most workers believe scaling requires driving. Curiously, however, Zakharov and Zaslavskiy[4] have demonstrated that the (7/2) exponent is an *exact result* within the formalism, *in the absence of driving or dissipation.* **The question emerges: how can all this be self-consistent?** How do fluctuations in the windfield drive the wavefield to achieve a Gaussian distribution, and a spectral exponent consistent with a) the average windfield; b) absence of an average windfield; and c) the Kolmogorov hypothesis? Please see Ref. 19 for speculations which I hope to develop in future work.

ACKNOWLEDGMENTS

This work was supported in part through ONR Grant No. N00014-92-J-1266. I thank Mr. Feng Yu for preparing the new figures and computing the new exponents.

REFERENCES

1. K. Hasselmann, *J. Fluid Mech.* **12**, 481 (1962).
2. V. Ye Zakharov and N. N. Filonenko, *Dokl. Akad. Nauk SSSR* **170**, 1292 (1966).
3. O. M. Phillips, **The Dynamics of the Upper Ocean**, 2nd ed. (Cambridge, 1977), Chap. 4.
4. V. Ye. Zakharov and M. M. Zaslavskiy, *Izv. Atm. Oc. Phys.* **18**, 747 (1982).
5. S. A. Kitaigorodskii, *J. Phys. Oceanog.* **13**, 816 (1983).
6. V. E. Zakharov, *Basic Plasma Physics* Vol. 2, ed. A. A. Galeev and R. N. Sudan. (Elsevier, 1984) and references therein), Chap. 5.1.
7. O. M. Phillips, *J. Fluid Mech.* **156**, 505 (1985).
8. M. L. Banner, *J. Phys. Ocean.* **20**, 966 (1990) contains an excellent critical review.
9. S. A. Kitaigorodskii, *Johns Hopkins APL Technical Digest* **8**, 11 (1987).
10. W. J. Plant, *J. Geophys. Res.* **87**, 1961 (1982).
11. Proceedings of Marseilles Conference on Wave Dynamics, ed. M. Donelan (1993).
12. J. L. Hammack and D. M. Henderson, *Annu. Rev. Fluid Mech.* **25**, 55 (1993); see also H. C. Yuen and B. M. Lake, *Ad. Appl. Mech.* **22**, 67 (1982).
13. F. Family and T. Vicsek, *J. Phys.* **A18**, L75 (1985); M. Kardar, G. Parisi and Y.-X. Zhang, *Phys. Rev. Lett.* **56**, 889 (1986).
14. V. Ye. Zakharov and V. S. L'vov, *Izv. Vusov. Radiofiz.* **18**, 1970 (1975).
15. M. S. Longuet-Higgins, *Proc. R. Soc. Lond.* **A347**, 311 (1976); M. J. H. Fox, ibid., **348**, 467 (1976).
16. J. F. Willemsen, *Phys. Rev. Lett.* **67**, 1964 (1991).
17. J. F. Willemsen, "Dynamical Time Scales Generated by a Model Nonlinear Hamiltonian," *Phys. Rev. Lett.* 1172 (1993).
18. J. F. Willemsen, "Analysis of Compact Models for Nonlinear Gravity Wave Interactions," *Phys. Fluids* **A5**, 2775 (1993).
19. J. F. Willemsen, in *Nonlinear Dynamics of Ocean Waves*, ed. M. Shlesinger (World Scientific Pub., Singapore Press, 1992) pp. 55–74.

Fractals, Vol. 1, No. 3 (1993) 539–546

A FRACTAL MODEL FOR SEISMICITY AT IZU-TOKAI REGION, CENTRAL JAPAN

B. BODRI*
Geophys. Re. Group, Hung. Acad. Sci., c/o Geophys. Dept., Eötvös Univ., Budapest, Hungary

Abstract

Fractal approach has been applied to investigate regional seismicity at the Izu peninsula — Tokai area, Central Japan. The frequency-magnitude distribution of earthquakes, distribution of epicenters, origin times of earthquakes, the fracture fault system in the region have been considered, and the fractal dimensions corresponding to them were calculated. A good correspondence in the fractal dimension values was found.

The frequency-magnitude distribution in the area shows a fractal dimension of 1.28, whilst $D = 1.15 \pm 0.18$ is representative of the geometry of the distribution of earthquake epicenters. The fractal dimension of faults for the Izu peninsula is found to be 1.16 ± 0.04, and in the whole Izu-Tokai region, values $1.1 < D < 1.3$ are characteristic. The temporal distribution of earthquakes yields a fractal dimension of 0.51 ± 0.03, which indicates a relatively weak clustering of events in time. Independent autocorrelation analysis also shows that the earthquakes in the area of study occur to a large extent statistically independent.

The general conclusion is that crustal deformation in the Izu-Tokai region occurs on a scale-invariant matrix faults. The behavior of the system is controlled by a single parameter, the fractal of dimension.

*Mailing address: B. Bodri
Department of Geophysics
Eötvös University
Budapest, Ludovika tér 2.,
Hungary 1083

1. INTRODUCTION

Geophysical phenomena are extremely complex. Deterministic modeling of many of them appears either fundamentally impossible, or can be carried out only with limited accuracy. For the majority of processes, only statistical bounds on the probability of some characteristics are revealed. This is especially true for seismicity, which is basically chaotic and nonpredictable. It was recognized long ago that in many geographic regions, earthquakes have a well-defined statistical distribution. The number of earthquakes N with a magnitude greater than M satisfies the empirical Gutenberg-Richter relation:

$$\ln N = -bM + a\,, \tag{1}$$

where a and b are constants.

The definition of fractal distribution is that the number of objects N with a characteristic size greater than r satisfies the relation:

$$N = Cr^{-D}\,, \tag{2}$$

where D is the fractal dimension and C a constant of proportionality. The applicability of a fractal relation implies that the physical process responsible for its creation is scale invariant over the range of applicability of the relation.

Seismicity has recently been recognized to show fractal geometry. Aki[1] demonstrated that Eq. (1) is equivalent to the definition of a fractal distribution, so that the fractal dimension of seismic activity D is simply twice the b-value ($D = 2b$). Subsequently, it has been established in numerous studies that seismicity is scale invariant over a wide range of scales both regionally and on a world-wide basis (e.g. Ref. 2).

Fractal distribution of regional seismicity has opened a very productive way for the analysis of its features. Seismicity is a result of a very complex geological-geophysical process. The fractals are a method of correlating a variety of quantitative observations characterizing seismicity. In this work we use the fractal concept to better understand the relative importance of, and the interactions among, different factors in the seismicity generating process at Izu peninsula — Tokai district area, Central Honshu, Japan. The area of study is situated at the Pacific coast of Central Honshu between the meridians 137° and 139 °E and the parallels 34.5° and 35.5 °N (Fig. 1). Geologically, it constitutes a mosaic of blocks separated by tectonic lines or conjugate sets of active faults.

2. EARTHQUAKE DATA

As a data source on seismicity, we used the listings of the Seismological Bulletin of the Japan Meteorological Agency (JMA) for a period of 7 years from 1983 to 1989. Aftershock sequences, which are obviously correlated to the main shock were eliminated from the data. The intensities of earthquakes vary in the range from less than 1 (JMA-scale) to $M = 6.0$ (some earthquakes are of undefined intensity), the hypocenter depths falling between 0 and 349 km.

Earthquakes of medium size ($M < 3.0$–4.0) are most characteristic of the area of study. The probability of strong earthquakes occurring decreases steeply. The recurrence time between earthquakes already at $M \geq 4.0$ becomes, by more than an order of magnitude, longer than that for events with $M \geq 3.0$ (Fig. 2).

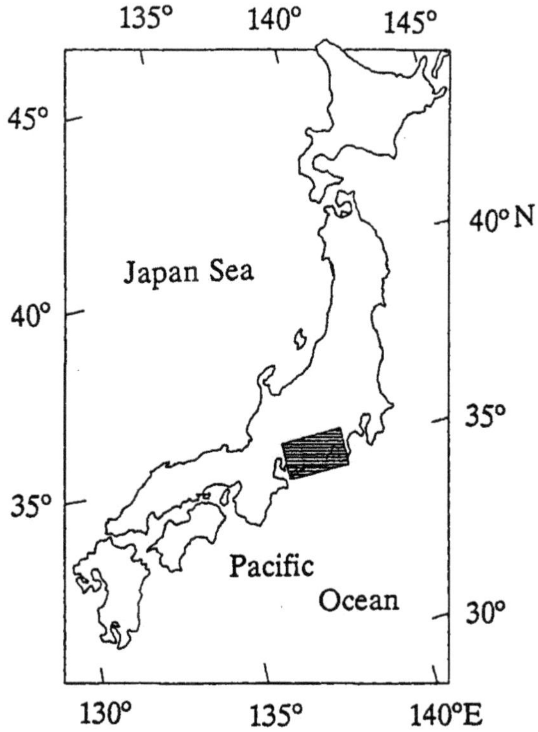

Fig. 1 Index map of Japan, showing (by hatching) the area of study.

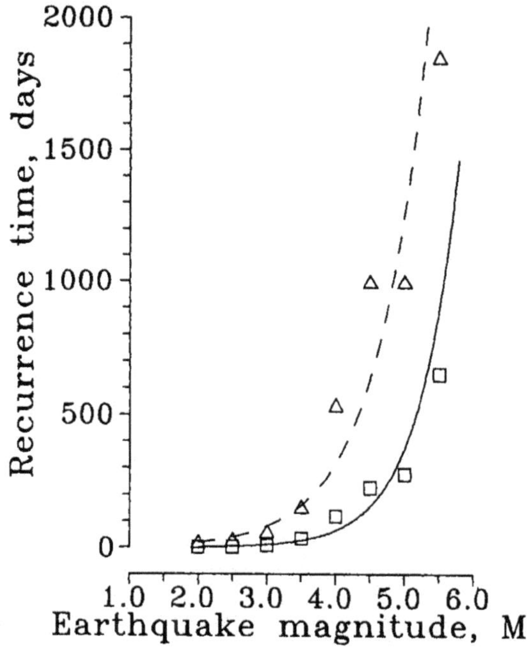

Fig. 2 Maximum (triangles and dashed line) and average (squares and solid line) recurrence times of earthquakes vs. magnitude for the Izu-Tokai area.

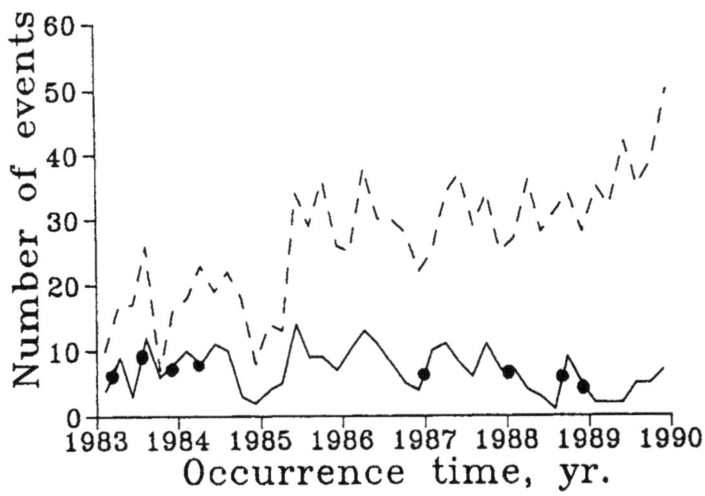

Fig. 3 Variation of the number of earthquakes occurring with 60-day time intervals over the considered years. Dashed line: earthquakes with $M \geq 2.0$. Solid line: earthquakes with $M \geq 3.0$. Dots indicate large ($M \geq 5.0$) earthquakes.

Figure 3 shows the variation of the number of earthquakes that have occurred in the area within 60-day intervals, as a function of time over the whole period considered. A clear trend in the number of events of $M \geq 2.0$ can be observed in the diagram. The number of events increases; this increase, however, is not to be considered as a result of increasing seismic activity, but reflects simply the progressive improvement in the techniques of the detection of weak earthquakes. The solid curve in Fig. 3, representing the number of earthquakes with $M \geq 3.0$, on the 5% significance level by Fisher's statistics does not show any trend. This means that the series of events with magnitudes $M \geq 3.0$ can be considered *stationary*. Therefore we have used $M = 3.0$ as the over limit magnitude, and all following considerations will be restricted to earthquakes of $M \geq 3.0$ only (a total of 291 such events occurred in the study area). The hypothesis about independence of the events with $M \geq 3.0$ in their sequence in time has been checked by calculating the autocorrelation function for the actual time intervals between the subsequent events. Autocorrelation shows values close to zero, thus the subsequent seismic events occur *independently* from one another. It was also examined if seismic events ($M \geq 3.0$) in the Izu-Tokai area show a Poisson distribution. We used the Kolmogorov-Smirnov statistics, according to which (at the 5% significance level) the sequence of earthquakes occurring in the Izu-Tokai area *does not show a Poisson distribution.* Consequently, one cannot use this statistics to obtain probability estimates for the characteristics of the distribution of earthquakes, and in attempts for predictions one has to rely upon empirical analysis of the available data.

3. TEMPORAL DISTRIBUTION OF EARTHQUAKES

Earthquakes can be considered as point events in space and time. The scale-invariant properties of their temporal distribution can be investigated on the basis of 1-D fractal cluster analysis (the Cantor dust model). The fractal dimension of temporal occurrences of earthquakes is obtained in this work by a time increments — counting algorithm. We divide the total investigated time interval into a series of progressively smaller intervals. The total time interval for which data is available is about 7 years. For intervals longer than 50 days, every interval contains an earthquake. When the intervals appear shorter than 2.5 days, only single events occur in each time interval, and a no clustering is observed. For lengths of time segments between these limits, a fractal clustering is observed which gives a fractal dimension $D = 0.51 \pm 0.03$ (Fig. 4). For the Cantor dust models, the fractal dimensions lie in the interval $0 < D < 1$ and indicate the strength of the clustering. The more isolated the clusters, the smaller is the value of D. The value that we obtain for D indicates the absence of a strong clustering. However, in the present case it is difficult to give a direct interpretation of the fractal dimension. The noticeable curvature of the data in Fig. 4 suggests a possible superposition of scale-invariant clustering and random events (for comparison see, e.g. Ref. 3).

4. SPATIAL DISTRIBUTION OF EARTHQUAKES

Considering the problem of fractal clustering in spatial dimensions, the scale-invariant properties of the epicenter distribution can be studied on the basis of 2-D fractal cluster analysis (the Sierpinski carpet model).

Figure 5 shows the epicenter distribution of earthquakes with magnitudes $M \geq 3.0$ in the study area. The fractal dimension of the spatial distribution of epicenters has been calculated by the well-known box-counting technique. Calculations were carried out separately for $1° \times 1°$ squares on the left (Tokai region) and the right (Izu peninsula area) sides of the study area. Clustering appears in the $\sim$ 2.5–25 km spatial interval. The fractal dimensions of these two particular areas show values $D_l = 1.33 \pm 0.18$ and $D_r = 0.96 \pm 0.17$, respectively. Such a difference indicates a stronger degree of clustering in the Izu peninsula area. We take the weighted mean value $D_{\rm wm} = 1.15 \pm 0.18$ as the characteristic fractal dimension for the epicenter distribution over the whole area (Fig. 6).

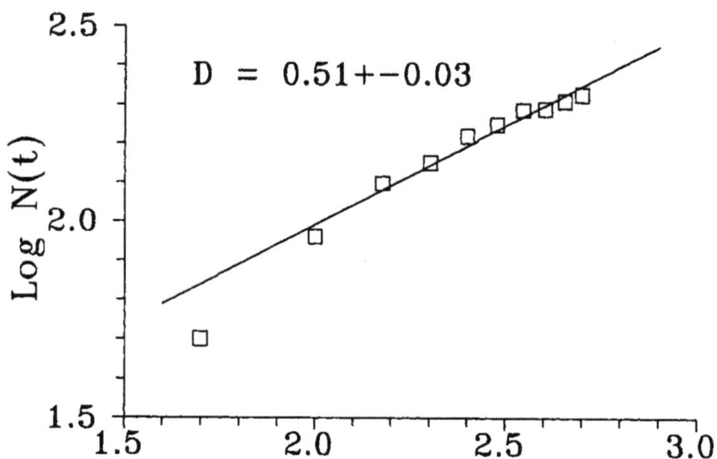

Fig. 4 Number of time intervals (presented by squares) that contain seismic event as a function of the total number of time intervals (with length t). T is the length of the whole time period over which data is considered. Solid line represents best fit to data. The lowest data point is not considered for fitting.

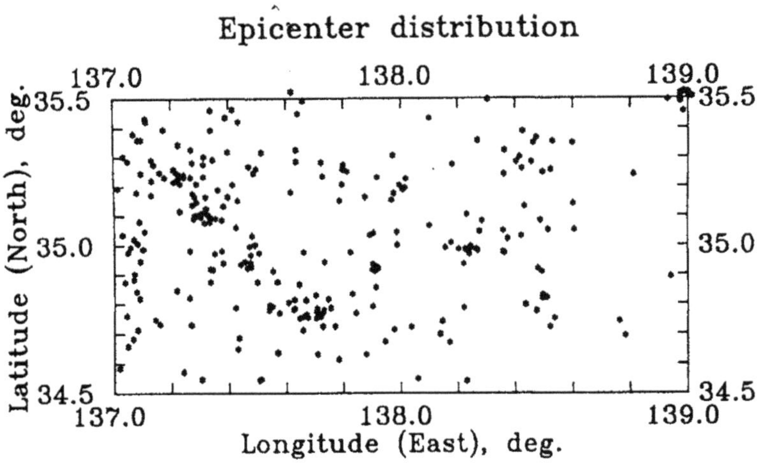

Fig. 5 Distribution of earthquake epicenters (for $M \geq 3.0$ only; their number = 291) in the study area for the period 1983–1989.

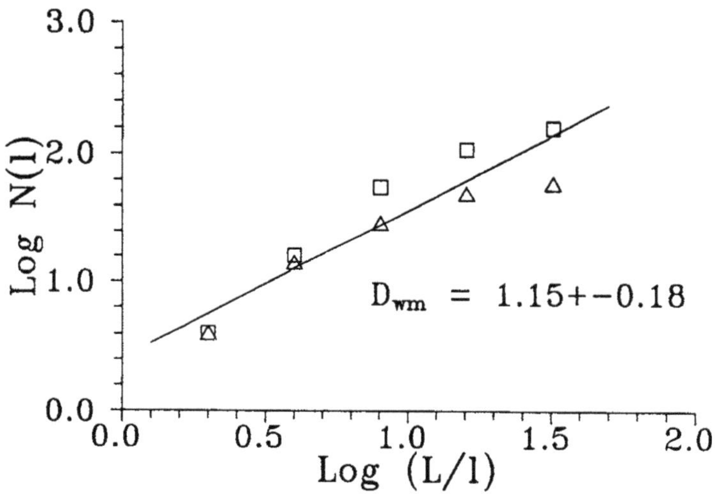

Fig. 6 Number of square boxes that contain epicenter as a function of the number of boxes along one side of the grid dividing the area. L is linear extension (~ 90 km) of the 1°*1° square area, l is a box side length. Squares: data for Tokai part; triangles: data for Izu part. Solid line: regression line with weighted mean coefficients from the two data sets.

5. THE FREQUENCY-MAGNITUDE DISTRIBUTION

The frequency-magnitude distribution of seismicity in the Izu-Tokai area is illustrated in Fig. 7, where the number of earthquakes with magnitudes greater than M is plotted as a function of M. In the magnitude range $3.0 \leq M < 6.0$, the data is in good agreement with Eq. (1) taking $b = 0.64 \pm 0.03$. This value for the b parameter corresponds to the fractal dimension $D = 1.28$.

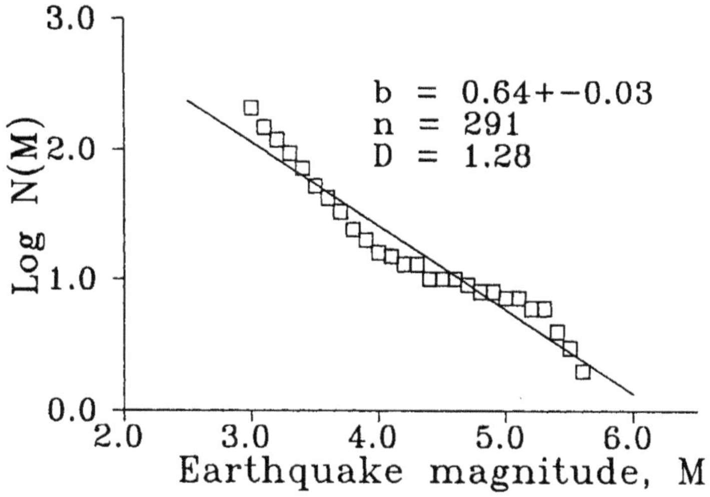

Fig. 7 Number of earthquakes that occurred in the study area for the period 1983–1989, with magnitudes greater than M as a function of M. n is total number of earthquakes by which the frequency-magnitude relation is obtained. Solid line with slope $-b$ represents best fit to data.

6. THE DISTRIBUTION OF CRUSTAL FAULTS

Seismicity in the study area is generally associated with a series of faults that have been identified from aerial photographs and partly from surface mapping. The fractal dimension of fault systems in Japan has been investigated by Hirata,[4] who considered data only on active faults that have moves during the Quaternary. Using the square box-counting algorithm he has shown that the spatial distribution of faults in the Izu-Tokai area is fractal at the scale $\sim$ 2–75 km, with a fractal dimension of $1.1 < D < 1.3$. Of the Izu peninsula, the region that has been investigated most intensively in Japan, on the same scale the fractal dimension $D = 1.16 \pm 0.04$ is characteristic.[4]

7. CONCLUSIONS

As shown above (for summary see Table 1), the temporal-spatial-magnitude patterns of earthquake occurrences in the 3.0–6.0 magnitude range at the Izu-Tokai region are fractals. The good coincidence of the fractal dimensions of spatial-magnitude observations in Table 1 implies that the process responsible for seismicity creation is scale invariant over the range of applicability of the obtained relations, and the behavior of the tectonic system in the study area is controlled by a single parameter, the D value.

Temporal distribution of earthquakes suggests that seismic events in the region do not exhibit significant clustering. They appear to a large extent statistically and apparently causally independent.

Active seismicity in the study area indicates a large regional stress level. However, as seen in Fig. 3, no significant drop in seismicity appears after large earthquakes. The shows that tectonic structure of the region is such that great earthquakes cannot reduce the stress level over an extended area, i.e., most of the regional stress reduces by small and medium-size earthquakes.

Table 1 Fractal Dimension of Different Observations Characterizing Seismicity in the Izu-Tokai Region

Observation		Fractal Dimension
Temporal distribution of earthquakes		0.51 ± 0.03
Spatial distribution in:	Tokai part	1.33 ± 0.18
	Izu part	0.96 ± 0.17
	whole area	$\mathbf{1.15 \pm 0.18}$
Frequency-magnitude distribution		1.28
Fault distribution in:	Izu peninsula	1.16 ± 0.04
	whole area	**1.1–1.13**

Generally, earthquakes are a major manifestation of fault motion. It can be concluded from Ref. 4 that crustal deformation in the Izu-Tokai region occurs on a scale-invariant matrix of faults. Such a fault geometry requires a substantial portion of the deformation in a fault system to occur on minor faults and not on the main faults.[5] This result is in good correspondence with our conclusion about the mechanism of tectonic stress sink in the area.

REFERENCES

1. K. Aki, "A Probabilistic Synthesis of Precursory Phenomena," in *Earthquake Prediction*, eds. D. W. Simpson and P. G. Richards (Amer. Geophys. Union, Washington, 1981) pp. 556–574.
2. D. L. Turcotte, "A Fractal Model for Crustal Deformation," *Tectonophysics* **132**, 261–269 (1986).
3. R. F. Smalley, Jr., J.-L. Chatelain, D. L. Turcotte and R. Prévot, "A Fractal Approach to the Clustering of Earthquakes: Applications to the Seismicity of the New Hebrides," *Bull. Seism. Soc. Amer.* **77**, 1368–1381 (1987).
4. T. Hirata, "Fractal Dimension of Ault Systems in Japan: Fractal Structure in Rock Fracture Geometry at Various Scales," *PAGEOPH* **131**, 157–170 (1989).
5. G. King, "The Accomodation of Large Strains in the Upper Lithosphere of the Earth and other Solids by Self-Similar Fault Systems: The Geometrical Origin of *B*-Value," *PAGEOPH* **121**, 761–815 (1983).

Fractals, Vol. 1, No. 3 (1993) 547–559

FRACTAL ANALYSES OF ANISOTROPIC FRACTURE SURFACES

B. L. COX and J. S. Y. WANG
Earth Sciences Division, Lawrence Berkeley Laboratory, Berkeley, California 94720, USA

Abstract

Natural surfaces of rock fractures often have anisotropic asperity distributions, especially for shear fractures or faults. The asperity distributions could be treated as self-affine fractals with directional dependent scaling in the plane of the rock surfaces. Different fractal analyses (divider, slit-island, variogram) are applied to surface distributions of asperity data (topography): (1) a granitic fracture from the Stripa mine in Sweden; (2) a faulted and geothermally altered fracture from Dixie Valley, Nevada, USA. The cutoff patterns (indicator maps) of the granitic fracture show a radial pattern, while those of the faulted fracture show a very anisotropic stretched pattern of shapes. Different cutoff patterns of the same surface generally yield the same fractal dimension with the slit-island technique. The slit-island technique assumes that the cut-off patterns are self-similar in the plane of the surface, with the perimeter versus area analyzed for the entire population of contours, regardless of aspect ratio. We measure the variance in the two coordinate directions as a function of perimeter/area ratio for the anisotropic fracture from Dixie Valley to determine a self-affine scaling ratio for the slit-island analysis. We compare this ratio with anisotropy ratios obtained from simulated flow models based on channeling of flow through the largest openings. The possible applications of fractal analyses to both the geometry and flow are evaluated.

1. INTRODUCTION

Naturally fractured rocks provide channels for fluid flow, which are important for the transport of both contaminants and economically valuable metals and hydrocarbons. Many researchers have measured geometry and fluid flow through laboratory-scale discrete fractures in rocks,[1] providing a means of investigating relationships between flow properties and fracture geometry. The measurement of fracture openings (apertures) is not easy, but has been accomplished by several different methods, including mechanical profilometry,[2] filling the fracture with a dyed silicone polymer cast and imaging the transmitted light through the cast,[3] and making an epoxy replica of the rock, filling the rock with a dyed fluid and imaging the fluid.[4] In this paper we analyze fracture aperture data obtained by imaging dyed fluid through epoxy replicas, as well as data obtained by mechanical profilometry. We analyze the fracture aperture patterns using fractal methods.

Fractal analysis is one means of characterizing the geometry of the fracture openings (aperture topography).[5–8] Previous fractal analysis of two different fractures by slit-island method showed that the slit-island method is insensitive to anisotropy in the plane of the fracture.[8] Faulted fractures show stretched patterns along the direction of fault movement, and are therefore anisotropic in that the aspect ratio of the patterns varies in the two coordinate directions of the fracture plane. In this paper we study (1) whether or not the anisotropic faulted Dixie Valley fracture would be self-affine in the plane of the fracture; and (2) whether the anisotropic geometry of the asperities would create a significant flow anisotropy.

Here, we describe the two fracture data sets and show the results of fractal analyses by the slit-island, divider, and variogram methods.[7] Then, we apply a variation of the method of Matsuchita and Ouchi[9] to the two different fractures. In this method, the standard deviation of the component vectors of different lengths of contours or profiles are plotted on a log-log plot against the total length of the segments. If the two directions are self-similar, the slope should be the same; if they are self-affine, the slopes differ. Finally, we apply a steady-state electrical analog flow simulation to the fractures in two orthogonal directions, to determine to what extent the anisotropy of the patterns affects the anisotropy of fluid flow.

2. FRACTURE APERTURE DATA

Fracture aperture measurements of two fractures were obtained by light transmission through dyed fluid in epoxy replicas of the two rock samples.[4] One of these is a granitic rock core drilled perpendicular to the fracture, obtained from the Stripa mine in Sweden (Stripa). The other fracture is from a highly altered gabbro obtained from a steeply dipping fault in Dixie Valley, Nevada, USA (Dixie Valley 1). The Dixie Valley fracture was also measured by a mechanical profilometer[2] (Dixie Valley 2). The two Dixie Valley fracture data sets were analyzed and compared. The statistical properties of the three data sets are shown in Table 1.

Surface topography of the absolute fracture openings (Fig. 1) shows that the natural fracture aperture surfaces are not isotropic. The Stripa fracture shows a warping along the y-coordinate direction but not in the x-direction, while the Dixie Valley fracture shows a wavy pattern in the x-coordinate direction, but not in the y-direction. Histograms of the aperture thicknesses[8] showed that the Stripa fracture is more skewed, while the Dixie Valley fracture is more peaked. The cutoff patterns (indicator maps) for the two fractures show a distinct radial pattern for the Stripa fracture, and a series of stretched patterns elongated in the faulting direction (Y-direction) for the Dixie Valley 1 fracture (Fig. 2).

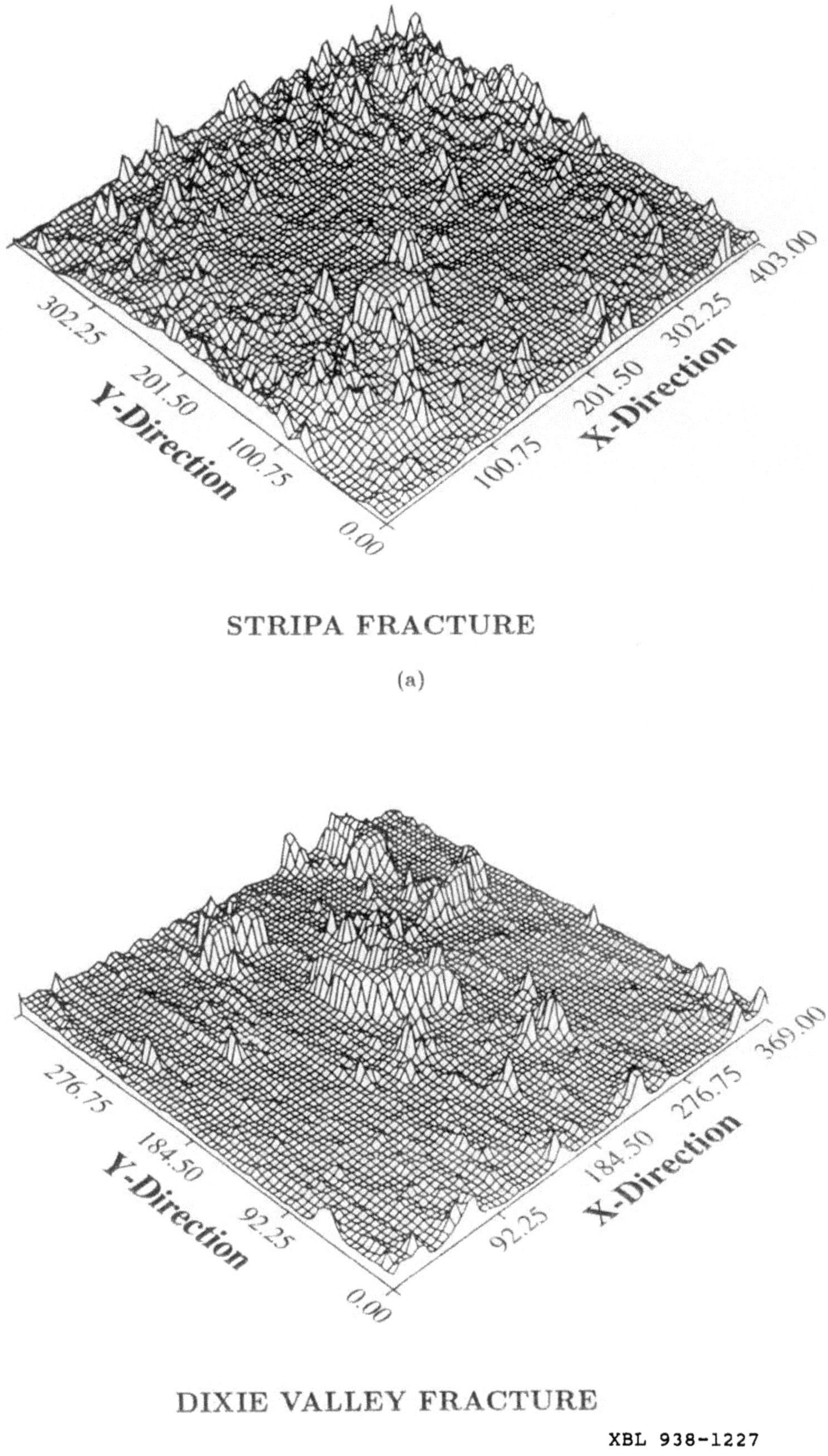

(b)

Fig. 1 Surface plots of fracture apertures for (a) Stripa, and (b) Dixie Valley 1 fractures. Vertical exaggeration is approximately 100 times.

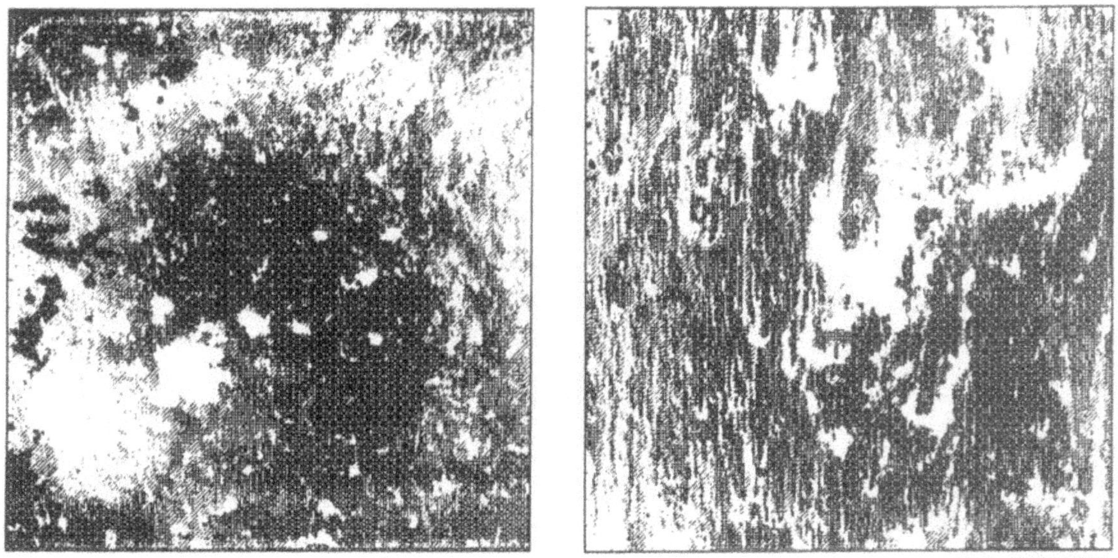

XBL 938-1228

(a) (b)

Fig. 2 Cutoff patterns for (a) Stripa, and (b) Dixie Valley 1 fracture apertures. White area corresponds to largest aperture cutoff.

Table 1 Statistics of Aperture Distributions

	Stripa	Dixie Valley 1	Dixie Valley 2
Length of Side	81 mm	74 mm	72 mm
Grid Spacing	0.2 mm	0.2 mm	0.5 mm
Number of Grid Points	403 × 403	369 × 369	144 × 144
Minimum Aperture	0.0 mm	0.0 mm	0.0 mm
Maximum Aperture	0.110 mm	0.153 mm	0.242 mm
Mean Aperture	0.037 mm	0.042 mm	0.041 mm
Standard Deviation	0.021	0.021	0.019
Covariance	0.50	0.58	0.46
Skewness	1.01	1.84	3.53
Kurtosis	0.95	5.05	18.4

3. FRACTAL ANALYSIS OF FRACTURE APERTURE PATTERNS

There are many different methods of measuring the fractal dimension, and we have used three of these (slit-island, divider, variogram) to measure the fractal dimension of the Stripa and Dixie Valley 1 fractures.[7,8]

3.1 Slit-Island Method

The slit-island method was first introduced by Mandelbrot et al.[10] who applied it to the fractal analysis of steel fractures. They filled steel fracture faces with nickel, then polished the fracture faces parallel to the fracture plane, so that approximately equal amounts of nickel and steel were evident on the polished fracture faces. This created a pattern of islands of nickel in a sea of steel, which they called "slit-islands". We applied this technique numerically to the fracture aperture data by selecting several cutoffs, creating surface contours which divided the surface into two kinds of shapes, white (islands) and black (sea). Both the perimeter and area are measured for each island, and the measurements are plotted on a log-log plot, with the perimeter on the x-axis and the area on the y-axis. The measurements follow a straight line, implying self-similarity to the scaling of the population of islands. The fractal dimension of the curves is $D_s = 2/\text{slope}$. The fractal dimension by the slit-island method was independent of the cutoff choice (Table 2). One of the cutoffs and slit-island plots for Dixie Valley is shown in Fig. 3. The fractal dimension of the Stripa contours is 1.3, while that of Dixie Valley is 1.4.

Table 2 Slit Island Analyses

Name	Cutoff, mm	#Islands	Area Percent	Total Perimeter	D_s
Stripa	0.027	536	36	5.178 m	1.31
Stripa	0.037	791	55	5.558 m	1.30
Stripa	0.050	919	72	5.294 m	1.33
Dixie Valley 1	0.030	383	28	5.994 m	1.39
Dixie Valley 1	0.040	698	53	6.656 m	1.40
Dixie Valley 1	0.050	897	71	5.208 m	1.36

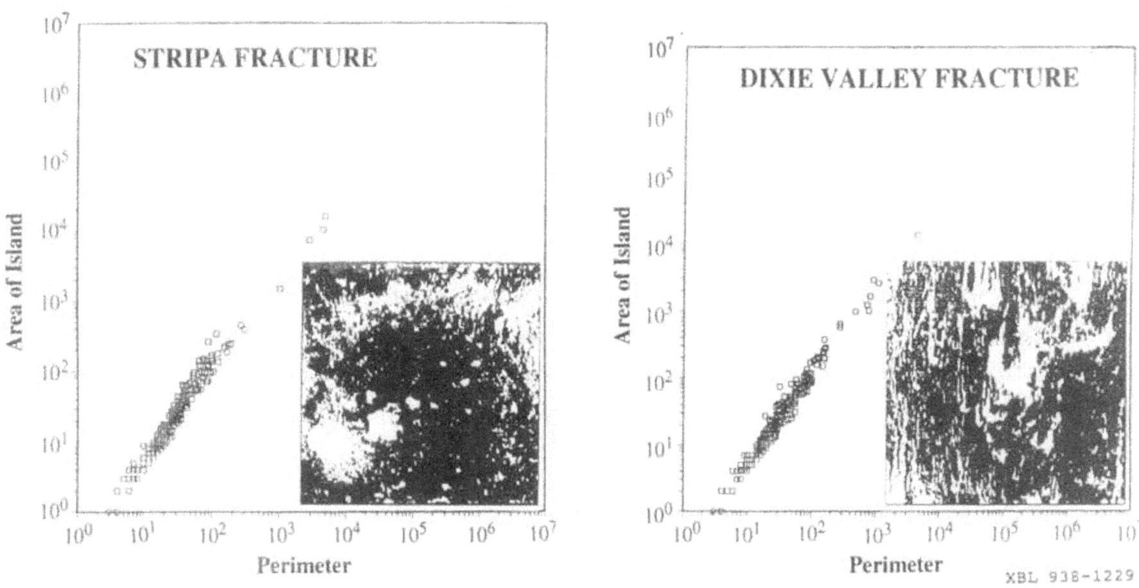

(a) (b)

Fig. 3 Slit-Island analyses for (a) Stripa, and (b) Dixie Valley 1 fractures. Cutoffs are 0.037 mm (a) and 0.040 mm (b). White areas (45% of (a); 47% of (b), represent apertures greater than cutoffs.

3.2 Divider and Variogram Methods

Profiles may be analyzed by the divider and variogram methods to determine the fractal dimension of the surface. However, this analysis is one-dimensional, in contrast to the slit-island technique, which is two-dimensional. The slope of the log-log plot of the profile length versus the sampling length was used to determine the fractal dimension by the divider method, where $D_d = 2 - \text{slope}$, and the profile length is the total length of the profile calculated along the sampling length intervals. The variogram fractal dimension was determined by plotting the semivariance versus the sampling distance on a log-log plot,[7,8] and using the relationship $D_v = (4 - \text{slope})/2$.

The divider and variogram fractal analyses are not equivalent, and the variograms tend to give higher fractal dimensions (D_v in Table 3). Both methods are sensitive to anisotropy, and we found some evidence of anisotropy in both fractures, although in the Dixie Valley fracture the anisotropy was more pronounced. In general, the divider method agreed with the slit-island method, while the variogram fractal dimension was much higher.

Table 3 Divider and Variogram Fractal Analyses

Name	Direction	D_d	D_v
Stripa	Y	1.36	1.82
Stripa	X	1.36	1.62
Dixie Valley 1	Y	1.36	1.81
Dixie Valley 1	X	1.46	1.89

4. ANISOTROPY/SELF-AFFINITY MEASUREMENTS

4.1 Anisotropy

Anisotropy of the fracture aperture patterns can be readily seen in the surface and cutoff plots (Figs. 1 and 2), as well as in the fractal analysis of profile lengths and variograms (Table 3). Another indication of anisotropy is the standard variogram plotted on linear axis. We observed that the linear part of the standard geostatistical variogram of the Dixie Valley fracture for the direction perpendicular to the fault movement was shifted along the semi-variance axis for the correlated apertures relative to the variogram for the direction parallel to the fault (Fig. 4). There was no offset in the standard variograms for Stripa.[8]

4.2 Self-Affinity

The concept of self-affinity was introduced by Mandelbrot,[11,12] and has been a very important concept for the determination of fractal scaling. Self-affinity is particularly important for topography, where the vertical scaling processes are generally much different than the horizontal scaling. It is necessary to check for self-affinity before determining a fractal dimension. Several methods for checking and correcting for self-affine scaling have been

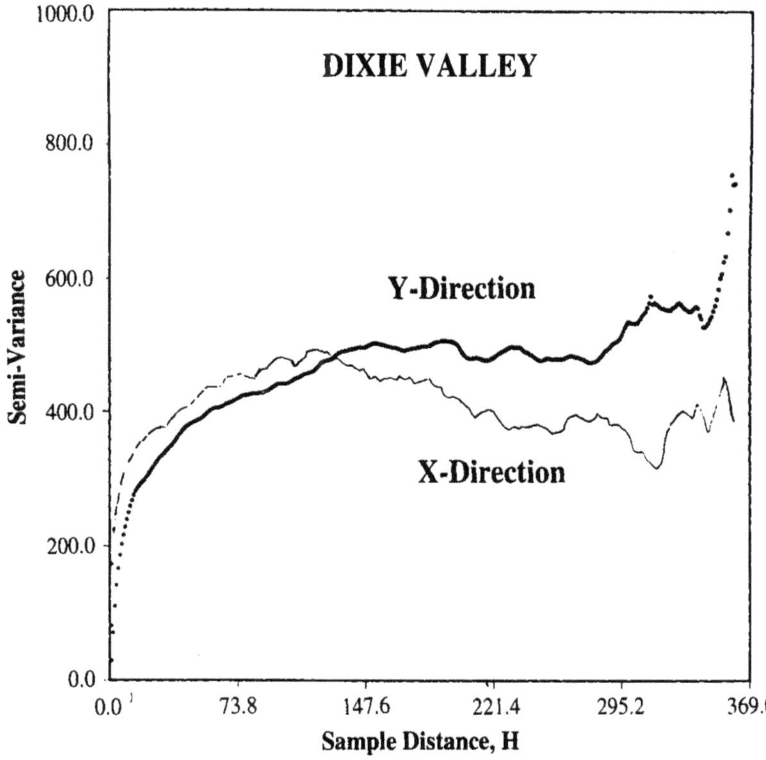

Fig. 4 Standard variograms for Dixie Valley 1, for profiles along the fault direction (y-direction), and transverse to fault direction (x-direction).

published. We adapted the method of Matsuchita and Ouchi[9] to the slit-island cutoffs by determining the standard deviation of the x and y coordinates defining the perimeter boundary for each island. These were then plotted on log-log graphs with the log perimeter on the x-axis (Fig. 5). The open circles represent the X-direction, and solid circles the Y-direction. The Stripa fracture showed little difference in the two directions, while the Dixie Valley fracture showed a definite offset. However, the slope is the same for the two directions, indicating that the island distribution was anisotropic, but not self-affine.

4.3 Flow Anisotropy

We used a two-dimensional steady-state electrical analog fluid flow model to determine the flow anisotropy of the fractures. The program we used for our simulation is described in Moreno et al.[13] The program solves for steady state flow from one constant potential head boundary to the opposite (parallel) boundary with a lower potential; the other two boundaries of the square fracture are no flow boundaries. The input for the program is the square grid of aperture values. These openings correspond to resistance values, where the resistance is inversely proportional to the cube of the aperture. The current between adjacent grid points varies according to the local apertures. Applying mass balance of current for each node, the field of unknown potentials at each node is solved using an iterative conjugate gradient matrix solver. From the potential at each node the flow between adjacent nodes is determined.

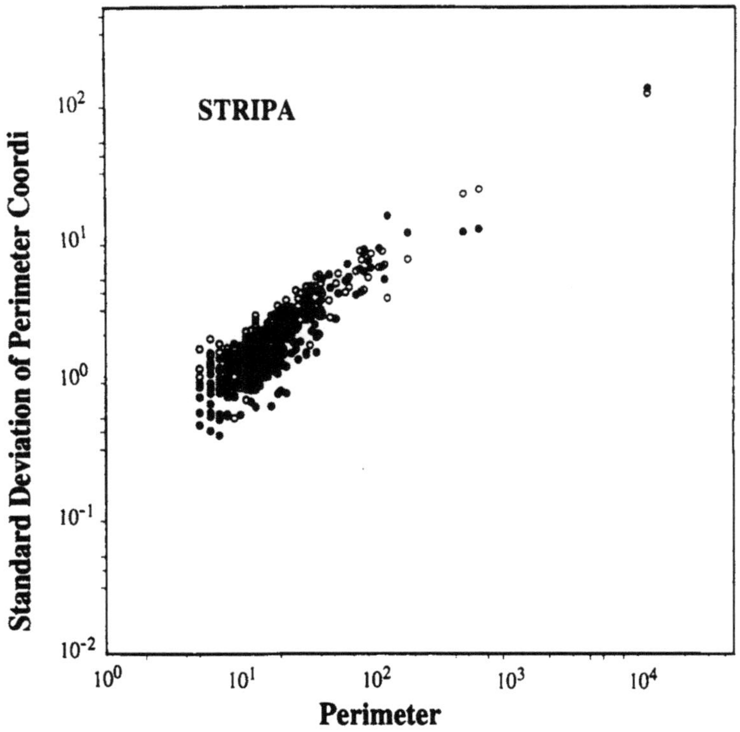

(a)

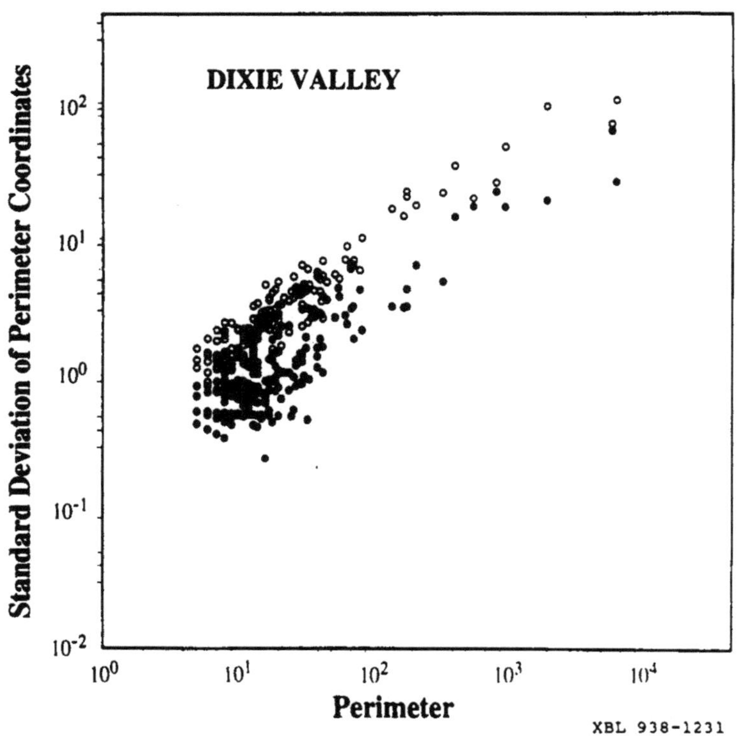

(b)

Fig. 5 Self-Affine analyses for (a) Stripa, and (b) Dixie Valley 1 islands for 0.037 mm (a) and 0.040 mm (b) cutoffs. Standard deviation (x-direction, filled circles; y-direction, open circles) is relative to all the perimeter points for each island.

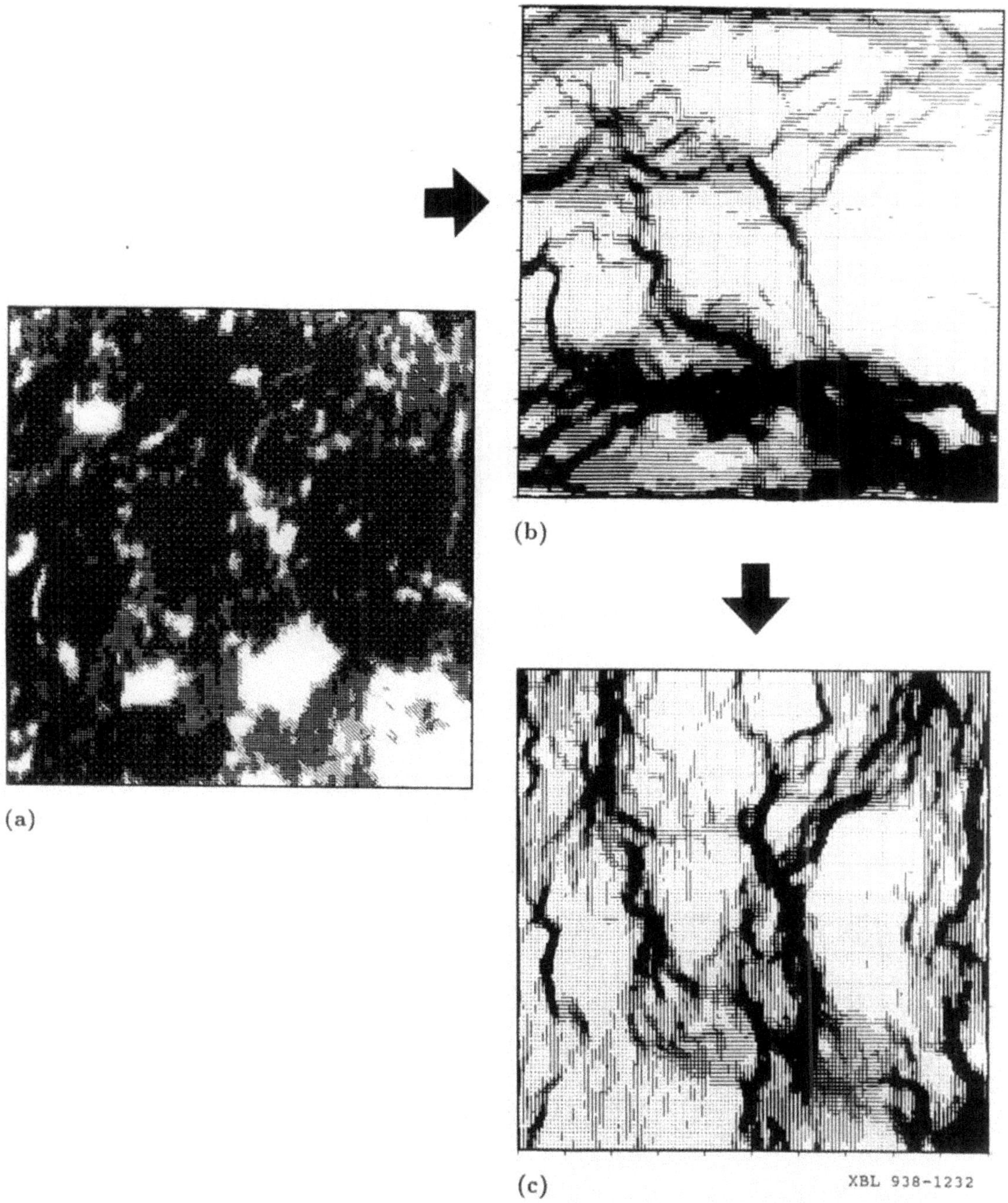

Fig. 6 **Fracture aperture (a) cutoff, and (b), (c) flow channels, for a 123 × 123 subsection of the Stripa fracture. Channel widths are proportional to the square root of the current for each individual plot.**

We have simulated flow through selected subregions of grid sizes up to 128 × 128. A 123 × 123 section of the Stripa data set was selected from the center of the 403 × 403 data set. An indicator map is shown in Fig. 6(a), where the white areas indicate the widest openings. The simulated flow channels through the fracture are shown in Figs. 6(b) and 6(c) for the two different directions. In the channel maps, the black lines correspond to the fluid flow, and the width of the lines is proportional to the square root of the current, and is scaled independently for each channel map. The anisotropy ratios of the total currents in the two directions are summarized in Table 4.

Table 4 Flow Anisotropy

	Stripa (Fig. 6)	Dixie Valley 1 (Fig. 8)	Dixie Valley 2 (Fig. 7)
Length of Side	24.6 mm	22.0 mm	64.0 mm
Grid Spacing	0.2 mm	0.2 mm	0.5 mm
Number of Grid Points	123 × 123	110 × 110	128 × 128
Flow Anisotropy Ratio	1.1	1.9	1.1

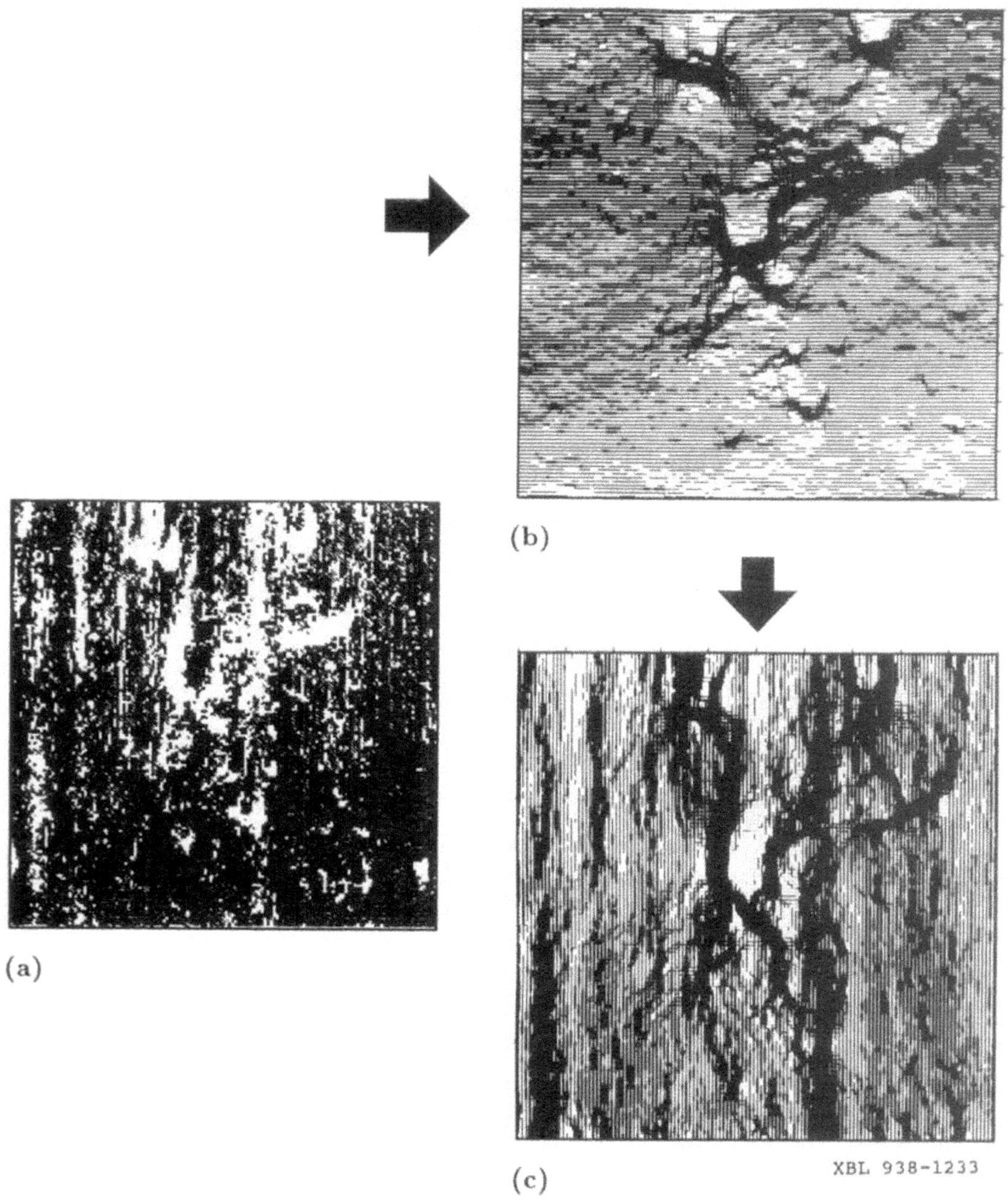

Fig. 7 Fracture aperture (a) cutoff, and (b), (c) flow channels, for a 128 × 128 subsection of the Dixie Valley 2 data set. Channel widths are proportional to the square root of the current for each individual plot.

A 128 × 128 section of the Dixie Valley 2 data set, representing approximately 80 percent of the total fracture, was tested for flow anisotropy. The aperture indicator map and channeling maps are shown in Fig. 7, and the flow anisotropy is shown in Table 4. A small (110 × 110) relatively uniform section of the Dixie Valley 1 data set was tested for flow anisotropy (Fig. 8; Table 4). Note that although the width of the channels transverse to the elongated direction are wider than those parallel to the elongations, the absolute flow through Fig. 8(c) is nearly twice that of Fig. 8(b).

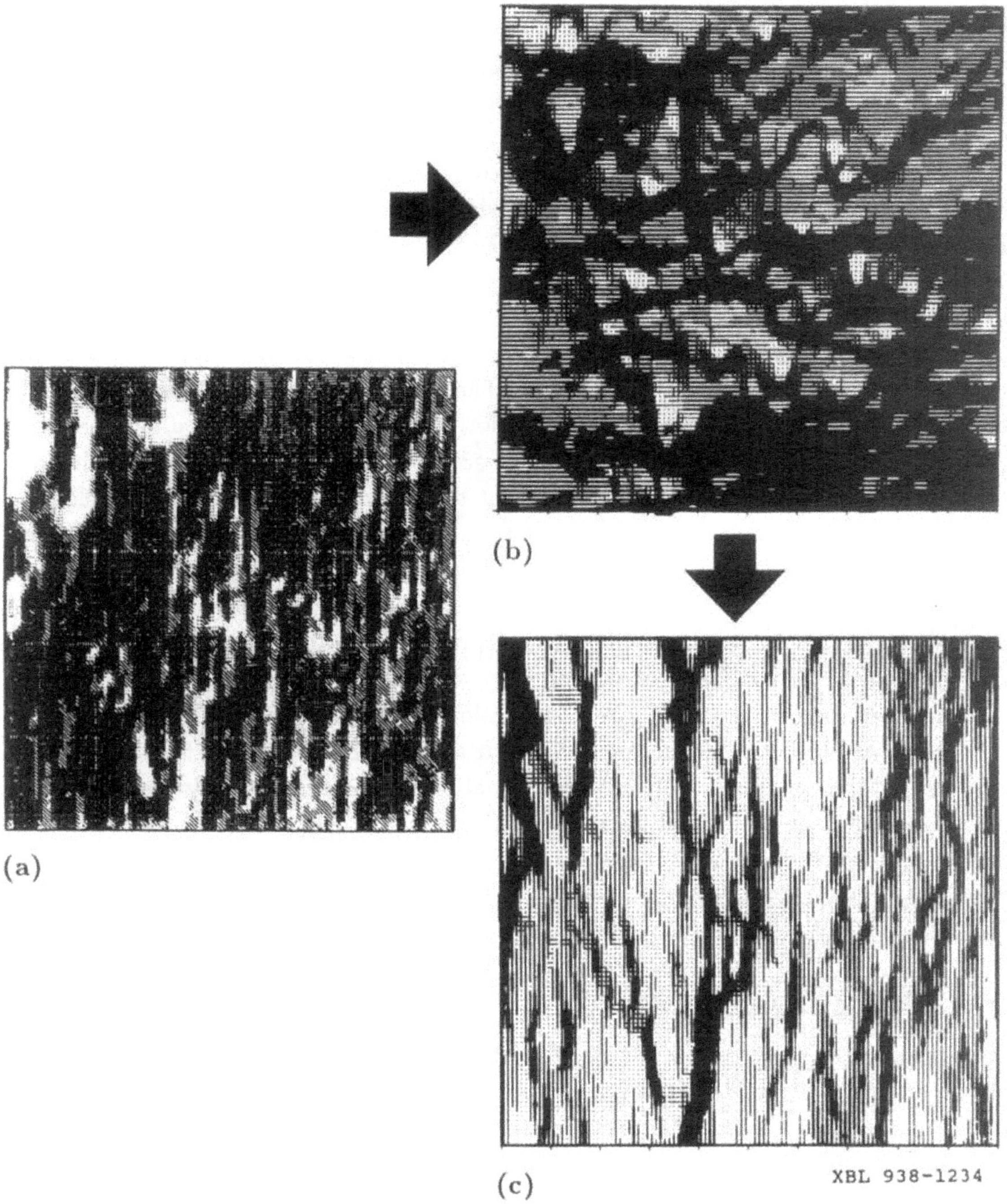

Fig. 8 Fracture aperture (a) cutoff, and (b), (c) flow channels, for a 110 × 110 subsection of the Dixie Valley 1 data set. Channel widths are proportional to the square root of the current for each individual plot.

5. DISCUSSION

The obvious anisotropic patterns of asperities in the faulted Dixie Valley fracture result in offsets in the semi-variogram, and variations in the fractal dimension with direction for the divider and variogram fractal analyses. However, the slit-island method is insensitive to anisotropy. If we plot the standard deviation of the component vectors of the perimeter of each island versus the length of the perimeter on a log-log graph, we see a definite difference in the amplitude of the standard deviations, but the slope is the same, indicating that the fracture pattern is not self-affine. This can be understood if one looks at the shapes of the patterns. Although these shapes have an aspect ratio different from 1, they preserve the same ratio no matter what size the shapes. These are self-similar shapes, but they are anisotropic. The processes that formed the aperture pattern were probably some combination of initial stress-induced fracturing followed by cycles of grinding, filling, and possibly precipitation and dissolution of minerals. These dynamic processes appear to be self-similar in the plane of the fracture.

The analysis of the perimeter versus area could be taken a step further. We analyzed the entire population of islands, and not individual islands. The fractal dimension of the entire patterns of contours is not necessarily the same as that of single contour lines.[14] We have performed the slit-island and self-affine analyses ignoring this possibility, and without considering possible sets of different shapes.

The anisotropy does not appear to be very significant for the two-dimensional steady-state electrical resistance model of fluid flow. The channeling patterns depend on direction that the flow traverses, but the ratios of the flows for the two directions differed less than 2 for the most anisotropic case (Dixie Valley 1 subset), and were barely distinguishable for the larger Dixie Valley 2 area. It appears that the size and distribution of the elongated shapes relative to the total size of the flow field play an important role. A small sample may show very strong anisotropy which will disappear if a larger region of the fracture is sampled. When the sample region is much larger than the average sizes of elongated islands the flow can easily circulate around the islands, resulting in flow fields with total flow rate insensitive to the local anisotropic islands.

In summary, we demonstrated several indications of anisotropy in the two fractures. The cutoff patterns of the asperities were self-similar and not self-affine. The flow anisotropy as indicated by the conductivity in two orthogonal directions was not very great. However, the channeling patterns differed in the two orthogonal directions for all three cases. These channel patterns indicate that transport of solutes and particles might differ significantly even though the flow rate might be nearly the same.

ACKNOWLEDGMENTS

The authors with to thank Dr. Robert Zimmerman and Dr. Yvonne Tsang for reviewing this paper, and to Mr. Frank Hale for helping with the flow simulations. We also appreciate helpful discussion with Dr. Tamas Vicsek. Work is performed under the auspices of the Director, Office of Civilian Radioactive Waste Management, Yucca Mountain Site Characterization Project Office, Regulatory and Site Evaluation Division of the U.S. Department of Energy through Contract Number DE-AC03-76SF0098.

REFERENCES

1. J. S. Y. Wang, "Flow and Transport in Fractured Rocks," *Reviews in Geophysics Supplement: U.S. National Report to International Union of Geodesy and Geophysics 1987–1990*, pp. 1987–1940 (April, 1991).
2. W. L. Power and T. E. Tullis, "The Contact Between Opposing Fault Surfaces at Dixie Valley, Nevada, and Implications for Fault Mechanics," *J. Geophys. Res.* **97**(NB11), 15425–15435 (1992).
3. B. L. Cox, K. Pruess and P. Persoff, "A Casting and Imaging Technique for Determining Void Geometry and Relative Permeability Behavior of a Single Fracture Specimen," *Proc. 15th Workshop Geothermal Reservoir Engineering*, pp. 173–178 (Stanford University, Stanford, California, 1990).
4. P. Persoff, K. Pruess and L. Myer, "Two-phase Flow Visualization and Relative Permeability Measurement in Transparent Replicas of Rough-walled Rock Fractures," *Proceedings, 16th Workshop on Geothermal Reservoir Engineering* (Stanford University, Stanford, California, 1991).
5. S. R. Brown, S. R., "A Note on the Description of Surface Roughness Using Fractal Dimension," *Geophys. Rev. Lett.* **14**(11), 1095–1098 (1987).
6. W. L. Power and T. E. Tullis, "Euclidean and Fractal Models for the Description of Rock Surface Roughness," *J. Geophys. Res.* **96**(NB1), 415–424 (1991).
7. B. L. Cox and J. S. Y. Wang, "Fractal Surfaces: Measurement and Applications in the Earth Sciences," *Fractals* **1**(1), 87–115 (1993).
8. B. L. Cox and J. S. Y. Wang, "Single Fracture Aperture Patterns: Characterization by Slit-island Fractal Analysis," *Proceedings, International High Level Radioactive Waste Management Conference* (Las Vegas, NV, 1993).
9. M. Matsuchita and S. Ouchi, "On the Self-affinity of Various Curves," *Physica* **D38**, 246–251 (1989).
10. B. B. Mandelbrot, D. E. Passoja and A. J. Paullay, A. J., "Fractal Character of Fracture Surfaces of Metals," *Nature* **308**, 721–722 (1984).
11. B. B. Mandelbrot, *The Fractal Geometry of Nature* (New York, W. H. Freeman, 1982).
12. B. B. Mandelbrot, "Self-affine Fractals and Fractal Dimension," *Physica Scripta* **32**, 257–260 (1985).
13. L. Moreno, Y. W. Tsang, C. F. Tsang, F. V. Hale and I. Neretnieks, "Flow and Tracer Transport in a Single Fracture: A Stochastic Model and its Relation to Some Field Observations," *Water Resources Research* **24**(12), 2033–2048 (1988).
14. M. Matsuchita, S. Ouchi and K. Honda, "On the Fractal Structure and Statistics of Contour Lines on a Self-affine Surface," *J. Phys. Soc. Japan* **60**(7), 2109–2112 (1991).

Fractals, Vol. 1, No. 3 (1993) 560–567

BI-MULTIFRACTAL ANALYSIS AND MULTI-AFFINE MODELING OF NON-STATIONARY GEOPHYSICAL PROCESSES, APPLICATION TO TURBULENCE AND CLOUDS

A. B. DAVIS, A. L. MARSHAK[†] and W. J. WISCOMBE
NASA Goddard Space Flight Center, Greenbelt, MD 20771, USA
[†]*SSAI, 5900 Princess Garden Parkway, Lanham, MD 20706, USA*

Abstract

In recent years, considerable progress has been achieved in the description of natural variability, largely due to the widespread use of scale-invariant concepts such as fractals and multifractals. In particular, this last concept has been used to clarify the fuzzy notion of "inhomogeneity" by introducing and quantifying the effects of intermittency. In this paper, we present a more comprehensive approach to multifractal data analysis and simulation that includes and combines the currently popular singularity analysis techniques with the more traditional approach based on structure functions. Being related to the new idea of "multi-affinity", these last statistics are regaining favor and constitute the proper framework to address the problem of quantifying and qualifying yet another outstanding fuzzy notion, that of "non-stationarity". This is an important step because non-stationary behavior is ubiquitous in Nature.

Using turbulence as an example, we also show how a unified multifractal formalism can help in extracting, from data alone, the "effective constitutive laws" that describe phenomenologically the nonlinearities of the macroscopic transport processes that shape the geophysical field represented by the dataset. Finally, we argue that the essential multifractality of any natural system can be captured on the "$q = 1$ multifractal plane" and describe ways in which it can be used in practical geophysical problems.

1. MOTIVATION AND OVERVIEW

Most data analysis techniques in usage (including the currently popular multifractal ones), make at least implicitly stationarity assumptions about the data, viewed as a representative realization of a stochastic process. Nature in the meantime is providing us with ample evidence of non-stationary behavior. In the following section, we show that "multi-affinity"[1] is the proper framework when dealing with non-stationary scale-invariant processes since directly related to structure functions. In Sec. 3 we survey multi-singularity analysis as a means to characterize quantitatively and qualitatively intermittency. The possibilities opened simply by connecting the two multi-scaling approaches are very promising (Sec. 4). Finally, we synthesize our outlook with the help of the "$q = 1$ multifractal plane".

2. MULTI-AFFINITY ANALYSIS, QUANTIFYING AND QUALIFYING NON-STATIONARITY

2.1 Scale-invariance, Stationarity and Stationary Increments

We take $\phi(x)$ to be a generic geophysical signal representative of either a time-series or a field assumed, for simplicity, to be a one-dimensional probing of some kind. More specifically, we are given $N+1$ real values over the interval $[0, L]$, sampled at rate $1/\ell$:

$$\phi(x), \quad \frac{x}{\ell} = 0, 1, \ldots, N\left(N = \frac{L}{\ell}\right). \tag{1}$$

In a typical situation, the above signal is highly fluctuating, "rough", possibly discontinuous. This is an indication that complicated, strongly nonlinear physical processes are at work generating $\phi(x)$, and the only symmetry we can reasonably posit for the system is scale-invariance. We will therefore use scale, denoted r, as a parameter in all of our statistics and seek statistically robust power-law behavior with respect to r. In sharp contrast with mathematical models, real world systems only scale over a finite range going from the "inner" (or "homogeneity" or "dissipation") scale η_ϕ to the "outer" (or "forcing" or "integral") scale R_ϕ. The instrumentally accessible portion of this range is $[\max\{\ell, \eta_\phi\}, \min\{L, R_\phi\}]$ but, for simplicity, we will take $\ell = \eta_\phi$ and $L = R_\phi$ throughout the following.

Our first task is to define "stationary" features of the dataset $\phi(x)$ to work on, with statistical stationarity meaning in essence "invariance under translation". The simplest scale-dependent statistic of a stationary process is its 2-point auto-correlation function:

$$G_\phi(r) = \langle \phi(x)\phi(x-r)\rangle \sim r^{-\mu_\phi}(\ell \le r \le L), \quad 0 < \mu_\phi < 1 \tag{2a}$$

where $\langle\cdot\rangle$ designates ensemble-averaging. The exponent μ_ϕ has a natural range: correlations are generally expected to decrease with increasing separation ($\mu_\phi > 0$) but too rapid decays ($\mu_\phi \ge 1$) are basically equivalent to "δ-correlations".

An even more popular 2nd order statistic is the energy spectrum $E_\phi(k)$:

$$E_\phi(k) = \sum_{\pm}\langle \hat\phi(\pm k)\hat\phi(\pm k)^*\rangle \sim k^{-\beta_\phi}\left(\frac{1}{L} \le k \le \frac{1}{2}\ell\right), \quad 0 < \beta_\phi = 1 - \mu_\phi < 1 \tag{2b}$$

where $\hat{}$ designates a Fourier transform. The two statistics are closely related since $G_\phi(r)$ and $E_\phi(k)$ are in Fourier duality (Wiener-Khintchine theorem). In scale-invariant situations, this leads to the connection between the two exponents in Eqs. (2). We are more interested in the limit, $\mu_0 \to 0$ ($\beta_\phi \to 1$), where we anticipate long-range correlations which are to be interpreted as a symptom of non-stationarity. We will focus more specifically on processes with $1 < \beta_\phi < 3$ which are said to have "stationary increments". Many turbulent[2–4] or otherwise natural[5,6] signals are in this class.

In the up-coming subsection, we characterize the multi-scaling properties of increments themselves over the full range of available scales, thus dealing with the non-stationarity directly. In Sec. 3, we adopt an altogether different strategy by first deriving from $\phi(x)$ another field which is itself stationary and can justifiably be investigated using methods requiring this property. This new field is required to contain the most interesting information on $\phi(x)$, specifically concerning its "intermittency".

2.2 Scaling Structure Functions and Special Cases

Define the increment $\Delta\phi(r;\, x) = \phi(x) - \phi(x-r)$ and consider the scaling of the structure functions:

$$\langle|\Delta\phi(r;\, x)|^p\rangle \sim r^{\zeta_\phi(p)}(\ell \le r \le L)\,, \quad -1 < p < \infty \tag{3}$$

where normalization requires that $\zeta_\phi(0) = 0$. For the lower limit on p and ways to elevate it, see Ref. 7. Within the framework of turbulence (revisited below), $\zeta_\phi(p)$ is given a formal multifractal interpretation in Ref. 8 by using its Legendre transform. The associated $D_\phi(h)$ "spectrum" is understood to be the fractal dimension of the subset of $[0,\, L]$ where the local Hölder exponent $h(x) = h$, this last exponent being defined in $|\Delta\phi(r;\, x)| \sim r^{h(x)}$.

Weakly variable increments ($\langle|\Delta\phi(r;\, x)|^p\rangle \approx \langle|\Delta\phi(r;\, x)|\rangle^p$) lead directly to $\zeta_\phi(p) = pH$, in other words, simple- or mono-scaling structure functions where $0 < H < 1$. The former ($H \to 0$) limit brings us back to stationary scaling processes. The latter ($H \to 1$) limit corresponds to almost everywhere differentiable processes; in particular, this case contains noiseless trends: $\phi(x) = ax + b$ where a and b are random variables (as long as $a \ne 0$, almost surely). If $\zeta_\phi(p)$ is not linear then it is necessarily concave ($\zeta''_\phi(p) < 0$) and $H_\phi(p) = \zeta_\phi(p)/p$ will be a decreasing hierarchy of exponents. Two values of p have received a lot of attention.

Firstly, there exists a Fourier duality for non-stationary processes between $E_\phi(k)$ and $\langle|\phi(x) - \phi(x-r)|^2\rangle$, often referred to as *the* structure function. For scaling processes, this leads to[9]:

$$\beta_\phi = 2H_\phi(2) + 1 \ge 1\,, \tag{4}$$

where the "=" is obtained in the stationary limit ($H_\phi(2) \to 0$). Beyond the opposite limit ($H_\phi(2) \to 1$), we find once differentiable functions ($\beta_\phi \ge 3$), hence $1 < \beta_\phi < 3$.

Secondly, one can relate the fractal dimension $D_{g(\phi)}$ of the graph $g(\phi)$ of $\phi(x)$, viewed as a self-affine[10] object in 2-space, to

$$H_1 = H_\phi(1) = 2 - D_{g(\phi)} \ge 0\,. \tag{5}$$

The "=" is obtained again in the stationary ($H_1 \to 0$) limit where $D_{g(\phi)} = 2$ (the graph fills space). In the opposite ($H_1 \to 1$) limit of almost everywhere differentiable processes, we retrieve graphs with $D_{g(\phi)} = 1$, akin to rectifiable curves. In other words, stationarity comes with more "roughness" and discontinuity, non-stationarity with more "smoothness" and continuity in scaling processes.

In contrast to the above discussed case of narrowly distributed increments ($H_\phi(p) \equiv H$), we can talk about "multi-affine" processes[1] in the more interesting cases where $H_\phi(p)$ is non-constant. We will retain H_1 (quite literally) as a first order quantifier of scale-invariant non-stationarity and the whole $H_\phi(p)$ hierarchy as a means to qualify it. We provide elsewhere an illustration of these ideas using "bounded" cascade models [Ref. 11 and references therein].

3. SINGULARITY ANALYSIS, QUANTIFYING AND QUALIFYING INTERMITTENCY

3.1 Defining a Related Measure

We wish to perform a singularity analysis of the now relatively standard type (based on a measure) and not like in the above (Hölderian) variant. We need to define a non-negative field related to $\phi(x)$ which should furthermore be stationary, having a well-behaved autocorrelation function and a spectral exponent which does not exceed one. A convenient choice is provided by:

$$\delta\phi(x) = |\Delta\phi(\ell;\, x)| = |\phi(x) - \phi(x-\ell)|\,, \quad \ell \le x \le L\,, \tag{6}$$

which can be described as the "absolute (small-scale) gradient field". Other options have been used.[12–14]

The next step is to degrade the resolution of the field defined in Ref. 6. Namely, we compute:

$$p_\phi(r;\, x) = \sum_{x'/\ell = x/\ell - r/\ell + 1}^{x'/\ell = x/\ell} \delta\phi(x')\,, \quad r \le x \le L(\ell \le r \le L)\,, \tag{7}$$

and let

$$\varepsilon_\phi(r;\, x) = \frac{p_\phi(r;\, x)}{r}\,. \tag{8}$$

3.2 Scaling Properties and Special Cases

Consider the scalings of $\langle \varepsilon_\phi(r;\, x)^q \rangle$ as parameterized in Ref. 15:

$$\langle \varepsilon_\phi(r;\, x)^q \rangle \sim r^{-K_\varepsilon(q)}(\ell \le r \le L)\,, \quad -\infty < q < \infty\,. \tag{9}$$

The exponent function $K_\varepsilon(q)$ is convex with predetermined values $K_\varepsilon(0) = K_\varepsilon(1) = 0$ and weak variability, $\langle p_\phi(r;\, x)^q \rangle \approx \langle p_\phi(r;\, x) \rangle^q$, yielding simply $K_\varepsilon(q) \equiv 0$. The multifractal significance of $K_\varepsilon(q)$ was established in Refs. 15 and 16 by showing that its Legendre transform yields the well-known singularity spectrum, $f_\varepsilon(\alpha)$.

Using the above parameterization, one can define the non-increasing hierarchy $D_\varepsilon(q) = 1 - K_\varepsilon(q)/(q-1)$ known as generalized or Reyni dimensions.[17,18] The single most important exponent in this whole approach is quite possibly:

$$C_1 = 1 - D_\varepsilon(1) = K'_\varepsilon(1) \ge 0\,, \tag{10}$$

the "information" codimension. It provides us with a straightforward measure of intermittency in the system, as determined by the deviation of $D_\varepsilon(q)$ or $K_\varepsilon(q)$ from a constant. The specific kind of intermittency is of course determined by the whole hierarchy $D_\varepsilon(q)$.

4. GEOPHYSICAL APPLICATIONS

4.1 Effective Constitutive Laws in Scale-invariant Format

It is legitimate to ask whether or not both multifractal statistics are independent of each other. Do $\zeta_\phi(p)$ and $K_\varepsilon(q)$ not convey, to some extent, the same information? We are convinced that the answer is yes, that the system is "physically" multifractal in a unique way. However, it is also clear that the two approaches capture only one facet each of this deep multifractal reality and that they are best viewed as complementary, at least in absence of any other knowledge.

Most geophysical fields of interest can be related to some globally conserved quantity (e.g. velocity and total kinetic energy). Any macroscopic (non-local) connection between a field $\phi(x)$ hence $\zeta_\phi(p)$, on the one hand, and its gradients $\delta\phi(x)$ hence $K_\varepsilon(q)$, on the other hand, is bound to be related to the transport of this quantity, from place-to-place and/or scale-to-scale.

4.2 The Case of Fully Developed Turbulence

One instance where an extra *a priori* scaling relation is available — largely thanks to dimensional analysis — is 3D turbulence. In this problem, we can take $\phi(x) = u(x = |\langle \mathbf{u} \rangle| t)$ using Taylor's frozen turbulence hypothesis in a flow with mean velocity $\langle \mathbf{u} \rangle$. Indeed, a deterministic (event-per-event) interpretation of Kolmogorov's famous (effective constitutive) relation[2]

$$\varepsilon_u(r; x) \approx \frac{|\Delta u(r; x)|^3}{r}, \quad r \leq x \leq L (\ell \leq r \leq L), \tag{11}$$

leads to

$$\zeta_u(p) = \frac{p}{3} - K_\varepsilon\left(q = \frac{p}{3}\right). \tag{12}$$

Note that $\zeta_u(0) = 0$ as required and that $\zeta_u(3) = 1$ as expected from Eq. (11). The $K_\varepsilon(p/3)$ term in Eq. (12) is the "intermittency" correction to the Kolmogorov's[2] theory for homogeneous turbulence ($H_u(p) \equiv 1/3$). Recently, Eq. (11) has been the subject of some debate (Ref. 19 and references therein) and the consensus seems to be that it is best interpreted in a statistical sense, that it applies extremely well on average but can fail completely for any given event.

Returning to the case of arbitrary geophysical data, we can turn the above questions around. Given enough high-quality data, can one extract from it such powerful relations as Eq. (12) or even Eq. (11) using scale-invariance analysis? To this effect, we suggest a merger of the two basic (uni-variate) multifractal statistics into a single bi-variate one. Consider the scaling of:

$$\langle |\Delta\phi(r; x)|^p \varepsilon_\phi(r; x)^q \rangle \sim r^{X_\phi(p,q)} (\ell \leq r \leq L), \quad -1 < p < \infty, \quad -\infty < q < \infty. \tag{13}$$

In the above, we have looked at $X_\phi(p, 0) = \zeta_\phi(p)$ and $X_\phi(0, q) = -K_\varepsilon(q)$. It is not hard to see that, in the unlikely situation where $|\Delta\phi(r; x)|$ and $\varepsilon_{\Delta\phi}(r; x)$ are independent random variables, we have $X_\phi(p, q) = \zeta_\phi(p) - K_\varepsilon(q)$. By way of contrast, the event-wise connection in Eq. (11) implies $X_u(p, q) = p/3 - K_\varepsilon(q + p/3)$ which in turn yields Eq. (12). So, in principle, a careful bi-multifractal analysis of data can lead to the determination of important scaling relations.

5. DISCUSSION AND CONCLUSIONS

In summary, we surveyed two multifractal data analysis techniques based on structure functions and singular measures, yielding respectively the exponent hierarchies $H(q)$ and

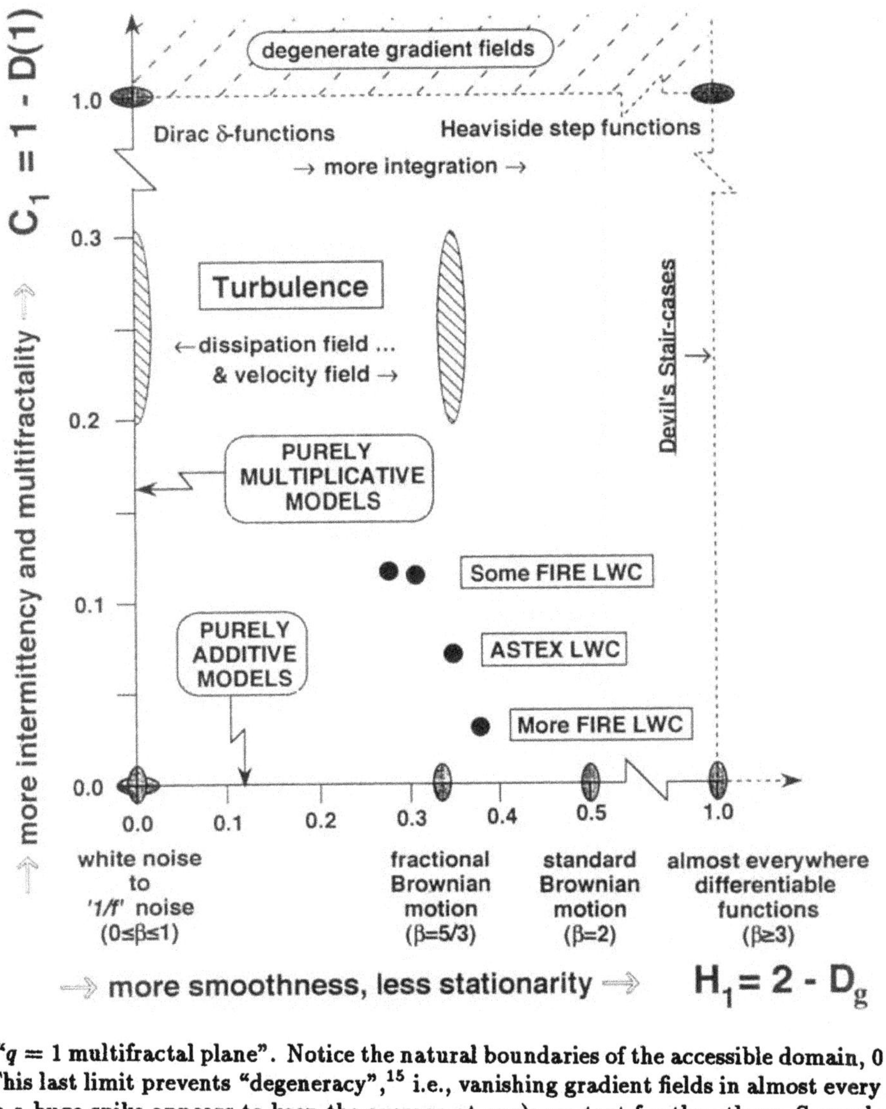

Fig. 1 The "$q = 1$ multifractal plane". Notice the natural boundaries of the accessible domain, $0 \leq H_1 \leq 1$, $0 \leq C_1 \leq 1$. This last limit prevents "degeneracy",[15] i.e., vanishing gradient fields in almost every realization (now-and-then a huge spike appears to keep the average at one); see text for the others. Several well-known models are found on these boundaries: "multiplicative" or turbulent cascade models, e.g. Ref. 9, "additive" models such as fractional Brownian motion and "Devil's stair-cases".[10] In contrast, typical geophysical signals tend to live inside this domain, as illustrated by four cloud liquid water content (LWC) datasets obtained during two different campaigns (FIRE, ASTEX). Inside we would also find the handful of multi-affine models found in the literature [Ref. 11 and references therein].

$D(q)$. The latter is now a standard tool for quantifying and qualifying "intermittency" and we argued that the former (multi-affine) technique can be used analogously for the presently more fuzzy notion of "non-stationary". We also merged the two approaches into a unified bi-multifractal one. Guiding ourselves with the well-studied case of 3D turbulence, we showed how this new concept can be used in a systematic quest for scale-invariant "effective constitutive laws." It is hoped that, from these, insight can be gained into the nonlinear physical mechanisms at work generating, in particular, the geophysical signal being processed.

In the meantime, we show in Fig. 1 some of the model- and data-based activity happening on the "$q = 1$ multifractal plane" where the coordinates are simply $H_1 = H(1)$ and $C_1 = 1 - D(1)$. In our view, the axes measure directly the degrees of non-stationarity (horizontally, H_1) and of intermittency (vertically, C_1) in the system. Non-vanishing intermittency leads to multi-scaling in singular measures (as a matter of definition) and also in structure functions (but in a non-trivial way). At this level of approximation, the rest — what precise flavor of multifractality is present — is a matter of detail. We advocate the

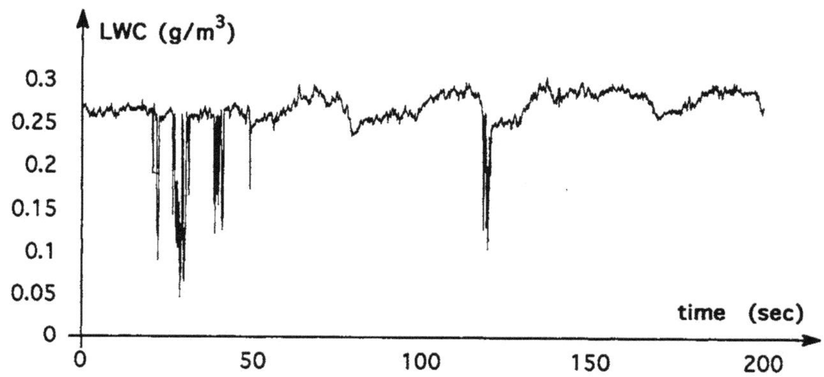

(a)

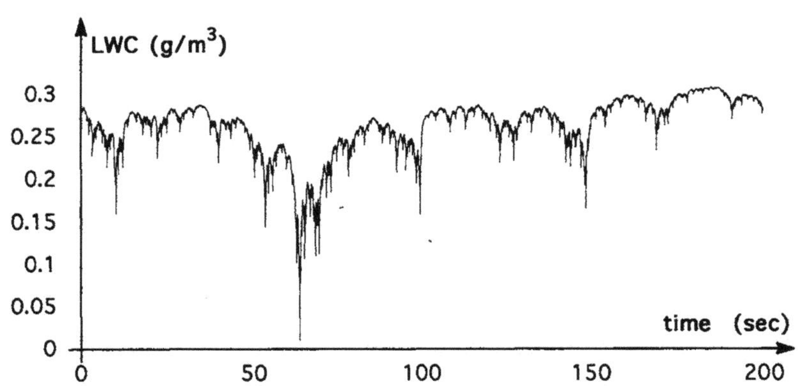

(b)

Fig. 2 (a) A sample of atmospheric liquid water density captured in a stratocumulus cloud deck. (b) A simple multi-affine stochastic model for the above, with comparable H_1 and C_1. More specifically, we generate a fractionally integrated standard cascade model,[15] and a linear transformation is then applied in order to fit the 1-point mean, variance and sign of skewness. This last procedure does not affect any of the exponents considered here since they are based on increments or on small scale differences.

use of such (H_1, C_1)-plots as simple universal tools in geophysical data analysis, whether it be for a classification study when several datasets are involved or as a way to determine the parameters of a stochastic model for a single dataset.[20] In Figs. 2(a) and 2(b) we illustrate this last exercise in the case of cloud structure. Figure 2(a) shows a typical *in situ* liquid water density probing described in more detail elsewhere[21] and in Fig. 2(b) we see a multi-affine two-parameter model.[15,20] We are confident that many other applications will be found.

ACKNOWLEDGMENTS

This work was supported by DoE's ARM project, grant DE-A105-90ER61069 to NASA Goddard Space Flight Center. We thank A. Arneodo, R. Cahalan, D. Lavallée, S. Lovejoy, C. Meneveau, W. Ridgway, D. Schertzer, Y. Tessier and T. Warn for fruitful discussions.

REFERENCES

1. T. Vicsek and A.-L. Barabási, *J. Phys. A: Math. Gen.* **24**, L845 (1991).
2. A. N. Kolmogorov, *Dokl. Akad. Nauk SSSR* **30**, 4, 299 (1941).
3. S. Corrsin, *J. Appl. Phys.* **22**, 469 (1951).
4. A. Obukhov, *Izv. Akad. Nauk. SSSR, Ser. Geogr. i Geofiz.* **13**, 55 (1949).
5. R. F. Cahalan and J. B. Snider, *Remote Sens. Environ.* **28**, 95 (1989).
6. S. Lovejoy, D. Schertzer, P. Silas, Y. Tessier and D. Lavallée, *Ann. Geophysicae* **11**, 119 (1993).
7. J. F. Muzy, E. Bacry and A. Arneodo, *Phys. Rev.* **E47**, 875 (1993).
8. G. Parisi and U. Frisch, in *Turbulence and Predictability in Geophysical Fluid Dynamics*, eds. M. Ghil, R. Benzi and G. Parisi (Amsterdam, 1985) p. 84.
9. A. S. Monin and A. M. Yaglom, *Statistical Fluid Mechanics*, vol. 2, (MIT press, Boston, Mass, 1975).
10. B. B. Mandelbrot, *Fractals: Form, Chance and Dimension* (W. H. Freeman and Company, San Francisco, 1977).
11. A. Marshak, A. Davis, R. Cahalan and W. Wiscombe, Multi-singular and Multi-affine Properties of Bounded Cascade Models (this volume).
12. C. Meneveau and K. R. Sreenivasan, *Phys. Rev. Lett.* **59**, 13, 1424 (1987).
13. F. Schmitt, D. Lavallée and D. Schertzer, *Phys. Rev. Lett.* **68**, 3, 305 (1992).
14. Y. Tessier, S. Lovejoy and D. Schertzer, *J. Appl. Meteorol.* **14**, 160 (1993).
15. D. Schertzer and S. Lovejoy, *J. Geophys. Res.* **D92**, 9693 (1987).
16. C. T. Halsey, M. H. Hensen, L. P. Kadanoff, I. Procaccia and B. I. Shraiman, *Phys. Rev.* **A33**, 1141 (1986).
17. P. Grassberger, *Phys. Rev. Lett.* **97**, 227 (1983).
18. H. G. E. Hentschel and I. Procaccia, *Physica* **D8**, 435 (1983).
19. G. Stolovitzky, P. Kailasnath and K. R. Sreenivasan, *Phys. Rev. Lett.* **69**, 8, 1178 (1992).
20. A. Davis, A. Marshak and W. Wiscombe, Multifractal Analysis and Stochastic Simulation of Non-stationary Geophysical Processes (in preparation for *J. Geophys. Res.*).
21. A. Marshak, A. Davis, W. Wiscombe and R. Cahalan, The Multifractal Properties of Observed Liquid Water Content in Marine Stratocumulus (in preparation for *J. Geophys. Res.*).

Fractals, Vol. 1, No. 3 (1993) 568–575

ESTIMATION OF UNIVERSAL MULTIFRACTAL INDICES FOR ATMOSPHERIC TURBULENT VELOCITY FIELDS

F. SCHMITT and D. SCHERTZER
Laboratoire de Météorologie Dynamique, Boîte 99,
Université Paris 6, 4 place Jussieu,
75252 Paris cedex 05, France

S. LOVEJOY
Physics Dept., McGill University, 3600 University St.,
Montréal, Qué., H3A 2T8, Canada

Y. BRUNET
INRA Bioclimatologie, 71 av. Edouard-Bourleaux, BP 81,
33883 Villenave d'Ornon cedex, France

Abstract

We study wind turbulence with the help of universal multifractals, using atmospheric high resolution time series. We empirically determine the three universal indices (H, C_1, and α) which are sufficient to characterize the statistics of turbulence. The first, H, which characterizes the conservation of the field, is theoretically and empirically known to be $\approx 1/3$, while C_1 corresponds to the inhomogeneity of the mean field ($C_1 = 0$ for homogeneous fields, and $C_1 > 0$ for inhomogeneous and intermittent fields). The most important index is the Lévy index α corresponding to the degree of multifractality ($0 \leq \alpha \leq 2$, $\alpha = 0$ for a monofractal). The two latter indices are directly obtained by applying the double trace moment technique (DTM) on the turbulent field. Analyzing various atmospheric velocity measurements we obtain: $\alpha \approx 1.45 \pm 0.1$ and $C_1 \approx 0.25 \pm 0.1$. These results show that atmospheric turbulence has nearly the same multifractal behavior everywhere in the boundary layer, corresponding to unconditionally hard multifractal ($\alpha \geq 1$) processes. This describes the entire hierarchy of singularities of the Navier-Stokes equations.

1. INTRODUCTION

In the limit of high Reynolds number, turbulence is known to be scale invariant.[1,2] One of the points which has been important in characterizing fully developed turbulence is to know whether it belongs to universal classes which could describe any high Reynolds number turbulent flow.[3,4] We test here such a scale invariant model of turbulence, called the *universal multifractal model*,[5,6] using different atmospheric turbulent datasets. This model is determined by only three indices, which completely describe the scaling behavior of the turbulent fields. We compare the values of these indices for various atmospheric datasets, in order to confirm the hypothesis of their constancy.

2. THE DATA AND THEIR SPECTRA

The datasets are turbulent velocity measurements made in the atmospheric boundary layer: the most important one was taken near Montreal, at 3 meters from the ground, with a high-resolution hot wire anemometer, and the others with sonic anemometers in Paris, and in Bordeaux (see Table 1 for a description of the experimental conditions).

Table 1 Description of the Characteristics of Three Atmospheric Datasets

Dataset #	Anemometer	Sampling Frequency (Hz)	Place of Acquisition	Altitude (m)	Total Number of Data Points used	β
1	Hot wire	2000	Montreal	3	720 000	1.70
2	Sonic	200	Paris	20	25 000	1.72
3	Sonic	10	Bordeaux	25	360 000	1.70

Their wind energy spectrum follows a power law over a wide range of scales:

$$E_\nu(k) \approx k^{-\beta} \tag{1}$$

with the rate of acquisition, the range of scaling, and β given in Table 1. Figure 1 shows the power spectrum of the dataset #1. The values of the spectral slopes are very close to the Kolmogorov-Obukhov phenomenological derivations predicting $\beta = 5/3$.[1,2] We then assume the validity of the refined[a] similarity hypothesis[3]:

$$\Delta\nu(\ell) \approx \varepsilon_\ell^{1/3} \ell^H \tag{2}$$

where $\Delta\nu(\ell)$ is the wind shear $|\nu(\mathrm{x}+\ell) - \nu(\mathrm{x})|$ at scale ℓ, ε_ℓ is the local rate of energy transfer (from scales larger to scales smaller than ℓ). The index H is the mean deviation from conservation of the velocity shears, and is related to β by $\beta \cong 2H + 1$ (e.g. see Ref. 7).

[a]Kolmogorov 1941 assumed ε to be homogeneous in space, i.e., $\varepsilon = \bar{\varepsilon}$, whereas Eq. (2) takes into account the small scale intermittency of turbulence with the introduction of a very intermittent energy transfer field.

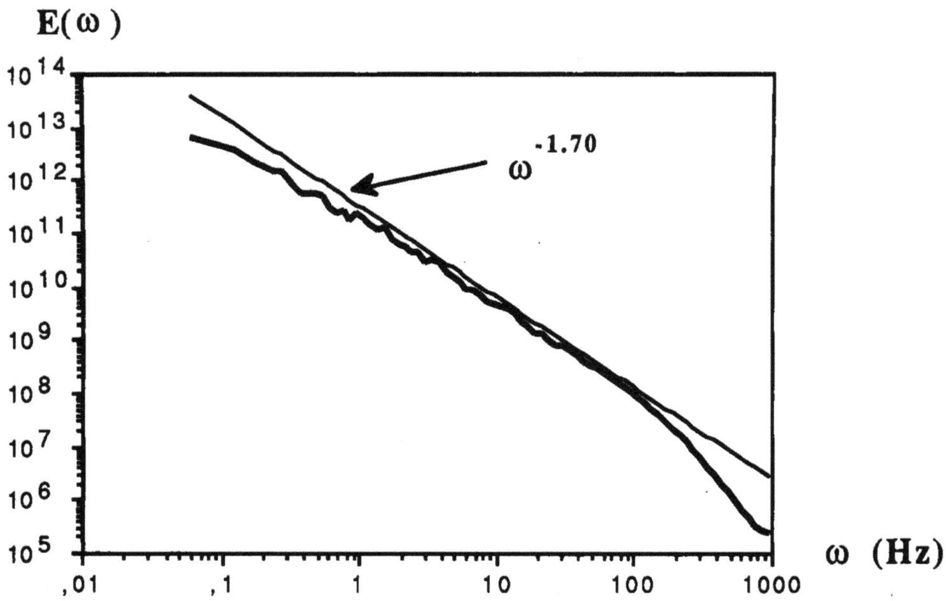

Fig. 1 The wind energy spectrum of atmospheric turbulence corresponding to dataset #1. For a wide range of scales, it follows a power law relation $E_\nu(k) \approx K^{-\beta}$ with $\beta \approx 1.68 \pm 0.05$ as predicted by Refs. 1 and 2.

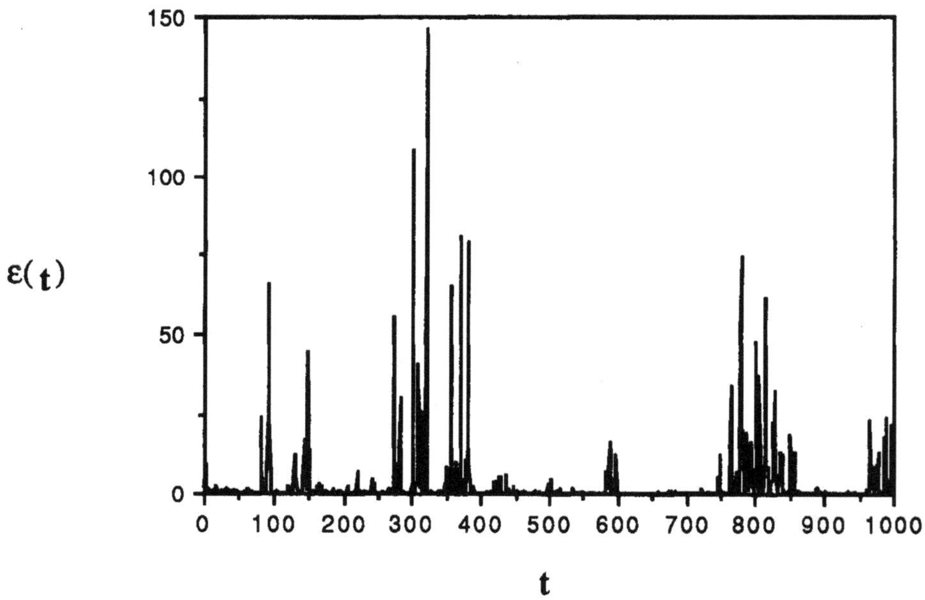

Fig. 2 A pattern of ε, the rate of energy transfer from large to small scales, for atmospheric turbulence: this is a very intermittent (and scaling) field.

This gives us $H = 0.33 \pm 0.03$ for the three datasets, as predicted by phenomenological models. In order to study the statistical properties of the very intermittent dissipation field ε_ℓ, we perform a fractional differentiation of order 1/3 of the wind field (i.e., a multiplication by $k^{1/3}$ in Fourier space,[6]) and then take the third power: the result is a very intermittent field, whose pattern can be seen in Fig. 2 for atmospheric measurements corresponding to dataset #1.

3. UNIVERSAL MULTIPLE SCALING OF THE FIELD ε

We have studied the energy flux ε using multifractal measures resulting from a multiplicative cascade process.[4,8–11] When the cascade has proceeded over a scale ratio $\lambda = \frac{L}{\ell}$ (the ratio of the largest scale of interest to the smallest scale) the density of the conserved energy flux has the singular behavior[5,12,13]:

$$\varepsilon_\lambda \approx \lambda^\gamma \tag{3}$$

When $\lambda \to \infty$ (or $\ell \to 0$), $\gamma > 0$ is an order of singularity. The statistical moments of this field will have the following scaling behavior[5]:

$$\langle(\varepsilon_\lambda)^q\rangle \approx \lambda^{K(q)} \tag{4}$$

where $\langle\ \rangle$ indicates ensemble averaging, and $K(q)$ is a nonlinear function[b] which characterizes all the statistics of the field.

For universality classes (the stable and attractive limit obtained when mixing cascade processes, see Refs. 5 and 6), the functions $K(q)$ depend only on two indices:

$$K(q) = \begin{cases} \dfrac{C_1}{\alpha - 1}(q^\alpha - q) & \alpha \neq 1 \\ C_1 q \ln(q) & \alpha = 1 \end{cases} \tag{5}$$

where the Lévy index α ($0 \leq \alpha \leq 2$) can be understood as an interpolation between the two extremes and well known cascade models of turbulence: the β-model ($\alpha = 0$)[4,8,9] and the lognormal model ($\alpha = 2$)[3,14,15]; and the second index corresponds to the mean inhomogeneity of the field: $C_1 = 0$ for homogeneous fields, and the larger C_1, the more the intermittent field. We have estimated these indices for wind tunnel and atmospheric energy fluxes, using a generalization of the (simple) scaling exponent $K(q)$.

4. ESTIMATION OF α AND C_1 WITH THE HELP OF THE DOUBLE TRACE MOMENT (DTM) ANALYSIS TECHNIQUE

The basic idea of the DTM technique[16,17] is to generalize the application of statistical methods to the quantity $(\varepsilon_\Lambda)^\eta$. This is done by taking the ηth power of ε_Λ at the scale ratio Λ (the outer or largest scale of interest to the smallest scale of homogeneity), and then studying its scaling behavior at decreasing values of the scale ratio $\lambda \leq \Lambda$:

$$\varepsilon^{(\eta)}_{\lambda,\Lambda} = \frac{(\varepsilon_\lambda)^\eta}{\langle(\varepsilon_\lambda)^\eta\rangle}\langle(\varepsilon_\Lambda)^\eta\rangle \tag{6}$$

The moments of this new field then have the following multiple scaling behavior:

$$\langle[\varepsilon^{(\eta)}_{\lambda,\Lambda}]^q\rangle \approx \lambda^{K(q,\eta)} \tag{7}$$

where $K(q, \eta)$ is the q, η double trace moment scaling exponent related to $K(q, 1) \equiv K(q)$ by [$\langle(\varepsilon_\Lambda)^\eta\rangle$ is a constant]:

$$K(q, \eta) = K(q\eta) - qK(\eta) \tag{8}$$

---

[b] $K(q)$ is related to the strange attractor notation[13] by: $K(q) = (q-1)D - \tau(q)$ where D is the dimension of the space.

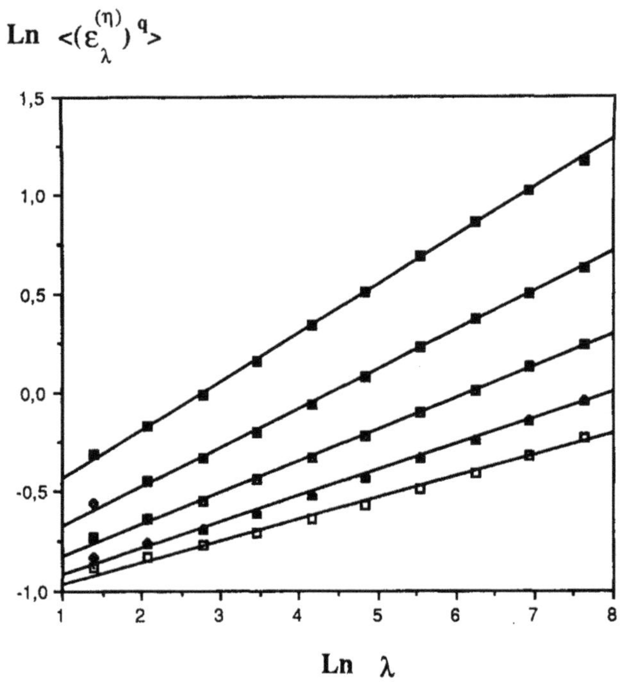

Fig. 3 A representation of $\langle[\varepsilon_{\lambda,\Lambda}^{(\eta)}]^q\rangle$, vs. λ in a log-log plot for atmospheric energy flux: the straight lines show that scaling of Eq. (4) is well respected. These straight lines correspond to $q = 2$ and $\eta = 0.43$, 0.49, 0.56, 0.65, 0.74 (from bottom to top).

It gives for universality classes:

$$K(q, \eta) = n^\alpha K(q) \tag{9}$$

By keeping q fixed (but different from the special values 0 or 1), the slope of $|K(q,\eta)|$ as a function of η on a log-log graph gives the value of the index α, which with the help of the intersection with the line ($\eta = 1$) yields C_1. Varying q then allows for a systematic verification of Eq. (9), and hence the universality hypothesis.

Figure 3 shows $\langle[\varepsilon_{\lambda,\Lambda}^{(\eta)}]^q\rangle$ vs. λ in a log-log plot for atmospheric energy flux corresponding to dataset #1: the straight lines show that the scaling of Eq. (7) is well respected, and their slope are the estimates of $|K(q, \eta)|$. Figure 4 shows the curves $\log|K(q, \eta)|$ vs. $\log(\eta)$ for different values of the parameter q, for atmospheric energy flux corresponding to dataset #1: the straight lines show that Eq. (9) is well respected for a wide range of η-values, and their slopes give the estimates shown in Table 2.

Table 2 Estimates of the Indices α and C_1 obtained for the Different Datasets presented in Table 1, showing that their Values are likely to be Steady in the Atmosphere (at least in the boundary layer)

Dataset	Maximum Scale Ratio	α	C_1
1	2048	1.50 ± 0.05	0.25 ± 0.05
2	1024	1.45 ± 0.1	0.29 ± 0.1
3	512	1.4 ± 0.1	0.24 ± 0.1

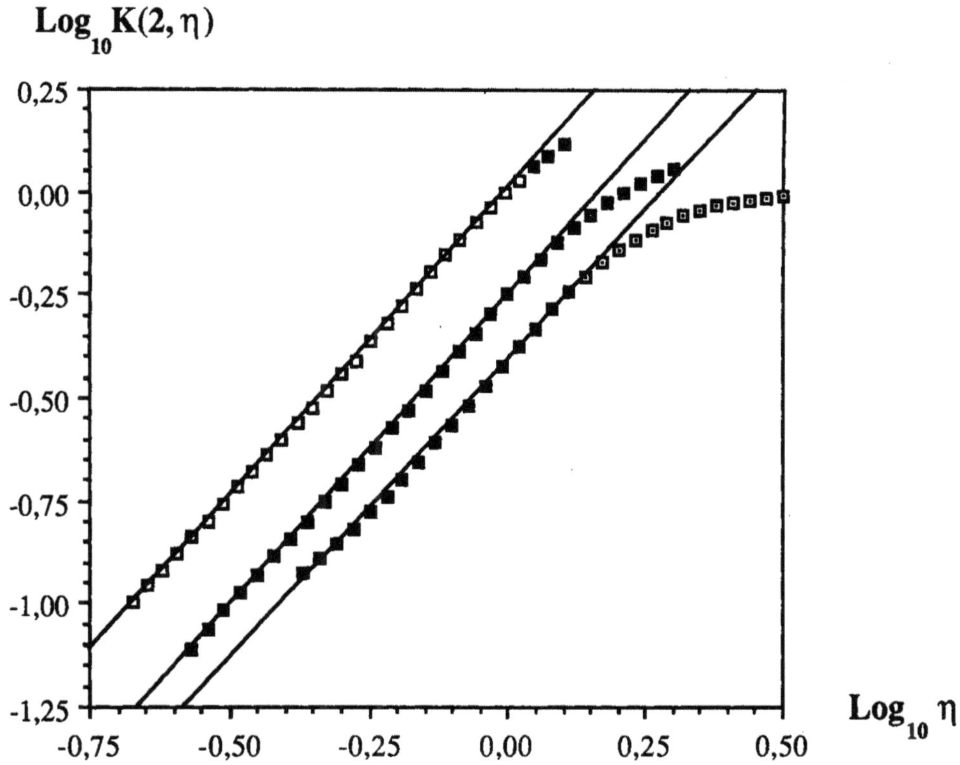

Fig. 4 The curves $\mathrm{Log}|K(q,\eta)|$ vs $\mathrm{Log}(\eta)$ for $q = 2$, 2.5 and 3 (from bottom to top) for atmospheric turbulence. The straight lines show that Eq. (9) is well respected for a wide range of η- values, and their slopes give the following estimates: $\alpha \approx 1.50 \pm 0.05$ for atmospheric turbulence corresponding to dataset #1. We can then estimate $C_1 \approx 0.25 \pm 0.05$.

According to error estimates presented in Table 2, we propose the following values for the indices α and C_1 in the atmospheric boundary layer: $\alpha \approx 1.45 \pm 0.1$ and $C_1 \approx 0.25 \pm 0.1$. We note that these values are in the same range that estimates obtained for wind tunnel turbulence[18–20]: $\alpha \approx 1.3 \pm 0.1$ and $C_1 \approx 0.25 \pm 0.05$. These values of α and C_1 show that turbulence is a hard multifractal process[18,21–23]: because $\alpha > 1$, divergence of moments[4,10,20,24] is expected to occur.

5. MOMENTS OF THE ENERGY FLUXES AND STRUCTURE FUNCTIONS OF THE WIND FIELD

With the help of these results, we compare empirical statistical moments with $K(q)$ given by Eq. (5), in Fig. 5: for moments order $q < 3.0 \pm 0.5$[c] the empirical and "theoretical" scaling exponents are in excellent agreement. This critical moment[21,24] corresponds either to a second order phase transition analogue (a maximum moment computable due to the finite size of the dataset), or to a first order phase transition[27] analogue

[c]Equations (2) and (4) give the scaling velocity structure functions: $\langle[\delta v(l)]^q\rangle \approx \lambda^{-\zeta(q)} \approx \langle(\varepsilon_\lambda)^{q/3}\rangle\lambda^{-Hq}$, which gives the relation $\zeta(q) = Hq - K(\frac{q}{3})$. Thus, in the case of database #1, it is useful to study the structure function $\zeta(q)$ only up to moment order $3q_{max} \approx 9.0 \pm 1.5$, unlike Ref. 25, which computed structure functions up to order 18, and Ref. 26 which used the latter structure function to evaluate α.

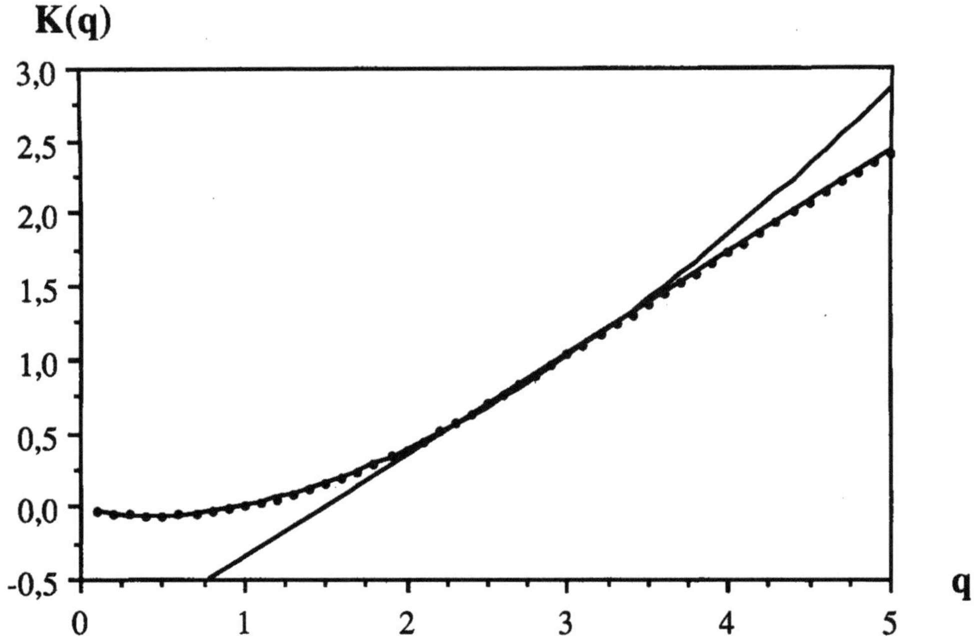

Fig. 5 A comparison between empirical statistical moments (squares) with $K(q)$ given by Eq. (4) (universality hypothesis, continuous line); for moments order $q < 3.0 \pm 0.5$ for atmospheric turbulence the empirical and "theoretical" scaling exponents are in excellent agreement. For larger moments, an empirical linear behavior is observed which can be explained by multifractal phase transitions.

(corresponding to a critical order of divergence of moments[d]). This question is studied elsewhere.[28]

6. CONCLUSION

Using various atmospheric datasets, we obtain similar values of the three indices (H, C_1, α) which completely characterize the statistics of turbulence: $H = 0.33 \pm 0.03$, $C_1 \approx 0.25 \pm 0.1$ and $\alpha \approx 1.45 \pm 0.1$ for the three atmospheric datasets studied. These results show that turbulence is a "hard" multifractal process ($\alpha \geq 1$), and that it belongs to theoretically predicted universality classes.

ACKNOWLEDGMENTS

We acknowledge C. Hooge, Ph. Ladoy, D. Lavallée and Y. Tessier for fruitful discussions. We are grateful to P. Schuepp, M. Duncan, F. Fabry, B. Katz and S. Buino for helping to collect the atmospheric data in Montreal and the Atmospheric Radiation Measurement program contract #DE-FG03-90ER61062 for partial financial support.

[d]This divergence of moments is expected to occur because of some very violent "head" small scale singularities, which are not smoothed by integration: if the number of samples increases, more and more of such singularities are encountered, and the statistical moments diverges.

REFERENCES

1. A. N. Kolmogorov, *CR Acad. Sci. USSR* **30**, 299 (1941).
2. A. Obukhov, *Izv. Akad. Nauk. S.S.S.R.* (Geogr. I. Jeofiz., 1949), Chap. 13, p. 55.
3. A. N. Kolmogorov, *J. Fluid Mech.* **13**, 82 (1962).
4. B. Mandelbrot, *J. Fluid Mech.* **62**, 331 (1974).
5. D. Schertzer and S. Lovejoy, *J. Geophys. Res.* **92**, 9693 (1987).
6. D. Schertzer and S. Lovejoy, in *Scaling, Fractals and Nonlinear Variability in Geophysics*, eds. D. Schertzer and S. Lovejoy (Kluwer, Holland, 1991) p. 41.
7. A. S. Monin and A. M. Yaglom, in *Statistical Fluid Mechanics: Mechanics of Turbulence*, ed. J. Lumley (MIT Press, 1975).
8. E. A. Novikov and R. Stewart, *Izv. Akad. Nauk. SSSR*, (Ser. I. Geofiz., 1964), Chap. 3, p. 408.
9. U. Frisch, P. L. Sulem and M. Nelkin, *J. Fluid Mech.* **87**, 719 (1978).
10. D. Schertzer and S. Lovejoy, in *U.I.T.A.M. Symp. on Turbulence and Chaotic Phenomena in Fluids*, Preprint vol., pp. 141–144, 1983.
11. C. Meneveau and K. R. Sreenivasan, *Phys. Rev. Lett.* **59**, 1424 (1987).
12. G. Parisi and U. Frisch, in *Turbulence and Predictability in Geophysical Fluid Dynamics and Climate Dynamics*, eds. M. Ghil, R. Benzi and G. Parisi (North Holland, Amsterdam, 1985).
13. T. C. Halsey, M. H. Jensen, L. P. Kadanoff, I. Procaccia and B. Shraiman, *Phys. Rev.* **A33**, 1141 (1986).
14. A. Obukhov, *J. Fluid Mech.* **13**, 77 (1962).
15. A. M. Yaglom, *Sov. Phys. Dokl* **2**, 26 (1966).
16. D. Lavallée, Ph.D. Thesis, University McGill, Montréal, Canada (1991).
17. D. Lavallée, S. Lovejoy, D. Schertzer and Ph. Ladoy, in *Fractals in Geography*, eds. L. de Cola and N. Lam (in press).
18. D. Schertzer, S. Lovejoy, D. Lavallée and F. Schmitt, *Nonlinear Dynamics of Structures*, eds. Sagdeev et al. (World Scientific, Singapore, 1991) p. 213.
19. F. Schmitt, D. Lavallée, S. Lovejoy, D. Schertzer and C. Hooge, *C.R. Acad. Sci. Paris* t.**314** série II, 749 (1992).
20. F. Schmitt, D. Lavallée, D. Schertzer and S. Lovejoy, *Phys. Rev. Lett.* **68**, 305 (1992).
21. D. Schertzer and S. Lovejoy, *Physica* **A185**, 187 (1992).
22. S. Lovejoy and D. Schertzer, *J. Geophys. Res.* **95**, 2021 (1990).
23. S. Lovejoy and D. Schertzer, *Scaling, Fractals and Nonlinear Variability in Geophysics*, eds. D. Schertzer and S. Lovejoy (Kluwer, Holland, 1991) p. 111.
24. D. Schertzer and S. Lovejoy, *Fractals: Physical Origin and Consequences*, ed. L. Pietronero (Plenum, New York, 1989) p. 49.
25. F. Anselmet, Y. Gagne, E. J. Hopfinger and R. A. Antonia, *J. Fluid Mech.* **140**, 63 (1984).
26. S. Kida, *J. Phys. Soc. Japan* **60**, 5 (1991).
27. D. Schertzer and S. Lovejoy, *Fractals in the Natural and Applied Sciences*, Elsevier (1993), in press.
28. F. Schmitt, D. Schertzer, G. Brethenoux, S. Lovejoy and Y. Brunet, *C. R. Acad. Sci. Paris* (1993), submitted.

Fractals, Vol. 1, No. 3 (1993) 576–584

A MINIMUM ENERGY DISSIPATION MODEL FOR RIVER NETWORKS AND THEIR ASSOCIATED TOPOGRAPHIES

T. SUN, P. MEAKIN and T. JØSSANG
Physics Department, Oslo University,
P. O. Box 1048 Blindern, 0316 Oslo, Norway
Center for Advanced Study
at The Norwegian Academy of Science and Letter
P. O. Box 7606 Skillebekk, 0205 Oslo, Norway

Abstract

The model for the river networks presented here is based on minimum energy dissipation principles. The foundation for this model is the empirical relationship $s \sim Q^{\alpha}$ between the link slope s in channel networks and the mean annual discharge Q in that link. The associated landscapes were constructed using a range of values for the exponent α. The surfaces appear to be more complex than simple self-affine fractals. The boundaries of drainage basins covering the entire drainage area were found to have an effective fractal dimension about 1.10 for all values of α in the range $-1 < \alpha < 0$. A universal power-law size (area) distribution is also found for the drainage basins obtained from this minimum energy dissipation model.

1. INTRODUCTION

The structure of natural drainage networks has, for a long time, been of considerable interest to hydrologists and geomorphologists. However, no concensus has yet developed concerning the relationship between the structure of natural channel networks and their associated landscapes. Rivers have frequently been cited as familiar examples of natural fractal structures,[1] and landscapes are often used as an example of self-affine fractals.[2] But the way in which their fractal scaling structures relate to each other remains poorly understood. Here we introduce a model for river networks and their associated landscapes, which reflects the close relationship between them.

The model introduced here is based on the assumption that an open system with constant energy injections tends to form a structure that minimizes the total energy dissipation in the system.[3,4] Drainage networks that are embedded in a drainage area subject to a uniform energy input from precipitation are a familiar example of an open system that may form a structure that minimizes the energy dissipations in the whole network.[5–7] The precipitated water flows downhill and dissipates its potential energy. By minimizing the total energy dissipation in the drainage area, the optimal structure of drainage networks and their associated landscapes are obtained.

The foundation for this model is the empirical relationship[8] $s \sim Q^\alpha$ between the link slope s in channel networks and the mean annual discharge Q in that link, together with the field observation that the velocity of water flow is almost constant everywhere in a channel network.[9] The constant velocity of water flow means that the contribution of kinetic energy to the energy dissipation in the stream is negligible. If P_i denotes the energy dissipation in link i of length L_i, Q_i is the mean discharge in link i, and s_i is the slope in that link, then the energy dissipation in the link can be written as:

$$\begin{aligned} P_i &= Q_i L_i s_i \\ &= Q_i^{1+\alpha} L_i \end{aligned} \tag{1}$$

Here the relationship:

$$s_i \sim Q_i^\alpha \tag{2}$$

has been applied. Thus the total energy dissipation in the whole drainage network can be expressed as:

$$P = \sum_i P_i = \sum_i Q_i^{1+\alpha} L_i \tag{3}$$

where the index i sums over all the links. The landscape is associated with the drainage network by:

$$h_i = \sum_j s_j L_j = \sum_j Q_j^\alpha L_j \tag{4}$$

Here h_i is the elevation of site i on the landscape and the index j labels all of the links connecting the site i to the outlet on the boundary of A, where the height is 0 (sea level) via the path of steepest descent. The flow Q_i is calculated as:

$$Q_i = \sum_j Q_j + 1 \tag{5}$$

where it is assumed that the rain falls uniformly on the whole the area A and there are no losses due to evaporation, subsurface flow, etc. The summation is over all the tributaries of i that have flow Q_j into site i. The addition of a unit flow to $\sum_j Q_j$ represents the precipitation onto the ith site.

In the computer simulations, the optimal structure for the drainage networks that minimizes the value of P, in Eq. (3), is calculated first and then the landscape is constructed according to Eq. (4). A triangular lattice (grid) is superimposed on the area A. The lengths L_j are set to 1 because on a triangular lattice, all the links (bonds) have the same length. The "simulated annealing" method[10] is used to calculate the optimal structure of the drainage networks. The "cost function" is the energy dissipation given in Eq. (3). The network modification rules and annealing schedule that were used by Sun et al.[7] have also been used here. The initial configuration, needed for simulated annealing, is obtained by performing a branched Eden growth simulation.[11] Loops are avoided in the model drainage networks since they are rarely found in nature.

The value of α was given by Leopold[8] as -0.49. However, a large range of values has been observed in natural river networks.[12–14] Six different values of α ($\alpha = -0.75, -0.625, -0.5, -0.325, -0.25$ and -0.125), were tested in the study. As an example, Figs. 1(a)–1(d) illustrate a minimum energy dissipation drainage network structure, a picture of its associated landscape, the contours of the landscape, and the boundaries of the drainage basins that cover the drainage area A for $\alpha = -0.5$.

The surfaces of the optimal drainage areas appear rough, like most geographical surfaces in nature. These surfaces were characterized by calculating the height difference correlation function $C(r)$ given by:

$$C(r) = \langle [h(x) - h(x+r)]^2 \rangle \tag{6}$$

For a self-affine fractal $C(r)$ is an algebraic function of r ($C(r) \sim r^{2H}$) and the exponent H is the Hurst exponent.[2,15] Log-log plots of this correlation function for the minimum energy

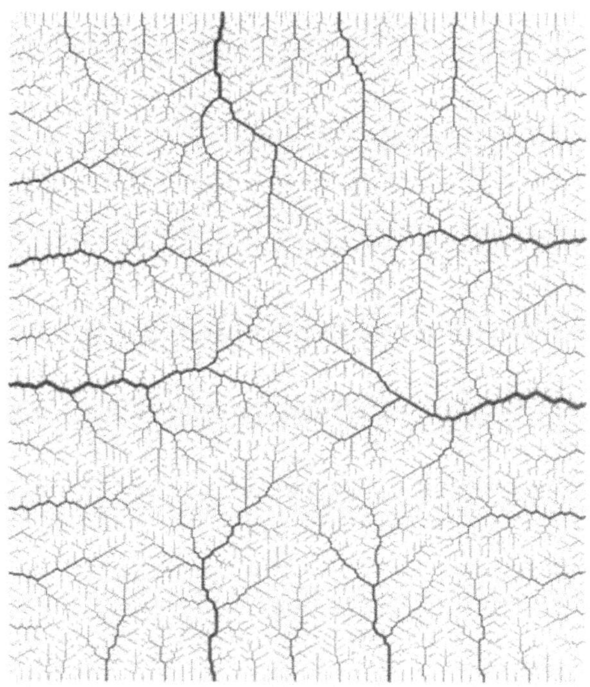

(a)

Figs. 1(a)–1(d) The patterns obtained from the minimum energy dissipation model for river networks and its associated landscapes with $\alpha = -0.5$. In Fig. 1(a), the structure of the drainage networks is displayed. The thickness of each segment in the figure is proportional to the square-root of the amount of flow it carries. The arrows in the figure indicate the direction of the flow on the sites. A "3 dimensional" picture of the corresponding landscape is shown in Fig. 1(b). The heights are indicated by different colors (red, yellow, green, blue and violet in order of decreasing height). The contours of this landscape are displayed in Fig. 1(c). The relative height difference (normalized by the greatest height in the whole landscape) between two thin contour lines is 0.0769, while it is 4 times as large ($0.0769 \times 4 = 0.308$) between the thick contour lines. Figure 1(d) shows the boundaries of all the drainage basins that cover the drainage area A.

(b)

(c)

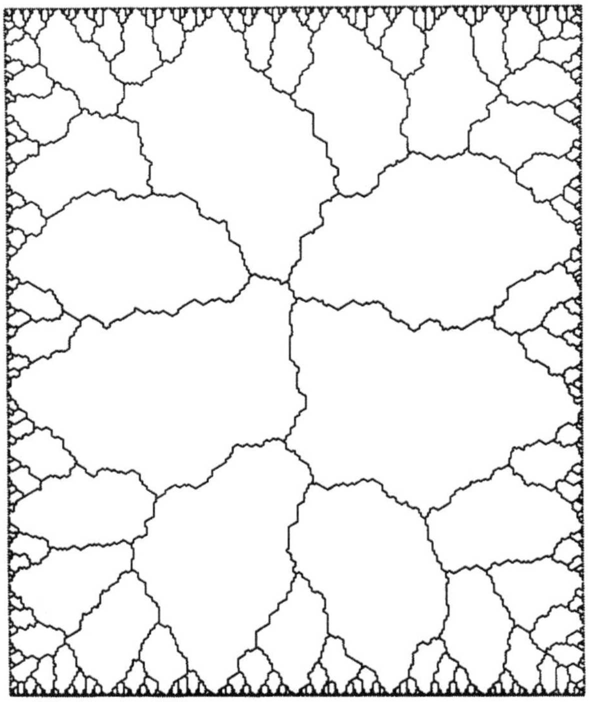

(d)

Fig. 1 (*Continued*)

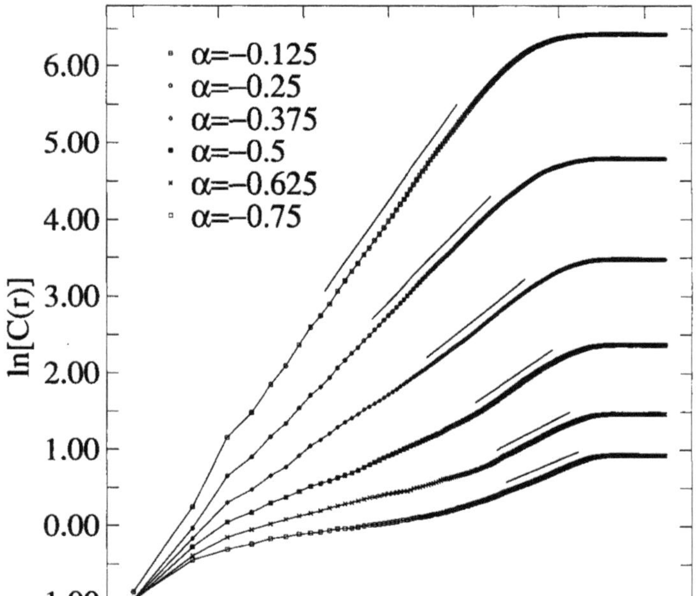

Fig. 2 The height difference correlation functions $C(r)$ of the minimum energy dissipation surfaces are shown as log-log plots in this figure for all six values of α. The six straight fitting lines in the figure have slopes of 1.57, 1.15, 0.887, 0.758, 0.581 and 0.471 corresponding to $\alpha = -0.125, -0.25, -0.375, -0.5, -0.625$ and -0.75, respectively.

dissipation surfaces obtained using different values of α are shown on Figure 2. However, the form of the correlation functions for the drainage basins appear to be more complex than a simple power law. For large values of α (for example, $\alpha = -0.125$, -0.25 and -0.375), most of the data in the log-log plots can be represented by a straight line corresponding to a power-law form for $C(r)$. But when α becomes smaller, the correlation functions become more complex. The effective "Hurst exponents" for the optimal drainage basins 0.785, 0.575, 0.444, 0.379, 0.290 and 0.236 were obtained for the six values for α listed above. Compared with Hurst exponents ranging from 0.55 to 0.75 for natural landscapes,[16] the surfaces of the minimum energy dissipation drainage networks obtained using small values of α, appear rougher. To compare with the optimal landscape illustrated in Fig. 1(b), Fig. 3 shows another optimal landscape generated with larger α ($a = -0.25$). It can be seen that the landscape illustrated in Fig. 3 is more realistic than that in Fig. 1(b).

The fractal dimension D of a contour (entire iso-level set) of a simple self-affine fractal can be related to its Hurst exponent H by[15]:

$$D = 2 - H \tag{7}$$

For a simple self-affine fractal, all the contours have the same fractal dimension, independent of the height at which the contour is taken. The values obtained for the effective fractal dimensions of contours at different height for different values of α are shown in Fig. 4.

Fig. 3 Picture of a landscape associated with the minimum energy dissipation drainage networks. The value of α used in this figure is -0.25. The heights are indicated by different colors (red, yellow, green, blue and violet in order of decreasing height).

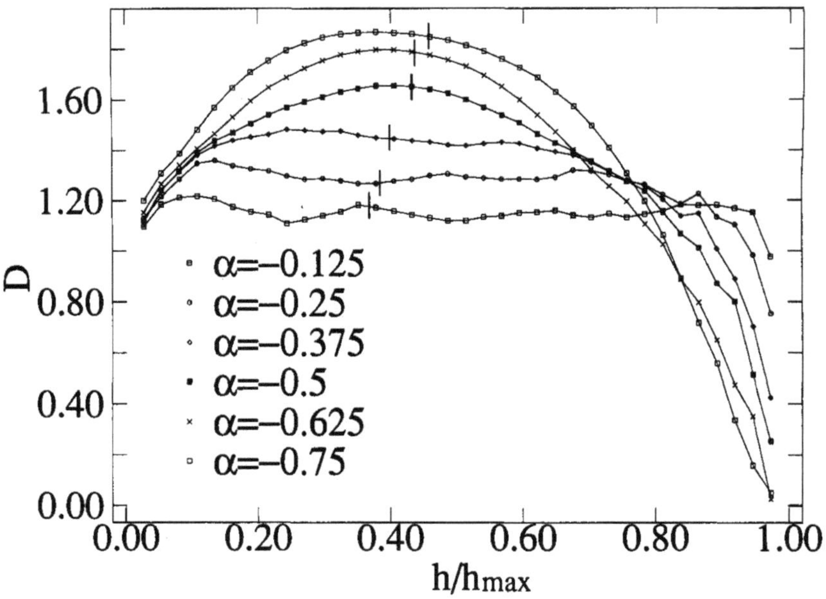

Fig. 4 The effective fractal dimensions of the contours of the minimum energy dissipation surfaces obtained by box-counting analysis are plotted as a function of the relative height (normalized by the highest height in the whole correspondence landscape) of the contours for all the six different values of α. The short vertical lines indicate the positions of the relative average height of each surface.

It can be seen from the figure that the surface of optimal drainage basins with $\alpha = -0.375$, -0.25 and -0.125, have a relatively larger number of contours with more or less the same fractal dimension. When α is small ($|\alpha|$ is large), only a few contours concentrated in a narrow range near the average height have similar fractal dimensions. This result agrees with the previous result from the height-height correlation measurement. For $\alpha = -0.125$, the fractal dimension of contours calculated from Eq. (11) is $D = 2 - H = 2 - 0.785 = 1.215$, which is very close to the value indicated in Fig. 4.

It has been shown[7] that optimal drainage networks have a power-law distribution of basins sizes covering the whole drainage area. Figure 5 shows the cumulative size distribution of the basins covering the area A for different values of the exponent α. All the data sets can be fitted quite well by straight lines on the log-log plot with the same slope of -0.50. This means that the number of the basins of size A decreases algebraically with increasing A with the same universal exponent for different values of α.

A power-law relationship between the length p of basin boundaries and their size (area) A has also been found $p \sim A^{D_p/2}$. Figure 6 shows the dependence of $\ln(p)$ on $\ln(A)$ obtained from the optimization model with all 6 values for α. The straight line in Fig. 5 has the slope 0.55. This implies that the boundaries of the optimal drainage basins have a fractal dimension of $D = 0.55 \times 2 = 1.10$. Meakin et al. have shown[17] that the boundaries for river networks generated by a "coalescing self-avoiding random walk model" has a fractal dimension of 1.25. Field observations by S. P. Breyer and R. S. Snow[18] on 12 river basins, whose area ranges from 150 km^2 to 50 000 km^2 in the United States, show that the boundaries of drainage basins have effective fractal dimensions between 1.06 and 1.12. So like real river basins in nature, optimal river basins tend to have more "smooth" boundaries than the basins obtained by simple statistical models.

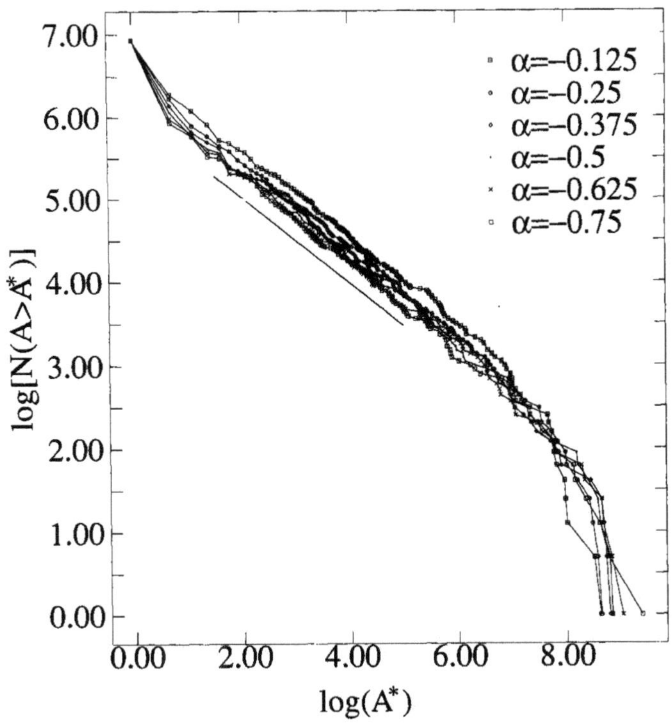

Fig. 5 The dependence of $N(A > A^*)$, the number of the basins that have areas larger than A^*, on the area A^* for all six values of α. As indicated in the figure, all the data sets can be fitted to a straight line with slope of about -0.50.

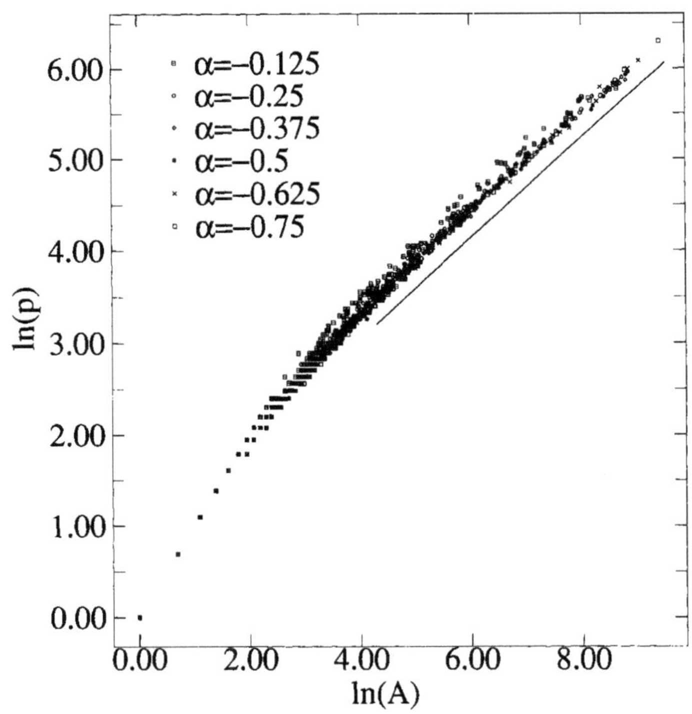

Fig. 6 The dependence of the length of the minimum energy dissipation drainage basins perimeters p on their areas A for $\alpha = -0.125$, -0.25, -0.375, -0.5, -0.625 and -0.75. Except for the basins with very small size (area A), all the data sets can be represented by the same straight line with slope of about 0.55 as indicated in the figure.

A model for river networks and their associated landscapes has been studied. The model is based on the minimum energy dissipation principle. In many respects, the various scaling properties obtained from this model are similar to those observed in nature. The scaling properties of the associated optimal landscape appear to be more sensitive to changes of the parameter, α, in the model than the scaling properties of the associated river networks.

REFERENCES

1. D. G. Tarboton, R. L. Bras and I. Rodriguez-Iturbe, *Water Res. Res.* **24**, 1317 (1988).
2. M. Matsushita and S. Ouchi, *Physica* **D38**, 246 (1989).
3. P. S. Stevens, *Patterns in Nature* (Little, Brown, Boston, Mass., 1974).
4. I. Prigogine, *From Being to Becoming — Time and Complexity in the Physical Science* (W. H. Freeman and Company, 1980).
5. A. Rinaldo, I. Rodriguez-Iturbe, R. Rigon, R. L. Bras, E. Ijjasz-Vasquez and A. Marani, *Water Res. Res.* **28**, 2183–2195 (1992).
6. T. Sun, P. Meakin and T. Jøssang, *A Mathematical Model for River Basin Formation — A Minimum Energy Dissipation Approach* (in preparation).
7. T. Sun, P. Meakin and T. Jøssang, *The Topograph of Optimal Drainage Basins* (in preparation).
8. L. B. Leopold and T. Jr. Maddock, *U.S. Geol. Survey Prof. Paper* **252**, 56 (1953).
9. L. B. Leopold, *Am. Jour. Sci.* **251**, 606–624 (1953).
10. S. Kirkpatrick, C. D. Gelatt, M. P. Vecchi, *Science* **220**, 671–680 (1983).
11. P. Meakin, *Physica Scripta* **45**, 69–74 (1992).
12. M. G. Wolman, *U.S. Geological Survey Prof. Paper* **272**, 56 (1955).
13. L. B. Leopold and J. P. Miller, *U.S. Geol. Survey Prof. Paper* **A282**, 1–36 (1956).
14. J. J. Flint, *Water Res. Res.* **10(5)**, 969–973 (1974).
15. P. Meakin, *The Growth of Rough Surfaces and Interfaces* (1993).
16. D. L. Turcotte, *Fractals and Chaos in Geology and Geophysics* (Cambridge University Press, 1992).
17. P. Meakin, J. Feder and T. Jøssang, *Physica* **A176**, 409–429 (1991).
18. S. P. Breyer and R. S. Snow *Geomorphology* **5**, 143–157 (1992).

Fractals, Vol. 1, No. 3 (1993) 585–593

NATURAL GOLD DENDRITES FROM HYDROTHERMAL Au-Ag DEPOSITS: CHARACTERISTICS AND COMPUTER SIMULATIONS

PAUL A. SCHOENLY and JAMES A. SAUNDERS
Department of Geology, Auburn University, Auburn, AL 36849-5305, USA

Abstract

Epithermal vein gold deposits in the western United States locally contain bonanza zones with extraordinary concentrations of gold in banded ores consisting of alternating gold-rich and barren silica bands. Gold-rich bands commonly consist of coalescing or isolated gold dendrites, which occur in a matrix of variably preserved colloidal silica. Most dendrites appear to have grown outward from near-vertical vein walls and have a rough radial symmetry. SEM images of dendrite surfaces suggest that they formed by the aggregation of spherical particles in the 10–100 nm range. Fractal dimension (D_f) of true two-dimensional slices through the dendrites typically are in the 1.6–1.7 range, which compares favorably with the theoretical D_f value of ~ 1.7 for 2-dimensional diffusion-limited colloid aggregation (DLA). Monte Carlo computer simulations best reproduce natural dendrite morphologies when particle randomness is large, sticking efficiency = 1, and when particles are required to contact 3 adjacent particles on the dendrite as a prerequisite for aggregation. Thus, the physical transport and aggregation of gold colloids offers an attractive explanation for the origin of these enigmatic super high-grade gold ores.

1. INTRODUCTION

Shallow hydrothermal ("epithermal") vein gold deposits hosted by volcanic rocks are fossil equivalents of the upper part of present day geothermal systems. In geothermal systems of this variety, deeply circulating groundwaters are heated by magmas, initiating large convection cells. Water ultimately returns to the near surface in upflow zones, which are equivalent to veins in the fossil geothermal systems. In the geothermal systems that produced gold deposits, magmas contributed gases such as HCl, CO_2 and H_2S to the circulating water, and probably gold and other metals as well. H_2S causes gold to be quite soluble in hydrothermal solutions; loss of H_2S causes gold precipitation. As the hydrothermal solutions move toward the surface, pressure decreases, which commonly leads to solution boiling. Boiling preferentially removes such volatiles as H_2S from the liquid fraction, which results in gold precipitation. Furthermore, silica (SiO_2), adularia ($KAlSi_3O_8$), and calcite ($CaCO_3$) are also deposited due to other boiling-induced changes such as temperature drop and increase in solution pH.

Here we document gold textures from the boiling zone of one of the highest-grade gold veins ever mined — the Sleeper deposit, Nevada, USA. Gold exhibits fractal structures implying they formed by colloid aggregation as a result of random collisions of gold particles with the vein walls, which indicates deposition under highly turbulent conditions. We believe these features to be the first example of natural 3-dimensional colloid aggregates.

2. MORPHOLOGY OF GOLD AND SILICA

Gold in the high-grade veins of the Sleeper deposit[1] occurs as three dimensional dendritic structures intergrown with fine-grained silica, both of which were deposited on a substrate of silica deposited previously on the vein wall. Gold in the richest bands occur as "forests" of dendrites composed primarily of large and dense dendrites with smaller dendrites occurring in between. Dendrite forests are extensive in both the length and width over the deposition surface, and individual dendrites occur irrespective of the surface orientation of fluid flow direction. Scanning electron microscope images (Fig. 1) show the wide range of scales through which the gold dendrites are self-affine.

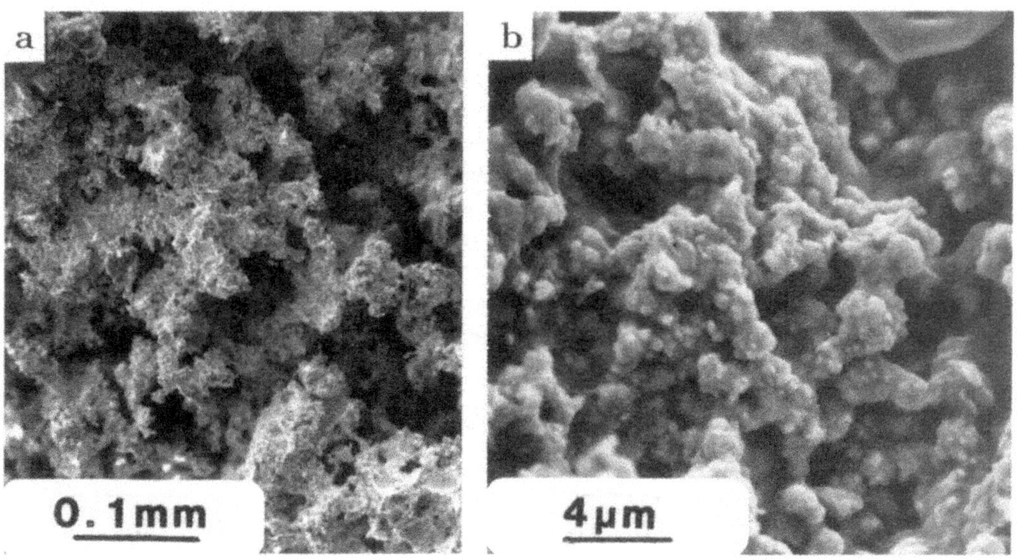

Fig. 1 SEM images of gold dendrite surfaces, where silica has been removed with hydrofluoric acid, under low (a), and high (b), magnification. Despite the wide difference in magnification between the two images, the gold structures still display a remarkable degree of self-affinity. The images are oriented roughly parallel to the vein wall.

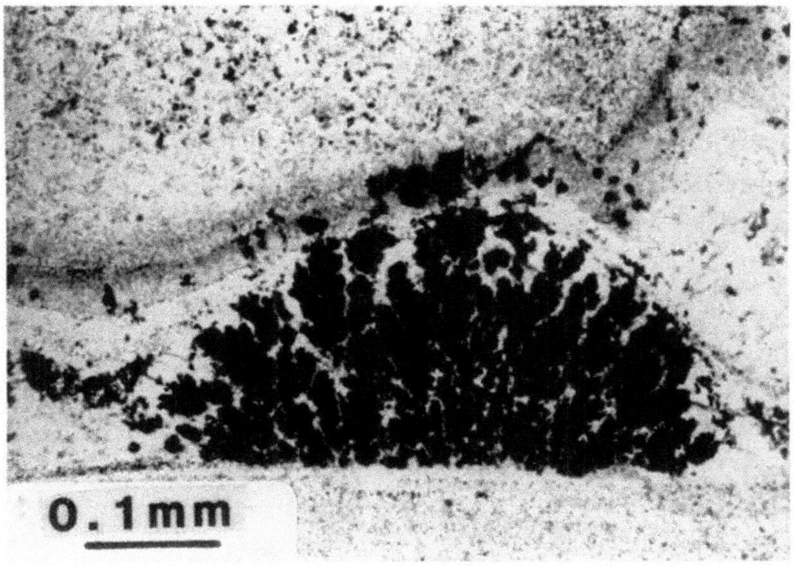

Fig. 2 Photomicrograph (transmitted light) of a radial gold dendrite (black) in a matrix of opaline silica (gray) from the Sleeper deposit, Nevada. The image represents a 2-dimensional projection of a $\sim 30\mu m$-thick slice through the dendrite.

Gold structures may be divided into three modes of occurrence on the basis of textures as seen by microscopic observations (in transmitted and reflected light) of thin sections through the dendrites. The most common mode of forests are those in which the individual dendrites are restricted in their lateral distribution by competition with other neighboring gold dendrites. In this case, individual dendrites apparently grew orthogonally out from the vein walls (depositional substrate) as a result of competition with neighboring dendrites. As a unit, the forests are dense and free space between the individual dendrites is minimal. The second most common forest type is typified by the occurrence of radial dendrites (Fig. 2). In these dendrites, particle distribution is less restricted parallel to the vein wall, and the resulting dendrites are radial and hemispheroidal in shape. Individual radial dendrites are larger than the dendrites in the restricted-growth forests. The forests containing radial dendrites locally may also have areas of restricted-growth dendrites. Dendrites from the deepest parts of veins, which formed at higher temperatures, apparently exhibit some crystallographic control on dendrite formation. These dendrites have branches emanating from linear (in two dimensions) trunks at 45° angles, with branches retaining the characteristic irregular shapes seen at shallower levels. We interpret this style of dendrite formation as similar to snowflake growth, where both random particle accretion and crystallography control growth. Not all gold present in the ores occurs as dendrites; lower grade ores contain disseminated gold grains in finer-grained silica.

Silica in the ores occurs as variably recrystallized amorphous colloidal silica (opal), with opal commonly preserved in gold-rich bands. Within gold forests, colloidal silica apparently filled the fjords between branches of gold and did not form separate dendrites. Infilling by silica apparently provided some structural support for the growing gold dendrites. Silica occurs as dense layers of opal often recrystallized to quartz in gold-poor bands,[1] and exhibits rolling and globular surfaces. Rarely, relict dendrites of silica are preserved.

3. FRACTAL DIMENSION OF GOLD DENDRITES

Photomicrographs of restricted-growth and radial dendrites were digitized to provide numerical input for image analysis. From this data, the height-density distribution and the density-density correlation functions were determined. The height-density distribution (Fig. 3) measures the density of structures perpendicular to the vein wall by counting the number of particles N at a distance H from the growth surface using the equation:

$$\rho(h) = \sum_{x} \frac{\rho(h, x)}{L} \tag{1}$$

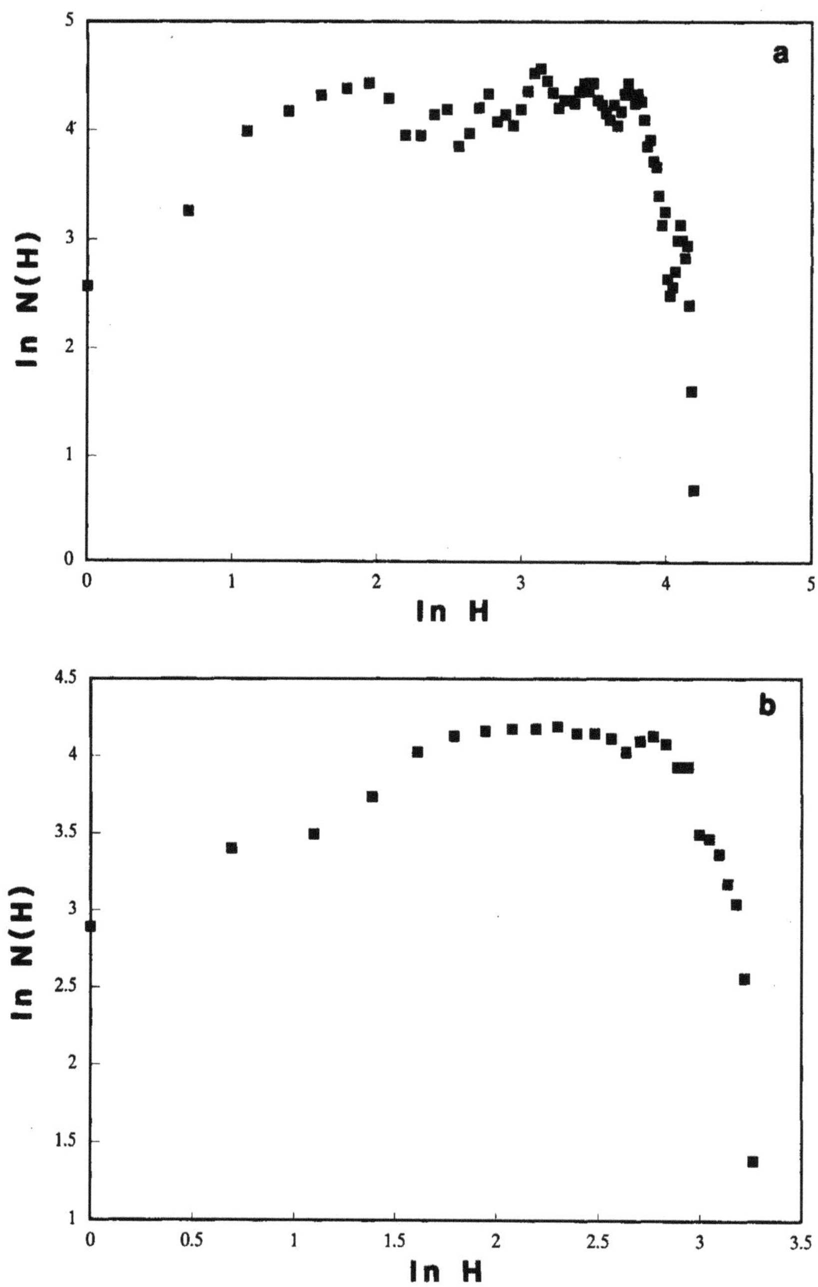

Fig. 3 Height-density distribution log-log graphs for the radial (a), and restricted growth dendrites (b).

where L represents the particles particle length of the depositional substrate, $\rho(h, x)$ is the probability of finding a particle at that height, and $\rho(h)$ is the summation of these normalized $\rho(h, x)$ values for each individual height.[2] The density-density correlation function (Fig. 4) measures the lateral distribution of the particles with the equation:

$$c_h(x) = \sum_{x'} \frac{\rho(h, x + x')\rho(h, x')}{L} \tag{2}$$

where L is the total number of particles along the depositional substrate and ρ is the

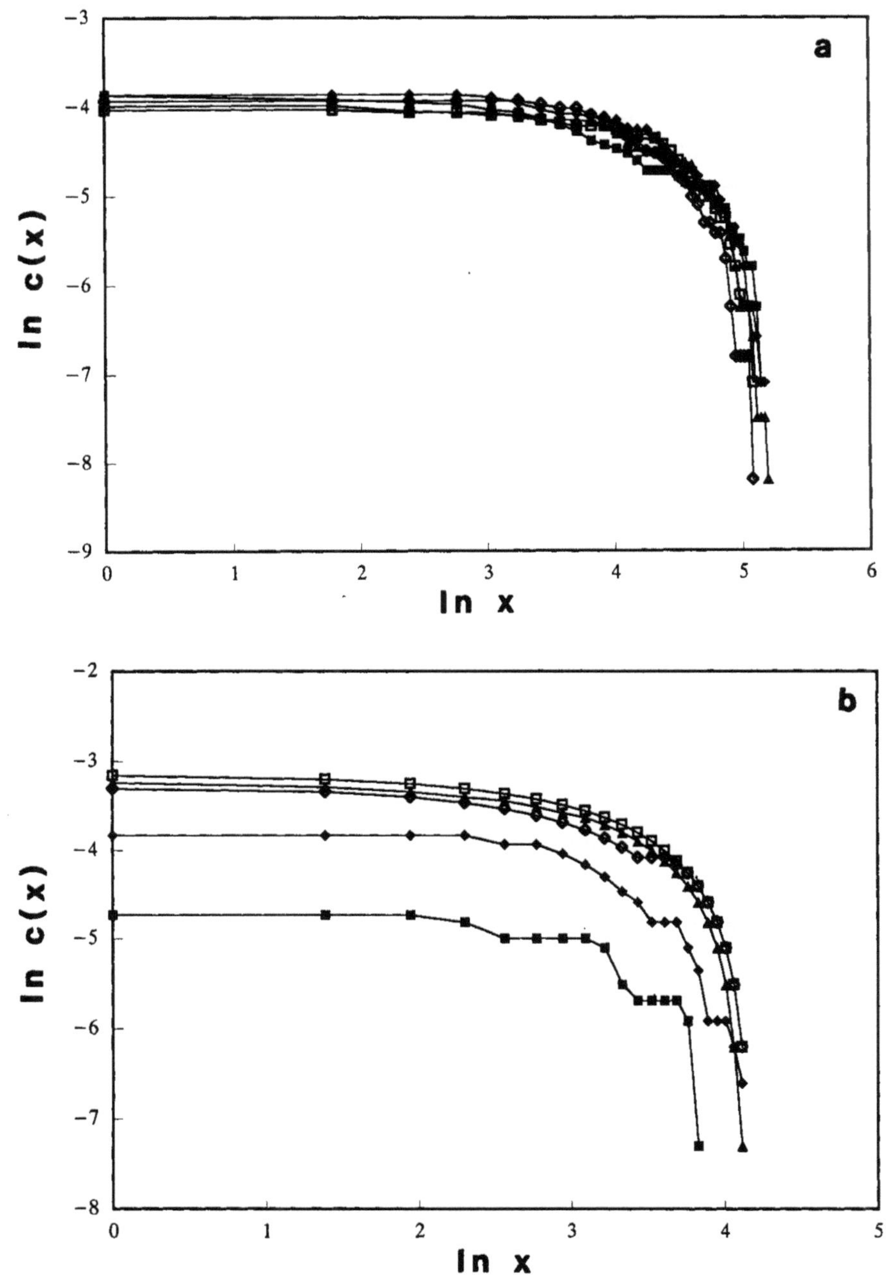

Fig. 4 Density-density correlation function log-log graphs for the radial (a), and restricted (b), growth dendrites.

probability of finding a particle at the prescribed lateral and height coordinates.[3] Both the restricted-growth and radial dendrites display similar height-density distributions, whereas slight differences are evident in the density-density correlation graphs for the radial and restricted-growth dendrites.

From the height-density distribution the logarithmic density ratios for the examples of gold dendrites were calculated in directions both parallel and perpendicular to the principal growth direction.[2] These ratios were used to calculate the fractal dimension with the equation:

$$D_f = 1 + \frac{1 - \nu_{\parallel}}{\nu_{\perp}} \tag{3}$$

where $\nu_{\parallel}$ and $\nu_{\perp}$ are the logarithmic density ratios, which are scaling factors related to dendrite asymmetry.[3] Fractal dimensions calculated using Eq. (3) are 1.65 for a true 2-dimensional slice (as seen in reflected light) through the radial dendrite, and 1.86 for partial projections (as seen in transmitted light through $\sim 30\ \mu$m thick slices) of restricted-growth dendrites (Table 1). Two-dimensional computer simulations have shown that diffusion-limited deposition produces forests with fractal dimensions approximately equal to 1.57,[2] whereas ideal DLA structures have fractal dimensions of 1.70–1.72.[4]

Table 1 Scaling Exponents ($\nu_{\perp}, \nu_{\parallel}$) and Effective Fractal Dimensions for Radial and Restricted-growth Dendrites

	$\nu_{\perp}$	$\nu_{\parallel}$	D_f
Radial	0.746	0.511	1.65
Restricted	0.630	0.460	1.86

4. INTERACTION BETWEEN GOLD AND SILICA COLLOIDS

Gold and silica apparently were deposited simultaneously throughout the formation of individual gold-rich bands within the high-grade vein of the Sleeper deposit.[1] Particles apparently traveled up the open vein structure at high velocities and with highly turbulent movement. Turbulence apparently resulted as a consequence of solution boiling and interaction with the highly irregular vein surface.[1] Throughout the movement up the vein, gold and silica colloids would have apparently experienced a series of collisions, which may have resulted in minor clustering of individual gold particles but not silica. This clustering process has been shown in laboratory experiments to be a major aggregation mechanism for gold colloids,[5,6] but generally leads to very tenuous aggregate structures. Silica, which has a lower sticking efficiency, probably does not bond with gold or other silica particles without some difficulty. Even though gold has a high sticking efficiency, the shielding by silica, along with repulsive surface electrostatic forces between individual gold particles, may have prevented immediate aggregation during transport. Thus a limited amount of cluster aggregation may have occurred during transport, and the process does not appear to have been the predominant process responsible for producing the relatively dense dendrites deposited on the vein walls.

5. COMPUTER SIMULATIONS

Computer simulations have been used to evaluate possible *in-situ* depositional processes for gold and silica. The interpreted highly turbulent movement of the fluid carrying the silica and gold colloids allows the modeling of dendrite formation through the use of the random walk method originally proposed by Witten and Sander.[7] In our simulation, two particle types were used to represent silica and gold colloids. Silica was modeled using a sticking efficiency significantly less than that of the gold particles. Various proportions of gold and silica were tested; the program successfully simulated the dendritic gold textures when the gold: silica ratio was approximately > 3:1. This proportion was necessary to form gold structures that were cohesive in form despite the presence of silica during the gold deposition process. At this gold: silica ratio, silica infills the fjords between branches of gold dendrites. Decreasing the sticking efficiency of silica further can reproduce the observed gold dendrites at lower gold: silica ratios.

Textural observations suggest that silica played only a minor role in determining the ultimate gold dendrite morphology, so our modeling efforts focused on simulating the dendrites using single particle modeling techniques. The most successful approach to simulating 2-dimensional images of natural dendrites allowed particles to move randomly only a set number of steps from a large starting radius. At the end of the particle's movement, the particle was still displayed even if it had not reached the surface of the growing dendrite. This approach differs from classic simulation techniques,[7] where the non-aggregated particles are removed from the simulation. However, leaving the non-aggregated particles in the simulation accounts for the possibility that gold particles are locally imbedded in the surrounding silica. A starting radius of 100 and movement limit of 500 steps most accurately approximated the cross-section photographs of the gold dendrites. To best simulate the dense nature of the dendrite branches, a variable was introduced that limited aggregation to sites on the dendrite where more than one adjacent pixels were lit.[8] Best results were obtained when a particle was required to contact 3 adjacent particles on the dendrite for aggregation (Fig. 5).

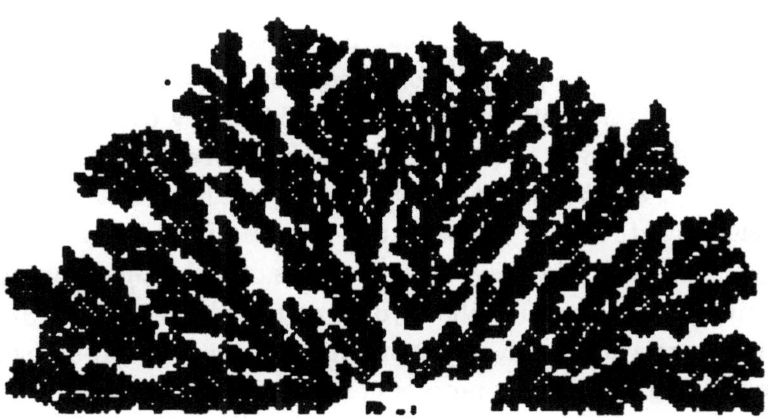

Fig. 5 Computer-generated radial dendrite approximately composed of 10 000 particles which simulates a two-dimensional projection of a three-dimensional radial gold dendrite (cf. Fig. 2). This simulation used a modification of a model proposed by Vicsek[8] in which particle aggregation is possible only on growth sites with a set number of occupied sites. In this simulation, three adjacent particles were required for aggregation to occur.

6. DISCUSSION

The height-density distribution and density-density correlation functions differ from results of the same measurements for computer-generated diffusion-limited deposition forests presented by Vicsek[3] and Meakin et al.[2] In both the restricted growth and radial dendrites in the Sleeper ores, height-density gradually increases with greater height only to decay rapidly as the tips of the dendrites are approached (Fig. 3). This observed increase in density is atypical of "normal" diffusion-limited deposition forests that typically are more dense at the base and decrease in density with increasing height. The density-density correlations are atypical of diffusion-limited deposition as well, because no minima for x is defined. However, they show marked conformity to similar measurements of DLA clusters.[3] This similarity is unusual because in DLA clusters, particles originate from a circular particle source area, whereas deposition along a vein surface suggests more linear ballistic deposition would be expected. This unexpected result might be due to localized turbulence, which could cause particles to originate from a circular source area, leading to radial dendrite formation. The restricted-growth dendrites display textures more characteristic of a particle source area typical of diffusion-limited deposition, yet the dendrites that compose the forests are fairly uniform in height and lateral distribution. This anomaly might be due to rapid annealing and segregation of gold particles aggregated at the inferred depositional temperature of $\sim 150°C$.[1] Thus, annealing might have obscured terminated basal growth sites typical of diffusion-limited deposition forests. Restricted-growth dendrite uniformity might also be a result of processes at surface interfaces and post-depositional relaxation.[7] Furthermore, fluid shearing has also been shown to lead to significant restructuring of colloidal aggregates,[9,10] and this process may have been important in the formation of natural dendrites deposited in a rapidly moving, turbulent fluid traveling in a vein structure.

We have observed dendritic gold textures in other similar bonanza grade deposits, such as National, Midas, and Jumbo (Nevada) and Republic (Washington). This suggests that the gold textures observed at the Sleeper deposit may be typical of high-grades ore zones within volcanic-hosted epithermal deposits, thus implying that diffusion-limited deposition of gold colloids may be a universal process in these systems.

7. CONCLUSIONS

On the basis of computer simulations of dendrite forms and through thin section and SEM observations, the processes responsible for deposition of gold and silica colloids may be evaluated. Because of its high sticking efficiency, gold largely controlled the ultimate morphology of gold dendrites even though silica may have been much more abundant in the hydrothermal fluid. Particle movement was highly random as a result of the localized turbulent movement of the fluid. Cross-section images of dendrites reveal that they consist of highly winding and relatively thick branches, which apparently formed primarily by diffusion-limited deposition. Results to date at the Sleeper deposit indicate that dendritic gold textures are widespread, suggesting that colloid aggregation by diffusion-limited deposition was a major ore-forming process in the bonanza ore zones.

ACKNOWLEDGMENTS

This research was supported by grants from U.S. National Science Foundation (EAR-91171713) and the AMAX Foundation.

REFERENCES

1. J. A. Saunders, *Geology* **18**, 757–760 (1990).
2. P. Meakin, J. Kertesz and T. Vicsek, *J. Phys. A: Math. Gen.* **21**, 1271–1281 (1988).
3. T. Vicsek, *Fractal Growth Phenomena* (World Scientific, Singapore, 1992).
4. P. Meakin, *Phys. Rev.* **A27**, 1495–1505 (1983).
5. D. A. Weitz and M. Olivera, *Phys. Rev. Lett.* **52**, 1433–1436 (1984).
6. M. Y. Lin et al., *Nature* **339**, 360–362 (1989).
7. T. A. Witten and L. M. Sander, *Phys. Rev. Lett.* **47**, 1400–1403 (1981).
8. T. Vicsek, *Phys. Rev. Lett.* **53**, 2281–2284 (1984).
9. M. Y. Lin et al., *J. Coll. Interface Sci.* **137**, 263–280 (1990).
10. V. J. Oles, *Coll. Interface Sci.* **154**, 351–358 (1992).

Fractals, Vol. 1, No. 3 (1993) 594–600

NUMERICAL DETERMINATION OF FRACTAL AND ENTROPY CHARACTERISTICS OF THE SEISMIC PROCESS ON THE TERRITORY OF ARMENIA

A. Yu. SHAHVERDIAN and N. Z. AKOPOV
Yerevan Physics Institute, Alikhanian Brothers Str 2, 375 036, Yerevan, Republic of Armenia

Abstract

An application of certain methods of the theory of dynamical chaos to researching of the seismic process is considered. By means of computer-assisted processing of the earthquakes catalogue the numerical values of the fractal, information and correlation dimensions for this process on the territory of Armenia are determined. Using the numerical method, the inequality $K > 0$ for the Kolmogorov-Sinay entropy K for seismic process is examined.

1. INTRODUCTION

Under the assumption that two different earthquakes cannot occur simultaneously, we can identify the seismic event with m-dimensional vector $\mathbf{x} = (x^1, \ldots, x^m)$ $(= \mathbf{x}(t))$ components of which are the main parameters of event, e.g. it may be the geographical coordinates of epicenter, energetic order, etc. In such a way, we can research (may be diverting from the natural time scale) the evolution of the seismic process, analysing the motion of the point $\mathbf{x}_n = \mathbf{x}(t_n)$ (t_n — the coming time of the event) in the bounded phase space.

In the theory of dynamic (or deterministic) chaos, discrete dynamical systems determined by a nonlinear mapping are considered. It is shown that in many cases, the trajectories of these systems demonstrate the complex unpredictable behavior. For classification dynamics of these systems Renyi[1] entropy dimensions D_q ($q \geq 0$) are generally used and various methods of its determination are considered. In the case of numerical methods, the explicit analytical form of the given mapping is used only for generating by computer the large number of its iterations, after which the numbers D_q (by programming) are calculated. As a matter of fact the constants D_q for logistic and Henon's mapping are calculated in this way in Refs. 2 and 3. Such a method is also accepted in some softwares[4] for nonlinear systems. In this paper, we apply certain methods of this theory to the previously given array of empirical data (earthquakes catalogue) but we are not interested in the analytical form of the physical low which generates these data. By means of computer-assisted processing, we estimate the numerical values of the 3 fundamental quantities D_q ($q = 0, 1, 2$) for the trajectory $T = (\mathbf{x}_n)_{n=1}^{N}$ (N is the number of records from catalogue). In the present context, they measure the degree of irregularity of the seismic process on the researched territory. The other quantity considered in this paper is the Kolmogorov-Sinai (KS-) entropy K. Our aim here is the numerical examination of the inequality $K > 0$, which represents the strong mathematical expression of chaotic property.

Some works in this area should be noted: Kagan and Knopoff[5] have calculated D_0 for epicenters of all earthquakes; Okubo and Aki[6] have obtained D_0 for the whole complex of the faults in California; Ref. 7 is devoted to the fractal objects in geophysics; while the existence of the low-dimensional chaos in certain seismic time series is reported in Ref. 8. Analogy between the behavior of the logistic mapping and some problems of meteorology is noted in Ref. 9, while a model of the seismic process with many degrees of freedom and with zero Lyapunov exponent is suggested in Ref. 10.

2. SOME AUXILIARY MATHEMATICAL FACTS

This section discusses some concepts and known results which are necessary for formulating our problems; for details see Refs. 2 and 11.

Let $t = (\mathbf{x}_n)_{n=1}^{\infty}$ be a trajectory of the discrete dynamical system in the d-dimensional phase space which we assume to be divided in the d-dimensional hypercubic bins with edge length r. In Sec. 3, the numerical values of the following 3 quantities are estimated:

$$D_0 = \lim_{r \to 0} \frac{\ln M(r)}{\ln \frac{1}{r}} \qquad \text{— fractal (capacity) dimension} \tag{1}$$

$$D_1 = \lim_{r \to 0} \frac{\sum_{i=1}^{M(r)} p_i \ln p_i}{\ln r} \qquad \text{— information (entropy) dimension} \tag{2}$$

$$D_2 = \lim_{r \to 0} \frac{\ln \sum_{i=1}^{M(r)} p_i^2}{\ln r} \qquad \text{— correlation dimension} \tag{3}$$

here p_i is the probability for T to fall in the i-th bin [$p_i > 0$ for $1 \leq i \leq M(r)$]. D_0 characterizes the extent of filling of phase space with a trajectory; D_1 determines the growth order of uncertainty contained in the seismic process which is dependent on the increase of the forecasting precision; D_2 is connected with the autocorrelation function of seismic time series $\mathbf{x}(t_n)$.[12]

The other characteristic estimated in Sec. 3 is the KS-entropy K,[2,11] which determines the mean rate of H information lost about the position of point in the phase space. In Sec. 3, H is an information on the epicenter's geometrical position (case T_2) and information on the epicenter's position and the energetic class of the seismic event (case T_3). The following inequality[2] holds:

$$K \geq K_2 = \lim_{r\to 0}\lim_{n\to\infty} K_2(r, n), \quad K_2(r, n) = \ln\left(\frac{C_n(r)}{C_{n+1}(r)}\right) \tag{4}$$

where C_n is the generalized Grassberger-Procaccia correlation integral:

$$C_n(r) = \lim_{N\to\infty} \frac{1}{N^2}\sum_{i,j}\theta\left(r - \left(\sum_{m=0}^{n-1} |\mathbf{x}_{i+m} - \mathbf{x}_{j+m}|^2\right)^{1/2}\right)$$

and θ is Heaviside function. Hence the inequality $K_2 > 0$ is a sufficient condition for the existence of chaos.

3. MAIN RESULTS AND DISCUSSIONS

We consider 2 close problems — planar and spaced. In the first case, calculations are made for trajectory $T_2 = (\mathbf{x}_n)_{n=1}^N$, $\mathbf{x}_n = (x_n^1, x_n^2)$; in second case — for $T_3 = (\mathbf{x}_n)_{n=1}^N$, $\mathbf{x}_n = (x_n^1, x_n^2, x_n^3)$. Here x_n^1, x_n^2 are the geographical coordinates (latitude φ and longitude λ respectively) of epicenter, x_n^3 is the energetic class C (instead of C one can consider the earthquake's magnitude M which is connected with C by linear relationship) of the event.

On the Fortran-77 four not large programs "Dimension 2(3)", "Entropy 2(3)" (the numbers indicate the trajectory's dimension) were written. The calculations of constants D_q are made for the region $38° \leq \varphi \leq 42°$, $42° \leq \lambda \leq 48°$ containing the territory of Armenia and for all earthquakes registered during 1962–1992 for which $5.5 \leq C \leq 11.5$ (number of records $N \cong 14\,500$). To estimate KS-entropy for this region, the restrictions $7.5 \leq C \leq 13.5$ or $7 \leq C \leq 13$ ($N \cong 9500$ and $N \cong 10\,500$ respectively) have been applied. Such restrictions are done only with the purpose of reducing the calculating time. We also changed the natural spatial scale into the more usual one — by homotetical transformation, the phase space is mapped on the rectangle $[0, 2/3] \times [0, 1]$ for a 2-dimensional case, and on the parallelepiped $[0, 2/3] \times [0, 1] \times [0, 1]$ for a 3-dimensional case.

For the dimensions D_q determination, we follow the traditional (cellular) scheme[2,7,11]; the phase space is divided on the S bins with linear size r; for each bin with number i the number S_i of attendances at the i-th bin by trajectory T is calculated, and by means of values $p_i = S_i/S$ and Eqs. (1)–(3), the dimensions D_q are calculated. Regarding the estimates for the errors of calculations, see Ref. 7.

Note that the catalogue used by us with each event contains its for- and aftershock records (and for these reason in Eq. (5) is a sign of inequality). In Figs. 1 and 2, the existence of a trend is distinctly seen even in the case of the (so-called "non-random") catalogue.

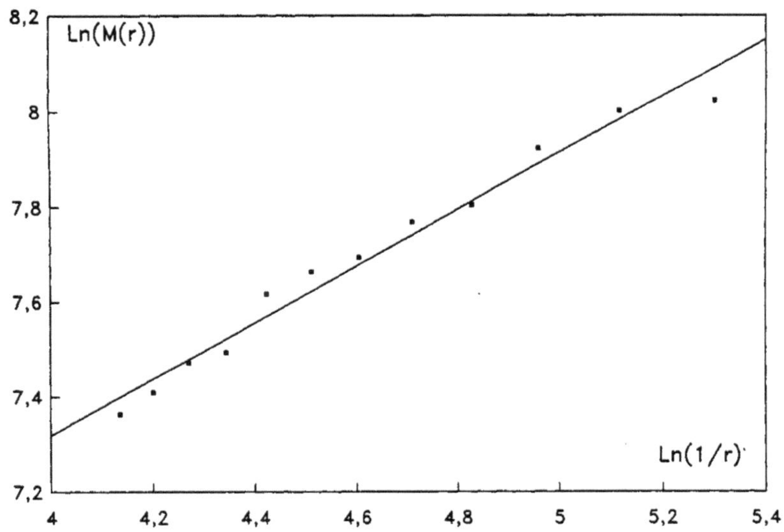

Fig. 1 Dependence of r on the number $M(r)$ covering T_2 nonempty bins assigned on doubly logarithmic scale. Straight line is the trend for this dependence ($N \cong 14\ 500$).

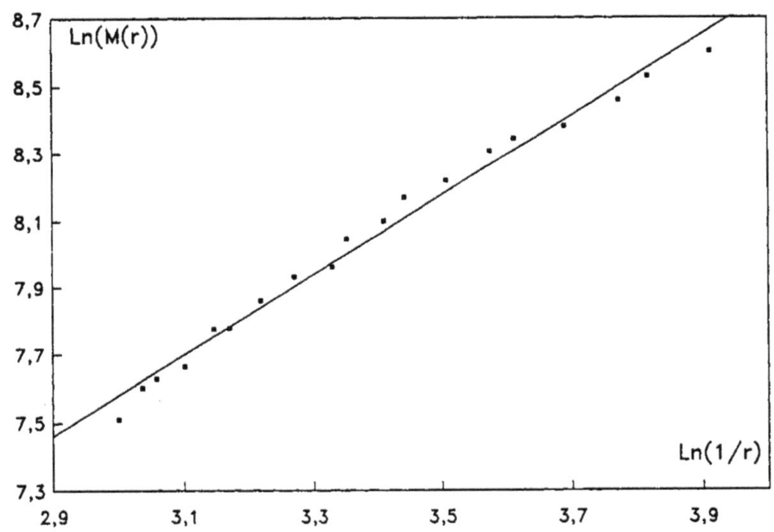

Fig. 2 Dependence of r on the number $M(r)$ covering T_3 nonempty bins assigned on doubly logarithmic scale. Straight line is the trend for this dependence ($N \cong 14\ 500$).

From Figs. 3 and 4, one can see that for each $q = 0, 1, 2$, the numerical values $D_q(r)$ [these are the functions after sign 'lim' in Eqs. (1)–(3)] tend to certain fractional numbers D_q with decreasing r. This means that we deal with strange attractors T_2 and T_3. Regretfully restrictions of the time do not let us keep track of its thin self-similar structure. The lower limits for r are bounded by the number $r = 0.005$, on account of the fact that the processing data had only 2 nonzero signs after the decimal point. For this reason we have not the possibility to execute the limit operation in Eqs. (1)–(3). The alternative is the consideration of some local (for certain interval values of r) (see Ref. 7) characteristics which are defined by substitution of the limit operation in Eqs. (1)–(3) by the averaging over enough numbers of r (note that these mean values do not necessarily coincide with the limiting

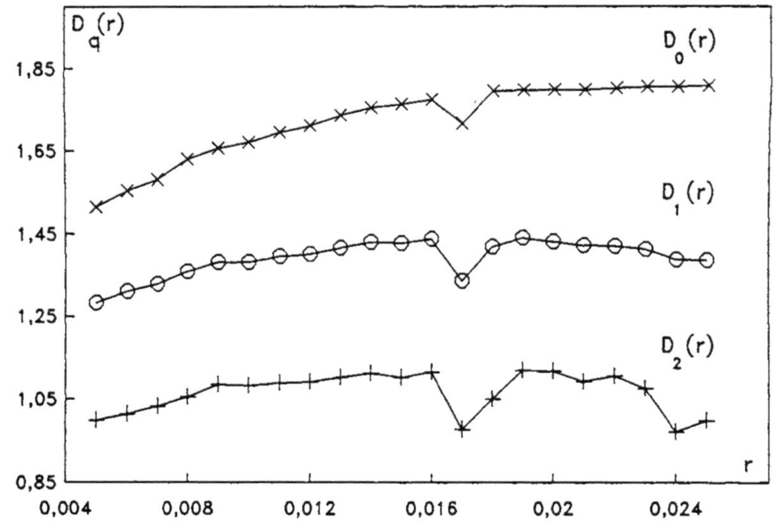

Fig. 3 The values $D_q(r)$ from Eqs. (1)–(3) tend to certain fractional numbers D with the decrease of r; 2-dimensional case ($N \cong 14\ 500$).

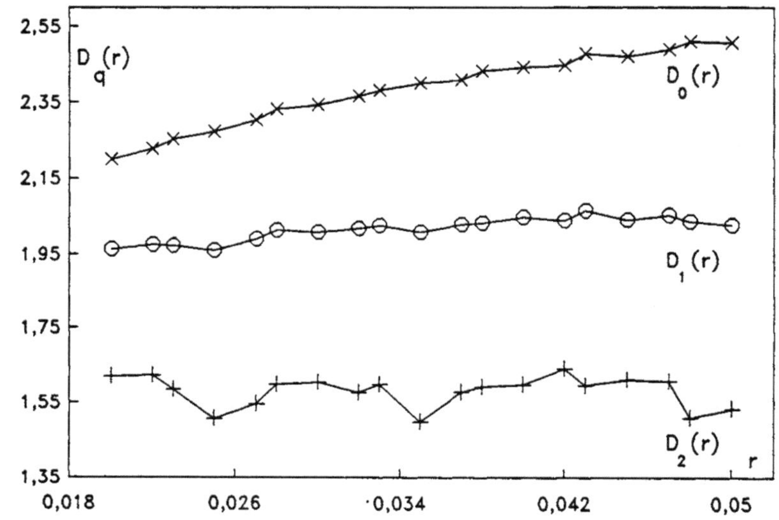

Fig. 4 The values $D_q(r)$ from Eqs. (1)–(3) tend to certain fractional numbers D_q with the decrease of r; 3-dimensional case ($N \cong 14\ 500$).

value D_q). We have obtained the numerical values sought-for dimensions — for the region $38° \leq \varphi \leq 42°$, $42° \leq \lambda \leq 48°$ and $5.5 \leq C \leq 11.5$, the following estimates hold:

$$\begin{aligned} &\text{for the attractor } T_2 : D_0 \leq 1.66, D_1 \cong 1.37, D_2 \cong 1.07 (0.005 \leq r \leq 0.015) \\ &\text{for the attractor } T_3 : D_0 \leq 2.26, D_1 \cong 1.98, D_2 \cong 1.58 (0.02 \leq r \leq 0.03) \end{aligned} \tag{5}$$

For the estimation of the KS-entropy, we proceed from Eq. (4). Such algorithms have the advantage that the calculation time depends very weakly on the bin linear size. But they resquare the order N^2 executing the same type operations; for faster algorithms, see Ref. 11.

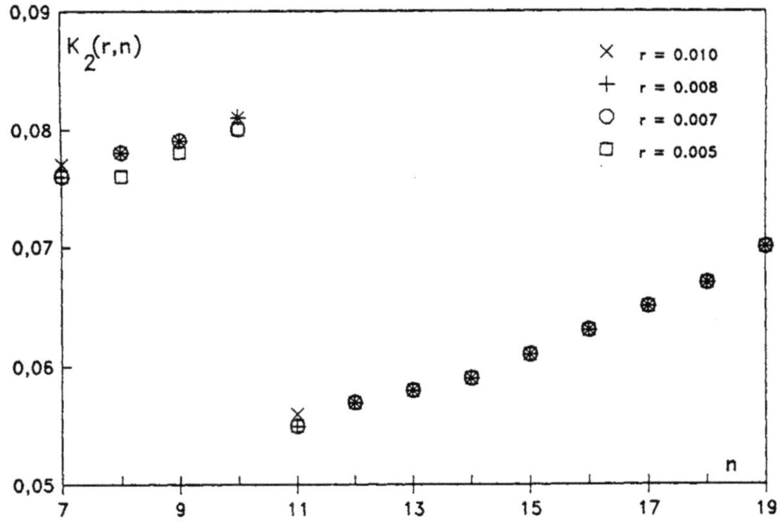

Fig. 5 Function $K_2(r, n)$ from Eq. (4) is separated from zero. First 6 members are greater than 1 and are not shown. For $12 \leq n \leq 19$, we obtain identical values for different r ($N \cong 9500$).

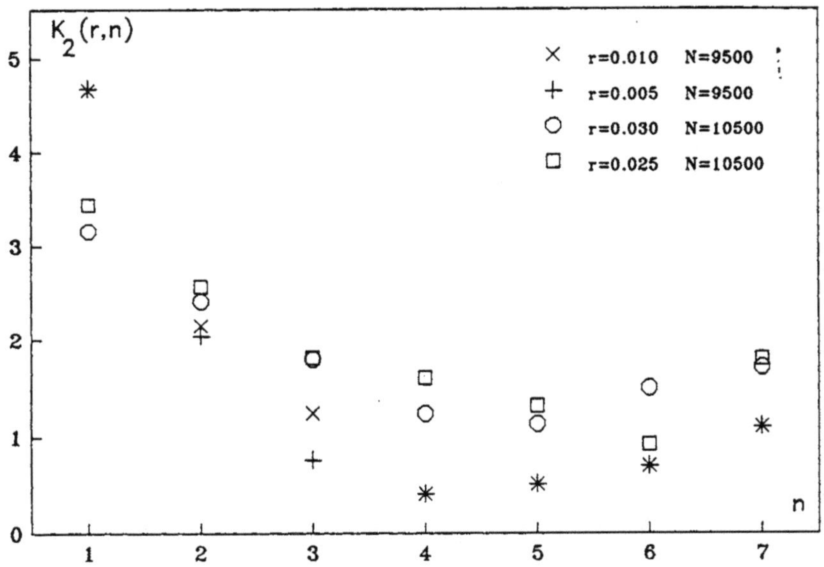

Fig. 6 The values of the first 7 members of the sequence $K_2(r, n)$ for 3-dimensional trajectory.

Examples for obtaining values are given in the Figs. 5 and 6; about 11 and 12.5 hours of computational time for the fast-working computer IBM PC 386-DX have been resquared.

Results in Fig. 6 are given only for 7 values of n from Eq. (4) and they must be considered as not so convincing (but we hope that it will stimulate efforts to obtain more precise estimates for 3-dimensional trajectories). This was mainly due to the deficiency of using the catalogue in which the values of energetic class were given with insufficient accuracy. Note also that to obtain good estimates for a 3-dimensional case, apparently, much more input data are required than in a 2-dimensional case.

In contrast to it, the results in Fig. 5 seriously testify to the inequality $K > 0$, and that once again confirm the chaotic property of attractor T_2. That means that the problem of the earthquakes forecasting must be considered within the framework of the theory of dynamic chaos. Thereafter, for example, the deterministic forecasting period is upper-bounded by the number τ_d (compare Ref. 9), and the long range forecasting on the time $t > \tau_d$ must necessarily be formulated in theoretic-probability terms. The other conclusion is the logarithmic dependence of the mean-predicting time τ_m on the accuracy of r forecasting — it has been followed from Farmer's formula[2]: $\tau_m = const.(1/K)Ln(1/r)$.

Note that the present results are the first approach and do not seek to be definitive. For example, one can obtain more exactly the estimates for D_q on account of good approaching of the geographical borders for researched territory — it will not increase the calculation time essentially. The method for construction of the empirical function of the probability[a] distributions and calculation of the invariant measure for chaotic systems is described in Ref. 11. Probably these questions will be considered in the next work.

REFERENCES

1. A. Renyi, *Probability Theory* (North-Holland, Amsterdam, 1970).
2. H. G. Shuster, *Deterministic Chaos* (Physik-Verlag, Weinheim, 1984).
3. M. Henon, *Comm. Math. Phys.* **50**, 69 (1976).
4. T. C. Parker, L. O. Chua, *Proc. IEEE* **75**, N8 (1987).
5. Y. Y. Kagan, L. Knopoff, *Geoph. J. of the Royal Astr. Soc.* **62** (1980).
6. P. Okubo, K. Aki, *J. Geoph. Res.* **92**, NB1 (1987).
7. M. B. Geilikman and V. F. Pisarenko, in *The Discrete Properties of the Geophysical Medium*, eds. M. A. Sadovskiy and O. Yu. Shmidt (Earth's Phys. Inst. Ac. Sci. USSR, Moscow, 1989).
8. V. A. Mukhamedov, B. S. Karriev and E. G. Kanel, in *Geodesy-Seismology: Deformation and Forecasting* (Int. Symp., Yerevan, 1989, The Abstracts, Yerevan, 1989).
9. E. Lorenz, in *Problems and Prospects in Long and Medium Forecasting*, eds. D. M. Burridge and E. Kallen (Springer-Verlag, Berlin, 1984).
10. P. Bak and K. Chen, in *Fractals and their Uses in Earth Sciences* (Geolog. Soc. of American Book Series, 1990).
11. F. C. Moon, *Chaotic Vibration* (John Wiley & Sons, N.Y., 1987).
12. P. Grassberger and I. Procaccia, *Phys. Rev. Lett.* **50** (1983).
13. T. Rikitake, *Earthquake Prediction* (Elsevier, Amsterdam, 1976).

[a]The other method for calculation of the seismic events probabilities, based on the Gutenberg-Richter low, is given in Ref. 13; here in the analytical relationships (see Chap. 15, Sec. 1.2) an error is made which can lead to rough mistakes.

Fractals, Vol. 1, No. 3 (1993) 601–605

SOME LONG-RUN PROPERTIES OF CLIMATIC RECORDS

L. BODRI*
Geophys. Res. Group, Hung. Acad. Sci., c/o Geophys. Dept., Eötvös Univ., Budapest, Hungary

Abstract

Fractal analysis allows for a mathematical comparison of climatic changes obtained from a variety of observations and recorded at different time scales. Analysis of the oxygen isotope curve of Pacific core V28–239 indicated fractal geometry of the oxygen isotope record for time scales 5000 to 2 000 000 years, with a fractal dimension of 1.22. On a time scale of some hundred years, tree ring index values yield an average fractal dimension of 1.32 ± 0.02. For annual precipitation records from between 1817 and 1963 at nine major cities in the United States, a mean fractal dimension of 1.26 ± 0.03 was found. Analysis of the annual mean surface air temperatures at seven meteorological stations in Hungary for the period of 1901–1991, indicates that the considered temperatures are fractals with a mean fractal dimension of 1.23 ± 0.01. For the global surface air temperature change estimated from meteorological station records from 1880 to 1985, we derive in the present study a fractal dimension of 1.21.

It is reasonable therefore to assume that climatic changes are characterized by one general fractal dimension over the spectral range 10 to 10^6 years. Records with such values of fractal dimension have some long-term persistence: even observations sufficiently distant from each other are not completely independent. If these conclusions become confirmed through analysis of a wider set of climatic records, long-run climatic prediction (in statistical sense) on different time scales will appear feasible.

*Correspondence to: L. Bodri
Department of Geophysics
Eötvös University
Budapest, Ludovika tér 2.,
Hungary 1083

1. INTRODUCTION

It appears typical to divide each branch of science into separate fields. Each of these fields, even if to a certain degree there is overlap of one field with another, can be studied alone. This approach has proved very successful, although one may meet with difficulties when, in the frames of a general conception, one sets to investigate mutual interaction of fields. The hypothesis of a scale invariance provides a possibility to make comparisons of the results of the studies of different fields.

Many physical processes on the Earth appear scale invariant and may be studied within the framework of fractal geometry (e.g. Ref. 1). The definition of a fractal distribution[2] is that the number of objects N with a characteristic size greater than r scales with the relation:

$$N = Cr^{-D}, \tag{1}$$

where D is the fractal dimension and C is a constant of proportionality. Mathematical representation [Eq. (1)] could be valid over an infinite range and yields a possibility to characterize phenomena by a single parameter, the fractal dimension. However, for any physical application, there will be an upper and a lower limit on the applicability of the fractal distribution. If scale invariance extends over a sufficient range of length scales, the fractal distribution provides a useful description of the applicable statistical distribution.

In this paper we consider the fractal geometry of different weather-climate records.

2. FRACTAL GEOMETRY FOR CLIMATIC RECORDS

In meteorology, it is especially characteristic to range the object of study of this science into distinct fields. The distinction is made mostly by time scale. Meteorologists choose to examine fluctuations with a time scale of days or weeks; "macrometeorologists" concentrate on periods from months to few years; climatologists deal with centuries; and paleoclimatologists are interested in longer time scales. Usually each kind of fluctuation is analyzed separately, although the different fields undoubtedly interfere with each other. The available records of meteorological-paleoclimatic changes come from a variety of data sources, are expressed in quite different units, and seldom directly express climate changes (e.g. growth rings). In such a case fractal analysis, which allows comparison of irregularity characteristics of records over a wide range of scales and is independent of the units of measure, provides an extremely successful tool for comparison.

Mandelbrot and Wallis[3] were the first to apply fractal analysis to some climatic records. Rescaled range analysis (R/S analysis) of the dendroclimatologic data (tree ring index values) yields an average fractal dimension of 1.32 ± 0.02. We have calculated this number from 16 values of D (see Ref. 3, Table 1), varying in the range 1.22–1.42. Most of the records were taken from forest-border sites in the southwestern USA and cover the period from the 13th to the 20th centuries. Since different tree species have their own characteristic responses to climatic changes, we have taken into account only records for Douglas fir and Ponderosa pine which behave much alike and can be used together. Besides, their growth ring widths have no autocorrelation from one year to the next. In other words, their tree ring width characterizes only the weather of the given year.[4]

Investigating annual precipitation from between 1817 and 1963 for nine major cities in the United States, the same authors concluded that regional precipitation was a fractal with mean fractal dimension of 1.26 ± 0.03 (the range of variations: 1.12–1.36).

Another kind of climatic data on a similar time scale has been considered in a study by Bodri.[5] This author investigated the annual mean surface air temperatures at seven meteorological stations in Hungary for the period of 1901–1991, and has shown that the considered temperatures were fractals with a mean fractal dimension of 1.23 ± 0.01 (with an interval of variations from 1.19 to 1.28).

Fluegeman and Snow[6] have determined the fractal dimension of long-range paleoclimatic data: oxygen isotope ratio in carbonates deposited on the ocean bed. Oxygen isotope ratios in the ocean bed carbonate remains of benthic species depend on the temperatures of the water in which these organisms lived, and thus directly indicate the prevailing temperatures of the sea floor (e.g. Ref. 4). In the work above, the oxygen isotope curve of the Pacific core V28–239 has been investigated. The results of the analysis indicated fractal geometry of the oxygen isotope record for time scales 5000 (sample interval) to 2 000 000 years, with fractal dimension equal to 1.22. The authors considered this value as characteristic of global climate change over the last 2 m.y. as expressed in changing δO^{18} ratios.

Below we present the results of an R/S analysis of global surface air temperature change estimated from meteorological station records for the period 1880–1985. Surface air temperature has been measured at a large number of meteorological stations for the past century. Hansen and Lebedeff[7] have combined observational data from more than 2500 stations in a way which is desired to provide accurate long-term variations. They find that meaningful global temperature change can be obtained for the past century, despite the fact that the meteorological stations are confined mainly to continental and island locations. Figure 1 shows the global surface air temperature change as constructed from the Hansen and Lebedeff[7] data. The results indicate a global warming of about 0.5 K in the past century. This linear trend was eliminated from data before R/S analysis. A detailed description of the analysis procedure is presented in Ref. 5. Figure 2 illustrates the variations of R/S values vs. time span. The slope H was estimated from the time spans of 5, 7, 10, 20, 40 and 70 years.[3] It equals 0.79 ± 0.03, which yields a fractal dimension $D = 2 - H = 1.21$.

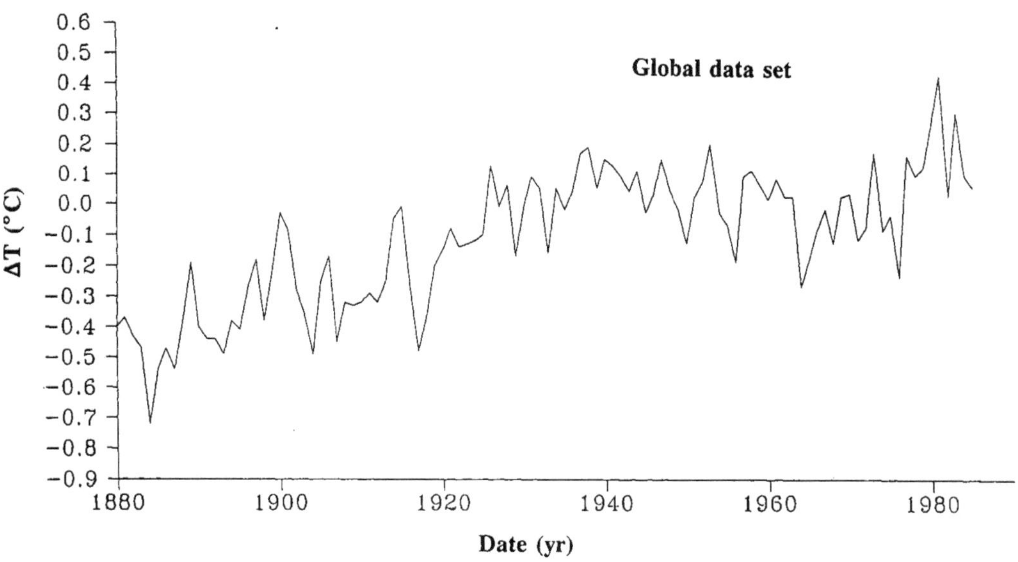

Fig. 1 Global surface air temperature change estimated from meteorological station records (by data from Ref. 7).

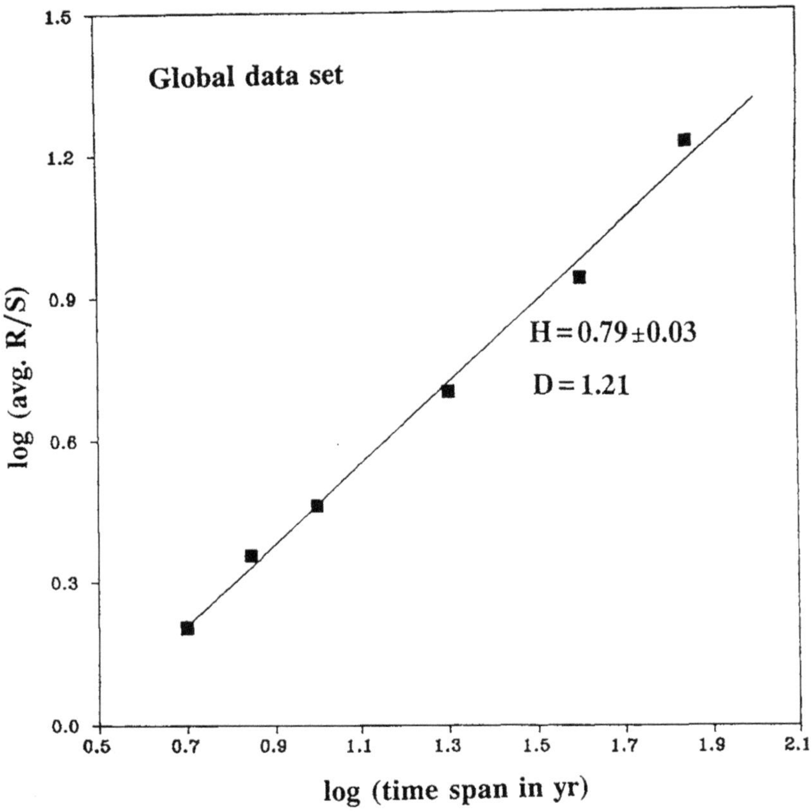

Fig. 2 Plot of average rescaled (R/S) values (marked by square boxes) vs. time spans for global surface air temperature change. Plotted line represents a linear regression line for time spans of 5, 7, 10, 20, 40 and 70 years.

Table 1 **Fractal Dimension of Different Climatic Records**

Type of Record	Time Interval of Record (yr)	Sample Interval (yr)	Fractal Dimension
Global surface air temperature change	105	1	1.21
Annual surface air temperatures	90	1	1.23 ± 0.01
Annual precipitation	~ 150	1	1.26 ± 0.03
Tree ring indices	650–300	1	1.32 ± 0.02
Oxygen isotope ratio	2^*10^6	5^*10^3	1.22

Such a value of H ($H > 0.5$) suggests the long-term persistence of climatic trends through time.[2]

3. CONCLUSIONS

Table 1 lists the D-values of different climatic records that have been analyzed to date from the point of view of their fractal geometric features. The good agreement of fractal dimensions from the time scales of 1–100 years to scales of 5000 to 2 000 000 years, suggests the inference that the irregularity characteristics of a variety of climatic records are very similar. If extensive further studies of the different weather-paleoclimatic phenomena will support this conclusion, one may propose the obvious hypothesis that this similarity reflects the uniform interaction of the climate-forming processes and the materials being affected. The reason for this general relationship can only be speculated upon. Nevertheless, such investigations will provide insight into the systematic nature of climatic records at different times scales. Unique fractal geometry of climatic records would permit interpolation of climatic changes from one scale to the next. It would appear realistic to reconstruct the entire power spectrum of the climatic changes from discrete data sets and with some additional information.

REFERENCES

1. D. L. Turcotte, "Fractals in Geology and Geophysics," *PAGEOPH* **131**, 171–196 (1989).
2. B. B. Mandelbrot, *The Fractal Geometry of Nature* (Freeman and Co., New York, 1982), p. 468.
3. B. B. Mandelbrot and J. R. Wallis, "Some Long-run Properties of Geophysical Records," *Water Resources Res.* **5**, 321–340 (1969).
4. H. H. Lamb, *Climate: Present, Past and Future. Vol. 2: Climatic History and the Future* (Methuen and Co. Ltd., London, 1977), p. 835.
5. L. Bodri, "Fractal Analysis of Climatic Data: Mean Annual Temperature Records in Hungary," *Climatic Change* (1993) (in press).
6. R. H. Fluegeman and R. S. Snow, "Fractal Analysis of Long-range Paleoclimatic Data: Oxygen Isotope Record of Pacific Core V28–239," *PAGEOPH* **131**, 307–313 (1989).
7. J. Hansen and S. Lebedeff, "Global Trends of Measured Surface Air Temperature," *J. Geophys. Res.* **92**, 13,345–13,372 (1987).

Fractals, Vol. 1, No. 3 (1993) 606–616

UNDERSTANDING COMPLEXITY IN THE STRUCTURE OF RAINFALL

BALAJI RAJAGOPALAN and DAVID G. TARBOTON
Utah Water Research Laboratory, Dept. of Civil Engg.
Utah State University, Logan, Utah – 84322-8200, USA

Abstract

This paper aims at understanding the structure of high resolution storm rainfall rates. Scale invariance property has been observed from the distribution function. Departures from simple scaling (i.e., multiscaling) have been noted, suggesting a cascade phenomenon. Close agreement of theoretical and estimated singularity spectra indicates the plausibility of modeling rainfall process by random cascades. Connections between multiscaling and the singularity spectrum have been identified.

1. INTRODUCTION

Study of the rainfall process and its scaling properties, both spatial and temporal is of great importance in hydrometrology. The traditional approach is to analyze the statistical dependence structure of the precipitation process by stochastic modeling, which has resulted in a large number of stochastic models. The statistical dependence structure has been shown by Rodriguez-Iturbe et al.[1] to be closely tied to the scales at which descriptions are sought. Consequently, the parameter estimation of any stochastic model used for modeling the rainfall process is also linked to the scales of the measured rainfall. This poses problems for the definition and identification of universal models for the rainfall occurrence process. In order to be able to understand the rainfall process over a wide range of scales, scale considerations are very important while modeling the rainfall process in time/space.

Phone: 801-750-3172
Fax: 801-750-3663
Email: rbala@pub.uwrl.usu.edu

Lovejoy and Schertzer[2] introduced the ideas of scale invariance and fractals into hydrologic rainfall modeling through evidence that rainfall may be scaling. Schertzer and Lovejoy[3] note that the rainfall process possesses no characteristic scale, i.e., the process can be said to be scale invariant. Rodriguez-Iturbe et al.[4] and Sharifi et al.[5] have studied storm rainfall in detail and have found evidence of deterministic chaos in the process, indicating that the underlying dynamics is deterministic. These serve as motivating factors for exploring the scaling properties to better understand the structure present in the rainfall process.

We present the analysis of storm rainfall rates, towards possible modeling by a simple random cascade. Simple and multiscaling properties have been studied. Singularity spectra have been estimated using data driven methods and connections have been established with multiscaling. Theoretical singularity spectrum from a multiplicative random cascade has been fit to the estimated spectrum. Close agreement between both the spectra provides justification for modeling the rainfall process with a random cascade. Data sets are first described, analysis of scale invariance is next presented. Probings into the scaling properties of the moments are then discussed. Singularity spectrum analysis and their concurrence with moment scaling are next presented. Discussion of the results concludes the paper.

2. DATA DESCRIPTION

Two very high resolution rainfall rate data sets were used in this study. Data from a storm in Indiana (USA) was collected using a distrometer (a LASER instrument) by Dr. Huang. Rainfall amounts each ten seconds are determined from drop sizes and counts of the number of drops in each size class over a 50 cm^2 area. In this paper the storm of October 9, 1990 was studied. Figure 2(a) shows the data. The other data set analyzed was the Boston storm of October 25, 1980. Rainfall depth was accumulated every fifteen seconds by a high resolution tipping bucket rain gauge located at MIT (USA). Figure 1(a) shows this data. The Boston data was the same data as used by Rodriguez-Iturbe et al.[4] in their search for deterministic chaos. They analyzed the correlation dimension of various embedding dimensions and concluded that this data suggested the existence of a low dimensional attractor.

3. EVIDENCE FOR SCALING

Rodriguez-Iturbe[6] note the intermittency and strong dependence of statistical moments on a few data values for the Boston storm data. Similar behavior was seen in the Indiana storm data. This is evidence that scale dependence may be important.

Lovejoy and Schertzer[2] studied scaling in rainfall rate using radar measured data by looking at the tails of the exceedence probability distribution function. They note that

$$\text{Prob}(|\Delta R| > \Delta r) \sim \Delta r^{-\beta} \tag{1}$$

In other words, the probability that the change of rainfall flux ΔR between two instants separated by a time ΔT is larger than any value Δr, and is proportional to $\Delta r^{-\beta}$. Equation (1) is a power-law relation and plots as a straight line in double logarithmic paper, with β the slope of the line.

We did a similar analysis. Here $\Delta R(\Delta T)(= R(t_2) - R(t_1))$ is the difference between successive depths of rainfall that accumulates in interval $\Delta T(= t_2 - t_1)$. ΔT is called the time of aggregation. Four times of aggregation were chosen for the present study. For the

Boston storm the ΔTs are 15 sec., 30 sec., 60 sec. and 120 sec., while for the Indiana storm they are 10 sec., 20 sec., 40 sec. and 80 sec. Figures 1(b) and 2(b) are the plots of Prob($|\Delta R| > \Delta r$) vs. Δr on a logarithmic scale for various ΔTs for the Boston and Indiana storm, respectively. The Weibull plotting position formula[7] was used. The striking feature of the plots is that, doubling ΔT does not change the separation between the lines and the tails are practically parallel. This is suggestive of self-similarity.

Self-similarity and scaling in space-time rainfall is discussed bys Waymire and Gupta.[8] They define a process $X(t)$ as simple scaling if upon rescaling $t \rightarrow \lambda t$, and defining:

$$X_\lambda(t) = X(\lambda t), \quad \lambda > 0 \tag{2}$$

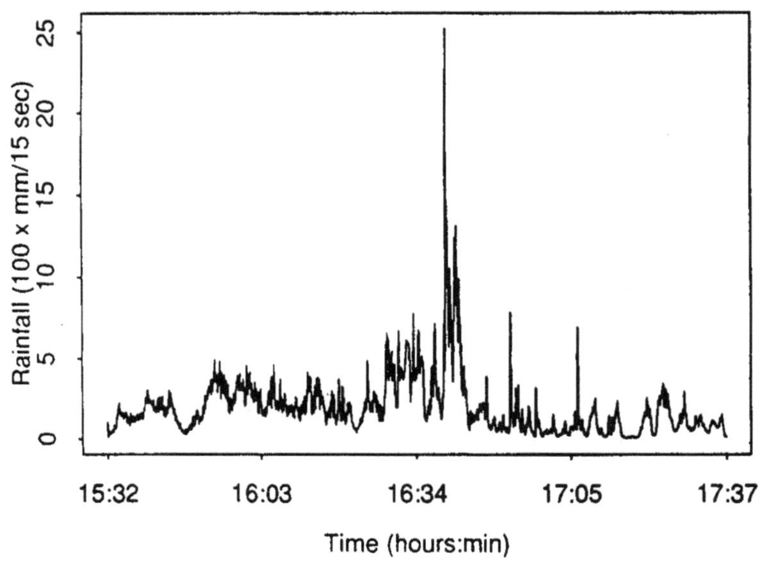

(a)

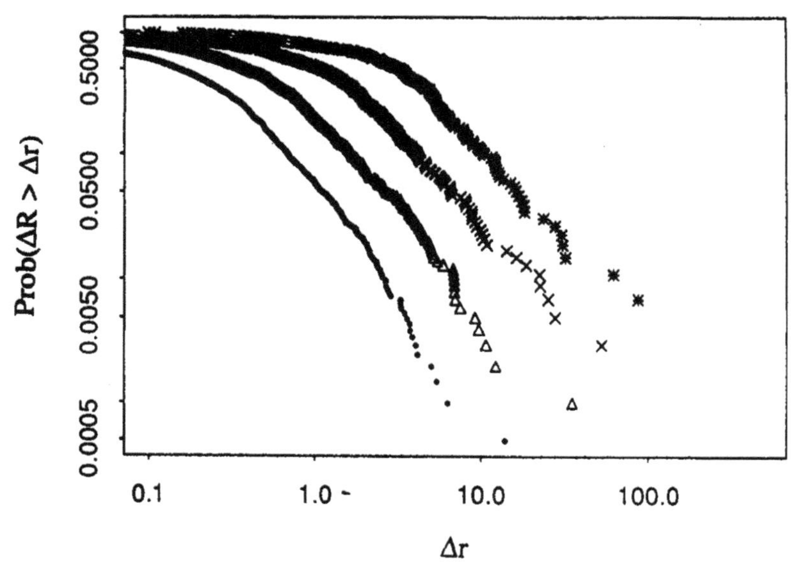

(b)

Figs. 1(a)–1(d) (a) Time series plot of the October 25, 1980 storm in Boston. (b) Probability distributions of rainfall fluctuations for the October 25, 1980, storm in Boston, plotted on log-log scale.

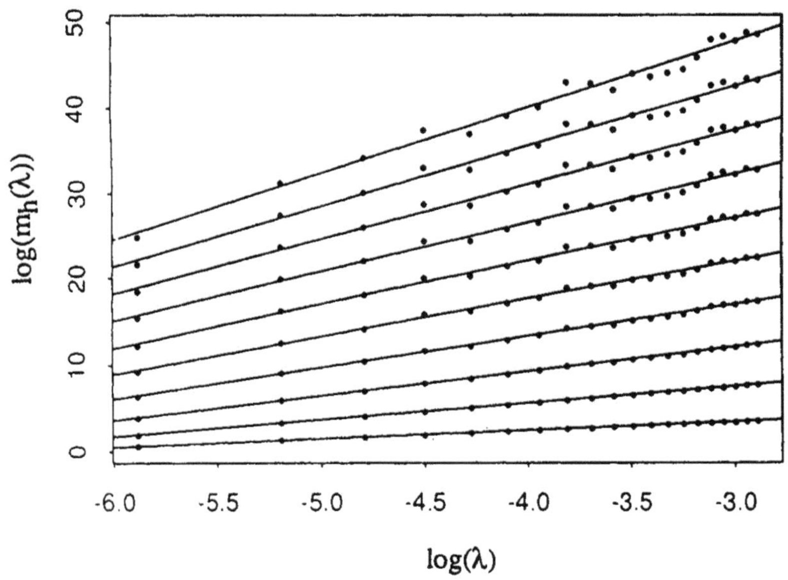

(c)

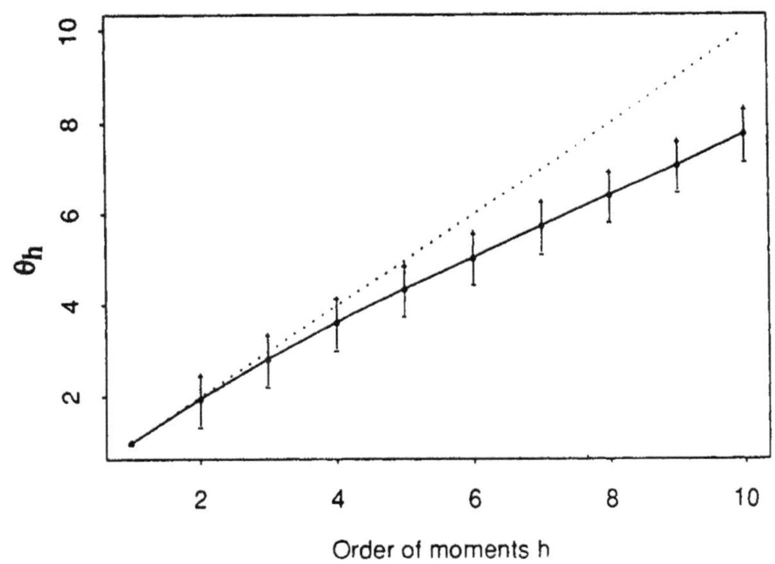

(d)

Fig. 1 (*Continued*)
(c) $\log(m_h(\lambda))$ vs. $\log(\lambda)$. Plotted for first ten moments ($h = 1, \ldots, 10$), in ascending order from the bottom. (d) Departures from simple scaling in the growth of slopes with respect to order of the moments. The dotted line indicates simple scaling, solid line is the estimate from the data. The vertical bars are 95% confidence bands, estimated as two times the standard error of the regression slope from Fig. 1(c).

the following relation holds,

$$X_\lambda(t) \stackrel{d}{=} \lambda^\theta X(t) \tag{3}$$

θ can be either positive or negative and is called the scaling exponent, $\stackrel{d}{=}$ denotes identity in statistical distribution.

Simple scaling implies that there is symmetry across scales in the irregularities of the process, reflecting an underlying structural self-similarity which can be characterized by *fractal dimension* D.[9] The exponent β and the fractal dimension D are related as[6]:

$$D = 2 - 1/\beta \tag{4}$$

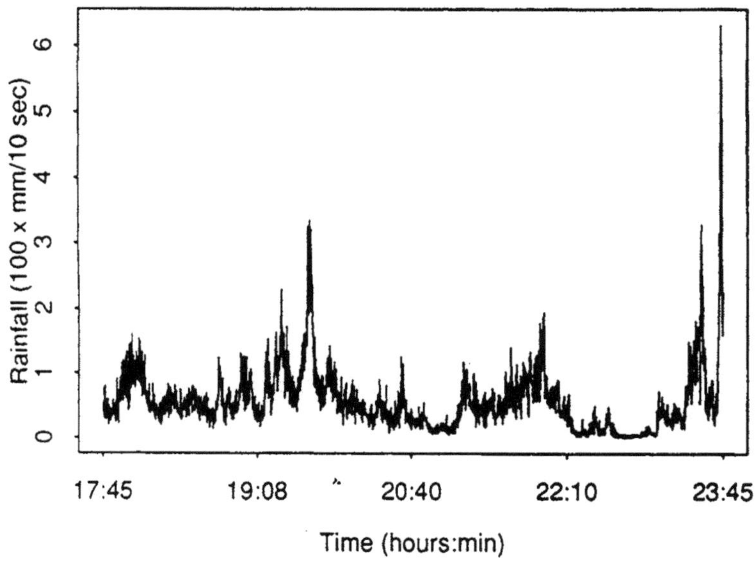

(a)

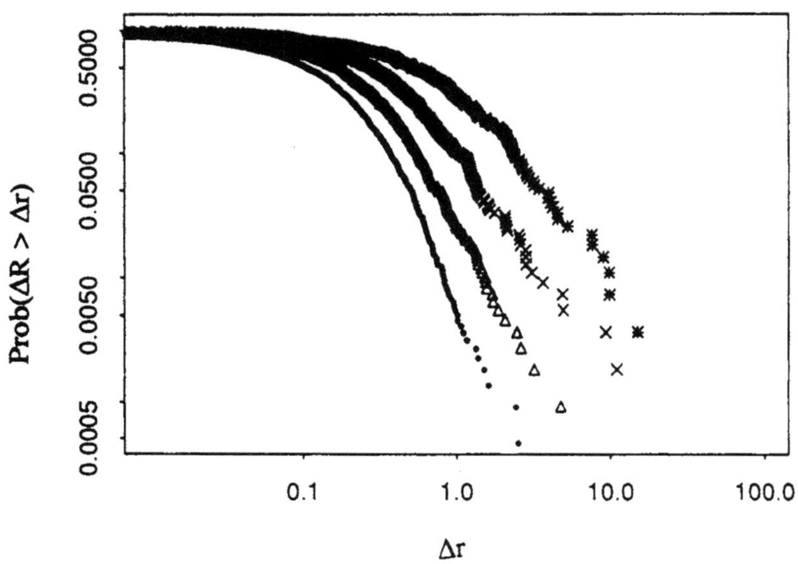

(b)

Figs. 2(a)–2(d) **(a)** Time series plot of the October 9, 1990 storm in Indiana. **(b)** Probability distributions of rainfall fluctuations for the October 9, 1990, storm in Indiana, plotted on a log-log scale.

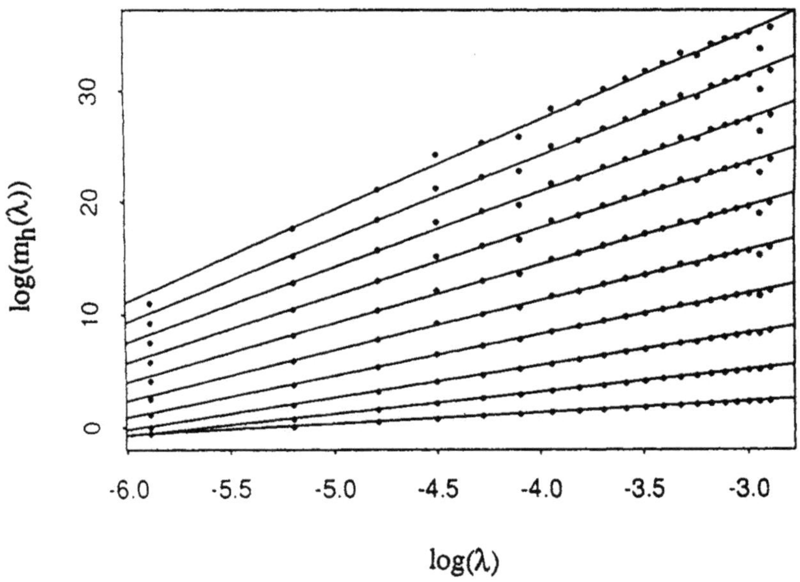

(c)

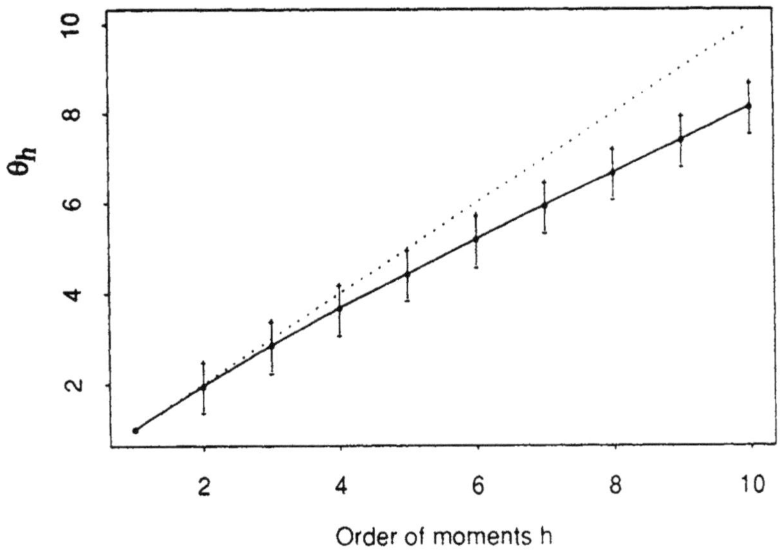

(d)

Fig. 2 (*Continued*)
(c) $\log(m_h(\lambda))$ vs. $\log(\lambda)$. Plotted for first ten moments ($h = 1, \ldots, 10$), in ascending order from the bottom. (d) Departures from simple scaling in the growth of slopes with respect to order of the moments. The dotted line indicates simple scaling, solid line is the estimate from the data. The vertical bars are 95% confidence bands, estimated as for Fig. 1(d).

4. SCALING OF MOMENTS

Properties of random variables are frequently studied through the analysis of moments. Gupta and Waymire[8] have studied the moment scaling of spatial rainfall process and have developed a class of multiplicative processes in light of their observations. Here, we investigated the moment scaling of temporal rainfall process and our results support their findings.

If the process $X_\lambda(t)$ has finite moments $E[X_\lambda^h(t)]$ of order h and if it is simple scaling, then we can write:

$$E[X_\lambda^h(t)] = \lambda^{h\theta} E[X^h(t)] \tag{5}$$

By taking log on both sides we get,

$$\ln\left(E[X_\lambda^h(t)]\right) = h\theta \log(\lambda) + \log\left(E[X^h(t)]\right) \tag{6}$$

Denoting, $m_h(\lambda) = E[X_\theta^h(t)]$, and $m_h(1) = E[X^h(t)]$, we can write Eq. (6) as:

$$\ln(m_h(\lambda)) = h\theta \ln(\lambda) + \ln(m_h(1)) \tag{7}$$

This log-linear scaling of the moments is of the general form:

$$\ln(m_h(\lambda)) = a_h + \theta_h \ln(\lambda) \tag{8}$$

With $\theta_h = h\theta$, the scaling exponent corresponding to moment h. Scaling only of the moments (as opposed to pdf) is called wide sense scaling by Gupta and Waymire.[8] One can imagine (and observe) that moment scaling is possible with θ_h nonlinear in h. This form of scaling is more general than simple scaling and is called multiscaling. The moment scaling exponent θ_h is related to the mass exponent $\tau(q)$ and singularity spectrum $f(\alpha)$ more commonly used to characterize multiscaling. These relationships are discussed below.

First we analyzed the moment scaling properties of the storm data. $m_h(\lambda)$ vs. λ is plotted on a log-log plot for each order of moment h. In this case, the first ten moments were evaluated. Figures 1(c) and 2(c) indicate this. The slope θ_h for each h is estimated via regression. Figures 1(d) and 2(d) are the plots of θ_h vs. h. The dotted line is the linearity of slope change assuming simple scaling ($\theta_h = h\theta$), while the solid line is the one obtained from the data analysis as described above.

From Figs. 1(c) and 2(c), it can be seen that the log-log linearity of moment scaling is preserved. However from Figs. 1(d) and 2(d) the linearity of slope change displays concave departures from simple scaling. Gupta and Waymire[8] show that by virtue of concavity only scale magnification ($\lambda < 1$) is possible. They[8] suggest this is a consequence of "cascading down" of some large-scale flux to successively smaller scales. This feature is similar to the cascading of the energy flux in fully developed turbulent flows. This suggests that the rainfall process can be modeled using cascade processes. This is further investigated in the next section.

5. MULTIFRACTALS

Physical systems that exhibit chaotic behavior are common in nature. Such systems lose information exponentially fast; as a result it is possible to follow and predict their motion in any detail only for short time scales. Chhabra et al.[10] argue that, to describe their long term dynamical behavior, one must resort to suitable statistical descriptions and one such description is the multifractal formalism. They say that the multifractal formalism comes from the fact that highly nonuniform probability distributions arising from the nonuniformity of the system often possess rich **scaling properties** including that of self-similarity.

In particular, if we cover the support of the measure with boxes of size λ and define $P_i(\lambda)$ to be the fraction of integrated measure in the ith box, then we can define an exponent (singularity strength) α_i by:

$$P_i(\lambda) \sim \lambda^{\alpha_i} \tag{9}$$

If we count the number of boxes $N(\alpha)$ where the fraction P_i has singularity strength between α and $\alpha + d\alpha$, then $f(\alpha)$ can be loosely defined as the fractal dimension of the set of boxes with singularity strength α by,

$$N(\alpha) \sim \lambda^{-f(\alpha)} \tag{10}$$

This leads to the notion that a fractal measure can be comprised of interwoven sets of singularities of strength α, each characterized by its own fractal dimension $f(\alpha)$. Such measures are termed multifractals.[11] The plot of α vs. $f(\alpha)$ is called the singularity spectrum.

The scaling exponent for the qth moments of the measure, provides an alternative description of the singular measure. Using the same notations as above, series of exponents parametrized by q can be defined as:

$$\sum_i P_i^q(\lambda) \sim \lambda^{-(q-1)D_q} = \lambda^{-\tau(q)} \tag{11}$$

D_q is called the generalized dimension; these characterize the nonuniformity of the measure: positive q's accentuate the denser regions and negative q's accentuate the rare ones. For special values of q, D_q can be realized as the dimension of a special set. which supports a particular part of the measure. For example, $D_{q=0}$ is the dimension of support of the measure, which is also the maximum of the singularity spectrum. $(q-1)D_q$ is defined as $\tau(q)$, also called the mass exponent. $f(\alpha)$, α, $\tau(q)$, q are related through a Legendre transformation. For details, the reader is referred to Chhabra et al.,[10] Feder,[12] Chhabra and Jensen.[13] The singularity spectrum provides a precise and intuitive description of the multifractal measure in terms of interwoven sets, with differing singularity strengths α, whose fractal dimension is $f(\alpha)$.

We next establish connections between multiscaling of moments described in the previous section and multifractals discussed above. It can be seen from Eq. (8) that:

$$m_h(\lambda) \sim \lambda^{\theta_h} \tag{12}$$

$m_h(\lambda)$ is the moment of order h at aggregation scale λ. Here h plays the same role as q. Now recognize that the fraction of rainfall in an interval i with length λ is $P_i(\lambda) = R_i(\lambda)/T$ where T is the total storm rainfall amount. Also recognize that the number of intervals of length λ used in the calculated of moments $N(\lambda) \sim \lambda^{-1}$. Therefore,

$$m_h(\lambda) = \frac{1}{N(\lambda)} \sum_i R_i^h(\lambda) = \frac{1}{N(\lambda)T} \sum_i P_i^h(\lambda) \sim \lambda^{1-\tau(q)} \tag{13}$$

Comparing to Eq. (12), it follows that

$$\theta_h = 1 - \tau(q) \tag{14}$$

Remember here that $h = q$. The $f(\alpha)$ curved can be computed using the Legendre transform of the $\tau(q)$ curve, or directly using Chhabra and Jensen's[13] method. It was observed that the method did not work unless we used box sizes λ, that were factors of length of the data.

The $f(\alpha)$ spectrum estimated using the direct method of Chhabra and Jensen,[13] is shown in Figs. 3(a) and 4(a), plotted as points. As moment scaling analysis suggested a cascade phenomena, we estimated the singularity spectrum of the best fitting random cascade, using closed form equations for random binomial cascades provided in Halsey et al.[14] These are plotted as dotted lines in Figs. 3(a) and 4(a). The estimated and theoretical spectra are in reasonable agreement; however, one could choose more complicated

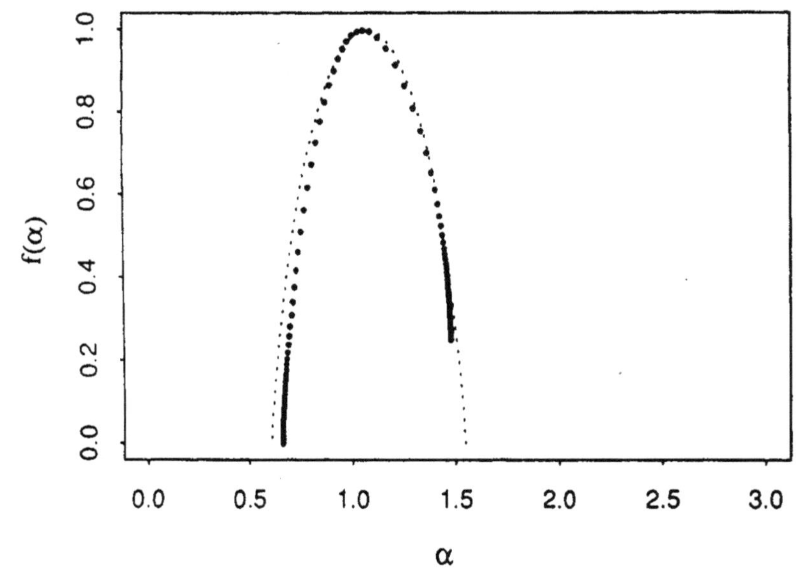

(a)

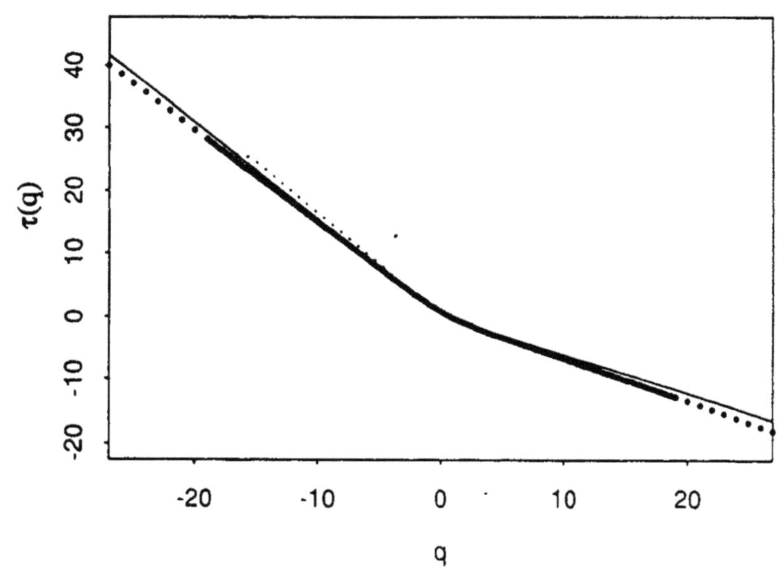

(b)

Fig. 3 (a) Singularity spectrum for the Boston storm. Points are estimated from the data using Chhabra and Jensen's method. The dotted line is the theoretical spectrum for a random binomial cascade with $p = 0.6575$. (b) $\tau(q)$ vs. q plot for the Boston storm. The points are estimated from the data. The dotted line is estimated from multiscaling [Eq. (14)]. The solid line is the curve for a random binomial cascade with $p = 0.6575$.

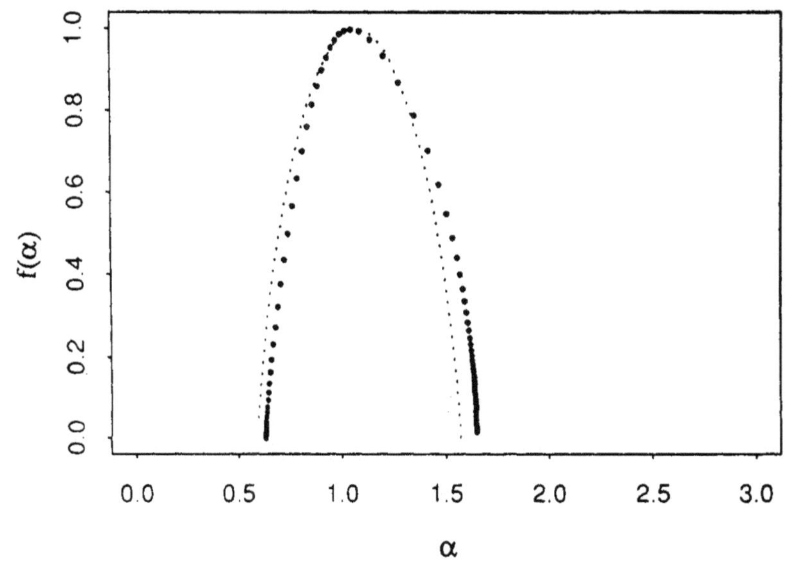

(a)

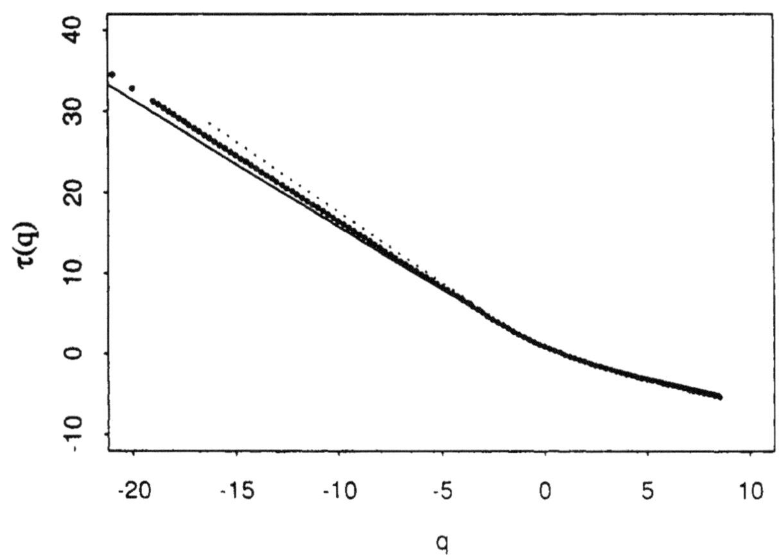

(b)

Fig. 4 **(a) Singularity spectrum for the Indiana storm. Points are estimated from the data using Chhabra and Jensen's method. The dotted line is the theoretical spectrum for a random binomial cascade with $p = 0.6636$. (b) $\tau(q)$ vs. q plot for the Indiana storm. The points are estimated from the data. The dotted line is estimated from multiscaling [Eq. (14)]. The solid line is the curve for a random binomial cascade with $p = 0.6636$.**

cascade models. Finally in Figs. 3(b) and 4(b), $\tau(q)$ curves calculated using the direct method of Chhabra and Jensen,[13] moment scaling [Eq. (14)] and from the fitted random cascade are compared. Closeness of all three lends confidence to the fact that the rainfall process can be modeled by cascade processes. This suggests that convections developed at larger time scales cascade down to smaller time scales. Meneveau and Sreenivasan[15] have developed a binomial cascade model (with $p = 0.7$) for fully developed turbulence. The plausibility of modeling the rainfall process by a random cascade with a similar parameter p, suggests ties between rainfall process and atmospheric turbulence.

After the appropriate cascade model is obtained, it could be used to simulate the rainfall process. Thus, we have a simple cascade model for the observed storms, which has the ability to reproduce the scaling properties of the process.

6. DISCUSSION

The analysis of storm data sets indicate that the seemingly irregular patterns exhibited by the storm rainfall, has a structure embedded in it that binds the process across a wide range of scales. The probability distributions show that the rainfall processes observed are scaling. The moment scaling analysis suggests multiscaling and as a result a cascading phenomenon. The establishment of connection between multiscaling of moments and the singularity spectrum, through $\tau(q)$ [Eq. (14)] implies that the two different analyses provide qualitatively, similar information. This lends confidence to the inferences. The close agreement of $\tau(q)$ curves estimated from all three approaches, and the singularity spectra both of the estimated and theoretical, suggest the presence of a cascading phenomena.

ACKNOWLEDGMENTS

Thanks are due to Dr. Chi-Hua Huang, (USDA-RAS-MWA, National Soil Erosion Research Laboratory, West Lafayette, IN) for providing the high resolution storm data sets of Indiana, and also Dr. Shatiq Islam (Univ. of Cincinatti) for providing the Boston storm data set. We are grateful for discussions with Tom Over.

REFERENCES

1. Rodriguez-Iturbe, V. K. Gupta and Ed Waymire, *Water Resources Research*, **18**(5), 1451 (1982).
2. S. Lovejoy and D. Schertzer, *Water Resources Research* **21**(8), 1233 (1985).
3. D. Schertzer and S. Lovejoy, *Journal of Geophysical Research* **92**(D8), 9693 (1987).
4. Rodriguez Iturbe, B. F. De Power, M. B. Sharifi and K. P. Georgakakos, *Water Resources Research* **25**(7), 1667 (1989).
5. M. B. Sharifi, K. P. Georgakakos and I. Rodriguez-Iturbe, *Journal of the Atmospheric Sciences* **47**(7), 888 (1990).
6. Rodriguez-Iturbe, *Transport Processes and the Hydrological Cycle*, (Venice, 1992), p. 240.
7. Chow, *Bulletin, Eng. Experiment Station*, (Univ. of Ill., Urbana), **414**, 26 (1953).
8. V. K. Gupta and Ed. Waymire, *Journal of Geophysical Research* **95**(D8), 1999 (1990).
9. B. B. Mandelbrot, *The Fractal Geometry of Nature* (Freeman and Co., New York, 1982).
10. A. Chhabra, C. Meneveau, R. V. Jensen and K. R. Sreenivasan, *Phys. Rev.* **A40**(9), 5284 (1989).
11. Vicsek, *Fractal Growth Phenomena*, (World Scientific, Singapore, 1989).
12. Feder Jens, *Fractals* (Plenum Press, New York, 1989).
13. A. Chhabra and R. V. Jensen, *Phys. Rev. Lett.* **62**, 1327 (1989).
14. T. C. Halsey, M. H. Jensen, L. P. Kadanoff, I. Procaccia and B. I. Shraiman, *Physical Review* **A33**(2), 1141 (1986).
15. C. Meneveau and K. R. Sreenivasan, *Phys. Rev. Lett.* **59**(13), 1424 (1987).

Methods

INTRODUCTION

Fractal geometry has provided us with a mathematical characterization of intrinsically irregular structures with self-similar properties. This possibility has extended the range of application of scientific methods to a broad class of phenomena that are very common in nature but have not been previously considered due to the lack of appropriate mathematical methods. This new situation has led to a change of perspective in Natural Sciences and the intrinsic irregularity of self-similar structures can now be considered as a fundamental concept for the understanding of the properties of complex systems.

The simple concept of Fractal Dimension consists in a generalization of the Euclidean dimension in the metric sense. It corresponds essentially to the exponent that characterizes the relation between mass and length-scale. This exponent describes the decay of the conditional density (or the conditional average density) from every point of the system. In this respect every point is associated with a singular behavior that is characterized by the same exponent for all points.

The power-law behavior refers to the convolution of the real correlations of a system that usually shows fluctuations or periodic oscillations in logarithmic space. These oscillations are typical of deterministic fractals (like the Sierpinski gasket) for which self-similarity is only valid for a finite set of scale transformations. It is possible to characterize these oscillations with an imaginary part of the fractal dimension but this possibility has received, up to now, little attention. In general the oscillations are eliminated in a large random fractal or by taking the average of many systems.

An important element in dealing with real finite systems consists in the identification of the lower and upper cutoffs that will limit the range of validity of the self-similar properties. These effects lead to important crossover phenomena that can complicate the extraction of the correct scaling behavior. An example of the problems induced by finite size effects can be found in the paper of Aguilar et al.

The fractal dimension provides a fundamental characterization of the metric properties of the system but it corresponds to a very limited information. Two systems can be quite different and still have the same fractal dimension. For example, the properties of connectivity that relate to the topological characteristics of the system require the introduction of different concepts like the fracton or spectral dimension.

Even for the metric properties, however, the situation can be more complex than the one corresponding to a simple fractal. For example, the nature of self-similarity can have a spectrum of possibilities. In this case it is necessary to generalize the analysis in terms of Multifractals. The paper by Beghdadi et al. discusses a new approach to characterize complex systems in which multifractality is complemented by the concept of configurational entropy.

A different type of complication may be due to the fact that scaling is anisotropic for different directions. This leads to the concept of self-affinity that is essential for the study of irregular surfaces and interfaces.

Apart from physical structures these methods have been used with success for the storage and transmission of pictures and information. An example of this is the paper by Koike.

Often self-similar properties are due to a cascade or multiplicative process that represents the generator of the complex properties of the structure. It is important to identify this generator because it represents an important element to understand the nature of the phenomenon in order to formulate a microscopic theory. The usual methods of fractal analysis cannot provide any information on this point because they focus on the global scaling properties. An attempt in this direction is instead provided by the Wavelet analysis that is a sort of local Fourier transform aimed at the identification of the local fragmentation processes of the structure at different scales (see the paper by Arneodo et al.). In this respect, it should be mentioned that the usual Fourier transform is not particularly useful for fractal structures because the most interesting correlation properties are related to the phases and not just to the amplitudes of the various Fourier components.

During the past ten years, all these methods have been applied to a variety of experimental data and computer simulations providing a new deep understanding of the properties of complex natural phenomena. With regard to the analysis of experiments, I have been involved in the application of these new methods to the analysis of galaxy correlations. The results have shown that the usual assumption of homogeneity that was usually made in this field is actually inconsistent and that the distribution of visible matter in the universe is fractal and multifractal up to the present observational limits.

However, the most popular playground for all these concepts has been the model of Diffusion Limited Aggregation (DLA) that represents the prototypical fractal growth model based on a well defined physical phenomenon. These detailed studies of DLA clusters have demonstrated the power of these methods but also their limitations and point to the need for further progress in the methods of analysis of complex structures.

For example, it is clear that DLA clusters are in some sense self-similar, but the precise determination of the fractal dimension depends on the geometry of the clusters (radial or cylindrical) and on the method of analysis. These small but persistent discrepancies cannot be explained in terms of simple finite size effects. Also the multifractal analysis of the growth probabilities shows an anomalous behavior for the small probability region. Other unexpected scaling properties have been detected with the analysis of moments and other methods. The Wavelet analysis has provided interesting information on the nature of the fragmentation properties of the structure (tip splitting; see Arneodo et al.) but it cannot clarify at the moment the nature of these anomalies.

This situation makes clear that, despite all the methods mentioned above, we have not yet achieved a complete characterization of the scaling properties of DLA clusters. One possible suggestion for the origin of these complications may be the fact that these clusters have a growth direction that gives rise to scaling properties that are nonhomogeneous and anisotropic. These possibilities are not included in most of the usual methods of analysis.

With regard to Theoretical Methods to understand the origin of fractal structures in natural phenomena, the situation has evolved in the following way. Considering again the case of DLA as an example, there have been many attempts to apply the usual methods of equilibrium statistical mechanics like the Renormalization Group (RG). These approaches have met with fundamental difficulties because, despite some analogies with critical phenomena, irreversible fractal growth is a new type of phenomenon for which new theoretical concepts are necessary.

The main characteristics of these phenomena are:

- The dynamics is irreversible and one cannot assign a statistical weight to a configuration without taking into account of its history.
- The fractal structure corresponds to the attractor of the dynamics and it should be related to an attractive fixed point.
- The long range interaction induced by the Laplace equation.

It is therefore essential to focus explicitly on the dynamical process and on the identification of the scale invariant dynamics. This is the new perspective of the Fixed Scale Transformation approach (see Pietronero) that allows a reasonably systematic treatment of these problems, in particular with respect to their self-organized nature, and the analytical calculation of the fractal dimension.

***by* L. Pietronero**

Fractals, Vol. 1, No. 3 (1993) 629–649

BEYOND CLASSICAL MULTIFRACTAL ANALYSIS USING WAVELETS: UNCOVERING A MULTIPLICATIVE PROCESS HIDDEN IN THE GEOMETRICAL COMPLEXITY OF DIFFUSION LIMITED AGGREGATES

A. ARNEODO, F. ARGOUL, J. F. MUZY and M. TABARD
Centre de Recherche Paul Pascal, Avenue Schweitzer, 33600 Pessac, France

E. BACRY
Université de Paris VII, UFR de Mathématiques, Tour 45-55, 2 Place Jussieu, 75251 Paris Cedex 05, France

Abstract

We emphasize the wavelet transform as a very promising tool for solving the inverse fractal problem. We show that a dynamical system which leaves invariant a fractal object can be uncovered from the space-scale arrangement of its wavelet transform modulus maxima. We illustrate our theoretical considerations on pedagogical examples including Bernoulli invariant measures of linear and nonlinear expanding Markov maps as well as the invariant measure of period-doubling dynamical systems at the onset of chaos. We apply this wavelet based technique to analyze the fractal properties of DLA azimuthal Cantor sets defined by intersecting the inner frozen region of large mass off-lattice DLA clusters with a circle. This study clearly reveals the existence of an underlying multiplicative process that is likely to account for the Fibonacci structural ordering recently discovered in the apparently disordered arborescent DLA morphology. The statistical relevance of the golden mean arithmetic to the fractal hierarchy of the DLA azimuthal Cantor sets is demonstrated.

1. INTRODUCTION

The diffusion-limited aggregation (DLA) model introduced by Witten and Sander[1] about a decade ago has become the paradigm model for diffusion-controlled pattern forming phenomena.[1-3] This prototype model mimics two-dimensional Laplacian growth processes according to the following algorithm: particles originating from far away are added, one at a time, to a growing cluster via random walk trajectories in the plane. Extensive on-lattice and off-lattice computer simulations have produced complex branched fractals that bear a striking resemblance to the tenuous tree-like structures observed in viscous fingering, electrodeposition, bacterial and neuronal growths.[1-11] The appealing simplicity of the DLA model and its relevance to various experimental situations have stimulated considerable experimental, numerical and theoretical interest.[1-11] But having regard to the efforts spent, the progress in capturing the screening mechanisms that govern DLA growth has been very limited. Actually, only a little is known about the ramified DLA morphology which is still very mysterious to many extents. In particular, we do not know whether some structural order is hidden in the apparently disordered geometry of DLA clusters. More generally, we still appear to be quite far from a physical understanding of Laplacian growth phenomena. This explains why, after more that ten years of extensive inquiry, the DLA model remains one of the most exciting theoretical challenge in the physics of structure formation.

One of the main obstacles to theoretical progress lies in the lack of structural characterization of the growing clusters. Most of the previous studies have mainly focused on the multifractal analysis of either the DLA geometry or the growth probability distribution along its boundary.[1-3,8] While the later displays unambiguous multifractal properties[3] (the harmonic measure of DLA clusters is extremely non-uniform), the statistical self-similarity of the DLA morphology is still a very much debated question.[12-14] Recent geometrical analysis of large mass off-lattice DLA clusters (of sizes up to 30 millions particles) suggests the possibility that plane DLA clusters might be multifractal[15]; according to B. B. Mandelbrot[16] they might even not be self-similar at all. However, the statistical homogeneity of the inner "frozen" region (namely the perimeter sites that are unlikely to grow further) of rather large size DLA clusters (up to 10^6 particles) is rather well admitted since, within the numerical uncertainty, the generalized fractal dimensions are all equal[17]: $D_q = 1.61 \pm 0.03$. (Note that the data do not exclude some weak multifractal departure from statistical homogeneity.[18]) This means that the mass locally behaves as a power-law of the length scale with a scaling exponent $\alpha \sim 1.61$, which is independent of the point chosen on the cluster. But the generalized fractal dimensions D_q and the $f(\alpha)$ singularity spectrum are thermodynamical functions that provide only "macroscopic" statistical information about the self-similarity properties of fractal objects. The incompleteness of the multifractal description mainly lies in the fact that, to some extent, the information concerning the hierarchical architecture of these objects has been filtered (averaged) out.

To achieve a more elaborate structural analysis, we have recently advocated[19] the use of the continuous wavelet transform[20-23] (WT). This technique has proved to be well adapted to the large hierarchy of scales involved in fractal patterns.[24-29] When using this mathematical microscope to explore the intricate DLA morphology, we have discovered the existence of Fibonacci sequences in the internal fractal branchings (in the inner "extinct" region) of large mass off-lattice DLA clusters.[17,30,31] This analysis also reveals that this underlying hierarchy is likely to be intimately related to a predominant structural five-fold symmetry. Our aim here is to establish the statistical relevance of the golden mean arithmetic to the structural fractal ordering of DLA clusters. Our investigation mainly concerns one-dimensional cuts of large-mass off-lattice aggregates. This is the prize we have to pay for our demonstration to be quantitative. Then, one can use the 1D continuous wavelet transform which has recently been shown to be a very promising tool for solving the inverse fractal problem.[32]

The paper is organized as follows. In Sec. 2, we describe a new technique based on the wavelet transform modulus maxima (WTMM) representation that allows us to extract, directly from the data, a dynamical system which leaves invariant a given fractal object embedded in a one-dimensional space. Several well-known examples of multifractal measures in the context of dynamical system theory are used to illustrate our purpose. In Sec. 3, we apply this wavelet based technique to the so-called "azimuthal Cantor set" defined by intersecting a DLA cluster, in its inner frozen region, with a circle. Several azimuthal Cantor sets, corresponding to a statistical sample of 50 off-lattice DLA clusters of mass $M = 10^6$, are analyzed. Therefore we expect the structural information extracted from our WT analysis to be pertinent to the statistical study of the DLA morphology. We conclude in Sec. 4.

2. SOLVING THE INVERSE FRACTAL PROBLEM WITH WAVELETS

Fractal and multifractal concepts[33–42] have proved very fruitful in the context of a statistical "thermodynamic" description of scale invariant objects. But there is a need to get deeper insight into the complexity of such objects and eventually to extract some "microscopic" information about their underlying hierarchical structure. In many cases, the self-similarity properties of fractal objects can be expressed in terms of a dynamical system which leaves the object invariant. The inverse problem consists in recovering this dynamical system (or its main characteristics) from the data representing the fractal object. This problem has been previously approached within the theory of Iterated Function Systems.[43–45] But the methods developed in this context are based on the search for a "best-fit" within a prescribed class of IFS attractors (mainly linear homogeneous attractors). In that sense, they approximate the self-similarity properties more than they reveal them. Here, we show that, in many situations, the space-scale representation of the wavelet transform of a fractal object can be used to extract some one-dimensional (1D) map which accounts for its construction process. These results have been announced in a previous short communication in Ref. 32.

The fractal objects we will use to carry out our demonstration are the invariant measures of "cookie-cutters". A cookie-cutter[46] is a map of $A = [0, 1]$ which is hyperbolic ($|T'| > 1$) and so that $T^{-1}(A)$ is a finite union of s disjoint subintervals $(A_k)_{1 \le k \le s}$ of A. For each k, $T_k = T|_{A_k}$ is a one to one map on A. An invariant measure μ associated to T is a measure which satisfies $\mu \circ T^{-1} = \mu$. We will suppose that μ is multiplicatively distributed on A:

$$\mu \circ T_k^{-1} = p_k \mu , \quad \forall k \in \{1, \dots, s\} \tag{1}$$

where $\sum p_k = 1$. These self-similar measures are also referred to as Bernoulli invariant measures of expanding Markov maps.[37] These measures have been the subject of considerable mathematical interest.[37,40,46] Practically, they have been widely used for modeling a large variety of highly irregular physical distributions; notable examples include strange repellers which characterize transient behavior of nonlinear dynamical systems[46] and the spatial distribution of the dissipation field in fully developed turbulent flows.[38,47]

The 1D continuous wavelet transform of a measure μ according to the analyzing wavelet ψ is defined as[25,26]:

$$W_\psi[\mu](b, a) = \int_A \psi\left(\frac{x-b}{a}\right) d\mu\,, \tag{2}$$

where $a \in \mathbb{R}^{+*}$ is the scale parameter and $b \in \mathbb{R}$ is the space parameter. Usually ψ is chosen to have some vanishing moments, up to a certain order, so that it is orthogonal to possible regular (i.e., polynomial) behavior of μ. In the particular case of invariant measures of cookie-cutters, there are no such behavior, so we will use a simple "smoothing wavelet"[39–42,48]: $\psi = \exp(-x^2)$. By combining Eqs. (1) and (2), a straightforward calculation at the first order in a ($a \ll 1$) leads to the following "self-similarity" relation[40–42]:

$$W_\psi[\mu](b, a) = \frac{1}{p_k} W_\psi[\mu](T_k^{-1}(b), T_k^{-1'}(b)a)\,, \quad \forall k \in 1, \ldots, s \tag{3}$$

where $T_k^{-1'}$ is the first derivative of T_k^{-1}. This relation can be interpreted as describing the self-similarity properties of the wavelet transform itself in the (b, a) half-plane.[24–29] Our goal is to study the self-similarity properties of μ through those of its wavelet transform $W_\psi[\mu]$. For that purpose, we are not going to deal with the whole wavelet transform but only with its restriction to the local maxima of its modulus,[49,50] i.e., the local maxima of $|W_\psi[\mu](x, a)|$ considered as a function of x. In fact, one can easily prove that the self-similarity relation [Eq. (3)] still holds when restricted to the set of modulus maxima of the wavelet transform. For more details, we refer the reader to our previous work in Ref. 40 and to a recent preprint by W. L. Hwang and S. Mallat,[51] where an alternative approach to recover the affine self-similarity parameters through a voting procedure based on Eq. (3) is reported.

2.1 Bernoulli Invariant Measures of Linear Expanding Markov Maps

In the case of linear cookie-cutters, the T_k^{-1}'s are linear, i.e., $T_k^{-1}(x) = r_k x + t_k$, where $r_k < 1$. Then the self-similarity relation (3) becomes:

$$W_\psi[\mu](b, a) = \frac{1}{p_k} W_\psi[\mu](r_k b + t_k, r_k a)\,, \quad \forall k \in 1, \ldots, s\,. \tag{4}$$

The meaning of this relation is illustrated on a particular example in Fig. 1. For the sake of simplicity, we choose $s = 2$, $p_1 = p_2 = 1/2$ and the T_k's to be linear: $T_1(x) = 5x/3$ and $T_2 = 5x - 4$. The corresponding invariant measure is shown in Fig. 1(a). The position of its WTMM in the space-scale (half-)plane is displayed in Fig. 1(b). One can see on this figure that the WTMM skeleton in the entire rectangle $[0, 1]\times]0, a_0]$ (a_0 is an appropriate coarsest scale which actually depends on the shape of the analyzing wavelet ψ), is similar to its parts that are respectively contained in the two rectangles delimited by the dashed lines, namely $[0, 3/5]\times]0, 3a_0/5]$ and $[4/5, 1]\times]0, a_0/5]$, up to a global rescaling of the wavelet transform.

Let us describe on this particular example our technique for recovering from the WTMM skeleton, the discrete (cookie-cutter) dynamical system T. The local maxima are lying on

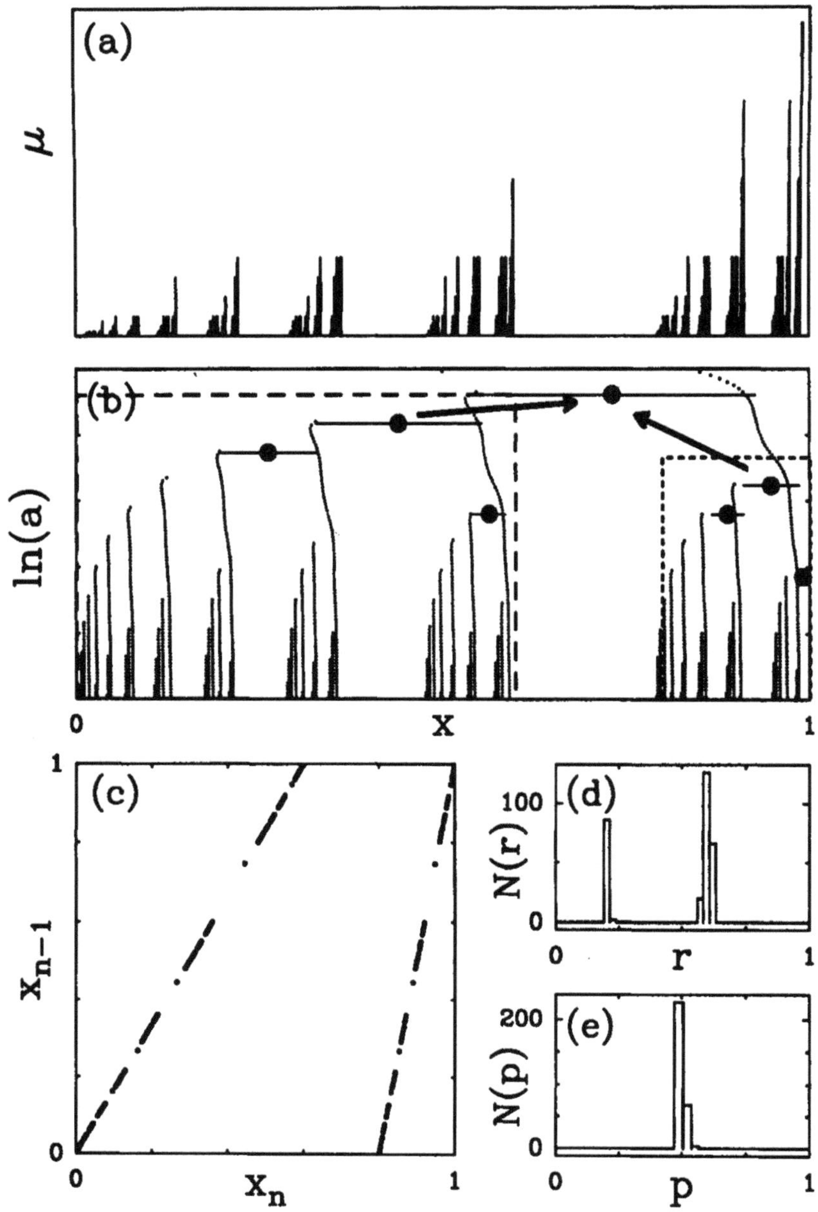

Fig. 1 (a) Invariant measure of the two branch cookie-cutter $T_1(x) = 5x/3$, $T_2(x) = 5x - 4$, distributed with equal weights $p_1 = p_2 = 0.5$ on the interval $[0, 1]$. (b) Position in the (x, a) half-plane of the WTMM of the measure shown in (a), using a gaussian analyzing function; large scales are at the top. According to the self-similarity relation [Eq. (4)], the maxima line arrangement in the two dashed rectangles is the same as in the original rectangle. The bifurcation points associated to each rectangle are represented by the symbols (•). Arrows indicate the matching of these bifurcation points according to the self-similarity relation [Eq. (4)]. (c) 1D map that represents the position x_{n-1} of an order $n-1$ bifurcation point vs. the position x_n of the associated order n bifurcation point following the tree matching defined in (b). The graph of this map corresponds exactly to the original cookie-cutter. (d) Histogram of scale ratios $r = a_n/a_{n-1}$ between the scales of two associated bifurcation points. (e) Histogram of amplitude ratios $p = |W_\psi[\mu](x_n, a_n)|/|W_\psi[\mu](x_{n-1}, a_{n-1})|$ computed from two associated bifurcation points.

connected curves usually referred to as "maxima lines".[49,50] We call "bifurcation point" any point on the space-scale plane located at a scale where a maxima line appears and which is equidistant to this line and to the closest longer line. The bifurcation points lie on a binary tree whose root is the bifurcation point at the coarsest scale. Each bifurcation point defines naturally a subtree which can be associated to a rectangle in the space-scale plane. This root corresponds to the original rectangle $[0, 1]\times]0, a_0]$, whereas its two sons correspond to reduced copies delimited by the dashed lines. As illustrated in Fig. 1(b), the self-similarity relation (4) amounts to matching the "root rectangle" with one of its "son rectangles", i.e., the whole tree with one of the subtrees. More generally, this relation associates any bifurcation point (x_n, a_n) of an order n subtree to its hierarchical homologous (x_{n-1}, a_{n-1}) of an order $n-1$ subtree. It follows from Eq. (4) that $x_n = r_k x_{n-1} + t_k$ and $a_n = r_k a_{n-1}$. Thus by plotting x_{n-1} vs. x_n, one can expect to recover the graph of the initial cookie-cutter T. This reconstructed 1D map is displayed in Fig. 1(c). As one can see, the linear branches T_1 and T_2 are remarkably well reproduced by the data. Let us point out that the non-uniform repartition of the data points on the theoretical branches results from the lacunarity of the measure induced by the "hole" between the branches T_1 and T_2 of the piece-wise linear map. In Fig. 1(d), we show the histogram of the (contracting) scale ratio values between the scales of two bifurcation points of successive generations: $r = a_n/a_{n-1}$, computed during investigation, systematically, of the WTMM skeleton in Fig. 1(b). As expected, it displays two peaks corresponding to the two slopes $r_1 = 3/5$ and $r_2 = 1/5$ of T_1^{-1} and T_2^{-1} respectively. Let us note that the peak corresponding to the smallest value of r is lower than the other one; this is a direct consequence of the finite cut-off used in our WT calculation at small scales. On a finite range of scales, the construction process involves fewer steps with the smallest scale ratio r_2 than steps with the largest one r_1. (The so-computed histogram can be artificially corrected in order to account for these finite size effects; its suffices to plot $N(r)\ln(1/r)$ instead of $N(r)$.) Figure 1(e) displays the histogram of amplitude ratio value $p = |W_\psi[\mu](x_n, a_n)|/|W_\psi[\mu](x_{n-1}, a_{n-1})|$. Up to the numerical uncertainty, this distribution appears to be a Dirac at $p = 0.5$ which indicates [Eq. (4)] that the weights [defined in Eq. (1)] are $p_1 = p_2 = 0.5$ or, in other words, that the measure is uniformly distributed on the cookie-cutter Cantor set.

Remark 1. Let us mention that the distribution $N(r)$ of scale ratios [Fig. 1(d)] is in a way redundant with the 1D map [Fig. 1(c)], since it is basically made up of two Diracs located at the inverse of the slopes of the two branches of this piece-wise linear map. On the contrary, the distribution $N(p)$ of amplitude ratios [Fig. 1(e)] brings a very important piece of information which is not present in the 1D map: the repartition of the weights at each construction step. In the case where this repartition is not uniform, we get a histogram $N(p)$ which no longer reduces to a single point $p = 1/2$ and one can furthermore study the joint law of p with r in order to find out the specific "rules" for associating a p with a r (see Fig. 10).

In the case where s is no longer equal to 2, one can easily adapt the technique by trying to match not only the root bifurcation point on its sons but also on its grandsons and so on .... For instance, in the case $s = 3$, we will match the root with one of its sons and with each of the two sons of its other son. The general algorithm that we have developed uses a "best matching" procedure that automatically "chooses" the matching which is the most consistent, e.g. such that the different derivatives of $W_\psi[\mu]$ follow the same self-similarity rule as $W_\psi[\mu]$. Thus the algorithm is not looking for a given number s of branches that the user would have guessed a priori; it automatically comes up with the "best" values of s. In Fig. 2 are shown the 1D map, the histogram of scale ratios and the histogram of amplitude ratios obtained for the Bernoulli measure generated with the following model parameters:

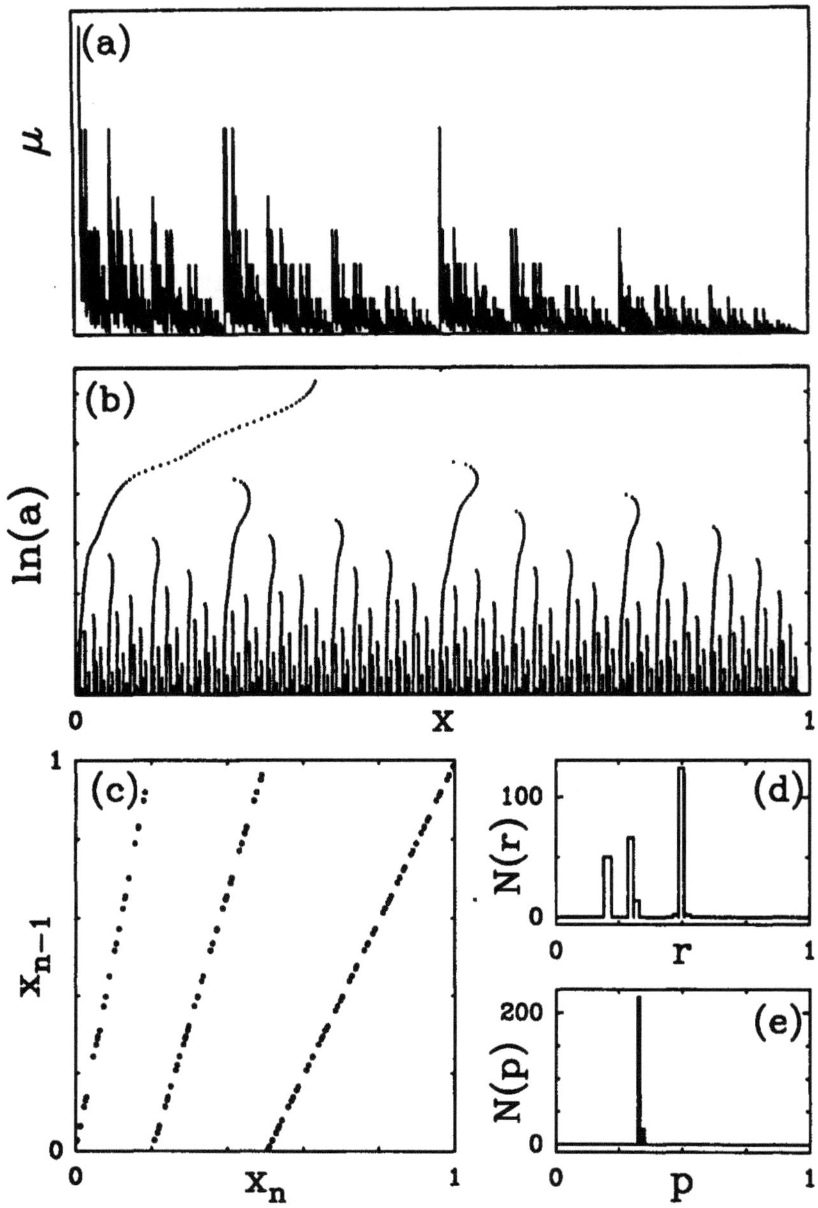

Fig. 2 (a) Invariant measure of the three branch cookie-cutter $T_1(x) = 5x$, $T_2(x) = 10x/3 - 2/3$, $T_3(x) = 2x - 1$, distributed with equal weights $p_1 = p_2 = p_3 = 1/3$ on the interval $[0,1]$. (b) WTMM skeleton computed with a gaussian analyzing function. (c) 1D map obtained with our tree matching algorithm (see text). (d) Histogram of scale ratios $r = a_n/a_{n-1}$. (e) Histogram of amplitude ratios $p = |W_\psi[\mu](x_n, a_n)|/|W_\psi[\mu](x_{n-1}, a_{n-1})|$.

$s = 3$, $p_1 = p_2 = p_3 = 1/3$ and $r_1 = 0.2$, $r_2 = 0.3$, $r_3 = 0.5$. All these values are very accurately recovered by our algorithm.

Remark 2. In this work, we only consider measures which do not involve any "memory effect" in their hierarchical structure, i.e., the successive (backward) iterations always consist in applying the same dynamical system T, independently of the previous iterations. However, in a certain way, a construction rule involving a finite memory can be accounted by increasing the number s of branches of a "no-memory" map T. As illustrated in Fig. 2, this class of dynamical systems is directly amenable to our WT algorithm procedure.

Nevertheless, it is important to emphasize that because of finite size effects, it is meaningless to look for some dynamical systems with a rather high number of branches; generally, there would not be enough scales in the data in order to ensure the theoretical validity of the outcoming discrete map.

2.2 Bernoulli Invariant Measures of Nonlinear Expanding Markov Maps

In the former examples, we have described our wavelet based technique to solve the inverse fractal problem for piece-wise linear cookie-cutters. Since locally in the space-scale plane, the self-similarity relation (3) looks like Eq. (4), we can apply exactly the same technique for nonlinear expanding maps. Let us point out that the hyperbolicity condition is a priori required for the first derivative of T_k^{-1} involved in the right-hand side of Eq. (3), to be finite. Figure 3 displays the 1D map extracted from the WTMM skeleton of the uniform Bernoulli measure associated to a nonlinear cookie-cutter made of two inverse hyperbolic tangent branches. Once again, the numerical results match perfectly the theoretical curve. In this case, the histogram of amplitude ratios $N(p)$ is still concentrated at a single point $p = 1/2$. But the histogram of scale ratios $N(r)$ involves more than simply two scale ratios as before [Fig. 1(d)], since the nonlinearity of the map implies that new scale ratios are actually operating at each construction step. This explains the broadening of the two peaks observed in Fig. 3(d). A careful analysis of the fine structure of this histogram would require the investigation of a large number of construction steps, but this is out of the scope of the present study.

2.3 Invariant Measures of Period-Doubling Systems at the Onset of Chaos

As a first application of our wavelet based technique to a physical problem, let us analyze the natural measure associated to the iteration of quadratic unimodal maps at the accumulation point of period-doublings. It is well-known[52] that the discrete time dynamical system: $x_{i+1} = f_R(x_i) = 1 - Rx_i^2$, exhibits, as the parameter R is increased, an infinite sequence of subharmonic bifurcations which accumulate at R_∞ where the system possesses a 2^∞ orbit. Beyond this critical value, the dynamics becomes chaotic. As independently emphasized by M. Feigenbaum[53] and by P. Coullet and C. Tresser,[54] there exists a deep analogy between this transition to chaos and second-order phase transitions in critical phenomena. At the critical value $R = R_\infty$, the map f_{R_∞} belongs to the stable manifold of the fixed point f^* of the renormalization operation[52–54]:

$$\mathcal{R}[f] = \frac{1}{f(1)}[f \circ f](f(1)x)\,. \tag{5}$$

The asymptotic behavior of the dynamics generated by f_{R_∞} is "universal" and corresponds to the one of f^* which is confined on a Cantor set. The natural invariant measure is defined on this Cantor set as the visiting probability of the orbit of $x = 0$. This measure is displayed in Fig. 4(a). Disregarding the fact that $f_{R_\infty}(x)$ is not hyperbolic at $x = 0$ ($f'_{R_\infty}(0) = 0$), we have carried out our WT analysis on the invariant interval $[f^*(1), 1]$. The results of this analysis are shown in Fig. 4. A well-defined 1D map with two distinct hyperbolic branches is numerically reconstructed in Fig. 4(c). A finer resolution computation would reveal that

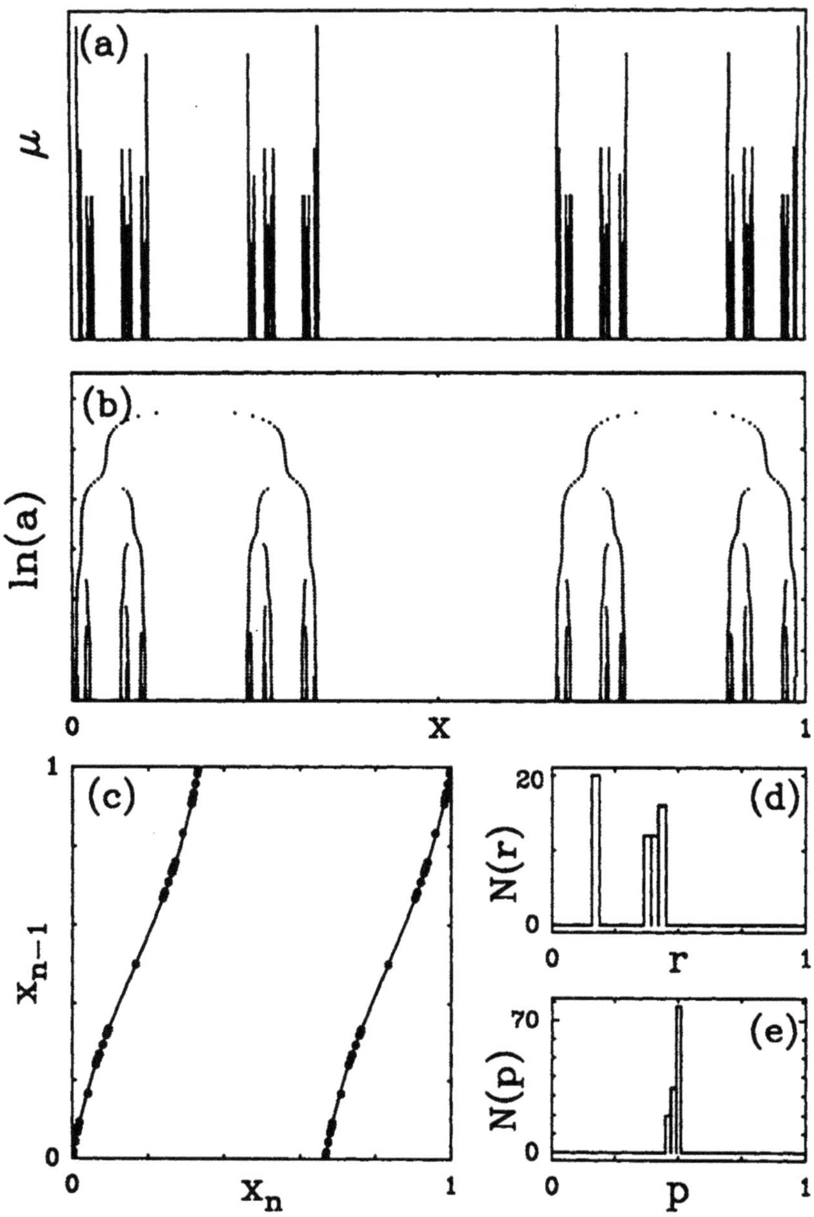

Fig. 3 (a) Invariant measure of a nonlinear cookie-cutter made of two inverse hyperbolic tangent branches. (b) WTMM skeleton computed with a gaussian analyzing function. (c) 1D map obtained with the same wavelet transform tree matching analysis as in Fig. 1; the original nonlinear dynamical system (solid lines) is recovered accurately. (d) Histogram of scale ratios $r = a_n/a_{n-1}$. (e) Histogram of amplitude ratios $p = |W_\psi[\mu](x_n, a_n)|/|W_\psi[\mu](x_{n-1}, a_{n-1})|$.

the left-hand branch is linear with a slope $-1/r = 1/f^*(1) \simeq -2.5$, whereas the right-hand one is nonlinear. A close inspection of the scaling ratio histogram in Fig. 4(d) confirms this observation. The amplitude ratio histogram computed in Fig. 4(e) displays a unique peak at $p = 1/2$, which suggests that the weights associated to the two branches of the 1D map are equal ($p_1 = p_2 = 1/2$). The period-doubling natural measure can thus be seen as the invariant measure of the cookie-cutter shown in Fig. 4(c) with uniform probability distribution. This result was derived theoretically by F. Ledrappier and M. Misiurewicz in Ref. 55; these authors proved that the invariant measure of $f^*(x)$ is the same as the one of

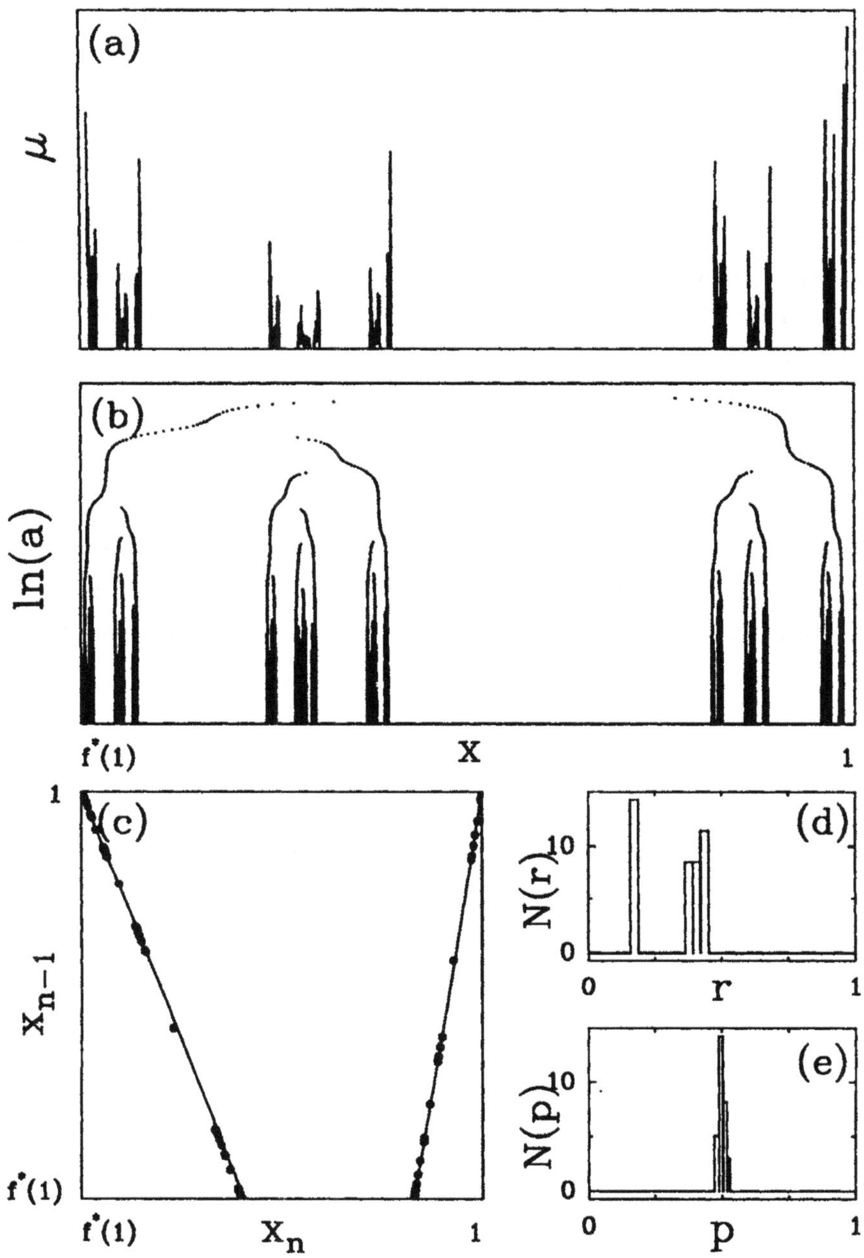

Fig. 4 (a) Invariant measure associated to the critical period-doubling dynamical system f^* (see text). (b) WTMM skeleton computed with a gaussian analyzing function. (c) 1D map obtained with our tree matching algorithm; the solid lines represent the theoretical prediction [Eq. (6)]. (d) Histogram of scale ratios $r = a_n/a_{n-1}$. (e) Histogram of amplitude ratios $p = |W_\psi[\mu](x_n, a_n)|/|W_\psi[\mu](x_{n-1}, a_{n-1})|$.

the cookie-cutter defined by:

$$T(x) = \begin{cases} x/f^*(1) & \text{for} \quad x \in [f^*(1), f^{*2}(1)]\,; \\ f^*(x)/f^*(1) & \phantom{\text{for}} \quad x \in [f^*(-f^*(1)), 1]\,, \end{cases} \tag{6}$$

This map is hyperbolic; it is represented by solid lines in Fig. 4(c). Our numerical data are in remarkable agreement with the theoretical prediction.

3. UNCOVERING A MULTIPLICATIVE PROCESS IN THE DLA AZIMUTHAL CANTOR SETS USING WAVELETS

A characteristic feature of diffusion-limited aggregation is the fact that most of the growth takes place in an "active" zone, near the outer radius of the cluster, which collects practically all the new particles. This active zone moves outward, leaving behind an "extinct" region[56–58] that can be considered as asymptotic in the sense that it is extraordinarily unlikely to be modified by further growth. This screening of the inner region by the tips is the basic reason for the fractal branching in DLA growth. To proceed to a quantitative fractal analysis of the DLA edifice, we thus need to investigate rather large DLA clusters so that their inner inactive regions contains several generations of branching. In Fig. 5(a), we show a 10^6 particle off-lattice aggregate generated using an efficient algorithm[17,30,31] which combines the simplicity of the off-lattice algorithm designed in Ref. 59, to the rapidity of on-lattice hierarchical algorithms.[60] Figure 5(b) illustrates the inner frozen region of this cluster as delimited by the circle drawn in Fig. 5(a); about 8×10^4 particles are contained in this disk of radius $R = 480$ particle sizes. When one restricts the multifractal analysis to this frozen region, the generalized fractal dimensions are found to be equal to the fractal dimension[17,30,31]:

$$D_q = D_F^{\mathrm{DLA}} = 1.61 \pm 0.03\,, \quad \forall q \tag{7}$$

Similar estimates are obtained when different clusters of the same size are analyzed. Moreover, consistently for each cluster, the numerical value for D_F^{DLA} in the inner extinct region is equal, up to the numerical uncertainty, to the fractal dimension of the entire aggregate.[17,30,31,61] This is not such a surprising result since recent numerical simulations have demonstrated that the subset of inaccessible sites of DLA clusters is a "fat fractal" that involves a finite proportion ($\sim 37\%$) of the total number of perimeter sites.[57,58]

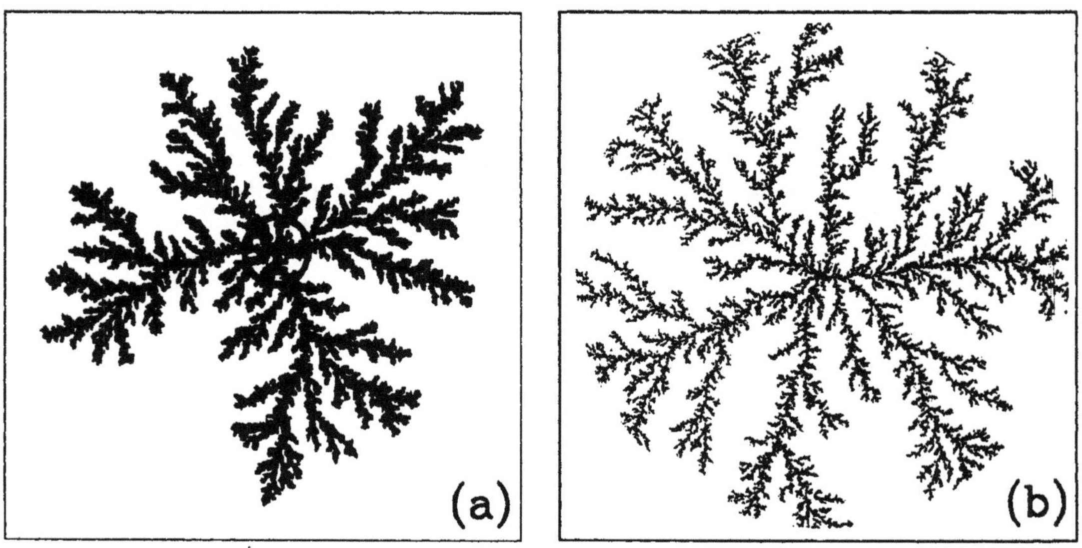

Fig. 5 (a) A 10^6 particle DLA cluster computed with an off-lattice random walker model. (b) The inner frozen region delimited by the circle sketched in (a); about 8×10^4 particles are contained in a disk of radius $R = 480$ particle sizes.

With the specific purpose of applying the tree matching WTMM technique described in Sec. 2, we will concentrate ourselves mainly on the analysis of the azimuthal Cantor set obtained by intersecting a DLA cluster with a circle of radius $R = 480$ particle sizes that somehow delimits its inner frozen region. (Note that this Cantor set is rigorously defined in the unphysical double limit $M \to +\infty$ and $R \to +\infty$, when taken in this particular order.) For this work, we have grown a total number of 50 off-lattice DLA clusters of mass $M = 10^6$ similar to the one in Fig. 5(a). Our demonstration will thus rely on a statistical sample made up of 50 DLA azimuthal Cantor sets. Further investigations for smaller and larger circle radius are currently in progress.

As a first unavoidable step, we have carried out a statistical multifractal analysis of our 50 DLA azimuthal Cantor sets. We have used both the classical box-counting technique and its recent generalization based on the WTMM method.[17,30,31] The fractal dimension is found to be equal to:

$$D_F^A = 0.63 \pm 0.03\,, \tag{8}$$

with a statistical uncertainty computed from the 50 different DLA realizations and not exclusively from the linear regression fit procedure as in Eq. (7). According to the Mandelbrot rule[62] for one-dimensional cuts of fractal sets, the fact that, up to the numerical uncertainty, $D_F^A = D_F^{\text{DLA}} - 1$ is a strong indication that DLA clusters are homogeneous fractal aggregates. A systematic analysis of the generalized fractal dimensions D_q^A of the azimuthal Cantor sets does not contradict this observation. To a good approximation, all the dimensions D_q^A are equal to the fractal dimension $D_F^A = D_{q=0}^A$. Let us mention, however, that the data obtained and the associated statistical error bars do not exclude some possible weak multifractal departure from homogeneity; they seem to be also compatible with a slight decrease of D_q^A as a function of q. The results of a study specially devoted to the multifractal analysis of DLA azimuthal Cantor sets in both the inner frozen region and the outer active region will be reported elsewhere.[18]

The WTMM representation of the azimuthal Cantor set of a 10^6 particle off-lattice DLA cluster is shown in Fig. 6. Actually, we have magnified three regions of the WTMM skeleton corresponding to three well-separated regions of the azimuthal Cantor set issued from three distinct main branches of the considered off-lattice cluster.[17] As in Sec. 2, the analyzing wavelet is simply the "smoothing" Gaussian function $\psi = e^{-x^2/2}$. Unlike the WTMM skeletons of the invariant repellers associated to discrete dynamical systems (Figs. 1–4), one does not see, at first sight, any conspicuous recursive structure in the bare WTMM unfolding of the DLA azimuthal Cantor sets in Fig. 6. One can, however, proceed to a systematic investigation of the value of the scale ratio between two successive bifurcations in the WTMM skeletons. The results of the statistical analysis of our 50 off-lattice DLA clusters are shown in Fig. 7. The histogram $N(r)$ of scale ratio values involves only the bifurcation points in those skeletons that occur above some lower scale. We have checked that this histogram is insensitive to the value of this lower cut-off, when varying it from rather large scales to scales of the order of a few particle sizes. This scale invariance strongly suggests that there is no correlation between successive generations of branching in the WTMM skeleton (no memory effects). Let us note that the scale ratio histogram in Fig. 7 displays a (unique) maximum at the value $r^* = 1/\lambda^* = 0.44 \pm 0.03$ ($\lambda^* = 2.2 \pm 0.2$). The generations of branching are thus expected to occur preferentially at scales $a_n = a_0 r^{*n} = a_0(2.2)^{-n}$, where a_0 is a macroscopic scale that is determined by the size

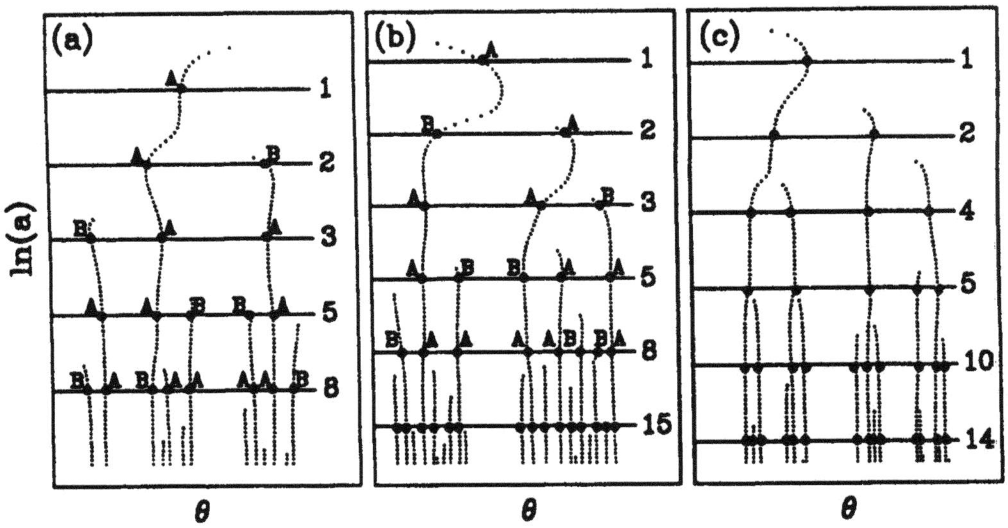

Fig. 6 (a) Enlargements of the WTMM skeleton of a DLA azimuthal Cantor set. These enlargements correspond to three distinct main branches of a 10^6 particle off-lattice DLA cluster. The horizontal lines mark the scale $a_n = a_0 r^{*n}$ with $r^{*-1} = 2.2$. In (a) and (b), the number of WTMM at each generation follows the Fibonacci series (10); moreover, a symbol A or B can be assigned to each of these maxima according to the Fibonacci recursive process (9). (c) Illustration of a local departure (defects) from the Fibonacci structural ordering.

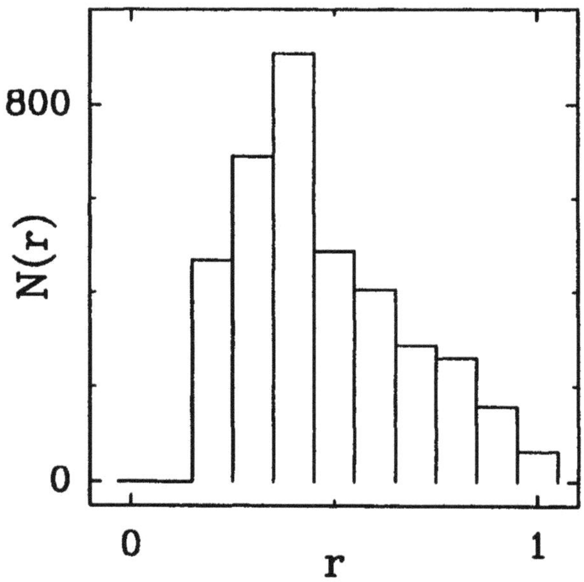

Fig. 7 Histogram of values of the scale ratio between two successive bifurcations in the WTMM skeletons of 50 DLA azimuthal Cantor sets. A single maximum is observed for $r^* = 1/\lambda^* \simeq 0.44 \pm 0.03$ ($\lambda^* \simeq 2.2 \pm 0.2$).

of the DLA branch under study. The horizontal lines in the (θ, a) half-plane in Fig. 6 are drawn as guide marks for those successive generations (in a statistical sense).

As seen in Fig. 6(a), the number of WTMM at each generation follows closely the Fibonacci series.[17] Let us recall that Fibonacci sequences are naturally generated by the recursive process[63]:

$$A \to AB\,, \quad B \to A\,. \tag{9}$$

If one starts with the species B at the generation $n = 0$, one gets A at the generation $n = 1$, and successively AB, ABA, $ABAAB$, $ABAABABA, \ldots$. The population F_n at the generation n can be deduced from the populations F_{n-1} and F_{n-2} at the two preceding generations, according to the iterative law:

$$F_n = F_{n-1} + F_{n-2}\,, \quad F_0 = F_1 = 1\,. \tag{10}$$

Note that F_{n-1} and F_{n-2} are also the respective populations of A and B at step n. A remarkable property of the Fibonacci series $\{F_n\} = \{1, 1, 2, 3, 5, 8, 13, 21, 34, \ldots\}$ is that the ratio of two consecutive Fibonacci numbers converges to the golden mean ϕ:

$$\lim_{n \to +\infty} \frac{F_{n+1}}{F_n} = \phi = \frac{1+\sqrt{5}}{2} = 1.618\cdots \tag{11}$$

As illustrated in Figs. 6(a) and 6(b), by assigning a symbol A or B to each maxima line issued from a bifurcation point, one obtains a coding of the WTMM skeleton that complies with the Fibonacci recursive law [Eq. (9)]. As indicated in Fig. 6(b) purposely, some departure from the Fibonacci ordering are usually observed at small scales, but this is not surprising since, at scales a of the order of a few particle sizes, the azimuthal Cantor sets are very sensitive to small changes in the radius R of the intersecting circle. According to Eq. (11), the branching ratio in the WTMM skeleton is likely to converge to the golden mean. Now if one uses the general formula established for Cantor sets that involves only one scale-factor, one gets the following estimate for the fractal dimension D_F^A of the DLA azimuthal Cantor sets[17,39–42]:

$$D_F^A = \frac{\ln \phi}{\ln \dfrac{1}{r^*}} \simeq \frac{\ln 1.62}{\ln 2.2} \simeq 0.61\,, \tag{12}$$

where we have identified the branching ratio and the scale ratio to the values that have been recorded the most frequently in our statistical study, namely ϕ and $1/r^*$ respectively. Let us note that this numerical value for D_F^A is in good agreement with our previous measurements in Eq. (8) based on box-counting techniques.

One can observe that the histogram in Fig. 7 is rather widely spread around $r^* \simeq 0.44$; this clearly indicates the existence of important fluctuations in the scale ratio value. As illustrated in Fig. 6(c), these fluctuations can produce some local departure from the Fibonacci structural ordering. A close examination of the WTMM skeletons of our set of DLA azimuthal Cantor sets reveals the presence of many of these structural defects and thus raises the question of the statistical pertinence of this Fibonaccian architecture.[17]

The WTMM based technique described in Sec. 2 provides a very attractive method[64] to look for some "mean 1D map" which could explain (and quantify) the presence of a predominant statistical Fibonaccian structural hierarchy in the DLA azimuthal Cantor sets. In order to carry out this analysis in a very efficient way, let us first proceed to a systematic investigation of the symbolic coding of the WTMM skeletons of our 50 DLA azimuthal Cantor sets. A close inspection of this coding reveals some randomness in the relative position of the symbols A and B at each bifurcation $A \to AB$. Apparently, B is likely to be found on the right or on the left of A: among the 1586 bifurcation points for which the coding has been achieved (note that the coding of the maxima lines issued from a bifurcation point requires the knowledge of the next bifurcation point for each of these two lines and thus cannot be processed at too small scales), we have recorded $N(A_L) = 747$ (47%) bifurcations from which A emerges on the left of B; and $N(A_R) = 839$ (53%) from which A emerges on the right. Moreover, the computation of the correlation matrix between two successive bifurcation points does not indicate any memory effect that would condition the relative position A and B from one bifurcation to the next: among the 1080 pairs of successive coded bifurcations, we have recorded $N(A_L, A_L) = 242$ (22.4%) pairs with both A on the left; $N(A_L, A_R) = 284$ (26.2%) pairs with an A on the right following an A on the left; $N(A_R, A_L) = 245$ (22.7%) pairs with an A on the left following an A on the right; and $N(A_R, A_R) = 309$ (28.6%) pairs with both A on the right. Therefore, within the statistical uncertainty, one cannot distinguish between the random occurrence of the symbols A and B at each bifurcation point in the WTMM skeleton of the DLA Cantor sets and a fair tossing coin.

This observation is of fundamental interest in order to adapt the tree matching WTMM technique described in the previous section to the presence of this statistical left-right symmetry at each bifurcation point of the WTMM skeleton of the DLA Cantor sets. We have thus modified the algorithm illustrated in Fig. 1 in such a way that its does not distinguish between a measure which is invariant under the map $T(x)$ on $[0,1]$ and a measure which is invariant under the map $\hat{T}(x) = 1-T(1-x)$ which is the symmetric of T when one flips the interval $[0,1]$ into $[1,0]$. More explicitly, when one proceeds to the tree matching described in Fig. 1(b), everytime A is found on the right of B, the algorithm automatically flips the whole subtree issued from this bifurcation point in such a way that the skeleton actually processed, is made up only of bifurcation points with an A emerging on the left. Then our tree-matching algorithm can be run to extract the map $T(x)$. The 1D map $\hat{T}(x)$ can equally be computed from our tree matching algorithm when implementing the opposite (arbitrary) choice of flipping whenever A is found on the left of B.

The 1D map $T(x)$ reconstructed from the analysis of as many as 50 off-lattice DLA clusters is shown in Fig. 8. In practice, we have identified 240 main branches overall in these 50 aggregates. The angular width of each of these branches has been normalized to 1 before computing the WTMM skeleton of the azimuthal Cantor sets. The data points obtained when scanning these skeletons with our (binary $s = 2$) tree matching algorithm obviously do not fall on a well-defined 1D map. But neither are they a scatter of rather uncorrelated points. The set of data points clearly separates into two distinct "noisy" branches. The solid lines in Fig. 8 correspond to the piece-wise linear 1D map:

$$T(x) = \begin{cases} \lambda_A^* x & \text{for} \quad x \in [0, r_A^*] \\ \lambda_B^*(x-1)+1 & \phantom{\text{for}} \quad x \in [1-r_B^*, 1] \end{cases} \tag{13}$$

where $\lambda_A^* \sim r_A^{*-1} \sim 2.2$ and $\lambda_B^* \sim r_B^{*-1} \sim \lambda_A^{*2} \sim 4.8$. This linear map does not come out from a best regression fit of the data (such a procedure would probably favor a fit with two slightly nonlinear branches). It is simply a 1D map made up of two linear branches whose slopes correspond to the inverse of the preferential scale ratios respectively found when the histogram shown in Fig. 7 is split into the two histograms represented in Figs. 9(a) and 9(c) respectively. These two histograms account for the scale ratio fluctuations observed in the WTMM skeletons when, at each bifurcation point, one computes separately the scale ratio obtained when following either the maxima line with symbol A or the one with symbol B.

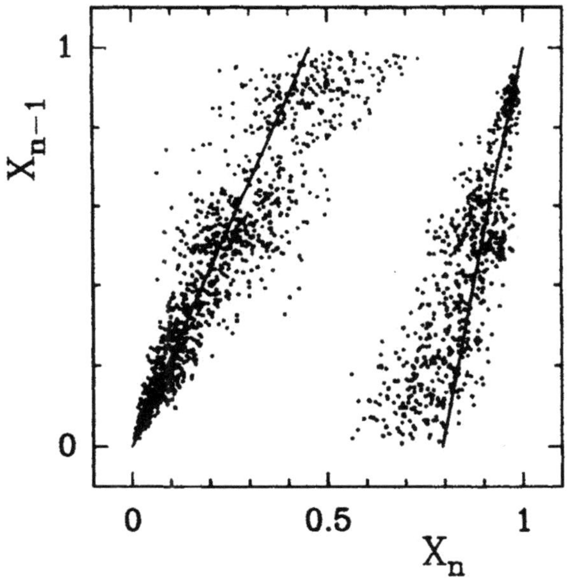

Fig. 8 1D map extracted from the WTMM skeletons of 50 DLA azimuthal Cantor sets using the tree matching algorithm described in the text. The solid lines correspond to the two branches of the linear cookie-cutter (13) with the respective slopes $\lambda_A^* = 2.2$ and $\lambda_B^* = \lambda_A^{*2} = 4.8$.

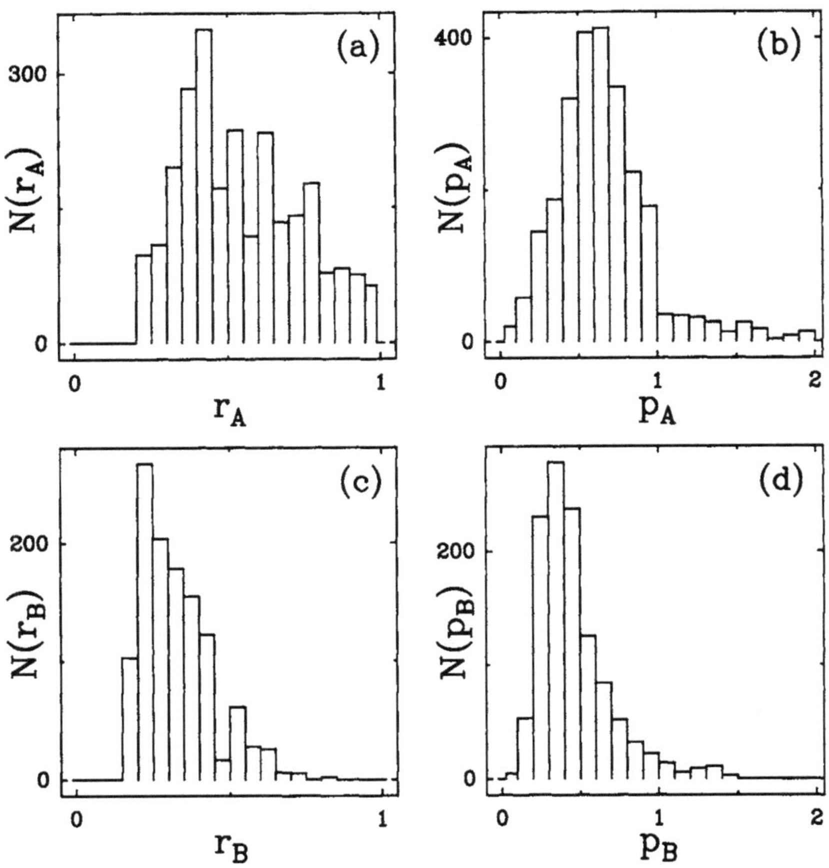

Fig. 9 Histograms of scale ratio values (a) $N(r_A)$, and (c) $N(r_B)$, as computed from each bifurcation point of the WTMM skeletons when following either the maxima line with symbol A or the one with symbol B towards the next bifurcation point at smaller scale (see Fig. 6). The corresponding histograms of amplitude ratio values $N(p_A)$ and $N(p_B)$ are shown in (b) and (d), respectively. Our statistical sample of 50 DLA azimuthal Cantor sets is the same as in Fig. 7.

These histograms are individually scale invariant. They both display a maximum for $r_A^* = 0.44 \pm 0.03$ and $r_B^* = 0.21 \pm 0.03$, respectively.

The fact that $\lambda_B^* \simeq \lambda_A^{*2}$ has a remarkable consequence on the piece-wise linear map [Eq. (13)], since it implies that the slope of one of its branches is likely to be the square of the slope of its other branch. A straightforward computation shows that if one assumes the equality $\lambda_B^* = \lambda_A^{*2}$, then the number of n-cylinders[37,46] (subintervals) of a given size λ_A^k generated by iterating backward T, is exactly the Fibonacci number F_k. A 1D map model as simple as the piece-wise linear map [Eq. (13)] therefore provides a rather natural understanding of the origin of the Fibonacci structural hierarchy discovered on individual realizations in Figs. 6(a) and 6(b). The concentration of data points around the solid lines in Fig. 8 can thus be regarded as a quantitative indication of the existence of a statistically predominant multiplicative process hidden in the fractal complexity of the DLA azimuthal Cantor sets. (One can even recognize, on each branch of this noisy 1D map, a rather non-uniform repartition of the data points that might be the clue for some lacunarity.) There is, however, some randomness in this multiplicative process since at each bifurcation point in the WTMM skeletons, one has to toss a coin in order to decide whether one iterates the map T^{-1} or its symmetric $\hat{T}^{-1}$. Moreover, the amount of spreading of the data points around this 1D map model, or in other words, how noisy is the reconstructed 1D map, can be seen conversely as the signature of the statistical importance of the structural defects to the Fibonacci fractal ordering.

Now, as far as the statistical homogeneity of the DLA azimuthal Cantor sets is concerned, one can easily convince oneselves that for the Bernoulli invariant measures of the piece-wise linear cookie-cutter model [Eq. (13)] to be homogeneous, the respective weights p_A and p_B, distributed multiplicatively at each iteration, have to satisfy the requirement $p_A^* = p_B^{*2}$. Since $p_A^* + p_B^* = 1$, one gets exactly $p_A^* = \phi^{-1}$ and $p_B^* = \phi^{-2}$, i.e., the ratio p_A/p_B is equal to the golden mean. This requirement is automatically satisfied if one considers that the measure of an interval, at a given resolution r_A^{*k}, is proportional to the number of n-cylinders of length r_A^{*k} that it contains. Thus, at the first stage of construction, the measures p_A and p_B of the left and right 1-cylinder are respectively proportional to the successive Fibonacci numbers F_{k-1} and F_{k-2}. In the limit $k \to \infty$, the ratio p_A/p_B converges to the golden mean, while individually p_A^* and p_B^* converge to $p_A^* = \phi^{-1} \simeq 0.618$ and $p_B^* = \phi^{-2} \simeq 0.382$. In Figs. 9(b) and 9(d) are respectively reported the histograms of amplitude ratios $N(p_A)$ and $N(p_B)$, extracted from our WTMM tree matching algorithm. Both histograms display a unique rather well-defined maximum in very good agreement with those expected values for p_A^* and p_B^*. This is a strong indication that DLA azimuthal Cantor sets are likely to be homogeneous fractals. The fact that the ratio of the integrals of the two histograms $N(p_A)$ and $N(p_B)$ (i.e., the ratio of the total numbers of data points in the left and in the right noisy branches of the reconstructed 1D map in Fig. 8), is also found close to the golden mean, provides additional evidence that the defects to the Fibonacci structural ordering do not seem to induce any significantly noticeable multifractal departure from statistical homogeneity.

But this might be a premature conclusion since the stochastic nature of the WTMM skeletons of the DLA azimuthal Cantor sets is far from being completely characterized by the histograms of scaling ratios and amplitude ratios shown in Figs. 7 and 9. The knowledge of various joint probability distributions is of fundamental interest for that purpose. We have investigated in Fig. 10 the correlations between the variables $\ln r$ and $\ln p$ at each bifurcation point of the WTMM skeletons. These two random variables appear to

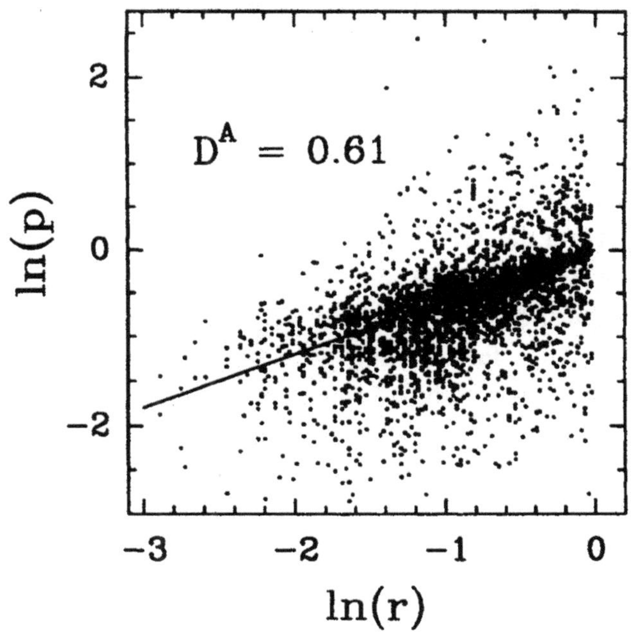

Fig. 10 ln p vs. ln r, where $r = a_n/a_{n-1}$ is the ratio between the scales of two bifurcation points of the WTMM skeletons that are associated by our tree matching algorithm (see text), while $p = |W_\psi[\mu](x_n, a_n)|/|W_\psi[\mu](x_{n-1}, a_{n-1})|$ is the corresponding amplitude ratio. A linear regression fit of the data provides a slope $\alpha = D_F^A = 0.61 \pm 0.03$, in good agreement with previous estimates of both the fractal dimension D_F^A [Eqs. (8) and (12)] and the local scaling exponent α of the 50 DLA azimuthal Cantor sets.

be strongly correlated since most of the data points are distributed around a line of slope $\alpha \simeq 0.61$. Similar results have been obtained by H. L. Hwang and S. Mallat in Ref. 51. This means that the amplitude ratio p and the scale ratio r are statistically related according to the law $p = Cr^{0.61}$. This result is in remarkable agreement with previous WT measurements[17,31] of the local scaling exponent $\alpha = D_F^A = 0.61 \pm 0.03$ of the DLA azimuthal Cantor sets. The scatter of points around the solid line in Fig. 10 is an indication that this scaling exponent might well not be unique. The computation of other (higher order) joint probability distributions is likely to provide a more refined analysis of the possible fluctuations of this scaling exponent. These fluctuations might quantitatively explain some weak multifractal departure from statistical homogeneity as previously noticed in our box-counting calculation of the generalized fractal dimensions D_q of the DLA azimuthal Cantor sets.[18]

4. DISCUSSION

In Ref. 3, H. E. Stanley et al. have clearly stated one of the main challenges raised by the puzzling DLA morphology: "... eventhough no two DLA's are identical ... nonetheless every DLA that we are likely to ever see has a generic form that even a child can recognize ...". We believe that the results reported in this work are a very attractive breakthrough in this spirit, since they provide statistical evidence for the existence of a multiplicative process hidden in the structural organization of DLA clusters. Actually, when exploring the WTMM skeletons of 50 DLA azimuthal Cantor sets with a tree matching algorithm specially devoted to solving the inverse fractal problem, we have revealed the pertinence of the piece-wise linear cookie-cutter T defined in Eq. (13), to account for the presence of a statistically predominant Fibonacci structural ordering. We have shown that there exist mainly two sources of randomness superimposed to this structural ordering. The first one is in the Fibonacci multiplicative process itself since at each construction step of the Cantor set, one has to toss a coin in order to decide whether one iterates T^{-1} or its symmetric $\hat{T}^{-1}$. The second one appears as intrinsic noise in the reconstructed 1D map and is likely to be the signature of structural defects to the Fibonacci fractal hierarchy.

This very promising step towards a comprehensive understanding of the DLA morphology perhaps raises more questions than it answers. Where does the preferential scale ratio $r^* \simeq 0.44$ come from? In a forthcoming publication, we will argue that this scale ratio can be derived from the linear stability theory of Saffman-Taylor viscous fingers grown in sector shape cells.[65] Is the noise observed in the reconstructed 1D map in Fig. 8 (or in other words the presence of structural defects) characteristic of Laplacian morphologies or is it simply the consequence of the noise which is intrinsic to the discrete DLA algorithm? To answer this question we have recently adapted the method of noise reduction[66] to our off-lattice DLA algorithm; we are currently running this rather time consuming algorithm. Is this Fibonacci structural ordering only present in the inner frozen region of the DLA aggregates or does it extend to the external active region? Does the branching process of the DLA cluster itself display this Fibonacci multiplicative structure, as suggested in our previous study in Refs. 17, 30 and 31? Is there any deep connection between this Fibonacci hierarchy and the statistical five-fold symmetry? Do the structural defects in the Fibonacci architecture correspond to some local five-fold symmetry breakings in the DLA fractal branching? To bring some answers to these interrogations we thus need to bridge the gap between the 1D WTMM analysis of the DLA azimuthal Cantor sets and the 2D WT analysis of the DLA clusters. The implementation of a tree matching algorithm that generalizes our wavelet based method described in Sec. 2, from 1D to 2D, is currently in progress. Further applications of this method to turbulent velocity signals, surface roughening and DNA "walks" nucleotide sequences, look very promising.

ACKNOWLEDGMENTS

This work was supported by the Direction des Etudes et Techniques under contract N°92/097 and by the Centre National des Etudes Spatiales under contrat N°92/0225.

REFERENCES

1. T. A. Witten and L. M. Sander, *Phys. Rev. Lett.* **47**, 1400 (1981); *Phys. Rev.* **B27**, 5686 (1983).
2. P. Meakin, in *Phase Transitions and Critical Phenomena*, eds. C. Domb and J. L. Lebowitz, Vol. 12 (Academic Press, New York, 1988), p. 355.
3. H. E. Stanley, A. Bunde, S. Havlin, J. Lee, E. Roman and S. Schwarzer, *Physica* **A168**, 23 (1990).
4. L. Pietronero and E. Tosatti, eds., *Fractals in Physics* (North-Holland, Amsterdam, 1986).
5. W. Guttinger and D. Dangelmayr, eds., *The Physics of Structure Formation* (Springer, Berlin, 1987).
6. H. E. Stanley and N. Ostrowsky, eds., *Random Fluctuations and Pattern Growth* (Kluwer, Dordrecht, 1988).
7. J. Feder, *Fractals* (Pergamon, New York, 1988).
8. T. Vicsek, *Fractal Growth Phenomena* (World Scientific, Singapore, 1989).
9. L. Pietronero, ed., *Fractal's Physical Origin and Properties* (Plemum, New York, 1989).

10. A. Aharony and J. Feder, eds., *Fractals in Physics*, Essays in honour of B. B. Mandelbrot, *Physica* **D38** (1989).
11. A. Bunde, ed., *Fractals and Disorder* (North-Holland, Amsterdam, 1992).
12. P. Meakin and S. Havlin, *Phys. Rev.* **A36**, 4428 (1987).
13. F. Argoul, A. Arneodo, G. Grasseau and H. L. Swinney, *Phys. Rev. Lett.* **61**, 2558 (1988); **63**, 1323 (1989).
14. G. Li, L. M. Sander and P. Meakin, *Phys. Rev. Lett.* **63**, 1322 (1989).
15. T. Vicsek, F. Family and P. Meakin, *Europhys. Lett.* **12**, 217 (1990).
16. B. B. Mandelbrot, in Ref. 11, p. 95.
17. A. Arneodo, F. Argoul, E. Bacry, J. F. Muzy and M. Tabard, *Phys. Rev. Lett.* **68**, 3456 (1992); in *Growth Patterns in Physical Sciences and Biology*, eds. J. M. Garcia-Ruiz, E. Louis and L. M. Sander (Plenum, New York, 1993), p. 191.
18. A. Arneodo, F. Argoul, E. Bacry, J. Elezgaray, J. F. Muzy and M. Tabard, in preparation.
19. F. Argoul, A. Arneodo, J. Elezgaray, G. Grasseau and R. Murenzi, *Phys. Lett.* **A135**, 327 (1989); *Phys. Rev.* **A41**, 5537 (1990).
20. J. M. Combes, A. Grossmann and P. Tchmitchian, eds., *Wavelets* (Springer, Berlin, 1988).
21. P. G. Lemarié, ed., *Les Ondelettes en 1989* (Springer, Berlin, 1990).
22. Y. Meyer, ed., *Wavelets and Their Applications* (Springer, Berlin, 1992).
23. Y. Meyer and S. Roques, eds., *Progress in Wavelet Analysis and Applications* (Editions Frontières, Paris, 1993).
24. M. Holschneider, *J. Stat. Phys.* **50**, 963 (1988); Thesis, University of Aix-Marseille II (1988).
25. A. Arneodo, G. Grasseau and M. Holschneider, *Phys. Rev. Lett.* **61**, 2281 (1988); in Ref. 20, p. 182.
26. A. Arneodo, F. Argoul, J. Elezgaray and G. Grasseau, in *Nonlinear Dynamics*, ed. G. Turchetti (World Scientific, Singapore, 1989), p. 130.
27. F. Argoul, A. Arneodo, G. Grasseau, Y. Gagne, E. Hopfinger and U. Frisch, *Nature* **338**, 52 (1989).
28. A. Arneodo, F. Argoul and G. Grasseau, *Lect. Notes in Maths* **148**, 125 (1990).
29. A. Arneodo, F. Argoul, E. Bacry, J. Elezgaray, E. Freysz, G. Grasseau, J. F. Muzy and B. Pouligny, in Ref. 22, p. 286.
30. A. Arneodo, F. Argoul, J. F. Muzy and M. Tabard, *Phys. Lett.* **A171**, 31 (1992); *Physica* **A188**, 217 (1992).
31. A. Arneodo, F. Argoul, E. Bacry, J. Elezgaray, J. F. Muzy and M. Tabard, in Ref. 23, p. 21.
32. A. Arneodo, E. Bacry and J. F. Muzy, "Solving the Inverse Fractal Problem from Wavelet Analysis," CRPP preprint (1993).
33. B. B. Mandelbrot, *The Fractal Geometry of Nature* (Freeman, San Francisco, 1982).
34. P. Grassberger and I. Procaccia, *Physica* **D13**, 34 (1984).
35. G. Parisi and U. Frisch, in *Turbulence and Predictability in Geophysical Fluid Dynamics and Climate Dynamics*, eds. M. Ghil, R. Benzi and G. Parisi (North-Holland, Amsterdam, 1985), p. 84.
36. T. C. Halsey, M. H. Jensen, L. P. Kadanoff, I. Procaccia and B. I. Shraiman, *Phys. Rev.* **A33**, 1141 (1986).
37. P. Collet, J. Lebowitz and A. Porzio, *J. Stat. Phys.* **47**, 609 (1987).
38. G. Paladin and A. Vulpiani, *Phys. Rep.* **156**, 148 (1987).
39. J. F. Muzy, E. Bacry and A. Arneodo, *Phys. Rev. Lett.* **67**, 3515 (1991); *Phys. Rev.* **E47**, 875 (1993); in Ref. 23, p. 323.
40. E. Bacry, J. F. Muzy and A. Arneodo, *J. Stat. Phys.* **70**, 635 (1993).
41. E. Bacry, Thesis, University of Paris VII (1992).
42. J. F. Muzy, Thesis, University of Nice Sophia Antipolis (1993).
43. M. F. Barnsley and S..G. Demko, *Proc. Roy. Soc. London* **A399**, 243 (1985).
44. M. F. Barnsley, *Fractals Everywhere* (Academic Press, New York, 1988).
45. C. R. Handy and G. Mantica, *Physica* **D43**, 17 (1990).
46. D. Rand, *Ergod. Th. and Dynam. Sys.* **9**, 527 (1989).
47. C. Meneveau and K. R. Sreenivasan, *Phys. Rev. Lett.* **59**, 1424 (1987); *J. Fluid. Mech.* **224**, 429 (1991).
48. J. M. Ghez and S. Vaienti, *Nonlinearity* **5**, 772 and 791 (1992).

49. S. Mallat and S. Zhong, *IEEE Trans. on Pattern Analysis and Machine Intelligence* **14**, 710 (1992).
50. S. Mallat and W. L Hwang, *IEEE Trans. on Information Theory* **38**, 617 (1992).
51. W. L. Hwang and S. Mallat, "Characterization of self-similar multifractals with wavelet maxima," *New York University Technical Report No. 641* (Computer Science Department, 1993).
52. P. Collet and J. P. Eckmann, *Iterated Maps of the Interval and Dynamical Systems* (Birkhauser, Boston, 1980).
53. M. J. Feigenbaum, *J. Stat. Phys.* **19**, 25 (1978); **21**, 669 (1979).
54. P. Coullet and C. Tresser, *J. Physique, Colloq.* **39**, C5 (1978); C. R. Acad. Sci. **287**, 577 (1978).
55. F. Ledrappier and M. Misiurewicz, *Ergodic. Th. and Dyn. Syst.* **5**, 595 (1985).
56. M. Plischke and Z. Rácz, *Phys. Rev. Lett.* **53**, 415 (1984).
57. F. Argoul, A. Arneodo, J. Elezgaray and G. Grasseau, in *Measure of Complexity and Chaos*, eds. N. B. Abraham, A. M. Albano, A. Passamante and P. E. Rapp (Plenum, New York, 1989), p. 433.
58. C. Amitrano, P. Meakin and H. E. Stanley, *Phys. Rev.* **A40**, 1713 (1989).
59. B. Derrida, V. Hakim and J. Vannimenus, *Phys. Rev.* **A43**, 888 (1991).
60. R. C. Ball and R. M. Brady, *J. Phys.* **A18**, L809 (1985).
61. P. Ossadnik, *Phys. Rev.* **A45**, 1058 (1992).
62. B. B. Mandelbrot, *J. Stat. Phys.* **34**, 895 (1984).
63. T. H. Garland, *Fascinating Fibonacci: Mystery and Magic in Numbers* (Dayle Seymour, Palo Alto, 1987).
64. A. Arneodo, F. Argoul, E. Bacry, J. F. Muzy and M. Tabard, "Uncovering a multiplicative process in one-dimensional cuts of diffusion-limited aggregates with wavelets," preprint (1993).
65. R. Combescot and M. Ben Amar, *Phys. Rev. Lett.* **67**, 453 (1991).
66. J. Nittmann and H. E. Stanley, *Nature* **321**, 663 (1986).

Fractals, Vol. 1, No. 3 (1993) 650–662

THE FIXED SCALE TRANSFORMATION: STATUS AND PERSPECTIVES

L. PIETRONERO
Dipartimento di Fisica, Università di Roma "La Sapienza", P. le A. Moro 2, 00185 Roma, Italy

Abstract

Irreversible fractal growth models like DLA and DBM have confronted us with theoretical problems of a new type that cannot be described in terms of the standard concepts like field theory and the renormalization group. The Fixed Scale Transformation is a theoretical scheme of a new type that is able to treat these problems in a reasonably systematic way. The idea is to focus on the dynamics at a given scale and to compute accurately the correlations at this scale by suitable lattice path integrals. The use of scale invariant growth rules then allows the generalization of these correlations to coarse-grained cells of any size and therefore to obtain the fractal dimension. We summarize the present status of the FST approach by focusing on the most recent results about the scale invariant dynamics of DLA/DBM. The possible extensions to other problems like the sand pile model (self-organized-criticality) and simplified models of turbulence will also be considered.

1. THE GENERAL PROBLEM

After the introduction of fractal geometry,[1] the prototype fractal growth models based on well defined physical processes are Diffusion Limited Aggregation[2] (DLA) and the Dielectric Breakdown Model[3] (DBM). These models have been generalized or extended to a variety of other models[4] and they represent the central problem for the understanding of fractal growth.

Despite some apparent analogies with the usual critical phenomena, DLA and DBM are profoundly different from the problems of equilibrium statistical mechanics. The attempts to describe them using the usual theoretical concepts, like field theory and the renormalization group (RG), have been rather problematic and did not lead very far. One of the key points is that these models are intrinsically irreversible and the *time evolution* has to be considered explicitly. It is not possible to assign a statistical weight to a configuration without considering its dynamical evolution.

In principle, one would like to compute the correlation function $G(\bar{r}_1, \bar{r}_2, t)$ that gives the probability of occupation of the points $\bar{r}_1$ and $\bar{r}_2$ after a time t. This time is defined by the total number of particles or bonds added to the structure. Actually the fractal properties become well defined only asymptotically with respect to two limits:

1. *Asymptotic time limit* ($t \to \infty$). This means that one should consider regions of the system that are very far from the growing interface and that will not be modified any more by further growth (freezing condition).
2. *Large scale limit* ($|\bar{r}_1 - \bar{r}_2| \to \infty$). This implies that one should consider distances at which eventual scale invariant properties become well defined. For a fractal structure, we expect therefore the following behavior:

$$\lim_{r=|\bar{r}_1-\bar{r}_2|\to\infty} \lim_{t\to\infty} G(\bar{r}_1, \bar{r}_2; t) = g(\bar{r}_1, \bar{r}_2) \cong |\bar{r}_1 - \bar{r}_2|^{-(d-D)} \tag{1}$$

where D is the fractal dimension and d is the embedding Euclidean dimension.

From a theoretical point of view, the problem is to describe the asymptotic behavior with respect to time for space variables that become renormalized by a factor to ($r' = r/b$). One should then be able to control the three limits: $t \to \infty$, $r \to \infty$, $b \to \infty$, with the condition that $r' = r/b =$ constant.

The attempts to formulate this problem in a field theory framework have met with fundamental difficulties.[5] If the free field is a random walk, the interaction due to the effect of Laplace equation, becomes unrenormalizable in any dimension. This corresponds to the fact that, in problems like DLA and DBM the interaction is always relevant and there is no upper critical dimension. The application of standard concepts of Real Space Renormalization Group (RSRG) have also been rather problematic because, by integrating over the degrees of freedom inside a given box, it is not possible to implement the asymptotic time limit (freezing) condition.[5] Also, in the interesting Nagatani approach,[6] it is not possible to distinguish between fractal and non-fractal structures.

Given this situation we have tried to formulate a theoretical framework of a novel type, the Fixed Scale Transformation (FST) that focuses explicitly on the dynamical evolution.[7] The FST method has been improved and clarified in the past few years and it is now able to treat in a reasonably systematic way the problems of irreversible fractal growth. A detailed discussion of the FST method can be found in Ref. 5. Here we present a brief summary of the situation focusing in particular on the most recent developments and on the possible extensions to other problems.

2. FST STRATEGY

The basic point of the FST method is to treat separately the two limits $t \to \infty$ and $b \to \infty$. For the asymptotic time limit, it is very important to describe accurately the time convergence of the asymptotic structure. This is crucial in order to be able to distinguish between fractal and non-fractal structures. For example, the Eden model leads asymptotically to compact structure but its convergency is rather slow. So it is essential to achieve a systematic treatment of this convergence in order to understand whether the structure is fractal or compact.[7]

The second problem ($b \to \infty$) refers to the renormalization of the dynamics (growth process) for which long times are not so crucial. This will be done by focusing mainly on the largest growth probabilities that correspond essentially to short time events. However, the fixed point asymptotic growth rules defined in this way are then used also for long times in the FST transformation. In this way, it is possible to compute the fractal dimension analytically for a variety of growth models with an accuracy of the order of a few per cent.

Let us first consider the problem of the asymptotic time limit. We start by defining the nearest neighbors ($n.n.$) pair correlations at the minimal scale and secondly, we will extend this result to all scales and relate it to the fractal dimension.

Consider for example the original DBM growth rules. In order to define the $n.n.$ correlation we have to consider a pair of sites and study the possible configurations of these pairs that occur in the asymptotic structure generated by the growth process. We are interested in conditional probabilities, so one of the sites will certainly be occupied (black). This leads to only two possibilities: a configuration of type 1 with an occupied (black) and an empty (white) site, and a configuration of type 2 with both sites occupied (black). The probabilities for the occurrence of these configurations are defined as C_1 and C_2 respectively.

In order to compute these probabilities it is convenient to consider the probability M_{ij} ($i, j = 1, 2$) that a pair configuration of type i is followed, in the growth direction (k), by a pair configuration of type j (Fig. 1). This leads to a transfer matrix problem:

$$\begin{pmatrix} C_1^{k+1} \\ C_2^{k+1} \end{pmatrix} = \begin{pmatrix} M_{12} & M_{21} \\ M_{12} & M_{22} \end{pmatrix} \begin{pmatrix} C_1^k \\ C_2^k \end{pmatrix} \tag{2}$$

whose fixed point is given by:

$$C_1^* = \left(1 + \frac{M_{12}}{M_{21}}\right)^{-1} ; \quad C_2^* = 1 - C_1^* . \tag{3}$$

The matrix elements M_{ij} can be computed by a *lattice path integral* over the possible growth processes that correspond to the configurations i and j. How to do this in practice can be found in Refs. 5 and 7. Note that the configurations considered should correspond to the asymptotic (frozen) structure. This implies that the lattice path integrals should be extended until the growing interface is far enough from these configurations, so that they can be considered as asymptotic (long time limit).

In order to make this calculation quantitatively accurate, it is important to consider also the possible environments (boundary conditions) outside the growing column defined by the dashed lines in Fig. 1. This implies that the matrix elements M_{ij} should be replaced by a weighted average[5,7]:

$$M_{ij} = \sum_n P(\lambda_n) M_{ij}(\lambda_n) \tag{4}$$

where λ_n is a parameter that characterizes the various possible boundary conditions and $P(\lambda_n)$ is the corresponding probability distribution that will depend on (C_1, C_2). This leads to the more complex nonlinear transformation:

$$\begin{pmatrix} C_1^{k+1} \\ C_2^{k+1} \end{pmatrix} = \sum_n P(\lambda_n, C_1^k C_2^k) \begin{pmatrix} M_{11}(\lambda_n) & M_{21}(\lambda_n) \\ M_{12}(\lambda_n) & M_{22}(\lambda_n) \end{pmatrix} \begin{pmatrix} C_1^k \\ C_2^k \end{pmatrix} \tag{5}$$

Its fixed point can also be found analytically using suitable approximation scheme.[5,7]

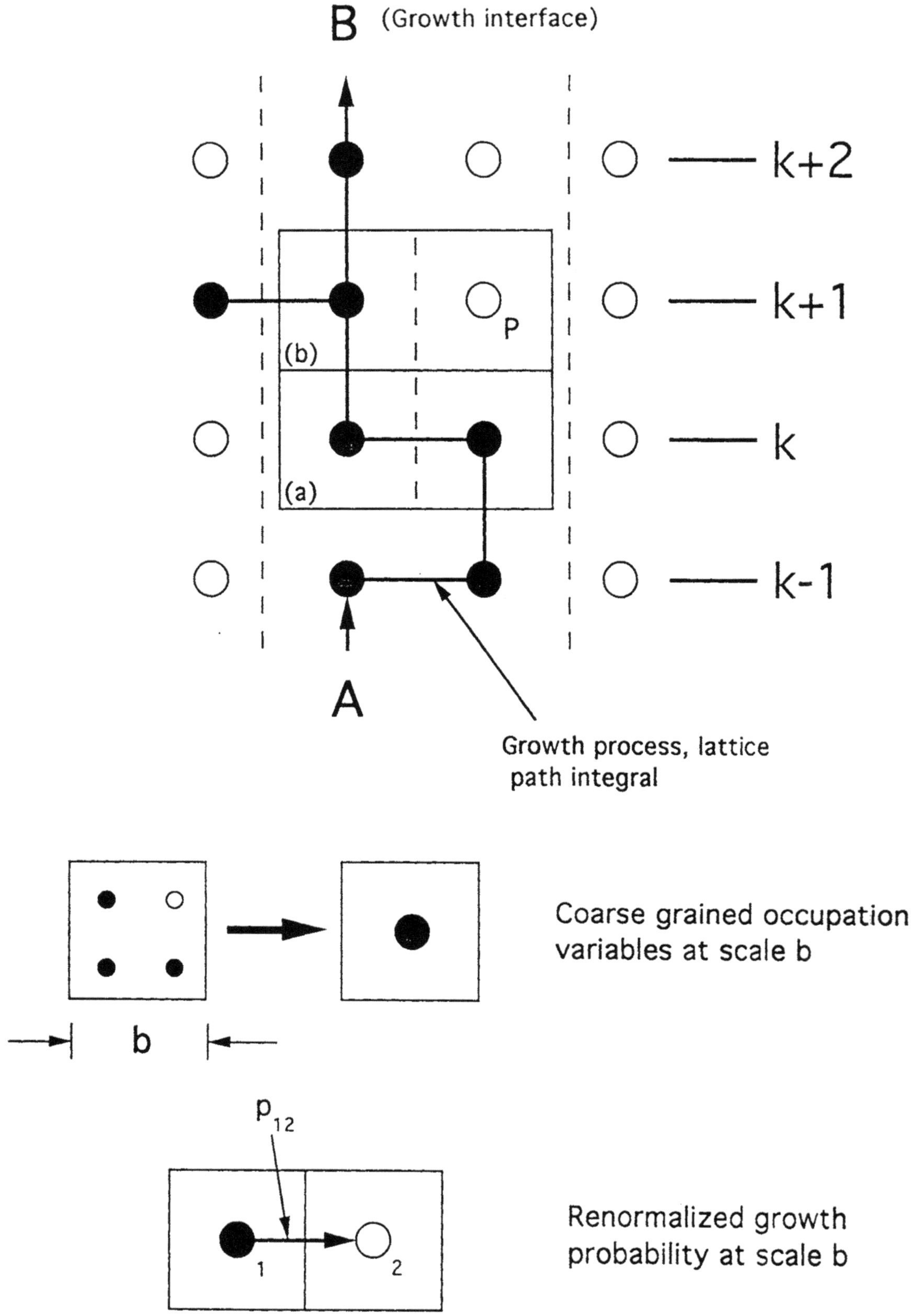

Fig. 1 A crucial concept in the FST method is the nearest neighbor correlation between cells of a given scale b. One can study this problem by considering the probability that a cell of type i (**a**) is followed, in the growth direction (k), by a cell of type j (**b**). These probabilities are defined by lattice path integrals corresponding to the growth dynamics at a given scale (**b**). If one uses a scale invariant growth dynamics, the abovè correlations can be extended to coarse-grained cells of any size and the fractal dimension can be computed.

The possibility of including explicitly the fluctuations of boundary conditions makes the FST appreciably more accurate than other real space methods like RSRG. This advantage is due to the fact that in the FST one does not change scale and in the end it reduces to the analysis of the occupation probability of a single site. For a technical discussion see Refs. 5 and 7.

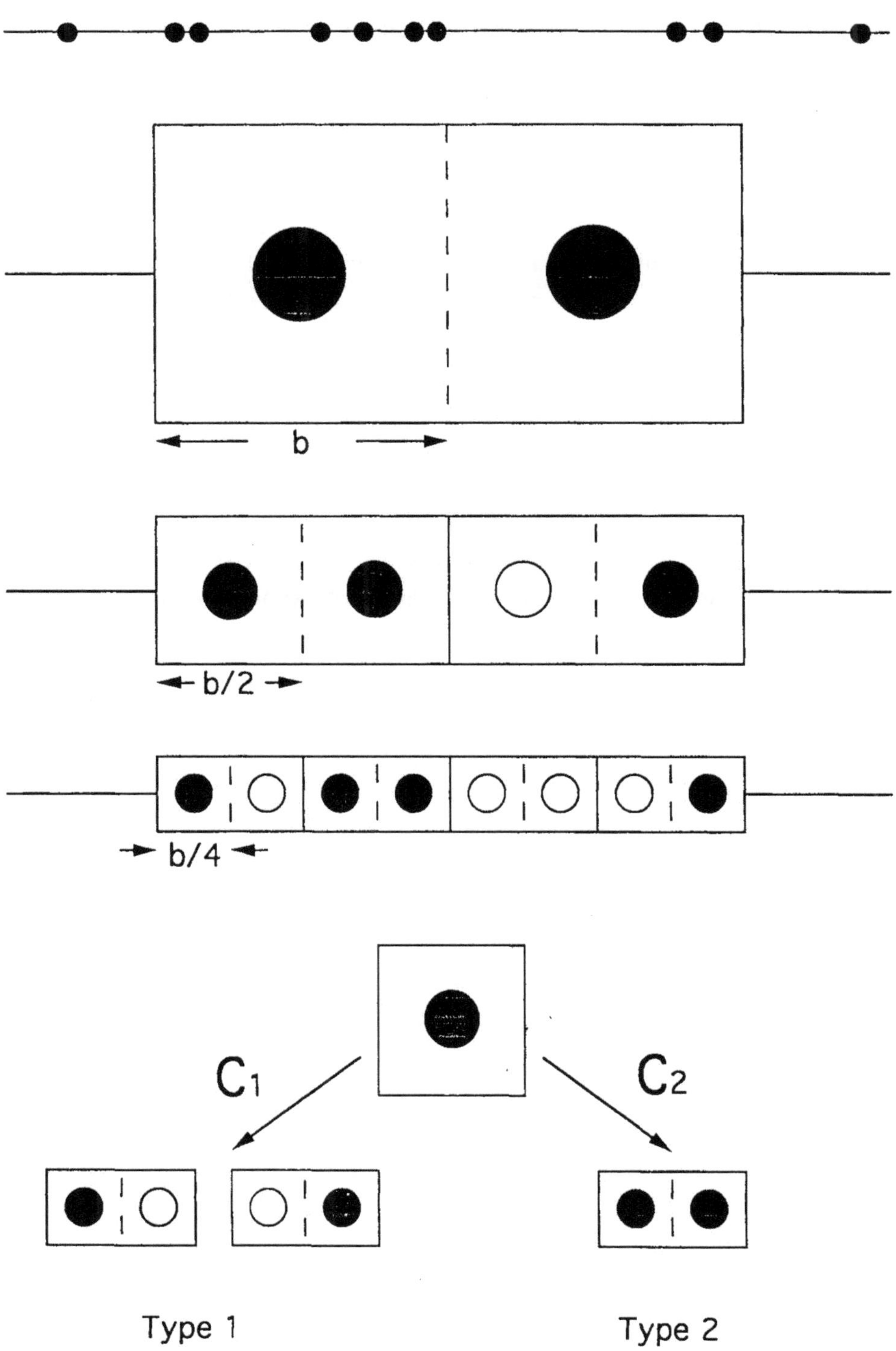

Fig. 2 Box covering process for the set of points defined by the intersection of the structure with a line perpendicular to the growth direction.

Strictly speaking, the FST scheme we have discussed until now allows only the computation of the *n.n.* pair correlations at the minimal scale. However, if it is possible to interpret our "sites" as coarse-grained cells and to use *growth rules* that are *scale invariant*, then the fixed point probabilities (C_1^*, C_2^*) would characterize the correlations between coarse-grained cells of any size (Fig. 1). This would allow us to perform the large scale limit ($b \to \infty$) because long range correlations would be described by the *n.n.* pair correlations of large cells.

In such a situation, our pairs of cells of variable sizes can be shown to correspond to the box covering process of an intersection of the structure with a line perpendicular to the growth direction (Fig. 2). In this way, the probability D of the structure[5,7]:

$$D = 1 + \frac{\ln(C_1^* + 2C_2^*)}{\ln 2} . \tag{6}$$

In order to perform the important step corresponding to the large scale limit one should be able to control how the growth rules change under a scale transformation (Fig. 3). DLA and DBM are intrinsically critical in the sense that their dynamics evolves into the scale invariant one without tuning any parameter. However, the lack of a characteristic length in the original (small scale) growth rules does not guarantee that these hold at a generic scale. Therefore, the questions of the universality and scale invariant properties of DLA and DBM are related to their effective asymptotic dynamics. In addition, the knowledge of this effective dynamics is one of the key points in the understanding of *why these models give rise to fractal structures.* This problem is in general very complex due to the large number (in principle infinite) of parameters that would appear in the renormalized dynamics. In practice, however, one can fix a subset of parameters and study their evolution under scale change. In view of these *a priori* restrictions this procedure is not very systematic. However, it allows an understanding of some general symmetry properties that belong to the scale invariant dynamics[8]: (i) growth probabilities should be defined per bond. If one starts with

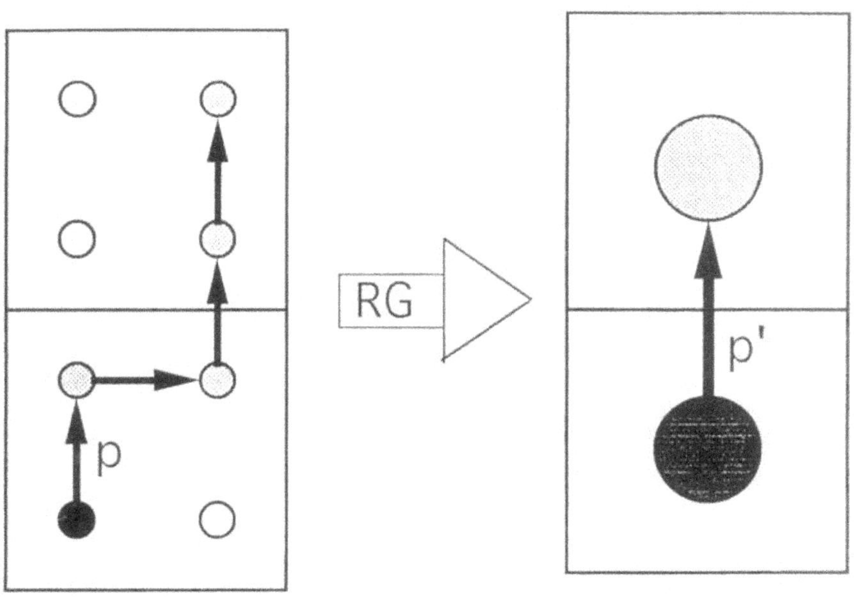

Fig. 3 Simple scheme that shows the nature of the renormalization process for the growth rules.

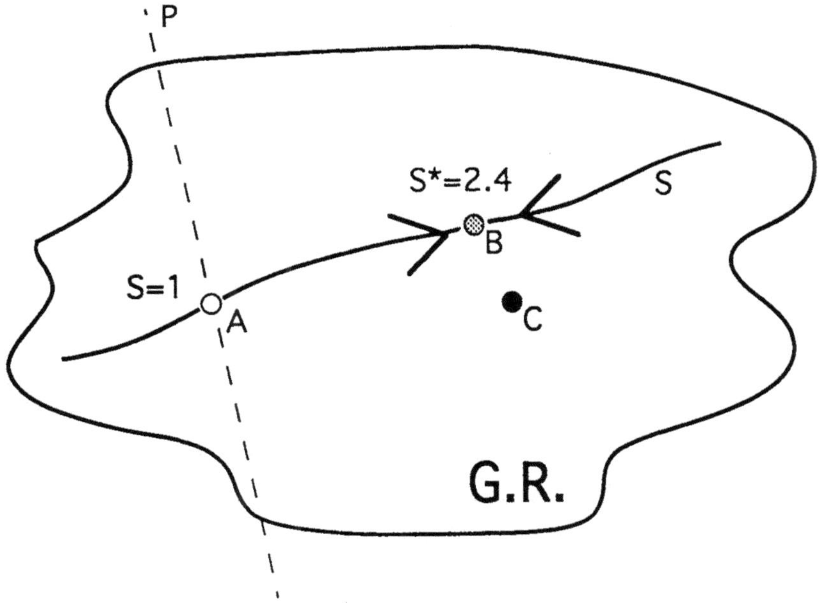

Fig. 4 Schematic picture that illustrates the space of growth rules (G.R.). The point A corresponds to the original DLA/DBM growth rules. Under an ideal renormalization this point would flow to point C that represents the "exact" scale invariant growth rules. Our scheme consists in renormalizing the growth rules along the line defined by the noise reduction parameter S. The idea is that the fixed point obtained in this way (point B) is a good approximation to the "exact" one (point C). The line P corresponds to a generalization of the growth rules that goes outside the correct basin of attraction.

site probabilities, these evolve into bond type under scale transformation; (ii) An eventual probability assigned to a diagonal bond disappears under scale change.

Within the complex space of growth rules (Fig. 4), the FST framework points, however, to the essential concept: Fractals can be generated only if screening persists in the scale invariant regime.[5,7] Note that the presence of screening due to the Laplace equation in the original growth rules does not guarantee that a similar effect persists for coarse-grained variables (Fig. 3). In fact, if one studies the growth rules for a coarse-grained cell, a problem of noise reduction naturally appears. The larger the cell, the larger will be the number of particles (bonds) necessary to span it. Naively therefore one could expect that, asymptotically, the effective noise reduction parameter (S) could diverge. This would then eliminate screening effects and lead to a compact structure. Therefore the key feature of the asymptotic growth rules is the identification of the *fixed point noise reduction parameter.*

The scale transformation from the noise reduction at scale b ($S = S(b)$) to that at scale $b' = 2b$ ($S' = S(b')$) can be written as[9,10]:

$$\frac{1}{S'} = \frac{1}{\langle N(S)\rangle} + \frac{\langle \delta N(S)^2\rangle}{\langle N(S)^2\rangle} \tag{7}$$

where $N(S)$ is the number of particles (bonds) that have to be added to span a coarse-grained cell of size b. It is possible to express the average of this number in the following way[10]:

$$\langle N(S)\rangle = \sum_{\alpha} W^{(\alpha)}(S) \sum_{i} P_i^{(\alpha)}(S) N_i^{(\alpha)}(S) \tag{8}$$

where $W^{(\alpha)}$ is the weight of each starting occupation configuration (α) for the cell that will be renormalized. The index i refers to a particular path whose total probability is given by

$P_i^{(\alpha)}$ and $N_i^{(\alpha)}$ is the corresponding total number of particles. An analogous equation can be written for $\langle N^2(S)\rangle$.

In this way, we obtain a renormalization scheme for the noise reduction parameter S.[10] This allows solving of the problem of the large scale limit ($b \to \infty$). This renormalization scheme is essentially defined by the largest probabilities which in turn correspond to small time effects. The idea is that long time effects are crucial for the asymptotic occupation of the sites but not for the identification of the scale invariant growth rules. These, in turn, are determined by short time effects but are also defined for long times so that they can then be used in the FST transformation to treat the long time effects.[5] The main results of this renormalization scheme[10] are that the fixed point noise reduction parameter turns out to be of the order of unity ($S^* = 2.4$) with an *attractive fixed point* (Fig. 4).

These results clarify therefore the *self-critical* nature of DLA patterns, in contrast, for example, to percolation in which the fixed point p_c is repulsive. This is due to the fact that, under scale change, noise is automatically generated by the dynamics of the system because there are always paths of different lengths that can span a given cell.[10]

In addition, the fact that the value of S^* is close to 1 shows that *screening is asymptotically preserved* and that the minimal scale growth rules are already rather close to the asymptotic ones. This allows us to understand why the initial FST studies of DLA,[7] in which the minimal scale growth rules were used, give indeed reasonable results for the fractal dimension.

The nature of this problem is therefore that the crucial point is to focus on the dynamics ($t \to \infty$ limit). The scale invariance of the growth rules seems to be somewhat less critical, as long as they possess the correct symmetry structure. In this respect, irreversible fractal growth models are quite different from the usual problems of equilibrium statistical mechanics.

3. SUMMARY OF RESULTS

In Tables 1 to 4, we summarize the main results that have been obtained with the FST method. The complete list of references for all the data presented in these table can be found in Ref. 5.

Table 1 corresponds to the application of the FST to fractal structures generated by standard equilibrium problems in statistical mechanics. For example, for the percolating cluster, one has first to interpret the problem in terms of fractal growth. It can be shown that the identification of the scale invariant dynamics corresponds to the determination of the critical probability p_c. From this non-universal critical parameter and the lattice topology, the FST allows the computation directly of the fractal dimension D, a universal exponent.[11] In this respect, the FST method is fundamentally different from the renormalization group even for standard problems. For the Ising and Potts models, the scale invariant "fictitious" dynamics is defined by the scale invariant structure of the Hamiltonian and the critical temperature T_c, while for polymers and lattice animals it corresponds to the value of the critical fugacity K_c. The results of this table are therefore a test of the FST methods in problems in which the result is known. It should be noted however that the FST results for the Potts clusters with $q = 2$; 3[12] were obtained before the conjecture of the exact results.[13]

In Table 2, we report the FST results to problems that are intrinsically dynamical and for which the standard theoretical concepts have already become problematic.

Table 1 Fractal dimension of structures generated by equilibrium critical phenomena computed with the FST method

Equilibrium Hamiltonian Models at Critical Point			
Model	D_{exact}	$D_{simulation}$	D_{FST} **(best estimate)**
Percolation ($d = 2$)			
Percolating cluster:			
square lattice	$91/48 = 1.8958$		1.8914 (test for
triangular lattice			1.8710 universality)
Backbone		1.647 (0.001)	1.60
Chemical distance		1.17 (0.01)	1.17
Ising and Potts Models ($d = 2$)			
Clusters:			
$q = 2$ (Ising)	$187/96 = 1.9479$		1.9570
$q = 3$	$153/80 = 1.9125$		1.9181
$q = 4$	$15/8 = 1.875$		1.8841
Droplets:			
$q = 2$ (Ising)	$15/8 = 1.875$		1.8414
$q = 3$	$28/15 = 1.8666$		1.7817(*)
$q = 4$	$15/8 = 1.875$		1.7525(*)
Self Avoiding Polymer ($\nu = 1/D$)			
$d = 2$	0.75		0.745
$d = 3$		0.58 (0.01)	0.592
Lattice Animals ($\nu = 1/D$)			
$d = 2$		0.641 (0.001)	0.644

(*) = simplified scheme

In Table 3, we report the application of more and more refined FST schemes to DLA and DBM in two dimensions. It should be noted that for these problems, the simulations give slightly different results depending on the method of analysis and on the geometry of the structures, radial or cylindrical.[14] Possibly these problems are due to the anisotropy and eventual self-affinity of these structures and it would be desirable to clarify these points with more sophisticated methods of analysis. In any case, these problems should not affect the dimension of the intersection of the structure with a line perpendicular to the growth direction, i.e., the property directly addressed by the FST approach. In case of anisotropy and self-affinity, the problem is then how to relate the dimension of the

Table 2 **Fractal Dimensions for Simple Dynamical Models Computed with the FST method**

	Simple Dynamical Models		
Model	D_{exact}	$D_{simulation}$ or Series	D_{FST} (best estimate)
Directed Percolation ($d = 2$)			
Transverse dimension	$591/790 = 0.7481$ (conjectured)	0.748	0.7466
Invasion Percolation ($d = 2$)	$91/48 = 1.8958$ (conjectured)		1.8872
I.P. with trapping		1.82	1.8196
Self Organized Criticality ($d = 2$)			
Bulk properties of clusters		2.0 (0.1)	1.9936

intersection to that of the whole structure. At the moment, we have treated this problem in the simplest way but more refined methods can be developed if necessary.

The first set of the FST results refer to the simple version in which the growth rules are not renormalized but have already the correct symmetry with the probability defined per bond (DBM). The last result refers, instead, to the FST with renormalized growth rules ($S^* = 2.4$). The result $D = 1.71$ should be considered as an upper limit because in this last calculation the effect of empty configurations has not been included. In addition, the open-closed approximation scheme to treat the fluctuations of boundary conditions corresponds probably to an upper limit for D.

In Table 4, we report the FST results for most of the irreversible fractal growth problems that have been studied in detail by extensive simulation. It is especially interesting to note that in the case of Cluster-Cluster aggregation the application of standard theoretical concepts like field theory or RSRG is not even conceivable, while the FST approach can be extended quite naturally to these problems.[15]

4. CONCLUSION AND PERSPECTIVES

The FST method is a theoretical framework especially suitable to treat irreversible fractal growth models.[5] The basic idea is to split the $t \to \infty$ limit (freezing condition) and the $b \to \infty$ limit (scale invariant dynamics). By focusing on a "fixed scale", one studies the fixed point with respect to the dynamical evolution of the system. This leads to an accurate calculation of the nearest neighbors correlations at this scale. If one uses scale invariant dynamics, these correlations become the same for coarse-grained variables of any size and

Table 3 Detailed FST analysis for DLA/DBM in two dimensions. FST Schemes of Increasing Complexity are considered. The final result $D = 1.71$ should be considered as an upper limit for the reasons discussed in the text

DLA/DBM in 2d: FST Detailed Analysis	
Method	**Fractal Dimension**
Computer Simulations	
Circular geometry:	
mass-length relation	1.715
density-density correlation	1.67 (0.01)
box counting	1.60–1.65
Cylinder Geometry:	
Box counting	1.60–1.65
FST with Small Scale Growth Dynamics:	
Closed boundary conditions	1.55 (lower bound)
Open-closed scheme	1.6406 (reference value for further improvements)
Inclusion of empty configurations	1.6242 (a)
Scale Invariant Dynamics (Spontaneous proliferation with respect to the renormalization of noise reduction S)	Self-Organization: Attractive Fixed Point Fixed point values: $S^* = 2.4$ (The corresponding growth probabilities are close to the those of the small scale dynamics ($S = 1$))
FST with Scale Invariant Dynamics ($S = 2.4$)	
Open-closed scheme	1.71 (a)

(a): should be considered with respect to the open-closed scheme.

correspond to the generator of a box covering process from which the fractal dimension can be computed. The scheme is rather accurate and flexible as shown by the numerous applications discussed.

The next step would be to try to apply these concepts to different problems like the renormalization of the sand pile dynamics[16] in order to obtain the exponents that characterize the mass distribution and the temporal evolution. Applications to simplified cellular automata models of turbulent behavior should also be possible.

Table 4 FST Results for the most studied Irreversible Fractal Growth Models. Note that the problem of Cluster-Cluster aggregation cannot even be formulated within the standard theoretical framework while the FST generalization to this model is straightforward

Fractal Growth Models		
Model	$D_{simulation}$	D_{FST} (Best Estimate)
Eden Model ($d = 2$) (bulk dimension)	2.0 (0.1)	2.0 (extrapolation, slow convergency
DLA/DBM ($d = 2$)		
Cluster:		
DLA (DBM, Eta = 1)	1.60–1.71 (for details see Table 3)	1.64–1.71 (for details see Table 3)
DBM, Eta = 2	1.40–1.45	1.43
DBM, Eta = 0.5	1.85–1.90	1.78 (slow convergency, close to Eden limit)
Backbone	1.00 (0.05)	1.04
DLA/DBM ($d = 3$)		
Cluster:		
DLA (DBM, Eta = 1)	2.5 (0.1)	2.49
DBM, Eta = 2	2.13 (0.1)	2.17
DBM, Eta = 3	1.89 (0.1)	1.91
Cluster-Cluster Aggregation		
$d = 2$	1.45 (0.05)	1.39
$d = 3$	1.80 (0.05)	1.90
Fracture Model ($d = 2$)	1.52 (0.04)	1.560

REFERENCES

1. B. B. Mandelbrot, *The Fractal Geometry of Nature* (Freeman, San Francisco, 1982).
2. T. A. Witten and L. Sander, *Phys. Rev. Lett.* **47**, 1400 (1981).
3. L. Niemeyer, L. Pietronero and H. J. Wiesmann, *Phys. Rev. Lett.* **52**, 1038 (1984).
4. T. Vicsek, *Fractal Growth Phenomena* (World Scientific, Singapore, 1992).
5. A. Erzan, L. Pietronero and A. Vespignani: "The Fixed Scale Transformation Approach to Fractal Growth," submitted to *Rev. Mod. Phys.* (1993).
6. T. Nagatani, *J. Phys.* **A20**, L381 (1987); *Phys. Rev.* **A36**, 5812 (1987).
7. L. Pietronero, A. Erzan and C. Evertsz, *Phys. Rev. Lett.* **61**, 861 (1988); *Physica* **A151**, 207 (1988).
8. R. De Angelis, M. Marsili, L. Pietronero, A. Vespignani and H. J. Wiesmann, *Europhys. Lett.* **16**, 417 (1991); see also A. Vespignani and L. Pietronero, this volume pp. 1002–1007.
9. P. W. Barker and R. C. Ball, *Phys. Rev.* **A42**, 6289 (1990).
10. R. Cafiero, L. Pietronero and A. Vespignani, *Phys. Rev. Lett.* **70**, 3939 (1993).

11. L. Pietronero and A. Stella, *Physica* **A170**, 64 (1990).
12. A. Erzan and L. Pietronero, *J. Phys.* **A24**, 1875 (1991).
13. C. Vanderzande, *Physica* **A185**, 235 (1992).
14. L. Pietronero, in *Growth and Form, Nonlinear Aspects*, eds. M. Ben Amar, P. Pelcé and P. Tabeling (NATO ASI Ser. B, Vol. 276, Plenum, New York, 1991), p. 345.
15. L. Pietronero, S. Sidoretti and A. Vespignani, *J. of Fractals*, **1**, 41 (1993).
16. P. Bak, C. Tang and K. Wiesenfeld, *Phys. Rev. Lett.* **59**, 381 (1987).

Fractals, Vol. 1, No. 3 (1993) 671–679

ENTROPIC AND MULTIFRACTAL ANALYSIS OF DISORDERED MORPHOLOGIES

A. BEGHDADI
L.P.M.T.M., C.N.R.S., Institut Galilée, Université Paris Nord, 93430 Villetaneuse, France

C. ANDRAUD, J. LAFAIT and J. PEIRO
Laboratoire d'Optique des Solides, Université Pierre et Marie Curie, C.N.R.S., 4, Place Jussieu, 75252 Paris Cedex 05, France

M. PERREAU
L.E.M., C.N.R.S./O.N.E.R.A., BP 102, 92322 Châtillon Cedex, France

Abstract

We propose the configuration entropy as an efficient tool of characterization of the disorder of random morphologies and as a pertinent morphological parameter for describing the optical properties. When increasing the size of observation of an image, it undergoes a maximum at a characteristic length which is the optimum length at which the image must be observed to get the maximum information. When applied to computer simulated images, the configuration entropy is more powerful, less ambiguous and less sensitive to the finite size of images than the generalized fractal dimension.

1. INTRODUCTION

Morphology governs all the physical properties of heterogeneous materials. This is particularly the case for the optical properties of granular metal films.[1] The problem of relating quantitatively the morphology and the optical properties under the form of parameters in a model is not yet solved. Some attempts were made with success to explain the characteristic extra absorption occurring around the percolation in these granular metals.[2,3]

They use characteristic parameters involved in scaling laws at percolation: metal fraction p, correlation length ξ, critical exponents, fractal dimension D_F of the infinite cluster, and the fractal dimension of its backbone which can be extracted from a statistical geometrical analysis of the morphology.

We have, however, some examples of granular metal layers deposited under different conditions (rate of deposition, temperature of the substrate, ...) which exhibit notably different optical properties, although having the same metal fraction, and fractal dimension and a flat multifractal spectrum. These parameters: metal fraction, fractal dimension (and doubtless multifractal spectrum) are thus not enough to describe the optical properties. This is the reason why we have explored another way of description of the disorder related to thermodynamical concepts and theory of information: the configuration entropy.[4]

The entropy of a physical system has often been described as a measure of randomness in the structure of the system. Moreover, the entropy measures the lack of information[5] or the degree of uncertainty in the realization of an event in a given experiment. Since the pioneer work of Shannon[6] in information theory, the entropy concept is widely used in many fields.[7–9]

2. CONFIGURATION ENTROPY

Shannon defined the entropy of a system as a function of the probability of occurrence of different states of the system. If a system has n different possible states with probability of occurrence p_i, $i = 1, 2, \ldots, n$, then the gain in information from the occurrence of the event i is defined as:

$$I_i = \ln \frac{1}{p_i} \tag{1}$$

The expected value of such a gain in information is defined as the entropy H of the system:

$$H = \sum_{i=1}^{n} p_i I_i \qquad \text{or} \qquad H = -\sum_{i=1}^{n} p_i \ln p_i \tag{2}$$

It can be noted that since H is a function of the probability p_i associated with a given source of symbols or a random variable, one can derive an entropy for each defined source. In the present study, the system is a binary image representing a two-phase material and the elementary source symbol is the pixel intensity which has two possible values 1 or 0 ("black" and "white" representation).

The image is analyzed through an elementary sliding cell of size $l \times l$. For a given position of the cell C_{ij}, we count the number of active pixels (pixels having value 1) contained in C_{ij}. Once the whole image is examined, we compute $N_k(l)$: number of cells containing k active pixels, and $N(l)$: total number of visited cells. Thus, the probability associated with the state of the cell is defined as follows:

$$P_k(l) = \frac{N_k(l)}{N(l)} \tag{3}$$

The corresponding entropy is:

$$H(l) = -\sum_{k=0}^{l^2} p_k(l) \ln p_k(l) \tag{4}$$

Obviously, H is a measure of the degree of uncertainty in the realization of a given state of the cell of size $l \times l$. Using the Lagrange multiplier method, one can easily show that $H(l)$ is maximized under the constraint:

$$\sum_{k=0}^{l^2} p_k(l) = 1 \tag{5}$$

when the different cell configurations are equally probable, i.e., when

$$p_k(l) = \frac{1}{l^2+1} \tag{6}$$

for every k. To examine the image structure at different length scale, the cell size is increased from $l = 1$ to $l = L/2$. To compare the entropy values obtained for different l values one has to normalise $H(l)$. We thus compute for each l value the normalized entropy defined by:

$$H^*(l) = \frac{H(l)}{H_{\max}(l)} \tag{7}$$

where

$$H_{\max}(l) = \ln(l^2+1) \tag{8}$$

$H^*(l)$ undergoes a maximum at a characteristic size l^* that we call the entropy optimum length. The box of size $l^* \times l^*$ presents a maximum of disorder and is the optimum area at which the image must be observed and eventually characterized with the aim of optical studies. We have also proved that the configuration entropy increases with the cluster ramification and the system lacunarity.

3. FRACTALITY, MULTIFRACTALITY AND ENTROPY

It has been shown by many authors that many physical systems consisting of objects having complex shape can be described in terms of fractal geometry.[10–14] Since the pioneer works of Mandelbrot, the characterization of random or statistical fractals has been increasingly developed and refined with the introduction of multifractal formalism.[15–19]

The concept of multifractal is related to the properties of the distribution of the mass associated with a measure defined on the object.[17–20] The object can be regarded in this approach as a family of different homogeneous fractal sets on which the measure has a given singularity. To analyze the multifractality of a given object, the support of the measure is covered with boxes of size $l \times l$ and the probability $p_i(l)$, which is the integrated measure, is computed in each box. The generalized dimension defined by:

$$D_q = \frac{1}{q-1} \lim_{1 \to 0} \frac{\ln \sum_i p_i^q(l)}{\ln l} \tag{9}$$

characterizes the non-uniformity of the measure.[19]

In practice, Eq. (9) cannot be directly used to compute the generalized dimension. Indeed, the exact value of the generalized dimension is only obtained when the box size tends to zero. Vicsek et al.[20–22] proposed practical methods to compute the generalized dimension. They must assume the box size to be much greater than the smallest elementary cell in the system and much smaller than the system size. They have thus shown that the standard methods for determining fractal dimensions have to be applied with some caution. The box counting method is only efficient when the cluster size is large enough and in the sand box method the results depend on the selected centers of the analysis boxes.

Recently, Blacher et al.[23] used the approach of Chhabra et al.[19] to compute the Hausdorff dimension of simulated clusters. Chhabra et al. define a family of normalized measures as follows:

$$\mu_i(q, l) = \frac{(p_i(l))^q}{\sum_j (p_j(l))^q} \tag{10}$$

Thus, the Hausdorff dimension is given by:

$$f(q) = \lim_{l \to 0} \frac{\sum_i \mu_i(q, l) \ln \mu_i(q, l)}{\ln l} \tag{11}$$

It can be noted that for $q = 1$, the numerator is a quantity similar to the one used in our approach [Eq. (4)], namely, the configuration entropy. For this reason, the corresponding dimension D_1 is called the information dimension. In our approach, the maximum value of entropy allows the measurement of the optimum box size corresponding to the highest degree of disorder in the system under observation. In the multifractal analysis, the maximum value of $f(q)$, for $q = 0$, gives the Hausdorff dimension of the object of the measure, namely D_0.

The main difficulty in using the multifractal formalism lies in the fact that the ideal limit $l = 0$ cannot be reached in practice. Furthermore, one usually needs only a very limited subset of the spectrum $f(q)$ in order to describe the important properties of the associated measure. Only the cases $q = 0$, $q = 1$ and $q = 2$ could be unambiguously relied on to make a physical interpretation of the system. Merits and limitations of the multifractal analysis have been discussed by Aharony.[24] In the same way, Chhabra et al.[19] pointed out the risks in the estimation of the Hausdorff dimension and cited different sources of errors in some physical cases. The difficulties arising in practice are due to the fact that the relevant quantities used in the multifractal concept are estimated asymptotically. In our approach, however, there is no ambiguity in the computation of our relevant quantities. Moreover, the measure of the correlation length ξ marking the limit of the scaling behavior is determined with poor accuracy using the fractal formalism.[25] On the other hand, the optimum length characterizing the maximum of disorder is unambiguously determined with our entropic approach.

The main purpose of the present study is not to compare in detail our approach to the fractal and multifractal analysis but to propose another powerful tool to measure the disorder in systems having a complex structure, and to show that, in a usual case, this tool is efficient and less ambiguous than these analysis. Furthermore, the characteristic length defined through the entropy concept could be used for optical characterization of two-phase materials.

4. APPLICATION TO RANDOM MORPHOLOGIES

We now demonstrate the efficiency of the configuration entropy as compared to the fractal dimension in two limit cases:

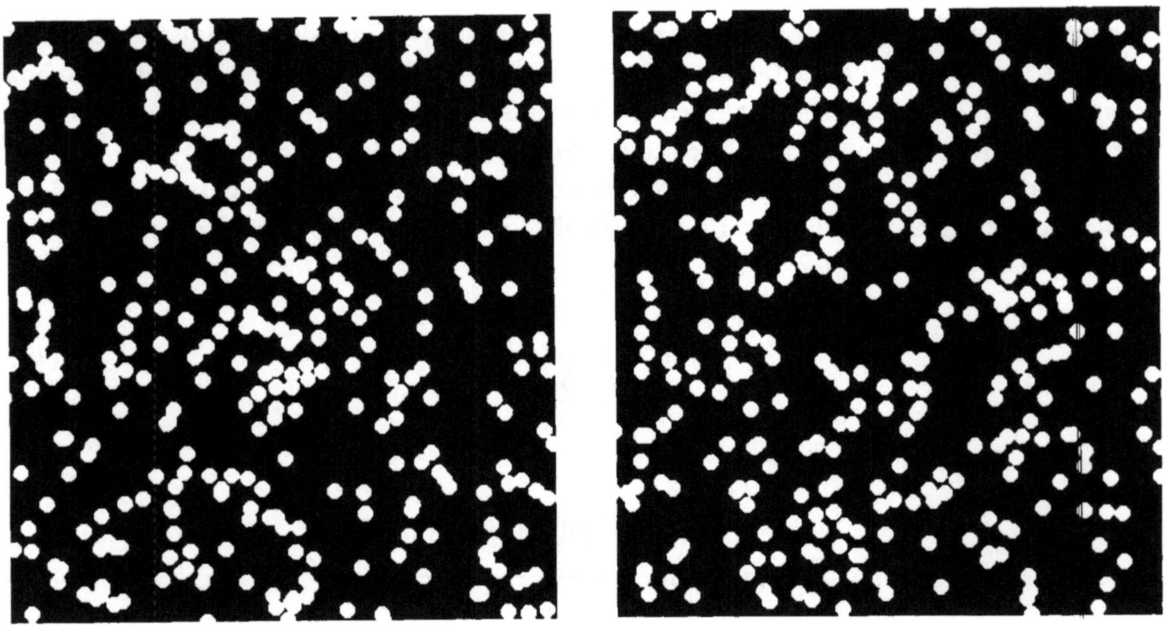

(a) (b)

Fig. 1 Computer simulated random morphologies obtained by random throws of overlapping disks of size 11 pixels: **(a)** First throw; **(b)** Second throw. (Image size 400 * 400 pixels.)

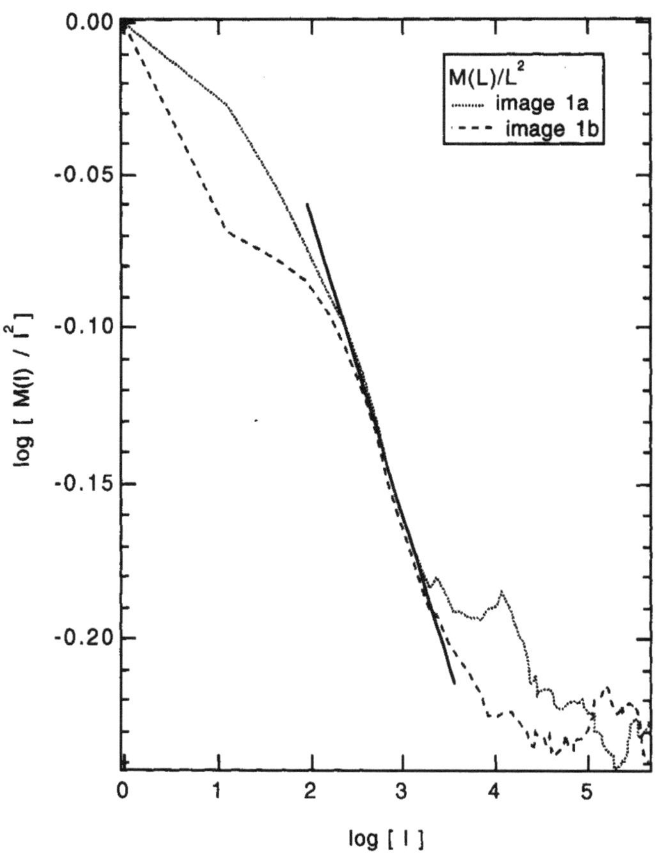

Fig. 2 Log-log plot of the black mass contained in a square of size l in images 1(a) and 1(b) (sand box method). The linear regression in the scaling domain is indicated. Slope ≈ -0.1; $D_F \approx 1.9$.

(1) Two very close images of the same size (400*400) composed of penetrating disks of the same size and the same concentration 0.20. The images differ only in the fact that they have been obtained after two different sequences of random trials. They can hardly be differentiated by visual observation [Figs. 1(a) and 1(b)]. The fractal dimension of the complementary set of the disks in the image is determined by the classical sand box method.[20,21] The scaling region clearly shows up in Fig. 2. The fractal dimension D_F thus deduced is 1.90 for both images. The finite size of the images and the lacunarity reduce the accuracy of the linear regression in the scaling region. For the same reasons, the average behavior following the scaling region merges into fluctuations. It is thus impossible to correctly evaluate the correlation length marking the transition between these two regimes. We then used the sand box method for the calculation of the generalized fractal dimension of order q: D_q, related to the multifractal spectrum via Legendre transform.[20,21] Both images do not present any multifractality, the straight line characterizing the different orders being indiscernible. The classical geometrical parameters of both images are thus equal.

The configuration entropy as calculated by our method presented above, exhibits for its parts some differences when applied to these images (Fig. 3). First, the two entropy curves are clearly discernible; second, both curves undergo a maximum $H^*_{\max}$ for an optimum length l_0. The small differences in these values, as shown in Fig. 3, are nevertheless significant

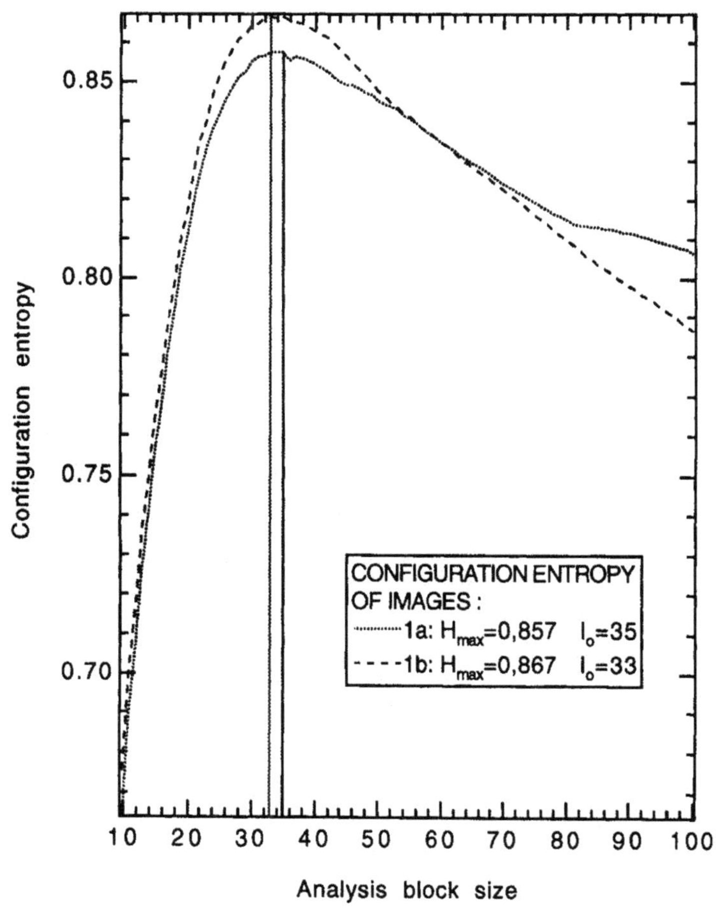

Fig. 3 Configuration entropy of images 1(a) and 1(b). Entropy maxima and related optimal lengths are indicated.

because of the high accuracy of their determination. It can be concluded that, in this limit case of quasi similar images, the configuration entropy succeeds in emphasizing an undeniable difference whereas other classical tools fail.

(2) We now envisage the analytical generation of two images for which one knows exactly the fractal dimension. These images are square Sierpinski carpets from which some pixels have been randomly removed at each step of iteration. n, being the size of the starting pattern, p, the number of blocks to remain in this starting pattern and k, the number of iterations: the size of the image is $n^{k*}n^k$ and its fractal dimension is $D_F = \log(p)/\log(n)$. Both images have the same concentration: 0.23 and their fractal dimension is: $D_{Fa} = 1.43$; $D_{Fb} = 1.73$.

Because of the low number of iterations, it is impossible to determine the fractal dimension by the sand box method. Nevertheless, the configuration entropy gives non-ambiguous results as shown in Figs. 4(a) and 4(b).

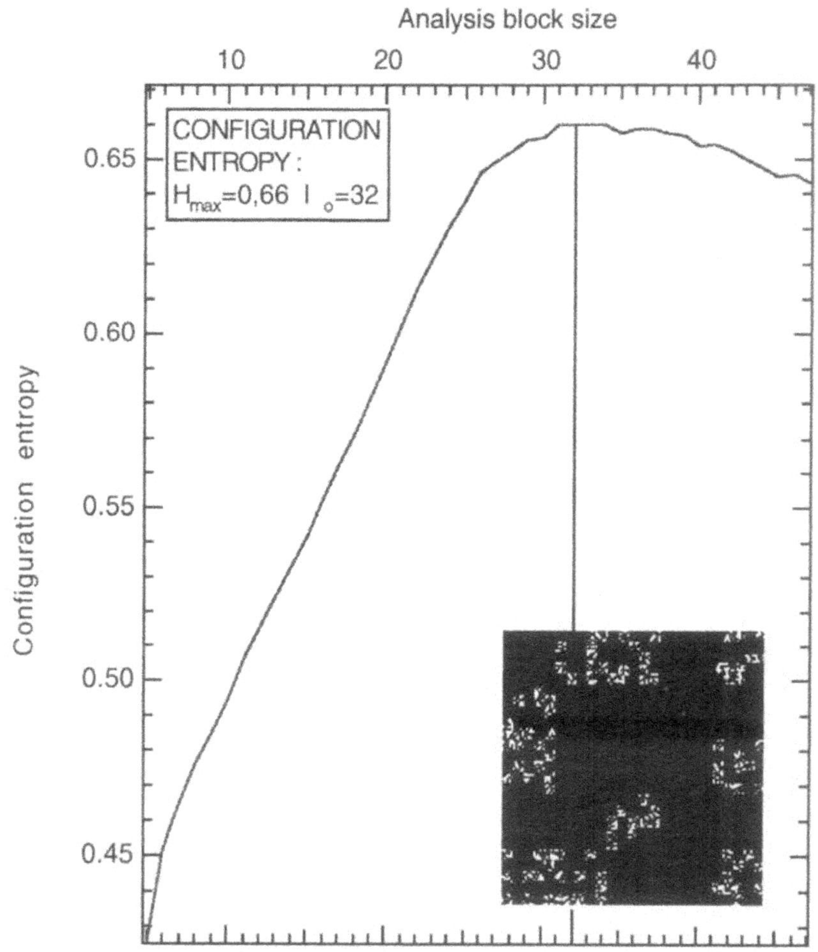

(a)

Fig. 4 Computer simulated morphologies obtained by randomly removing sites on a square Sierpinski carpet with starting pattern size n, number of remaining sites in this pattern p, number of iterations k, and the corresponding entropy.

(a) $n = 5$ $p = 10$ $k = 3$ size $= 125 * 125$ pixels $D_{Fa} = 1.43$

(b) $n = 6$ $p = 22$ $k = 3$ size $= 216 * 216$ pixels $D_{Fb} = 1.73$

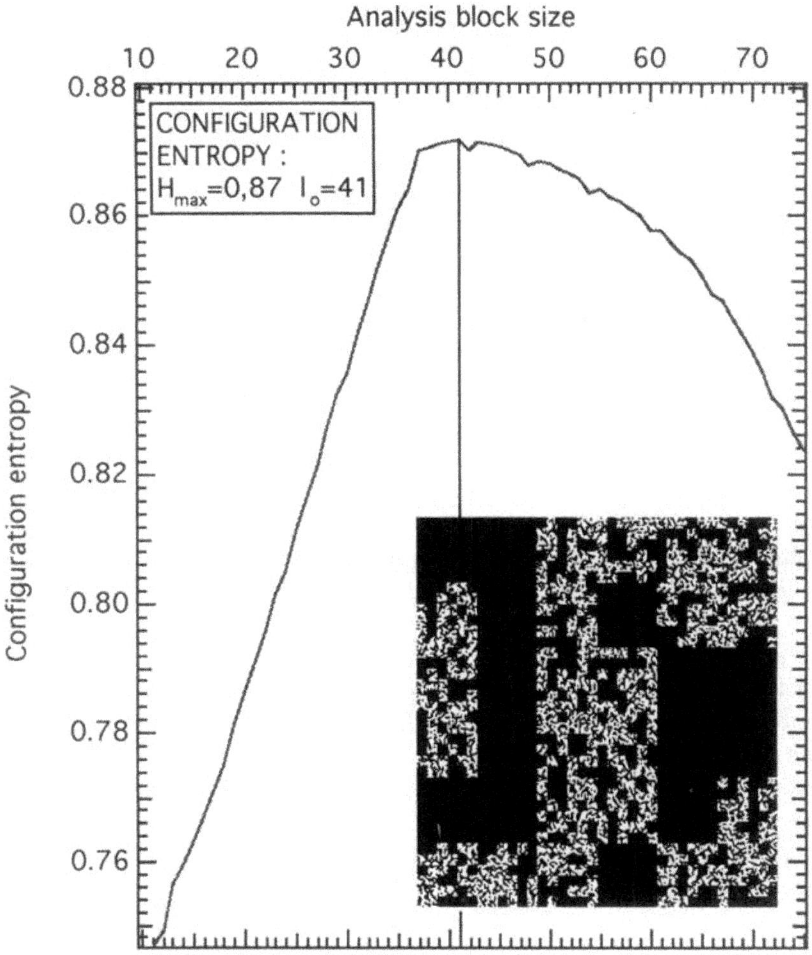

(b)

Fig. 4 (*Continued*)

In these two extreme examples, the classical geometrical parameters are either indiscernible or cannot be determined geometrically, whereas the configuration entropy is always unambiguous: well defined distinct maxima and of different optimum lengths.

5. CONCLUSION

We have shown in these limit cases, that the configuration entropy is a powerful tool well adapted to the characterization of disorder and able to separate out very close morphologies when fractal and multifractal analysis fails. In any case, this tool never gives ambiguous results and is less sensitive to the finite size of images, in contrast to fractal and multifractal analysis. The next but difficult step will be to incorporate this parameter in a quantitative model of the optical properties of these random morphologies.

REFERENCES

1. P. Gadenne, A. Beghdadi and J. Lafait, *Optics Commun.* **65**, 17 (1988).
2. Y. Yagil, M. Yosefin, D. J. Bergman, G. Deutscher and P. Gadenne, *Phys. Rev.* **B43**, 1342 (1991).

3. T. Robin and B. Souillard, *Physica* **A193**, 79 (1993).
4. H. O. Peitgen and D. Saupe, *The Science of Fractal Images* (Berlin, 1988).
5. L. Brillouin, *Science an Information Theory* (Academic Press Inc., New York, 1956).
6. C. E. Shannon, *Bell. Syst. Tech. J.* **210**, 623 (1948).
7. B. R. Frieden, *J. Opt. Soc. Amer.* **62**, 511 (1972).
8. S. F. Gull and T. J. Newton, *Applied Optics* **25**, 156 (1986).
9. A. Mohammad-Djafari and G. Demoment, *Applied Optics* **26**, 1745 (1987).
10. B. B. Mandelbrot, *The Fractal Geometry of Nature* (W. H. Freeman, New York, 1982).
11. R. F. Voss, R. B. Laibowitz and E. I. Allessandrini, *Phys. Rev. Lett.* **49**(19), 1441 (1982).
12. R. Orbach, *Science* **231**, 814 (1986).
13. J. F. Gouyet, *Physique et structures fractales* (Masson, Paris, 1992).
14. L. Pietronero, *Physica* **A144**, 257 (1987).
15. B. B. Mandelbrot, *J. Fluid. Mech.* **62**, 331 (1974).
16. G. Paladin and A. Vulpiani, *Phys. Rep.* **156**, 147 (1987).
17. M. H. Jensen, L. P. Kadanoff and A. Libchaber, *Phys. Rev. Lett.* **55**, 2798 (1985).
18. T. C. Halsey, M. H. Jensen, L. P. Kadanoff, I. Procaccia and B. I. Shraiman, *Phys. Rev.* **A33**, 1141 (1986).
19. A. B. Chhabra, C. Meneveau, R. V. Jensen and K. R. Sreenivasan, *Phys. Rev.* **A40**, 5284 (1989).
20. T. Vicsek, *Fractal Growth Phenomena*, 2nd ed. (World Scientific Publishing Co, Singapore, 1992).
21. T. Vicsek, *Physica* **A168**, 490 (1990).
22. T. Tel, A. Fülöp and T. Vicsek, *Physica* **A159**, 155 (1989).
23. S. Blacher, F. Brouers and R. Van Dyck, to appear in *Physica A* (1993).
24. A. Aharony, *Physica* **A168**, 479 (1990).
25. A. Beghdadi, J. Lafait, P. Gadenne, A. Constans and O. Bouet, *Acta Stereol* **6/III**, 809 (1987).

Physics

INTRODUCTION

When Benoit Mandelbrot discovered the generality of fractal geometry in nature, he suggested physics as only one of the fields where his concepts should find applications. There are several reasons why fractals became popular first among physicists. Beyond physicists' usual sensibility to general, universal concepts, the idea of scaling has made in the seventies a victory march through the whole condensed matter physics due to the renormalization group theory. The time was just ripe for fractals in physics. The first examples were in fact taken from the so-called geometric phase transitions like percolation or polymer models. During this period some skeptics (including myself) thought that fractals are just another name of the game (looking for exponents), maybe with a bit more emphasis on the geometric aspect.

The sign, that something more is going on, came for many of us through Witten and Sander's model of diffusion limited aggregation. This was obviously an invention triggered by the fractal concept. Intensive studies of growth phenomena followed and by now much information has accumulated in this field. Most of the results are related to computers which have gone through an enormous development during the last decades. The previously unimaginable memory, speed and graphics capacities have certainly contributed to the boom in the field of fractals. Computational Physics, the new and gradually equally important branch besides experimental and theoretical physics, has grown simultaneously with the knowledge about fractals.

Due to the often beautiful experimental and computer generated patterns, the new aspect of fractal aesthetics emerged in addition to the purely scientific one. Some consider this as a gain, others are suspicious about it. I think that in one way or another aesthetics has always been present in science.

It was at the time when multifractality became popular — much later after being discovered by Mandelbrot — that most physicists realized the impact of the fractal approach. By now, I believe, it has become evident that this concept means much more than an appropriate language to describe similar features in diverse branches of natural sciences (and possibly beyond). The fact that it has helped us to get insight into connections between seemingly totally different phenomena like turbulence and aggregation, or rough surfaces and time dependent fluctuations, would already justify the success fractals. But this concept also generated new fields of research which lead to important results. The reader may find some of them in this section of the present book.

A few papers in this volume are devoted to such classic fractal problems as percolation or to the discussion of vibrational properties of fractal networks. We have a coherent picture of these phenomena which, however, have to be refined again and again.

The papers in this section deal, to a considerable extent, with interesting new experiments as well as new theoretical approaches to kinetic surface roughening. This is an example where the interest in fractals motivated research has finally led to a deeper understanding of the physics in different areas such as fluid displacement, epitaxial growth or even the motion of magnetic flux lines. The scenario described by the so-called Kardar-Parisi-Zhang equation seems to be only one of the basic possibilities. The studies done in recent years revealed the importance of the form of the noise on the kinetics and the morphology of rough surfaces.

Another type of growth phenomenon is unstable fractal pattern growth including Laplacian and related growth problems. Here much is known but almost nothing has been solved. It seems to be typical in modern science that researchers leave unsolved problems and look for new fields where possibly faster success can be achieved. This unfortunate development, possibly related to the funding system, is perhaps responsible for the underrepresentation of these basic fractal fields in this volume. On the other hand, it is nice to see how fast the fractal concept spreads and reaches such terrains like fracture formation, a field of extreme technological importance. It is already a significant contribution that fractality adds new points of view to the classification of fracture patterns — it would be a real breakthrough if it contributed to the understanding of some of the mechanisms.

Some years ago, Per Bak and his collaborators asked the simple and very physical question: What is the origin of the general occurrence of the (spatial and temporal) scale invariance in nature. They argued that a general tendency of slowly driven non-equilibrium systems, manifested in what they called "Self-Organized Criticality" (SOC), is responsible for this phenomenon and they suggested simple "sand pile" models to demonstrate their concepts.

As usual, the problem posed turned out to be more complicated than at the first sight, and the investigation of SOC has become an interesting part of fractal research.

The concept of SOC not only stimulated intensive studies of diverse cellular automata including traffic models, it is also at least in part "responsible" for the increasing interest in the physics of granular media. Sand is an extremely interesting object for a statistical physicist! It is a really complex system with many particles, each having many degrees of freedom. No wonder that interesting scaling behavior has been found in these systems ....

This book contains papers about percolation, pattern growth, rough surfaces, self-organized criticality, traffic models, granular and porous media, crack propagation and astrophysical application of fractals. However, some of the areas of active research are not included, such as dynamical systems, turbulence, shape of wave functions, etc. These lists already indicate that the pioneers' period in fractals is over. The fractal concept has become an established part of our knowledge, it added to our understanding of nature and it provides us with a tool for approaching new phenomena.

by **J. Kertész**

Fractals, Vol. 1, No. 4 (1993) 745–752

LARGE SCALE STRUCTURE OF INTERFACES: AN INVERSE METHOD

CHI-HANG LAM
Applied Physics Department, Yale University,
Box 2155 Yale Station, New Haven, CT 06520, USA

LEONARD M. SANDER
H.M. Randall Laboratory of Physics, The University of Michigan,
Ann Arbor, MI 48109-1120, USA

Abstract

We propose an inverse method to extract effective couplings and the renormalization group flow for growing interfaces. We apply it to discrete surface growth models in the Kardar-Parisi-Zhang universality class in $1+1$ dimensions and obtain the first measurement of a universal coupling constant. We consider interfaces not in the steady state. It may also be applicable to analyze experimental data and for other forms of interface growth.

1. INTRODUCTION

The growth of disorderly interfaces far from equilibrium has been intensely studied in recent years, and a number of remarkable advances have been made.[1,2] Examples include the Edward-Wilkinson (EW) equation describing sedimentation,[3] the Kardar-Parisi-Zhang (KPZ) equation describing vapor deposition and directed polymers,[4] and the diffusion-limited aggregation model.[5] In all of these cases the interface is specified by a set of kinetic rules. It is driven by noise to a steady state which has scaling features, and these features are seldom obvious from the microscopics of the rules. The most spectacular example of this is DLA where the large-scale structure appears to be at least approximately fractal, and no theory is known to predict this fact from the simple model.

A particularly interesting example is the Kuramoto-Svinshinsky equation of flame-front theory.[6] In this case, the interface is in fact specified by a deterministic equation which has an instability leading to cellular structures on the interface. It has recently been shown that on large scales, the correlations of the interface act exactly as if they were governed by the noise-driven KPZ equation. Again, there is scaling structure on these scales which is not in the least obvious from short-range considerations.

We have proposed[8] an inverse method to find essentially complete information about interface growth processes and to explicitly reveal the scaling. It closely resembles the spirit of Ma's Monte Carlo renormalization group, which computes numerically the effective block spin Hamiltonian from equilibrium spin configurations.[9] Our method can be considered as the dynamical and continuum counterpart and is applicable to both equilibrium and nonequilibrium systems. We can calculate directly the evolution equation for a given surface from simulation or possibly experimental data. The renormalization group flow of the equation can be obtained with *all* its parameters computed. We will review the method in this paper and give a new application to interfaces not in the steady state.

The method is as follows[8]: A time sequence of snapshots of a surface captures its dynamical properties up to the corresponding resolution. It is usually not difficult to guess a general candidate equation for the evolution containing all the possible terms consistent with symmetry. Our task is to discriminate among them. Given a snapshot of the surface, realizations of the equation with different coefficients give different predictions about the next snapshot. The correct equation is the one which, on the average, gives the best predictions over a large number of trials. It can thus be obtained by minimizing the errors of the predictions. Any terms with zero coefficients are proved to be absent.

We illustrated the method by working through applications to growth models in the KPZ universality class in 1+1 dimensions. The height-height correlation function of these surfaces in the steady state follows the scaling form: $\langle |h(x,t) - h(0,t')|^2\rangle \sim |x|^{2\alpha} f[|t-t'|/|x|^z]$. The roughness exponent α and the dynamical exponent z characterize the universality class and are thought to be the same for many microscopic models. Though these examples have been extensively studied previously, this novel technique is powerful enough to yield entirely new and important results as well as reproduce the known ones. We briefly review the method here.

The evolution of the surfaces we studied is known to be given by the KPZ equation[4]:

$$\frac{\partial h}{\partial t} = c + \nu \nabla^2 h + \frac{\lambda}{2}(\nabla h)^2 + \eta(x,t) \tag{1}$$

where the noise η has a Gaussian distribution and mean 0 and a correlator given by:

$$\langle \eta(x,t)\eta(x',t')\rangle = 2D\delta(x-x')\delta(t-t') \tag{2}$$

In addition, there is a lower wavelength cutoff, or spatial resolution, l. From dynamical renormalization group (DRG) calculations, for a given surface, ν and D depend on the resolution l of observation and, asymptotically for large l, we have[10,4]:

$$\nu(l) \sim l^\alpha \quad D(l) \sim l^\alpha \tag{3}$$

The parameters λ and $D(l)/\nu(l)$ are independent of l due to symmetry[11] and a fluctuation dissipation theorem[10] respectively. Both of them have been measured numerically[12–14] and, for some particular models, can be obtained analytically.[13,15]

We take the candidate evolution equation to be in a generalized form:

$$\frac{\partial h}{\partial t} = c + \nu\partial^2 h + \frac{\lambda}{2}(\partial h)^2 + a_4\partial^4 h + a_5\partial^2(\partial h)^2 + a_6\partial h\partial^3 h + a_7\partial^6 h + \cdots + \eta \tag{4}$$

where all the differentiations on the right are with respect to x. Discretizing time, we have:

$$\frac{\Delta h(x,\, t)}{\Delta t} \simeq \mathbf{a}\cdot\mathbf{H}(x,\, t) + \eta(x,\, t) \tag{5}$$

where $\mathbf{a} = (c,\, \nu,\, \frac{\lambda}{2},\, a_4, \ldots)$ and $\mathbf{H}(x,\, t) = (1,\, \partial^2 h,\, (\partial h)^2,\, \partial^4 h, \ldots)$ are vectors. We find the parameter vector $\mathbf{a}$ so that the equation best describes the snapshots. To measure $\mathbf{H}(x,\, t)$ numerically, we first simulate a given discrete model to obtain an initial discrete surface $h_d(x,\, t)$. We coarse-grain it by truncating the Fourier components of wavelength smaller than l. From the coarse-grained surface, we get $\mathbf{H}(x,\, t)$ which thus depends on l. The differentiations and multiplications involved are done in the Fourier and real spaces respectively. An additional coarse-graining is then carried out to maintain the cutoff which is shifted by the nonlinearities. We obtain m realizations of $h_d(x,\, t+\Delta t)$ by further growing the *same* surface $h_d(x,\, t)$ m times. Averaging gives the mean discrete profile at time $t + \Delta t$ from which we obtain the coarse-grained mean surface advance $\langle \Delta h(x,\, t)\rangle$, which depends on both the spatial and temporal resolutions l and Δt. For large m, Eq. (5) averages to:

$$\frac{\langle \Delta h(x,\, t)\rangle}{\Delta t} \simeq \mathbf{a}\cdot\mathbf{H}(x,\, t) \tag{6}$$

The fitting can be done most conveniently with discrete data set. We look at N discrete points $(x_i,\, t)$ each separated spatially by l and denote $\mathbf{H}_i = \mathbf{H}(x_i,\, t)$ and $\langle \Delta h_i\rangle = \langle \Delta h(x_i,\, t)\rangle$. To enhance the statistics, the data set can be taken from M realizations of the pair $H(x,\, t)$ and $\langle \Delta h(x,\, t)\rangle$ generated from the corresponding M realizations of $h_d(x,\, t)$. Using a least squares fitting procedure, the mean square deviation of Eq. (6) from equality is:

$$\mathcal{D} = \frac{1}{N}\sum_{i=1}^{N}\left[\frac{\langle \Delta h_i\rangle}{\Delta t} - \mathbf{a}\cdot\mathbf{H}_i\right]^2 \tag{7}$$

Minimizing $\mathcal{D}$ with respect to $\mathbf{a}$ gives the main equation of our inverse method, from which $\mathbf{a}$ can be found:

$$\mathbf{A}\mathbf{a} = \mathbf{b} \tag{8}$$

where the matrix $\mathbf{A}$ and the vector $\mathbf{b}$ are defined by:

$$\mathbf{A} = \frac{1}{N}\sum_{i=1}^{N}\mathbf{H}_i\otimes\mathbf{H}_i \quad \mathbf{b} = \frac{1}{N}\sum_{i=1}^{N}\frac{\langle \Delta h_i\rangle}{\Delta t}\mathbf{H}_i \tag{9}$$

The symbol $\otimes$ denotes the vector outer product. Straightforward algebra gives the noise correlator as:

$$D = \frac{\Delta t l \mathcal{D}'}{2} \tag{10}$$

where $\mathcal{D}'$ is defined like $\mathcal{D}$ in Eq. (7) with $\langle \Delta h_i\rangle$ replaced by an unaveraged Δh_i. We have assumed the δ-correlations in Eq. (2). We have now obtained expressions for all the parameters in the continuum formulation.

We applied the method to discrete models in the KPZ universality class. Two of them are single step models[17] in which the nearest-neighbor height differences of the surface is constrained to be ± 1. A particle of height 2 can be added to the aggregate only if the resulting step heights obey the constraint. The first model (SS I) is the original single step model, where there is one growth attempt at a random site per site per unit time.[17] The other one (SS II) adopts sublattice updating. During each time step, growth is attempted independently with 50% probability at all odd sites and then at all even sites.[15] We also simulated 2 variants of the ballistic deposition model with sublattice updating (BD I and BD II) with growth rules given by $h(x, t+1) = \max\{h(x, t)+\eta(x, t), h(x-1, t), h(x+1, t)\}$ and $h(x, t+1) = \frac{1}{2}[max\{h(x, t), h(x+1, t), h(x-1, t)\} + h(x, t)] + \eta(x, t)$[16] respectively, where the η's are independent uniform random deviates.

For details of the fitting procedure see Ref. 8. The resulting measured model parameters for the SS I model are shown in Fig. 1. The statistical errors are small except noticeably

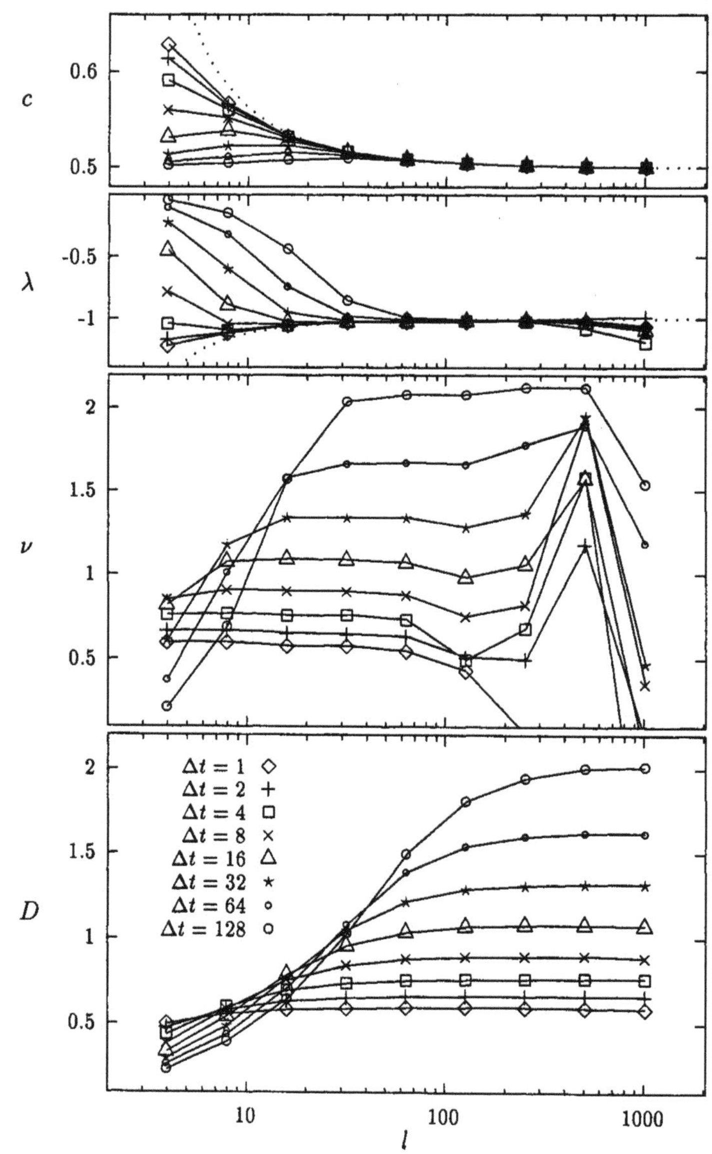

Fig. 1 Measured continuum parameters c, λ, ν and D as functions of the resolutions l and Δt for the SS I model.

Table 1 Numerical Values of λ, ν and D for $\Delta t = 128$, which give D/ν and $\bar{\lambda}$. Bracketed values are exact results

Model	λ	$\nu/\Delta t^{1/3}$	$D/\Delta t^{1/3}$	D/ν	$\bar{\lambda}^2$
SS I	−1.00 (−1.00)	0.413	0.399	0.966 (1.000)	5.66
SS II	−0.712 (−0.707)	0.254	0.236	0.929 (0.971)	5.68
BD I	1.86	0.253	0.0325	0.128	5.30
BD II	1.64	0.265	0.048	0.181	5.45

for ν at large l where the magnitude of this term is small compared with the others. For $l^z \lesssim \Delta t$, where $z = \frac{3}{2}$ in $1+1$ dimensions, all the parameters are affected by the finite time difference approximation. From Fig. 1, both c and λ converge to the expected exact values with the "finite size corrections" $\frac{1}{2}[1-(l-1)^{-1}]^{-1}$ and $-[1-(l-1)^{-1}]^{-1}$ respectively (dotted lines).[18] The parameters ν and D admit renormalization and are much more interesting. Both ν and D converge as l increases and become functions of Δt only. Table 1 lists the numerical values of ν, D and D/ν at $\Delta t = 128$ for large l together with λ. Bracketed values are exact results.[13,15] The errors are estimated to be around 2 to 5%. The values of D/ν measured agree reasonably with the exact results but are systematically smaller because D, which is related to the error in the fitting, is underestimated due to the finite sample size. The agreement supports the δ correlation of the renormalized noise assumed in deriving Eq. (10).

Since the DRG calculations look at the long time behavior,[10] while we have finite observation time Δt, the asymptotic form in Eq. (3) is generalized to:

$$\nu(l, \Delta t) = l^\alpha f_1\left(\frac{\Delta t}{l^z}\right) = \Delta t^\beta f_2\left(\frac{\Delta t}{l^z}\right) \tag{11}$$

and similarly for D, where the early time exponent $\beta = \frac{\alpha}{z} = \frac{1}{3}$. The convergence of ν and D, which manifests itself as $\lim_{y\to 0} f_2(y) \to$ constant, corresponds to the fact that Δt controls the extent of the renormalization for sufficiently large l. This will be explained further later. With increasing Δt, the asymptotic scaling form [Eq. (11)] is expected and is demonstrated through the reasonable data collapse for ν at $\Delta t \gtrsim 32$ for the SS I model in Fig. 2. The result for D is similar. The success of the data collapse demonstrates that our method yields the correct exponents. By examining curves similar to Fig. 2 for different exponents, we can show that we must have $\beta = 0.33 \pm 0.01$ and $z = 1.50 \pm 0.04$.

For models in this class, asymptotic scaling is approached when a dimensionless coupling constant $\bar{\lambda}^2(\Lambda) = \lambda^2 D/\nu^3 \Lambda$ flows to a finite universal fixed point upon renormalization in $1+1$ dimensions, where $\Lambda = \pi/l$ is the upper momentum cutoff. Now the extent of renormalization is controlled by Δt and Λ should be replaced by an effective cutoff $c(\lambda^2 D/\nu)^{-1/3}\Delta t^{-1/z}$, where c is a universal constant and the remaining proportionality constants are from dimensional analysis. This leads to a redefined coupling constant $\bar{\lambda}^2(\Delta t) = [(\lambda^4 D^2/\nu^5)\Delta t]^{2/3}$. The numerical values are listed in Table 1. We obtained $\bar{\lambda}^2(\infty) = 5.5 \pm 0.2$, which is the first numerical estimate of the universal fixed point.

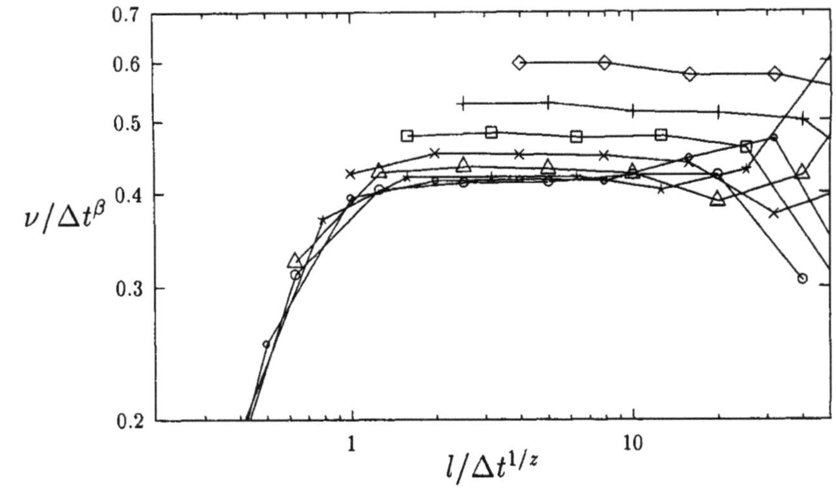

Fig. 2 Data collapse of the values of ν from Fig. 1.

Up until the present, we have been focusing on the evolution of surfaces in the steady state. This is also the regime to which renormalization group techniques were directly applied.[10] The early development of initially flat surfaces is less well understood. Most exact results become invalid. For example, the fluctuation dissipation theorem, which we needed[8] to prove that renormalized parameters should converge for large l, does not hold[10] and the probability functional of the surface is unknown analytically.

We applied the inverse method to unsaturated surfaces generated with the single step model (SS I). The simulation is similar to the previous case. However, instead of fully developed surfaces, $h_d(x,t)$ were prepared by depositing $t = 200$ layers of particles onto an initially flat substrate, which is, more precisely, zig-zag substrates following the ± 1 step height constraint. Their roughness is saturated only locally for segments shorter than the correlation length $\xi \simeq t^{1/z} \simeq 34$, where $z = \frac{3}{2}$. Figure 3 shows the computed parameters. For spatial resolution $l \ll \xi$, the parameters are the same as those in Fig. 1 for saturated surfaces. For $l \gtrsim \xi$, the renormalized coefficients are different in general. This is because, in the renormalization group sense, the short wavelength modes being integrated out do not follow the steady state distribution. They perturb the evolution of the remaining modes differently and contribute different renormalizations.

Since the fluctuation dissipation theorem does not hold, our parameters do not have to converge for large l. As shown in Fig. 3, ν turns negative for large l. The $\nu\nabla^2 h$ term then corresponds to an instability which can be understood as follows. It is known that the nonlinear coupling of the λ term speeds up the roughening of flat surfaces. This is because the r.m.s. surface width W scales with time like $W \sim t^{\beta}$, where the early time exponent $\beta = 1/3$ for the KPZ case is bigger than $\beta = 1/4$ for the Edward-Wilkinson case. After the short wavelength modes are integrated out, their contribution to the roughening of the remaining modes due to the nonlinear coupling have to be represented in the renormalized parameters. This amounts to an instability of the long wavelength modes. The same conclusion can also be obtained from a microscopic point of view. Note that a locally smoother (rougher) segments of the surface grow faster (slower), where we assumed negative λ but the argument is similar for positive λ. Hence the spatial variations in the local roughness are able to induce long wavelength fluctuations. Moreover, a particular pattern

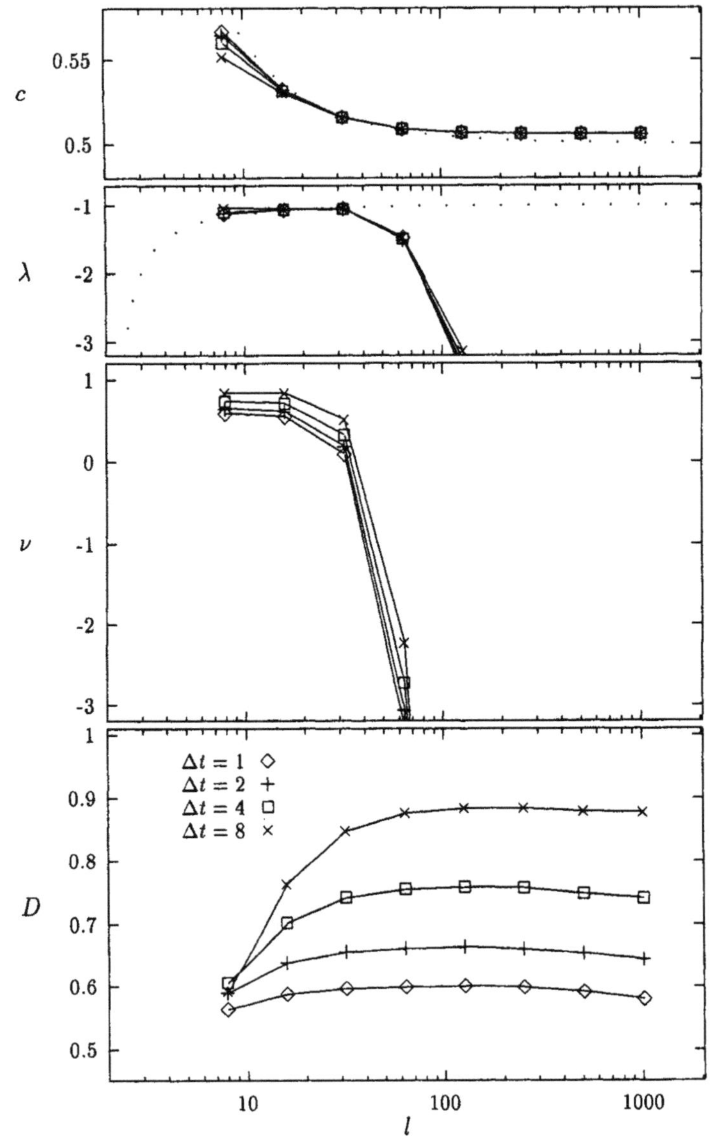

Fig. 3 **Measured continuum parameters c, λ, ν and D as functions of the resolutions l and Δt for the SS I model in the transient regime.**

of roughening persists for the period of time when the variations in the local roughness remain practically unchanged. This persistence allows further advance of already advanced local segments and is precisely an instability.

For the previous steady state case, λ is not renormalized due to a symmetry of the KPZ equation with respect to an infinitesimal tilt.[11] However, since we now consider the early development of a flat surface, the substrate orientation breaks the symmetry. This permits renormalization of λ as is evident in Fig. 3. The averaged growth speed c in Fig. 3 does not converge to the exact value $\frac{1}{2}$ for steady state surfaces. This is because a flatter surface is, on average, smoother and grows faster for the single step model with negative λ. The noise intensity D is indistinguishable in the two cases within our numerical accuracy.

We have demonstrated that when the spatial resolution l is comparable to the correlation length ξ, very different values of the parameters are obtained. In many cases, $l \ll \xi$ is desired since it is when the standard KPZ description holds. Then the preparation of fully roughened surfaces ($\xi = \infty$) is important. In our investigation[8] of $1+1$ dimensional surfaces in the KPZ class, they were generated directly according to the known distribution. In general, the distribution is unknown and deriving efficient algorithms to approximate the steady state surfaces is an important task.

We should note that the effective instability which we tracked in this simple case is rather like that of the Kuramoto-Sivashinsky case where a negative ν is eventually converted to a positive one after the surface develops enough. We think that an application to this case would be of particular interest.

Another case which we intend to study is the DLA model which has been a deep puzzle for a decade. In this case, even the formulation of a mean-field theory has proven to be a major challenge.[19] We can imagine trying a continuum formulation in that case as well, and trying to fit to data from extensive simulations which are now easily done.

In conclusion, our inverse method shows great promise for analysis of interface growth. Previous applications yielded the first computation of *all* the parameters in the KPZ equation at large scales. This is the most direct numerical verification of the applicability of the equation. The finiteness of the universal coupling constant has been proved numerically and its value computed. We have shown here that the interface not in the steady state acts as if it is unstable. The method should have very wide applicability.

ACKNOWLEDGMENTS

We thank D. S. Graff, F. Guinea, D. A. Kessler, D. Ko, O. Pla and R. B. Stinchcombe for interesting discussions. This work is supported by NSF Grant DMR91-17249.

REFERENCES

1. F. Family and T. Vicsek, *Dynamics of Fractal Surfaces* (World Scientific Pub. Co., 1991).
2. L. M. Sander, in *Solids Far From Equilibrium*, ed. C. Godreche (Cambridge, 1992).
3. S. F. Edwards and D. R. Wilkinson, *Proc. R. Soc. Lond.* **A381**, 17 (1982).
4. M. Kardar, G. Parisi and Y.-C. Zhang, *Phys. Rev. Lett.* **56**, 889 (1986).
5. T. A. Witten and L. M. Sander, *Phys. Rev. Lett.* **47**, 1400 (1981); *Phys. Rev.* **B27**, 2586 (1983).
6. Y. Kuramoto and T. Tsuzuki, *Prog. Theo. Phys.* **55**, 356 (1976); G. I. Sivashinsky, *Acta Astronautica* **6**, 569 (1979).
7. V. Yakhot, *Phys. Rev.* **A24**, 642 (1981), S. Zaleski, *Physica* **D34**, 427 (1989), C. Jayaprakash, F. Hayot and R. Pandit, *Phys. Rev. Lett.* **71**, 561 (1993).
8. C.-H. Lam and Leonard M. Sander, *Phys. Rev. Lett.* **71**, 561 (1993).
9. S. Ma, *Phys. Rev. Lett.* **37**, 461 (1976).
10. D. Forster, D. R. Nelson and M. J. Stephen, *Phys. Rev.* **A16**, 732 (1977).
11. E. Medina, T. Hwa, M. Kardar and Y.-C. Zhang, *Phys. Rev.* **A39**, 3053 (1989).
12. D. E. Wolf and L.-H. Tang, *Phys. Rev. Lett.* **65**, 1591 (1990).
13. J. Krug, P. Meakin and T. Halpin-Healy, *Phys. Rev.* **A45**, 638 (1992).
14. J. G. Amar and F. Family, *Phys. Rev.* **A45**, 5378 (1992).
15. L.-H. Tang, N. M. Forrest and D. E. Wolf, *Phys. Rev.* **A45**, 7162 (1992).
16. C.-H. Lam and L. M. Sander, *Phys. Rev. Lett.* **69**, 3338 (1992).
17. P. Meakin, P. Ramanlal, L. M. Sander and R. C. Ball, *Phys. Rev.* **A34**, 5091 (1986).
18. J. Krug and P. Meakin, *J. Phys.* **A23**, 987 (1990).
19. T. A. Witten and L. M. Sander, *Phys. Rev.* **B27**, 2586 (1983); R. Ball, M. Nauenberg and T. Witten, *Phys. Rev.* **A29**, 2017 (1984); E. Brener, H. Levine and Y. Tu, *Phys. Rev. Lett.* **66**, 1978 (1991).

Fractals, Vol. 1, No. 4 (1993) 753–766

THE MORPHOLOGY AND EVOLUTION OF THE SURFACE IN EPITAXIAL AND THIN FILM GROWTH: A CONTINUUM MODEL WITH SURFACE DIFFUSION

FEREYDOON FAMILY and JACQUES G. AMAR
Department of Physics, Emory University, Atlanta, GA 30322, USA

Abstract

A number of discrete models as well as continuum equations have been proposed for describing epitaxial and thin film growth. We have shown that there exists a macroscopic groove instability in many of these models. This unphysical feature in the continuum equations arises from the truncation or linearization of the diffusion operator along the surface. A similar artifact occurs in the discrete models, because in these models adatoms only diffuse horizontally and must take an unphysical vertical jump at step edges. We have proposed and studied a continuum equation for epitaxial and thin-film growth in which the full diffusion along the surface is taken into account. The results of the solutions of this continuum equation, for the growth and the morphology of the surface, are in excellent agreement with recent low temperature molecular-beam epitaxy and ion-sputtering experiments. In particular, we find that at late times dynamic scaling breaks down and the surface is no longer a self-affine fractal. The surface develops a characteristic morphology whose dependence on deposition rate and surface diffusion is similar to that found in experiments.

1. INTRODUCTION

Growth techniques such as molecular beam epitaxy (MBE) and sputtering are widely used in materials processing.[1] In recent years, a number of nonequilibrium growth models have been proposed and investigated[2] in order to describe MBE and other deposition processes.

The goal of these studies is to determine the relevant dynamical mechanisms which control the Kinetic and the morphology of the surface in thin film growth. Much of these studies are based on the investigations of the dynamic scaling behavior of the surface[3] which depends sensitively on the details of various surface relaxation and roughening processes.[2] In particular, assuming an initially flat interface, the scaling of the interface width $w(L, t)$ on length scale L at time t is expected to be of the form,[3] $w(L, t) = L^{\alpha} f(t/L^{\alpha/\beta})$ with $f(x) \sim x^{\beta}$ for $x \gg 1$ and $f(x) \to const$ for $x \ll 1$, where α is the roughness exponent and β is the growth exponent. Therefore, the dynamic scaling exponents α and β can be used in order to determine which of the many different competing surface processes are the dominant ones in a particular type of growth process.

The evolution and the morphology of a growing surface[2] generally depends on an interplay between the deposition noise and various surface relaxation mechanisms such as evaporation-condensation and diffusion of adatoms on the surface.[4–6] Since desorption is negligible in MBE and vacuum deposition techniques, surface diffusion has been assumed to be the dominant relaxation process in various models that have been proposed[6–15] for describing thin film and epitaxial growth.

We have studied a number of continuum[6–8,10] and discrete models[7,8,11] for thin film growth. We have found[16] that there exists a macroscopic groove instability in many of these models. This unphysical feature in the continuum equations arises from the truncation or linearization of the diffusion operator along the surface.[11,16] A similar artifact occurs in the discrete models, because in these models adatoms only diffuse horizontally and must take an unphysical vertical jump at step edges.[13,14,16] Therefore, we have proposed and studied a continuum equation for epitaxial and thin-film growth in which the full diffusion along the surface is taken into account.[17] We review the results of the solutions of this continuum equation and compare them with MBE[18–21] and ion-sputtering experiment.[22] We find that the growth and the morphology of the surface in our model are in excellent agreement with recent experiments. In particular, we find that at late times dynamic scaling breaks down and the surface is no longer a self-affine fractal. The surface develops a characteristic morphology whose dependence on deposition rate and surface diffusion is similar to that found in experiments.

2. GROOVE INSTABILITY

We have found that there exists a macroscopic grooved phase in both the truncated continuum equations[7–9] and the discrete "horizontal diffusion" models for which $\alpha \geq 1$.[7,8,10–12] This can be shown analytically for the linear surface diffusion equation,[6–8]

$$\frac{\partial h}{\partial t} = -\kappa \nabla^4 h + \eta \tag{1}$$

where $h(\mathbf{x}, t)$ is the height of the $(d-1)$ dimensional surface, κ is the surface mobility and the noise $\eta(\mathbf{x}, t)$ is assumed to be delta-correlated in time and space with

$$\langle \eta(\mathbf{x}, t)\eta(\mathbf{x}', t')\rangle = 2D_R \delta^{d-1}(\mathbf{x} - \mathbf{x}')\delta(t - t'), \tag{2}$$

The noise strength D_R is proportional to the deposition rate R and we have neglected the contribution due to thermal (conserved) noise. The exact solution for the height-difference squared correlation function,

$$G_2(x;\,t) = \langle [h(x;\,t) - h(0;\,t)]^2 \rangle\,, \tag{3}$$

may be written as (starting from a flat interface at $t = 0$)[16]

$$G(x;\,t) = \frac{2D}{\pi\kappa} \int_{\Lambda_L}^{\Lambda_a} \frac{dk}{k^{6-d}} (1 - \cos(kx))(1 - e^{-2\kappa k^4 t}) \tag{4}$$

where the cutoffs are $\Lambda_L \sim 1/L$ and $\Lambda_a \sim 1/a$ where a is the short length-scale cutoff and L is the system size. For $t/L^4 \gg 1$, the exponential may be neglected, and for the case $x/L \ll 1 \ll x/a$, limits of the integral may be taken to be 0 and ∞, assuming that the integral converges. One thus has $G_2(x;\,\infty) \sim x^{2\alpha} \sim x^{5-d}$ in agreement with the exact expression $\alpha = (d-5)/2$. For $d = 2$ and $d = 3$, however, the integral diverges at $u = 0$ so that this is not the case. In particular, for $d = 2$ one obtains, $G_2(x;\,\infty) \sim Lx^2$ while for $d = 3$ one obtains $G_2(x;\,\infty) \sim x^2 \ln(L/x)$. Thus, for $d = 2$ and $d = 3$ the scaling of the correlation function $G_2(x;\infty)$ disagrees with the known scaling of the saturation width as a function of system size, which is already a sign that the surface at saturation is not self-affine. Our results indicate that in $d = 2$, the surface at saturation consists essentially of linear pieces with slope of order $\sqrt{G_2(x;\,\infty)}/x = \pm\sqrt{L}$. The average slope may also be directly calculated in $d = 2$ as,

$$\langle (\nabla h)^2 \rangle \sim \int_{\Lambda_L}^{\Lambda_a} \frac{dk}{k^{4-d}} (1 - e^{-2\kappa k^4 t}) \sim L \quad \text{as} \quad t \to \infty\,. \tag{5}$$

Assuming a single groove whose width is the system size L then the saturation width scales as $L * \sqrt{L} = L^{3/2}$, in agreement with the known result for the linear equation (1). In $d = 3$ one obtains $\langle (\nabla h)^2 \rangle \sim \ln(L)$ at saturation and $\sim \ln(t)$ at early time.

The fact that there is a single groove at saturation for the linear equation may be seen more clearly by considering the correlation function of the slope $m = \nabla h$ at saturation,

$$G_{\Delta m}(x) = \langle [m(0) - m(x)]^2 \rangle \sim x \text{ in } d = 2 \ (\sim \ln(x) \text{ in } d = 3) \quad \text{for } x/L \ll 1 \tag{6}$$

Thus, in $d = 2$ one has $G_{\Delta m}(x)/\langle m^2 \rangle = x/L$ ($= \ln(x)/\ln(L)$ in $d = 3$), which implies an orientational persistence length at saturation $\xi_P \sim L$. This corresponds to a single macroscopic groove which spans the length of the system.

We have tested our analytical picture by numerically integrating Eq. (1) in $d = 2$ and 3, using a finite-difference method on a lattice ($\kappa = 1$, $D_R = 0.1$, lattice spacing $\Delta x = 1$, and time-step $\Delta t = 0.01$) starting from a flat interface at $t = 0$.[16] Figure 1(a) shows the development of the grooves at the early stages of the growth for a system of size $L = 128$ in $d = 2$, whereas Fig. 1(b) shows the final single groove at saturation for the same system. Figure 2 shows the surface obtained from numerical integration of Eq. (1) in $d = 3$ for a system of size 32×32 at late time. We again see that the steady-state surface appears to consist essentially of a single groove, although $d = 3$ is somewhat marginal.

In order to investigate the effect of nonlinearities on the groove instability, we have also numerically solved the nonlinear equation,[10]

$$\frac{\partial h}{\partial t} = -\kappa \nabla^4 h + \lambda_2 \nabla^2 (\nabla h)^2 + \eta \tag{7}$$

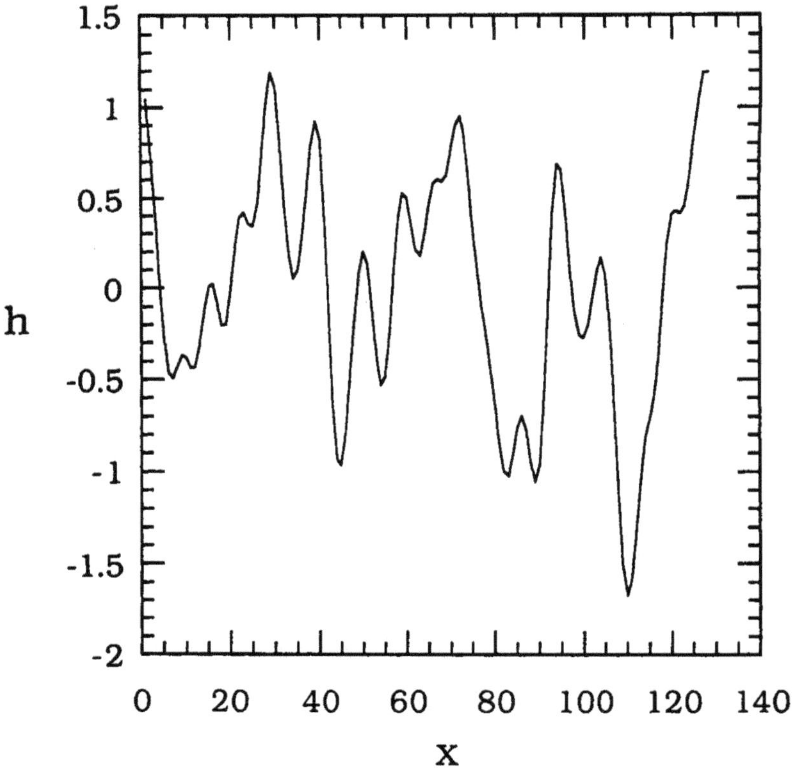

(a)

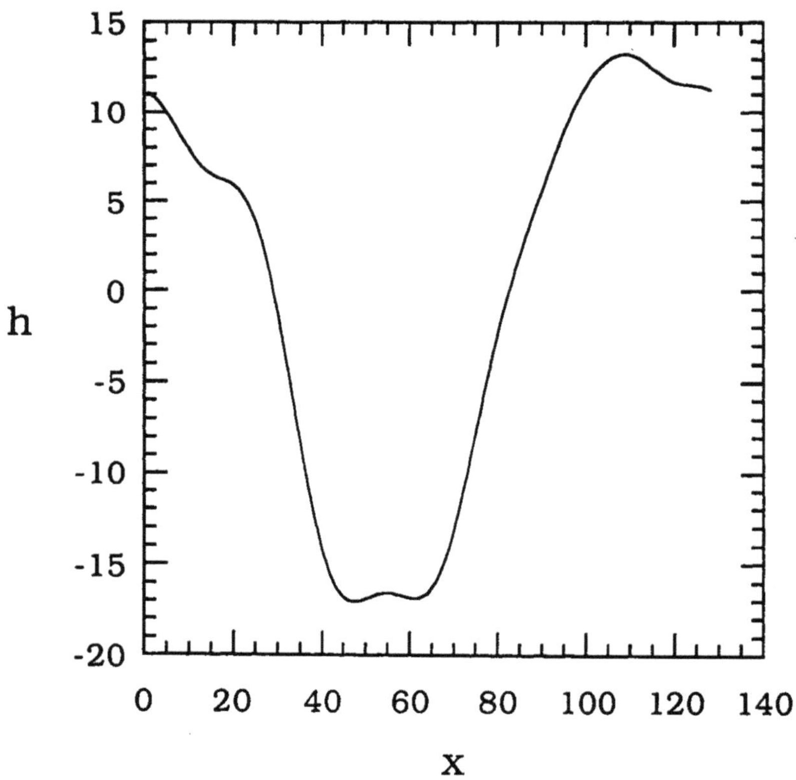

(b)

Fig. 1 (a) Interface profile obtained from numerical simulation of Eq. (1) with $D_R = 0.1$, $\kappa = 1.0$, $L = 128$, at time $t = 10.0$, (b) Snapshot of single groove formed at late time ($t = 100\,000$) in the same model.

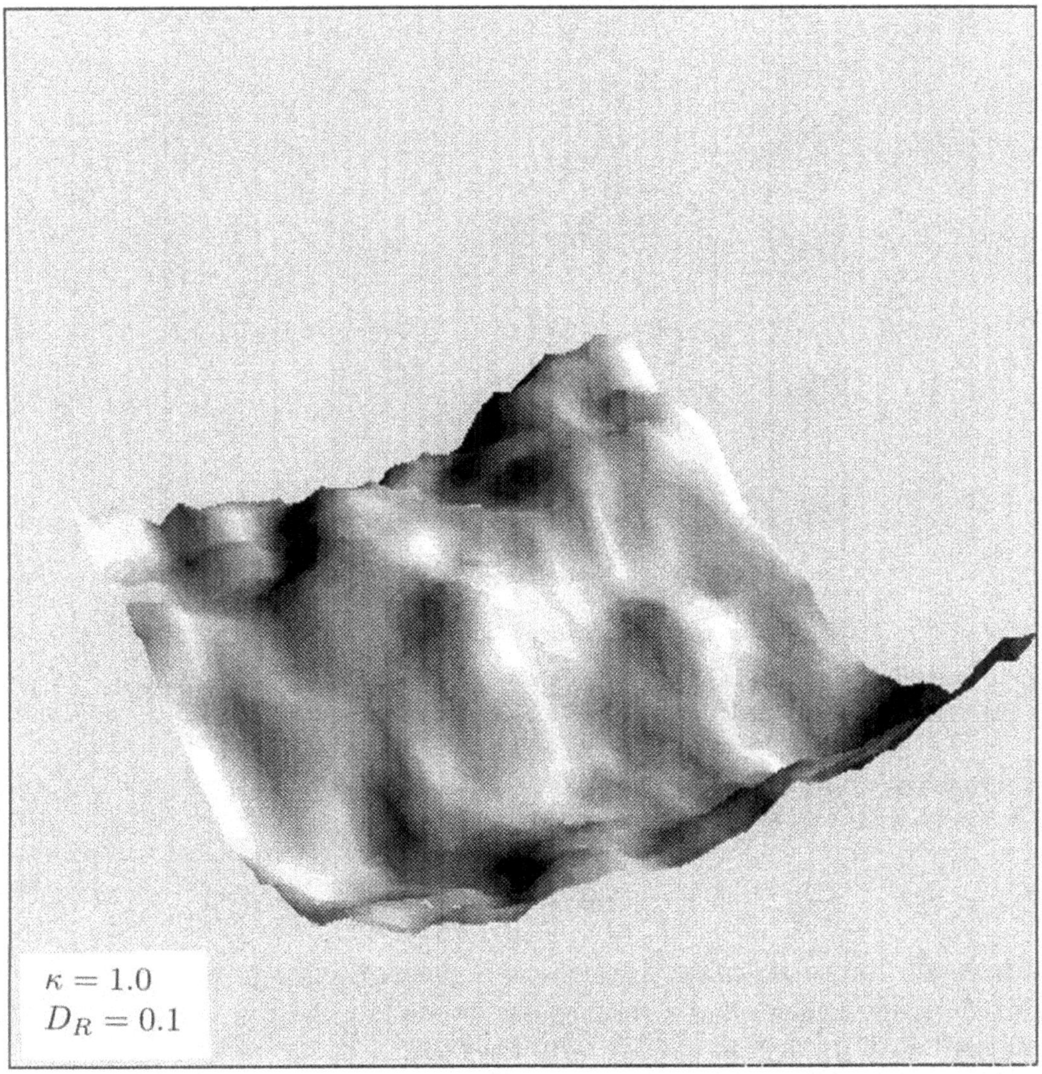

Fig. 2 Picture of interface obtained from numerical simulation of Eq. (1) in $d = 3$ with $L = 32$ showing a "groove" at late time $t = 10\,000$.

for which it is known that $\alpha = (5-d)/3$. In $d = 2$, $\alpha = 1$ which is marginal. By numerically solving Eq. (7), we have found evidence that in the steady-state there exists a groove as well in this case.[16] This shows that the grooved phase is not an artifact of the linear models, but it can also exist in the nonlinear Eq. (7) and the corresponding discrete models. In fact, the groove instability is expected to occur for any model with roughness exponent $\alpha \geq 1$.

The existence of the grooved phase in Eqs. (1) and (7) may be traced to the fact that these equations were derived by expanding the full diffusion operator in powers of ∇h and keeping the lowest order term. From a renormalization group point of view, this would have been an appropriate approximation if in the long time limit and large length scales the higher order terms were negligible. However, since in these models $\alpha \geq 1$, all higher order terms are of the same order and are relevant.[11,16,17] Therefore, truncation of the full diffusion operator is not justified in the case of growth with surface diffusion. In fact, in the next section, we show that the continuum model with full surface diffusion[17] gives quite different results from the truncated equations.

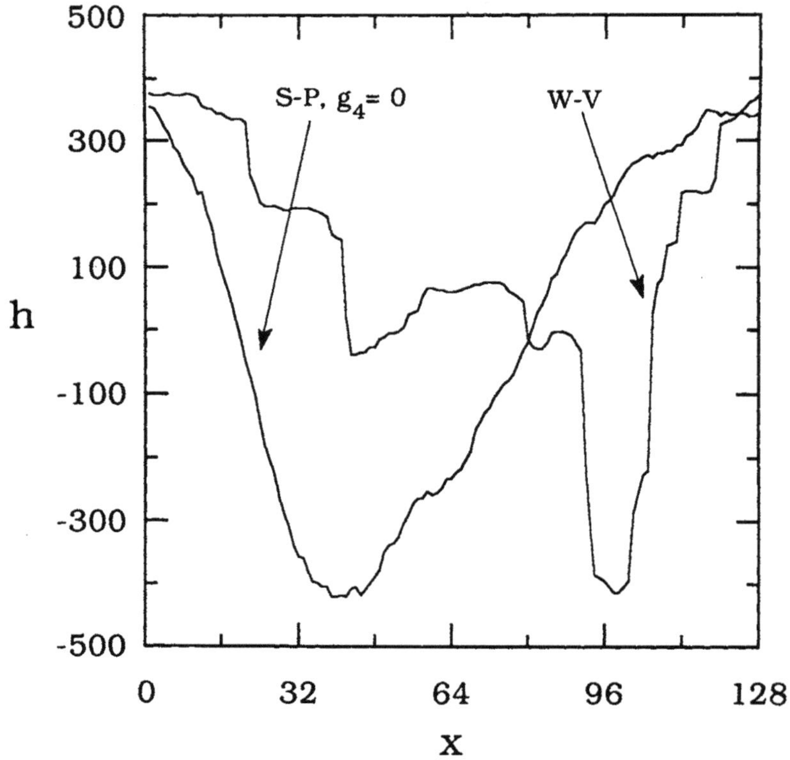

Fig. 3 Snapshots of interface profile for Siegert-Plischke model (SP) at $t = 128 \times 10^6$ in $d = 2$ with $g_4 = 0.0$, $\tau = 0.1$, $\beta J = 0.01$, and $L = 128$, and for Wolf-Villain model (WV) with $L = 128$ at $t = 3.2 \times 10^6$.

We have also investigated the existence of a grooved phase in some of the discrete models.[7,11] Figure 3 shows that a macroscopic groove is formed in the Siegert and Plischke model[11] with $g_4 = 0$, in analogy with the linear equation result. However, due to the presence of a small surface tension term,[14,23,24] the Wolf-Villain model[7] is in a different universality class and although the surface has large fluctuations (see Fig. 3), it does not form a macroscopic groove as in the other models. The unphysical feature that is common to all the discrete models with surface diffusion is that they are basically "horizontal diffusion" models[13,14] in the sense that adatoms can only diffuse horizontally. To avoid forming overhangs, an atom that reaches a step edge is not allowed to diffuse vertically and it must instantaneously jump vertically. This is the origin of the groove instability in these models. This unphysical simplification[13,14] is needed in the discrete models in order to avoid overhangs, but it clearly leads to the formation of the grooved phase at late times.

Thus, a proper model of growth with surface diffusion must take the full diffusion of adatoms along the surface into account. In the next section, we describe the model that we have developed and studied[17] in which diffusion of atoms along the surface is the dominant relaxation mechanism.

3. CONTINUUM MODEL OF SURFACE GROWTH

In a surface growth process in which the dominant relaxation is by diffusion of atoms along the surface, the number of atoms is conserved.[4-6] This means that the normal velocity of the interface obeys a continuity equation,

$$v_n(\mathbf{x}, t) = \frac{1}{\sqrt{g}}\frac{\partial h}{\partial t} = -\Omega\nabla\cdot\mathbf{j}(\mathbf{x}, t) + \eta(\mathbf{x}, t) \tag{8}$$

where $g = 1 + (\nabla h)^2$, Ω is the atomic volume, and the noise η satisfies Eq. (2) where $D_R = \Omega R/2$ is proportional to the deposition rate R. As originally proposed by Mullins,[5] we assume that the particle current $\mathbf{j}$ is driven by the gradient of a surface chemical potential, $\mathbf{j} = -\Lambda\nabla\mu$, where the chemical potential μ is given by $\mu = \Omega\frac{\delta F}{\delta h}$. The Ginzburg-Landau functional F may be taken as the free energy of the drumhead model,

$$F = \mathcal{H}_{DH} = \int d^{d-1}x\sigma\sqrt{g}\,, \tag{9}$$

where σ is the surface tension. Therefore the continuum equation describing surface diffusion is given by

$$\frac{\partial h}{\partial t} = \Omega\nabla\sqrt{g}\,\Delta_s\frac{\delta F}{\delta h} + \eta\,, \tag{10}$$

where

$$\Delta_s = \frac{1}{\sqrt{g}}\partial_i\left\{\sqrt{g}\left[\delta_{ij} - \frac{1}{g}h_ih_j\right]\partial_j\right\}, \tag{11}$$

is the Laplace-Beltrami operator Δ_s along the surface. Previous expressions for the Laplacian operator along the surface[25] contained a large number of terms and were not amenable to numerical or analytical investigations. We have derived[17] a more physical and computationally more convenient form for the Laplace-Beltrami operator so that we can rewrite Eq. (10) as,

$$\frac{\partial h}{\partial t} = -\kappa\left[\frac{1}{\sqrt{g}}\Re M - M(\nabla h\cdot\nabla M)\right] + \eta \tag{12}$$

where M is the mean curvature (sum of two principal curvatures),

$$M = -\frac{1}{\sigma}\frac{\delta F}{\delta h} = g^{-3/2}\Re h\,, \quad \text{and} \quad \Re = \sum_{ij} G_{ij}\partial_{ij}\,, \tag{13}$$

with $G_{ij} = g\delta_{ij} - h_ih_j$, so that in 3-dimensions we have,

$$\Re = (1 + h_y^2)\partial_{xx} + (1 + h_x^2)\partial_{yy} - 2h_xh_y\partial_{xy}\,, \tag{14}$$

and $\kappa = \Omega\Lambda\sigma$. Assuming the Nernst-Einstein relation gives $\kappa = \frac{\rho_s D_s \Omega^2\sigma}{k_BT}$, where ρ_s is the surface density of diffusion atoms and D_s is the surface diffusion constant (which depends on temperature).

We have integrated Eq. (12) numerically on a two-dimensional ($L \times L$) grid with $\Delta x = 1$ for a variety of different values of κ and D_R, with lattice sizes ranging from $L = 128$ to 256 and time step ranging from $\Delta t = 0.0025$ to 0.01.[17] In our integration method, we calculated the mean curvature M at each point using the second-derivatives of the height in the $\Re$ operator. The $\Re$ operator was then applied a second time in a two-step process. As a test of our algorithm, we first studied the case of a one-dimensional interface and obtained $\beta \simeq 0.5$, in excellent agreement with the results of Golubovic and Karunasiri.[26]

Figure 4(a) shows a plot of our results in $d = 3$ for the interface width w as a function of time t for four different values of the deposition rate R ($D_R = 0.2 - 0.5$) for $\kappa = 1$. We

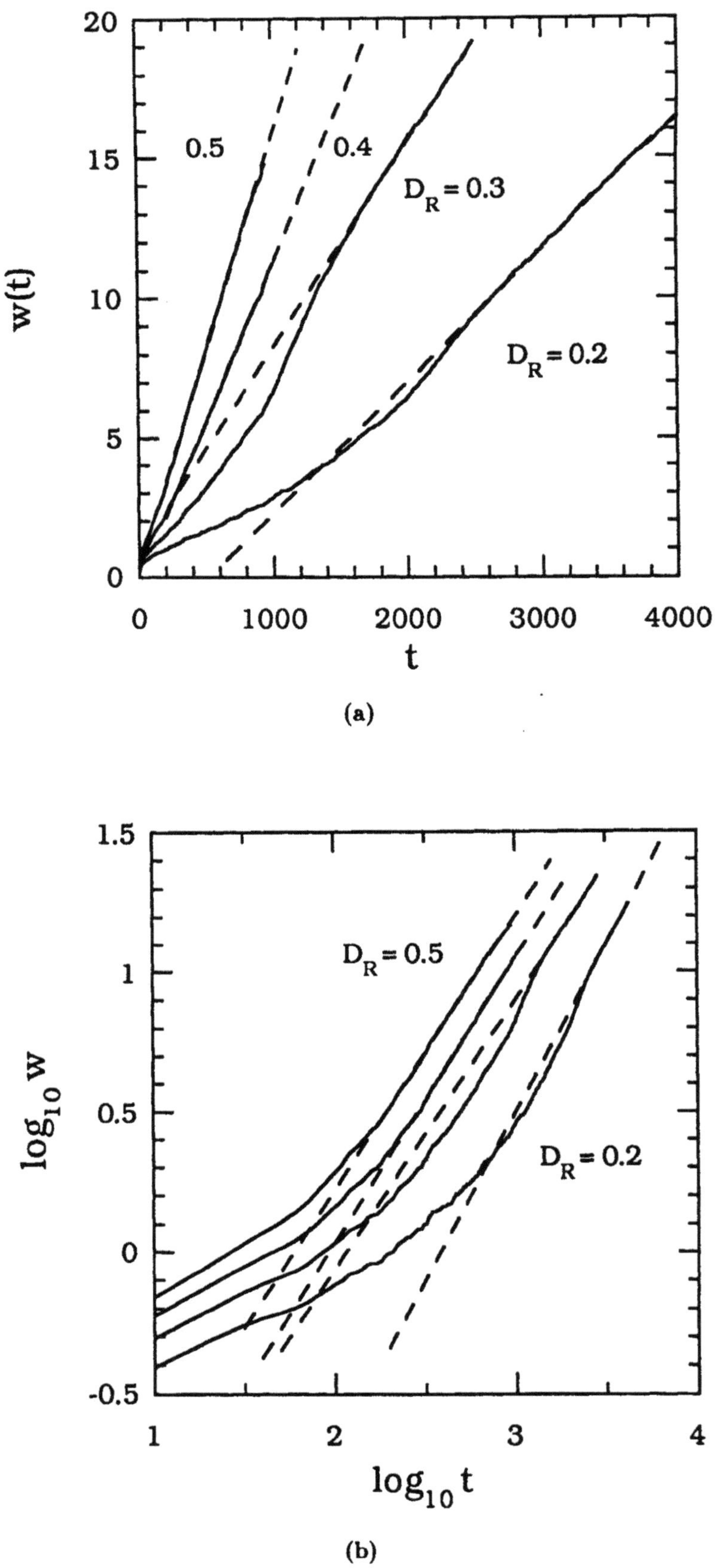

(a)

(b)

Fig. 4 (a) Plot of surface width as a function of time for $\kappa = 1$ and $D_R = 0.2 - 0.5$. Dashed lines show linear fits at late times. (b) Log-log plot of same data showing crossover from $\beta = 0.25$ behavior to linear behavior at late times.

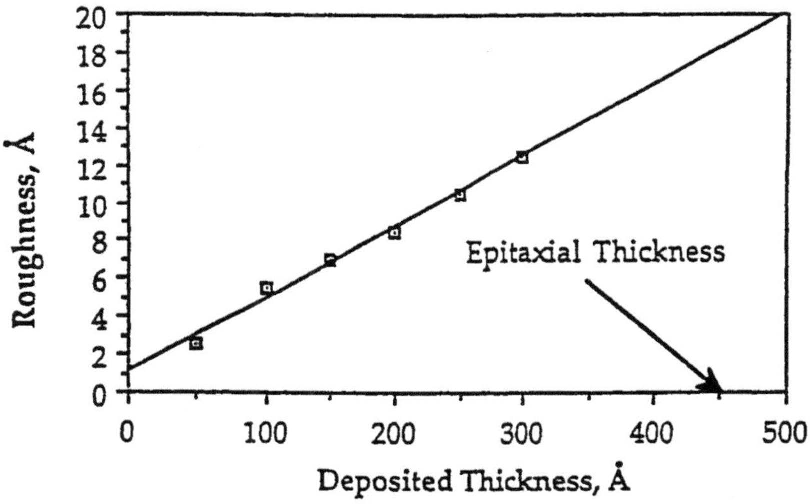

Fig. 5 Plot of the width as a function of the deposition height (time) in the low temperature MBE experiment of Eaglesham and Gilmer[19] shows the linear behavior which is also observed in our model (see Fig. 4). This data taken at low temperatures corresponds to the high D_R limit of our model indicating rapid linear growth of the width with time.

found similar results for higher values of κ as well.[17] The linear behavior at late times is consistent with the experimental results of Eaglesham[19] for low temperature MBE growth of Si and other semiconductor films. For comparison, one of Eaglesham's graphs[19] which shows the growth of the width of a Si film as a function of the deposition height (which corresponds to time) is shown in Fig. 5. This data clearly confirms our result that in the low temperature MBE, where the surface mobility is low and the deposition rate (or noise strength D_R) is high, the linear growth is reached very quickly. The linear growth of the width with time has also been observed by Williams et al.[22] in their ion-sputtering experiments on graphite. Figure 4(b) shows a log-log plot of the same data showing the crossover from growth with an exponent $\beta = 0.25$ at very early times, which is consistent with the linear equation (1), to linear growth behavior.

Figure 6 shows a typical log-log plot of $G_2(r)$ for $D_R = 0.4$ at two different times after the linear behavior has set in. The slope indicates that $\alpha \simeq 0.75$. For other values of κ and D_R the value of α ranges from $\alpha = 0.70$ to 0.85 at late times. However, the correlation function $G_2(r)$ does not scale with time in the normal manner, because the range over which scaling is observed does not increase at late times, while its magnitude increases approximately as t^2. This suggests that the surface consists of mountains (and valleys) which have a typical width ξ which does not increase at late times, but whose height increases linearly.

In order to further study the dependence of ξ on deposition and diffusion rate, we have also studied[17] the circularly-averaged height-height correlation function,

$$G(r) = \langle \tilde{h}(\mathbf{0})\tilde{h}(\mathbf{r})\rangle , \tag{15}$$

where $\tilde{h}(\mathbf{r}, t) = h(\mathbf{r}, t) - \langle h(\mathbf{r}, t)\rangle_r$, and the structure function,

$$S(k) = \left\langle \left| \sum_j e^{i\mathbf{k}\cdot\mathbf{r}_j}\tilde{h}(\mathbf{r}_j)\right|^2 \right\rangle . \tag{16}$$

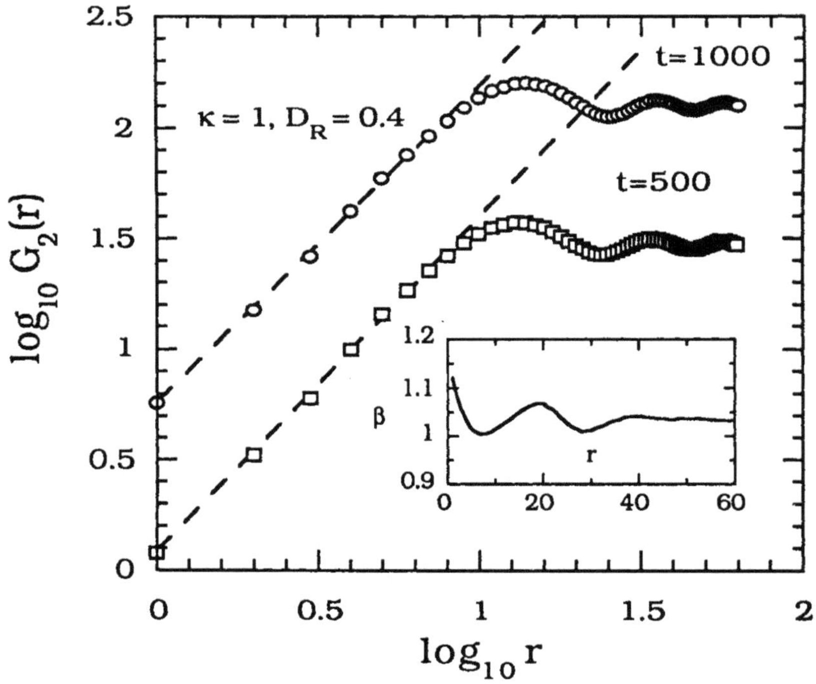

Fig. 6 Log-log plot of $G_2(r)$ for $\kappa = 1.0$ and $D_R = 0.4$ at $t_1 = 500$ and $t_2 = 1000$.

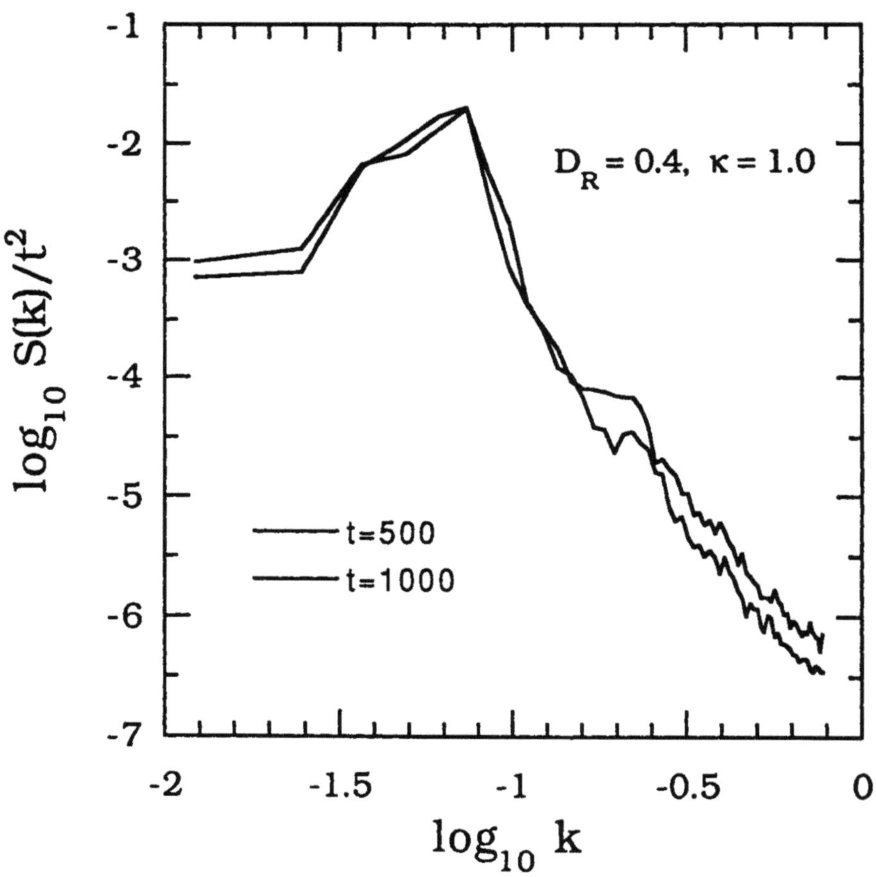

Fig. 7 Plot of the circularly averaged structure factor $S(k)$ shows a broad peak similar to the one observed in the experiments of William et al.[22] indicating the existence of a characteristic transverse length scale ξ.

As expected, we find that ξ increases with increasing surface diffusion κ and decreases with increasing deposition rate D_R. This is demonstrated in Fig. 7 for the circularly-averaged structure factor showing a broad peak indicating the existence of a characteristic transverse length scale, similar to the experimental results of Williams et al.[22]

In order to compare the morphology of the surface in our model with experiments, in Fig. 8 we show pictures of the surface at late times for two different deposition rates ($D_R = 0.2$ and 0.4), together with the STM topograph of Ar ion sputtered graphite.[22] The surfaces in both our model and the experiment of Williams et al.[22] consist of a set of valleys and mountains whose height is increasing in time. These irregularities grow until a crossover time, determined by the competition between the deposition flux and the diffusion constant, after which the correlation length ξ no longer increases and linear growth of the interface width sets in. We have studied[17] the dependence of this crossover time on the noise strength D_R and surface diffusion κ, and we find that it increases much more rapidly with decreasing deposition rate than with increasing κ. We have also calculated the critical aspect ratio $r_c = w_c/\xi$ at the point at which linear growth of the interface width sets in. For $D_R = 0.4 - 0.5$, we find $r_c \simeq 0.24$ ($\kappa = 1 - 2$) while for $D_R = 0.2, 0.3$ ($\kappa = 1$), $r_c \simeq 0.6 - 0.7$. We note that at late times, dynamic scaling breaks down because the surface width does not saturate. The reason is that the diffusion is parallel to the surface and the relevant length scale is no longer the horizontal width of the system but rather the total length along the surface.

(a)

Figs. 8(a)–8(c) Pictures of surface at late times for $L = 128$ with $\kappa = 1.0$ (a) $D_R = 0.2$ at $t = 2000$.

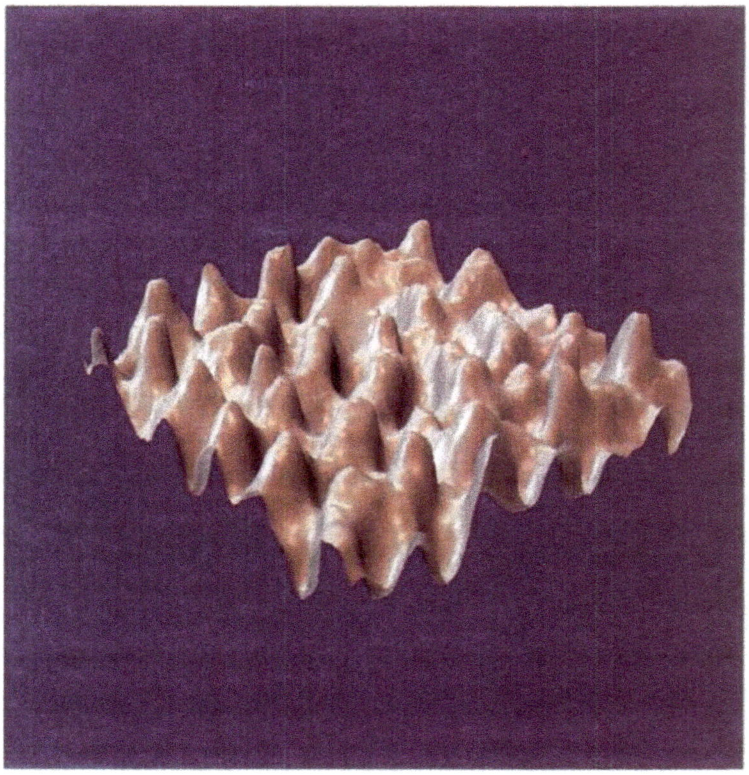

(b)

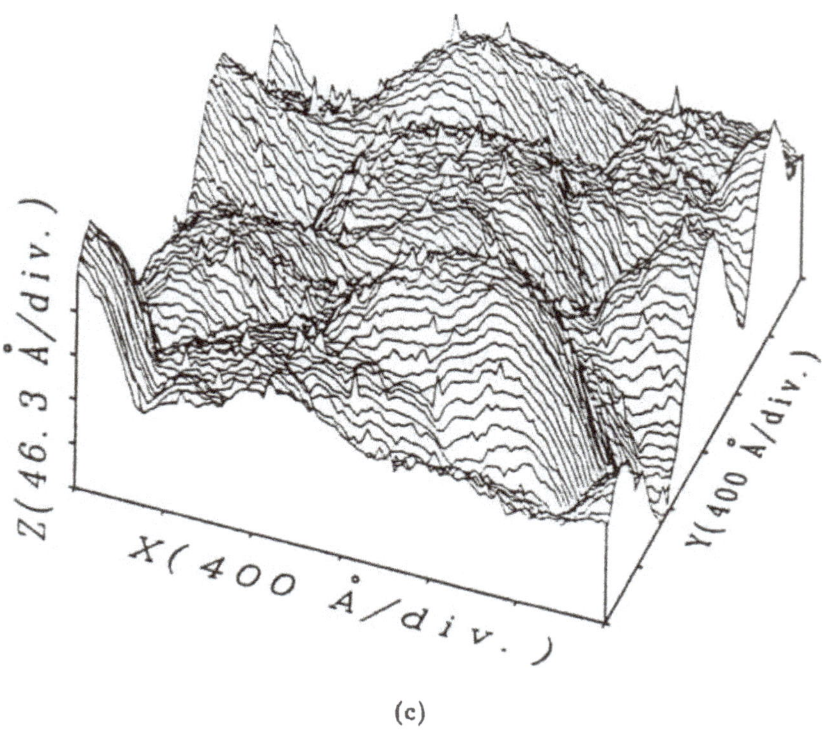

(c)

Fig. 8 (*Continued*)
(b) $D_R = 0.4$ at $t = 500$. **(c)** Experimental STM topograph of graphite from Williams et al.[22] shows strong similarity to our surface morphologies shown in (a) and (b).

4. CONCLUSIONS

We have discussed the existence of a groove instability in continuum equations and some of the discrete models of surface growth.[16] The truncation of the full continuum equation for surface diffusion[11] and vertical jumps in the "horizontal diffusion" models[13,14] are responsible for the development of a macroscopic groove in late stages of growth in these models.[16] This implies that the full diffusion along the surface must be taken into account in a proper model of growth with surface diffusion. In this context we discussed a continuum equation model,[17] which incorporates the full diffusion of atoms along the surface. The results of this model are qualitatively and quantitatively different from the previous models. In terms of comparison with experiments, our model suggests that at temperatures where surface diffusion is large, we expect the width to increase as $t^{1/4}$ and the roughness exponent α, obtained from the correlation function, to be close to 1 before the surface features have fully developed. Such behavior has previously been observed in the experiments of Chiarello et al.[27] ($\alpha = 0.95 \pm 0.1$ for Au at 500 K), Chevrier et al.[20] ($\beta \simeq 0.25$), and He et al.[21] ($\beta = 0.22 \pm 0.02$ and $\alpha = 0.79 \pm 0.05$). However, at lower temperatures or high deposition rates, the nonlinearities due to full diffusion along the surface become important at early times and may lead to linear growth of the surface width as observed by Eaglesham[19] and by Williams et al.[22] We note that the data of Chevrier et al.[20] also appear to be crossing over to a higher value of β ($\beta \simeq 0.6$) at later times, consistent with our results. A similar crossover behavior as a function of temperature has also been observed in recent experiments by Tong et al.[28] on CuCl.

Finally, we note that had been an alternative explanation of the experiments of Eaglesham[19] had been given in terms of models with asymmetric hopping rates (due to step barriers).[19,29] In order to test these ideas we have carried out[30] large scale simulations of the solid-on-solid model of Eaglesham and Gilmer[19] with a step barrier of 0.1. Figure 9 shows the surface morphology at late times.[30] As can be seen from this figure, the surface

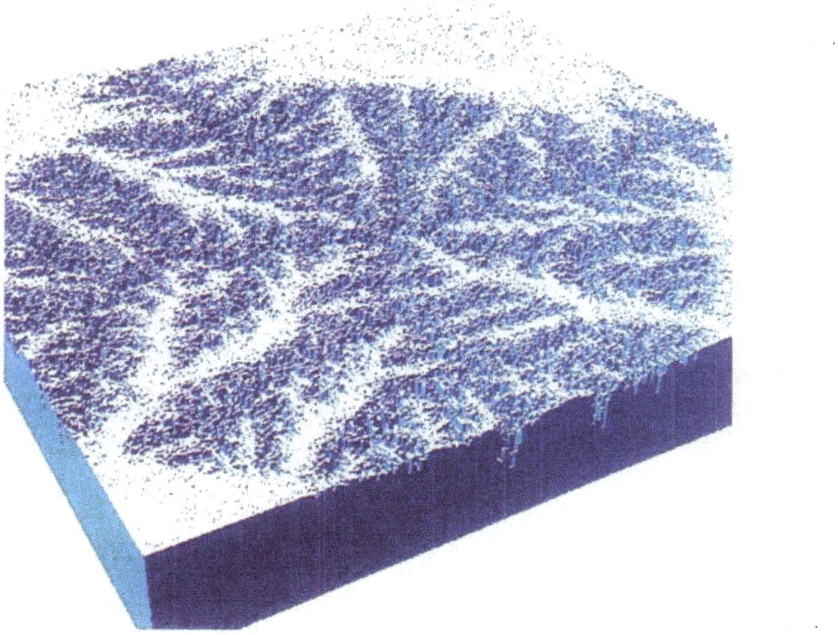

Fig. 9 The existence of a deep groove with large surface fluctuations is shown in this plot of the surface morphology in the step barrier model of Eaglesham and Gilmer.[19] The deepest point in the groove is at the center of the figure and is as deep as the height of the surface.

consists of a single deep groove with large surface fluctuations. Although this type of groove can give rise to a rapid (linear or higher than linear) growth of the width with time, it leads to a surface morphology that does not agree with experiments. It would be interesting to investigate the effect of an asymmetric barrier in our continuum model,[17] because the step barrier model of Eaglesham and Gilmer[19] is a "horizontal diffusion" model and that could be the reason for the existence of the unusual type of groove in their model.

ACKNOWLEDGMENTS

We would like to thank Pui-Man Lam for his contributions in the study of groove instability, and Dan Platt and Ken Lao for their help with graphics, David Eaglesham for discussions of his experiments as well as for allowing us to reprint one of his experimental graphs here, and Stan Williams for permitting us to reproduce the STM picture of the surface morphology from his experiment. This work was supported by the National Science Foundation grant DMR-9214308 and by the Office of Naval Research.

REFERENCES

1. J. A. Thornton, *Ann. Rev. Mater. Sci.* **7**, 239 (1977).
2. F. Family and T. Vicsek, eds., *Dynamics of Fractal Surfaces*, World Scientific, Singapore (1991).
3. F. Family and T. Vicsek, *J. Phys.* **A18**, L75 (1985).
4. C. Herring, in *The Physics of Powder Metallurgy*, ed. W. E. Kingston (McGraw-Hill, NY, 1951).
5. W. W. Mullins, *J. Appl. Phys.* **28**, 333 (1957); *J. Appl. Phys.* **30**, 77 (1959).
6. J. Villain, *J. Phys.* (France) **I1**, 19 (1991).
7. D. E. Wolf and J. Villain, *Europhys. Lett.* **13**, 389 (1990).
8. S. Das Sarma and P. Tamborenea, *Phys. Rev. Lett.* **66**, 325 (1991).
9. L. Golubovic and R. Bruinsma, *Phys. Rev. Lett.* **66**, 321 (1991).
10. Z. W. Lai and S. Das Sarma, *Phys. Rev. Lett.* **66**, 2348 (1991).
11. M. Siegert and M. Plischke, *Phys. Rev. Lett.* **68**, 2035 (1992).
12. M. Wilby, D. D. Vvedensky and A. Zangwill, *Phys. Rev.* **B46**, 12896 (1992).
13. H. Yan, *Phys. Rev. Lett.* **68**, 3048 (1992).
14. D. A. Kessler, H. Levine and L. M. Sander, *Phys. Rev. Lett.* **69**, 100 (1992).
15. See T. Shitara, D. D. Vvedensky, M. R. Wilby, J. Zhang, J. H. Neave and B. A. Joyce, *Phys. Rev.* **B46**, 6815 (1992) and references therein.
16. J. G. Amar, P.-M. Lam and F. Family, *Phys. Rev.* **E47**, 3242 (1993).
17. J. G. Amar and F. Family, preprint (1993).
18. D. J. Eaglesham, H.-J. Gossmann and M. Cerullo, *Phys. Rev. Lett.* **65**, 1227 (1990).
19. D. J. Eaglesham and G. H. Gilmer, in *Surface disordering: Growth, roughening and phase transitions*, eds. by R. Jullien, J. Kertesz, P. Meakin and D. E. Wolf (Nova, NY, 1993); D. J. Eaglesham, private communication.
20. J. Chevrier, V. Le Thanh, R. Buys and J. Derrien, *Europhys. Lett.* **16**, 737 (1991).
21. Y.-L. He, H.-N. Yang, T.-M. Lu and G.-C. Wang, *Phys. Rev. Lett.* **69**, 3770 (1992).
22. R. S. Williams, R. Bruinsma and J. Rudnick, in *Evolution of Surface and Thin-Film Microstructure, Fall 1992 MRS Proceedings Series*, eds. H. A. Atwater, M. Grabow, E. Chason and Max Legally (Mat. Res. Soc., Pittsburgh, PA, 1992); E. A. Eklund, R. Bruinsma, J. Rudnick and R. S. Williams, *Phys. Rev. Lett.* **67**, 1759 (1991).
23. D. A. Kessler and B. G. Orr, preprint (1993).
24. J. Krug, M. Plischke and M. Siegert, preprint (1993).
25. B. J. Palmer and R. G. Gordon, *Thin Solid Films* **158**, 313 (1988).
26. L. Golubovic and R. P. U. Karunasiri, *Phys. Rev. Lett.* **66**, 3156 (1991).
27. R. Chiarello, V. Panella, J. Krim and C. Thompson, *Phys. Rev. Lett.* **67**, 3408 (1991).
28. W. M. Tong, R. S. Williams and M. S. Anderson, private communication.
29. D. D. Vvedensky, A. Zangwill, C. N. Luse and M. R. Wilby, preprint (1993).
30. K. Lao, F. Family and J. G. Amar, preprint (1993).

Fractals, Vol. 1, No. 4 (1993) 767–783

GRANULAR COCKTAIL ROTATED AND SHAKEN

G. BAUMANN, E. JOBS and D. E. WOLF
Theoretische Physik, FB 10, Universität-GH-Duisburg,
D-47048 Duisburg, Germany

Abstract

In order to study scaling phenomena in granular materials, simple extensions of a model originally proposed by Jullien, Meakin and Pavlovitch (*Phys. Rev. Lett.* **69**, 640 (1992)) are developed. It is shown that the model reproduces phenomena like the heap formation due to friction at the container walls during vertical shaking, as well as size segregation in a rotating drum. The geometrical mechanisms underlying these phenomena are discussed, as well as preliminary results about the critical behavior.

1. INTRODUCTION

Natural fractals,[1] the geometrical counterpart of any critical phenomenon, most often emerge in driven systems. By this we mean systems kept in or approaching a nonequilibrium steady state. In this paper, we investigate scaling behavior in driven granular media like sand. The notion of a granular medium means a many-particle system in which each of the grains has a macroscopically large number of internal degrees of freedom responsible for the dissipation necessary for any driven system to counterbalance the energy input.

In 1992, Jullien, Meakin and Pavlovitch (JMP)[2] introduced a simple model for a shaken granular medium. It is based on a growth model which has been studied a lot in the context of kinetic roughening. In the continuous version of this model, one imagines that spheres are deposited one by one along randomly chosen trajectories perpendicular to a substrate. When a sphere touches the surface or another sphere for the first time, it takes the path of steepest descent until it reaches a local minimum. In this minimum, it gets stuck forever. Because inertia and elasticity are neglected, bouncing of the spheres as well as later rearrangements of the settled ones, upon the impact of an incoming sphere, are forbidden. These conditions could be regarded as the limit of extremely large static friction and low energy restitution coefficient (inelastic collisions).

The new idea of Jullien, Meakin and Pavlovitch was to list all the spheres of the deposit in order of increasing heights of their centers. Then they redeposite them over and over again in the order of the current list to simulate a vertical shaking process. The implicit assumptions are that the packing comes to rest after each shake and that the acceleration is big enough to lead to a temporary vertical dilation of the system which allows the lower particles to settle while the higher ones are still in free flight.

This model reproduces a well-known phenomenon, the size segregation (i.e., clustering of grains of equal size) during vertical shaking: in a mixture of large and small spheres the former move up to the surface. The simulations[2–4] predicted a critical ratio of the radii which depends on the angle of repose: only if the sizes of the particles are different enough, the big spheres raise continuously no matter how small the amplitude of the shaking. If the sizes of the particles are too similar, segregation changes its character to an intermittent process, where the amplitude of the shaking and random contributions to the motion become essential. For the two-dimensional case this was recently confirmed experimentally by Duran et al.[4]

The simulations by JMP were not the first which explained the size segregation by vertical shaking as a local geometrical effect. The extensive Monte Carlo simulations of Rosato et al.[5,6] already described essentially all the properties of the intermittent, amplitude-dependent segregation regime correctly. Studies paying more attention to the other parameters of the problem, like the ones by Metha and Barker[7] and molecular dynamics simulations[8] are not in conflict with these findings.

The main advantage of the JMP model compared to the other simulations is its simplicity: one only needs to determine the positions where the relaxing sphere changes its contacts with other spheres. These positions are given by geometrical equations and are easy to calculate. In order to find the final position of a sphere, typically two intermediate positions have to be calculated. The algorithm is very fast so that one is able to study a great number of particles (10^6 and more). It would be hardly possible to study the critical properties close to the segregation threshold, e.g. in molecular dynamics.

The JMP model is certainly an oversimplification of the real processes. It corresponds to extreme physical conditions, and the question arises, whether in this limit it could serve as a useful reference model for the dynamics of granular media in general. Therefore we studied if the JMP-model can be easily extended to account for phenomena like the size segregation in rotating drums (Sec. II) and the heap formation during vertical shaking (Sec. III). Section IV contains our conclusions and an outlook.

No theory of granular media which convinces by its predictive power has yet been established. The modeling very much resembles the mixing of a cocktail, putting together and modifying whatever ingredients are handy and which appeal to the bar mixer's intuition. The physicist's taste however is only met if he gets new insights into the nature of granular media, and if the model stands the experimental test. A reference model that meets these criteria could be a significant progress towards a systematic theory.

2. SEGREGATION IN A ROTATING DRUM

2.1 Experiment

The mixing or demixing of solid particles in a drum which rotates around a horizontal axis is a common industrial process. In spite of the considerable number of scientific papers published about it, the knowledge has remained more or less empirical. The basic mechanisms are not understood. Many parameters are involved in the process such as: size, shape, surface roughness, material properties of the grains, friction forces, velocity, geometry and filling of the drum. Some of them should be more important than others, and one would like to extract an idealized model capturing the essential physics of this process.

Since Oyama's pioneering experiments in 1939,[9] it has been known that particles of different size tend to demix rather than to mix in a rotating drum "mixer". An initial homogeneous mixture of grains of two sizes shows two different kinds of segregation: in the radial direction, perpendicular to the rotation axis, a core of smaller particles forms, and in the direction parallel to the rotation axis bands of alternating particle size occur. The phenomena are qualitatively different, if the drum is more than half filled, than for half or less filling.[10]

More recent experiments of D. Bideau et al.[11] show that the radial segregation is not restricted to three dimensions (Fig. 1). In two dimensions the segregation happens already in the first revolution of the drum, whereas band segregation in three dimensions is observed only after a greater number of revolutions.

2.2 Implementation

With a modified JMP-model, we were able to simulate the two dimensional experiments. The main modification we introduced is a rigid rotation of the whole packing by a small angle $\Delta\Phi$ before each relaxation (update) of the system. In order to prevent the disks slipping back into their old positions and ensure that small and large disks at the wall exchange places, we had to make the wall "sticky" close to the bottom. Physically this corresponds to the fact that for small slopes the static friction at the wall is larger than the tangential component of the gravitation force.

After the rigid rotation, all disks are allowed to relax to the nearest local minimum sequentially, in the order of increasing heights of their centers. As the steric hindrance of higher particles is ignored during the relaxation, this seems a bit artificial: in contrast to shaking, which can be viewed as a periodic dis- and reassembling of the configuration, rotation hardly leads to sufficiently large temporary gaps to allow this kind of relaxation. We shall show, however, that in the most interesting cases the system develops a structure in which the relaxation happens predominantly via surface avalanches and block slip motion, both realistic processes. Unphysical restructuring was observed only in exceptional cases during our computer simulations and should not change the dynamics significantly.

A possible exception is the uppermost region of the disk configuration which is influenced by the upper half of the wall. In every update of the configuration, each disk follows at first a fictitious vertical trajectory down from a higher place above the drum. At the first contact with the previously fallen disks, it may happen that it overlaps with the wall. In this case it is put to a point above the disk which was touched first, where it is in contact with the wall, too, and is allowed to relax from there into the closest local minimum. This causes some irregular motion in the upper region of the drum.

(a)

(b)

Fig. 1 Experimental results of F. Cantelaube-Lebec and D. Bideau, a system at the beginning and after one revolution.

2.3 Qualitative Comparison with the Experiment

Here we only consider the case of less than half filling of the drum. Figure 2 shows a system in its initial configuration and after one revolution. In accordance with the experiments of D. Bideau et al., we obtain radial segregation already in the first revolution of the drum, and the size, shape and position of the segregation cloud are similar. We were able to reproduce structural properties of the configuration such as the dense packing and the layerwise ordering of the large disks along the wall and the disordered region they form in the upper part.

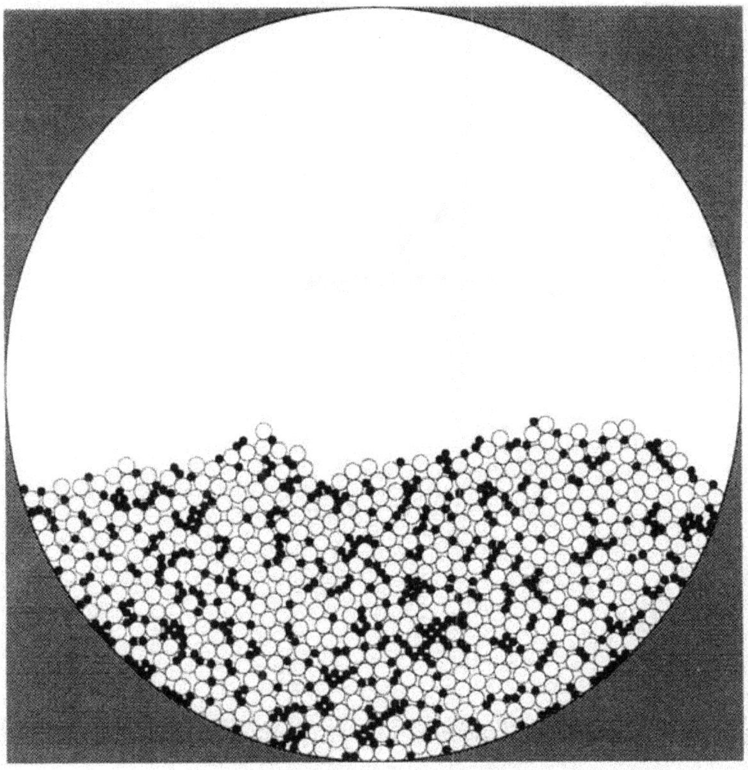

(a)

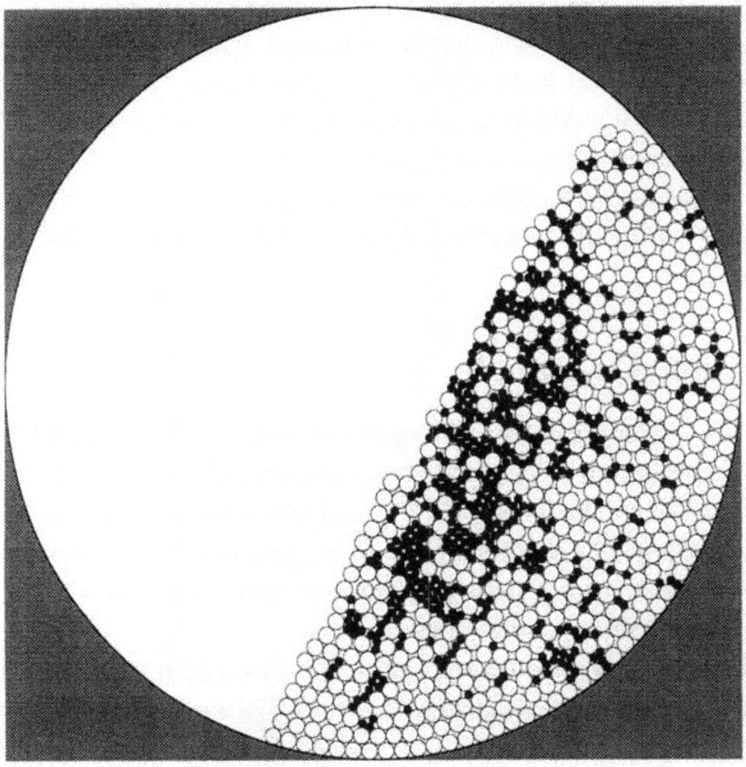

(b)

Fig. 2 Simulation of 1000 disks, 35% small, radius of the small disks $r = 0.5$, radius of the drum $R = 50$ (in units of the radius of large disks); (a) initial configuration, (b) after one revolution.

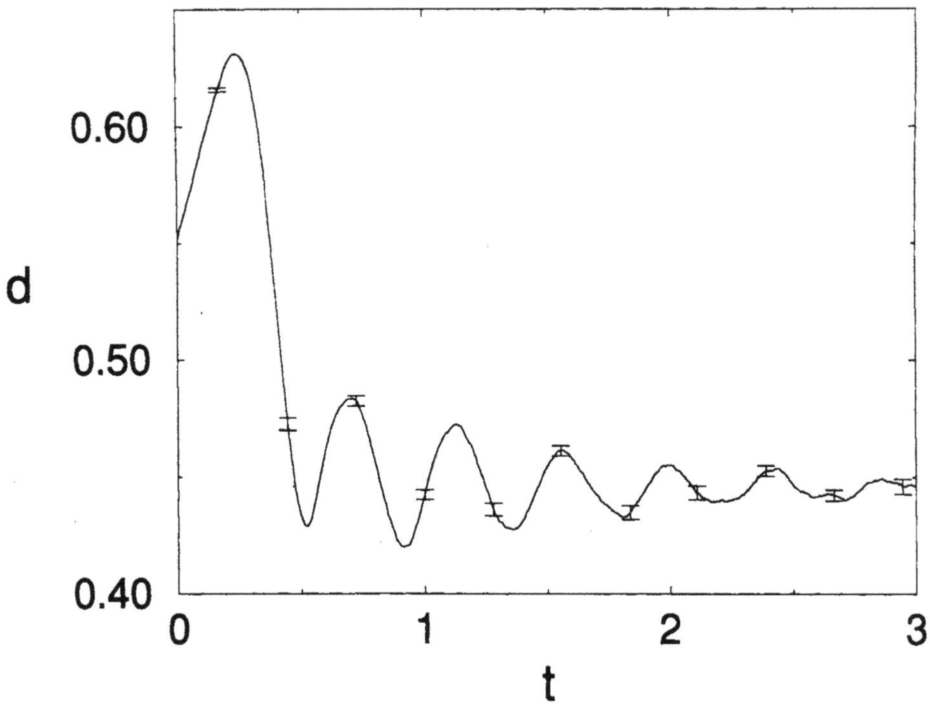

Fig. 3 Average distance d of the small disks from their center of mass in units of the radius of the drum, R, vs. time t measured as number of revolutions; simulation parameters as in Fig. 2.

As a measure of segregation we calculated the average distance of the small particles from their center of mass. This quantity is largest for the initial random distribution and decreases when the small disks cluster together. Figure 3 shows the typical behavior averaged over 50 runs with different initial configurations: the distance of the small particles from their center of mass decreases essentially within one revolution to its asymptotic value around which it has damped oscillations with ca. half a revolution of the drum as period (the drum is about half filled so that a rotation by 180° means a complete turnover of the whole configuration).

2.4 A Critical Ratio of Radii

The mechanism for segregation by shaking is different than in a rotating drum. During vertical shaking the large particles segregate to the surface, because the cyclic loosening of the whole packing allows smaller disks to slip underneath the larger ones. By contrast, the rotating drum transports all disks, small and large, upwards. The larger ones roll down the surface whereas smaller ones get stuck in niches on their way. This explains why the small disks are concentrated in the middle region.

The niches between large disks have a maximum size, which depends on their configuration. Hence, to get stuck in these niches, the small disks must not exceed a certain size. So we expect a critical ratio of the small to the large disk radii. Only for a ratio smaller than the critical value r_c will one find segregation. In our simulations we find segregation if the radius of the small disks is $r = 0.5$ (Fig. 2), but no segregation for $r = 0.9$ (in units of the large disk radius).

It is plausible that r is the only relevant parameter for segregation. The radius of the large disks and the period of the rotation are chosen as unit length and as unit time, respectively, throughout this paper. The other parameters of the model, the number of the small disks, the number of the large disks and the diameter of the drum, should not be crucial for segregation. A fifth parameter is the angle increment $\Delta\Phi$ (or time step). We do not expect it to be important, provided it is chosen small enough.

2.5 Segregation Scenario for a Test Particle

If the above mechanism is right, segregation in a rotating drum is a purely kinetic effect which does not require many interacting small disks; on the contrary: their excluded volume is an unnecessary complication if one wants to evaluate the segregation quantitatively. For example, Fig. 3 would look different if the number of small disks increases. It is an open problem to find a proper normalization of the segregation measure such that it only depends on the relevant parameter r.

We try to avoid this problem by studying the motion of a single small test particle. The number of large particles and the radius of the drum will be chosen big enough that they do not influence the results any more. Then the only parameter of the model is the radius r of the small disk.

What does "segregation" mean if there is only one small disk? We expect the following scenario: For $r < r_c$, the small disk will pass a trajectory similar to a spiral into a fixpoint which in general will not coincide with the center of mass of all the particles in the drum. The fixpoint should essentially correspond to the center of mass of the segregation cluster, if there were many small disks. On the other hand, for $r > r_c$, no such fixpoint should exist. The trajectory of the small disk should continue to visit a large part of the drum in an irregular fashion, instead.

It turns out, however, that this scenario is too simple. The simulations with a test particle are hampered by a new complication, a structural self-organization of the monodisperse system (all disks of equal size) which has to be discussed first.

2.6 The Monodisperse System

Starting from a random packing of the disks, the rotation leads to an ordered, "crystalline" structure with distinct domains separated by defect lines, and with an amorphous zone in the upper region [Fig. 4(a)]. This structure forms within less than half a revolution and then essentially reproduces itself in every step, apart from the surface configuration, occasional shifts of the defect lines by one lattice constant, and the amorphous structure.

Figure 4(b) illustrates our statement above that the structural self-organization of the system prevents unphysical motions which would require large gaps between the particles. Following the motion of each disk during 1/4 of a revolution shows the existence of a block in the center, close to the surface of the packing, which does not take part in the rotation. If it is twisted too much, it slips back along the defect lines, in contrast to the outer parts of the packing which are transported to the surface in concentric shells, and there roll down to the bottom. The "erosion" zone at the surface taking part in the surface avalanches has a thickness of 2–3 diameters of the disks in our simulation.

The localized motion of the grains in the slipping block does not imply that its structure is static: the defect lines inside the central block do move, and the whole network can be significantly deformed in spite of the absence of large displacements of the discs. The twist-induced dilation will turn out to be important for the motion of a small test particle put into this system.

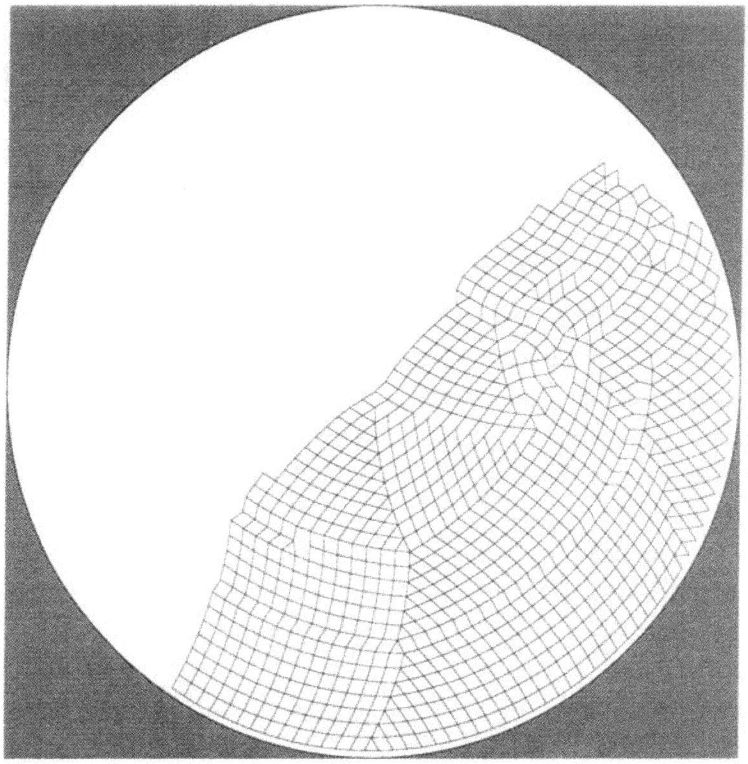

(a)

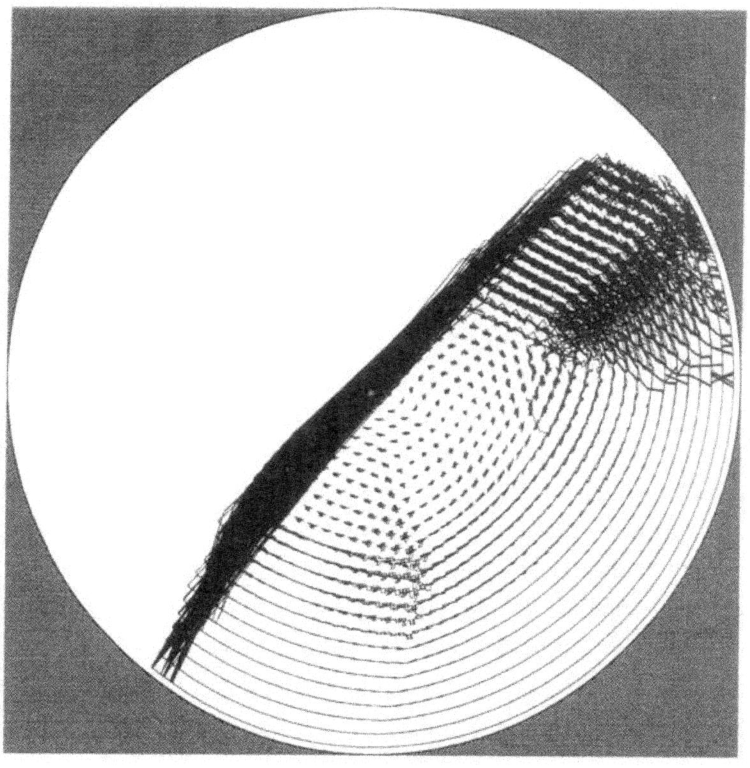

(b)

Fig. 4 (a) Lattice structure of the steady state of a monodisperse system, bonds connect the centers of disks which are in contact; (b) trajectories of the disks, central block slips back along the defect lines shown in (a).

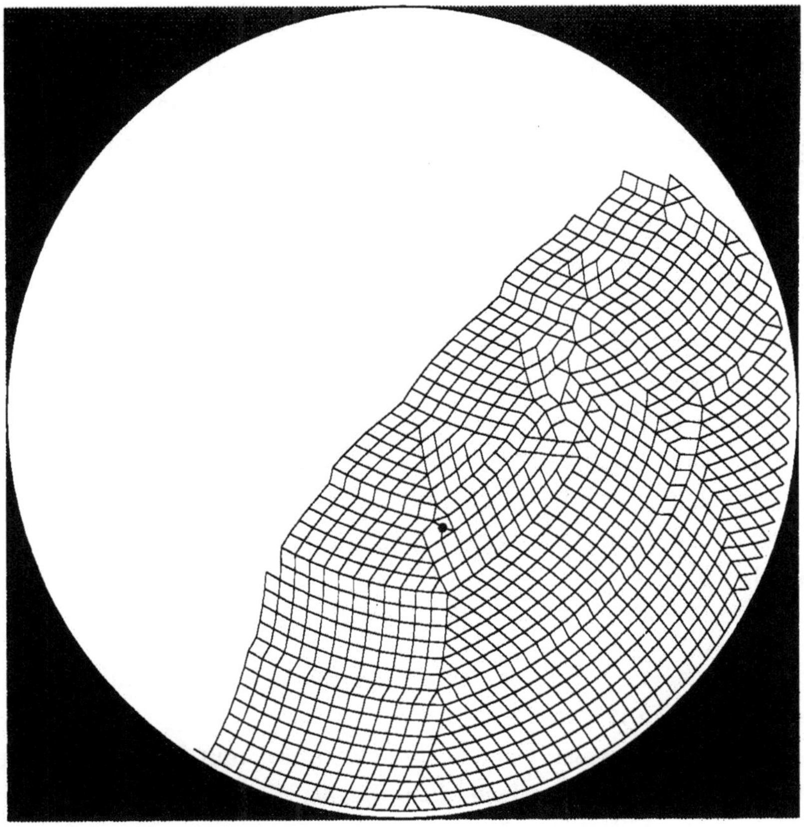

Fig. 5 Same as Fig. 4(a), but with a test particle of radius 0.5, indicated by the circle, at its final position.

2.7 The Monodisperse System with a Test Particle

One of the most prominent features of the motion of a single small disk in a system of large disks is its slowness close to the slipping central block. Even in the erosion zone it stays much longer than the large disks, because it gets stuck in the niches, and the avalanches simply sweep over it.

The second important observation is that in contrast to the large disks, the test particle may penetrate into the slipping block. However, once inside, it cannot escape any more, but may still move down along a lattice plane with small slope until it reaches a defect line going upwards. At this place it gets stuck forever. The twist-induced dilation of the block should play a major role for this motion. Of course, the presence of the small disk changes the structure of the slipping block, especially by introducing new defect lines due to the displacement of the large disks above it (Fig. 5).

3. HEAPING DUE TO VERTICAL SHAKING

3.1 Experiment

In 1992 Clément, Duran and Rajchenbach did a series of experiments in order to investigate how a two dimensional granular medium behaves when shaken vertically.[12] They filled the gap between two coaxial cylinders, i.e., a two-dimensional box with periodic boundary conditions, with beads of equal size. The diameter of the beads was almost equal to the distance between the two cylinders. The size of the box was such that the beads arranged themselves in a dense packing. Inserting a stick with the bead diameter between the cylinders, a situation equivalent to a planar two-dimensional box with side walls could be studied. Two material parameters were varied: the energy restitution coefficient and the bead-bead friction.

They observed two phenomena which exclude each other: surface fluidization and heaping. Surface fluidization requires a high coefficient of restitution and a frequency low enough to allow for an efficient energy transfer from the bottom to the surface by shock waves. Heaping is induced by the side walls and requires a strong bead-bead friction which must not be rendered ineffective at the surface by fluidization.

The heaping starts at the sidewalls where two small bumps are formed. They grow and, keeping their slope, their maxima move towards the center of the system as shown in Fig. 6. Finally the two heaps merge to build a steady surface shape with a maximum in the middle.

Measurements of the distance between a heap and the closest wall, Δ, showed a logarithmic time dependence[12]:

$$\Delta \sim \ln(\alpha t)\,. \tag{1}$$

Different accelerations yield different slopes in the semilogarithmic plot, the larger the maximal acceleration, the larger the slope.

Overlayed pictures of successive shakes give an important hint on the mechanism. One can see thin active zones at the surface of the heaps. These are the only parts of the heaps where a restructuring takes place. Small rolls are transporting the matter from the point where the heap meets the wall to the top of the heap. The probability of the occurrence of these rolls will determine the flux to the top of the heap. If one could calculate this probability it should be possible to explain the whole process quantitatively.

It should be kept in mind that this picture is only right for shaking with small acceleration (larger than gravitational acceleration, though). For larger accelerations, it was observed that significant parts of the bead packing are displaced and long range slips occur.

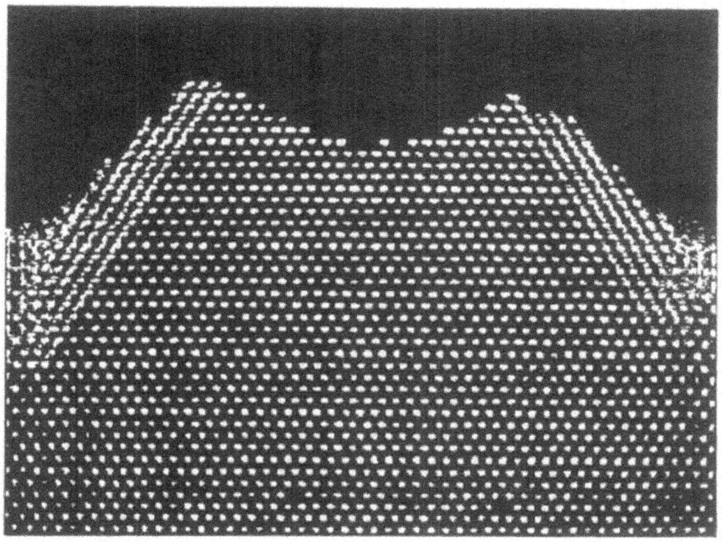

Fig. 6 Picture of the experiment (from Ref. 12).

3.2 Implementation

The inner part of our program is a suitably adapted implementation of the JMP-algorithm.[2] As mentioned in the introduction, it corresponds to the limit of high bead-bead friction and low restitution coefficient. Therefore it should be possible to reproduce the heaping observed in the experiment. For this purpose two problems had to be solved: How to incorporate the walls, and how to account for the difference between shaking with small and large acceleration?

The wall friction is simulated in the following way: as in the JMP model, the particles are sorted in the order in which they will arrive at the bottom after the next shake. The friction with the side walls accelerates the adjacent particles downwards so that they arrive prior to the inner ones which were at the same height after the last shake. This picture is confirmed by molecular dynamics simulations done by Jysoo Lee.[13] Therefore we sort the particles according to increasing effective heights:

$$\tilde{y} = y - F(x, y), \tag{2}$$

where the function F decreases towards the center.

The y-dependence of F takes care of the second problem. Our physical picture is the following: any restructuring requires some dilation of the dense packing. The experiments of Clément et al. show that the dilatancy depends on the acceleration, as mentioned at the end of the last section: for small acceleration the dilatancy and hence the restructuring are concentrated within a narrow region at the surface, whereas for large acceleration it extends further down into the bulk of the packing. The JMP model does not contain any parameter like the acceleration, but the inhomogeneous dilatancy can easily be accounted for by effectively suppressing any restructuring far away from the surface. In our program this is achieved by letting the function F decay from the surface into the bulk, so that the wall friction has no effect deep in the bulk.

We used the function:

$$F(x, y) = E \cdot \exp \frac{-(x-1)}{D} \cdot \frac{1-(h(x)-y)}{G} \tag{3}$$

for $h(x) - y \leq G$ and $F = 0$ otherwise. Here, $1 \leq x \leq L/2$ is the distance of a particle center from the closest side wall (all radii are equal 1) and y its height measured from the bottom of the box. $h(x)$ is the local height of the surface defined by the highest line that connects the centers of spheres in contact.

D, E and G are parameters. D is the characteristic length over which particles feel the wall friction. G is the width of the dilation zone close to the surface in which the wall friction can cause the displacements of spheres directly. Displaced spheres can push higher spheres out of their former positions so that inside a heap restructuring occurs also in a larger distance from the surface than G. The strength E of the bonus for particles adjacent to the wall determines if there is any restructuring at all, as will be shown below.

Obviously these ideas are not restricted to monodisperse systems. However, as we want to compare with the experiments, we specially adapted the JMP algorithm to the monodisperse situation with triangular packing. Then the final position of the spheres are lattice positions, thus much easier to calculate. A particular advantage of this approach is that rounding errors of the particle coordinates can be eliminated. Spheres with equal effective y-coordinate [Eq. (2)] are allowed to relax in a random sequence.

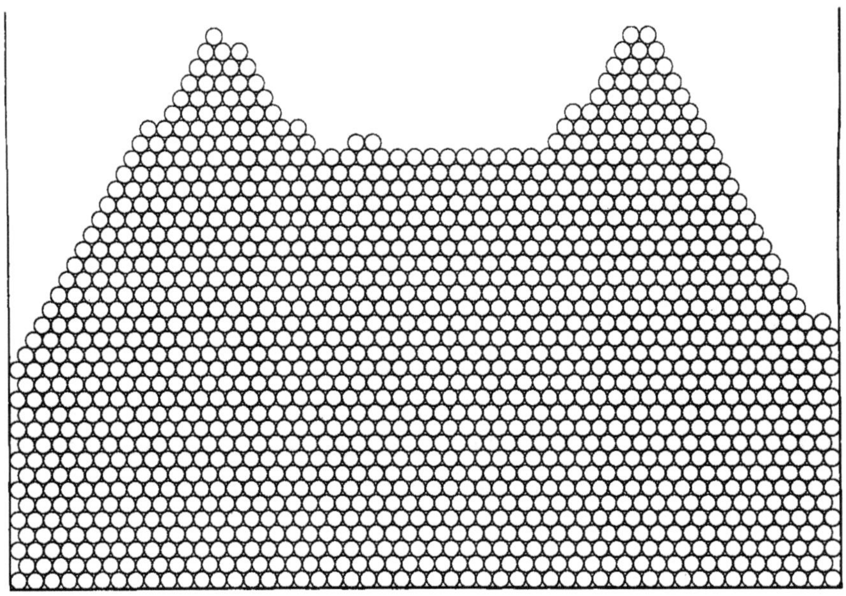

Fig. 7 State of simulation after 500 shakes.

3.3 Results and Comparison with the Experiment

The qualitative results of our simulations, Fig. 7, agree with the experiment. To our knowledge this is the first simulation which reproduces the two heaps occurring at the walls and wandering to the middle, where they merge. In our model the slope of the heaps is always close to 60°, larger than in the experiments due to the extremely large static friction and small restitution coefficients which suppress avalanches.

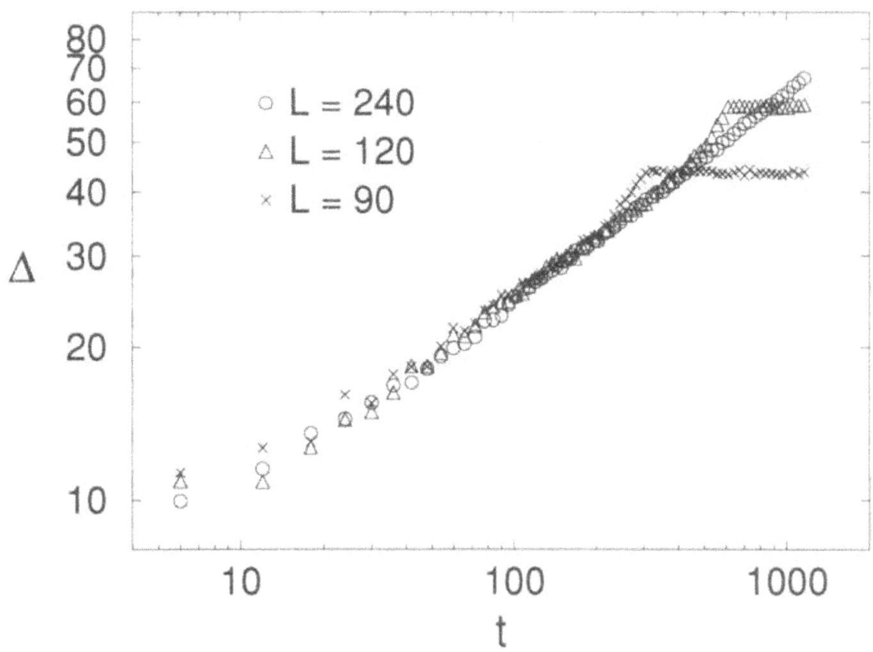

Fig. 8 Power law and finite size effect for small systems; $E = 2.9$, $D = 0.8$, $G = 10.0$.

As one can see in Fig. 8, the distance Δ between the heaps and the walls obeys a power law as a function of time with an effective exponent which, depending on the parameters, varies between 0.3 and 0.5. This is not in accordance with the logarithmic behavior seen experimentally. Below we shall show that random contributions to the motion of the beads remove this discrepancy. Also visible is the finite size effect for small systems of length L: when the two heaps begin to merge, at $\Delta \approx 0.35L$, the dynamics changes. The closer the heaps are in this stage, the less matter is needed to fill the space between them and therefore — if the mechanism of mass transport does not change — the heaps will grow faster than expected from the power law.

A simple argument explains why a power law can be expected and why the exponent should be less than 1/2. If we assume a constant flux from the foot to the upper part of the heap, then its area grows linearly in time and therefore the distance to the wall should behave like $t^{1/2}$. Our analytical considerations in the next section indicate that the flux decreases the larger the heap is. Therefore the exponent will be smaller.

If the function $F(x, y)$ extends over too large a distance D for a given strength E, no heaping is observed. The physical reason is that the shearing of order E/D becomes too weak. The phase boundary between the heaping and the non-heaping phase can be calculated analytically:

$$E(D) = \begin{cases} \dfrac{\sqrt{3}}{1 - e^{-1/D} \cdot \dfrac{1-\sqrt{3}}{G}} & \text{for } G > \sqrt{3} \\ \sqrt{3} & \text{for } G \leq \sqrt{3} \end{cases} \tag{4}$$

This formula is obtained by assuming that a restructuring occurs only, if one sphere of layer n of the dense packing settles prior to a neighboring one which was in layer $n-1$ after the

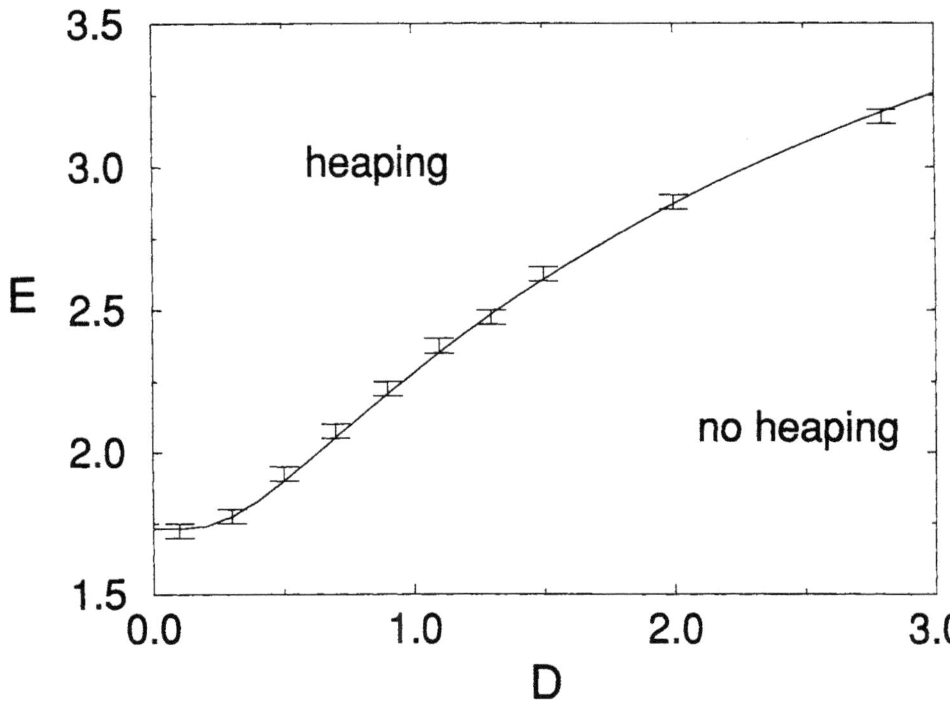

Fig. 9 Theoretically predicted phase boundary and experimental results; $G = 5$.

previous shake. Figure 9 shows the theoretical curve and some values determined from the simulation. The agreement is better than 3%.

Why does one not observe a logarithmic law? A very suggestive analogy can be drawn to the size segregation during shaking as described in the introduction. There the large particle raises continuously in a surrounding of sufficiently small ones. However, if the sizes are too similar, it gets frozen into a static configuration, which can be broken up intermittently by the inclusion of some randomness. In the case of the heaping, our simulations with sufficiently strong shearing show a continuous build-up of the heaps. This is in contrast to the experimental finding that the "block motion appears stochastically in time with a duration and a waiting time much larger than the period of excitation",[12] in other words it is intermittent. This suggests that the experiment is done in a regime where our simulation gives no heaping because of the absence of random motion.

Therefore we added a random contribution uniformly distributed between −0.2 and +0.2 to both x and y in every shake. Figures 10(a) and (b) show the results for parameters, where no heaping was obtained previously, plotted double-logarithmically and semi-logarithmically. Now the surface forms heaps, but longer simulations are needed to decide unambiguously whether the time dependence is really logarithmic. At the present stage the results do not rule out our conjecture that a transition is obtained between a logarithmic heaping for weak shearing as observed in the experiment and an algebraic heaping for strong shearing.

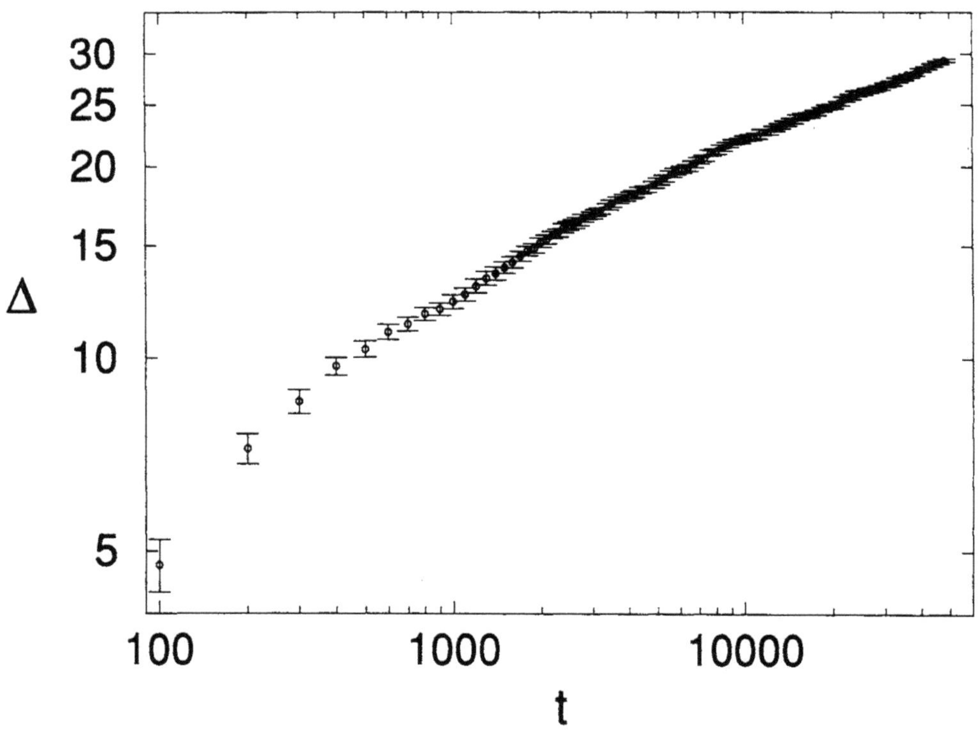

(a)

Fig. 10 Results of a simulation with randomness, $E = 2.5$, $D = 2.0$, $G = 5.0$, in a log-log plot (a), and in a log-lin plot (b).

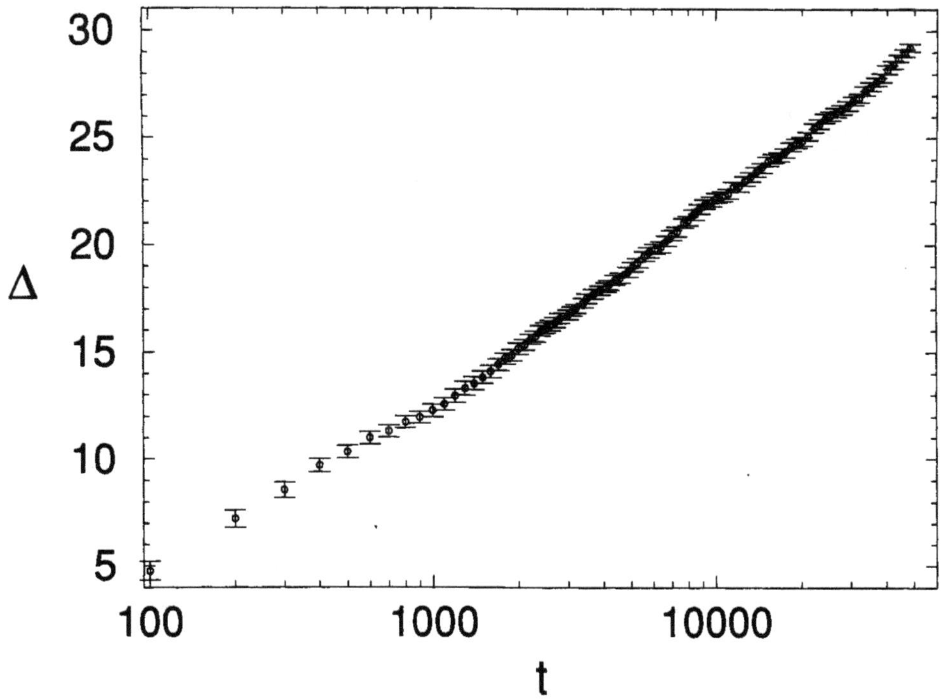

(b)

Fig. 10 (*Continued*)

3.4 Basic Mechanism

The analytic description of the model is still in progress but there are preliminary results that shed some light on the underlying heaping mechanism. As in our simulation the system returns to a triangular dense packing after each shake, all particle movements can be described by lattice vectors. We only consider realistic parameter values such that for each shake a particle returns either to its previous or to a nearest neighbor position, with the exception that it rolls down the surface, which is the only case, when it is displaced by more than a lattice constant. If the final position was occupied previously, the corresponding particle must be pushed aside to one of its nearest neighbor positions and so on, leading to a chain of displacement vectors. One can distinguish two different classes of chains, examples of which are given in Fig. 11:

- *Pure convection events.* The chain is closed, therefore the net flux is zero; no restructuring has occurred.
- *Heaping convection events.* The chain is not closed, but starts and ends on the surface; the particle at the beginning of the chain was virtually moved to the end of the chain.

A chain can only be open if between its start and its end point there was a local maximum or at least one point with slope zero before the update. This shows that the longer the incline the longer the open chains must be, and the higher the probability for a chain to be closed. This explains why the exponent in Fig. 8 is smaller than 1/2.

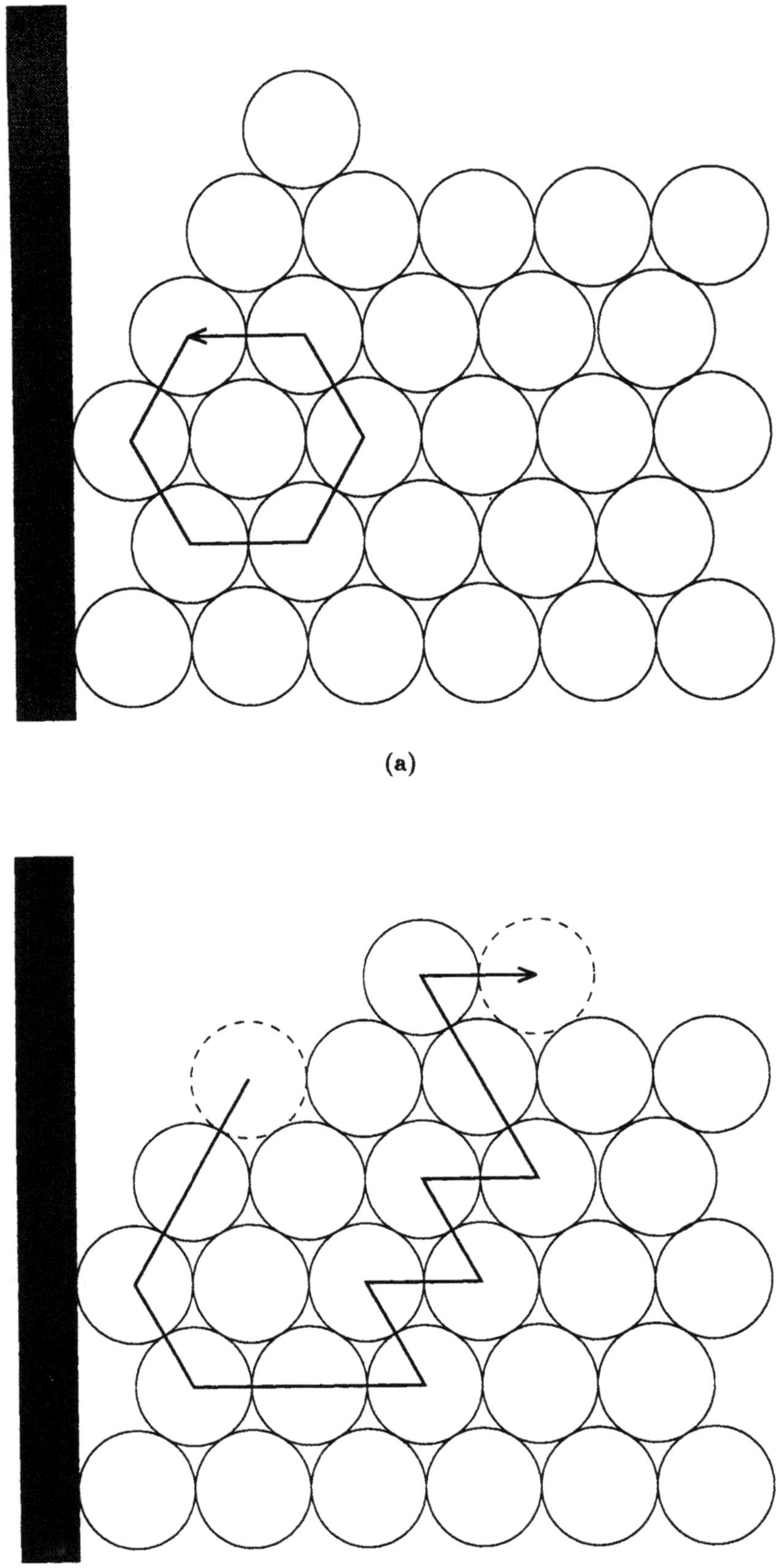

(b)

Fig. 11 Two typical displacement chains close to the left side wall, as recorded in the simulation; **(a)** closed path, no net flux; **(b)** open path, net flux of one particle.

4. CONCLUSION

The main conclusion supported by our simulations is that the JMP-model can describe several different kinetic phenomena in granular matter, despite the diversity of the underlying geometrical mechanisms. Moreover, due to its simplicity it lends itself to analytical treatment in special cases as indicated in the last section. Therefore it is a suitable reference model which may give deeper insight into a wide range of kinetic phenomena with a structural origin in granular materials. However, all dynamic effects involving kinetic or elastic energies, such as fluidization, remain elusive. They have to be treated by molecular dynamics simulations, for example.

Our computer simulations yield several predictions which should be checked experimentally such as the self-organized structure in a rotated monodisperse system and the existence of a threshold for the wall friction above which the heaping is no longer logarithmic but obeys a power law with an exponent smaller than 1/2. The investigation of the critical exponents for the rotating drum as well as for the heaping are now in progress, and it would be interesting to see more experiments with which the simulations and the analytical results could be compared.

5. ACKNOWLEDGMENTS

We like to thank Hans Herrmann, Remi Jullien, János Kertész and Jysoo Lee for many interesting discussions, as well as the groups of Daniel Bideau and Jacques Duran for information on their experiments prior to publication. This work was performed within SFB 341 of the Deutsche Forschungsgemeinschaft.

REFERENCES

1. B. B. Mandelbrot, *The Fractal Geometry of Nature* (Freeman, San Francisco, 1982).
2. R. Jullien, P. Meakin and A. Pavlovitch, *Phys. Rev. Lett.* **69**, 640 (1992).
3. R. Jullien, P. Meakin and A. Pavlovitch, *Europhys. Lett.* **22**, 523 (1993).
4. J. Duran, J. Rajchenbach and E. Clément, *Phys. Rev. Lett.* **70**, 2431 (1993).
5. A. Rosato, F. Prinz, K. J. Strandburg and R. Swendsen, *Powder Technol.* **49**, 59 (1986).
6. A. Rosato, K. J. Strandburg, F. Prinz and R. H. Swendsen, *Phys. Rev. Lett.* **58**, 1038 (1987).
7. G. C. Barker, A. Mehta and M. J. Grimson, *Phys. Rev. Lett.* **70**, 2194 (1993).
8. H. J. Herrmann, T. Pöschel, and G. H. Ristow: *Granular Materials*, video film, HLRZ, Forschungszentrum Jülich, 1992.
9. Y. Oyama, *Bull. Inst. Phys. Chem. Res., Tokyo* **18**, 600 (1939).
10. D. S. Cahn, D. W. Fuerstenau, T. W. Healy, R. Hogg and H. E. Rose, *Nature* **209**, 494 (1966).
11. D. Bideau, private communication.
12. E. Clément, J. Duran and J. Rajchenbach, *Phys. Rev. Lett.* **69**, 1189 (1992).
13. J. Lee, preprint.

Fractals, Vol. 1, No. 4 (1993) 784–794

KINETIC SURFACE ROUGHENING AND MOLECULAR BEAM EPITAXY

S. DAS SARMA
Department of Physics, University of Maryland,
College Park, Maryland 20742-4111, USA

Abstract

We review recent developments in our understanding of Molecular Beam Epitaxy as a kinetically rough growth phenomenon. It is argued that while the most general growth conditions lead to generic growth universality, actual growth conditions allow a complex interplay of several different dynamic universality classes producing rich crossover behavior determined by growth temperature, incident flux rate, and local solid state physics and chemistry of the growing material. Possible coarse-grained continuum growth equations which may be applicable to Molecular Beam Epitaxy are discussed.

1. INTRODUCTION

In growth via Molecular Beam Epitaxy (MBE, we use the abbreviation MBE to denote "molecular beam epitaxial" as well), a controlled flux of atoms (or, molecules) are deposited in ultra high vacuum normal to a substrate which is held at a constant temperature.[1] By suitably varying the incident atomic flux and the substrate temperature, one grows a thin film of the incident material on the substrate. While the average thickness of the growing film is determined by the average incident atomic deposition rate (we ignore any desorption of the incident atoms, which is an excellent approximation under the usual MBE growth conditions), kinetic roughness invariably builds up in the growing film due to the shot noise in the incident atomic beam. In this paper, we give a brief review of our current theoretical understanding of MBE growth from the viewpoint of the kinetic roughening phenomenon with emphasis on recent developments.

Kinetic roughening in a dynamically growing thin film via the vapor deposition MBE process is assumed to obey dynamical scaling[2] characterized by two independent critical exponents z (the dynamical exponent), and either β (the growth exponent) or α (the roughness exponent) with $z = \alpha/\beta$. The kinetic surface roughness $W(L, t)$, which is defined as the root mean square fluctuation in the thickness of the growing film, is thought to obey the dynamic scaling hypothesis[2]:

$$W(L,t) \sim L^{\alpha} f(t/L^{z}), \tag{1}$$

where L is the lateral width of the substrate, t is the growth time, and $f(x)$ is the dynamic scaling function with the following asymptotic properties:

$$\begin{aligned} f(x) &= 1 \quad \text{for} \quad t \gg L^{z}, \\ &= x^{\beta} \quad \text{for} \quad t \ll L^{z}. \end{aligned} \tag{2}$$

Thus, the kinetic roughness W for a substrate of size L grows as $W(L \to \infty) \sim t^{\beta}$ for $t \ll L^{z}$, and, as $W(t \to \infty) \sim L^{\alpha}$ for $t \gg L^{z}$. Note that the lateral correlation length for kinetic roughness $\xi \sim t^{1/z}$ and, W saturates for $\xi \gg L$ as the lateral correlation spreads over the whole system. A knowledge of the critical exponents z, α, β (and the scaling function f) solves the specific dynamical growth problem from the nonequilibrium statistical mechanics viewpoint. More specifically, a dynamical universality class is defined by a specific set of values of the critical exponents α, β, (and, z) and, just as in ordinary (static) critical phenomena, seemingly very different dynamical growth processes may have the same values of α and β and thus belong to the same universality class.

An issue that has attracted considerable recent interest[3–25] is the determination of the universality class of MBE growth. Note that strictly speaking, MBE growth violates the basic premise of the dynamic scaling hypothesis which requires that there be no characteristic lateral length scale in the problem (other than the system size L). In MBE, there is a temperature dependent diffusion length (l) which determines the lateral length over which an incident atom diffuses on the surface before incorporation into the growing film.[14,15] (Equivalently, one could think of the inverse of the diffusion rate or the atomic hopping rate as a characteristic time in the MBE growth problem.) When $l \sim L$, dynamic scaling hypothesis breaks down and the effective growth exponent vanishes purely due to finite size effects.[15,16] On the other hand, for $l \ll L$, dynamic scaling should hold for MBE growth. This is accomplished by having either a low temperature and/or a high deposition rate. Dynamic scaling should be asymptotically valid in the limit of very high deposition rates. Note that in the limit of a large diffusion length ($l \gg L$) or a high hopping rate, the system tends to locally equilibrate in between two successive deposition events and, thus, kinetic roughening which is a far from equilibrium phenomenon does not play an important role. Another significant point is that the relevant system size L for MBE growth is not always the lateral size of the whole substrate. Only on a purely singular flat surface L will be the size of the whole substrate. Usually the substrate is vicinal with flat terraces separated by steps. Then the system size L denotes the lateral size of the flat terraces because once $l \gtrsim L$, step flow growth takes over, and kinetic roughness plays only a minor role. The condition $l(T) \geq L$ defines the temperature[16] at which step flow growth mode takes over, and above this epitaxial growth temperature, MBE growth produces smooth and reasonably flat thin films. For studying kinetic roughness one needs to stay below this step flow region and, therefore, the growth temperature should be low and the substrate should be a symmetric singular surface.

Monitoring $W(L, t)$, the dynamical surface roughness of the growing film, in this kinetically rough (i.e., low temperature) regime enables[21-25] one to obtain α and β, and determine the universality class of MBE growth. Note that the nature of the MBE growth process invariably leads to the critical exponents β, α, and z to be temperature dependent (and, eventually in the high temperature region, finite size and equilibrium effects combine to invariably produce $\beta(T)$, $\alpha(T) \to 0$). Thus, crossover effects are always present in MBE growth studied as a dynamical critical phenomenon. In particular, at very low temperatures, when atomic hopping and diffusion are exponentially suppressed, lateral dynamical correlation length ξ is very small ($\xi \ll a$, where a is the lattice spacing) and therefore, MBE growth is dominated by the random deposition process, leading to $\beta = 1/2$, which is characteristic of the Poisson growth associated with the purely random deposition process, in the initial stages of growth. Nontrivial kinetic surface roughening effects, therefore, manifest themselves in MBE growth at intermediate temperatures — the temperature must not be too high for step-flow growth to take over nor too low for purely random growth to occur.

MBE growth is a discrete phenomenon with the growth properties determined by local atomistic rules controlling kinetic processes such as deposition, diffusion, relaxation, attachment to or detachment from steps, incorporation, evaporation, etc. It is possible to numerically simulate[1,16,18,20] MBE growth quite successfully by using local atomistic kinetic rules on lattice models. For a global theoretical understanding of the kinetic roughening process, however, one wants to employ a coarse-grained continuum description[2] (presumably using suitable partial differential equations) for the dynamical height variable $h(\underset{\sim}{r}, t)$ where h is the height (or, the thickness) of the growing interface at the substrate position $\underset{\sim}{r}$ and time t. Thus, $(\underset{\sim}{r}, h)$ is a d-dimensional dynamical vector defining the growing interface on a $d-1$ dimensional substrate. For the usual MBE growth, $d = 2+1$, denoting growth on a 2-dimensional substrate. The hope is to discover the continuum equation which describes (at least, the asymptotic critical properties of) MBE growth in a coarse-grained fashion. We emphasize that, similar to the situations in many other problems in physics, there is no general proof that such a single-valued, coarse-grained continuum description in terms of the variable $h(\underset{\sim}{r}, t)$ must necessarily exist for all conceivable sets of local rules defining kinetic growth. There are, however, several discrete kinetic growth models[2] whose asymptotic critical properties are well described by simple stochastic differential equations in the coarse-grained variable $h(\underset{\sim}{r}, t)$. For the rest of our discussion, we assume that the average surface height $\langle h \rangle$ has been subtracted from the variable h in its definition so that $h(\underset{\sim}{r}, t)$ denotes only the local surface height fluctuation.

In the absence of desorption and formation of bulk vacancies or surface overhangs (i.e., in the solid-on-solid model with no evaporation), there must be a conserved surface mass current $\underset{\sim}{j}$ in the MBE growth problem, leading immediately[6] to the current continuity equation for the coarse-grained height variable h:

$$\frac{\partial h}{\partial t} = -\nabla \cdot \underset{\sim}{j} + \eta \,, \tag{3}$$

where η is the random shot noise associated with the deposition process which is chosen to be an unconserved white noise (in space and time) throughout this paper. Under the most general conditions which are consistent with the symmetry of the problem, Eq. (3) leads to the following conserved continuum equation for the solid-on-solid (SOS) growth in the leading order:

$$\frac{\partial h}{\partial t} = \nu_2 \nabla^2 h - \nu_4 \nabla^4 h + \lambda_{42} \nabla^2 (\nabla h)^2 + \lambda_{43} \nabla (\nabla h)^3 + \eta \,, \tag{4}$$

where ν_2, ν_4, λ_{42}, λ_{43}, the coefficients of the various gradient terms as shown, are unknown within the continuum kinetic roughening theory (and, should be determined from a detailed microscopic description of the growth process). We emphasize that up to the fourth order in the gradient expansion, Eq. (4) contains all possible terms consistent with the symmetry of the problem and the condition of conserved growth. One, therefore, feels that, if a continuum, single-valued, stable description of conserved MBE growth is possible, then Eq. (4) must be it, at least in the leading order of gradient expansion. We mention that in the context of nonequilibrium growth, the first term (i.e., the $\nabla^2 h$ term) on the right-hand side of Eq. (4), was introduced[26] by Edwards and Wilkinson; the second term (i.e., the $\nabla^4 h$ term) was considered simultaneously by Wolf and Villain,[3] and by Das Sarma and Tamborenea[4] (it is also the leading order equilibrium mass diffusion current originally studied[27] by Herring, and by Mullins a long time ago); the third term (i.e., the nonlinear $\nabla^2(\nabla h)^2$ term) was introduced[6,13] by Lai and Das Sarma, and by Villain; and the last term (i.e., the nonlinear $\nabla(\nabla h)^3$ term) was introduced by Lai and Das Sarma..[6]

It is trivially obvious that the asymptotic large-scale long-time critical properties of Eq. (4) will be totally dominated by the Edwards-Wilkinson $\nabla^2 h$ term because it vanishes most slowly at long wavelengths. It is, however, not obvious that this term must necessarily be present in all situations and, if it is present, what its strength is. In particular, if both the $\nabla^2 h$ and $\nabla^4 h$ terms are present, the crossover length scale at which the Edwards-Wilkinson term starts becoming important is given by $[\nu_4/\nu_2]^{1/2}$ which can be very large indeed if $\nu_4 \gg \nu_2$. If the crossover length scale is much larger than the system size, then the Edwards-Wilkinson (EW) universality may not show up. Even for large systems, the lateral correlation length $\xi(t)$ must be of the order of this crossover length for the EW universality to manifest itself, which could take very long times of the order of $t \sim (\nu_4/\nu_2)^{z/2}$ if $\nu_2 \ll \nu_4$. Two situations where the EW term is known to be important are the presence of desorption (which introduces a local relaxation term described by $\nabla^2 h$), and the presence of gravitation or some similar relaxation term which tends to minimize the local height of the deposited particle. Thus, an attractive substrate or any other mechanism which biases downward hops of surface atoms (i.e., towards the substrate or decreasing h) over up-hops (i.e., away from the substrate or increasing local h), will necessarily introduce the EW term, and, at least asymptotically, lead to EW universality. This is borne out by actual simulations in all situations *except one* (the 1+ model[11] as introduced by Das Sarma and Tamborenea[4]). The crossover to EW universality is, however, extremely slow in many situations, making it unclear whether it is always experimentally observable. The EW equation being a linear equation is easily solved by a Fourier transformation leading to the EW critical exponents:

$$\beta = (3-d)/4, \quad \alpha = (3-d)/2\,; \quad z = 2\,. \tag{5}$$

Thus, EW universality leads to only logarithmic roughness in the physically most interesting $d = 2+1$ dimensions. We know of no direct experimental evidence supporting such logarithmic kinetic roughening in MBE growth — at high temperatures, MBE growth is smooth with very little surface roughening, but, as emphasized before, we attribute this smooth growth to finite-size effects and step-flow growth mode. In the low temperature kinetically rough growth regime, the surface roughness of MBE grown films is usually substantial and unlikely to be only logarithmic. Thus, ν_2, if present, is usually small in magnitude.

The linear $\nabla^4 h$ term in the growth equation is easily solved to produce the critical exponents:

$$\beta = (5-d)/8; \quad \alpha = (5-d)/2; \quad z = 4. \tag{6}$$

This term has been associated with surface diffusion because it is the leading term in the chemical potential gradient driven (near equilibrium) mass diffusion on a surface as studied by Mullins[27] a long time ago. In the MBE growth problem, one is very far from equilibrium, and there is no chemical potential in the usual sense and, therefore, the connection between the $\nu_4\nabla^4 h$ term in Eq. (4) and true (nonequilibrium) surface diffusion is unclear. But, the critical exponents corresponding to Eq. (6) show up in a number of discrete growth simulations[3,4,11,16–18] involving atomic hopping to the sites with the maximum local coordination numbers (i.e., maximum local chemical bonding). One, therefore, identifies the $\nu_4\nabla^4 h$ term as arising in the growth equation from the effect of surface diffusion of the deposited atom to the site with the maximum local bonding. The problem is that the sites with the maximum local bonding in a global (or, absolute) sense are obviously the trapping sites (i.e., a surface vacancy surrounded by atoms on all sides — an empty surface site with all its nearest lateral neighbor sites filled) which are necessarily also the local height minima. (Note that this is true only in a global sense — most often the sites with maximum local bonding, such as the kink sites, are *not* local height minima.) Thus, it seems that any surface diffusion-driven relaxation to local maximally coordinated sites must invariably also have a $\nabla^2 h$ term. Numerical simulations have not yet completely clarified this issue.[28,29] In particular, the original models introduced by Das Sarma and Tamborenea[4] (the 1+ model[11]), and, by Wolf and Villain[2] (the $d2+$ model[11]) seem to show only the $\nabla^4 h$ critical exponents given by Eq. (6), with no sign of crossover to EW behavior in the largest simulations carried out so far.[28,29] This is one of the more important unsolved puzzles in the study of MBE growth as a kinetic roughening phenomenon. There are four possibilities for these "anomalous" 1+ and $d2+$ models: (i) The crossover to EW universality is extremely slow and will eventually show up as computation power increases; (ii) for reasons not understood at the present time, the $\nu_2\nabla^2 h$ term (as well as the fourth order nonlinear terms) is identically zero in the growth equation; (iii) the coefficients ν_2, ν_4 etc. are time-dependent as the surface evolves dynamically and, therefore, even though the ν_2 term is formally present it does not show up in the exponents (of course, the time-dependence has to be very special for this to occur); (iv) dynamic scaling fails for these models. There is strong indication[28–30] that the nature of dynamic scaling is very different in these models (1+ and $d2+$)[11] compared with other kinetic roughening models. In particular, the exponents α and β one gets by studying $W(L, t)$ are substantially different from those (α' and β') obtained from a study of the asymptotic properties of the height-height correlation function $G(x, t)$ of the system.[28,29] This is bizarre and has been referred[28] to as "anomalous" dynamic scaling. In addition, the scaling function $f(x)$ of Eqs. (1) and (2) in these models does not approach[28] a constant (as in Eq. (2)) for $x(= t/L^z) \gg 1$, but behaves as $f(x \gg 1) \sim x[(\ln x)^{-1}]$ for $d = 2[3]$ dimensions. The fact that the critical exponent α connected with these models is large ($\alpha \simeq 3/2$, 1 in $d = 2, 3$) implies that the associated kinetic roughness is not self-affine, which may lead to these pathologies. In addition, the structure factor (the equal time Fourier transform of $G(x, t)$) shows an anomalous system size dependence[29] in these models. These issues need clarification.

The fourth-order nonlinearities in Eq. (4) are, of course, more relevant (in the renormalization group sense) than the linear $\nu_4\nabla^4 h$ term as can be easily seen by a power counting analysis (the fourth order nonlinearities $\nabla^2(\nabla h)^2$ and $\nabla(\nabla h)^3$ are, however, less relevant than the linear second order $\nabla^2 h$ term). They have been discussed using a renormalization group analysis[6] by Lai and Das Sarma. A simple dimensional analysis gives:

$$\nabla^2(\nabla h)^2 : \quad \beta = (5-d)/(7+d); \quad \alpha = (5-d)/3; \quad z = (7+d)/3. \tag{7}$$

$$\nabla(\nabla h)^3 : \quad \beta = (5-d)/2(3+d); \quad \alpha = (5-d)/4; \quad z = (3+d)/2. \tag{8}$$

Dimensional analysis also shows that the $\nabla(\nabla h)^3$ nonlinearity is more relevant than the $\nabla^2(\nabla h)^2$ term in the renormalization group sense. The result [(Eq. (7)] for the critical exponents of the $\nabla^2(\nabla h)^2$ term, which was originally obtained[6] via a one loop calculation, is actually exact to all orders[31] because the noise is unrenormalized and the vertex diagrams do not introduce any logarithmic corrections (i.e., the interaction strength remains local). The result corresponding to Eq. (8) for the critical exponents of $\nabla(\nabla h)^3$ nonlinearity is, however, not expected[31] to be exact because there are obvious logarithmic vertex corrections. An RPA treatment[31] summing the most divergent diagrams leads to a screening out of the logarithmic divergence and, therefore, any source of difficulty is at the four loop order. We expect[31] corrections (if any) to the exponents of Eq. (8) for the $\nabla(\nabla h)^3$ nonlinearity to be small.

The physical mechanisms responsible for the occurrence of the fourth order nonlinear terms in the growth equation are not explicitly clear at this stage. What is known is that simulations[11,17,18,32] of atomistic hopping which treat up- and down-hopping on an equal footing (i.e., both are equally likely) produce critical exponents consistent with Eq. (7), leading one to conclude that the nonequilibrium $\nabla^2(\nabla h)^2$ term shows up in the presence of atomistic up-hopping. This is consistent with the original finding[6] of Lai and Das Sarma who showed that allowing up-hopping in a stochastic growth model led to $\beta \simeq 1/3$ and $\alpha \simeq 1$ in $d = 1+1$, consistent with the $\nabla^2(\nabla h)^2$ nonlinearity. (We emphasize that in the presence of down-hopping alone, all SOS conserved models are expected[28,29] to eventually crossover to the EW universality, except for the 1+ and $d2+$ models which so far have shown only the linear $\nabla^4 h$ exponents, but with anomalous dynamic scaling behavior.) It is worthwhile to mention here that if down-hopping is suppressed in a growth model (the so-called[33] "Schwoebel effect"), one gets an instability because there is a local bias towards increasing h — in such a situation dynamic scaling is not valid and critical exponents are not a meaningful concept. Villain has argued[13] that such asymmetric diffusion barriers which suppress down-hopping lead to a $\nabla^2 h$ term in the growth equation with a negative coefficient (i.e., $\nu_2 < 0$) leading to an instability, just as an asymmetric barrier favoring down-hopping (i.e., suppressing up-hopping) produces the $\nu_2\nabla^2 h$ term with $\nu_2 > 0$ leading to the EW asymptotic behavior. We believe that when the diffusion barrier is symmetric between up- and down-hops, the universality class may correspond to the $\nabla^2(\nabla h)^2$ term, provided, of course, the conserved SOS conditions are obeyed in the growth process. The more relevant fourth order nonlinearity, the $\nabla(\nabla h)^3$ term, is problematic from a physical viewpoint because it is difficult[6] to identify physical mechanisms which produce this term during growth. Numerical studies show the unambiguous presence[11,34] of this nonlinearity in $d = 2+1$-dimensional stochastic simulations, particularly in situations where the eventual crossover is to the EW behavior. We believe that this term arises as a higher order process to the EW term, and may dominate the transient behavior of initial stages (~ a few thousand layers) of three dimensional conserved MBE growth until the EW crossover sets in.

To understand this connection between the $\nabla^2 h$ term and the $\nabla(\nabla h)^3$ term in the growth equation, we consider the standard surface tension driven interface free energy which is described by the Hamiltonian:

$$H \sim \int d^{d-1}r\sqrt{1+(\nabla h)^2} \sim \int d^{d-1}r[a|\nabla h|^2 + b|\nabla h|^4 + \cdots]. \tag{9}$$

The corresponding dynamics could be described by the Langevin equation[6]:

$$\frac{\partial h}{\partial t} = -\Gamma \frac{\delta H}{\delta h} + f_1 , \tag{10}$$

where f_1 is any manifestly nonequilibrium part of the growth equation which cannot be derived from a Hamiltonian. Using the standard surface free energy Hamiltonian defined by Eq. (9), we see that the a and the b terms in H lead immediately to the $\nabla^2 h$ and the $\nabla(\nabla h)^3$ terms respectively in the growth equation. In this sense, the $\nabla(\nabla h)^3$ nonlinearity can be thought of as a higher order correction to the EW term in the Langevin dynamics of the dynamically evolving interface. It is, therefore, likely that the presence of $\nabla(\nabla h)^3$ exponents in growth simulations should alert one to the possibility of an eventual crossover to EW universality, as has recently been verified[28,29] in the $d4+$ model of Das Sarma and Ghaisas,[11] where the initial transient regime corresponding to the $\nabla(\nabla h)^3$ nonlinearity lasts[34] over three orders of magnitudes of growth time, eventually crossing over to the expected EW universality.[29]

The Hamiltonian-based Langevin equation approach also allows an intuitive classification of the $\nabla^4 h$ term in the growth equation, which arises naturally in a curvature-driven surface free energy of the form:

$$H \sim \int d^{d-1}r|\nabla^2 h|^2 , \tag{11}$$

which, coupled with Eq. (10), immediately leads to the $\nabla^4 h$ term in the growth equation. Kim and Das Sarma have recently shown[35] that growth under the restricted curvature condition (which should be controlled by a Hamiltonian of the form of Eq. (11)) leads directly to the $\nabla^4 h$ universality. A curvature-driven free energy, as in Eq. (11), has been much studied recently[36] in the context of tethered membranes and crumpling transitions where the usual surface tension term (Eq. (9)) in the free energy is negligibly small compared with the bending energy which is characterized by the local curvature. The Langevin dynamics of such tethered membranes naturally leads to the $\nabla^4 h$ term in the growth equation without any $\nabla^2 h$ terms.

We note, however, that the nonlinear term $\nabla^2(\nabla h)^2$ in the growth equation does not arise from any Hamiltonian and is a purely nonequilibrium effect (contained in the functional f_1 in the generalized Langevin dynamics of Eq. (10)). Such nonequilibrium terms, which cannot be derived on the basis of an underlying Hamiltonian, are very important in the generic kinetic roughening model, and the celebrated KPZ nonlinearity[37,2] is such a nonequilibrium term in the second order growth equation.

The generic model for kinetically rough growth relaxes the SOS approximation of the conserved MBE growth considered so far in this paper, and allows desorption as well as formation of overhangs and vacancies in the growing deposit. This leads to the well-known Kardar-Parisi-Zhang (KPZ) equation[37] which has been much discussed in the literature[2] and is given by:

$$\frac{\partial h}{\partial t} = \nu_2 \nabla^2 h + \lambda_2 (\nabla h)^2 + \eta \,, \tag{12}$$

where the nonlinear $\lambda_2(\nabla h)^2$ term, which cannot be written in the $\underset{\sim}{\nabla} \cdot \underset{\sim}{j}$ form, breaks the current conservation condition of Eq. (4). The KPZ nonlinearity allows ballistic deposition and lateral sideways growth, both involving formation of vacancies and overhangs in the growing film. Desorption also leads to the KPZ nonlinearity (even within an SOS model, e.g. RSOS model[38] of Kim and Kosterlitz) as the EW $\nabla^2 h$ linear term gets renormalized. We emphasize that the $(\nabla h)^2$ KPZ nonlinearity cannot be derived from a Hamiltonian and is a generic example of the importance of manifestly nonequilibrium effects in kinetically rough growth.

Much has been written[2] about the KPZ equation. For our purpose, the most important point to emphasize is that if the KPZ nonlinearity is present, it defines the asymptotic universality class of the model because it is the most relevant term in the renormalization group sense. Thus, the presence of any overhangs or vacancies or desorption will immediately make the asymptotic universality class of MBE growth the KPZ universality. This has recently been verified[32] in large scale realistic atomistic simulations of MBE growth involving ballistic deposition and Arrhenius activated hopping. The question still remains, however, whether the eventual crossover to KPZ universality is preceded by any non-trivial transient growth regime belonging to any of the other universality classes discussed in this paper. This question takes on particular significance because a number of recent experiments[24,25] claim observation of critical exponents in MBE growth which correspond to the $\nabla^2(\nabla h)^2$ universality.

We briefly summarize here the results[32] of our recent large-scale MBE growth simulations which partially answer this question. These stochastic Monte Carlo simulations employ random deposition and Arrhenius activated hopping, explicitly obeying the principle of detailed balance locally (with the activation energy being dependent on the bonding configuration of the initial site only).[1,20] We find that when up- and down-hopping are both equivalently allowed without any bias, the initial growth regime corresponds to $\nabla^2(\nabla h)^2$ universality, eventually crossing over to the KPZ universality if overhangs and vacancies are allowed to form either during deposition or during hopping. (We do not allow any evaporation of atoms from the growing surface, which is consistent with the usual MBE growth conditions.) At high enough temperatures, MBE growth simulations follow EW behavior even in the ballistic deposition case, because there are very few vacancies and overhangs in the deposit due to very rapid atomic hopping, and the eventual crossover to KPZ behavior moves to very long times. When up-hopping is suppressed, both Das Sarma–Tamborenea (i.e., $\nabla^4 h$) and Edwards-Wilkinson (i.e., $\nabla^2 h$) universalities may show up in the initial transient regime (depending on the growth temperature) before eventual crossover to KPZ universality. If up-hopping is suppressed in an SOS growth simulation, the eventual asymptotic crossover is always to the EW universality, but very long transient regimes are invariably present characterized by critical exponents corresponding to $\nabla(\nabla h)^3$ or $\nabla^4 h$ universalities in $d = 3$ or 2 dimensions, respectively. Our preliminary simulations with an asymmetric Schwoebel barrier show the expected instability with no well-defined β or α (i.e., no dynamic scaling). More numerical work is clearly needed to establish the crossover behavior of MBE growth.

Other than the complicated crossover behavior discussed above, an important finding of our recent simulations[32] is that the crossover to the KPZ universality is invariably triggered by a large scale and rapid formation of defects in the deposit. In this transient regime of defect formation, dynamic scaling does not hold and $\beta_{\text{effective}} \approx 1$. Beyond this defect-forming transient regime, growth is described by the KPZ universality whereas before this transient, growth is better described by one of the other (SOS) models as discussed above. The onset of this defect formation moves to later times for higher temperatures in an (almost) exponential fashion and, before this transient region, defect formation is slow with a very low defect density in the growing deposit. We believe that the SOS model may be a reasonable description for MBE growth below this defect-forming transient regime with true KPZ universality taking over beyond this regime.

Before concluding we comment on the recent suggestion[8,39] that MBE growth may be describable by the complete surface diffusion equation (involving the Laplace-Beltrami operator along the surface for the local curvature) to which the $\nu_4 \nabla^4 h$ term is the leading order approximation in the small slope limit. The argument has been that the growth equation $\frac{\partial h}{\partial t} = -\nu_4 \nabla^4 h + \eta$ leads to $\alpha = 3/2$ in $d = 2(3)$, which manifestly violates the small slope approximation and, therefore, all nonlinear terms in the full surface diffusion equation involving higher powers of $|\nabla h|$ become relevant (marginal) $d = 2(3)$ in this super-rough ($\alpha \geq 1$) growth situation. A direct numerical integration[39] of the full diffusion equation leads to $\beta_{\text{effective}} \approx 1$ and the associated failure of dynamical scaling. We feel that it is unlikely that such a full diffusion equation has any physical relevance for MBE growth which is a far-from-equilibrium process, and the near-equilibrium surface diffusion equation does not, in our opinion, catch the physics of MBE growth. The surface diffusion process associated with the Arrhenius activated hopping at the MBE growth front is unlikely to be simply describable by the nonlinear surface diffusion equation. We have a second mathematical and formal reason for believing the irrelevance of the nonlinear surface diffusion equation to MBE growth — nonlinearities such as $\nabla^2(\nabla h)^2$ and $\nabla(\nabla h)^3$ (and the linear $\nabla^2 h$) terms, which we have argued may play significant roles in the crossover behavior of MBE growth, are more relevant from a renormalization group viewpoint than all the nonlinear surface diffusion terms and, therefore, Eq. (4) is likely to be a more appropriate continuum description for conserved SOS MBE growth than the full nonlinear surface diffusion equation. In any case, the presence of any overhangs and vacancies drives MBE growth into KPZ universality, making this discussion of the applicability of the full surface diffusion equation to the MBE process quite academic. We feel that the most appropriate continuum growth equation[40] which generically describes MBE growth is obtained by adding the $\lambda_2(\nabla h)^2$ KPZ nonlinearity to the right-hand side of Eq. (4) and keeping in mind that ν_2 could be negative in the presence of an asymmetric diffusion barrier which suppresses down-hopping. The actual evaluation of the various coefficients (i.e., ν_2, λ_2, ν_4, λ_{42} and λ_{43}) in this growth equation requires a full understanding of the underlying microscopic solid state physics and chemistry controlling growth of the specific material, and remains a formidable challenge for the theorists. We emphasize that, in principle, it is possible for these coefficients to be time-dependent as the growing surface evolves dynamically. In that case, the crossover behavior of MBE growth may turn out to be hopelessly complicated.

We emphasize that equilibrium crystal growth considerations[41] have been left out of the work reviewed here because our interest is explicitly in the nonequilibrium kinetic surface roughening of MBE growth. Equilibrium effects[41] lead to various growth modes such as three-dimensional islanding, two-dimensional layer-by-layer, and a mixed mode. Our interest here is in the self-affine fractal aspects of kinetically rough growth and, as such, we have concentrated on renormalization group dynamical scaling behavior. It should also be emphasized that real MBE growth, particularly for technologically important semiconductor materials, is extremely complicated with actual crystal geometry, surface reconstruction, chemical reactions, and extrinsic effects such as impurities playing possible crucial roles. It is not clear at this stage how much of the nonequilibrium statistical mechanics considerations of this review apply to *real* MBE growth — we feel that some of the basic ideas and concepts developed here should be useful in understanding real MBE growth, particularly at lower temperatures and higher deposition rates where kinetic roughening effects are important. In this respect, it is encouraging that dynamical scaling aspects of MBE growth are currently being investigated experimentally.[21-25] Much more experimental and theoretical work is needed before we can quantitatively understand MBE as a nonequilibrium statistical mechanical phenomenon.

ACKNOWLEDGMENTS

The author thanks S. V. Ghaisas, J. M. Kim, C. Lanczycki, and P. I. Tamborenea for helpful discussions. The author's work on MBE growth is supported by the US-ONR.

REFERENCES

1. S. Das Sarma, *J. Vac. Sci. Tech.* **A8**, 2714 (1990); **B10**, 1695 (1992); references therein.
2. F. Family and T. Vicsek, eds., *Dynamics of Fractal Surfaces*, (World Scientific, Singapore, 1991) and references therein.
3. D. E. Wolf and J. Villain, *Europhys. Lett.* **13**, 389 (1990).
4. S. Das Sarma and P. I. Tamborenea, *Phys. Rev. Lett.* **66**, 325 (1991).
5. L. Golubovic and R. Bruinsma, *Phys. Rev. Lett.* **66**, 321 (1991).
6. Z. W. Lai and S. Das Sarma, *Phys. Rev. Lett.* **66**, 2348 (1991).
7. L. H. Tang and T. Natterman, *Phys. Rev. Lett.* **66**, 2899 (1991).
8. M. Siegert and M. Plischke, *Phys. Rev. Lett.* **68**, 2035 (1992).
9. H. Yan, *Phys. Rev. Lett.* **68**, 3048 (1992).
10. D. A. Kessler, H. Levine and L. M. Sander, *Phys. Rev. Lett.* **69**, 100 (1992).
11. S. Das Sarma and S. V. Ghaisas, *Phys. Rev. Lett.* **69**, 3762 (1992) and **71**, 2510 (1993).
12. J. Krug, M. Plischke and M. Siegert, *Phys. Rev. Lett.* **70**, 3271 (1993).
13. J. Villain, *J. Phys. I (France)* **1**, 19 (1991).
14. S. V. Ghaisas and S. Das Sarma, *Phys. Rev.* **B46**, 7308 (1992).
15. S. Das Sarma, Z. W. Lai, and P. I. Tamborenea, *Surf. Sci. Lett.* **268**, L311 (1992).
16. S. Das Sarma and P. I. Tamborenea, *Phys. Rev.* **B46**, 1925 (1992); P. I. Tamborenea and S. Das Sarma, *Phys. Rev.* **E48**, 2599 (1993).
17. M. Kotrla, A. C. Levi and P. Smilaner, *Europhys. Lett.* **20**, 25 (1992).
18. D. Wilby, D. D. Vvedensky and A. Zangwill, *Phys. Rev.* **B46**, 12896 (1992).
19. L. Golubovic and R. P. U. Karunasiri, *Phys. Rev. Lett.* **66**, 3156 (1991).
20. I. K. Marmorkos and S. Das Sarma, *Phys. Rev.* **B45**, 11262 (1992); *Surf. Sci. Lett.* **237**, L411 (1990).
21. D. J. Eaglesham, H. J. Gossmann and M. Cerullo, *Phys. Rev. Lett.* **65**, 1227 (1990).
22. J. Chevrier, V. L. Tranh and J. Derrien, *Europhys. Lett.* **16**, 737 (1991).
23. Y. N. Yang, Y. S. Luo and J. H. Weaver, *Phys. Rev.* **B45**, 13803 (1992).
24. Y. L. He, H. N. Yang, T. M. Lu and G. C. Wang, *Phys. Rev. Lett.* **69**, 3770 (1992).
25. M. A. Cotta, R. A. Hamm, T. W. Staley, S. N. G. Chu, R. L. Harriott, M. B. Panish and H. Temkin, *Phys. Rev. Lett.* **70**, 4106 (1993).
26. S. F. Edwards and D. R. Wilkinson, *Proc. Roy. Soc. London* **A381**, 17 (1982).
27. W. W. Mullins, *J. Appl. Phys.* **30**, 77 (1959); C. Herring, in *The Physics of Powder Metallurgy*, ed. W. E. Kingston (McGraw-Hill, New York 1951), p. 143.

28. S. Das Sarma, S. V. Ghaisas and J. M. Kim, Phys. Rev. E, January 1994 (in press).
29. M. Plischke, J. D. Shore, M. Schroeder, M. Siegert and D. E. Wolf, preprint (1993), and private communication.
30. J. G. Amar, P. M. Lam and F. Family, *Phys. Rev.* **E47**, 3242 (1993).
31. S. Das Sarma and J. K. Bhattacharjee, to be published.
32. S. Das Sarma, C. Lanczycki, S. V. Ghaisas and J. M. Kim, preprint (1993); C. Lanczycki and S. Das Sarma, unpublished.
33. G. Ehrlich and F. G. Huda, *J. Chem. Phys.* **44**, 1039 (1966); R. L. Schwoebel and E. J. Shipsey, *J. Appl. Phys.* **37**, 3682 (1966); R. L. Schwoebel, *J. Appl. Phys.* **40**, 614 (1969); S. C. Wang and G. Ehrlich, *Phys. Rev. Lett.* **70**, 41 (1993).
34. S. V. Ghaisas and S. Das Sarma, unpublished.
35. J. M. Kim and S. Das Sarma, *Phys. Rev.* **E48**, 2575 (1993). See also, J. M. Kim and S. Das Sarma (preprint, 1993) — in this paper, local discrete growth rules are developed on a lattice which correspond *exactly* to the $\nabla^2 h$, $\nabla^4 h$, and $\nabla^2(\nabla h)^2$ continuum universalities, depending on the specific choice of the discrete rules. The connection between these discrete rules and MBE growth process remains unclear at the present time.
36. D. Nelson, in *Statistical Mechanics of Membranes and Surfaces*, eds. D. Nelson, T. Piran and S. Weinberg (World Scientific, Singapore, 1989), p. 137 and references therein.
37. M. Kardar, G. Parisi and Y. C. Zhang, *Phys. Rev. Lett.* **56**, 889 (1986).
38. J. M. Kim and M. Kosterlitz, *Phys. Rev. Lett.* **62**, 2289 (1989).
39. J. G. Amar and F. Family, preprint (1993).
40. In general, one can also add the symmetry-allowed non-conserving fourth order nonlinearities to the growth equation. These are $\lambda_{44}(\nabla^2 h)^2$, $\lambda'_{44}(\nabla h)^4$, $\lambda''_{44}(\nabla^2 h)(\nabla h)^2$, etc. We refrain from doing that because we cannot imagine a physical scenario where these terms could be present even in the absence of the lower order KPZ nonlinearity $\lambda_2(\nabla h)^2$. Once the KPZ nonlinearity (or, for that matter, the EW linear term) is present, these higher-order nonlinearities become irrelevant. We should mention that among all these fourth order nonlinear terms, the $(\nabla h)^4$ term is the most relevent from a renormalization group viewpoint.
41. J. A. Venables, G. D. Spiller and M. Hanbücken, *Rep. Prog. Phys.* **47**, 399 (1984).

Fractals, Vol. 1, No. 4 (1993) 795–805

EXAMPLES OF FRACTALS IN SOIL MECHANICS

HANS J. HERRMANN*, MUHAMMAD SAHIMI*,† and FRANK TZSCHICHHOLZ*,‡
*HLRZ, KFA Jülich, Postfach 1913, 52425 Jülich Germany
†Dept. of Chemical Engineering
Univ. of Southern California, Los Angeles, CA 90089-1211
‡Institut für Werkstofforschung
GKSS-Forschungszentrum, Postfach 1160, 21502 Geesthacht, Germany

Abstract

Models will be presented for fractal structures appearing naturally in soils. On the one hand, we discuss the opening of brittle media via hydraulic fracturing at constant pressure using a square lattice beam model with disorder. We consider the case in which only beams under tension can break, and discuss under which conditions the resulting cracks may develop fractal patterns. The stress field of the fractal cracks is visualized by photoelastic fringes. Then we present a modelization for a fluid penetrating under a pressure gradient into a fractal crack which it is itself opening. To do this, we investigate invasion percolation fingers in a quenched medium in which the randomness has a gradient corresponding to the density of microcracks that arise in a self-organized way around a large crack.

1. INTRODUCTION

In soil mechanics fractals can appear in various contexts. One example is hydraulic fracturing which is used systematically to improve oil recovery and has considerable importance for the functioning of geothermal wells.[1] An incompressible fluid, in general water, is pushed with pressure deep inside the ground by injecting it into a deep perforation. The fluid penetrates into the soil by opening long cracks that radially go from the injection hole into the material and eventually displace some more viscous fluids that are present in the ground.

Our interest will focus on the shape of the cracks and on the fluid finger penetrating into the crack.

Controlled experiments have been performed recently in two dimensional geometries.[2] Water or air is pushed from above into the center of a circular Hele Shaw cell filled with clay. While for slow motions the clay behaves like a fluid for high injection pressures it fractures like a solid (viscoelastic fluid). In this fracturing regime, the resulting cracks display a ramified structure and seem to obey self-similarity with a fractal dimension of 1.4–1.5. Compared to the fractal structures observed in viscous fingering (or diffusion-limited aggregation (DLA)), not only is the fractal dimension lower but also the angles between branches are nearly three times larger (i.e., about 90°) and no tip splitting is observed. These interesting observations remain largely unexplained.

While for viscous fingering the underlying instabilities (Saffman-Taylor, tip splitting, side branching) have been heavily investigated[3] and numerical clusters of tens of millions of particles have been generated,[4] the case of fracture is still not yet very well understood. A recent stability analysis of the shape of a circular hole with internal pressure was compared with the one of a perforated stretched membrane, and it was shown that contrary to common belief, there is a large difference between the two cases.[5] The origin of this is the nonlinear dependence of the growth velocity of the crack surface due to the threshold imposed by the cohesion force that must be overcome to break the material. For viscous fingering, these differences do not exist in the limit of large clusters.[6]

The breaking of a material from a microcrack was first simulated by Louis and Guinea[7] using a triangular network of springs and stretching it radially on the outer boundary into the six directions of a hexagon. They found fractal cracks with a fractal dimension close to that of DLA. These findings were studied subsequently by various groups.[8] It became clear that the patterns depend very much on the applied external displacement (shear, uni-axial, radial). Since networks of longitudinal springs need more extended structures to transfer momenta, it turned out to be more efficient to use a beam model[9] which automatically takes into account internal rotations and bending. For more details on this work we refer to Ref. 10.

In all previous work the imposed displacements only were applied at the external surface of the system. In hydraulic fracturing, however, the imposed load is a pressure that acts along the entire (inner) surface of the crack perpendicular to the surface. In this way, the point of application of the imposed load varies during the growth of the crack, which makes the situation more complex. We will present numerical results for such a situation using a beam model.[11]

The penetration of fluids into the soil[12] and the creation and propagation of major cracks influence each other. This coupling between fluid transport and cracking has nearly never been attacked theoretically although it seems to be crucial in applications of hydraulic fracturing. In Ref. 14, an invasion percolation model has been proposed for a non-wetting fluid in which a gradient is introduced by assuming quenched random permeabilities z_i distributed as $P(z_i) \propto r^{-\alpha}$, r being the distance from the injection point. This graded distribution of random permeabilities models is to some degree a pressure gradient. A numerical simulation of that model showed that for positive α the clusters are fractal with a fractal dimension of $d_f = 1.44 \pm 0.02$, independent of α. We will present a similar model[13] introducing a second gradient corresponding to the effect of the cloud of microcracks that is known to arise in a self-organized way with a density decaying like a power-law in the distance from the central crack.[10]

In the following section, we will describe the fracture model and its result.[11] In Sec. 3, we present the model for fluid invasion.[13] In the last section we conclude.

2. RUPTURE MODEL ON A SQUARE LATTICE

Let us consider the beam model (as defined on p. 232 of Ref. 10) on a square lattice of linear size L. Each of the lattice sites i carries three real variables: the two translational displacements x_i and y_i and a rotational angle φ_i. Neighboring sites are rigidly connected by elastic beams of length l. The beams are assumed to have the same shape and the same elastic behavior, governed by three material dependent constants $a = l/(EA)$, $b = l/(GA)$ and $c = l^3/(EI)$. Here E and G are the Young and shear moduli, A the area of the beam section and I the moment of inertia for flexion. We used for all simulations the values $a = 1.0$, $b = 0.0017$ and $c = 8.6$. The constant pressure has a value of 0.01. When a site is rotated ($\varphi_i \neq 0$), then the beams must bend accordingly and always tangentially form 90° angles with each other. In this way, local momenta are taken into account.

Since rocks typically crack under tension and to a much lesser degree under compression, we will assume that only beams that are under tension are allowed to break, i.e., to be irreversibly removed from the force and momentum balance equations. For example, for a horizontal beam between sites i and j, we have for the longitudinal force acting at site j: $F_j = \alpha(x_i - x_j)$, for the shear force: $S_j = \beta(y_i - y_j) + \frac{\beta}{2}l(\varphi_i + \varphi_j)$, and for the flexural torque at site j: $M_j = \frac{\beta}{2}l(y_i - y_j + l\varphi_j) + \delta l^2(\varphi_i - \varphi_j)$, using $\alpha = 1/a$, $\beta = 1(b + c/12)$ and $\delta = \beta(b/c + 1/3)$.

At the place into which the incompressible fluid is injected (center of the lattice) one vertical beam connecting sites i and j is removed. Since we want to simulate the loading of a crack by the injected fluid, a pair of opposite forces (dipole) is applied at the sites i and j pointing into the elastic bulk. The acting pressure p is just the force divided by the beam length l.

Each time a beam is removed, a new force dipole is applied at its corresponding neighboring sites i and j changing the previous balance of forces. So, each time a beam is broken, we calculate again the internal equilibrium of forces. This is done in our case using a conjugate gradient with a precision of $\varepsilon = 10^{-10}$ (see Eq. (47) in Ref. 15). After the equilibrium displacement field $\mathbf{u}_i$ corresponding to the new boundary conditions is determined, one will decide which is the next beam to be broken, i.e., we assume that the physical process of stress redistribution is much faster than the actual fracture process.

We will consider in the breaking procedure only beams along the surface of the inner hole. In that way only a single crack is generated. More generally, one could allow breaking also bonds not belonging to the crack surface. Then one would also form a cloud of microcracks around the main crack.[10] This leads to effects which will be considered in the next section.

For each of the beams on the surface, the force f_{ij} acting along its axis is determined and only if this f_{ij} is positive, i.e., under tension, the beam is eligible to be broken. In addition, since we are interested in random structures, we need to incorporate some stochastic mechanism or disorder into the model. The disorder may be either quenched, i.e., the beams have fixed (random) breaking thresholds; or it may be annealed, i.e., the next broken beam is selected randomly according to some probability distribution. In the following, we will use an annealed model but in real soils the disorder is rather quenched. A quenched case will be considered in the next section.

Among the eligible beams one is chosen with a probability p_{ij}, which is determined in analogy to dielectric breakdown[10] to be:

$$p_{ij} \propto \begin{cases} f_{ij}^{\eta} & \text{if } f_{ij} > f_{\text{coh}} \\ 0 & \text{otherwise} \end{cases} \tag{1}$$

where f_{coh} is the cohesion force. For most of our simulations, we have chosen $f_{\text{coh}} = 0$. Here, $0 \leq \eta < \infty$ is a phenomenological parameter determined by the *local* breaking mechanism of the material since a first principle mesoscopic breaking rule is not known. If $\eta \to \infty$, only the most stretched beam breaks, i.e., the one with the largest f_{ij}. The opposite case corresponds to $\eta = 0$, where all eligible beams can break with the same probability.

For the Laplacian case, the value $\eta = 1$ corresponds to DLA with a fractal dimension $d_f \cong 1.7$ in two dimensions.[16] For the present beam model, however, topologically linear cracks develop ($d_f = 1$) for $\eta = 1$. Similar results are obtained for $\eta > 1$. This indicates that the distribution of p_{ij} is much narrower than for DLA and essentially only the most stretched beams (located at the tips of the crack) break.

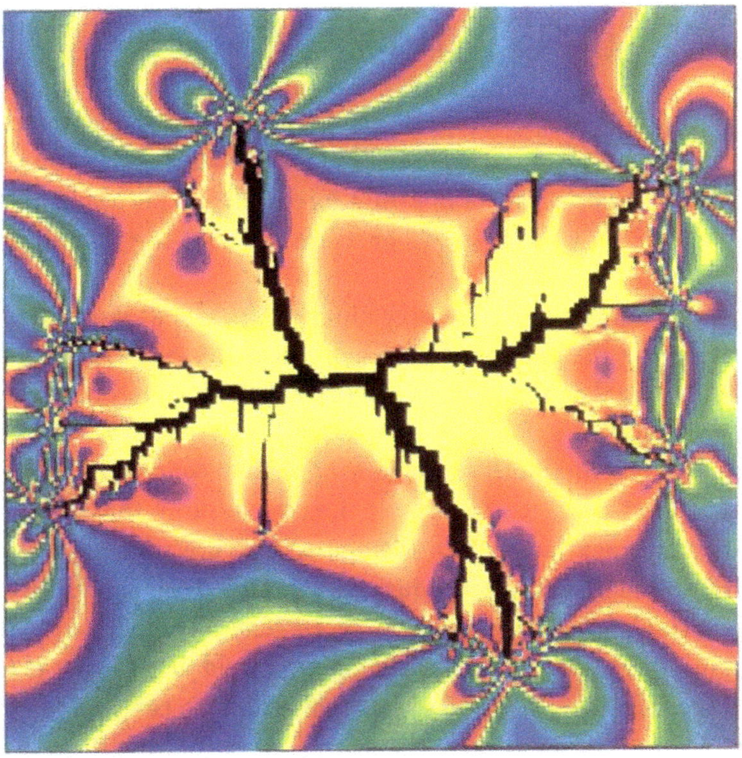

(a)

Fig. 1 A typical crack obtained with a beam model for hydraulic fracturing at constant pressure when $\eta = 0$, on a square lattice of (a) 150 × 150, and (b) 250 × 250 sites with (a) free, and (b) helical external boundary conditions. The cracks consists of (a) 680 and (b) 2200 broken beams, and only beams under tension are allowed to break. The different colors represent the difference of the diagonal elements of the hydrostatic stress field. The full range for the stress field is linearly mapped onto ten color-circle cycles starting at magenta going through blue, cyan, green, yellow and ending at red. The calculations for (b) took 500 hours on an IBM RS/6000-320H workstation.

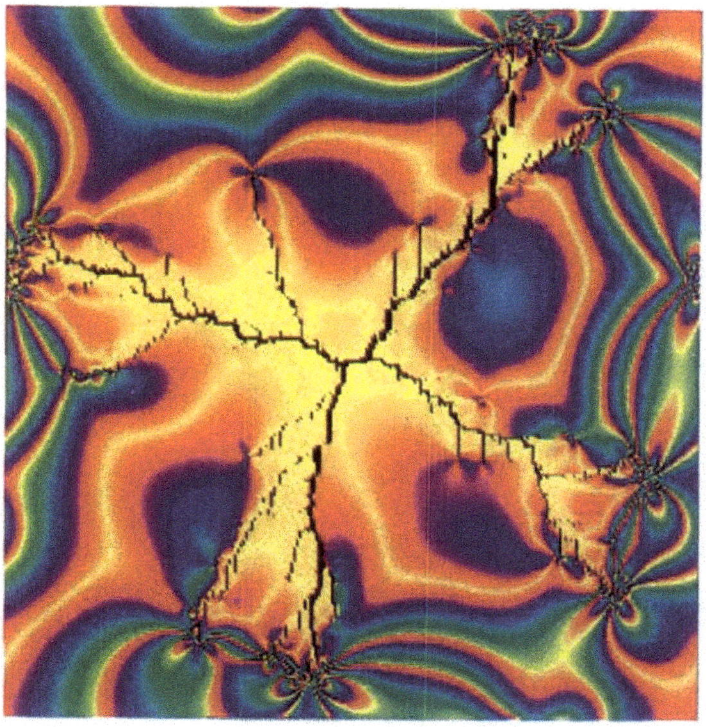

(b)

Fig. 1 (*Continued*)

More interesting is the case $\eta = 0$, i.e., in which all beams under tension can be selected for breaking with the same probability. For the Laplacian case, one obtains compact clusters of spherical shape (Eden clusters) and fluctuations occur only at their surface.[16] In the present case, the resulting cracks (shown in Fig. 1(a)) display ramifications which persist up to scales comparable to the system size L, and it seems plausible that they might be fractal.

The colors visualize the stress field through the photoelastic interface fringes, i.e., the image one would see experimentally if one passes polarized light through a plexiglass plate having that stress field. In fact, we show the difference between the two diagonal elements of the stress tensor modulated by the color spectrum such that ten nodes cover linearly the entire range. One sees that the strongest gradients occur around the crack tips and that the regions between crack branches are under virtually constant compression of magnitude p. Clearly the opening of the crack is larger on certain main cracks than on the side branches. Although only weakly visible, the system of Fig. 1(a) is no perfect square anymore but has small bulges due to the free boundaries. The ramified cracks seen in Fig. 1 much resemble crack patterns seen in nature and it seems plausible that the cracks might be fractal.

In Fig. 1(b) we see a crack generated with helical external boundaries in both directions, i.e., spanned on a torus. This is the largest crack simulation ever performed with the beam model. The crack opening is clearly weaker here than in Fig. 1(a). This comes from the fact that the helical boundaries act like a closed system, as if a certain pressure is exerted from outside. Figure 2 shows the evolution of the growth for the crack of Fig. 1(b). One clearly sees that screening stops the growth of the smaller clusters, like in DLA.

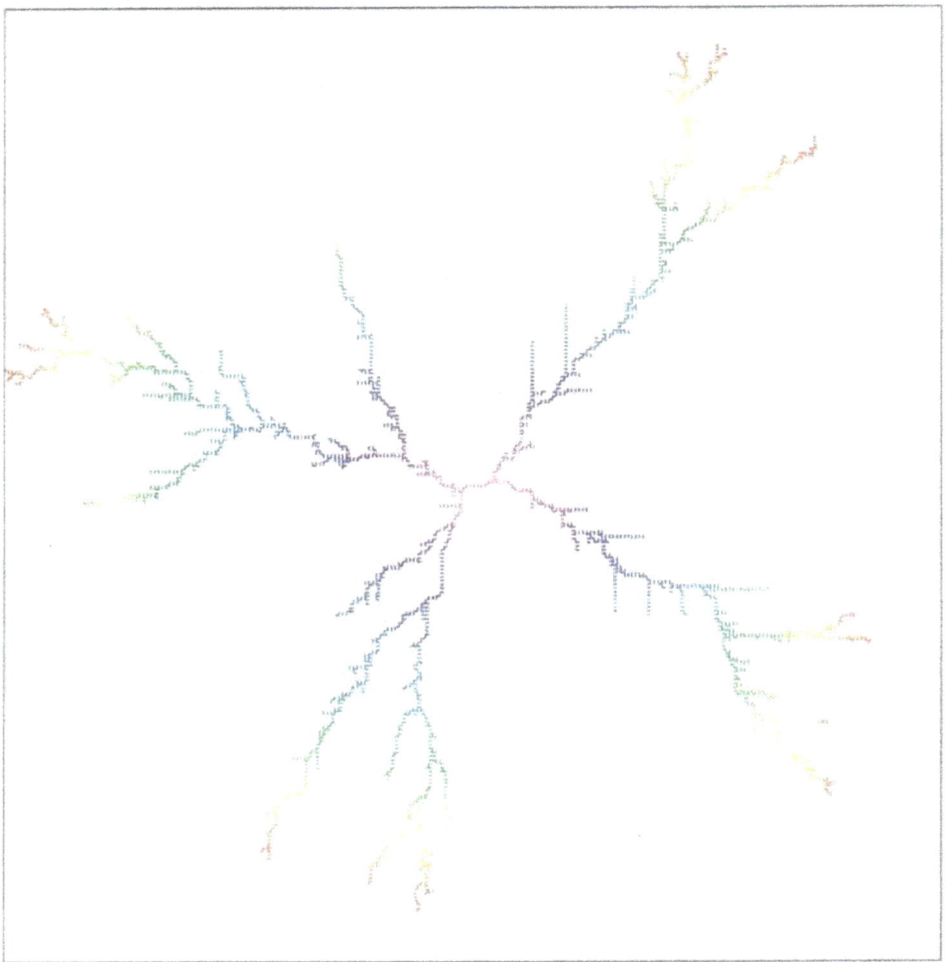

Fig. 2 Time evolution of the crack growth for the crack of Fig. 1(b). The colors indicate sites that were grown in a certain time interval.

To characterize quantitatively the shape of the cracks obtained when $\eta = 0$, we plot in Fig. 3 the number N of broken beams against the radius of gyration R_G of the crack. For a single crack the radius of gyration is defined as:

$$R_G^2(N) = \frac{1}{N}\sum_{i=1}^{N}(\mathbf{r}_i - \mathbf{r}_0)^2\,, \qquad \mathbf{r}_0 = \frac{1}{N}\sum_{i=1}^{N}\mathbf{r}_i\,, \tag{2}$$

with $\mathbf{r}_i$ being the position vector of the ith broken beam with respect to the undistorted lattice. The results shown in the log-log plot of Fig. 3 are for the case of free external boundaries. We see a power-law behavior $N \sim R_G^{d_f}$ over nearly two decades of the slope which gives a fractal dimension of $d_f = 1.56 \pm 0.05$. Similar results were obtained for the case of periodic boundary conditions. The fact that the clusters are fractal already for $\eta = 0$, is rather unexpected in the context of viscous fingering where $\eta = 0$ corresponds to the Eden model and the generated clusters are compact.

For a finite cohesion threshold force $f_{\text{coh}} > 0$ and $\eta = 0$, the cracks tend to grow more anisotropically. Also, a finite ratio f_{coh}/p introduces an additional length scale in the system.[17]

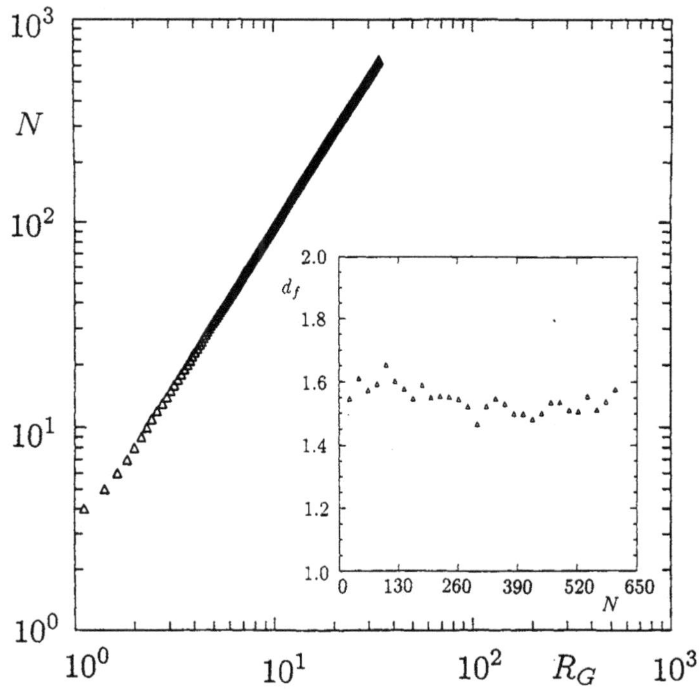

Fig. 3 Plot of the number of broken beams N as a function of the radius of gyration R_G of the cracks obtained for $\eta = 0$ and free boundary conditions. Averages are over 27 samples. The inset shows the successive slopes d_f of the data, and indicates an average value $d_f = 1.56 \pm 0.05$.

Our simulations are performed at constant pressure and we can measure the volume V of the crack opening which corresponds to the amount of fluid that has penetrated into the soil. We measure this volume through the actual displacements of the lattice sites as the crack grows. We define the volume elements ΔV_{ij} connecting sites i and j as $\Delta V_{ij} = (x_i - x_j)l$ for horizontal broken beams (l is the beam length). Similar expressions hold for vertical broken beams. The total crack opening volume V is just the sum of all the volume elements ΔV_{ij}.

In Fig. 4, we show the crack volume V for free and periodic boundary conditions as a function of the radius of gyration R_G in a log-log plot. The slope gives an exponent consistent with two, i.e., the spatial dimension. This agrees with the observation that a finite amount of fluid actually enters the hole. When the crack approaches the boundary, the curve has an artificial increase in slope due to boundary effects. For comparison, we also measured the volume of a deterministic straight crack as shown in Fig. 4. For such a crack in an infinite plate, it is well known from elasticity theory that the crack opening volume is proportional to the squared crack length (in two dimensions).[18] This shows that although the crack surface (number of broken beams) is fractal, the volume of the crack is not, *independent* of the crack structure and the external boundary conditions.

3. FLUID PENETRATION INTO A CRACK

To model a non-wetting fluid entering a damaging soil by invasion percolation, we consider a square lattice of horizontal length L_1 and vertical height L_2 and with periodic boundaries in the vertical direction. At line j_0 on the left boundary we inject the fluid, i.e., we occupy this site. Then on each site of the first two columns, a random permeability is placed

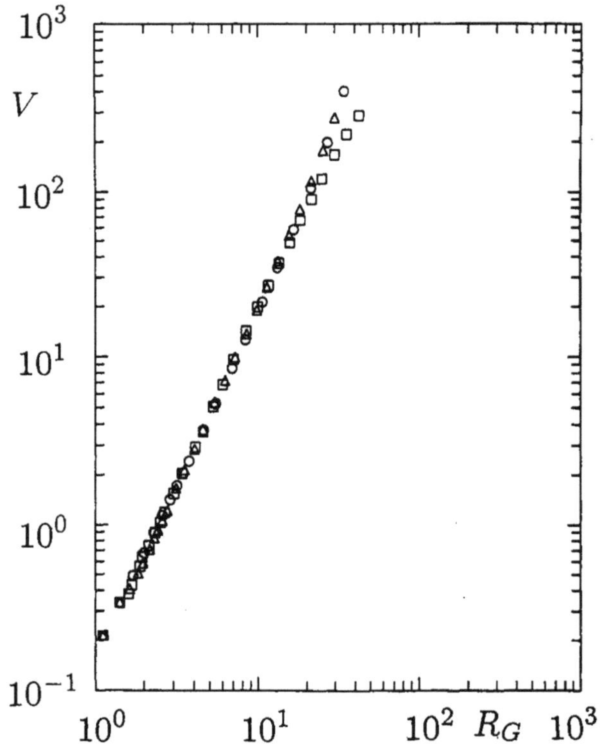

Fig. 4 Plot of the crack opening volume V as a function of the radius of gyration R_G, obtained by averaging over 12 samples using free boundary conditions (triangles) and averaging 4 samples using periodic boundary conditions (squares). For comparison, we also show the results for a straight crack (circles) obtained in the case $\eta \to \infty$ under free boundary conditions.

distributed according to $P(z_i) \propto |\text{mod}((j-j_0), \frac{L_2}{2})|^\beta$ where j is the line number of point i. Now the usual invasion percolation algorithm is applied: Among all the sites on the surface of the finger, i.e., all the sites adjacent to the occupied sites, the one having the smallest permeability value is chosen and occupied. Each time the finger advances by one column, the (random) permeabilities of the following column are chosen according to the distribution:

$$P(z_i) \propto \left|\text{mod}\left((j-j_t), \frac{L_2}{2}\right)\right|^2 \times (k-1)^{-\alpha} \tag{3}$$

where k is the number of the column and j_t is the number of the line at which the finger just advanced by a new column. When the finger touches the right boundary, the process is stopped.

In the described model, the noise is quenched as once the random permeabilities are defined on a site, they will not change anymore. For $\beta = 0$, we recover the model described in Ref. 14. If $\beta > 0$, the vertical width of the finger at a given column is confined by the power-law gradient of permeabilities so that if L_2 is not chosen too small, the finger can wind around the tube formed by the periodic boundary conditions making it insensitive to finite size effects in the vertical direction.

Next let us present our numerical results for the case in which the quenched noise is given by Eq. (3). In Fig. 5, we see the invaded clusters for various parameters. If $\beta = 2$, the invading finger is very straight [Fig. 5(a)] and for decreasing β [Figs. 5(b) and 5(c)],

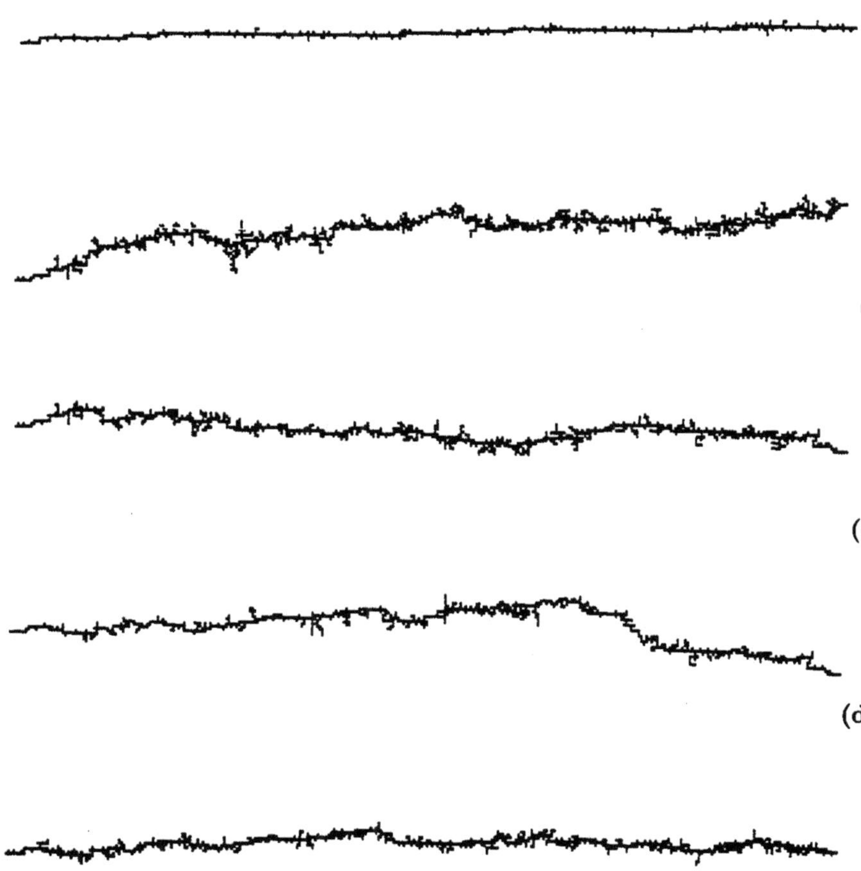

(a) (b) (c) (d) (e)

Fig. 5 Invading fingers when they percolate through a system of $L_1 = 320\,(L_2 = 41)$ for (a) $\beta = 2$, $\alpha = 1$; (b) $\beta = 0.5$, $\alpha = 1$; (c) $\beta = 0.7$, $\alpha = 1$; (d) $\beta = 0.7$, $\alpha = 4$ and (e) $\alpha = 1$ and an exponential distribution of the form $e^{0.5r}$ in the transverse direction.

it has much denser sidebranches and stronger deviations from the line j_0 = const given by the original injection point. The effect of α, on the contrary, is rather weak as seen when comparing Figs. 5(c) and 5(d). It is interesting to note that by taking an exponential distribution instead of a power law in the transverse direction, no qualitative difference is noted as can be seen from Fig. 5(e).

We define the width w of the invasion cluster as:

$$w = \frac{1}{N_s} \sum_{1}^{N_s} |j - j_0| \tag{4}$$

where N_s is the total number of sites in the finger. Then w is averaged over 1000 different samples. In Fig. 6, we see a log-log plot of the width w as a function of L_1. In all cases the data fall asymptotically on straight lines of slope a little below 0.5. Since the curves, however, have a slight upward curvature the resulting roughness exponent ζ defined through

$$w \sim L_1^{\zeta} \tag{5}$$

could well be 1/2. From Fig. 6, we also see that the detailed distribution of quenched permeabilities in the transverse direction has no noticeable effect on the roughening exponent. Even an exponential distribution gives the same result.

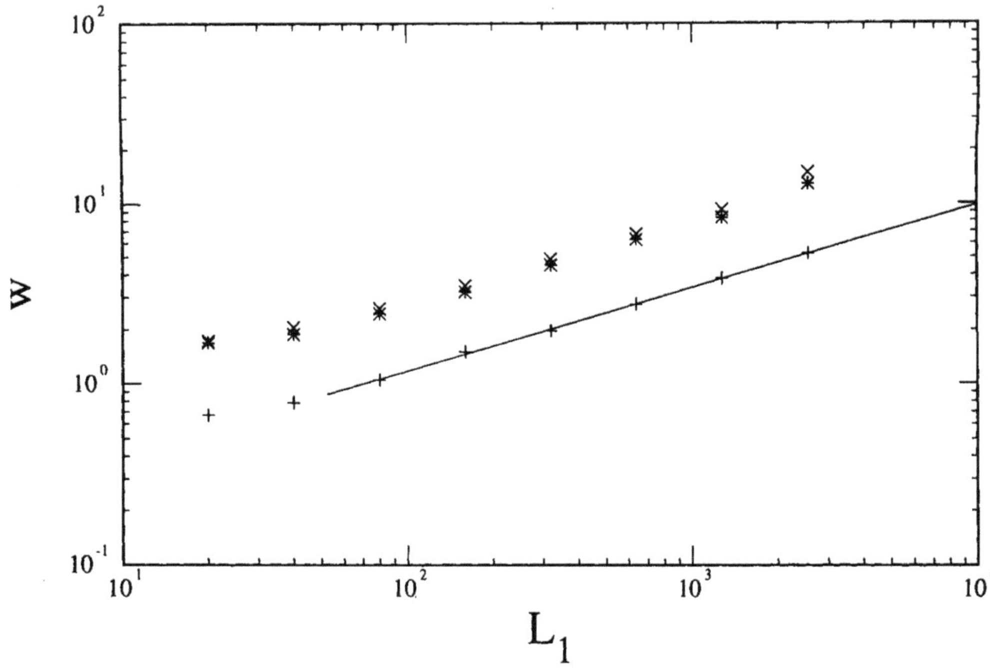

Fig. 6 Log-log plot of the width w of the finger as a function of its length L_1 for $L_2 = 23$, $\alpha = 1$ and $\beta = 1$ ($*$), $\beta = 2$ ($+$) and an exponential distribution $e^{0.5r}$ ($\times$). The straight line is a linear fit through the last six data points and has a slope of 0.46.

4. CONCLUSION

We have presented a discrete fracture model in two dimensions adapted to the conditions of hydraulic cracking of soils as typically used for instance to create additional heat exchange surface for geothermal energy recovery. Only tensile forces break the solid and the heterogeneities are considered to be dominant. For $\eta = 0$, the crack patterns are fractal with a fractal dimension of 1.56 ± 0.05. Our results reproduce some of the features observed in the 2d experiments of Orléans for high injection rates and rigid pastes[2]: The crack patterns are more kinky than DLA and the fractal dimensions agree within their error bars. The case $\eta = 0$ seems reasonable to describe a medium that has a very weak cohesion and is dominated by local heterogeneities as one would expect for the viscoelastic pastes used in Ref. 2.

Still our model should be considered as yet quite preliminary concerning the understanding of real industrial hydrostatic fracturing. On the one hand, real soils are three dimensional and in fact we do not know whether in three dimensions the patterns are finger-like as in two dimensions or rather consisting of randomly attached fracture planes. The latter case seems more likely although it has yet never been seen numerically up to now. Also the role of the heterogeneities must be investigated more closely. In soils, the disorder is quenched, i.e., the randomness (in breaking threshold, modulus, etc) are assigned before cracking starts. We have used "annealed" randomness: random numbers are used during the breaking process in order to select beams with a certain probability.

Our invasion percolation model with quenched disorder yields as result that the penetrating fluid essentially performs a random walk in the transverse direction ($\zeta = \frac{1}{2}$). The confining distribution given by the power-law distribution of microcracks only has an effect on the amplitude of the width of the clusters but not on the roughening exponent.

Other important factors that should be taken into account when trying to make the model more realistic are plasticity and hydrodynamic effects in the fluid, stress corrosion and short time effects, like shock waves.

REFERENCES

1. H. Takahashi and H. Abé, in *Fracture Mechanics of Rock*, ed. B. K. Atkinson (Academic Press, London, 1987), p. 241.
2. H. Van Damme, E. Alsac and C. Laroe, *C. R. Acad. Sci. Série II* **309**, 11 (1989); H. Van Damme, in *The Fractal Approach to Heterogeneous Chemistry*, ed. D. Avnir (Wiley, New York, 1989) p. 199; E. Lemaire, P. Levitz, G. Daccord, and H. Van Damme, *Phys. Rev. Lett.* **67**, 2009 (1991).
3. D. A. Kessler, J. Koplik and H. Levine, *Adv. Phys.* **37**, 255 (1988).
4. P. Ossadnik, *Physica* **A176**, 454 (1991).
5. H. J. Herrmann and J. Kertész, *Physica* **A178**, 227 (1991).
6. H. J. Herrmann, in *Proceedings of the NATO Workshop "Growth Patterns in Physical Sciences and Biology"* (1992).
7. E. Louis and F. Guinea, *Europhys. Lett.* **3**, 871 (1987); E. Louis, F. Guinea and F. Flores, in *Fractals in Physics*, eds. L. Pietronero and E. Tossatti (Elsevier, Amsterdam, 1986).
8. P. Meakin, G. Li, L. M. Sander, E. Louis and F. Guinea, *J. Phys.* **A22**, 1393 (1989); E. L. Hinrichsen, A. Hansen and S. Roux, *Europhys. Lett.* **8**, 747 (1988).
9. S. Roux and E. Guyon, *J. Physique Lett.* **46**, L999 (1985).
10. H. J. Herrmann and S. Roux, eds. *Statistical Models for the Fracture of Disordered Media* (North Holland, Amsterdam, 1990).
11. F. Tzschichholz, H. J. Herrmann, H. E. Roman and M. Pfuff, preprint.
12. M. Sahimi, *Rev. Mod. Phys.*, in press.
13. H. J. Herrmann and M. Sahimi, preprint.
14. L. de Arcangelis and H. J. Herrmann, *J. Phys.* **A23**, L923 (1990).
15. G. G. Batrouni and A. Hansen, *J. Stat. Phys.* **52**, 747 (1988).
16. L. Niemeyer, L. Pietronero and H. J. Wiesmann, *Phys. Rev. Lett.* **52**, 1033 (1984).
17. B. R. Lawn and T. R. Wilshaw, *Fracture of Brittle Solids* (Cambridge University Press, London, 1975).
18. R. G. Hahn, *Bruchmechanik* (Teubner, Stuttgart, 1976).

Fractals, Vol. 1, No. 4 (1993) 806–817

VIBRATIONAL PROBLEM OF FRACTAL NETWORKS

TSUNEYOSHI NAKAYAMA
Department of Applied Physics, Hokkaido University,
Sapporo 060, Japan

Abstract

This paper describes the recent progress in our understanding of the dynamical properties of fractal percolating networks via large-scale computer simulations.

1. INTRODUCTION

The realization that solving the problem of diffusion was equivalent to a solution of the scalar elastic vibration problem led Alexander and Orbach[1] to evaluate the density of states (DOS) for vibrations of fractal networks (fractons) with the introduction of the spectral or fracton dimension $\tilde{d}$. Since that time, a great deal of activities have been concentrated on understanding the nature of fractons, their spatial and dynamical properties and their relationship to physical observables. In addition, the dynamics of fractal networks has been used as a model to aid the understanding of thermodynamic and transport properties of random physical systems.

Numerical simulations have proven to be a valuable tool for obtaining insight into the dynamics of fractal networks. They do more than simply verify physical assumptions, pointing the way to a new qualitative understanding of the nature of excitations of strongly random structures. As the size of supercomputers increases, especially with the advent of array-processors, the size of fractal networks which could be simulated grows enormously, yielding a wealth of new results and insights. Examples are the calculation of the density of states (DOS) for fractons; the asymptotic form of the fracton wavefunction; and matrix elements for inelastic light scattering and the dynamical structure factor $S(q,\omega)$.[2] In this paper, I describe the results of large-scale simulations on the dynamical properties of random fractals (percolating networks).

2. FRACTON DYNAMICS OF ELASTIC NETWORKS WITH SCALAR DISPLACEMENTS

2.1 Density of States and Spectral Dimension

The spectral dimension $\tilde{d}$ is the key dimension to describe the dynamics of fractal networks in addition to the fractal dimension D_f. D_f describes how the mass of a geometrical object depends on length scale, whereas the spectral dimension $\tilde{d}$ characterizes the dynamics of fractal systems. The spectral dimension is related to the exponents characterizing the static structure (β, ν) and the conductivity exponent μ, and is given as:

$$\tilde{d} = \frac{2D_f}{2 + \dfrac{\mu - \beta}{\nu}}, \tag{1}$$

where β and ν are the exponents for the percolation probability and the correlation length, respectively. Nakayama[3] has pointed out that $\tilde{d}$'s of $d = 2 \sim 4$ percolating networks are determined accurately from the calculations of the densities of states (DOS), and equivalently to determine the conductivity exponent μ from the following relation obtained from Eq. (1):

$$\mu = \beta + 2\nu \frac{D_f}{\tilde{d} - 1}. \tag{2}$$

Let us show the calculated results of the DOS's for $d = 2 \sim 4$ elastic percolating networks. We consider a bond-percolating (BP) network consisting of N particles with unit mass and linear springs connecting two atoms in the nearest neighbor. The displacement u_i at the site i has only one component. In Fig. 1, filled squares indicate the calculated DOS per atom for $d = 2$ BP network at $p_c = 0.5$, formed on a 1100×1100 square lattice ($N = 657\,426$) with periodic boundary conditions. The spectral dimension $\tilde{d}$ is determined as $\tilde{d} = 1.325 \pm 0.002$; the data of $d = 2$ BP are given in Fig. 1. The DOS for $d = 3$ BP-networks are also presented in Fig. 1 by filled triangles (middle). These data show the averaged DOS over three samples at the percolation threshold $p_c (= 0.249)$. The networks, formed on $100 \times 100 \times 100$ cubic lattices, have 155 385, 114 303 and 143 026 atoms, respectively. The spectral dimension $\tilde{d}$ is determined as $\tilde{d} = 1.317 \pm 0.003$ by least squares fitting using the data of Fig. 1.

The DOS of $d = 4$ BP networks at $p_c = 0.160$, formed on $30 \times 30 \times 30 \times 30$ "quartic" lattices, are shown in Fig. 1 by filled circles, which are obtained by averaging over 15 samples. The network sizes are $N = 8410 \sim 64\,648$. The DOS in the frequency region $0.12 < \omega < 0.9$ clearly shows a power law as in the $d = 3$ case. The spectral dimension $\tilde{d}$ is estimated as $\tilde{d} = 1.31 \pm 0.03$ from least squares fitting using the data of Fig. 1. Using these estimated values for $\tilde{d}$ in Eq. (2), we can determine the values of the conductivity exponents as $\mu = 1.288$ for $d = 2$, $\mu = 2.02$ for $d = 3$ and $\mu = 2.45$ for $d = 4$ percolating networks, respectively. The value μ for $d = 2$ is obtained using the exact values of $\beta = 5/36$ and $\nu = 4/3$ ($D_f = d - \beta/\nu$) in Eq. (2). This indicates that $\mu = 1.288$ for $d = 2$ is quite accurate. The values μ for $d = 3$ and $d = 4$ are evaluated using the values obtained by

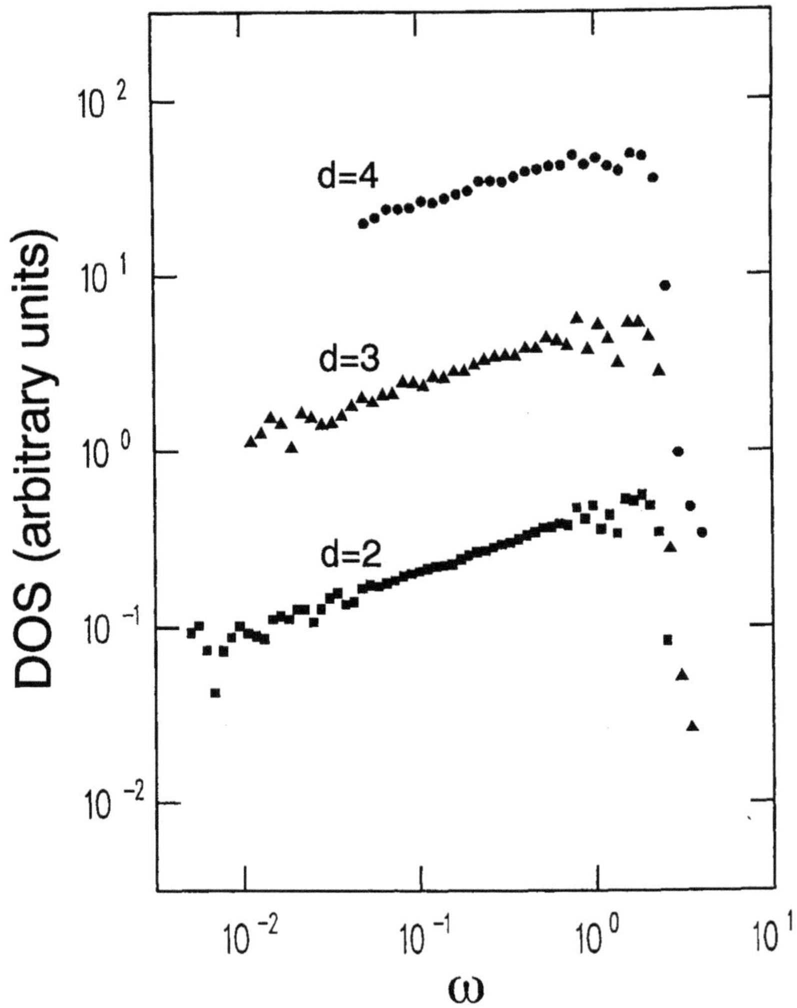

Fig. 1 The DOS per atom for $d = 2 - 4$ BP networks at $p = p_c$. The angular frequency ω is obtained in units of mass $m = 1$ and force constant $K = 1$.[3]

Monte Carlo calculations[4] in Eq. (2), namely $\beta = 0.43$ and $\nu = 0.88$ for $d = 3$, and $\beta = 0.65$ and $\nu = 0.68$ for $d = 4$.

2.2 Mode-Patterns of Scalar Fractons

Let us show a mode-pattern of fracton which was obtained by applying the forced oscillator method.[5] The network at $p_c = 0.5$ was formed on a 200×200 square lattice. Since the average frequency difference $\Delta\omega$ between eigenfrequencies is estimated as $\sim 10^{-4}$, the frequency resolution should be as small as 10^{-6}. Figure 2(a) shows the mode-pattern of a fracton (snap shot) excited on a $d = 2$ network, where the eigenmode belongs to the eigenfrequency $\omega = 0.049975168739345$. To clarify the details more directly, the cross section of amplitudes of the mode-pattern is shown in Fig. 2(b). Figure 2(b) exhibits the cross section of the vibrational amplitudes for the fracton mode along the white line in Fig. 2(a). It should be emphasized that the forced oscillator method[5] makes it possible to figure out the time development of a fracton eigenmode. Figure 2(b) shows the time development of a vibrating fracton from the top to the bottom with the phase difference of $\pi/3$. The filled circles and the open circles represent the occupied sites and the vacant sites in the percolating network, respectively. One sees that the fracton core (the red part of the upper figure in the middle with large amplitudes) possesses very clear boundaries for the edges of the excitation, almost of steplike character, with a long tail in the direction of the weak segments. This is in contrast to the case of homogeneously extended modes (phonons) in which the change of their amplitudes is correlated smoothly over a long distance. Displacements of atoms in "dead ends" and/or "blobs" move in phase, and the vibrational amplitudes fall off sharply at their edges. In addition, it is interesting to note that the tail (the portion spreading from the center of the figure in the right-below direction) extends over a very large distance with phase changes.

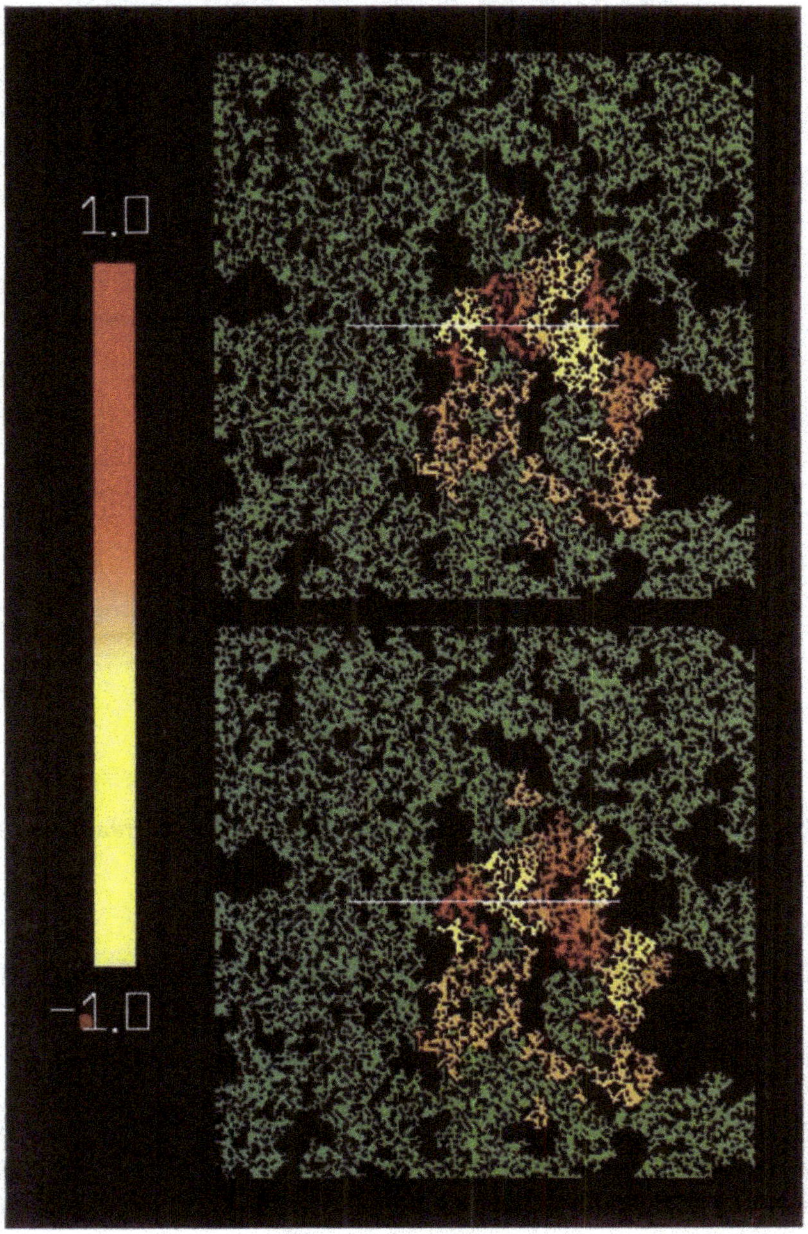

Fig. 2 **(a)** The snapshot of a fracton excited on $d = 2$ BP network at $p = p_c$ formed on 200×200 square lattice (see text). Colours indicate the amplitudes normalized by the maximum one. The upper and bottom figures are out-of-phase.

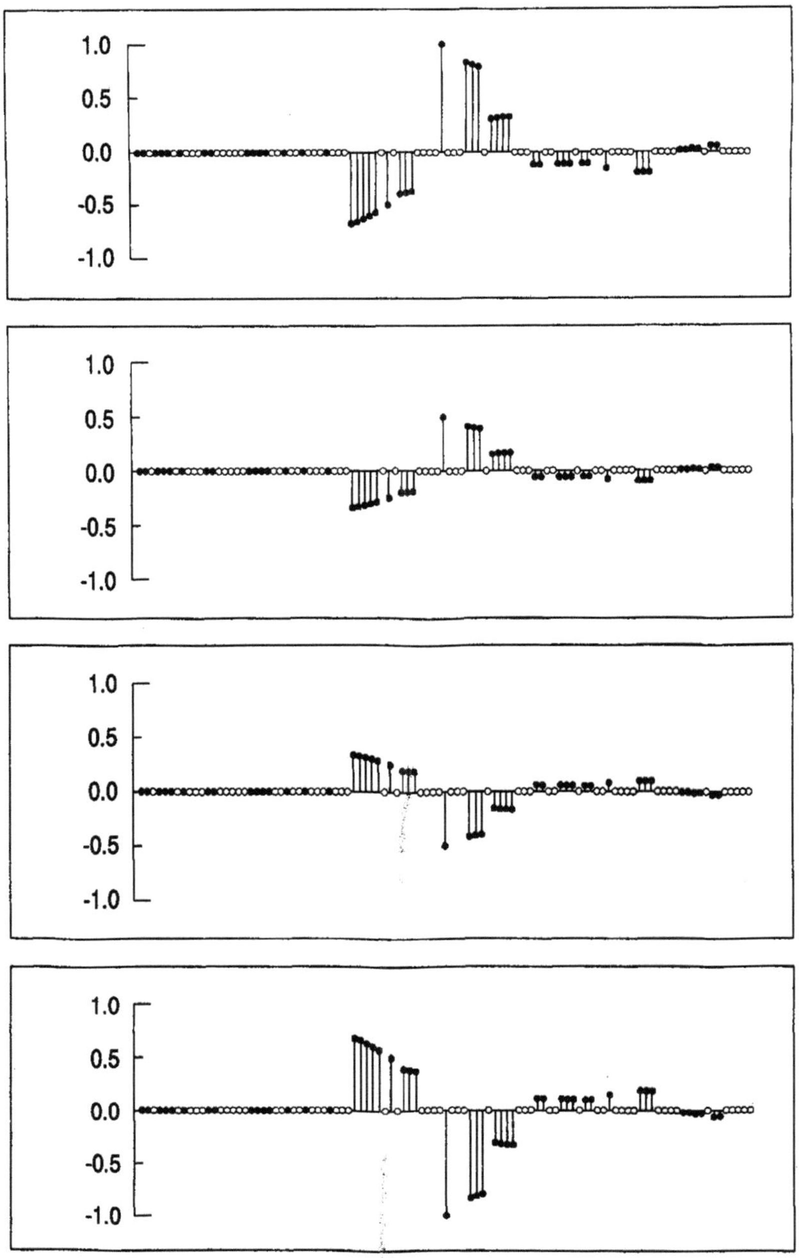

Fig. 2 (b) Cross section of a vibrating fracton along the white line shown in Fig. 2 (see text).

3. FRACTAL NETWORKS WITH VECTOR INTERACTIONS

3.1 Density of States

In realistic systems, the vector nature of atomic displacements becomes crucial. Percolating networks with rotationally-invariant vector-elastic-forces have different critical exponents for elastic moduli from those of scalar interactions. The same corresponding relations as with the scalar displacement (the mapping of the diffusion problem onto the vibrational problem) is more difficult because of the significant additional complication of the vector nature of the displacements. The Hamiltonian, taking vector displacements into account, is given by:

$$\mathcal{H} = \frac{1}{2}\sum_i m_i(\dot{\mathbf{u}}_i)^2 + \frac{1}{2}\alpha \sum_{\langle ij \rangle} K_{ij}[(\mathbf{u}_i - \mathbf{u}_j)\cdot \mathbf{r}_{ij}]^2 + \frac{1}{2}\beta \sum_{\langle ijk \rangle} K_{ij}K_{jk}(\delta\theta_{ijk})^2 . \qquad (3)$$

Here $\mathbf{u}_i$ is the vector displacement of the ith atom with unit mass ($m = 1$), $\mathbf{r}_{ij}$ the unit-vector between nearest neighbors $\langle ij \rangle$, and $\delta\theta_{ijk}$ the small change in angle between bonds $\langle ij \rangle$ and $\langle ik \rangle$ due to the displacements of atoms. The K_{ij} takes the value unity if both sites i and j are occupied by atoms; otherwise, $K_{ij} = 0$ and α and β are the bond-stretching and the bond-bending force constants, respectively. The rigidity threshold of this system is identical to the percolating threshold p_c.

There exist two kinds of characteristic length, $\xi(p)$ and l_c, for percolation networks with vector elasticity.[6,7] The length $\xi(p)$ scaled as $|p - p_c|^{-\nu}$ is the correlation length, namely the crossover-length scale from homogeneous to fractal structure. The mechanical length l_c determines the crossover length scale below which bond-stretching motion is energetically favorable and above which the bond-bending becomes dominant. This crossover length l_c depends only on the force constants α and β, namely $l_c \propto \sqrt{\beta/\alpha}$.[6] One can connect the characteristic lengths $\xi(p)$ and l_c with two characteristic frequencies, ω_ξ and ω_{l_c}, respectively.

The simulation results for the DOS of large-scale percolating networks with vector displacements are given in Fig. 3. Let us first discuss the crossover behavior of the DOS from bending to stretching fractons. For this purpose, the situation is considered where the bond-bending force constant β is larger than the stretching one α, so that the characteristic frequency ω_{l_c} is much smaller than the Debye cut-off frequency ω_D. The network was prepared at $p = p_c$ so that the condition $l_c < \xi(p)$ always holds.[8] The calculated DOS is shown

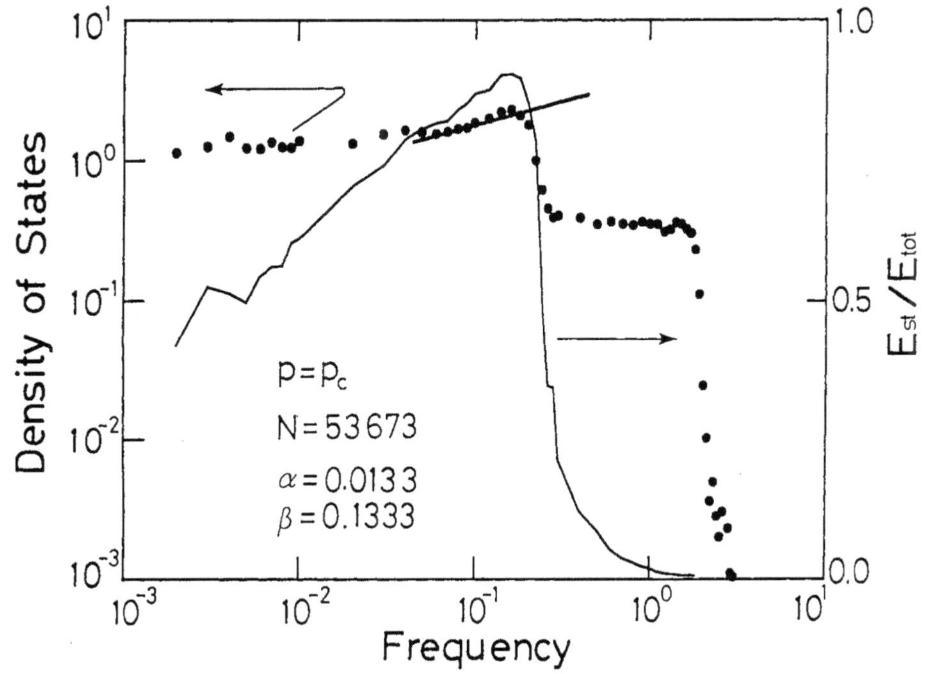

Fig. 3 The DOS for a network with vector displacements. The network is formed on a 500 × 500 square lattice at the percolation threshold p_c with 53 673 sites. The straight line through filled circles is only a guide to the eye for the law $\mathcal{D}(\omega) \propto \omega^{1/3}$. The solid curve indicates the ratio of potential energy E_{st}/E_{tot} as a function of frequency.[8]

in Fig. 3, where the percolating network formed on a 500×500 square lattice has 53 673 sites, and the force constants α and β are taken as 0.0133 and 0.133, respectively. This network has a cut-off frequency $\omega_D = 2.0784$. The step-like decrease of the states in the high-frequency regime in Fig. 3 indicates that the stretching motions are not excited above some frequency ω_0, whose value is determined by the force constant α. For the present case $\alpha = 0.0133$ and ω_0 is estimated to be 0.2309 from the relation $\omega_0 = 2\sqrt{\alpha/m}$. This value coincides with the observed value in Fig. 3.

In order to clarify the individual contributions from the bending or stretching motions, the ratio of the potential energies [Eq. (3)] was calculated as a function of frequency ω. The solid curve in Fig. 3 shows the ratio between the potential energy attributed to the stretching motion (E_{st}) and the total potential energy (E_{tot}) obtained by substituting the displacements of calculated eigenmodes into the potential energy of Eq. (3). The crossover frequency ω_{l_c} is estimated from the condition $E_{st}(\omega)/E_{tot}(\omega) = 1/2$, leading to $\omega_{l_c} \cong 0.005$. The DOS in the vicinity of ω_{l_c} is independent of frequency. The crossover region from bending to stretching fractons extends over at least two orders of magnitude in frequency. This is because the ratio $E_{st}(\omega)/E_{tot}(\omega)$ increases logarithmically, as shown in Fig. 3.

In order to clarify the contribution of bending fractons, the DOS's for the case $\beta/\alpha = 0.05$ have been calculated. Inelastic light scattering experiments can probe bending fractons due to polarization conditions.[9] These sets of force constants allow exclusive examination of the DOS for the bending-fracton regime because the stretching fracton regime is shifted into the high-frequency region. Figure 4 shows the results for the DOS for $d = 2$ BP at $p = 0.5 = p_c$ (open circles) and $p = 0.55$ (filled circles), which are formed on 800×800 lattice. The sets

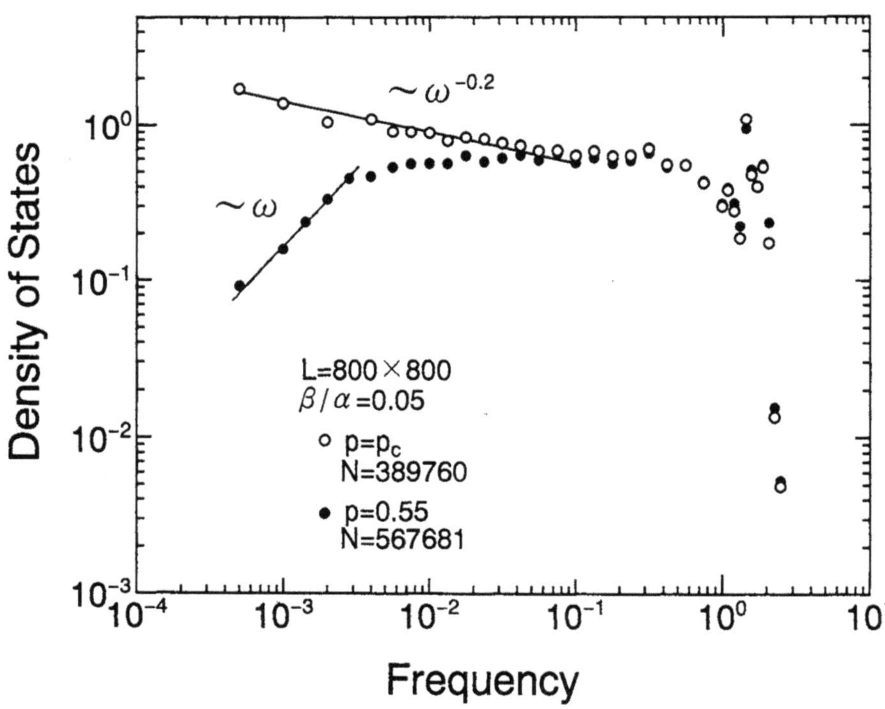

Fig. 4 The DOS for a network with vector displacements. Networks have the same structure as that of Fig. 3. Filled circles indicate the result for $\alpha = 1.0$ and $\beta = 0.01$. Open circles are for $\alpha = \beta = 0.12$. The straight lines in the l.h.s. is drawn by least squares fitting indicating the law proportional to $\omega^{\tilde{d}_b - 1}$ with $\tilde{d}_b = 0.79$. The straight line in the r.h.s. indicates the DOS proportional to $\omega^{1/3}$.[8]

of force constants $[\alpha, \beta]$ were taken as [1.0, 0.05]. The DOS for $p = p_c$, given by open circles in Fig. 4, weakly diverges as $\omega \to 0$ in accord with the theory by Feng.[6] The value of the bending-fracton dimension $\tilde{d}_b$ obtained by a least squares fitting from Fig. 4 is $\tilde{d}_b = 0.8$. The straight lines on the l.h.s. through the filled and open circles in Fig. 4 are drawn according to $\omega^{\tilde{d}_b - 1}$ with $\tilde{d}_b = 0.8$. This value agrees well with the predicted value ($\tilde{d}_b = 0.78$), in contrast with the case of stretching fractons.

It is interesting to see the behavior of the DOS at the phonon-bending fracton crossover. The filled circles are calculated results for $d = 2$ BP at $p = 0.55$. One can see, from the data given in Fig. 4, a clear crossover from phonons to bending fractons in the vicinity of $\omega \simeq 4 \times 10^{-3}$. There is no noticeable hump around the crossover as found in Fig. 4. Similar behavior is found for the case of scalar fractons.[10] Furthermore, the DOS in the bending fracton regime does not weakly diverge (see the frequency regime above $\omega \simeq 4 \times 10^{-3}$ in Fig. 4).

3.2 Mode Pattern of a Bending Fracton

Let us show the mode-pattern of a bending fracton as in the case for a fracton with scalar displacements described in Sec. 2.1. The forced oscillator method is applied to illustrate

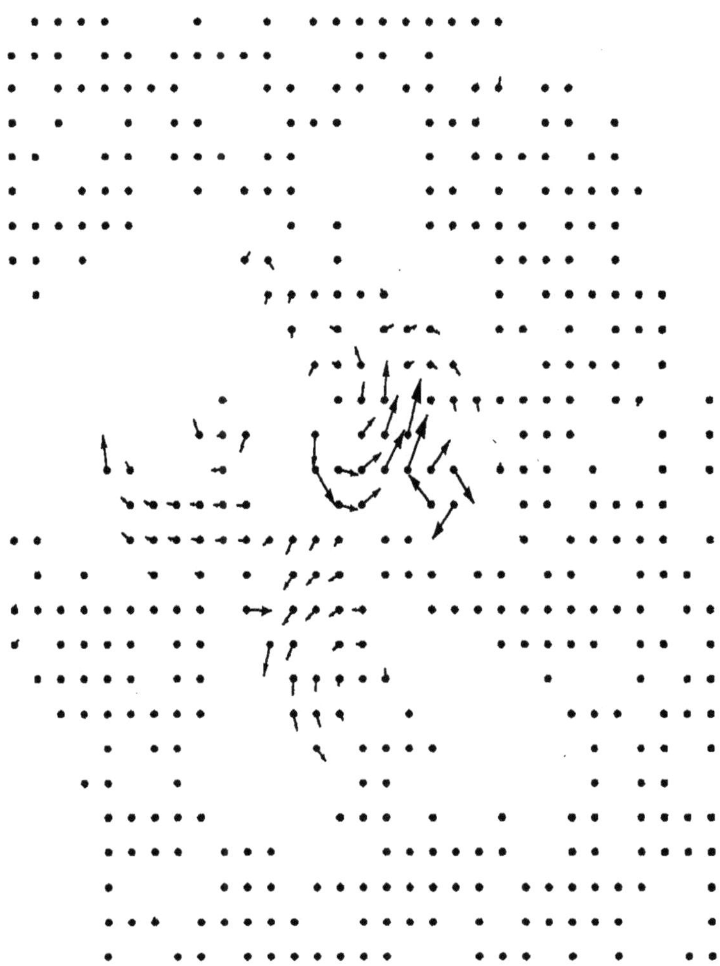

Fig. 5 The snapshot of the bending fracton eigenmode with $\omega = 0.074288$.

the mode-pattern of a bending fracton. Networks ($d = 2$) at p_c was formed on a 50×50 [$N = 1213$] square lattice.

Figure 5 shows the bending fracton with $\omega = 0.0742488$ (an eigenmode with bending displacements). The set of the force constants ($\alpha = 1.0$ and $\beta = 0.01$) is taken in order to see the bending motion. One sees from Fig. 5 that the bending fracton really executes rotational motions, whereas the rotational motions are correlated over a large distance with phase changes.

4. SPECTRAL DIMENSION FOR ANTIFERROMAGNETS

4.1 Density of States

This section presents simulated results on the DOS's for percolating antiferromagnetic networks ($d = 2-4$). This belongs to a different universality class than that for scalar elasticity. It is found that the value of the spectral dimension $\tilde{d}_{AF}$ for antiferromagnetic fractons takes a value very close to unity independent of the Euclidean dimension d.[11]

The linearized equations of motion for spin deviations S_i^+ from the perfect antiferromagnetic Néel order are expressed in unit of $2S/\hbar = 1$ by:

$$i\frac{\partial S_i^+}{\partial t} = \sigma_i \sum_i J_{ij}(S_i^+ + S_j^+), \tag{4}$$

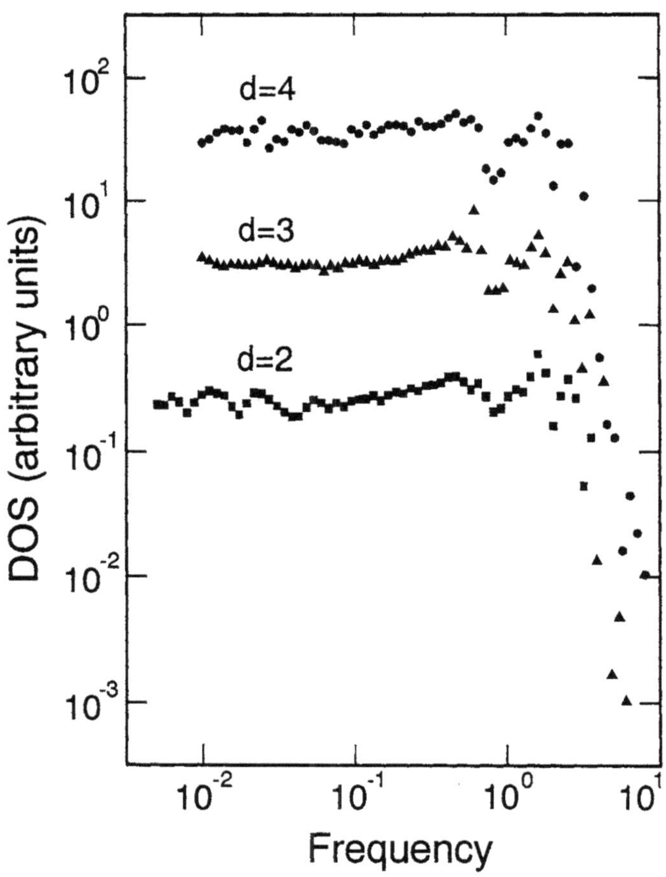

Fig. 6 The DOS per one spin for $d = 2$ (squares), $d = 3$ (triangles), and $d = 4$ (circles) BP antiferromagnets at $p = p_c$.

where $S_i^+ = S_i^x + iS_j^y$, and J_{ij} are chosen such that $J_{ij} = 1$ if both sites i and j are occupied and $J_{ij} = 0$ otherwise. In Eq. (4), σ_i is $+1$ for the site i belonging to the A-sublattice and -1 for the B-sublattice. These equations belong to a different class of equations of motion from that for ferromagnetic spin-waves or elastic vibrations with scalar displacements due to the factor σ_i in addition to the sign of the second term of the r.h.s. of Eq. (4). This indicates that the universality class of percolating antiferromagnets is different.

The results of the DOS's of spin-waves in bond percolating (BP) antiferromagnets at p_c are shown in Fig. 6. Filled squares in Fig. 6 represent the results for $d = 2$ percolating antiferromagnets, in which 11 BP networks have been prepared on 1000×1000 square lattices at $p_c(= 0.5)$ with periodic boundary conditions. The largest network has 605 544 spins. It is remarkable that the almost constant DOS for $\omega \ll 1$ is found in Fig. 6. The least squares fitting for filled squares in Fig. 6 leads to $\mathcal{D}(\omega) \propto \omega^{\tilde{d}_{\rm AF}-1}$, with $\tilde{d}_{\rm AF} = 0.97 \pm 0.06$. The DOS for $d = 3$ percolating antiferromagnets at $p_c(= 0.2488)$ are given in Fig. 6 by

Fig. 7 (a) The snapshot of the antiferromagnetic fracton with $\omega = 0.0511637$. Coloring are the same with that given in Fig. 2.

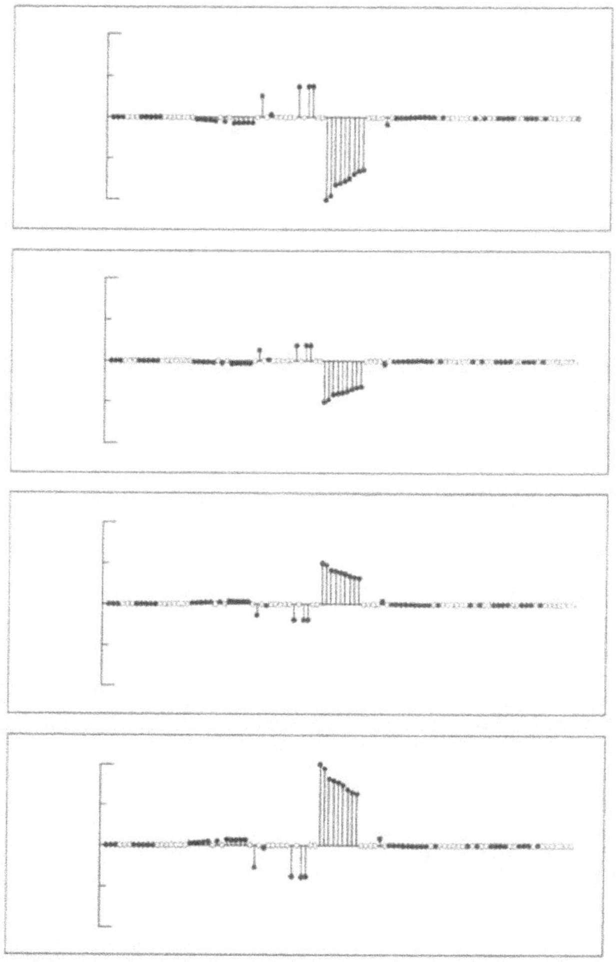

Fig. 7 (b) Cross section of a rotating AF fracton along the white line shown in Fig. 7(a). The data are plotted as if amplitudes of B-lattice spins take the same directions with those for A-lattice spins.

filled triangles, respectively. The BP networks of 13 realizations (the largest network has 177 886 spins) are formed on $96 \times 96 \times 96$ cubic lattices with periodic boundary conditions for all directions. The calculated DOS at low frequencies shows the constant behavior as a function of frequency as well as the case of $d = 2$ percolating magnets. The value of $\tilde{d}_{\rm AF} = 0.97 \pm 0.03$ is obtained by least squares fitting from data in Fig. 6. The DOS of spin-waves in $d = 4$ percolating networks are given by filled circles in Fig. 6, respectively. The BP networks of 12 realizations (the largest network has 26 060 spins) are formed on $28 \times 28 \times 28 \times 28$ hypercubic lattices at $p_c(= 0.160)$. From the data of the DOS, we have obtained $\tilde{d}_{\rm AF} = 0.98 \pm 0.09$ for $d = 4$ antiferromagnetic fractons.

Our remarkable findings ($\tilde{d}$ independent of Euclidean dimension d) can be interpreted in terms of a dynamic scaling argument.[11]

4.2 Mode Pattern of Antiferromagnetic Fracton

Figure 7(a) shows a mode-pattern of antiferromagnetic (AF) fracton which was obtained by applying the forced oscillator method. The network at $p_c = 0.5$ was formed on a 200×200 lattice. The site number is $N = 18\,890$. The color picture of the mode pattern of an

AF fracton (snap shot) on a $d = 2$ network is shown in Fig. 7(a), where the eigenmode belongs to an angular frequency $\omega = 0.0511637$. The colorings are the same as in the case of Fig. 2(a). One sees that the mode pattern is almost the same as the case for an elastic network except for a small localization length [see Fig. 7(b)]. This suggests that at very high frequency, AF fractons do not reflect the fractality of the system.

5. CONCLUSIONS

I have described the recent progress in our understanding of the dynamic properties of elastic and antiferromagnetic percolating networks via large-scale simulations. It should be emphasized that large-scale simulations have surely confirmed to a great extent the early scaling arguments and also stimulated new theoretical developments.[2]

ACKNOWLEDGMENTS

I would like to thank my colleagues K. Yakubo and T. Terao for fruitful discussions. S. Alexander and R. Obach should be acknowledged for stimulating discussions during the course of our works. This work was supported in part by a Grant-in-Aid of Scientific Research of the Special Project, "Computational Physics as New Frontiers in Condensed Matter Research," of the Japan Ministry of Education, Science and Culture.

REFERENCES

1. S. Alexander and R. Orbach, *J. Phys. (Paris) Lett.* **43**, L625 (1982).
2. For a review, see, for example, T. Nakayama, K. Yakubo and R. Orbach (to be published in *Rev. Mod. Phys.*).
3. T. Nakayama, *Physica* **A191**, 386 (1992).
4. P. Grassberger, *J. Phys.* **A19**, 1681 (1986).
5. K. Yakubo, T. Nakayama and H. J. Maris, *J. Phys. Soc. Jpn.* **60**, 3249 (1991).
6. S. Feng, *Phys. Rev.* **B32**, 5793 (1985).
7. I. Webman and G. S. Grest, *Phys. Rev.* **B31**, 1689 (1985).
8. K. Yakubo, K. Takasugi and T. Nakayama, *J. Phys. Soc. Jpn.* **59**, 1909 (1990).
9. Y. Tsujimi, E. Courtens, J. Pelous and R. Vacher, *Phys. Rev. Lett.* **60**, 2757 (1988).
10. K. Yakubo and T. Nakayama, *Phys. Rev.* **B36**, 8933 (1987).
11. K. Yakubo, T. Terao and T. Nakayama, *J. Phys. Soc. Jpn.* **62**, 2200 (1993).

Fractals, Vol. 1, No. 4 (1993) 818–826

ANOMALOUS INTERFACE ROUGHENING: THE ROLE OF A GRADIENT IN THE DENSITY OF PINNING SITES

L. A. N. AMARAL,* A.-L. BARABÁSI,* S. V. BULDYREV,*
S. HAVLIN,*,† and H. E. STANLEY*
*Center for Polymer Studies and Dept. of Physics,
Boson University, Boston, MA 02215 USA
†Dept. of Physics, Bar-Ilan University, Ramat Gan, ISRAEL

Abstract

We study the effect on interface roughening of a gradient ∇p in the density of pinning sites p. We identify a new correlation length, ξ, which is a function of ∇p : $\xi \sim (\nabla p)^{-\gamma/\alpha}$, where $\alpha = \nu_\perp/\nu_\parallel$ is the roughness exponent, and $\gamma = \nu_\perp/(1+\nu_\perp)$. The exponents $\nu_\perp$ and $\nu_\parallel$ characterize the transverse and longitudinal correlation lengths. To investigate the effect of ∇p on the scaling properties of the interface in $(1+1)$ and $(2+1)$ dimensions, we calculate the critical concentration, p_c, and the exponents γ and α from which $\nu_\perp$ and $\nu_\parallel$ can be determined. Our results are in qualitative agreement with some of the features of imbibition experiments.

1. INTRODUCTION

Recently the growth of rough interfaces has attracted great interest, partly fueled by the broad interdisciplinary aspects of the subject.[1–5] Early studies[1,2,5] showed that the *rms* width of an interface, w, scales with time, t, and system size, L, as:

$$w(L,t) \equiv \langle [h(x,t) - \bar{h}(t)]^2 \rangle^{1/2} \sim L^\alpha f\left(\frac{t}{L^z}\right), \tag{1}$$

where $h(x, t)$ is the height of the interface and the function $f(x)$ satisfies $f(x \ll 1) \sim x^\beta$ and $f(x \gg 1) \sim const.$ The exponents α and $z \equiv \alpha/\beta$ are called the roughness and dynamical exponents, respectively.

Several authors[6–12] have recently suggested that for the case of an interface growing in a disordered medium, the roughening is caused by a quenched noise which is a function of the height of the interface but not of time. This proposed quenched noise not only enabled researchers to explain the results of some imbibition experiments in $(1+1)$ and $(2+1)$ dimensions,[7,8] but also resulted in the mapping of the growth of the interface to directed percolation and directed surfaces.[6–9]

In Ref. 7 (and similarly in Ref. 6) the interface stops growing, i.e., becomes pinned, when it encounters a directed percolating cluster of *pinning sites*. This fact allows the determination of the roughness exponent in $(1+1)$ dimensions. In directed percolation, the characteristic lengths in the directions normal and parallel to the interface scale as[13]:

$$\xi_\perp \sim |p_c - p|^{-\nu_\perp}, \quad \xi_\parallel \sim |p_c - p|^{-\nu_\parallel}. \tag{2}$$

The width of the pinned interface is proportional to $\xi_\perp$. When a directed path percolates the system, for $L \ll \xi_\parallel$, we expect[6,7]:

$$w \sim \xi_\perp \sim \xi_\parallel^{\nu_\perp/\nu_\parallel} \sim L^{\nu_\perp/\nu_\parallel}. \tag{3}$$

Thus,

$$\alpha = \nu_\perp/\nu_\parallel. \tag{4}$$

From the calculated values of $\nu_\perp = 1.097 \pm 0.001$ and $\nu_\parallel = 1.733 \pm 0.001$,[14,15] Eq. (4) predicts for the roughness exponent $\alpha = 0.633 \pm 0.001$.

In Ref. 7, a simple set of imbibition experiments was presented. Paper towels were dipped into a reservoir filled with various colored liquids (e.g. coffee, ink) and the propagating wetting front observed. The wetting front reaches a certain height above the level of the liquid and stops propagating when the evaporation of the liquid induces the *pinning* of the interface by the inhomogeneities of the paper. The rough boundary between colored and uncolored areas is digitized and a roughness exponent $\alpha \cong 0.63$ was found,[7] in agreement with the predictions of directed percolation.

However, three experimental features remain unexplained.

(i) The scaling behavior was observed in Ref. 7 for length scales inferior to 1 cm only, much smaller than the system size;

(ii) Although the model of Ref. 7 correctly reproduces the anomalously large value of α observed in some experiments,[7] it does not make clear which mechanism drives the experimental system towards the critical probability;

(iii) The experiments show that the same fluid in the same paper reaches a higher level if the evaporation from the surface of the paper is reduced by enclosing the whole apparatus in a box or when the viscosity of the fluid is reduced by decreasing the concentration of coffee or ink (see Fig. 1).

To better understand the above experimental features, we offer the following considerations. It is reasonable to assume that the amount of liquid per unit area, $\rho(h)$, in the paper decreases with height, h, by the amount of evaporated liquid:

$$\frac{\partial \rho(h)}{\partial h} = -\frac{a}{v}, \tag{5}$$

Fig. 1 The completely stopped interfaces in imbibition experiments with coffee and paper towels: **(a)** high evaporation rate, **(b)** medium evaporation rate, and **(c)** low evaporation rate.

where a is the evaporation rate (the amount of fluid that evaporates from unit area in unit time) and v is the velocity of the fluid particles in the paper. Assuming that a and v are *constants*, we can conclude that ρ is a linear function of the height:

$$\rho(h) = \rho(0) - \frac{a}{v}h\,. \tag{6}$$

The fluid in the paper propagates through fibers randomly distributed and connected. We assume that the fluid will penetrate from one fiber to the next only if there is a certain amount of accumulated liquid in the wet fiber. Since the amount of this accumulation depends on the properties of the connections between the fibers — which are very heterogeneous — the concentration of the *pinning connections* between fibers increases with the decrease of the amount of water in the paper, $\rho(h)$. The critical threshold p_c corresponds to the average height reached by the liquid in the paper during the final stage of the experiment when the interface ceases to propagate.

Since in the model the concentration of *pinning connections* between fibers corresponds to the density of pinning sites, the probability of a site being blocked should increase with height. If we assume this increase to happen with a constant gradient, $\nabla p > 0$, then in the experiment ∇p should be inversely proportional to the height of the completely stopped interface. Due to the gradient, a new correlation length ξ is found, in analogy with the theory of Sapoval, Rosso and Gouyet on gradient percolation.[16–18]. This correlation length can be related to the upper cutoff of scaling behavior in the experiments of Ref. 7.

2. THE MODEL

Consider a square lattice and distribute randomly interface pinning forces in the sites. For each site we compare the pinning force with the interface driving force. If the former is larger, the site is considered blocked (pinning site), otherwise it is empty, in which case the site can be invaded during the growth. By varying the driving force with the height we can produce a gradient in the density of pinning sites.

We start from a flat interface on the bottom edge of a lattice (with periodic boundary conditions), and allow the interface to grow by invasion of empty sites. In the growth process we move along all columns, from left to right, one after the other (see Fig. 3). For each column we first try to grow it, from its top position, one site up. If this is not possible (because that site is blocked), we try to grow it to the left and/or to the right *if and only*

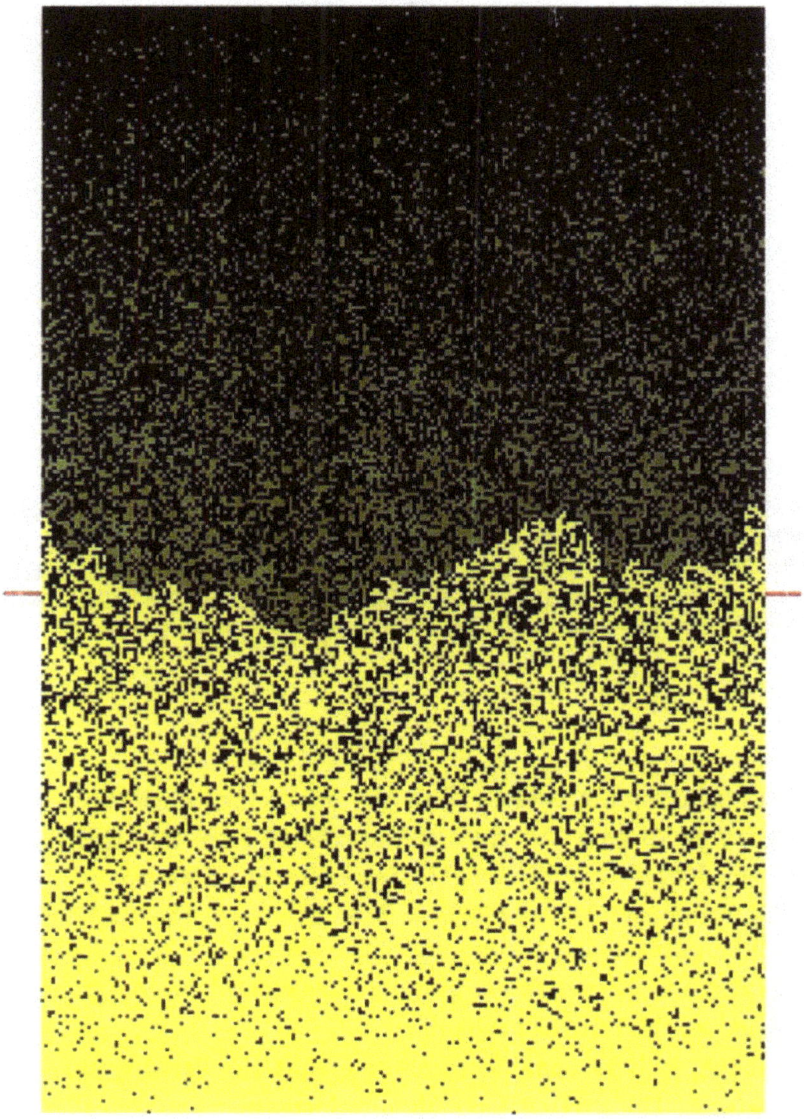

Fig. 2 We show here the pinned interface for a system with a constant gradient. Black dots represent pinning sites, green dots empty sites and yellow dots invaded sites. The red line shows the position of the critical probability. We can see that the interface stops when it reaches a percolating cluster of pinning sites, what occurs for $p = p_c$.

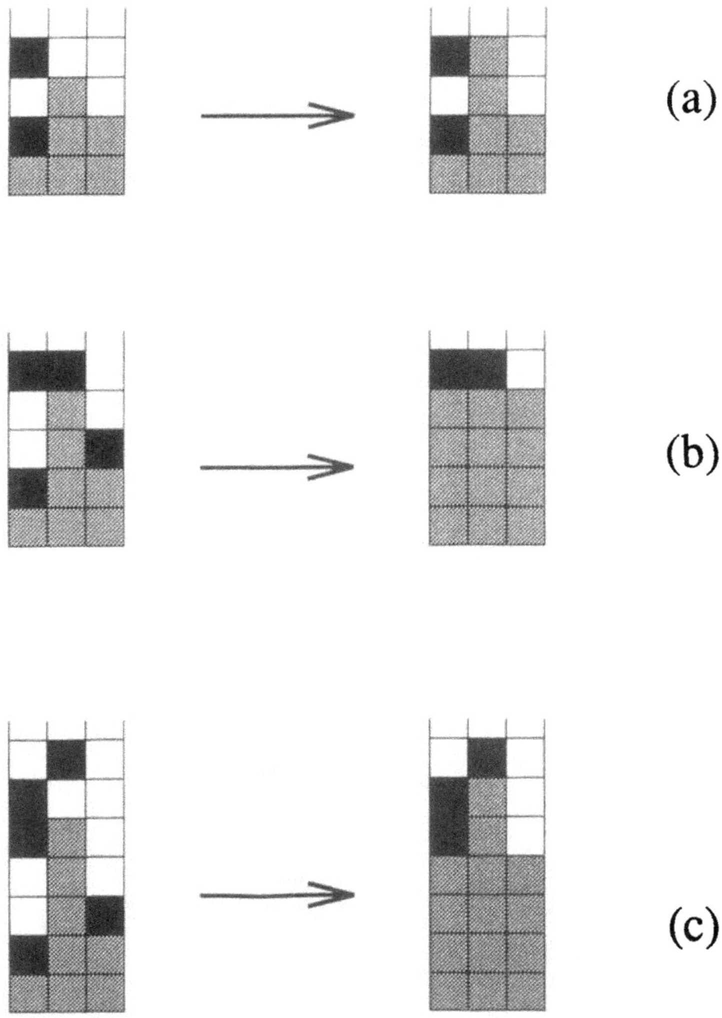

Fig. 3 Show are three cases of growth for the middle column with the conventions: grey sites are invaded, white sites are empty, and black sites are blocked (pinning) sites. In case (a) the top position grows one site up. In case (b) the top position grows both to the right and to the left, and the height of the neighbor column is updated accordingly. In case (c) the top position grows up and the second position from the top grows both to the left and to the right; the height of the neighbor columns is updated.

if the neighboring columns are shorter than the one being considered *and if* that site is empty. We keep trying to grow the sides of the column from all other positions, beneath the top one, that are higher than the neighboring columns. If we succeed, we set the height of the neighboring column to be the height of that site, since the model does not allow for overhangs. The time is increased by one unit once we reach the last column. We can see in Fig. 2, the variation in the density of pinning sites caused by the gradient and that the interface gets pinned at the critical probability.

As discussed above, it is reasonable to assume that the driving force depends on the height of the interface. The driving force should decrease monotonically with the height. This means that the density of pinning sites will increase monotonically. For example, in imbibition experiments the density of fluid, $\rho(h)$, decreases because of evaporation, and thus the driving force should also decrease.

The simplest choice [see Eqs. (5) and (6)] for the functional form of the driving force $f(h)$ is a linear function,

$$f(h) = f_0 - (\nabla p)h\,. \tag{7}$$

With this form of $f(h)$, we get the following density of pinning sites, $p(h)$:

$$p(h) = p_0 + (\nabla p)h\,. \tag{8}$$

Finding a suitable physical justification for this particular choice is difficult since the microscopic processes responsible for the interface growth and roughening are extremely complex. However, we shall find that the value of $p_0(< p_c)$ does not affect the final width, which further justifies the assumption that only the variation in the region close to p_c is important for the properties of the pinned interface.

3. RESULTS

3.1 Interface Growth in (1 + 1) Dimensions

The presence of the gradient in the density of blocked cells changes the width (see Fig. 4) of the pinned interface and its scaling form. Our results for the pinned interface show that the saturated width behaves as $w \sim \ell^{\alpha}$ for $\ell \ll \xi$ and as $w \sim (\nabla p)^{\gamma}$ for $\ell \gg \xi$, where $\xi = \xi(\nabla p)$ is the correlation length. We find:

$$\alpha = 0.63 \pm 0.01\,, \quad \gamma = 0.52 \pm 0.01\,. \tag{9}$$

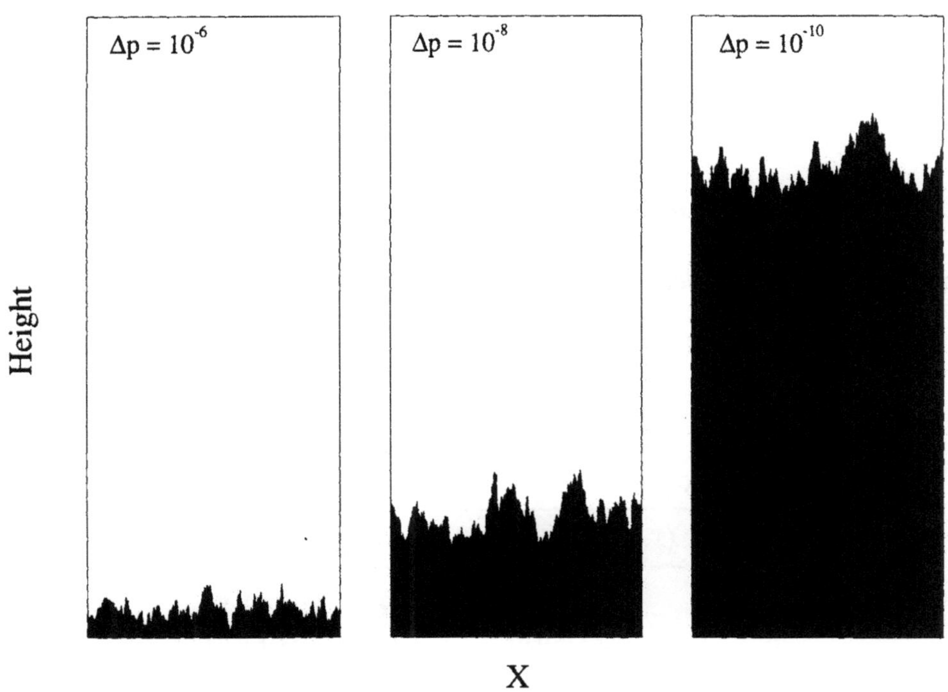

Fig. 4 Contour of completely pinned interfaces for three values of gradient: $\nabla p = 2^{-6}$, 2^{-8} and 2^{-10}. We can observe the increase in the final heights of the interface with the decrease of the gradient. For the widths, we can see an increase from the larger gradient to the other values but not between the smaller values. The reason for this lies in the fact that the system size is smaller than the correlation length ξ.

Moreover, we find our data to scale as:

$$w(\ell, \nabla p) \sim \ell^\alpha f\left(\frac{\ell}{(\nabla p)^{-\gamma/\alpha}}\right), \tag{10}$$

where the function $f(x)$ satisfies $f(x \ll 1) \sim const$ and $f(x \gg 1) \sim x^{-\alpha}$.

The value of α can be understood from the mapping to directed percolation, since the conditions for a complete pinning of the interface do not change from the model of Ref. 7. The presence of the gradient, which introduces a new length scale in the direction normal to the interface, is not important for $\ell \ll \xi$. On the other hand, for $\ell \gg \xi$, this second length scale will cause an early breakdown of the scaling behavior. By an argument similar to the one developed in Ref. 16 (see also Refs. 19 and 20) we can relate the exponent γ to $\nu_\perp$,

$$\gamma = \frac{\nu_\perp}{1+\nu_\perp}. \tag{11}$$

This relation predicts $\gamma \cong 0.523$ in $(1 + 1)$ dimension, in very good agreement with our result [Eq. (9)].

3.2 Interface Growth in $(2+1)$ Dimensions

The $(2 + 1)$ dimensional model can be mapped to directed surfaces,[8,9,21] a percolation problem that has not been thoroughly investigated. A study along the lines described above for $(1 + 1)$ dimensions allows us to estimate the values of the critical exponents and critical probability for that problem.

From our data we find:

$$\alpha = 0.43 \pm 0.03, \quad \gamma = 0.32 \pm 0.01. \tag{12}$$

From these results, we calculate the values of the transverse and longitudinal correlation exponents for the directed surfaces problem,

$$\nu_\perp = 0.47 \pm 0.02, \quad \nu_\parallel = 1.1 \pm 0.1. \tag{13}$$

Note that these results represent more accurate values for the exponents than those found earlier.[8]

Table 1 Our Results for the Critical Probability and Exponents

	$(1+1)$ dimensions	$(2+1)$ dimensions
p_c	0.47 ± 0.03	0.75 ± 0.03
α	0.63 ± 0.01	0.43 ± 0.03
γ	0.52 ± 0.01	0.32 ± 0.01
$\nu_\perp$	1.09 ± 0.01	0.47 ± 0.02
$\nu_\parallel$	1.73 ± 0.02	1.1 ± 0.1

4. CONCLUSIONS

We used the growth rule proposed in Ref. 7 to simulate the motion of the wet interface. The difference is that while in Ref. 7 the pinning sites were randomly distributed with a constant probability, p, in our model this probability changes with height according to Eq. (8). While the model of Ref. 7 needs to be tuned to reach the critical state at $p = p_c$, in the gradient model the system evolves naturally to the pinned critical state. However, since the gradient introduces a correlation length in the transverse direction we would hesitate to refer to this *self-tuning* ability of the model as self-organized criticality.[22]

An important conclusion from this work is the ability of our model to reproduce the important features of imbibition experiments. We observe both in simulations and experiments a dependence of the final height of the interface on the value of the gradient. We also find that a gradient in the density of pinning sites leads to the introduction of a new correlation length, in qualitative agreement with experiments. We verified the dependence of that correlation length on the gradient and its effect on the scaling properties of the interface. For the saturated pinned interface, we find good scaling of our data both in $(1+1)$ and $(2+1)$ dimensions. This allow us to estimate, with good accuracy, the critical exponents and the critical probability both for directed percolation and directed surfaces.

There are still a number of open questions related to the present model and the general problem of nonequilibrium roughening with quenched noise. A first question is the behavior near the pinning threshold. As we approach p_c, the interface velocity approachs zero as:

$$v \sim \left(1 - \frac{p}{p_c}\right)^{\zeta}. \tag{14}$$

Numerical integration of the KPZ equation with quenched noise suggests an exponent $\zeta = 0.64 \pm 0.08$.[23] In the presence of a gradient in the pinning force, this behavior is no longer valid. Recently, related models *without threshold* have been introduced,[24,25] which generate an interface with the same roughness exponent as in our model. It is very interesting that these models result in a nontrivial temporal multi-affinity,[26] while no spatial multi-affinity[27] has been observed. The question of the existence of this nontrivial multi-scaling in our model is still open.

It is important to understand the discrepancy between the different reported results of the roughness exponent. In the present model, since the pinning is due to the directed percolation cluster determined by the disorder, the roughness exponent is $\alpha \cong 0.63$. However different results have been reported from numerical integration of the KPZ equation (see Ref. 4) with quenched noise, resulting in $\alpha = 0.71 \pm 0.08$.[23] Renormalization group analysis leads to $\alpha = 1$,[28] while related scaling arguments gave 3/4.[23] So far an understanding of these discrepancies is not available.

After this work was completed we received a theoretical analysis by Olami *et al.*[29] that studies the behavior of the $(1+1)$ dimension model at criticality ($p = p_c$) and relates all the scaling exponents to a single one, α.

ACKNOWLEDGMENTS

We thank S. Schwarzer for his help with the preparation of the manuscript. L. A. N. Amaral acknowledges a scholarship from Junta Nacional de Investigação Científica e Tecnológica. S. Havlin acknowledges partial support from the Bi National US-Israel Foundation. The Center for Polymer Studies is supported by the National Science Foundation.

REFERENCES

1. T. Vicsek, *Fractal Growth Phenomena*, 2nd ed. (World Scientific, Singapore, 1992).
2. F. Family and T. Vicsek, eds., *Dynamics of Fractal Surfaces* (World Scientific, Singapore, 1991).
3. J. Krug and H. Spohn, in *Solids Far from Equilibrium: Growth, Morphology and Defects*, ed. C. Godrèche (Cambridge University Press, Cambridge, 1991).
4. M. Kardar, G. Parisi and Y.-C. Zhang, *Phys. Rev. Lett.* **56**, 889 (1986).
5. F. Family and T. Vicsek, *J. Phys.* **A18**, L75 (1985).
6. L.-H. Tang and H. Leschhorn, *Phys. Rev.* **A45**, R8309 (1992).
7. S. V. Buldyrev, A.-L. Barabási, S. Havlin, F. Caserta, H. E. Stanley and T. Vicsek, *Phys. Rev.* **A45**, R8313 (1992).
8. S. V. Buldyrev, A.-L. Barabási, S. Havlin, J. Kertész, H. E. Stanley and H. S. Xenias, *Physica* **A191**, 220 (1992).
9. A.-L. Barabási, S. V. Buldyrev, S. Havlin, G. Huber, H. E. Stanley and T. Vicsek, in *Imbibition in Porous Media: Experiment and Theory*, Les Houches Workshop, eds. R. Jullien, J. Kertész, P. Meakin and D. E. Wolf (Nova Science, New York, 1992).
10. G. Parisi, *Europhys. Lett.* **17**, 673 (1992).
11. N. Martys, M. Cieplak and M. O. Robbins, *Phys. Rev. Lett.* **66**, 1058 (1991).
12. D. Kessler, H. Levine and Y. Tu, *Phys. Rev.* **A43**, 4551 (1991).
13. A. Bunde and S. Havlin, eds., *Fractals and Disordered Systems* (Springer, Heidelberg, 1991).
14. J. W. Essam, K. De'Bell, J. Adler and F. M. Bhatti, *Phys. Rev.* **B33**, 1982 (1986).
15. J. W. Essam, A. J. Guttmann and K. De'Bell, *J. Phys.* **A21**, 3815 (1988).
16. B. Sapoval, M. Rosso and J. F. Gouyet, *J. Physique Lett.* **46**, L149 (1985).
17. M. Rosso, J. F. Gouyet and B. Sapoval, *Phys. Rev.* **B32**, 6053 (1985).
18. M. Rosso, J. F. Gouyet and B. Sapoval, *Phys. Rev. Lett.* **57**, 3195 (1986).
19. A. Hansen, T. Aukrust, J. M. Houlrik and I. Webman, *J. Phys. A: Math. Gen.* **23**, L145 (1990).
20. A. Hansen and J. M. Houlrik, *J. Phys. A: Math. Gen.* **24**, 2377 (1991).
21. S. Havlin, A.-L. Barabási, S. V. Buldyrev, C. K. Peng, M. Schwartz, H. E. Stanley and T. Vicsek, in *Growth Patterns in Physical Sciences and Biology* (Plenum, New York, 1992).
22. P. Bak, C. Tang and K. Wiesenfeld, *Phys. Rev. Lett.* **59**, 381 (1987).
23. Z. Csahók, K. Honda and T. Vicsek, *J. Phys.* **A26**, L171 (1993).
24. K. Sneppen, *Phys. Rev. Lett.* **69**, 3539 (1992); K. Sneppen and M. H. Jensen, *Phys. Rev. Lett.* **71**, 101 (1993).
25. J. Kertész, V. K. Horváth and F. Weber, *Fractals* **1**, 67 (1992).
26. A.-L. Barabási, *J. Phys. A: Math. Gen.* **24**, L1013 (1991).
27. A.-L. Barabási and T. Vicsek, *Phys. Rev.* **A24**, 2730 (1991); A.-L. Barabási, P. Szépfalusy and T. Vicsek, *Physica* **A178**, 17 (1991); A.-L. Barabási, R. Bourbonnais, M. Jensen, J. Kertész, T. Vicsek and Y.-C. Zhang, *Phys. Rev.* **A45**, R6951 (1992).
28. O. Narayan and D. S. Fisher, in press (1993).
29. Z. Olami, I. Procaccia and R. Zeitak, submitted to *Phys. Rev. E* (1993).

Fractals, Vol. 1, No. 4 (1993) 827–839

SURFACE ROUGHENING WITH QUENCHED DISORDER IN *d*-DIMENSIONS

SERGEY V. BULDYREV*, SHLOMO HAVLIN*,†, JANOS KERTÉSZ‡
ARKADY SHEHTER† and H. EUGENE STANLEY*
**Center for Polymer Studies and Department of Physics, Boston University, Boston, MA 02215, USA*
†*Department of Physics, Bar Ilan University, Ramat Gan, Israel*
‡*Institute of Physics, Technical University of Budapest, II-1521 Budapest XI, Hungary*

Abstract

We review recent numerical simulations of several models of interface growth in d-dimensional media with quenched disorder. These models belong to the universality class of anisotropic diode-resistor percolation networks. The values of the roughness exponent $\alpha = 0.63 \pm 0.01$ ($d = 1+1$) and $\alpha = 0.48 \pm 0.02$ ($d = 2+1$) are in good agreement with our recent experiments. We study also the diode-resistor percolation on a Cayley tree. We find that $P_\infty \sim \exp(-A/\sqrt{p_c - p})$, thus suggesting that the critical exponent for $P_\infty \sim (p_c - p)^{\beta_p}$, $\beta_p = \infty$ and that the upper critical dimension in this problem is $d = d_c = \infty$. Other critical exponents on the Cayley tree are: $\tau = 3$, $\nu_\parallel = \nu_\perp = \gamma = \sigma = 0$. The exponents related to roughness are: $\alpha = \beta = 0$, $z = 2$.

1. INTRODUCTION

The growth of rough interfaces in random media is a topic of current interdisciplinary interest.[1–5] For the most part, two types of models have been applied to interface-roughening phenomena:

(a) nonlinear Langevin-type equations — such as the KPZ equation[6] — resulting in *self-affine* interfaces, and
(b) spreading and invasion percolation models,[7] generating *self-similar* interfaces.

The self-affine interface (a) can be characterized by the rms surface width

$$w(\ell, t) \equiv \langle [h(x, t) - \langle h(x, t)\rangle]^2 \rangle^{1/2} . \tag{1}$$

Here $h(x, t)$ is the surface height at time t, and the angular brackets denote the average over x belonging to a $d-1$-dimensional hypercube of size ℓ^{d-1} in the horizontal cross section perpendicular to the direction of growth.

An alternative and equivalent quantity is the height-height correlation function $c(\ell, t)$.[2] The scaling exponents obtained from w and c are believed to be identical, so we use them interchangeably.

Analysis of the KPZ equation implies the scaling law[1–4]:

$$w(\ell, t) \sim c(\ell, t) \sim \ell^{\alpha} f\left(\frac{t}{\ell^z}\right) , \tag{2a}$$

where

$$f(u) \sim \begin{cases} u^{\beta} & u \ll 1 \\ \text{const} & u \gg 1 , \end{cases} \tag{2b}$$

and

$$z = \frac{\alpha}{\beta} . \tag{2c}$$

Of course, the scaling function $f(u)$ is different for w and for c. The roughness exponent is $\alpha = 1/2$ and the dynamical exponent is $\beta = 1/3$ for $d = 1 + 1$.[6] Numerical studies of the KPZ equation give $\alpha \approx 2/(d + 2)$ in $d \geq 2$.[8–10]

On the other hand, approach (b) (percolation-type models) produces *self-similar* interfaces[7] for which $w(\ell, \infty) \sim \ell^{\alpha}$ with $\alpha = 1$ and $\beta = 1$. For many phenomena in $d = 1 + 1$ — from bacterial growth[11] and viscous flows[12,13] to the wetting,[13–15] and burning[16] of paper — *self-affine* surfaces are found with anomalous exponents α and β significantly larger than the KPZ values but less than 1. Recent experimental data in $d = 2 + 1$ also show anomalously large values of α, e.g. for mountain surfaces $\alpha = 0.58$,[17,18] for wetting of porous media $\alpha = 0.5$,[19] and for ion beam erosion of metal surfaces $\alpha = 0.53$.[20]

An essential feature of many of these systems is that the noise is quenched, i.e., noise is a function of the interface position, as opposed to the uncorrelated noise of the KPZ equation. A critical question, then, is the scaling properties of systems with quenched noise. In particular, what phases, if any, are there which exhibit self-affine growth with $\alpha > \alpha_{\mathrm{KPZ}}$?

A complete answer to this question is not yet available. Recently, we and others have introduced an interesting class of models which exhibit new scaling behavior related to directed percolation. Another line of investigation[21–23] appears to indicate a direct transition in some models from a random-field Ising model phase with $\alpha = (5-d)/3$ to KPZ behavior. However, more recent work[24] suggests that this result might be due to the neglect of a KPZ-type nonlinearity in the dynamics. As we shall see, it is possible that the nonlinear equation of Csahók et al.[24] is in the same universality class as our models. Clarifying the relationship of these and other models[25,26] is a topic that clearly merits more study.

As a step in answering this question, we develop and generalize several models[13,14,27–31] of spreading percolation with anisotropy in the growth direction which belong to a new universality class characterized by a self-affine interface with the roughness exponents given in Table 1. We explore the relation of this universality class to the reversed percolation transition point in the diode-resistor percolation network,[28–30] at which the backflow current emerges in the direction opposite to the direction of diodes, when the concentration of resistors q approaches some critical value $q_c = 1 - p_c$ ($p = 1 - q$ is the concentration of diodes). In $d = 1 + 1$, this transition point is dual to the directed percolation point[28–30] and the roughness exponent α of the interface between blocked and unblocked diodes can be expressed in terms of correlation exponents of directed percolation.

In $d > 2$, little is known about the critical properties of this transition. Recently, several theoretical conjectures have appeared concerning the values of the critical exponents of the surface roughening models with depinning transition caused by the quenched noise for continuum systems.[24,25] The numerical data for the continuum models in $d = 1+1$ strongly support the argument that these models belong to the same universality class as discrete models described below. However, according to Refs. 24 and 25, $d = 5$ should be the upper critical dimension above which $\alpha = \beta = 0$. Our numerical results do not support these conjectures. To understand the problem of surface roughening in high dimensions, we studied the random diode-resistor network on a Cayley tree. Our results indicate that the critical dimension is $d_c = \infty$.

2. MODELS

The models A, B and C we have studied are straightforward generalizations to d dimensions of the $d = 1+1$ models defined in Refs. 13, 14, 27 and 28. In each model the porous media is simulated by a cubic lattice with certain fraction p of randomly blocked cells which represent the inhomogeneities of the media (see Fig. 1). The horizontal cross section of the lattice is the $d-1$ dimensional hypercube of volume L^{d-1} with periodic boundaries. Every lattice cell can be wet or dry. At $t = 0$, all lattice cells are dry except those with vertical coordinate $h \leq 0$. At every time step, we simultaneously examine all dry cells on the wet-dry interface and decide whether each of these cells should become wet on the next time step. The decision concerning each cell is taken according to the deterministic rules specific to each model. In analogy with spreading percolation,[2,32] we define shell t as the set of cells that become wet on time step t.

In Model A, each cell adjacent to the interface becomes wet if (i) it is unblocked, or if (ii) it lies below the unblocked cell, adjacent to the interface.

In Model B: (i) the unblocked cell becomes wet if its nearest neighbor from below is wet; (ii) the blocked cell becomes wet if its nearest neighbor from below is wet and if the height of the highest wet cell in one of the nearest neighboring columns is larger than the height of the cell under consideration.

In Model C, rule (i) is the same as in Model B, but rule (ii) is slightly different. The blocked cell with the coordinates $(x_1, x_2, \ldots, x_{d-1}, h)$ becomes wet if its nearest

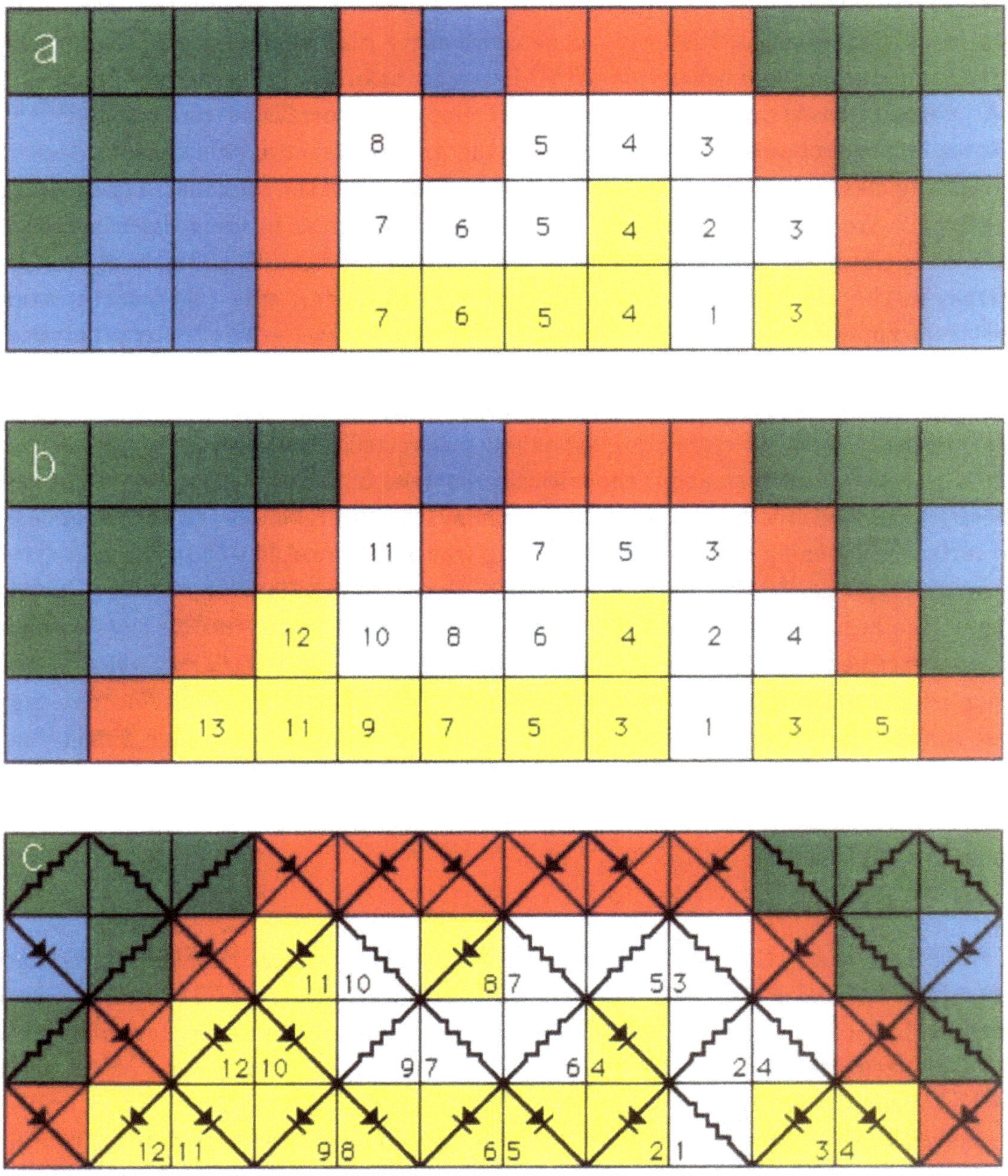

Fig. 1 Explanation of the surface growth models A, B and C in $d = 1 + 1$. Cells are randomly blocked with probability p (shown in blue) or unblocked with probability $1 - p$ (shown in green). Wet cells are shown in white (those that were unblocked before becoming wet), or yellow (those that were blocked). Wet cells are marked by numbers indicating time step at which they have become wet. Those blocked cells, that form the spanning path of directed percolation that has stopped the growth, are shown in red. **(a)** Model A, proposed by Buldyrev et al.[13-15] ($p_c = 0.4698$); **(b)** Model B proposed by Tang and Leschhorn[27] ($p_c = 0.5390$); **(c)** Model C proposed by Dhar et al.[28] ($p_c = 0.6447$). The configuration of blocked cells is the same for all three models, so the largest cluster of model C corresponds to the largest p_c. In model C diodes correspond to blocked cells, resistors to the unblocked cells. The order of wetting differs from the original definition of Ref. 28, in order to prevent overhangs in the moving parts of the interface. These changes, however, do not affect the shape of the pinned interface, which in both cases is the path of the bond directed percolation of diodes on the dual lattice.

neighbor from below is wet and at least one of the cells with the coordinates $(x_1+\Delta x_1,\ x_2+\Delta x_2,\ldots,\ x_{d-1}+\Delta x_{d-1},\ h+\Delta h)$ is wet, where the increments Δx_i and Δh obey the following constraints: $\Delta h \geq 0$, $|\Delta x_i| \leq 1$; if $\Delta h > 0$ then it is sufficient that $|\Delta x_i| \leq 1$; if $\Delta h = 0$ then $\Delta x_i = 0$ or $\Delta x_i = (-1)^{h+x_i}$. Model C is in fact equivalent to the diode-resistor network in which each blocked cell corresponds to a diode, and each unblocked cell to a resistor connecting a pair of the opposite vertices of the cell. Each diode or resistor has orientation $(\Delta x_1, \Delta x_2, \ldots, \Delta x_{d-1}, -1)$, where $\Delta x_i = -(-1)^{x_i+h}$ are determined by the coordinates of the cell [see Fig. 1(c)].

Model A is analogous to spreading percolation [rule (i)] with erosion of overhangs [rule (ii)]. In Models B and C, the height of each column at each time step can increase only by one cell. In Model A, the erosion of overhangs corresponds to much faster local growth. Also, in Models B and C, the maximal difference in heights of the two neighboring columns is 2, while in Model A, it can be any number.

Model A was suggested by Buldyrev et al.,[13-15] while Model B was originally proposed by Tang and Leschhorn[27] and can be considered as the discretized version of the Langevin equation with quenched noise and is very close to the models defined by Csahók et al.[24] and Parisi.[25] Comparison of the time development of all three models is shown in Fig. 1 for the same $d = 1+1$ configuration of blocked cells. The growth of each model is stopped by the spanning paths of directed percolation of the blocked cells with different definitions of connectivity (shown as a red fence of blocked cells). In Model A the path has five choices (North, South, East, North-East and South-East); in Model B three choices (East, North-East and South-East); and in Model C only two choices (North-East and South-East). This results in different percolation thresholds: $p_c^A = 0.4698$, $p_c^B = 0.539$, $p_c^C = 0.6447$. Numbers on the cells show the time at which they become wet. These models have different finite size corrections to scaling, but the asymptotic behavior is characterized by the same set of exponents.

3. DIMENSION 1 + 1: THEORY AND SIMULATIONS

In this section, we review theory and simulations for $d = 1+1$.[13-15] When the probability of blocked cells p is close to p_c, the growth is halted in many places by the paths of a directed percolation cluster. Each path can be characterized by two correlation lengths, $\xi_\perp$ and $\xi_\parallel$. When the path is spanning, i.e., when the growth is stopped completely, $\xi_\parallel$ is equal to the system size L, and $\xi_\perp$ is proportional to the width of the interface w. It is known from the theory of directed percolation[32,33] that the correlation lengths diverge in the vicinity of p_c as:

$$\xi_\perp \sim |p-p_c|^{-\nu_\perp}\,, \quad \xi_\parallel \sim |p-p_c|^{-\nu_\parallel}\,, \tag{3a}$$

where $\nu_\parallel \simeq 1.733$, and $\nu_\perp \simeq 1.097$.[32] Thus,

$$w \sim \xi_\perp \sim L^{\nu_\perp/\nu_\parallel} \equiv L^\alpha\,, \tag{3b}$$

where

$$\alpha = \nu_\perp/\nu_\parallel \simeq 0.633 \pm 0.001\,. \tag{3c}$$

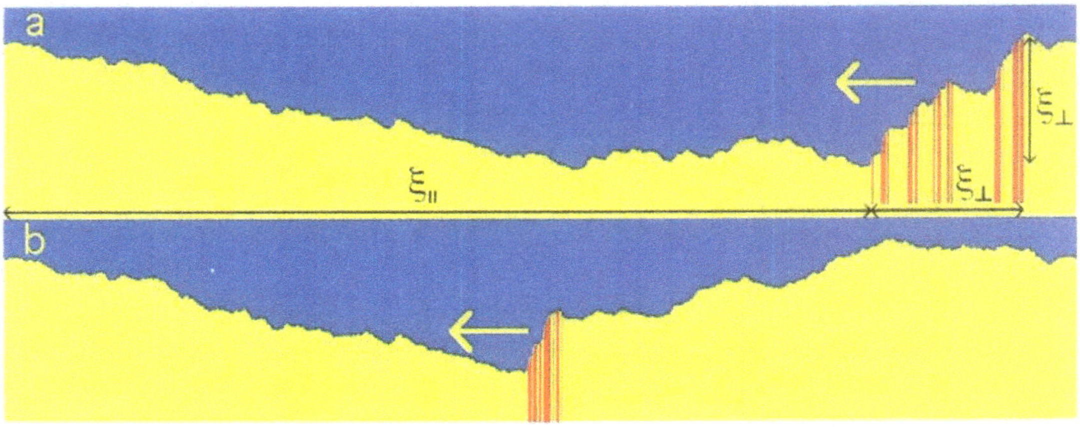

Fig. 2 Two successive snapshots of the interface of Model A still evolving near its pinning threshold ($p = 0.4698$). System size is 2^{12}, but the calculation are made for $L \leq 2^{17}$. Upper snapshot taken at $t = 18\,000$, lower at $t = 19\,500$. Yellow color indicates wet area, blue color indicates dry area, and red color indicates "live" columns, i.e., columns that contain cells which become wet at the current time step. Arrows show the direction of the propagation of steep eroding slopes, which has approximately equal horizontal and vertical dimensions both of the order of $\xi_\perp$. The horizontal dimension of the long blocked region is of the order of $\xi_\parallel$, while its vertical dimension is of the order of $\xi_\perp$.

Thus the spanning path of directed percolation describes the final state of the paper wetting experiment,[13–15] where the wetting front is completely pinned by inhomogeneities in the paper. The theoretical value of α [Eq. 3(c)] is in excellent agreement with both our simulations and our experiments.[13–15]

A natural question is "What is the *dynamics* of wetting?" To answer this, we study the dynamical behavior of the models below and above p_c. Figure 2 shows a snapshot of the wetting front as it continues to propagate in the $(1+1)$-dimensional media when $p < p_c$. Large sections of the interface are already pinned and the growth is occurring only in columns that contain unblocked cells on the wet boundary (shown in red). The average horizontal size of pinned sections is $\xi_\parallel$, while the average vertical size of pinned sections is $\xi_\perp$. The moving parts have constant steep slope [in Model B it is exactly equal to 2], and their vertical and horizontal dimensions are proportional to $\xi_\perp$. The height-height correlation function $c(\ell, t)$, calculated for such an interface for $\ell \simeq \xi_\perp$ is of the order of $\ell^{(3\alpha-1)/2\alpha} = \ell^{0.71}$. When $\xi_\perp$ is very large, the effective value of the exponent α for $l < \xi_\perp$ is close to 1. The effective roughness exponent of this *moving* interface — calculated directly from our numerical data near criticality in $d = 1+1$ — is $\alpha_{\text{dyn}} = 0.70 \pm 0.05$, in good agreement with the above geometrical arguments. The growth is now mostly a fast erosion of steep slopes, propagating horizontally with constant speed. This observation implies that the dynamical exponent $z_{\text{dyn}} \equiv \alpha_{\text{dyn}}/\beta$ has a value close to one. Thus, $\beta \simeq \alpha_{\text{dyn}}$, in good agreement with our numerical results.[13–15] This large value of α_{dyn} may explain the large values 0.7–0.8 found in dynamical experiments.[11,12]

In the case $d = 2+1$ we find that for all studied models $\alpha = 0.48 \pm 0.02$, $\beta = 0.41 \pm 0.02$ and $z = 1.16 \pm 0.02$, while the results for p_c for the different models are shown in Table 1. In $d > 2$ the duality to directed percolation breaks down, and the growth of the surface is pinned by self-affine hypersurfaces, that can be characterized by horizontal and vertical correlation lengths $\xi_\parallel$ and $\xi_\perp$ (see Ref. 34 for an isotropic model of percolating hypersurfaces).

Table 1 **Critical Exponents and Percolation Thresholds p_c^A, p_c^B, p_c^C, for Models A, B and C**

d	$1+1$	$2+1$	$3+1$	$4+1$
α	0.63 ± 0.01	0.48 ± 0.03	0.38 ± 0.04	0.27 ± 0.05
β	0.63 ± 0.01	0.41 ± 0.02	0.28 ± 0.03	0.18 ± 0.03
z	1.01 ± 0.02	1.16 ± 0.03	1.32 ± 0.5	1.50 ± 0.10
τ_{surv}	1.46 ± 0.02	2.18 ± 0.03	2.54 ± 0.05	3.00 ± 0.20
δ	0.60 ± 0.03	1.14 ± 0.06	1.6 ± 0.1	1.9 ± 0.2
p_c^A	0.4698 ± 0.0002	0.7423 ± 0.0002	0.8423 ± 0.0002	0.889 ± 0.0005
p_c^B	0.5388 ± 0.0002	0.8009 ± 0.0002	0.8857 ± 0.0004	0.9240 ± 0.0005
p_c^C	0.6447 ± 0.0001	0.9340 ± 0.0003	0.9863 ± 0.0004	0.9970 ± 0.0005

Below criticality moving parts form steep circular terraces, surrounding pinned parts (see Fig. 4). However, the moving parts in $d > 1+1$ do not move along straight lines, but rather perform a kind of correlated random walk, which implies $z \geq 1$. At $d = d_c$, z should become 2, as for uncorrelated random walk. We find that even for $d = 5$, $z = 1.5 \pm 0.1 < 2$, suggesting that $d_c > 5$.

The effective roughness exponent found for the moving interface in $d = 2+1$ near criticality is about 0.52 ± 0.05 which is in good agreement with our experiments on wetting of $3d$ porous media.[19]

4. AVALANCHES AND FRACTAL DUST

Above p_c, the growth is stopped by the spanning path of a directed percolation cluster in $d = 1+1$, or by a self-affine surface in $d = 2+1$. However, we can modify our models and assume that even when the growth is completely stopped, the blocked cells on the interface may still erode — but at an infinitesimal rate. With this assumption, we can remove blocked cells at random when the interface is completely stopped. Each removal will produce an avalanche of growth which eventually will die out when the front reaches another directed spanning path, or directed surface of blocked cells (see Fig. 3).

The distribution of avalanche sizes $P(V)$ is found to be[15]:

$$P(V) \sim V^{-\tau_{aval}} F\left(\frac{V}{V_0}\right), \tag{4a}$$

where V is the number of sites removed in an avalanche, and $V_0 \sim \xi_{\parallel}^{d-1}\xi_{\perp}$ is the characteristic volume. The probability $P(V)$ is estimated to be the ratio of the number of avalanches of size V to the total number of avalanches.

In $d = 1+1$, the maximum linear extent of the avalanches (Fig. 3) in the longitudinal and transverse directions is found to scale with exponents:

$$\nu_{\parallel}^{aval} = 1.73 \pm 0.02\,, \qquad \nu_{\perp}^{aval} = 1.10 \pm 0.02 \tag{4b}$$

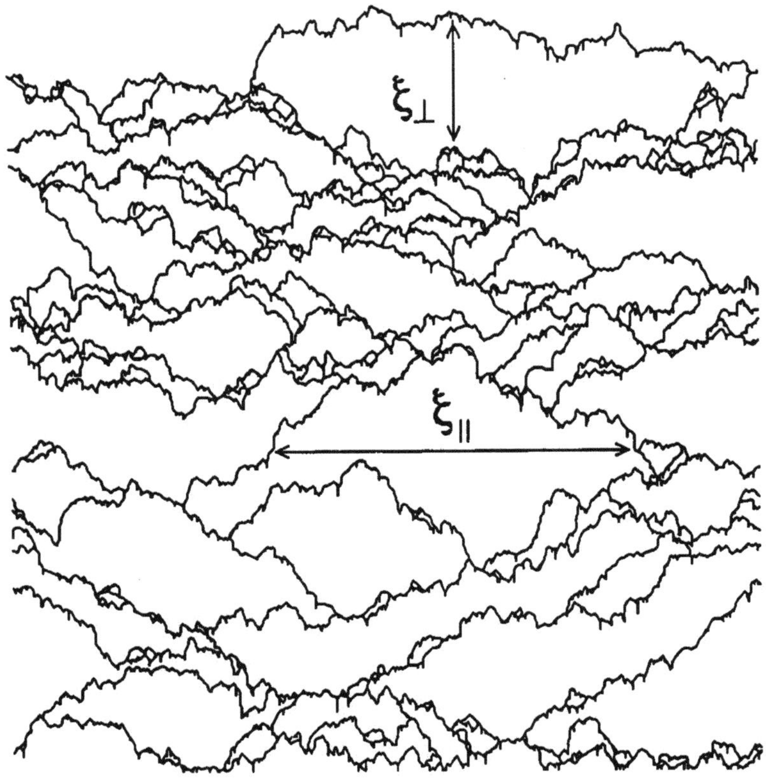

Fig. 3 Successive series of pinned interfaces of Model A, showing the boundaries of avalanches, produced by removing a randomly-chosen blocked cell from the previously-pinned interface. $L = 400$, $p = 0.5 > p_c$. Correlation lengths $\xi_{||}$ and $\xi_\perp$ are the typical sizes of the avalanches. (After Ref. 15.)

in excellent agreement with the correlation-length exponents of directed percolation. Moreover, we find[15,35]:

$$\tau_{\text{aval}} = 1.26 \pm 0.03\,. \tag{4c}$$

This result is in agreement with $\tau_{\text{aval}} = 2/(1+\alpha)$ found for $d = 1+1$ by Olami et al.[36]

From avalanche studies in $d = 2+1$ we find the correlation-length and roughness exponents to be $\nu_{||} = 1.06 \pm 0.1$, $\nu_\perp = 0.47 \pm 0.1$, which gives slightly smaller value of α than from analysis of height-height correlation function and width: $\alpha = \nu_\perp/\nu_{||} = 0.44 \pm 0.1$. This may be due to large error bars that are caused by comparatively small errors in the value of p_c.

An alternative way of producing avalanches is to start growth from a single unblocked cell at time $t = 0$ when the interface is flat (Fig. 4). Above p_c, the clusters of wet cells will be all pinned by the blocked cells. Below p_c some of these clusters will grow infinitely, but some will be stopped by the pinning surfaces.

In analogy with conventional percolation, the survival probability $P_{\text{surv}}(t)$ of the clusters for $t < t_0$ will decay as a power law, $P_{\text{surv}}(t) \sim t^{1-\tau_{\text{surv}}}$. Here t_0 is the characteristic time which is related to the correlation lengths: $\xi_{||} \sim t_0^{1/z}$, $\xi_\perp \sim t_0^{\beta}$. The exponent τ_{surv} is related to the τ_{aval} of Eq. (4):

$$(\tau_{\text{surv}} - 1) = \frac{(\tau_{\text{aval}} - 1)(d - 1 + \alpha)}{z}\,. \tag{5a}$$

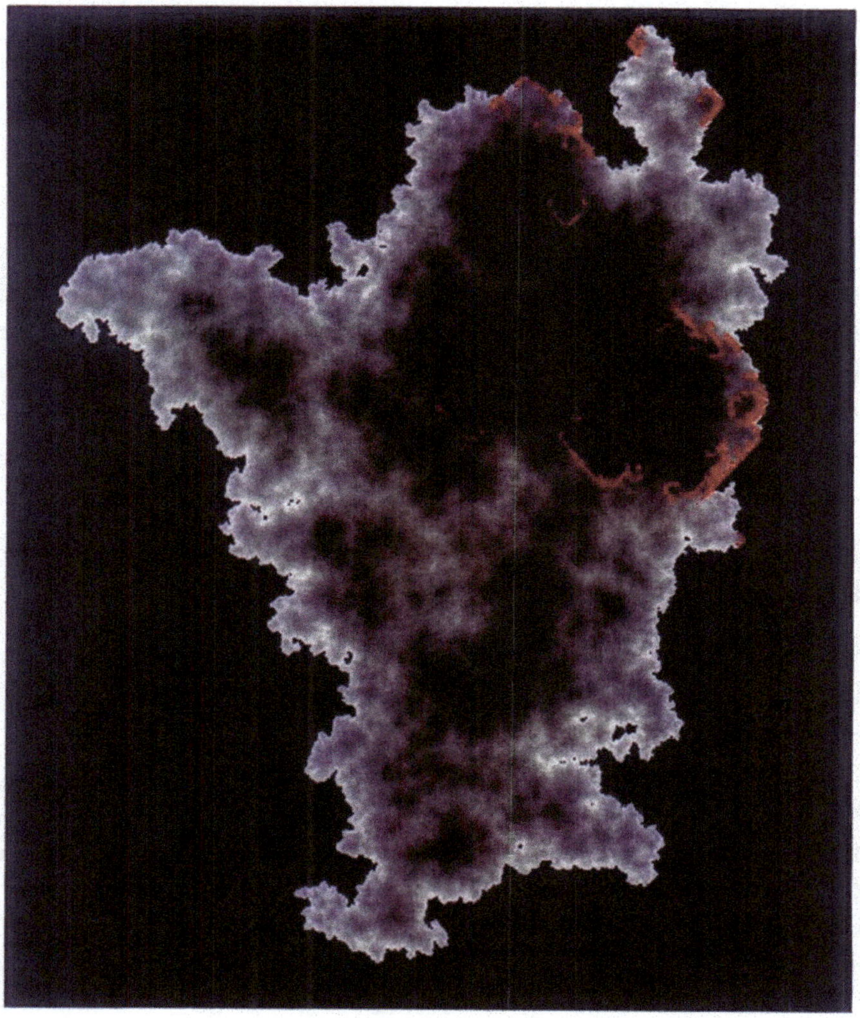

Fig. 4 Horizontal projection of the cluster of Model B ($p = 0.8009 = p_c$) which was started at the center of the screen 2^{10} time steps ago. The current diameter of the cluster is about 2^{10}. Blue area shows flat interface that is still dry since the beginning of the process. Darkest shades of gray corresponds to the largest heights of the interface. Red dots forming "fractal dust" indicate cells that become wet at the current time step.

For $t > t_0$, $P_{\rm surv}(t)$ either goes to zero exponentially for $p > p_c$ or approaches a constant value $P(\infty)$, the probability of an infinite cluster, for $p < p_c$. Thus, studying $P_{\rm surv}(t)$ provides a very accurate method of estimating p_c. The critical exponents can be derived from the parameters of the clusters exactly at p_c. One of these exponents is the exponent δ that characterizes the time dependence of the size of the percolation shell: $n(t) \sim t^{\delta}$. It is possible to connect this exponent with z and α:

$$\delta + 1 = \frac{d - 1 + \alpha}{z}. \tag{5b}$$

In $d = 1 + 1$, $z = 1$, so $\delta = \alpha = \beta$, which agrees with the simple geometrical picture (Fig. 2) that the projection of the shell corresponds to the length of the largest steep moving terrace, which scales as the vertical size of the whole system $w(t) \sim t^{\beta}$. We have studied the time dependence of the size of the shell $n(t) \sim t^{\delta}$ numerically. The results are in agreement with the above relations.

The projection of the shell forms a *fractal dust* (Figs. 2 and 4). In $d = 1 + 1$, when $z = 1$, the fractal dimension of this dust d_f is connected to the exponent of distribution of avalanches (i.e., blocked regions) through the relation:

$$d_f = \tau_{\text{surv}} - 1 = 0.46 \pm 0.02\,. \tag{6}$$

Note that $d_f < \delta$ which means that, in fact, fractal dust is packed in moving blocks (the largest moving block is of the order of the vertical system size of $w(t)$). These moving blocks behave like quasi-particles which are distributed in a fractal way with fractal dimension equal to d_f. Direct numerical studies of the correlation function of the dust support this point of view.

5. POSSIBLE GENERALIZATION OF THE DIODE-RESISTOR PROBLEM TO THE $d \to \infty$ LIMIT

The generalization of the random diode-resistor problem to infinite dimension ($d \to \infty$) is next considered. If our assumptions are correct, the model we study corresponds to the Cayley tree limit.[32,37]

Consider a Cayley tree with coordination number $z_c = 4$ [31] (Fig. 5). We assign an altitude $i \geq 0$ for each site on the Cayley tree. In this way, each site at level i is connected via two bonds to a higher level $i + 1$, and by two bonds to a lower level $i - 1$. The sites on the level $i = 0$ are connected only to the sites at level $i = 1$.

Next, with probability p a bond will be occupied by a diode pointing to a lower level or with probability $1 - p$ by a resistor. There exists a diode-resistor percolation threshold, $p = p_c$, above which there is no current between a site at level $i = 0$ and a site at level

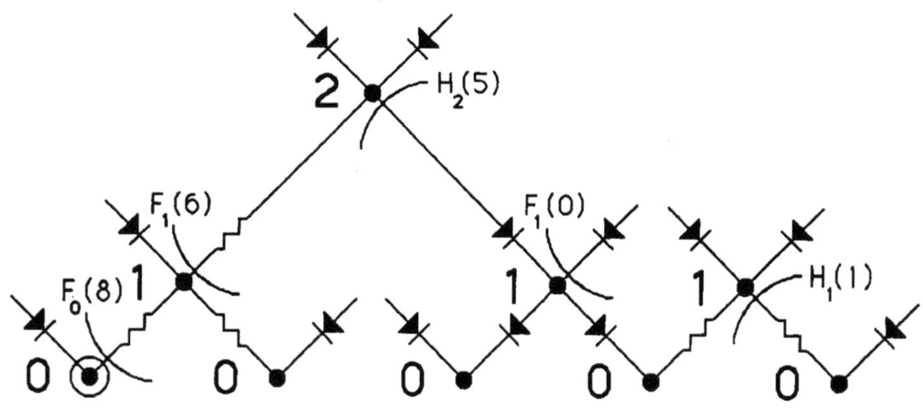

Fig. 5 A finite diode-resistor cluster of size $s = 9$ on the Cayley tree with coordinational number $z_c = 4$. The initial site is marked by a larger circle. Sites are arranged vertically according to the number of the level they belong to. The lowest level has the number equal to zero. In order to illustrate the concept of a branch, we show several examples of branches of various sizes s outgoing from sites at different levels i. These branches are indicated by arcs and probabilities $F_i(s)$, $H_i(s)$. For example, the right branch that is connected to the site at level $i = 2$ from below consist of $s = 5$ sites. Diodes that are connected to the cluster by their lower end only represent surface that stops percolation. These diodes are considered to be upper branches of zero mass, which always have probability $F_i(0) \equiv p$. Note that only 3 resistors (not 5 as shown) are necessary to create this cluster.

$i = +\infty$. Below p_c there is a finite probability that a site at level $i = 0$ will be connected with $+\infty$ by a path of resistors and diodes on which current can flow from $i = 0$ to $i = +\infty$. We define a cluster with respect to a given site at level i as the set of sites that can be reached by a current starting at this site. We also define a cluster branch to be a set of sites that are connected to a given site through one of the adjacent bonds.

To find p_c and the critical behavior we define the following quantities. Let $F_i(s)$ be the probability that a site at level i is connected via an upper bond to a branch of size s and $H_i(s)$ is the probability that a site at level i is connected via a lower bond to a branch of size s. Solving recursion relations for the corresponding generating functions, we find that the probability to be connected to the infinite cluster has the novel form[38]:

$$P_\infty \sim B \exp\left(-\frac{A}{\sqrt{p_c - p}}\right), \tag{7}$$

where $A = (\pi/3)\sqrt{2}\ln(3)$ for $z = 4$. From Eq. (7), it follows that the exponent β_p defined by $P_\infty \sim (p_c - p)^{\beta_p}$ has the value $\beta_p = \infty$ for the diode resistor on a Cayley tree.

We find the exact value of the critical threshold to be $p_c = 1 - 1/(z_c - 1)^2$. Furthermore, we find that the correlation function (the probability that a site on the level 0 is connected to a site on level h) decays at the critical point exponentially as $(z_c - 1)^{-h}$, which means that $\xi_\perp$ is finite and hence $\nu_\perp = 0$.

The logarithmic derivatives of the generating functions give the average sizes of branches that are connected to a site on level i from above and below respectively. Solving the corresponding recursion equations, we obtain that average size of the finite branch is finite below and above p_c, but has a first order discontinuity at $p = p_c$. This implies that $\gamma = 0$. We also find that the average cluster size, that reached level h, scales like $(z_c - 1)^h$, while the minimal cluster size at the level h scales like $(z_c/2)^h$. This means that the average number of summits of such a cluster scales like $(2 - 2/z_c)^h$. These summits perform a random walk in the horizontal dimension of the Cayley tree, which means that the horizontal radius of the cluster is the square root of the number of summits, which still grows exponentially with h, hence the roughness exponent is $\alpha = 0$. We find, however, that the horizontal characteristic length $\xi_\parallel$ remains finite at the critical point, hence $\nu_\parallel = 0$.[38]

Direct studies of the cluster size distribution show that the exponent of the cluster size distribution τ_{aval} is a monotonically increasing function of the diode probability p. Exactly at the critical point $F_0(s) = (s \ln s)^{-2}[C + O(1/\ln s)]$, which is consistent with the first order discontinuity of the average cluster size observed at the critical point. In the standard percolation notations, this means that $\tau = 3$, larger than $\tau = 5/2$ for the regular Cayley tree. Below the critical point $\tau_{\text{aval}}(p) < 2$, but there appears an abrupt exponential cutoff of the distribution at the cluster size $s_0 \sim \exp(A/\sqrt{p_c - p})$, implying $\sigma = 0$. The derivation of the above results will be published elsewhere.[38]

Using hyperscaling equation with the above values of the exponents $\nu_\perp + \nu_\parallel(d - 1) = \gamma + 2\beta_p$ we suggest that the critical dimension of diode-resistor problem is $d_c = \infty$.

6. DISCUSSION

We have studied several models of surface growth in the quenched disordered media near the pinning threshold. These models are in the universality class of the diode-resistor reverse percolation. The numerical results are in good agreement with anomalously large values of the critical exponents, obtained in many experiments in $d = 1 + 1$ and $d = 2 + 1$. The importance of quenched noise and pinning as a mechanism of surface roughening was suggested by several authors[24–26,39] who studied the continuum Langevin equations with quenched noise as models of surface growth. Our numerical results are in rather good agreement with the numerical results of those authors in $d = 1 + 1$, but differ in prediction of the behavior in high dimensions.

The behavior of the model on the Cayley tree suggests that the upper critical dimension is equal to infinity, in disagreement with Refs. 24 and 25. It should be mentioned that the invasion variant of the present models,[14] where at each time step a site with the smallest resistivity is taken out, was later interpreted as the model of self-organized criticality.[36,40,41] However, all the properties of the later model, including the distribution of avalanches, can be derived from the corresponding properties of the present models.

ACKNOWLEDGMENTS

We wish to thank A.-L. Barabási, G. Huber, D. Stauffer and T. Vicsek, for helpful discussions and NSF and OTKA for financial support.

REFERENCES

1. R. Jullien, J. Kertész, P. Meakin and D. Wolf, eds., *Surface Disordering: Growth, Roughening and Phase Transitions* (Nova Science, New York, 1992).
2. T. Vicsek, *Fractal Growth Phenomena*, 2nd ed. (World Scientific, Singapore, 1992).
3. J. Krug and H. Spohn, in *Solids Far From Equilibrium: Growth, Morphology and Defects*, ed. C. Godrèche (Cambridge Univ. Press, Cambridge, 1991).
4. P. Meakin, Phys. Reports (in press).
5. T. Halpin-Healey and Y.-C. Zhang, Phys. Reports (in press).
6. M. Kardar, G. Parisi and Y.-C. Zhang, *Phys. Rev. Lett.* **56**, 889 (1986); E. Medina, T. Hwa, M. Kardar and Y.-C. Zhang, *Phys. Rev.* **A39**, 3053 (1989).
7. N. Martys, M. Cieplak and M. O. Robbins, *Phys. Rev. Lett.* **66**, 1058 (1991); C. S. Nolle, B. Koiller, N. Martys, and M. O. Robbins, *Phys. Rev. Lett.* **71**, 2074 (1993).
8. J. M. Kim and J. M. Kosterlitz, *Phys. Rev. Lett.* **62**, 2289 (1989).
9. B. M. Forrest and L.-H. Tang, *Phys. Rev. Lett.* **64**, 1405 (1990).
10. K. Moser, J. Kertész and D. E. Wolf, *Physica* **A178**, 215 (1991).
11. T. Vicsek, M. Cserzö and V. K. Horváth, *Physica* **A167**, 315 (1990).
12. M. A. Rubio, C. A. Edwards, A. Dougherty and J. P. Gollub, *Phys. Rev. Lett.* **63**, 1685 (1990); V. K. Horváth, F. Family and T. Vicsek, *J. Phys.* **A24**, L25 (1991).
13. S. V. Buldyrev, A.-L. Barabási, F. Caserta, S. Havlin, H. E. Stanley and T. Vicsek, *Phys. Rev.* **A45**, R-8313 (1992).
14. S. Havlin, A.-L. Barabási, S. V. Buldyrev, C. K. Peng, M. Schwartz, H. E. Stanley and T. Vicsek, in *Growth Patterns in Physical Sciences and Biology*, eds. E. Louis, L. Sander and P. Meakin (Plenum, New York, 1992).
15. A.-L. Barabási, S. V. Buldyrev, S. Havlin, G. Huber, H. E. Stanley and T. Vicsek, in Ref. [1] (1992).
16. J. Zhang, Y.-C. Zhang, P. Alstrøm and M. T. Levinsen, in Ref. [1] (1992).
17. M. Matsushita and S. Ouchi, *Physica* **D38**, 246 (1989); G. Dietler and Y.-C. Zhang, *Physica* **A191**, 213 (1992).
18. A. Czirok, E. Somfai and T. Vicsek, *Phys. Rev. Lett.* **71**, 2154 (1993).
19. S. V. Buldyrev, A.-L. Barabási, S. Havlin, J. Kertész, H. E. Stanley and H. S. Xenias, *Physica* **A191**, 220 (1992).
20. J. Krim, I. Heyvaert, C. Van Haesendonck and Y. Bruyhseraede *Phys. Rev. Lett.* **70**, 57 (1993).
21. H. Ji and M. O. Robbins, *Phys. Rev.* **A44**, 2538 (1991).
22. B. Koiller, H. Ji, and M. O. Robbins, *Phys. Rev.* **B46**, 5258 (1992).

23. H. Ji and M. O. Robbins, *Phys. Rev.* **B46**, 14519 (1992).
24. Z. Csahók, K. Honda, E. Somfai, M. Vicsek and T. Vicsek *Physica* **A200**, 136 (1993).
25. G. Parisi, *Europhys. Lett.* **17**, 673 (1992).
26. T. Nattermann, S. Stepanow, L.-H. Tang and H. Leschhorn, *J. Phys. II France* **2**, 1483 (1992); O. Narayan and D. S. Fisher *Phys. Rev. B* (in press); M. H. Jensen and I. Procaccia (preprint).
27. L.-H. Tang and H. Leschhorn, *Phys. Rev.* **A45**, R-8309 (1992).
28. D. Dhar, M. Barma and M. K. Phani, *Phys. Rev. Lett.* **47**, 1238 (1981).
29. S. Redner, *Phys. Rev.* **B25**, 3242 (1982).
30. S. Redner, in *Percolation Structures and Processes*, eds. G. Deutscher, R. Zallen and J. Adler (A. Hilger, Bristol, 1983).
31. B. C. Harms and J. P. Straley *J. Phys.* **A15**, 1865 (1982).
32. A. Bunde and S. Havlin, eds., *Fractals and Disordered Systems* (Springer-Verlag, Heidelberg, 1991).
33. W. Kinzel, in *Percolation Structures and Processes*, eds. G. Deutscher, R. Zallen and J. Adler (A. Hilger, Bristol, 1983).
34. J. Kertész and H. J. Herrmann, *J. Phys.* **A18**, L1109 (1985).
35. S. V. Buldyrev, S. Havlin and H. E. Stanley, *Physica* **A200**, 200 (1993).
36. Z. Olami, I. Procaccia and R. Zeitak, submitted to *Phys. Rev. E* (1993).
37. D. Stauffer and A. Aharony, *Introduction to Percolation Theory* (Taylor and Francis, London, 1992).
38. S. V. Buldyrev, A.-L. Barabási, S. Havlin, J. Kertész, A. Shehter and H. E. Stanley (to be published).
39. D. A. Kessler, H. Levine and Y. Tu, *Phys. Rev.* **A43**, 4551 (1991).
40. K. Sneppen, *Phys. Rev. Lett.* **69**, 3539 (1992).
41. K. Sneppen and M. H. Jensen, *Phys. Rev. Lett.* **71**, 101 (1993).

Fractals, Vol. 1, No. 4 (1993) 840–845

EVIDENCE FOR MULTISCALING IN LARGE DLA CLUSTERS

C. AMITRANO and A. CONIGLIO
Dipartimento di Scienze Fisiche, Università di Napoli, Mostra d'Oltremare Pad. 19, 80125 Napoli, Italy

P. MEAKIN
Department of Physics, University of Oslo, Box 1048 Oslo 0316, Norway

M. ZANNETTI
Dipartimento di Fisica, Università di Salerno, 84081 Baronissi (SA), Italy

Abstract

We have measured the multiscaling dimension $D(x)$ on a set of large off-lattice DLA clusters of one million particles, finding results which strongly support multiscaling behavior. The multiscaling picture is also confirmed by the slow logarithmic decay in the fluctuations of the radius (relative to the cluster radius) as the size of clusters grow.

1. INTRODUCTION

More than ten years after Witten and Sanders[1] proposed their model for diffusion limited aggregation (DLA), no consensus has yet been reached concerning the value of the fractal dimensionality D in two dimensions. Accepted values of D range from $D \simeq 1.65$ in a strip with cylindrical geometry, to $D \simeq 1.71$ when the fractal dimensionality is calculated using the mass vs. radius relationship $M \sim R^D$ in a radial geometry. Though DLA is a critical phenomenon in the sense that there is an asymptotically divergent characteristic length scale, there are various differences from ordinary critical phenomena. For instance, DLA clusters show an interior frozen part and an external active periphery not present in other geometrical critical phenomena such as percolation. This feature of DLA is responsible for the poor performance of a real space renormalization group analysis[2] based on a spanning configuration in a cell, compared with analogous computations carried out in the percolation problem. In fact a different procedure for DLA in which the renormalization is limited to the mass in the frozen part gives much better results.[2] The distinction between a frozen and a growth region was also stressed by Pietronero et al.,[3] who have developed a theory of fractal growth which applies only to the frozen part. More recently, it has been proposed that planar DLA does not have a simple self-similar geometry.[4]

A more general concept of scale invariance was introduced[5] that may explain the various anomalies present in DLA and not in ordinary critical phenomena. This more general scale invariance leads to a multiscaling form for the density profile, which is the analog of the order parameter correlation function in critical phenomena. Defining the density profile as:

$$g(r,\, R)d^d r = dN$$

where dN is the number of particles in the volume $d^d r$ at a distance r from the origin and R is the radius of gyration, the following multiscaling form was proposed:

$$g(r,\, R) = r^{-d+D(r/R)} C \frac{r}{R} \tag{1}$$

where $C(x)$ is an amplitude that goes to zero rapidly for large x. Multiscaling contains a structure much richer than standard scaling. In the asymptotic regime, where r and R are both large with some fixed ratio $x = r/R$, the above form for the density profile predicts a different power-law decay for each value of x and consequently a different local fractal dimensionality $D(x)$ for each shell corresponding to that value of x. Specifically, moving from $x = 0$ toward increasingly larger values of x, $D(x)$ varies from the fractal dimensionality $D(0)$ of the inner frozen region to the fractal dimensionality of the outer shells in the growth region.

The proposed behavior implies that DLA satisfies symmetries more complex than those found in ordinary critical phenomena. With simple algebra it can be shown that: (i) under a transformation where lengths are rescaled by a factor l

$$r' = \frac{r}{l}$$

$$R' = \frac{R}{l}$$

the density profile [Eq. (1)] rescales as:

$$g(r',\, R') = l^{d-D(r/R)} g \frac{r}{R}$$

and that (ii) the above defined transformations have the group multiplication property. The density profile rescales by a factor which depends on x as opposed to standard scaling where $D(x) = D$ independent of x. Knowledge of the symmetry of a problem is of fundamental importance in developing a theory. In ordinary critical phenomena the renormalization group was based on standard scale invariance. Then, if multiscaling is correct for DLA a complete theory should be based on such a multiscaling property.

The multiscaling form [Eq. (1)] was tested in a previous paper.[6] There we reported on numerical calculations aimed at measuring $D(x)$ for DLA in two dimensions (see Table 1).

Table 1 Numerical value of $D(x)$ (A) for a set of 2000 square-lattice DLA clusters of 10^4 particles, (B) for a set of 200 off-lattice DLA clusters of 10^5 particles, and (C) for a set of 66 off-lattice DLA clusters of 10^6 particles. The error is estimated graphically, from the slopes of the log-log plots of the mass $M(r, R)$ as a function of r.

x	$D(x)$(A)	$D(x)$(B)	$D(x)$(C)
0.15	1.672 ± 0.01	1.684 ± 0.01	1.680 ± 0.01
0.25	1.676 ± 0.01	1.690 ± 0.01	1.694 ± 0.01
0.35	1.670 ± 0.01	1.691 ± 0.01	1.692 ± 0.01
0.45	1.680 ± 0.01	1.695 ± 0.01	1.700 ± 0.01
0.55	1.689 ± 0.01	1.701 ± 0.01	1.697 ± 0.01
0.65	1.696 ± 0.01	1.705 ± 0.01	1.703 ± 0.01
0.75	1.701 ± 0.01	1.706 ± 0.01	1.695 ± 0.01
0.85	1.722 ± 0.01	1.713 ± 0.01	1.712 ± 0.01
0.95	1.736 ± 0.01	1.723 ± 0.01	1.717 ± 0.01
1.05	1.752 ± 0.01	1.735 ± 0.01	1.718 ± 0.01
1.15	1.770 ± 0.01	1.748 ± 0.01	1.725 ± 0.01
1.25	1.765 ± 0.01	1.756 ± 0.01	1.742 ± 0.01
1.35	1.755 ± 0.01	1.765 ± 0.01	1.745 ± 0.01
1.45	1.744 ± 0.01	1.739 ± 0.02	1.741 ± 0.01
1.55	1.701 ± 0.02	1.704 ± 0.02	1.722 ± 0.02
1.65	1.611 ± 0.02	1.642 ± 0.03	1.667 ± 0.03
1.75	1.497 ± 0.02	1.530 ± 0.05	1.530 ± 0.04
1.85	1.310 ± 0.05	1.318 ± 0.05	1.320 ± 0.05

2. SIMULATIONS AND RESULTS

In a first set of calculations (A) 2000 DLA clusters of $N = 10^4$ particles were grown on the square lattice, while in a second set (B) 200 off-lattice DLA clusters of $N = 10^5$ particles were grown. For both data sets $D(x)$ was found to be roughly constant for small values of x, bending down at a value of x close to 1.5, with an overall behavior in agreement with the multiscaling picture. Recently, multiscaling was tested by another numerical calculation.[7] In this case, clusters of 10^5 and 10^6 particles were analyzed. While a $D(x)$ agreement with multiscaling was obtained from the 10^5 particles clusters, the one million particle clusters gave a roughly constant $D(x)$ indicating disappearance of multiscaling for very large clusters. Motivated by this result, we have repeated our own calculation using a set (C) of 66 of one million particles. Contrary to Ref. 7, our new data are consistent with our previous result even for one million particles and do confirm the multiscaling picture. The method that we have used is the same as before and is described in detail in our previous paper.[6] The plot of $D(x)$ for the one million particles is given in Fig. 1. To show the trend as a function of the number of particles, the data for the three sets of measurements have been plotted. The unphysical bump,[6] more pronounced in the set of 10^4 particles, flattens out as the number of particles increases.

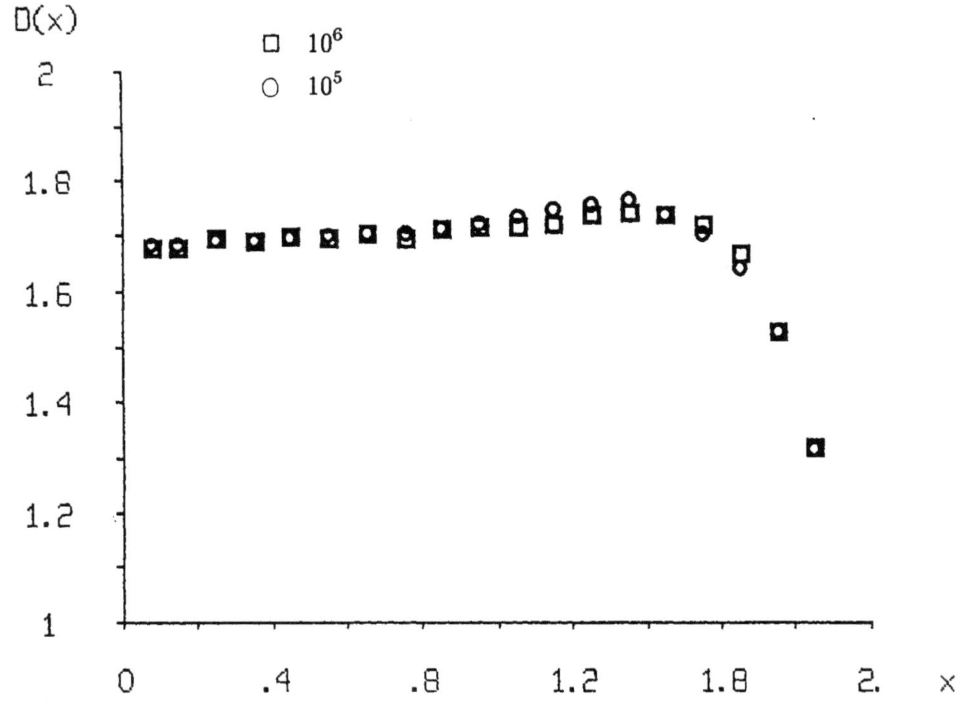

Fig. 1 Plot of $D(x)$ vs. x for (○) a set of 200 off-lattice DLA clusters of 10^5 particles, and (□) a set of 66 off-lattice DLA clusters of 10^6 particles.

It may be useful to characterize the behavior of $D(x)$ for small values of x. By assuming:

$$D(x) \cong D(0) - ax^{\beta} \tag{2}$$

for small x, we find $\beta \sim 10$ and $a \sim 1/2$. The multiscaling form of the density profile $g(r, R)$ implies a slow logarithmic convergence in the fluctuation of the radius defined by:

$$\sigma = \frac{\langle r^2 \rangle - \langle r \rangle^2}{\langle r \rangle^2}$$

where

$$\langle r^m \rangle = \frac{\int r^m g(r, R) d^d r}{\int g(r, R) d^d r}.$$

Using Eq. (2) and the above definitions from Eq. (1) we find:

$$\sigma = \frac{1}{D(0)[2 + D(0)]} + \frac{b}{(\ln R)^{1/\beta}} + O\left(\frac{1}{\ln R}\right) \tag{3}$$

where b is a constant. Thus multiscaling is revealed in the radius fluctuations by a slow logarithmic approach to a constant value as the radius of gyration grows. Our data are indeed consistent with this behavior as shown in Fig. 2.

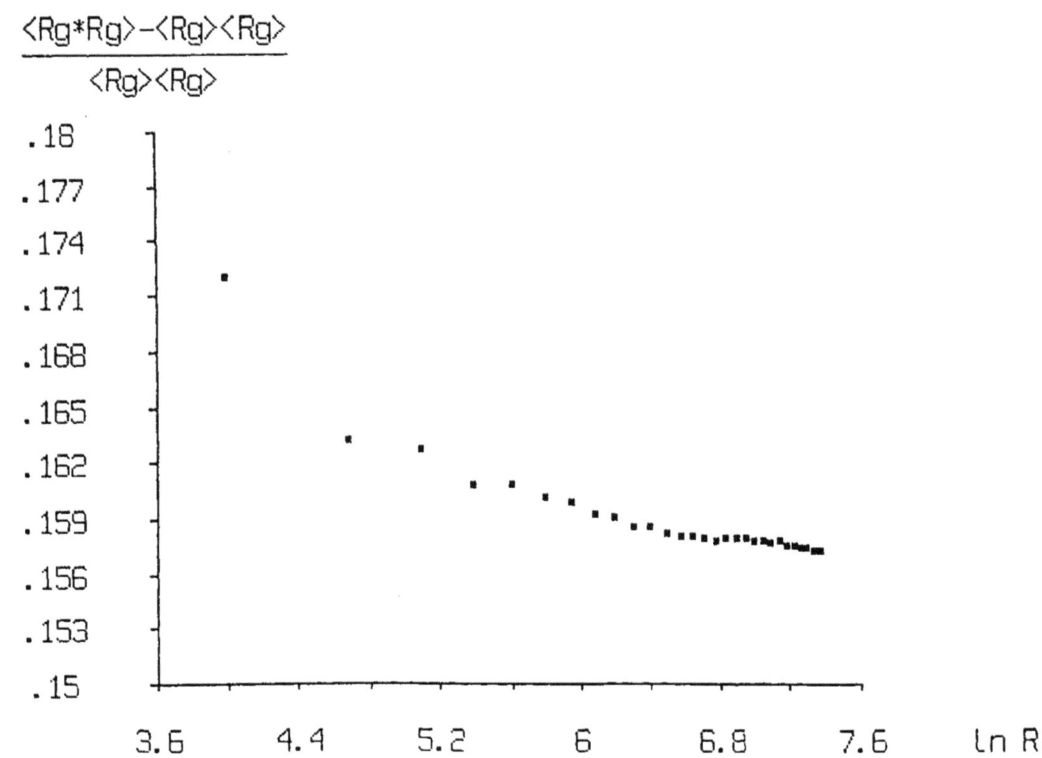

Fig. 2 Plot of $\sigma = \frac{\langle r^2\rangle - \langle r\rangle^2}{\langle r\rangle^2}$ vs. ln R for the set of 66 off-lattice DLA clusters of 10^6 particles.

3. DISCUSSION

The multiscaling form of the density profile [Eq. (1)] also implies multiscaling in the growth probability distribution $P(r, N)$. This quantity was introduced by Plischke and Racz[8] and is defined in such a way that $P(r, N)dr$ is the probability that the Nth particle added to the growing cluster sticks at a distance between r and $r + dr$ from the origin.

Since

$$\frac{dr^{d-1}g(r, R)}{dN \propto P(r, N)} \tag{4}$$

multiscaling in the density profile implies multiscaling in the $P(r, N)$. The fluctuation of the radius can also be defined using $P(r, N)$ instead of $g(r, R)$ to calculate the average $\langle r^m \rangle$.

It follows from Eq. (4) that σ has the form given in Eq. (3) with a slow logarithmic approach to a constant value. Plischke and Racz fitted their data for $P(r, N)$ with a Gaussian distribution centered around $r/R = 1$. If such a distribution is assumed, the only way to get multiscaling is by requiring that σ approaches zero logarithmically[5]

$$\sigma \sim \frac{c}{\ln R} \tag{5}$$

where c is a constant. However a density profile with the multiscaling form given by Eq. (1) where $D(x)$ is a decreasing function[6] of x and $C(0)$ finite, due to the relationship expressed in Eq. (4) is compatible neither with the Gaussian form of the density profile nor with

Eq. (5). In fact it gives a value for σ that slowly (logarithmically) approaches a constant value.

Very recently, Ossadnik[9] has reported new calculations on DLA clusters of one million particles and on a single cluster of 50 million particles. The result obtained from the one million particle clusters agrees with multiscaling (the discrepancy with the previous calculations reported in Ref. 7 was the consequence of a problem in the random number generator). Due to large fluctuations, the density profile results from the 50 million particle cluster are rather inconclusive while σ, defined by the growth probability distribution, converges slowly towards a constant, that is, consistent with multiscaling.

Finally, we emphasize that multiscaling is a distinct phenomenon from multifractality.[10] In the case of multifractality, the existence of infinitely many exponents is the consequence of the introduction of an appropriate measure on the fractal, and does not imply any condition on the symmetries.[11] For example, in percolation there is multifractality with respect to the voltage distribution[12] of the random resistor network; nevertheless the pair connectedness function at criticality, is invariant under standard scaling. This is the reason why standard renormalization group methods can be successfully applied to percolation.

REFERENCES

1. T. A. Witten and L. M. Sander, *Phys. Rev. Lett.* **47**, 1400 (1981).
2. C. Amitrano, A. Coniglio and F. di Liberto, *J. Phys.* **A21**, L201 (1988).
3. L. Pietronero, A. Erzan and C. Evertsz, *Phys. Rev. Lett.* **61**, 861 (1988).
4. B. B. Mandelbrot, *Physica* **A191**, 95 (1992).
5. A. Coniglio and M. Zannetti, *Physica* **A163**, 325 (1990).
6. C. Amitrano, A. Coniglio, P. Meakin and M. Zannetti, *Phys. Rev.* **B44**, 4974 (1991).
7. P. Ossadnik, *Physica* **A176**, 454 (1991).
8. M. Plischke and Z. Racz, *Phys. Rev. Lett.* **53**, 415 (1984).
9. P. Ossadnik, preprint.
10. A different case is when multifractality is obtained with respect to the mass distribution as found by T. Vicsek, F. Family and P. Meakin *Europhys. Lett.* **12**, 217 (1990). In this case, multifractality is directly related to a more complex geometry of the aggregate and is closer to multiscaling.
11. C. Amitrano, A. Coniglio and F. di Liberto, *Phys. Rev. Lett.* **57**, 1016 (1987); T. C. Halsey, M. H. Jensen, L. P. Kadanoff, I. Procaccia and B. I. Shraiman, *Phys. Rev.* **A33**, 1141 (1986).
12. L. De Arcangelis, S. Redner and A. Coniglio, *Phys. Rev.* **B31**, 4725 (1985).

Fractals, Vol. 1, No. 4 (1993) 846–859

SURFACTANT-MEDIATED SURFACE GROWTH: NONEQUILIBRIUM THEORY

ALBERT-LÁSZLÓ BARABÁSI*
Center for Polymer Studies and Department of Physics,
Boston University, Boston, MA 02215, USA

Abstract

A number of recent experiments have shown that surfactants can modify the growth mode of an epitaxial film, suppressing islanding and promoting layer-by-layer growth. Here a set of coupled equations are introduced to describe the coupling between a growing interface and a thin surfactant layer deposited on the top of the non-equilibrium surface. The equations are derived using the main experimentally backed characteristics of the system and basic symmetry principles. The system is studied using a dynamic-renormalization-group scheme, which provides scaling relations between the roughness exponents. It is found that the surfactant may drive the system into a novel phase, in which the surface roughness is negative, corresponding to a flat surface.

1. INTRODUCTION

Recently, there has been much theoretical interest in the statistical properties of nonequilibrium interfaces. Most of the growing interfaces naturally evolve into self-affine structures; the surface morphology and the dynamics of roughening exhibit simple scaling behavior despite the complicated nature of the growth process.[1–5] In particular, much attention has been focused on different models to describe thin-film growth by molecular-beam epitaxy (MBE).[6–14]

*Electronic address: alb@buphy.bu.edu

Under ideal MBE conditions, the primary relaxation mechanism is surface diffusion, which conserves the mass of the film. Experimentally, both lattice strain and surface free energy determine whether the film undergoes layer-by-layer growth, islanding, or layer-by-layer growth followed by islanding. In experiments involving growth of Ge on Si(100) surface, layer-by-layer growth is limited to 3–4 monolayers (ML) due to lattice mismatch between Si and Ge and is followed by formation of unstrained Ge islands. It was shown recently that islanding in the Ge/Si system can be suppressed effectively by use of a surfactant monolayer, *changing the growth mode* from island growth to layer-by-layer growth.[15] Suitable surfactants such as As and Sb strongly reduce the surface free energy of both Si and Ge surfaces and cause segregation at the surface during growth.

In this paper, we study the generic problem of nonequilibrium roughening of an interface covered by a thin surfactant layer (see Fig. 1). Building on experimental results and general symmetry principles, a set of nonequilibrium equations is proposed to describe the growth of an interface coupled to the fluctuations in the surfactant coverage. The analytic study of these equations indicates that the surfactant changes drastically the morphology of the interface in $2+1$ dimensions. In particular, the coupled system supports the existence of a novel phase characterized by negative roughness exponent, which can be identified with a morphologically flat surface. A summary of the main results were presented in an earlier publication.[16]

The paper is organized as follows: Sec. 2 presents a short review of the experimental studies. Section 3 introduces the key elements of the proposed nonequilibrium theory. Section 4 presents the main results of the dynamic renormalization group (DRG) analysis, and finally, Sec. 5 reflects on open problems and possible future developments.

2. EXPERIMENTAL STUDIES ON SURFACTANT MEDIATED GROWTH

The ultimate goal in crystal growth by MBE or other vapor-phase techniques is to control and influence the growth mode of a thin film. At high temperatures necessary to obtain sufficient mobility of the surface atoms, the growth mode is determined by the interface and surface free energies and the lattice strain. Lattice strain is specially relevant in heteroepitaxial growth, when attempt is made to combine different types of atoms in a layered structure.

Depending on the surface, interface and heteroepitaxial layer's free energy, three distinct growth modes can be observed. Technologically, it is most useful if the film grows in a layer-by-layer mode (Frank–Van der Merwe), when well controlled planar morphology is obtained. The deposited atoms diffuse on the surface and stick to the edge of nucleated islands. As a result the islands grows, finally covering the whole surface and completing the layer. On the top of the completed layer, new islands start to nucleate, and the previous process is repeated. The growth has an oscillatory character in time, which can be observed using reflection high-energy electron diffraction (RHEED) or other experimental techniques.

If the overlayer does not wet the surface, islanding is observed (Volmer–Weber mode), marked by the dumping of the intensity in RHEED measurements and absence of oscillations.

And finally, if the overlayer wets the surface, but the overlayer stress is unfavorable, the film might grow in a layer-by-layer mode, followed by islanding (Stranski–Krastanov mode).

One of the most widely studied heteroepitaxial structure is obtained by growth of Si on Ge, or Ge on Si. The Ge lattice is 4% larger than the Si lattice, thus generating considerable strain influencing the heteroepitaxial growth. As a result, Ge grows on a Si(100) lattice in the Stranski–Krastanov mode, while Si on Ge follows the Volmer–Weber model.

The lattice mismatch generates islanding after 3–4 monolayers of layer-by-layer growth during Ge deposition on a Si(100). Recently Copel, Reuter, Kaxiras and Tromp proposed the use of a surfactant monolayer of As to reduce the surface free energies and effectively alter the growth mode.[15]

The microscopic mechanism responsible for the unusual effect of the As on the growth process is partially understood. The As layer, with one extra valence electron per surface atom, fills the dangling bonds of the Si(100) surface, creating a stable termination. Furthermore, As segregates to the surface during growth. Si or Ge atoms deposited on the surface covered by an As monolayer rapidly exchange sites with the As and incorporate it into the subsurface. As a result, the heteroepitaxial structure incorporates negligible quantities of As.

Two main mechanisms were proposed to explain the effect of the surfactant on the growth process.[15,17] The first is a dynamic one, based on enhanced incorporation of the growth atoms. The As atoms drive any incoming Si or Ge atom to subsurface sites due to their ability to segregate easily. For a surfactant free surface, the deposited atoms diffuse to the surface, until they reach a step or a defect, where they stick. In contrast, with surfactant the freshly arrived atoms are driven into subsurface sites by an exchange mechanism with the As atoms, their diffusion being severely curtained. Thus in the presence of a surfactant, the adatom can be incorporated without a step or a defect.

The second mechanism uses first principle calculations to explain the effect of the As atoms on the stress distribution of the surface layers. A shortcoming of this mechanism is that it predicts only an increase in the epitaxial thickness before islanding appears, but does not account for the change in the growth mode. Experimental results indicate that after about 50 ML the strain is fully relieved, suppressing the driving force for island formation.

Although probably the combination of the two effects is responsible for the suppression of islanding, the nonequilibrium theory proposed in this paper is based on the first mechanism, providing a quantitative formulation of the dynamic phenomena occurring during surfactant mediated growth.

In addition to the investigations mentioned,[15,17] a number of subsequent experiments showed that surfactants can change the surface morphology in a wide variety of systems. It was found that both Sb and As can act efficiently as a surfactant for the Si/Ge system.[17] The effect of the surfactant on the lattice strain and on the appearance of dislocations was studied in details.[18–21] Low-energy electron microscopic observations were used to gain further understanding in the local exchange mechanism between Ge and surfactant. It has been argued that surface energy anisotropy, instead of surface energy, determines the changes in the growth mode of the Ge/Si system.[22]

Further experimental investigations found that Sb alters the growth of Ag on Ag(111).[23] Since submonolayer surfactant coverages were used, a new mechanism to explain these experiments was proposed. According to this, the Sb attaches to the edge of the islands and lowers the interlayer diffusion barrier of the diffusing adatoms. The Sb moves together with the edge of the growing islands and probably is segregated at the surface when the islands coalesce.

In subsequent experiments, antimony was found to change the structure of islands in Ge/Si growth,[24] and Te was used as surfactant to sustain layer-by-layer growth of InAs on GaAs(001).[25–27]

Since the system most investigated is the Ge/Si growth with As or Sb as surfactant, in what follows such a system involving surfactant mediated growth is discussed. Whether the proposed theory applies to all the experiments mentioned, or additional effects have to be considered, is an open question, which will not be addressed here.

3. NONEQUILIBRIUM THEORY

In order to construct a nonequilibrium theory to account for the nontrivial effect of the surfactant on the growth, we have to study separately the dynamics of the interface and surfactant, and then consider the possible coupling between the two quantities.

As mentioned above, under ideal MBE conditions, relaxation proceeds via surface diffusion. Atoms deposited on the surface diffuse until they find an energetically favorable position (usually near a step or a dislocation), where they stick mostly irreversibly. The diffusive dynamics is *conservative*, i.e., it does not change the mass of the film. The only change in the mass might come from deposition or desorption.

In contrast to the ideal MBE, there is experimental evidence that surfactant mediated growth of Ge on Si proceeds by highly local Ge incorporation with minimum surface diffusion.[30] Ge atoms that adhere to the As-capped surface rapidly exchange sites with the As atoms and incorporate into subsurface sites.

In the absence of surface diffusion, the growth equation may contain terms which violate mass conservation.[31] The simplest nonlinear growth equation with nonconserved dynamics was introduced by Kardar, Parisi, and Zhang (KPZ)[32]:

$$\partial_t h = \nu \nabla^2 h + \lambda (\nabla h)^2 + \eta \,. \tag{1}$$

Here $h(x,t)$ is the height of the interface in $d = d' + 1$ dimensions. The first term on the right hand side describes relaxation of the surface by a surface tension ν. The second term is the lowest order nonlinear term that can appear in the interface growth equation, and is related to lateral growth. $\eta(x,t)$ is a stochastic noise driving the growth; it can describe thermal and beam intensity fluctuations.

Equation (1) is the lowest order nonlinear equation compatible with the basic symmetries of a growing interface: it is isotropic in the substrate directions ($x \to -x$ transformation leaves the system invariant), and invariant to translation both in the substrate directions ($x \to x + a$) and in the growth direction ($h \to h + b$). But there is a broken up-down symmetry in h: the transformation $h \to -h$ does not leave the system invariant. The explanation to this broken symmetry is based on the existence of a preferred growth direction for the interface. In the absence of the nonlinear term λ this symmetry is obeyed as well. Another important property of this equation is that higher order nonlinear terms are irrelevant, i.e., they do not effect the growth exponents (to be defined later).

Additional terms in Eq. (1) will include the coupling to the surfactant fluctuations.

In describing the dynamics of the surfactant we shall choose as parameter the *width of the surfactant layer*, $v(x,t)$ (see Fig. 1). Throughout this paper, it is assumed that the surfactant layer is very thin, thus nonlocal effects do not contribute to the dynamics. The typical experimental coverage, which is the spatial average of $v(x,t)$, is around 1 ML. For coverages smaller than 1 ML, holes might appear in the surfactant layer. Since then it has been proposed that growth equations do not depend in an explicit form on the thickness of the layer, but only on its spatial derivative, and the system remains well defined even in the presence of such a holes.

An efficient surfactant must fulfill two criteria: it must be sufficiently mobile to avoid incorporation, and it must surface segregate. Careful experimental studies showed for the Ge/Si system that the bulk As concentration is less than 1%; thus the effect of As on growth is a surface phenomena.[15,17]

Neglecting the desorption of the surfactant atoms, the equation governing the surfactant kinetics obeys mass conservation.

This leads to the continuity equation:

$$\partial_t v = -\nabla \cdot \mathbf{j} + \eta' . \tag{2}$$

where η' is a conserved uncorrelated noise which incorporates the random local fluctuations of the surfactant, and $\mathbf{j}$ is the particle-number current density. The simplest linear equation with conserved dynamics correctly incorporating the effect of surface diffusion is[6]:

$$\partial_t v = -K\nabla^4 v + \eta' . \tag{3}$$

Equation (3) can be obtained from Eq. (2) by using a current $\mathbf{j} \sim \nabla\mu$, where μ is the local chemical potential on the interface. Considering $\mu \sim \nabla^2 v$, i.e., depends only on the local curvature of the thickness (describing local surfactant agglomerations), we obtain Eq. (3).

To account for the coupling between the growing surface and the surfactant, it is necessary to introduce additional terms in Eqs. (1) and (3). There are two main criteria which restrict our choice: The coupling terms must satisfy the symmetry conditions characteristic of the interface, and the obtained set of equations should be self-consistent, i.e., the resulting dynamics should not generate further nonlinear terms. In addition, the coupling terms included in Eq. (3) must obey the required mass conservation for the surfactant.

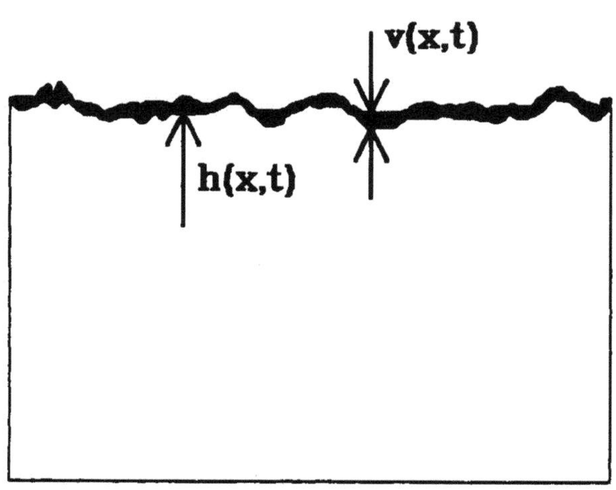

Fig. 1 Schematic illustration of the studied surfactant/surface system. The figure represents a cross section of the two dimensional surface of height $h(x,t)$ covered by a thin surfactant layer with thickness $v(x,t)$. A newly arriving atom penetrates the surfactant and is deposited on the top of the growing interface $h(x,t)$.

The simplest set of equations that satisfy the above condition is:

$$\partial_t h = \nu\nabla^2 h + \lambda(\nabla h)^2 + \beta(\nabla v)^2 + \eta_0 \tag{4a}$$

$$\partial_t v = -K\nabla^4 v + \gamma\nabla^2[(\nabla h)\cdot(\nabla v)] + \eta_1\,, \tag{4b}$$

where the noise terms η_0 and η_1 are assumed to be Gaussian distributed with zero mean and the following correlator:

$$\langle \eta_i(x,t)\eta_i(x',t')\rangle = \mathcal{D}_i\delta(x-x')\delta(t-t')\,. \tag{5}$$

Here

$$\mathcal{D}_0 = D_0 \tag{6}$$

and

$$\mathcal{D}_1 = -D_1\nabla^2 = D_2\nabla^4\,. \tag{7}$$

The D_2 terms is generated by D_0 and D_1 as will be shown below.

The generic nonlinear term $(\nabla v)^2$ in Eq. (4a) can be derived using symmetry principles. In Eq. (4b) the $\nabla^2[(\nabla h)\cdot(\nabla v)]$ term results from a current $\mathbf{j} = -\nabla[(\nabla h)\cdot(\nabla v)]$, and obeys mass conservation. Geometrical interpretation[9] of this term suggests that a positive γ drives the surfactant to cover uniformly the irregularities of the surface, i.e., enhances the wetting properties.[33] A negative γ has the opposite effect, assigning a non-wetting character to the surfactant. Since in experiments there is no evidence of surfactant agglomeration (non-wetting character), but it is energetically favorable to terminate the Ge layer with As atoms, we assume that the surfactant wets the surface, thus $\gamma > 0$.

The quantity of main interest is the dynamic scaling of the fluctuations characterized by the width[1]:

$$w_0^2(t,L) = \left\langle [h(x,t) - \bar{h}(t)]^2 \right\rangle = L^{2\chi_0} f(t/L^{z_0}) \tag{8}$$

where χ_0 is the roughness exponent for the interface $h(x,t)$, and the dynamic exponent z_0 describes the scaling of the relaxation times with the system size L; $\bar{h}(t)$ is the mean height of the interface at time t and the $\langle\ \rangle$ denotes ensemble average. The scaling function f has the properties:

$$f(u \to 0) \sim u^{2z_0/\chi_0} \tag{9}$$

and

$$f(u \to \infty) \sim \text{const.} \tag{10}$$

In a similar way one can define χ_0 and z_1 to characterize the fluctuations in the surfactant coverage $v(x,t)$.

4. ANALYTICAL STUDY

For $\beta = 0$, Eq. (4a) reduces to the KPZ equation.[1] For a one-dimensional interface the exponents can be obtained using DRG, resulting in the roughness exponent $\chi = 1/2$ and in the dynamic exponent $z = 3/2$. For higher dimensions, unfortunately no exact results are available. But due to the non-renormalization of the nonlinear term λ, the scaling relation $\chi + z = 2$ exists between the exponents, valid in any dimension. This reduces the number of independent exponents to one. A number of conjectures exist in the literature regarding the higher dimensional exponents, but so far none of them is proved. But numerical simulations on discrete models and direct integration of Eq. (1) helped to obtain reliable estimates for the exponents in higher dimensions as well. For the physically relevant dimension, $d = 2+1$, extensive numerical simulations give $\chi_0 = 0.385 \pm 0.005$ and $z_0 \simeq 1.6$.[34] Thus the interface is rough and the roughness increases with time as $\omega_0(t) \sim t^{\chi_0/z_0}$.

For $\gamma = 0$, Eq. (4b) is the fourth order linear diffusion equation with conserved noise [Eq. (3)], which can be solved exactly, resulting in $z_1 = 4$ and $\chi_1 = 0$.[38,39]

Thus neglecting the coupling terms, Eqs. (4a) and (4b) predict rather different values for z_i and the roughness exponents χ_i. To see how the couplings change the behavior, we have investigated Eq. (4) using a DRG scheme.

For this we rewrite Eq. (4) in its Fourier components:

$$\tilde{h}(k,\omega) = \tilde{\eta}_0(k,\omega)G_0(k,\omega) - \lambda G_0(k,\omega)\int\int d^dq d\Omega\, q(k-q)\tilde{h}(q,\Omega)\tilde{h}(k-q,\omega-\Omega)$$
$$- \beta G(k,\omega)\int\int d^dq d\Omega\, q(k-q)\tilde{v}(q,\Omega)\tilde{v}(k-q,\omega-\Omega) \qquad (11)$$

$$\tilde{v}(k,\omega) = \tilde{\eta}_1(k,\omega)G_1(k,\omega) + \gamma K^2 G_1(k,\omega)\int\int d^dq d\Omega\, q(k-q)\tilde{h}(q,\Omega)\tilde{v}(k-q,\omega-\Omega) \quad (12)$$

where $\tilde{\eta}_i(k,\omega)$, $\tilde{h}(k,\omega)$, and $\tilde{v}(k,\omega)$ are the Fourier components of the corresponding quantities and the correlators have the form:

$$G_0(k,\omega) = \frac{1}{\nu k^2 - i\omega} \qquad (13)$$

$$G_1(k,\omega) = \frac{-1}{Kk^4 + i\omega} \qquad (14)$$

During the DRG calculations, only one dynamic exponent $z = z_0 = z_1$ was used, valid if the Eqs. (4) do not decouple. Equations (11) and (12) are the starting point for the perturbative evaluation of $\tilde{h}(k,\omega)$ and $\tilde{v}(k,\omega)$. The basic diagrams are indicated in Fig. 2. The fast modes are integrated out in the momentum shell $e^{-l}\Lambda_0 \leq |k| \leq \Lambda_0$, and the variables are rescaled as $x \to e^l x$, $t \to e^{zl}t$, $h \to e^{\chi_0 l}h$, and $v \to e^{\chi_1 l}v$. The calculations have been performed up to one-loop order.

In what follows we shall skip most of the details of the calculation; the interested reader is referred to the literature.[40] We shall present only the main parts which are relevant to further arguments.

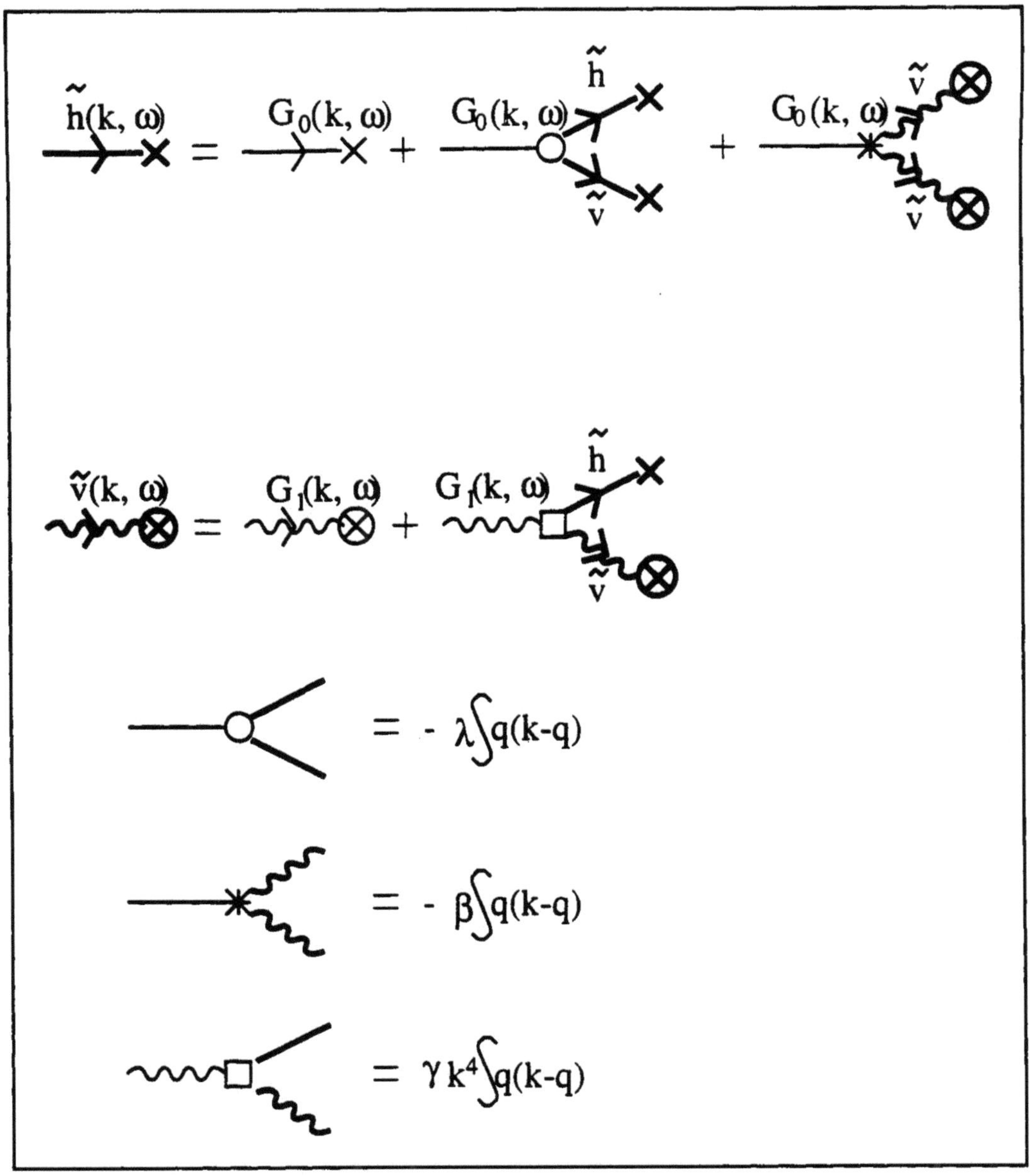

Fig. 2 Diagrammatic representation of the nonlinear integral equations [Eqs. (11) and (12)].

The first result is that the diagrams contributing to λ cancel each other, resulting in the flow equation:

$$\frac{d\lambda}{dl} = \lambda[z + \chi_0 - 2] \tag{15}$$

providing us with the scaling relation:

$$z + \chi_0 = 2 \, . \tag{16}$$

This relation is known to be the property of the KPZ equation and it is a consequence of Galilean invariance (GI). Since the DRG conserves the GI, this scaling law is expected to remain valid for all orders of the perturbation theory.

Fig. 3 The leading contribution to the effective noise spectral function. The encircled noise term corresponds to $\mathcal{D}_1$, and consists of D_1 and D_2 according to Eq. (7).

A second scaling relation can be obtained from the non-renormalization of the diffusion coefficient D_1:

$$\frac{dD_1}{dl} = D_1[z - d' - 2 - 2\chi_1], \tag{17}$$

resulting in:

$$z - 2\chi_1 - d' - 2 = 0. \tag{18}$$

The diagrams that contribute to D_1 (see Fig. 3) have a prefactor proportional to $\mathbf{k}^4$, thus they are irrelevant ($\mathbf{k}$ is the wave vector in the Fourier space). They in fact contribute to D_2, justifying its introduction in Eq. (7).

These two scaling relations already indicate that the coupled interface/surfactant system is qualitatively different from the uncoupled one, for a planar interface ($d' = 2$)[16,18] gives

$$\chi_0 + 2\chi_1 = -2, \tag{19}$$

which means that at least one of the exponents has to be negative.

A third scaling relation unfortunately is not available, but insight can be obtained from numerical integration of the flow equations obtained from the DRG. A correct flow must not scale the nonlinear terms β and γ to zero, which would decouple Eqs. (4a) and (4b). The finiteness of the nonlinear terms guarantee the validity of the scaling relations[16,18] as well. The integration showed the existence of the two main regimes:

(i) In the first regime one or both of the coupling terms (β, γ) scale to zero. In this case the two equations become completely (both coupling terms vanish) or partially (only one coupling term vanishes) decoupled, and the two equations might support different dynamic exponents z. The DRG scheme used is not reliable in this regime.

(ii) The presence of a strong coupling fixed point is expected when both of the nonlinear terms diverge. The integration shows that the *coupled phase* exists only for $z \geq 8/3$. The coupled phase is stable against small fluctuations in the coefficients and exists in a finite region of the parameter space. Since under experimental conditions, small fluctuations in the value of the control parameters are always expected, the stability of the system against them ensures the persistence of the coupled phase. But large deviations of the parameters introduce instabilities, which result in the breakdown of the smooth phase. This is in accord with the experimental observation, that surfactant induced layer-by-layer growth develops only under well controlled experimental conditions.

It is important to note that although there is no identifiable fixed point, in this phase the scaling relations[16,18] are exact. According to Eq. (16) for $z \geq 8/3$, the roughness exponent of the interface χ_0 is negative (see Fig. 4). With a negative roughness exponent, every noise-created irregularity is smoothed out by the growth dynamics and the resulting surface becomes flat. Thus the coupling of the surfactant to the growing interface results in the ***suppression of the surface roughness.*** This corresponds exactly to the experimentally observed behavior, i.e., the addition of the surfactant suppresses islanding, resulting in a morphological transition from rough (without surfactant) to flat (with surfactant) interface.

The roughness exponent of the surfactant from χ_1 [Eq. (18)] is negative if $z < 4$, while for $z > 4$ it becomes positive (see Fig. 4). In the Ge/Si system, for example, the As has a saturation coverage of 1 ML, which is independent of the system size and is governed only by the microscopic bonding of the As to the Ge dangling bonds. One expects no relevant fluctuations in the thickness of the coverage; this requires a negative roughness exponent for the surfactant and thus limits the dynamic exponent to values smaller than four.

The DRG analysis fails to provide the exact value of the dynamic exponent z. As in the case of many other growth phenomena, simple discrete models might be very helpful to obtain its value (see discussion later). Summarizing the results of the direct integration of the DRG equations, for $z > 8/3$, the existence of a strong coupling fixed point is observed, in which the interface roughness exponent is negative, corresponding to a flat phase. There is no upper bound in z for the existence of this phase, but physical considerations suggest that $z < 4$, in order to allow for the uniform surfactant coverage observed experimentally.

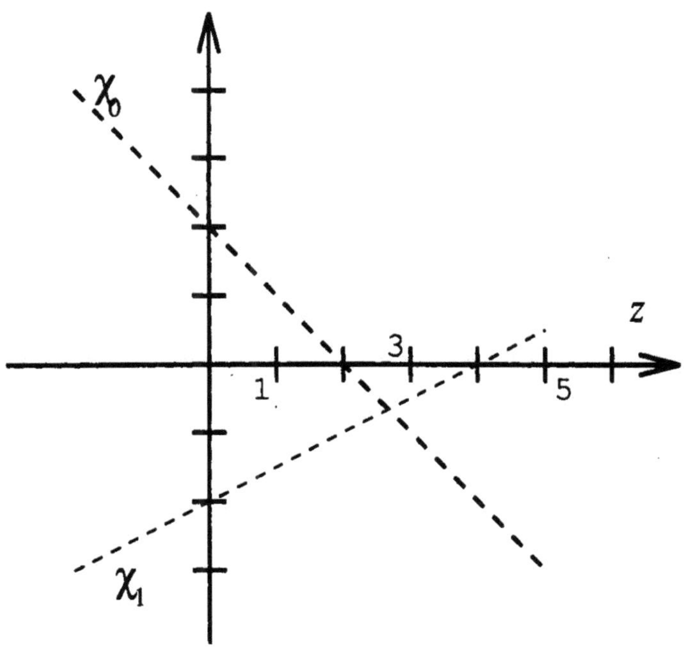

Fig. 4 The dependence of the roughness exponents χ_0 and χ_1 on the dynamic exponent z, according to the scaling relations Eqs. (16) and (18).

5. CONCLUSIONS AND FURTHER DEVELOPMENTS

In the previous sections, we introduced a set of coupled equations compatible with the basic symmetries and conservation laws of the surfactant/interface system studied experimentally. The main feature of these equations is that they predict a negative roughness exponent. We have argued that a negative roughness exponent describes a flat interface, in accord with the experimental observations. A natural question arises here: Is there any predicting power in this theory, or have experimental results merely been reproduced without generating further inquiries?

In this section, we examine the predictions made by the theory. The limits are presented as well: what are the physical ingredients we neglected, and whether and how could they be incorporated in a new theory along the presented lines.

As we have noted earlier, the analytic study does not provide us with the exact value of the exponents, but predicts that the dynamic exponent z lies in the narrow range between 8/3 and 4. If we could measure somehow the dynamic exponent z, the scaling relations would provide us with the other exponents. In fact, if one would be able to measure experimentally any of the exponents χ_i, β_i, or z, the other exponents could be obtained via Refs. 16 and 18.

The scaling theory[8–10] predicts that an originally flat interface becomes rough as a power law of time, $w \sim t^{\beta}$. Since in our case β is negative, an originally rough interface becomes smooth as a power law of time, until a limiting small roughness is reached. The only difference is in the system size dependence of the roughness: while in the usual growth models, the roughness increases as a power of L, in our case the interface is smooth, with a small thermal roughness w_0, independent of the system size. Thus a possible experimental check of the previous predictions would start from an initially rough interface and monitor directly the decrease of the roughness in time and fit the obtained curve with a power law. Previous experimental results indicated that it is possible to obtain the time dependence of quantities directly related to the surface roughness.[51–55] It would be interesting to see whether for the surfactant such a study could be carried out.

Such an experiment would result in the exponent β_0 for the interface ($\beta_0 = \chi_0/z$), from which using the scaling relations z and χ_0 could be determined. Hopefully the determined z would fall between the boundaries predicted by the theory.

Further test of the theory might come from direct numerical integration of the coupled Eqs. (4), with the aim to look for the coupled phase and obtain the value of the critical exponents. Integration proved to be successful in obtaining the exponents for the KPZ equation,[37] and for checking the DRG results for other coupled systems.[45]

Constructing and investigating discrete models in the same universality class as the studied continuum equations is another efficient and frequently very accurate way to obtain the scaling exponents.[48–50,35] For nonconserved coupled equations (see later) such models have been investigated[47] and gave results in accord with the DRG[47] and numerical integration.[45]

And finally let us mention some open problems related to the presented theory. It is important to note that when introducing Eq. (4), we did not use directly the existence of the strain which appears due to the lattice mismatch. Although an important problem,[56] a continuum description of strain-induced roughening is still missing. The proposed model [Eq. (4)] is expected to describe the *coupled* surfactant/interface system, but decoupling the surfactant does not necessary result in an equation describing heteroepitaxial islanding. Further studies are necessary to understand the microscopic (perhaps strain induced) origin of the nonlinear coupling terms.

In Eq. (4), the desorption of the surfactant atoms is neglected by considering that Eq. (4b) obeys mass conservation. Lifting the conservation law, Eq. (4b) should be replaced by a non-conservative equation. Such a system has recently been studied,[45,47] and it was found that in most cases the coupling does not change the KPZ scaling exponents. Enhancement of the exponents is possible *only* when the coupling is one-way, i.e., one of the equations decoupled from the other one is acting as a source of correlated noise.

Further, linear and/or nonlinear terms added to Eqs. (4) might influence the dynamics of the system. The goal here was to derive the *simplest* set of equations predicting the experimentally observed morphological phase transition; the study of other possible nonlinear terms and their relevance is left for future work.

Another shortcoming of the presented theory is that it does not predict oscillations in the interface roughness in the layer-by-layer growth regime, as is expected experimentally. This is due to the fact that the present continuum theory does not account for the *discreteness* of the lattice, responsible for the oscillations. But such a discrete pinning potential in principle can be introduced in Eqs. (4). The effect of such a pinning potential for both the conserved and nonconserved equation was studied in the literature.[57,58] It would be interesting to see how the coupling terms interact with the lattice potentials, and whether such a calculation leads to a coupled phase with periodic oscillations in time.

In conclusion, I have introduced a set of equations to describe the interaction of a growing surface with a surfactant. The main experimentally motivated requirements for Eqs. (4) were: (a) no surface diffusion of the newly landed adatoms; (b) conservative and diffusive surfactant dynamics, originating from neglecting incorporation and desorption of the surfactant during the growth process. The obtained equations indicate the existence of a coupled phase, in which two scaling relations between the three exponents are available. In this phase, the roughness exponent of the interface is negative, morphologically corresponding to a flat interface, as observed experimentally.

Moreover, Eqs. (4) serve as a good starting point for future studies of an interface coupled to a local *conservative* field, a problem of major interest in the context of recent efforts to understand the general properties of nonequilibrium stochastic systems.

6. ACKNOWLEDGMENTS

I wish to thank M. Gyure and E. Kaxiras for useful discussions and comments on the manuscript and H. E. Stanley for continuous encouragement and support. The Center for Polymer Studies is supported by National Science Foundation.

REFERENCES

1. T. Vicsek, *Fractal Growth Phenomena*, 2nd ed. (World Scientific, Singapore, 1992).
2. D. E. Wolf, in *Kinetics of Ordering and Growth at Surfaces*, ed. M. Lagally (Plenum, NY, 1990)
3. J. Krug and H. Spohn, in *Solids Far From Equilibrium: Growth, Morphology and Defects*, ed. C. Godrèche (Cambridge Univ. Press, Cambridge, England, 1991).
4. F. Family and T. Vicsek, *Dynamics of Fractal Surfaces* (World Scientific, Singapore, 1991).
5. *Surface Disordering: Growth, Roughening and Phase Transitions*, Proc. of the Les Houches Workshop 1992, eds. R. Jullien, J. Kertész, P. Meakin and D. E. Wolf (Nova Science, New York 1992).
6. D. E. Wolf and J. Villain, *Europhys. Lett.* **13**, 389 (1990).
7. J. Villain, *J. Phys.* **I1**, 19 (1991).
8. L. Golubović and R. Bruinsma, *Phys. Rev. Lett.* **66**, 321 (1991).

9. Z.-W. Lai and S. Das Sarma, *Phys. Rev. Lett.* **66**, 2348 (1991).
10. L.-H. Tang and T. Nattermann, *Phys. Rev. Lett.* **66**, 2899 (1991).
11. C. Roland and H. Guo, *Phys. Rev. Lett.* **66**, 2104 (1991).
12. D. E. Wolf, *Phys. Rev. Lett.* **67**, 1783 (1991).
13. P.-M. Lam and F. Family, *Phys. Rev.* **A44**, 4854 (1991).
14. J. Krug, M. Plischke and M. Siegert, *Phys. Rev. Lett.* **70**, 3271 (1993).
15. M. Copel, M. C. Reuter, E. Kaxiras and R. M. Tromp, *Phys. Rev. Lett.* **63**, 632 (1989).
16. A.-L. Barabási, *Phys. Rev. Lett.* **70**, 4102 (1993).
17. M. Copel, M. C. Reuter, M. Horn von Hoegen and R. M. Tromp, *Phys. Rev.* **B42**, 11682 (1990).
18. F. K. LeGoues, M. Copel and R. M. Tromp, *Phys. Rev. Lett.* **63**, 1826 (1989).
19. F. K. LeGoues, M. Copel and R. M. Tromp, *Phys. Rev.* **B42**, 11690 (1990).
20. M. Horn von Hoegen, F. K. LeGoues, M. Copel, M. C. Reuter and R. M. Tromp, *Phys. Rev. Lett.* **67**, 1130 (1991).
21. J. M. C. Thorton, A. A. Williams, J. E. Macdonald, R. G. Silfhout, J. F. van der Veen, M. Finney and C. Norris, *J. Vac. Sci. Technol.* **B9**, 2146 (1991).
22. D. J. Eaglesham, F. C. Underwald, and D. C. Jacobson, *Phys. Rev. Lett.* **70**, 966 (1993).
23. H. A. van der Vegt, H. M. van Pinxteren, M. Lohmeier, E. Vlieg, and J. M. C. Thorton, *Phys. Rev. Lett.* **68**, 3335 (1992).
24. H. J. Osten, J. Klatt, G. Lippert, B. Dietrich and E. Bugiel, *Phys. Rev. Lett.* **69**, 450 (1992).
25. N. Grandjean, J. Massies, and V. H. Etgens, *Phys. Rev. Lett.* **69**, 796 (1992).
26. C. W. Snyder and B. G. Orr, *Phys. Rev. Lett.* **70**, 1030 (1993).
27. N. Grandjean, J. Massies, *Phys. Rev. Lett.* **70**, 1031 (1993).
28. A recently proposed model [M. Sigert and M. Plischke, *Phys. Rev. Lett.* **68**, 2035 (1992)] predicted a smooth interface, but with grooved morphology, which is not observed in the surfactant system.
29. E. Kaxiras, *Europhys. Lett.* **21**, 685 (1993).
30. R. M. Tromp and M. C. Reuter, *Phys. Rev. Lett.* **68**, 954 (1992).
31. In fact, even for systems with surface diffusion but without mass conservation the long time behavior is described by the KPZ equation (Ref. 32). But if surface diffusion is absent, the growth equation can have only nonconservative dynamics, as is the case with the surfactant covered Ge interface.
32. M. Kardar, G. Parisi, and Y.-C. Zhang, *Phys. Rev. Lett.* **56**, 889 (1986).
33. P. G. de Gennes, *Rev. Mod. Phys.* **57**, 827 (1985).
34. For a detailed discussion on the different conjectures on the $d = 2+1$ KPZ exponents, and their numerical verification see Refs. 1, 35–37.
35. J. M. Kim and J. M. Kosterlitz, *Phys. Rev. Lett.* **62**, 2289 (1989).
36. B. M. Forrest and L.-H. Tang, *Phys. Rev. Lett.* **64**, 1405 (1990).
37. K. Moser, J. Kertész and D. E. Wolf, *Physica* **A178**, 215 (1991).
38. T. Sun, H. Guo and M. Grant, *Phys. Rev.* **A40**, 6763 (1989).
39. Z. Rácz, M. Siegert, D. Liu and M. Plischke, *Phys. Rev.* **A43**, 5275 (1991).
40. For the basic techniques of DRG see Refs. 41 and 42. Their application to nonequilibrium growth equations and related models can be found in Refs. 43 and 44. Coupled nonequilibrium equations investigated with DRG can be found in Refs. 47–47.
41. S. K. Ma, *Modern Theory of Critical Phenomena* (The Benjamin/Cummings Publishing Company, Reading, 1976).
42. P. C. Hohenberg and B. I. Halperin, *Rev. Mod. Phys.* **49**, 435 (1977).
43. D. Forster, D. Nelson and M. Stephen, *Phys. Rev.* **A16**, 732 (1977).
44. E. Medina, T. Hwa, M. Kardar and Y.-C. Zhang, *Phys. Rev.* **A39**, 3035 (1989).
45. D. Ertas and M. Kardar, *Phys. Rev. Lett.* **69**, 929 (1992).
46. D. Ertas and M. Kardar, *Phys. Rev. E* (in press).
47. A.-L. Barabási, *Phys. Rev.* **A46**, R2977 (1992).
48. S. Das Sarma and P. Tamborenea, *Phys. Rev. Lett.* **66**, 325 (1991).
49. D. A. Kessler, H. Levine and L. M. Sander, *Phys. Rev. Lett.* **69**, 100 (1992).
50. S. Das Sarma and S. V. Ghaisas, *Phys. Rev. Lett.* **69**, 3762 (1992).
51. J. Chevrier, V. Le Thanh, R. Buys and J. Derrien, *Europhys. Lett.* **16**, 737 (1991).
52. D. J. Miller, K. E. Gray, R. T. Kampwirth and J. M. Murduck, *Europhys. Lett.* **19**, 27 (1992).
53. Y.-L. He, H.-N. Yang, T.-M. Lu and G.-C. Wang, *Phys. Rev. Lett.* **69**, 3770 (1992).

54. H. You, R. P. Chiarello, H. K. Kim and K. G. Vandervorrt, *Phys. Rev. Lett.* **70**, 2900 (1993).
55. J.-K. Zuo and J. F. Wendelken, *Phys. Rev. Lett.* **70**, 1662 (1993).
56. B. G. Orr, D. Kessler, C. W. Snyder and L. Sander, *Europhys. Lett.* **19**, 33 (1992).
57. T. Hwa, M. Kardar and M. Paczuski, *Phys. Rev. Lett.* **66**, 441 (1991).
58. T. Sun, B. Morin and M. Grant in Ref. 5.

Fractals, Vol. 1, No. 4 (1993) 860–866

$1/f$ NOISE IN A TRAFFIC MODEL

MISAKO TAKAYASU
Research Institute for Fracture Technology, Faculty of Engineering, Tohoku University, Aramaki Aoba, Sendai 980, Japan

HIDEKI TAKAYASU
Graduate School of Information Sciences, Tohoku University, Aramaki Aoba, Sendai 980, Japan

Abstract

One-dimensional traffic flow is simulated by a cellular-automaton-type discrete model. As we increase the car density, the model shows a phase transition between a jam phase and a non-jam phase. By adding random perturbations we found a $1/f$ power spectrum in the jam phase, whereas a white noise is observed in the non-jam phase.

1. INTRODUCTION

One of the most interesting temporal fractals may be the so-called $1/f$ noises that are characterized by power spectra with infrared divergence. Although $1/f$ noises have been found almost everywhere, the basic mechanism is still mysterious. In this paper, we pay attention to the $1/f$ noise in the density fluctuation of highway traffic.[1]

Traffic flows have been numerically or theoretically analyzed mainly by using hydrodynamic models in which jams are represented by shock waves.[2] Recently, a new approach using the notions in statistical physics is attracting much interest. In the new approach, a traffic system is regarded as an interacting particle system which shows a phase transition between two phases; the low-density dynamical phase in which cars move smoothly at maximal speed, and the high-density jammed phase in which cars are either stopped or moving very slowly.[3]

The essential mechanism of this phase transition can be understood easily by a drastically simplified discrete model. The key point of jamming is quite simple that cars cannot run without sufficient spaces among them, namely, in a discrete model a car cannot move if the neighboring site in the moving direction is occupied. Biham et al. interpreted this mechanism in their cellular automaton models of traffic flow and analyzed the inherent dynamic phase transition.[3] Their evolution rule is identical to the cellular automaton of Wolfram's CA rule no. 184.[4] The same model was also studied in the context of surface growth[5] and lattice gas dynamics.[6]

The authors have introduced a similar discrete model of particle dynamics which is a revised version of the model of Biham et al. taking into account a new effect, delay in acceleration.[7] The system shows a hysteresis-like behavior that includes a bi-stability, and both continuous and discontinuous phase transitions. We have also found an indication of inverse-power type spectrum in the spatial density fluctuation by adding random perturbations.

In the next section, the model and results with deterministic rules are summarized. In the Sec. 3, we introduce the effect of random perturbations and show that the temporal density fluctuation observed at a site follows a white noise in the non-jam phase; however, in the jam-phase the fluctuation clearly shows an infra-red divergence with a $1/f$ type power spectrum. The last section is devoted to concluding remarks.

2. THE DETERMINISTIC MODEL

Let us start with the most basic model of Biham et al.[3] The model is defined on a one-dimensional lattice where at most one particle can shear a site. At every time step, each particle tries to move to the left neighboring site. This attempt fails if the destination site is already occupied; in that case, the particle is regarded as being involved in a jam. All particles simultaneously shift to the left if no pair of particles are shearing neighboring sites. Therefore, the maximum particle density without jam is 1/2. In a finite system with periodic boundary, the steady state mean velocity, $\langle v \rangle$, is determined only by the particle density, p, independent of the initial configurations. There remains no jam and $\langle v \rangle = 1$ for $p \leq 1/2$; on the other hand for $p > 1/2$ there exist jams. The mean velocity for $p > 1/2$ is solved analytically as $\langle v \rangle = (1-p)/p$.

Next, we add a new effect, the delay in acceleration, to make the system more interesting and realistic. Generally, cars running at a high speed can stop immediately by brakes, but it takes a much longer time to recover the speed by acceleration. In real traffic, not all cars on the same lane start running simultaneously as a signal changes, but a car accelerates after recognizing the front car's movement. This delay plays an important role in the formation of jams. Nagel and Schrekenberg introduced the delay in acceleration into a cellular automaton model by considering several modes for the car velocity. It has been reported that the density-velocity plots and space-time configuration agree nicely with real traffic data.[8] The effect of delay can be introduced to the simplest model in a different way by adding the new regulation: a particle at rest ($v = 0$) can move only when both the nearest and the next nearest neighboring sites in front are empty. As a result, each particle departing from a jam has two empty sites in front (see Fig. 1).

This modification does not affect running particles, so the configurations without jam remain unchanged. Namely, for $p \leq 1/2$, there exist initial configurations with all particles running stably keeping interparticle intervals longer than a site. On the other hand, at least a jam appears for $p > 1/2$, then the modification of delay in acceleration become effective. The mean velocity changes suddenly at the critical density, $p = 1/2$, as shown in Fig. 2(a).[7]

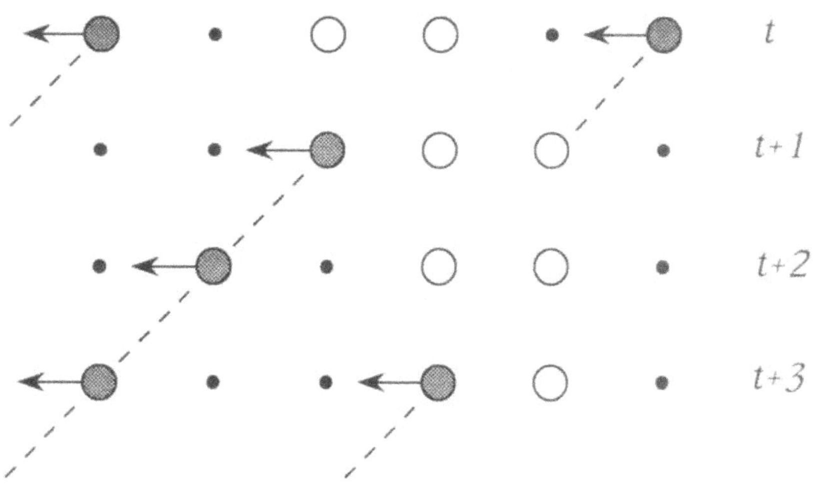

Fig. 1 Schematic picture of car's movement. Solid circles with arrows show moving cars with $v = 1$, and open circles show jamming cars with $v = 0$. Small solid circles show unoccupied lattice sites.

However, this is not the unique solution. If we start with the configuration that all particles are densely packed, then the mean velocity shows a continuous phase transition as shown in Fig. 2(b) with a non-differentiable point at $p = 1/3$. We can find a bi-stable state in the range $1/3 < p \leq 1/2$. From the standpoint of phase transition, the critical density between the jam phase and non-jam phase is $p_{c1} = 1/2$ in the case of Fig. 2(a) which is a kind of first order phase transition; and $p_{c2} = 1/3$ in the case of Fig. 2(b) which corresponds to a second order phase transition. Starting with an arbitrary initial condition, it can be shown that the density-velocity relation converges to one of the above two curves.

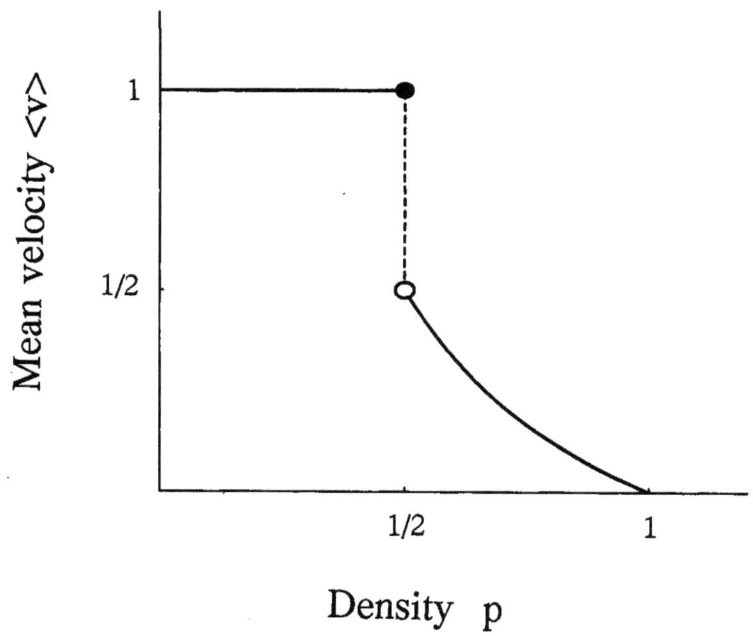

Fig. 2 Density-velocity diagrams depending on the initial conditions: (a) Starting from an initial condition that each particles keep regularly ordered intervals, namely, particles are placed on every other sites. The solid circle means including the point and the open circle means excluding the point.

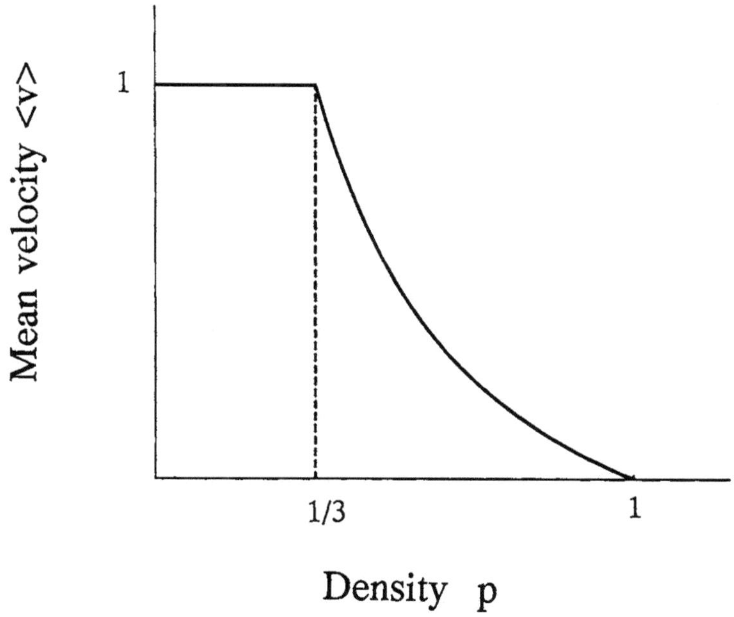

Fig. 2(b) Starting from a possibly maximum jamming initial condition that all particles are densely packed.

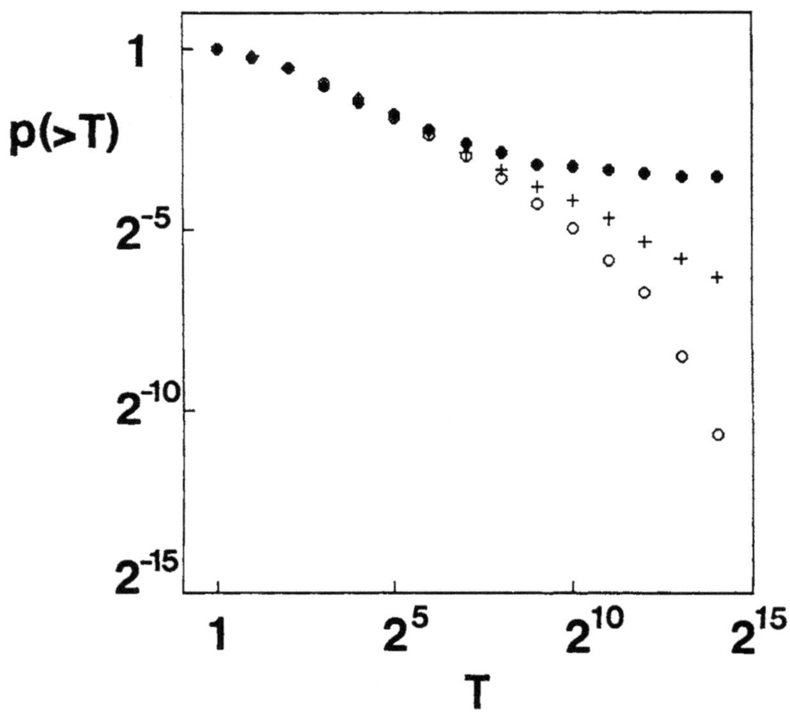

Fig. 3 The cumulative distribution of the total time on a log-log scale. The solid circles give the values for $p = 0.338$ (above the critical density), the crosses show the values for $p = 0.333$ (on the critical density), and the open circles correspond to $p = 0.330$ (below the critical density).

The continuous phase transition point is very much similar to the usual second order phase transitions in the sense that a quantity follows a power-law distribution. The quantity we pay attention to is a sum of jamming time, T, which is defined in the following way: assume that a car is at rest initially and we consider the jam triggered by this car. The ith car behind the trigger may be involved in the jam and stopped for T_i time steps ($T_i = 0$ if the jam is unsnarled before the ith car reaches). T is defined by the sum of T_i. The distribution of T starting with an independent random configuration is plotted in Fig. 3. As seen from this figure and also proved by theoretical arguments, the distribution function at the second order critical point is given by the following power law[7]:

$$P(> T) \propto T^{-1/2}. \tag{1}$$

The behaviors of the distribution below and above p_{c2} are also very similar to the cluster size distribution of the mean field percolation. Namely, the jam phase is characterized by the existence of infinite T_i, which corresponds to the percolated phase with an infinite size cluster.

3. RANDOM PERTURBATION AND $1/f$ NOISE

So far, we have seen the results of deterministic models. We now investigate the behavior of the system with a stochastic rule. The new rule is introduced as follows: at each time step cars move following the same rule as the rule with acceleration delay, but running cars stop spontaneously with certain probability χ. By this alternation, we obtain complicated space-time evolution patterns as typically shown in Fig. 4. Here the values of the parameters are $p = 0.5$ and $\chi = 0.5$. We can observe creation, annihilation and aggregation of jams.

The power spectrum of temporal density fluctuation at a site is observed in a simulation on 2400 sites with 4000 time steps and an average is taken over 10 realizations. As shown in Fig. 5, we have a $1/f$-type power spectrum in the jam phase, while in the non-jam phase,

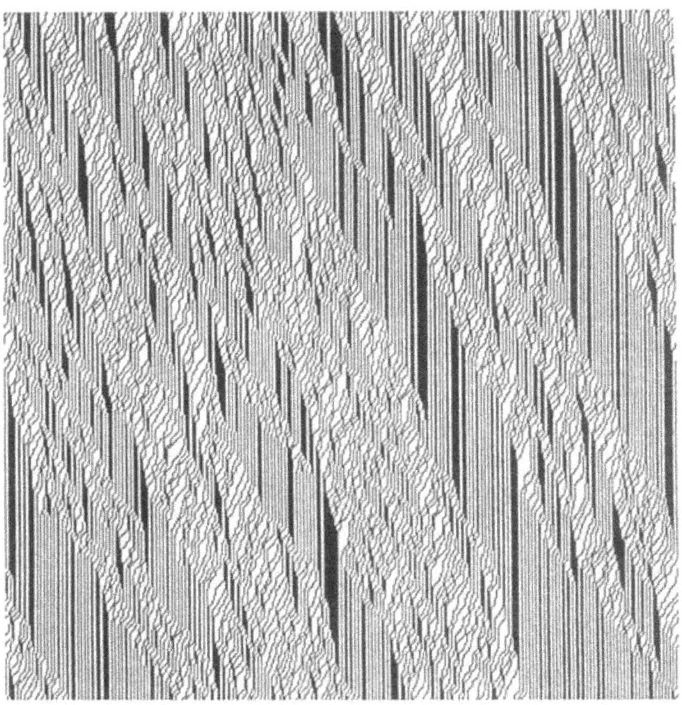

Fig. 4 A density profile for the stochastic model with delay in acceleration. The mean density is $p = 0.5$ and the spontaneous stopping rate is $\chi = 0.5$.

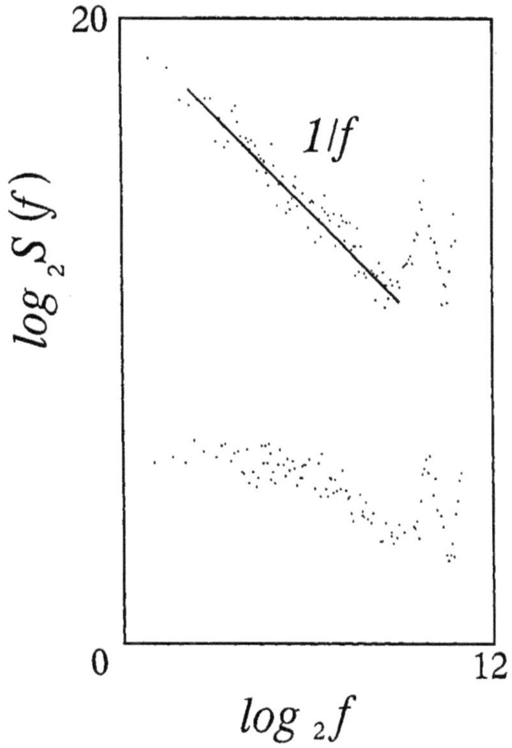

Fig. 5 Log-log plots of the temporal power spectra for the stochastic model of $p = 0.2$ (below), 0.5 (above, shifted upward for clear comparison) with $\chi = 0.5$. The straight guide line shows the slope for $1/f$.

the power spectrum is more like a white noise. The small peaks at large wave numbers are due to the characteristic of the model that jamming particles get out of the jam with two empty sites in front. The $1/f$ noise indicates in general that there exists a long time correlation. In our model, such long correlation may be supported by the jams with long life time. This estimation is consistent with the result that $1/f$ noise cannot be observed in the non-jam phase. Also, even though we are randomizing the density with relatively high probability, $\chi = 0.5$, we still have the long range correlation in a jam because of the delay in acceleration.

4. DISCUSSION AND SUMMARY

Jamming is a very universal phenomenon in systems composed of hard-core particles moving in one direction. Actually, Ristow and Herrmann recently confirmed by numerical simulation that the density fluctuations of granular material flowing out of a hopper also show approximately a $1/f$-type power spectrum.[9] They report that the effect of static friction is essential for the $1/f$ spectrum and the spectrum becomes clearer for smaller opening angle of the hopper. This is consistent with our results because the static friction delays acceleration and the decrease of opening angle makes the system one-dimensional. Our model clarified that the long time correlation in the $1/f$ behavior is due to the long life of jams. As we have seen in the deterministic model the mean life time of jam is infinite in the jam phase. In the non-jam phase, the mean life time is finite and consequently we observe the flat spectrum.

ACKNOWLEDGMENTS

The authors would like to thank Prof. H. Takahashi and Prof. A. Yu. Tretyakov for valuable discussions. M. T would like to acknowledge the support of grant from Japan Society for Promotion of Science.

REFERENCES

1. T. Musha and H. Higuchi, Japanese *J. Appl. Phys.* **15**, 1271 (1976).
2. See, e.g., W. Leutzbach, *Introduction to the Theory of Traffic Flow* (Springer-Verlag, Berlin, 1988); D. L. Gerlough and M. J. Huber, *Traffic Flow Theory* (National Research Council, Washington, D.C., 1975).
3. O. Biham, A. A. Middleton and D. Levin, *Phys. Rev.* **A46**, R6124 (1992).
4. S. Wolfram, *Theory and Applications of Cellular Automata* (World Scientific, Singapore, 1986).
5. J. Krug and H. Spohn, *Phys. Rev.* **A38**, 4271 (1988).
6. S. A. Janowsky and J. L. Lebowitz, *Phys. Rev.* **A38**, 4271 (1988).
7. M. Takayasu and H. Takayasu, preprint.
8. K. Nagel and M. Schrekenberg, *J. Phys. I France* **2**, 2221 (1992).
9. G. H. Ristow and H. J. Herrmann, preprint (1993).

Fractals, Vol. 1, No. 4 (1993) 867–874

THE ANISOTROPY OF ON-LATTICE SIMULATIONS OF AGGREGATE GROWTH

FERMÍN OTÁLORA and JUANMA GARCÍA-RUIZ
Instituto Andaluz de Geología Mediterránea, CSIC/Universidad de Granada, Campus Fuentenueva s/n, 18002, Granada (Spain)

Abstract

The origin of anisotropy in on-lattice simulations of cluster growth is investigated by quantifying the anisotropic behavior of the two processes involved in the simulation: transport and aggregation of nutrient particles. Three different anisotropic effects are described: anisotropic surface kinetics, anisotropic concentration profiles and anisotropic screening. The relevance of these sources of anisotropy in natural systems is discussed.

1. INTRODUCTION

On-lattice DLA simulations are influenced by the anisotropy of the underlying lattice.[1] A lot of work has been devoted to understanding the origin of such anisotropy, mainly from theoretical models,[2–4] but considerations on the importance of this observation for crystal growth are only pointed out as a secondary result of noise-reduced simulations.[5–8] The anisotropic surface tension of growing crystals has been proposed as the natural equivalent of this lattice-induced anisotropy, but small scale processes leading to macroscopic anisotropy during crystal growth remain uncorrelated with particle-level processes in cluster growth simulations. The aim of this work is to isolate the kind of small-scale anisotropic processes interplaying during on-lattice simulations and to estimate its relative importance in the morphogenesis of natural systems.

Aggregate growth simulations deal with two processes: bulk transport of a nutrient phase to the surface of the aggregate and on-surface aggregation kinetics. When these processes work on a lattice, anisotropic effects are expected to occur. In this work, we first enter upon the "structural" (i.e., crystallographic) anisotropy of aggregates where surface kinetics is taken into account (Sec. 2). In Sec. 3, anisotropic on-lattice probability distributions for random walkers are considered for the case of on-lattice transport. The last section contains a discussion of the anisotropic screening effects and the noise-reduced DLA in this framework, and includes finally some considerations on the relevance of these results to natural systems.

2. DIRECTIONAL BONDING STRUCTURAL ANISOTROPY

The DLA model[9] has often been modified to simulate aggregate growth in situations closer to equilibrium than the original model does.[10–14] These modifications commonly add to the model curvature dependent sticking probabilities and/or surface relaxation. As a consequence, growth probabilities are directly or indirectly controlled by the energetics of a directional (i.e., anisotropic) bonding configuration. Recently[14,15] we analyzed the morphology of aggregates grown using a modified DLA algorithm where surface relaxation

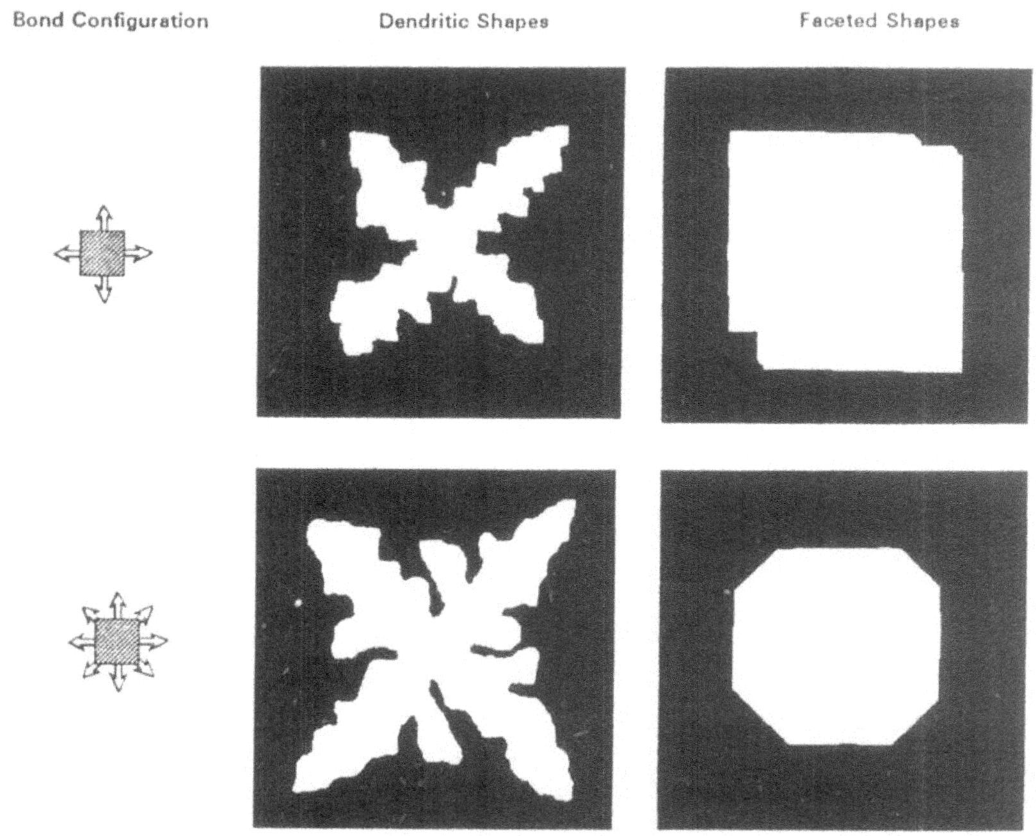

Fig. 1 Dendritic and faceted clusters grown by modified DLA computer simulations.[14–16] Dendritic and faceted crystals in the upper row are produced with a bond configuration consisting of four bonds in the direction of the main axis of lattice. Crystals in the lower row are produced using eight bonds, four along main lattice directions and four $\pi/4$ apart from there. Square shaped clusters ({10} faces) are obtained when bonding is allowed with nearest neighbors, while octagon ones ({10} and {11} faces) are characteristic of bonding with nearest and next nearest neighbors.

and energetically controlled growth probabilities are introduced. Two different bonding configurations are taken into account, one that includes the bonds of the first sphere of coordination in the Kossel crystal, and other including the bonds in the first and second spheres of coordination (Fig. 1). For the first case, the observed close to equilibrium growth morphology is a square cluster containing only {10} crystal faces, while for the second case, octagonal shapes containing both {10} and {11} faces are reported. These morphologies are controlled by the structural anisotropy of the cluster surface. It is possible to quantify such anisotropy by computing the orientation dependent surface energy of the aggregate. In order to perform this estimation, we compute the slope and unsaturated bond distribution on a set of crystal faces with different Miller indexes. By defining the specific surface energy for each such faces as:

$$\sigma = \sum N_i \Phi_i \tag{1}$$

where N_i is the number of unsaturated bonds per unit area with bond strength Φ_i lying in the i-th coordination sphere, we compute the equilibrium morphology of the cluster using the Wulff construction. In Fig. 2, Wulff plots (polar plot of the specific surface energy vs. face orientation) are shown for the nearest and next nearest bond configurations (the clusters represented in Fig. 1). The equilibrium morphology of crystals having these bond

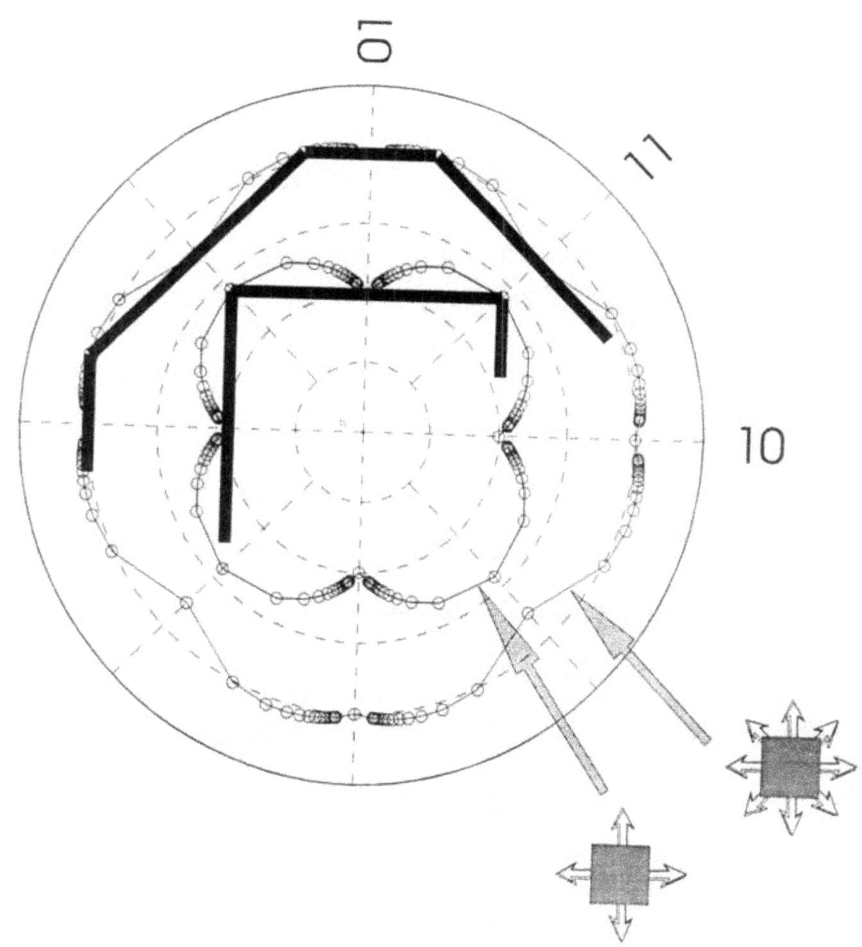

Fig. 2 Wulff plot of specific surface energy as a function of orientation. Surface energy is plotted for the two bonding configurations used to produce clusters in Fig. 1. Relative minima (valleys) in the plot correspond to stable faces in the equilibrium morphology.

configurations is made up of faces in the direction of local minima (valleys). Morphologies with square shape are expected to occur with the nearest bond configuration, while octagonal ones are the equilibrium morphology corresponding to the nearest and next nearest bond configuration. These structurally controlled morphologies computed from theoretical considerations are the same as those that we obtain in stochastic growth simulations (Fig. 1 and Refs. 14 and 15) showing the influence of structural control on cluster anisotropy.

3. PROBABILITY DISTRIBUTION OF ON-LATTICE RANDOM WALKERS. ANISOTROPIC TRANSPORT

When dealing with stochastic computer simulations, diffusion is simulated by using random walkers and assuming that their probability distribution fits a diffusive field as described by the Fick's equations, but the main deviation of the probability distributions of two (or more) dimensional on-lattice random walkers from Fick's equations is precisely a subtle kind of anisotropy. In particular, for the two dimensional case, the probability of a random walker being at position (x, y) after N steps can be computed as a function of the probability of a one-dimensional random walker of being at a position y after V ($V \leq N$) steps and the probability of a one-dimensional random walker of being at a position x after N–V steps; V representing the number of steps in the vertical direction (up plus down). The two dimensional probability distribution is:

$$P(N, x, y) = \left(\frac{1}{4}\right)^N \sum_{V=0}^{N} \frac{N!}{(N-V+x)!(N-V-x)!(V+y)!(V-y)!} \tag{2}$$

where the summation runs over the number V of steps taken in the vertical direction. This probability distribution is shown in Fig. 3(a). Figure 3(b) shows a stochastically computed distribution using 10^6 random walkers and Fig. 3(c) shows the off-lattice equivalent of Fig. 3(b). While the expected behavior of a simulated diffusion field seems like that in Fig. 3(c), the actual distribution is that of Fig. 3(b). A minor difference between distributions is that the on-lattice, probability is zero at sites with parity $(x+y)$ not equal to parity (N); this problem can be eliminated by averaging over two successive time steps. Nevertheless, the main difference is that the on-lattice distribution is no longer radially symmetric: in the areas of low probability (i.e., low concentration), probability is greater than zero in the main lattice directions at distances from origin where the probability become zero $\pi/4$ apart from lattice directions.

The coupling of this anisotropy with the boundary conditions imposed in aggregate growth simulations gives rise to anisotropic diffusion fields. In these simulations, the random walkers start its path on a circle that encloses the growing cluster and is assumed to be of homogeneous concentration. By adding the probability distributions centred in every site of this "generation circle", the on-lattice probability distribution of the whole field can be computed for a given time. Figure 4 shows the probability of a random walker starting its path on a generation circle with radius 120 pixels, of being on circles of radius 116, 112 and 104 pixels after 20 time steps. In spite of some noise induced by the discrete nature of the computations and the parity exclusions previously commented, clear trends are observed: the probability distribution shows maxima and minima in directions related to that of the underlying lattice. In addition, a shift in the maximum position is observed: maxima in the $r = 116$ curve correspond to minima in the $r = 104$ curve. This shift in the

(a)

(b)

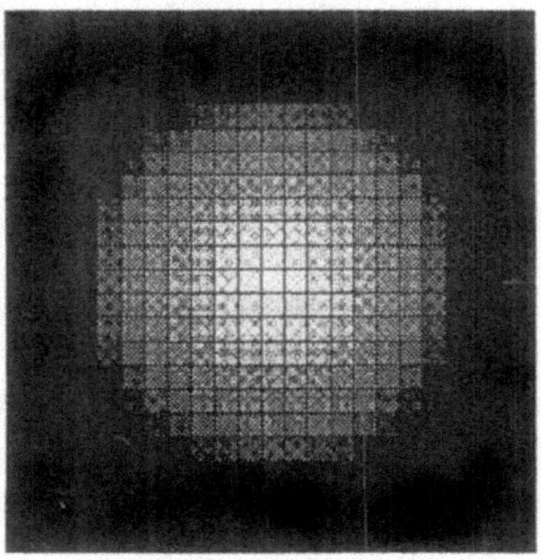

(c)

Fig. 3 Probability distributions for random walkers; **(a)** Exact distribution (Eq. 2); **(b)** Stochastically computed by tracking 10^6 on-lattice random walkers during 11 time steps; and **(c)** Off-lattice equivalent to (b).

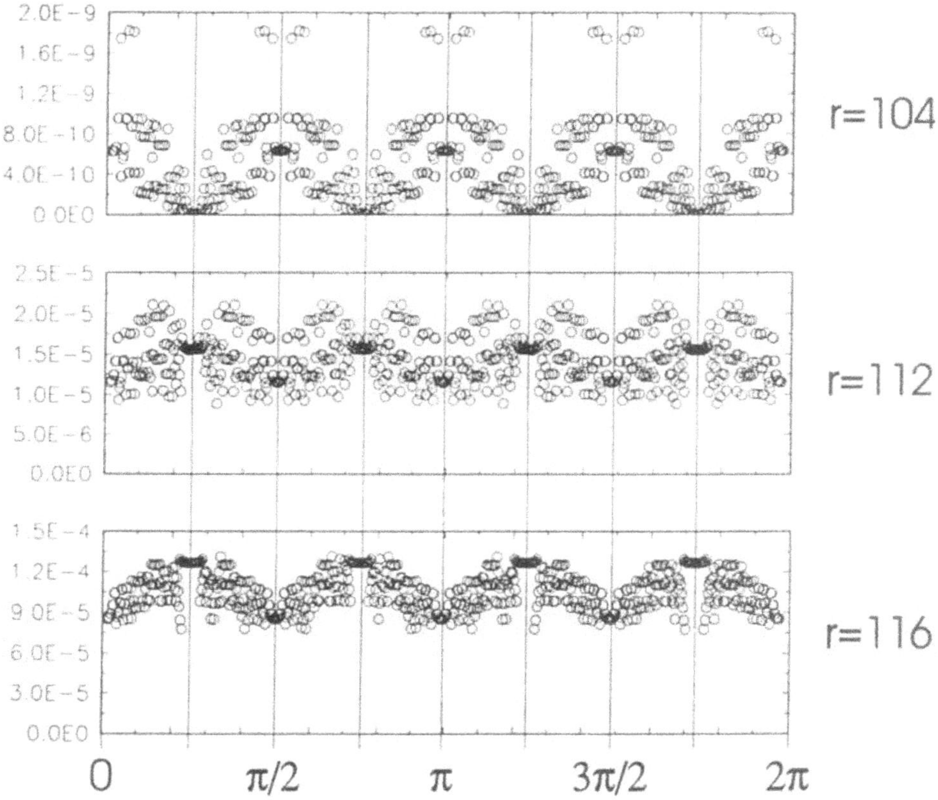

Fig. 4 Probability distribution taking into account boundary conditions. The lattice positions belonging to a "generation circle" with radius 120 units are computed. Exact probability distributions after $N = 20$ steps are centered in these positions, adding to its values where they superimpose. The value of every lattice position is divided by the number of sites making the circle to renormalize the whole probability distribution to 1. The three plots represent the probability values found in the intersection of three concentric circles and the generated field; the radius of the three circles is, respectively, 4, 8 and 16 units lower than the radius of the "generation circle" (approximately the distances between the maximum radius of the cluster and the generation circle in reported simulations).

maxima positions from the high concentration areas to the low concentration ones induces curved non-radial paths of preferred flux that has been previously reported in stochastic simulations.[16]

4. DISCUSSION

DLA is an unstable model for aggregate growth that produces open fractal structures.[9] This Saffman-Taylor-like instability is due to the selection of preferred growth positions at the tip of branches and the screening of inter-tip bays. Large DLA clusters evolve into diamond-like envelopes due to lattice anisotropy[17]; as this diamond shape grows, positions close to the centre of diamond sides become increasingly screened by the growth sites close to the corners of the diamond, leading to a cross-like structure reminiscent of dendritic crystal growth. The best known DLA-related algorithm producing anisotropic dendritic-like aggregates, noise-reduced DLA,[5-8] changes the aggregation rules of DLA: instead of aggregating a diffusing particle when it reaches a growth position, this event increments a local counter in the site position; when this counter comes up to a predefined value s, the site is occupied.

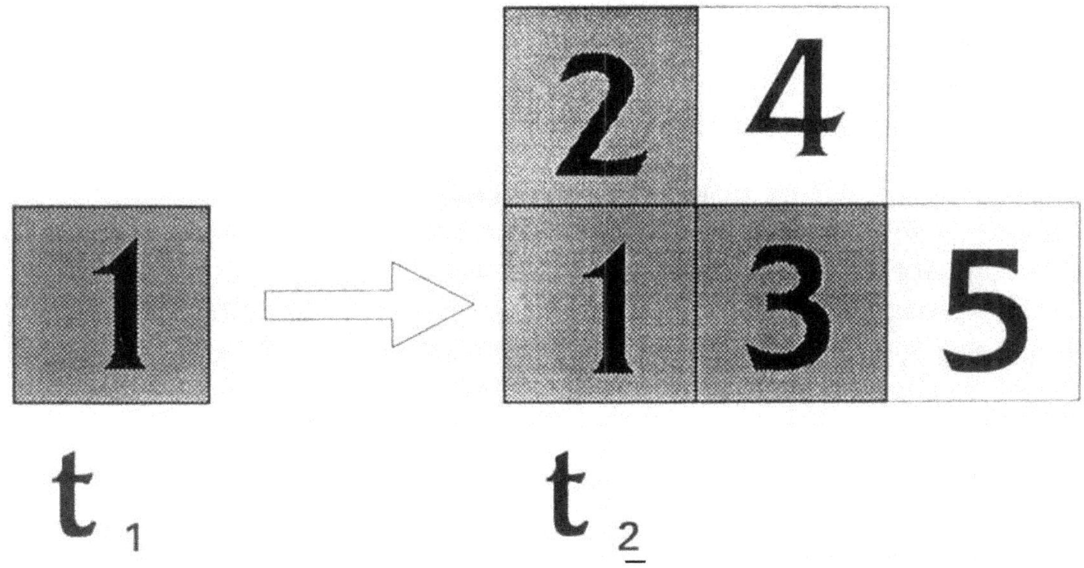

Fig. 5 Early state in the growth of a cluster. Two particles (2 and 3) have been added to the original seed (1). As a result, different surface sites have different growth probability, lower in screened bays than in exposed tips. Differences in bay/tip growth probabilities are exaggerated by noise reduction.

This algorithm is expected to allow the growth in a reasonable time of aggregates whose shapes approximate that of very large DLA clusters. Noise-reduced DLA deeply modifies the growth probability distribution over the interface sites. Let p_t and p_b be the growth probabilities of the tip (position 5) and bay (position 4) sites, respectively, in the simple situation illustrated in Fig. 5. Clearly $p_t > p_b$ because the tip position is more exposed to nutrient particles. Screening during DLA simulations is weakly anisotropic and stochastic noise obliterates it early during the simulation. If the configuration in Fig. 5 is now supposed to occur during a noise-reduced DLA simulation, the new probabilities p'_t and p'_b are related to p_t and p_b by $p'_t = p_t^s$ and $p'_b = p_b^s$ (where s is the value that the local counter should reach to be occupied), and $p'_t \gg p'_b$. This increment in the difference of tip and bay growth probabilities accelerates the previous transition from diamond to cross shapes, and is the origin of the anisotropy enhancement of Noise-reduced DLA simulations. This third anisotropic effect (anisotropic screening) is related to structural anisotropy, because it controls the distribution of screened and exposed growth position, and to anisotropic transport because the restrictions in the movement of random walkers control the geometry of screening.

Three different anisotropic effects have been described that inhabit on-lattice simulations of aggregate growth. Structural anisotropy is linked to the existence of an anisotropic bonding configuration, while anisotropic probability distributions are related to on-lattice transport, and anisotropically enhanced screening is a result of the interplay of both processes. In terms of relative importance, structural anisotropy is clearly the more important factor, because it controls the overall morphology of clusters; but when running increasingly out of equilibrium, anisotropic distributions and anisotropic screening start to play its role. Our preliminary results quantifying the relative importance of the three processes seem to support the fact that anisotropic screening is much more important in controlling the growth probabilities on the cluster than anisotropic transport. The anisotropy reported in on-lattice simulations of aggregate growth is not the result of a single nor simple process: on-lattice bonding and on-lattice transport interplay to produce anisotropic aggregates both being complex processes. Such complexity can be overcome by using off-lattice or semi-lattice models when simulating growth of isotropic or anisotropic aggregates, respectively, from an isotropic field; on the other hand, keeping this effects in simulations is an interesting way to make progresses in fields where transport and aggregation take place on a lattice, particularly for models where crystal lattices and crystal surfaces are involved. All the effects discussed in this work are real facts rather than artefacts in crystal growth experiments. For example, diffusion of adatoms and step advance on crystal surfaces are the two dimensional equivalent of two dimensional on-lattice DLA modifications, while solid state diffusion and solid state crystal growth (for example, solid state phase segregation) are the three dimensional analogue of these simulations. Therefore, on-lattice DLA is the better approach to emulate real phenomena such as mineral exolutions in crystalline media or step instabilities in crystal growth.

ACKNOWLEDGMENTS

Work performed under financial support from the Autonomous Government of Andalucia (Spain) and CICYT project PB89-0059.

REFERENCES

1. R. C. Ball and R. M. Brady, *J. Phys.* **A18**, L809 (1985).
2. L. A. Turkevich and H. Scher, *Phys. Rev. Lett.* **55**, 1026 (1985).
3. R. C. Ball, R. M. Brady, G. Rossi and B. R. Thompson, *Phys. Rev. Lett.* **55**, 1406 (1985).
4. T. C. Hasley, P. Meakin and I. Procaccia, *Phys. Rev. Lett.* **56**, 854 (1986).
5. J. Nittmann and H. E. Stanley, *Nature*, **321**, 141 (1986).
6. J. Kertesz and T. Vicsek, *J. Phys.* **A19**, L257 (1986).
7. R. C. Ball, *Physica* **A104**, 62 (1986).
8. J. Kertesz, T. Vicsek and P. Meakin, *Phys. Rev. Lett.* **57**, 3303 (1986).
9. T. A. Witten and L. M. Sander, *Phys. Rev. Lett.* **47**, 1400 (1981).
10. T. Vicsek, *Phys. Rev. Lett.* **53**, 2281 (1984).
11. Y. Saito and T. Ueta, *Phys. Rev.* **A40**, 3408 (1989).
12. J. R. Banavar, M. Kohmoto and J. Roberts, *Phys. Rev.* **A33**, 2065 (1986).
13. R. F. Xiao, J. I. D. Alexander and F. Rosenberger, *J. Crystal Growth*, **100**, 313 (1990).
14. J. García-Ruiz and F. Otálora, *Physica* **A178**, 415 (1991).
15. F. Otálora and J. M. García-Ruiz, *Crystal Properties and Preparation*, **36–38**, 686 (1991).
16. J. García-Ruiz and F. Otálora, *J. Crystal Growth*, **128**, 163 (1993).
17. R. C. Ball, M. J. Blunt and O. R. Spicack, in *Fractals in the Natural Sciences*, eds. M. Fleischmann, D. J. Tildesley and R. C. Ball, (Princenton University Press, 1990).

Fractals, Vol. 1, No. 4 (1993) 894–903

CHARACTERIZATION OF POROUS MEDIA WITH GEOMETRICAL MULTIFRACTALS

ANTOINE SAUCIER and JIRI MULLER
Institutt for Energiteknikk, Box 40, N-2007 Kjeller, Norway

Abstract

In a porous medium the local porosity often exhibits spatial variations. These variations can be characterized by a multifractal spectrum, as long as suitable scaling characteristics are present. We derive some of the properties of such geometrical multifractals, and discuss how the spatial distribution of a set is related to its multifractal spectrum.

1. INTRODUCTION

The concept of multifractal measure has recently been used for the geometrical characterization of the microstructure of disordered heterogeneous materials such as sedimentary rocks,[1–3] granular films,[4] polymer alloys,[4] paper formations,[5] diffusion limited aggregates[6] and electron deposition clusters.[7] Such characterizations can be useful for the investigation of relationships between the microstructure and the macroscopic physical properties of such materials.

In porous media the local porosity often exhibits spatial variations. These variations depend on the location and the size of the subregion considered within the porous medium, and they can be used for its detailed description.[8] If suitable scaling features are present, then the spatial variations of the local porosity can be characterized by a multifractal spectrum. The multifractal characterization of complex geometrical objects is based on the scaling behavior of a measure defined by the volume of a set, e.g. the pore-space volume in the case of porous media. A complex geometry is then characterized by a function $f(\alpha)$ — the multifractal spectrum — or equivalently by the generalized dimensions $D(q)$.

In practice, pictures of a porous medium are obtained at a finite resolution by digitizing microscope images of 2-dimensional thin sections.[2,3]

Geometrical multifractals, being defined by the volume of a set, are subject to specific constraints that are not obeyed by multifractals in general. The goal of this paper is to examine the effect of these constraints on the bounds of the coarse singularity index α and the generalized dimensions $D(q)$ (the effect of a finite resolution on these bounds is also discussed). We also show how the spatial distribution of a set is related to its multifractal spectrum.

2. PORE SPACE WITH MULTIFRACTAL PROPERTIES

2.1 Definitions

We consider a D-dimensional ($D = 1, 2$ or 3) bounded region $E \subseteq IR^D$ containing a porous medium with fixed porosity $0 < \phi_0 < 1$. For simplicity it is assumed that E is cubic and has size unity, so that the total pore-space volume equals the porosity. Within E the porosity may vary locally.

We define the measure of any domain $D \subseteq E$ to be the pore-space volume enclosed in D, normalized by the total pore-space volume contained in E.[a] The measure of a cubic box of size $0 \leq \delta \leq 1$ will be denoted by $p(\delta)$, and takes the form:

$$p(\delta) = \frac{\nu_p(\delta)}{\phi_0}, \tag{2.1}$$

where $\nu_p(\delta)$ denotes the pore-space volume contained in the box. Such measures, defined by the volume of a set — here the set being the pore-space — have the characteristic property of being homogeneous on their support. In other words, the mass density corresponding to the measure is constant and non-zero on the support (i.e., in the pore-space), and vanishes everywhere else (i.e., in the solid matrix).

Consider the measure $p_x(\delta)$ of a box centered about an arbitrary point x in E. As long as x does not lie exactly on the pore-space boundary, then for δ small enough the box will be either entirely within the pore-space, or entirely within the rock matrix.[b] In both cases, $p_x(\delta)$ will remain constant for small variations of x (i.e., $p_x(\delta) = 0$ in the rock matrix, and $p_x(\delta) = \delta^D/\phi_0$ in the pore-space). In that sense, measures based on the volume of a set are *locally homogeneous*. This implies, in particular, that the pointwise (or local) singularity index $\alpha(x)$ at point x,[10,11] defined by $\alpha(x) = \lim_{\delta\to 0} \ln(p_x(\delta))/\ln \delta$, is equal to D for any x in the pore-space.

The pore-space can be covered with a regular cubic grid of mesh δ. If $p_i(\delta)$ denotes the measure of the ith box of the pore-space covering, the moment generating function $\chi_q(\delta)$ is defined by:

$$\chi_q(\delta) = \sum_i [p_i(\delta)]^q, \tag{2.2a}$$

---

[a] A measure μ is a non-negative set function such that $\mu\{\emptyset\} = 0$ and $\mu\{\cup_i S_i\} = \sum_i \mu\{S_i\}$ for any sequence of disjoint sets $S_1, S_2, \ldots$. The measure considered in the text is the Lebesgue measure on $\mathbb{R}^D$.

[b] It is assumed here that both the pore-space and the solid matrix (minus its boundary) are open sets.[9]

where the summation extends over all the boxes of the pore-space covering,[11] and where q is a real parameter.

Consider a box-size $\delta_{\text{pore}} > 0$ sufficiently small for most boxes of the pore-space covering to be entirely within the pore space. The boxes inside the pore space have the same measure δ^D/ϕ_0, and for $\delta < \delta_{\text{pore}}$ there are approximately $\phi_0\delta^{-D}$ such boxes.[c] It follows from Eq. (2.2a) that the moment generating function takes the approximate form:

$$\chi_q(\delta) \sim \phi_0^{1-q}\delta^{D(q-1)} \tag{2.2b}$$

as $\delta \to 0$. $\chi_q(\delta)$ therefore scales with δ and the scaling exponent $\tau(q) = D(q-1)$ is linear in q. The "monofractal" behavior [Eq. (2.2b)] observed in the limit $\delta \to 0$ (the generalized dimensions $D(q) = \tau(q)/(q-1)$ are equal to D for all q) reflects the local homogeneity of the measure.

Next consider the case where the generating function $\chi_q(\delta)$ satisfies:

$$\chi_q(\delta) \sim \delta^{\tau(q)} \tag{2.3}$$

for all δ in some range of scales $\eta \leq \delta \leq \delta_0$, and where $\tau(q)$ is a *nonlinear* function of q. In this case, we shall say that the locally-homogeneous measure is multifractal. Here η and $\delta_0 \leq 1$ are respectively the inner and outer scales of the scaling range. It is emphasized that a multifractal behavior may exist only in some intermediate range of scales $\eta \leq \delta \leq \delta_0$ for which $\eta \gg \delta_{\text{pore}}$, since the monofractal behavior [Eq. (2.2b)] prevails as $\delta \to 0$.

The spatial variations of $p(\delta)$ reflect the spatial fluctuations of the pore-space geometry, which is the motivation for the term "geometrical multifractals." In the context of the geometrical characterization of fractal aggregates, geometrical multifractals have also been called "mass multifractals."[12,13] Since none of these terminologies reflects the main characteristic of these measures, which is to be locally homogeneous, we prefer in this paper to refer to them as "locally-homogeneous multifractals."

2.2 A Geometrical Constraint on the Measure μ

Locally homogeneous measures, which are determined by the pore-space volume via Eq. (2.1), are subject to the geometrical constraint:

$$\nu_p(\delta) \leq \delta^D . \tag{2.4}$$

Equation (2.4) must hold for all $\delta \geq 0$, and in particular for all δ in the scaling range $[\eta, \delta_0]$. Equation (2.4) follows naturally from the fact that the pore-space volume contained in a cube of size δ cannot exceed the total volume of this cube. With the definition (2.1), Eq. (2.4) takes the form:

$$p(\delta) \leq \frac{\delta^D}{\phi_0} . \tag{2.5}$$

[c]Let us denote by $N_B(\delta)$ the number of boxes of size δ needed to cover the pore-space. $N_B(\delta)\delta^D$ is an estimate of the volume ϕ_0 of the set at resolution δ. It is therefore expected that

$$N_B(\delta)\delta^D \to \phi_0$$

as $\delta \to 0$. Hence $N_B(\delta) \sim \phi_0\delta^{-D}$ as $\delta \to 0$.

The usual definition $p(\delta) = \delta^{\alpha}$ of the coarse (or coarse-grained) singularity index α,[11] replaced in Eq. (2.5), leads directly to:

$$\alpha \geq D - \frac{\ln \phi_0}{\ln \delta}, \tag{2.6}$$

that must hold for all δ.

Let us now make the hypothesis that the locally-homogeneous measure $p(\delta)$ is multifractal in the range $\delta \in [\eta, \delta_0]$, and denote by $\alpha_{\min}$ the minimum value of α. $\alpha_{\min}$ must satisfy Eq. (2.6) for all $\delta \in [\eta, \delta_0]$. Hence it satisfies:

$$\alpha_{\min} \geq \max_{\delta \in [\eta, \delta_0]} \frac{D - \ln \phi_0}{\ln \delta}. \tag{2.7}$$

The maximum in Eq. (2.7) occurs for $\delta = \eta$. Assuming $\eta \neq 0$, Eq. (2.7) yields:

$$\alpha_{\min} \geq D - \frac{\ln \phi_0}{\ln \eta}. \tag{2.8}$$

For locally-homogeneous multifractals, $\alpha_{\min}$ has therefore an infimum (i.e., maximum lower bound) that is determined by the dimension of space D, the total porosity ϕ_0 and the inner scale η. In the limit $\eta \to 0$, Eq. (2.8) yields $\alpha_{\min} \geq D$. For $\eta > 0$ and $0 < \phi_0 < 1$, Eq. (2.8) makes room for the occurrence of locally-homogeneous multifractals with $\alpha_{\min} < D$. For deterministic multifractals $\alpha_{\min}$ corresponds to the generalized dimension D_{∞}.[10] Since $D(q)$ is a monotonically decreasing function of q, it then follows from Eq. (2.8) that the generalized dimensions of a locally-homogeneous multifractal are also bounded by:

$$D(q) \geq D - \frac{\ln \phi_0}{\ln \eta}. \tag{2.9}$$

Note that Eq. (2.8) can also be rewritten in the form:

$$\phi_0 \leq \eta^{D - \alpha_{\min}}. \tag{2.10}$$

Consider a multifractal measure for which $\alpha_{\min} < \alpha(0) = D$, i.e., a multifractal with the usual "bell-shaped" $f(\alpha)$ function.[11] With $\alpha_{\min} < D$ fixed, Eq. (2.10) implies that $\phi_0 \to 0$ as $\eta \to 0$. Hence geometrical multifractals with wide scaling ranges must have small porosity.

3. EFFECT OF A FINITE RESOLUTION ON MEASURED SCALING PROPERTIES

3.1 General Considerations

In practice, the pore-space obtained for example by the digitization of a photograph is not known with an infinite resolution. We rather obtain a coarse-grained picture of the pore-space. The coarse-grained pore-space will typically be composed of a union of square "pixels" of size η_p, where η_p is the resolution or coarse-graining scale. As a result, the pore-space volume contained in a given box will be known only approximately, and we expect in general an effect of the coarse-graining on the measured scaling properties (e.g. range of the values of α, scaling exponents). The effect of the coarse-graining on the scaling properties will depend on the pore-space geometry, the coarse-graining process,[d] and the resolution. Unless all this information is known, the problem of comparing the scaling properties of the exact pore-space and of the coarse-grained pore-space remains ill-defined.

In the first step, we may however examine some of the simplest properties of the coarse-grained pore-space. In the following, we shall make the hypothesis that the coarse-grained pore-space also satisfies the scaling property:

$$\chi_q^*(\delta) \sim \delta^{\tau^*(q)} \tag{3.1}$$

for all $\delta \in [\eta^*, \delta_0^*]$. Such behavior has been observed for example on digitized 2-D pictures of thin sections of porous rocks.[1–3] The stars indicate here that the quantities $\chi_q^*(\delta)$, $\tau^*(q)$ and η^* have been obtained from the coarse-grained pore-space. It is emphasized that $\chi_q^*(\delta)$, $\tau^*(q)$ and η^* are not equal in general to the exact values $\chi_q(\delta)$, $\tau(q)$ and η. The scaling Eq. (3.1) of the coarse-grained sample may be inherited from the scaling Eq. (2.2) of the exact pore-space, and the exponents $\tau^*(q)$ may reflect for some q's the scaling properties of the exact underlying pore-space.

Note that the argument of Sec. 2.2 holds unchanged as long as all definitions are based on "star quantities" instead of exact quantities, i.e., on quantities estimated from the coarse-grained pore-space. It follows that the constraint Eq. (2.8) also holds for the coarse-grained pore-space, i.e.,

$$\alpha_{\min}^* \geq D - \frac{\ln \phi_0^*}{\ln \eta^*} . \tag{3.2}$$

3.2 An Upper Bound on α^* Due to the Resolution Limit

A limited resolution creates an additional constraint on α^*: the accessible singularity indices α^* will have a supremum (minimum upper bound). The largest values of α^* obtained from the multifractal analysis of the pore-space are determined by the smallest non-zero values of the measure. For a coarse-grained pore-space, this minimum value is clearly the volume of a single pixel. Hence we expect that for all δ:

$$v_p^*(\delta)|_{\text{non-zero}} \geq \eta_p^D . \tag{3.3}$$

Using again the definition (2.1) — with star quantities — and the convention $p^*(\delta) = \delta^{\alpha^*}$ leads directly to:

$$\alpha^*(\delta) \leq \ln \frac{\dfrac{\eta_p^D}{\phi_0^*}}{\ln \delta} . \tag{3.4}$$

[d]For example, in the case of digitized microscope pictures of porous rock thin sections, the coarse-graining process involves the combined properties of the microscope and of the thresholding required to produce black and white pixels (corresponding to pore-space and solid matrix respectively).

This coarse-grained measure is scaling [Eq. (3.1)]. Consequently, there exists a maximum value of α^*, denoted by $\alpha^*_{\max}$, that must satisfy Eq. (3.4) for all $\delta \in [\eta^*, \delta^*_0]$. $\alpha^*_{\max}$ therefore satisfies:

$$\alpha^*_{\max} \leq \min_{\delta \in [\eta^*, \delta^*_0]} \frac{\ln \frac{\eta_p^D}{\phi^*_0}}{\ln \delta} . \tag{3.5}$$

The minimum in Eq. (3.5) occurs for $\delta = \eta^*$, therefore:

$$\alpha^*_{\max} \leq D \frac{\ln \eta_p}{\ln \eta^*} - \frac{\ln \phi^*_0}{\ln \eta^*} . \tag{3.6}$$

The singularity indices estimated from the coarse-grained pore-space have therefore a supremum determined by the dimension D, the porosity ϕ^*_0, the inner scale η^* and the pixel size η_p. It is emphasized that the constraint (3.6) is only due to the resolution limit imposed by a non-zero pixel size η_p, and it should be noted that this constraint disappears in the limit $\eta_p \to 0$.

Taken simultaneously, Eqs. (3.2) and (3.6) lead to the global inequality:

$$D - \frac{\ln \phi^*_0}{\ln \eta^*} \leq \alpha^* \leq D \frac{\ln \eta_p}{\ln \eta^*} - \frac{\ln \phi^*_0}{\ln \eta^*} . \tag{3.7}$$

The bounds [Eq. (3.7)] were found to be in agreement with the experimental results obtained in Refs. 1 and 2 from digitized pictures of porous media. For deterministic multifractals $\alpha_{\min}$ and $\alpha_{\max}$ correspond to the generalized dimensions D_∞ and $D_{-\infty}$ respectively.[10] $D(q)$ being a monotonically decreasing function of q, it then follows from Eq. (3.7) that the generalized dimensions $D^*(q)$ measured on the coarse-grained pore space are bound by:

$$D - \frac{\ln \phi^*_0}{\ln \eta^*} \leq D^*(q) \leq D \frac{\ln \eta_p}{\ln \eta^*} - \frac{\ln \phi^*_0}{\ln \eta^*} . \tag{3.8}$$

It should be noted that the inequalities [Eqs. (3.7) and (3.8)] are more general than the bounds proposed previously in the literature.[14–16] It is emphasized that the results Eqs. (3.7) and (3.8) are not restricted to porous media, but apply to locally-homogeneous multifractals in general.

4. EXAMPLES OF GEOMETRICAL MULTIFRACTALS

A geometrical multifractal can be constructed with a multiplicative process.[11] These processes are simple models of multifractals that have been used in many fields of physics and geophysics. The process involves λ^D weights $w_i \geq 0$, $i = 1, \ldots, \lambda^D$, such that $\sum_i w_i = 1$. $\lambda = 2, 3, \ldots$ is an integer scale ratio. The first two steps of the construction process are illustrated in Fig. 1 for a 2-dimensional process with $\lambda = 2$. The unit square $B(\delta_0)$ of size $\delta_0 = 1$ is initially assigned a total measure ϕ_0 equal to the porosity. In the first construction step, $B(\delta_0)$ is divided into λ^D identical squares $B_i(\delta_1)$ of size $\delta_1 = \delta_0/\lambda$ (here $\delta_m = \delta_0\lambda^{-m}$) and each of these sub-squares is assigned a weight $\phi_0 w_i$, $i = 1, \ldots, \lambda^D$. The splitting process is then repeated on each sub-square: $B_i(\delta_1)$ is divided into λ^D squares of size δ_2 which

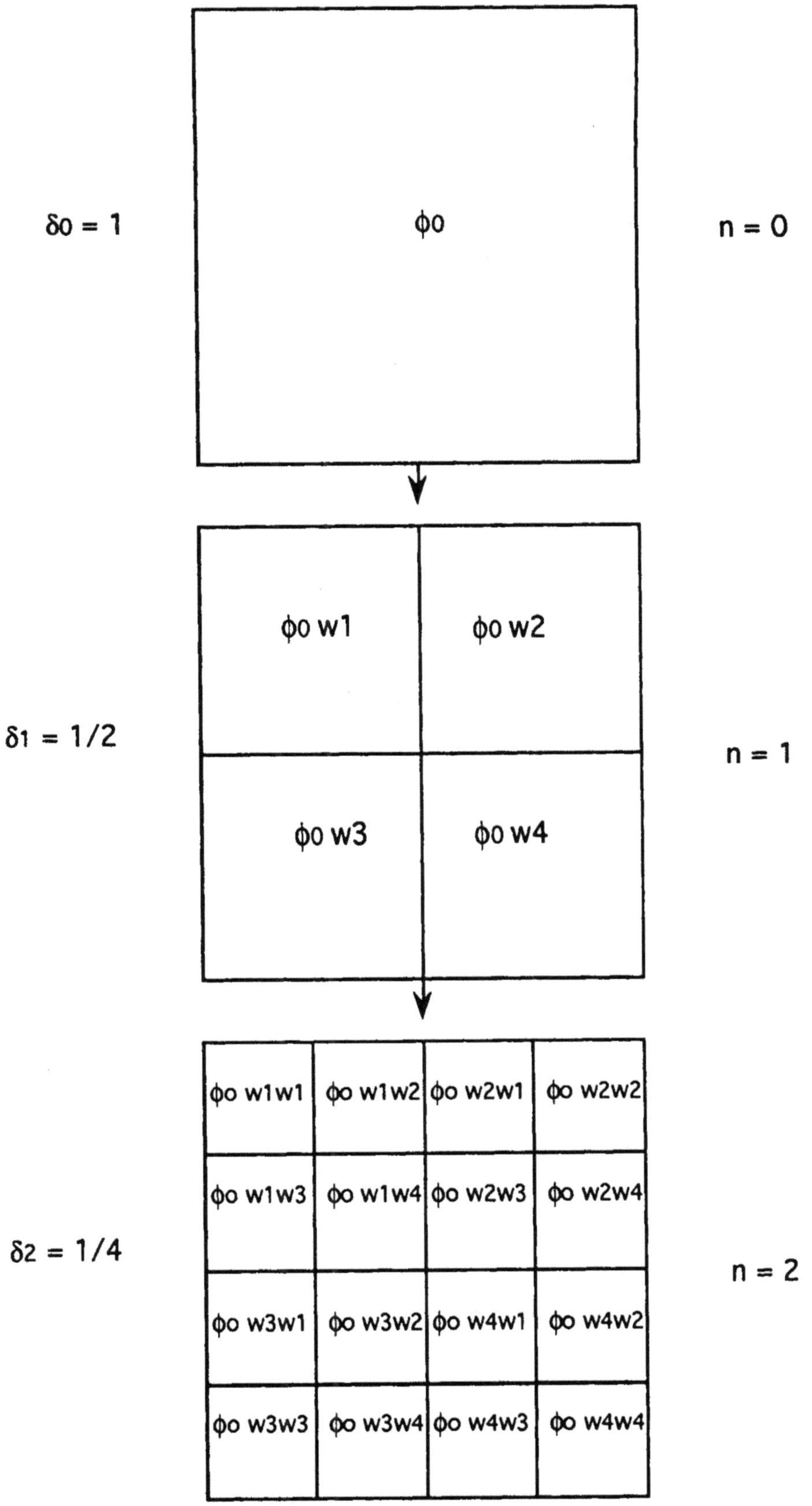

Fig. 1 The first two steps of construction of a deterministic two-dimensional multiplicative process, showing the squares $B(\delta_0)$, $B_i(\delta_1)$ and $B_j(\delta_2)$ of size δ_0, δ_1 and δ_2 respectively. Note that ϕ_0 is the total pore-space volume (equal to the porosity if $\delta_0 = 1$).

(a) (b)

Fig. 2 **(a)** A fairly homogeneous mass multifractal of porosity 30%; the generating function scales in the range $[1/2^4, 1]$. **(b)** A slightly clustered mass multifractal of porosity 30%; the generating function scales in the range $[1/2^4, 1]$.

are assigned the weights $\phi_0 w_i w_j$, $j = 1, \ldots, \lambda^D$. This process is continued n times down scale until a scale $\delta_n = \lambda^{-n}$ is reached. δ_n is the inner scale of the multiplicative process. The weights obtained at scale δ_n are interpreted to be the pore space volumes contained in the blocks $B_i(\delta_n)$. Note that the process is conservative, i.e., the total measure (equal to the pore-space volume) remains the same for all n.

For geometrical multifractals, the multiplicative process *cannot* be iterated *ad infinitum* because the measure of a block $B_i(\delta_n)$ may exceed its volume for n large enough. For multiplicative processes, the minimum value $\alpha_{\min}$ of α is related to the maximum value $w_{\max}$ of the multipliers by[11]:

$$\alpha_{\min} = -\log_\lambda(w_{\max}) \,. \tag{3.9}$$

Replacing this value of $\alpha_{\min}$ in the general constraint (2.8) with $\eta = \delta_n = \lambda^{-n}$ yields:

$$n \leq -\frac{\ln \phi_0}{\ln(\lambda^D w_{\max})} \,, \tag{3.10}$$

which is a restriction on the parameters $w_{\max}$, n and ϕ_0.

In the following 2-dimensional examples [Figs. 2(a) and 2(b)], we have plotted two sets having exactly the same porosity of 30%, the same scaling range for the generating function, but different multifractal spectra $f(\alpha)$ (Fig. 3). For both sets the pore-space in each grid block $B_i(\delta_n)$ was represented by a black square of area $p_i(\delta_n)$, located randomly within $B_i(\delta_n)$. The measures $p_i(\delta_n)$ were constructed with a multiplicative process with a scale ratio $\lambda = 2$, four iterations ($n = 4$) and with the multiplicative weights (w_1, w_2, w_3, w_4) equal to $(0.27, 0.2, 0.23, 0.3)$ and $(0.3, 0.1, 0.29, 0.31)$ for Figs. 2(a) and 2(b) respectively. These choices of parameters can be checked to satisfy the constraint (3.10). The generating functions of these measures scale exactly in the range $[1/2^4, 1]$,

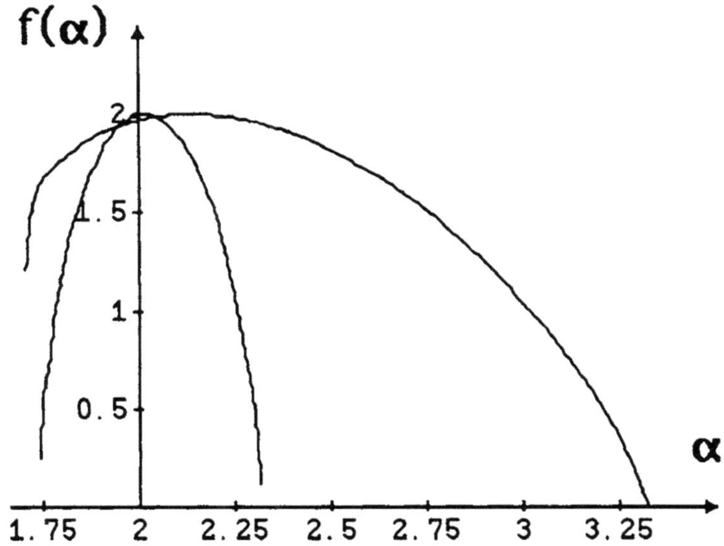

Fig. 3 Multifractal spectra of the mass multifractals of Figs. 2(a) and 2(b). The spectrum with a broad horizontal span corresponds to Fig. 2(b), i.e., the "clustered" set. The narrower spectrum corresponds to Fig. 2(a), i.e., to a relatively homogeneous set.

and the scaling is broken at smaller scales. In the scaling range, the mass exponents take the form[11]:

$$\tau(q) = -\log_2\{w_1^q + w_2^q + w_3^q + w_4^q\}. \tag{3.11}$$

The multifractal spectra $f(\alpha)$ of both measures (Fig. 3) were obtained with the usual parametric representation $\alpha = \tau'(q)$ and $f(\alpha) = q\tau'(q) - \tau(q)$.

The set of Fig. 2(a) exhibits a pore-space spread more homogeneously than the set of Fig. 2(b). This is partly because the difference between the minimum and maximum values of $p_i(\delta_n)$, equal to $\delta_n^{\alpha_{\max}}$ and $\delta_n^{\alpha_{\min}}$ respectively, is larger when $\alpha_{\max} - \alpha_{\min}$ is larger, i.e., when the multifractal spectrum is broader. It is also because the information dimension[e] is lower for the set of Fig. 2(b) than for the set of Fig. 2(a). Indeed, for multifractals most of the total measure — here the total pore-space volume — is concentrated on the blocks $B_i(\delta_n)$ such that $p_i(\delta_n) \approx \delta_n^{\alpha(1)}$, and $f(\alpha(1)) = \alpha(1)$ is the information dimension. The sets of Figs. 2(a) and 2(b) have $f(\alpha(1)) \approx 1.98$ and 1.89 respectively. Hence the pore-space volume of the set of Fig. 2(a) is concentrated on a nearly space-filling set, i.e., with $f(\alpha) \approx 2$, and therefore it has a relatively homogeneous appearance. By contrast, the pore-space volume of the set of Fig. 2(b) is concentrated on a set with lower dimension $f(\alpha)$, and therefore clustering is more visible (in this case, the set on which most of the measure concentrates is the quite visible subset of large black squares that resemble a Sierpinski carpet).

[e] For these multiplicative processes the information dimension is given by:

$$\alpha(1) = f(\alpha(1)) = -\frac{1}{\ln(2)} \sum_i w_i \ln(w_i).$$

REFERENCES

1. J. P. Hansen, J. L. McCauley, J. Muller and A. T. Skjeltorp, in *Random Fluctuations and Pattern Growth*, eds. H. E. Stanley and N. Ostrowsky (Kluwer Academic Publishers, 1988), p. 310.
2. J. Muller and J. L. McCauley, *Transport in Porous Media* **8**, 133 (1992).
3. J. Muller, *Physica* **A191**, 284 (1992).
4. S. Blacher, F. Brouers and G. Ananthakrishna, *Physica* **A185**, 28 (1992).
5. J. Muller, in *Wavelets, Fractals and Fourier Transforms: New Developments and New Applications*, eds. M. Farge, J. C. R. Hunt and J. C. Vassilicos (Oxford University Press, 1993), p. 397.
6. T. Vicsek. F. Family and P. Meakin, *Mat. Res. Soc. Extended Abstracts (EA-20)* (Materials Research Society, 1989), p. 253.
7. F. Argoul, A. Arneodo, G. Grasseau and H. L. Swinney, *Phys. Rev. Lett.* **61**, 2558 (1988).
8. R. Hilfer, *Phys. Rev.* **B45**, 7115 (1992).
9. K. Falconer, *Fractal Geometry — Mathematical Foundations and Applications* (J. Wiley & Sons, Chichester, New York, Toronto, 1990), p. 5.
10. T. C. Halsey, M. H. Jensen, L. P. Kadanoff, I. Procaccia and B. I. Shraiman, *Phys. Rev.* **A33**(2), 1141–1151 (1986).
11. C. J. G. Evertsz and B. B. Mandelbrot, in *Chaos and Fractals*, eds. H. O. Peitgen, H. Jürgen and D. Saupe (Springer Verlag, New York, 1992), pp. 849–969.
12. T. Tél and T. Vicsek, *J. Phys.* **A20**, L835 (1987).
13. T. Vicsek, *Fractal Growth Phenomena* (World Scientific, Singapore, 1989).
14. A. Aharony, *Physica* **A168**, 479 (1990).
15. J. L. McCauley, *Physica* **A187**, 18 (1992).
16. J. L. McCauley, *Technical Report IFE/KR/F92/17* (Institutt for energiteknikk, 1992).

Fractals, Vol. 1, No. 4 (1993) 904–907

CRITICAL EXPONENTS OF ELASTICITY IN A CONTINUUM PERCOLATION SYSTEM (INVERSE SWISS-CHEESE MODEL)

K. MARUYAMA, K. OKUMURA, H. YAMAUCHI and S. MIYAZIMA
Department of Engineering Physics, Chubu University,
Kasugai, Aichi 487, Japan

Abstract

Critical exponents of elastic constant of Inverse Swiss-cheese model (Continuum percolation problem) is experimentally investigated following our previous experimental studies on electric conductivity and permeability in a continuum percolation system.

1. INTRODUCTION

Differences in critical phenomenon between a conventional percolation and a continuum percolation model are revealed by several experiments concerned with mineralogy[1] and by theoretical investigation.[2] As for electric conductivity, the two-dimensional experiments for the conventional and continuum percolation models induced the same critical exponent 4/3.[2] It is in the three-dimensional system that we can see the difference in electrical conductivity exponents between the conventional and the continuum percolation systems.

The authors of this paper have performed a three-dimensional experiment of the continuum percolation in a laboratory. We measured the electric conductivity of the continuum percolation system and obtained the critical exponent 2.4 which is different from 1.9 for the three-dimensional convention percolation model.[3] Furthermore, we have discussed the elasticity of the continuum percolation model to reproduce the critical exponent 5.4, which is agreeable with Halperin et al.'s prediction of 4.1,[2] while that of the conventional model is 3.6. In this paper, we report on an experiment to investigate the elastic constant of a similar continuum percolation system, but based on an inverse Swiss-cheese model.

2. EXPERIMENTAL APPARATUS

In this paper we discuss the elastic constant in the continuum percolation system, that is, the inverse Swiss-cheese model. Our system consists of many rubber balls (16 mm diameter) in an acrylic tube with an inner diameter of 10 cm and a height of 50 cm, as shown in Fig. 1.

Here we pile up the rubber balls randomly in order to build up a string or random networks of balls by lifting up the balls. This construction is performed in a random network made of 0.2 mm copper wire which is almost completely plastic. Many rubber balls (the number is 250) are hung here and there in the random copper-wire network. At one side of the tube, a piston is set to press the rubber ball system, and at the other side a strain gauge on a 2 mm brass plate is used to measure the strain in the rubber ball system.

We measure the magnitude of force which travels through the rubber ball system in the acrylic tube. If an increase of the pressure Δp induces a decrease of the rubber system by length $\Delta \ell$, then it means that the system has an elastic constant $\Delta p/\Delta \ell$. Gradually pushing up the piston and measuring the pressure to push the rubber ball system, we obtain the decrease of the stress as shown in Fig. 2.

Fig. 1 The experimental apparatus prepared to measure the elastic constant of continuum percolation model.

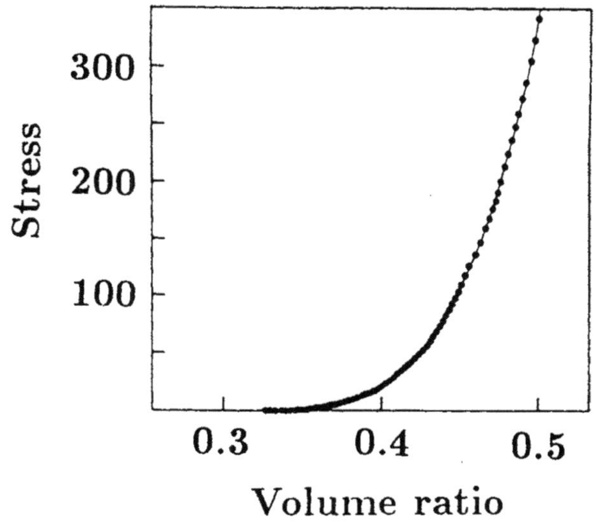

Fig. 2 The stress vs. the ratio of rubber volume to rubber volume + porous space.

3. CRITICAL EXPONENT

These data are plotted in a log-log plot diagram against $V - V_c$ in Fig. 3, where V is the volume of the rubber ball system including the free volume, and V_c is a critical volume and a parameter we change in order to obtain the straight portion in the curves in Fig. 3. The slope of the straight portion is the critical exponent, and we have obtained 5.4.

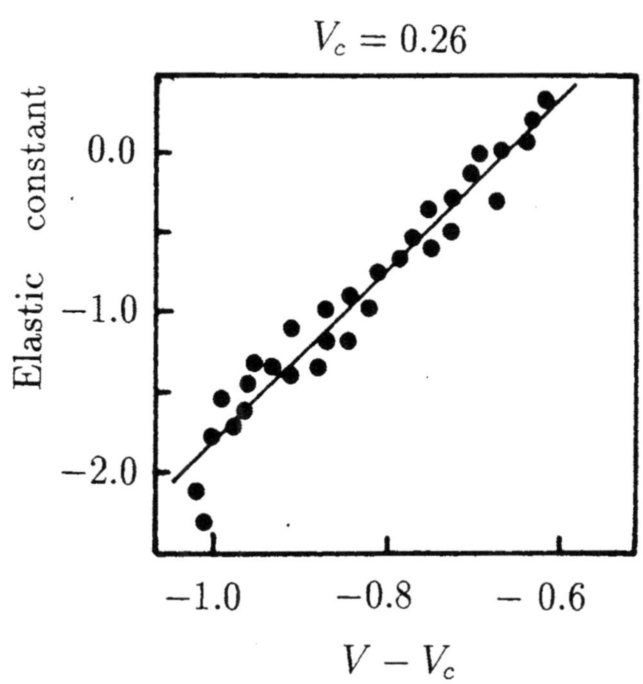

Fig. 3 Log-log plot of elastic constant vs. $(V - V_c)$, where the threshold volume V_c is determined so as to obtain a straight fitting line. The slope of the asymptotic line in this figure is 5.4.

4. RESULT AND DISCUSSION

The obtained critical exponent for the elastic constant is about 5.4 ± 0.6 which is close to the prediction, although the present percolation system is not exactly equivalent to the so-called inverse Swiss-cheese model. We put many rubber balls randomly into the acrylic tube. Even near the threshold value V_c, there are many skeleton structures which contribute to the elastic constant. When it approaches the threshold, i.e. by pushing up the piston, we obtained a phenomenon similar to the percolation criticality with the same critical exponent.

REFERENCES

1. P. Wong, J. Koplik and J. P. Tomanic, *Phys. Rev.* **B30**, 6606 (1984).
2. B. I. Halperin, S. Feng and P. N. Sen, *Phys. Rev. Lett.* **54**, 2391 (1985).
3. S. Miyazima, K. Maruyama and K. Okumura, *J. Phys. Soc. Jpn.* **60**, 2805 (1991).

Fractals, Vol. 1, No. 4 (1993) 908–916

ON THE DYNAMICAL EVOLUTION OF THE FRACTAL STRUCTURE OF INTERSTELLAR MEDIUM

Z. Y. YUE and B. ZHANG
I. Physikalisches Institut, Universität zu Köln, 50937 Köln, Germany
Department of Geophysics, Beijing University, Beijing, China
G. WINNEWISSER and J. STUTZKI
I. Physikalisches Institut, Universität zu Köln, 50937 Köln, Germany

Abstract

Two-dimensional compressible turbulence in a self-gravitating, magnetic interstellar medium is calculated as an initial value problem. It is shown that even if the initial density distribution is homogeneous and the initial velocity distribution contains only a few Fourier components, the nonlinear interaction among the Fourier components will generate more and more Fourier components and lead to a turbulent and fractal structure in the interstellar medium. The calculations are carried out for three different initial states. In order to see the time evolution, detailed density distributions and fractal dimensions of the density contours are calculated at three moments of time for each of the initial states. The results show that the fractal dimension remains almost the same (~1.4–1.5), although the detailed density distribution has changed considerably. The insensibility of the fractal dimension of density contours to both the initial conditions and the evolution time is in good agreement with observations of molecular clouds in the interstellar medium.

1. INTRODUCTION

Molecular clouds, as the main sites of star formation, have greatly attracted the attention of astrophysicists during the last two decades. All observational evidences indicate the fact that at all scales the structure of molecular clouds is not smooth, but clumpy and filamentary. (For detailed discussions, see Refs. 1–7.)

In our earlier paper,[8] we carried out two-dimensional hydrodynamic calculations for a self-gravitating, magnetic and compressible medium and concluded that the joint effect of the nonlinearity of the hydrodynamic equations and self-gravity is the mechanism for generating the observed complex structure of molecular clouds. In the present paper, we will study the dynamical evolution of the fractal structure of column density contours based on the same type of self-gravitating hydrodynamic calculations.

2. FORMULATION

In order to overcome the difficulty associated with the inclusion of the self-gravity, i.e., the Poisson equation, we formulate the problem in Fourier space by following the same procedure as in Ref. 8 and obtain the following equations governing the time evolution of the Fourier coefficients (An outline of the derivation of these equations will be given in Appendix A. For more details see Ref. 8):

$$\frac{d}{dt_*}\rho_{nm} = -\xi_{nm} \tag{1}$$

$$\frac{d}{dt_*}u_{nm} = -\eta_{nm} + n\left[1 - \left(\frac{\lambda_0}{\lambda_J}\right)^2 \frac{1}{n^2+m^2}\right]\rho_{nm} + \sqrt{2}\left(\frac{\lambda_0}{\lambda_J}\right)^2 v_{nm} \quad (n^2+m^2 \neq 0) \tag{2}$$

$$\frac{d}{dt_*}v_{nm} = -\zeta_{nm} + m\left[1 - \left(\frac{\lambda_0}{\lambda_J}\right)^2 \frac{1}{n^2+m^2}\right]\rho_{nm} - \sqrt{2}\left(\frac{\lambda_0}{\lambda_J}\right)^2 u_{nm} \quad (n^2+m^2 \neq 0) \tag{3}$$

where

$$\xi_{nm} = \frac{1}{2}\sum_{i,j} \rho_{ij}[n(u_{n+i,m+j} + u_{n-i,m-j}) + m(v_{n+i,m+j} + v_{n-i,m-j})] \tag{4}$$

$$\eta_{nm} = \sum_{i,j}[(n-i)u_{ij} + (m-j)v_{ij}]u_{n-i,m-j} \tag{5}$$

$$\zeta_{nm} = \sum_{i,j}[(n-i)u_{ij} + (m-j)v_{ij}]v_{n-i,m-j} \tag{6}$$

and $\sum_{i,j}$ is the summation over all the integer points in the (i, j) plane.

3. DYNAMICAL EVOLUTION OF THE DENSITY DISTRIBUTIONS AND THE FRACTAL DIMENSIONS OF THE DENSITY CONTOURS

We shall consider the following initial states of the Fourier coefficients which will be referred to as Case A, Case B and Case C respectively. In Cases A and B, the initial density is homogeneous, while in Case C, the initial density perturbation contains only one Fourier component. For each of the Cases A, B and C, the initial velocity contains only a few Fourier components. More specifically, these initial states are defined by their initial Fourier coefficients as follows:

Case A:

$$\rho_{00}(0) = 1\,, \qquad \rho_{nm}(0) = 0 \quad (\text{for } n^2 + m^2 \neq 0) \tag{7}$$

The initial Fourier coefficients of the velocity are defined by the following formulae:

$$u_{00}(0) = v_{00}(0) = 0 \tag{8}$$

$$u_{nm}(0) = v_{-m,n}(0) = \varepsilon_{nm}\frac{\lambda_0}{\lambda_J}\frac{(n-m)}{2(n^2+m^2)^{1/2}} \tag{9}$$

$$v_{nm}(0) = -u_{-m,n}(0) = \varepsilon_{nm}\frac{\lambda_0}{\lambda_J}\frac{(n-m)}{2(n^2+m^2)^{1/2}} \tag{10}$$

$$\frac{\lambda_0}{\lambda_J} = 1.2 \tag{11}$$

$$\begin{gathered}\varepsilon_{10} = -0.4\,, \quad \varepsilon_{20} = -0.2\,, \quad \varepsilon_{30} = -0.1\,, \quad \varepsilon_{11} = -0.3\,, \quad \varepsilon_{21} = -0.2\,, \\ \varepsilon_{nm} = 0 \quad (\text{for all other } (n,\, m))\end{gathered} \tag{12}$$

In Eqs. (9)–(11), λ_0 is a basic wavelength corresponding to the size of the cloud, and λ_J is the Jeans length based on the Alfven speed [see Eqs. (A6) and (A13) respectively].

Case B: It differs from Case A only by taking:

$$\varepsilon_{20} = 0.2\,, \quad \varepsilon_{21} = 0.2 \tag{13}$$

and

Case C: It differs from Case A only by taking:

$$\rho_{21}(0) = -0.05\,, \quad \frac{\lambda_0}{\lambda_J} = 1.0 \tag{14}$$

The density contour maps (with greyscales) at $t_* = 2.0$, $t_* = 3.0$, and $t_* = 4.0$ for each of the Cases A, B and C are shown in Figs. 1(a)–1(i), respectively, while the area-perimeter relations and the corresponding fractal dimensions of these density contour maps are shown in Figs. 2(a)–2(i) respectively. The fractal dimension is calculated from the perimeter-area relation[9]:

$$(\text{area})^{1/2} \propto (\text{perimeter})^{1/D} \tag{15}$$

where D is the fractal dimension.

4. CONCLUSIONS

By comparing Fig. 1 with Fig. 2, we see that although the detailed density distribution keeps changing with time (note that $\Delta t_* = 1$ corresponds to $\Delta t = 4.2 \times 10^5$ yr for Cases A and B, and to $\Delta t = 3.7 \times 10^5$ yr for Case C if the average number density of molecules $\bar{n}$ is taken to be 3×10^3 cm^{-3}), the fractal dimension of the density contour maps remains almost the same (~1.4–1.5) as long as the turbulence has been fully developed. (If the turbulence has not been fully developed, the definition of the fractal dimension D itself, in particular the derivation of D from Eq. (15), would not be meaningful.) The insensibility of the fractal dimension to both the initial conditions and the evolution time is in good agreement with observations. (For example, Scalo[3] found a fractal dimension $D = 1.4$ for the Taurus 100 μm column density image; Beech[10] obtained $D = 1.4$ for 24 Lynds clouds; Zimmermann[7] obtained $D = 1.45$ for ^{12}CO(1-0) map of NGC 1499, $D = 1.35$ for ^{12}CO(1-0) map of L1457, and $D = 1.51$ for ^{12}CO(2-1) map of L1457.)

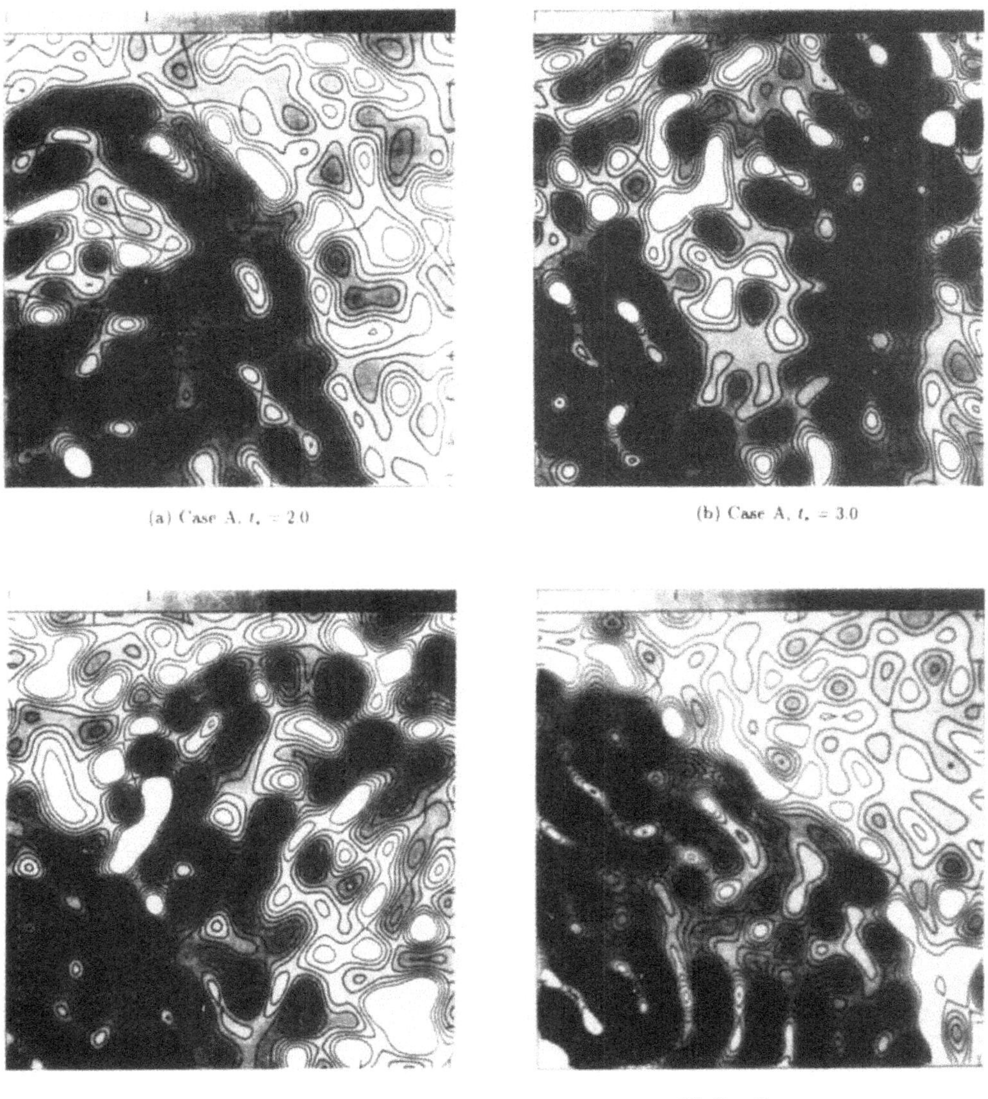

(a) Case A, $t_* = 2.0$ (b) Case A, $t_* = 3.0$

(c) Case A, $t_* = 4.0$ (d) Case B, $t_* = 2.0$

Figs. 1(a)–1(i) The column density contour maps for **(a)** Case A, $t_* = 2.0$, **(b)** Case A, $t_* = 3.0$, **(c)** Case A, $t_* = 4.0$, **(d)** Case B, $t_* = 2.0$.

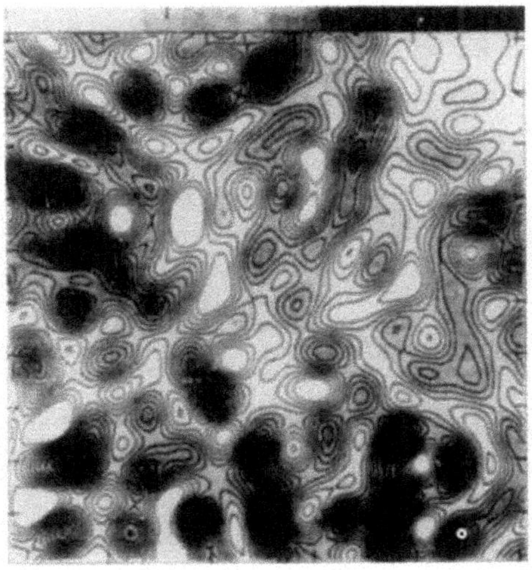

(e) Case B, $t_* = 3.0$

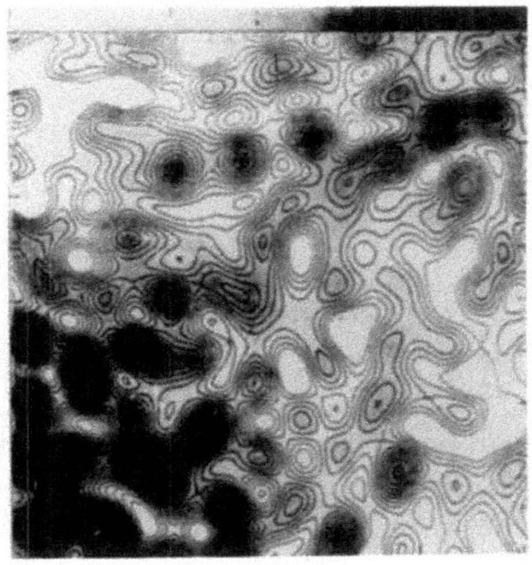

(f) Case B, $t_* = 4.0$

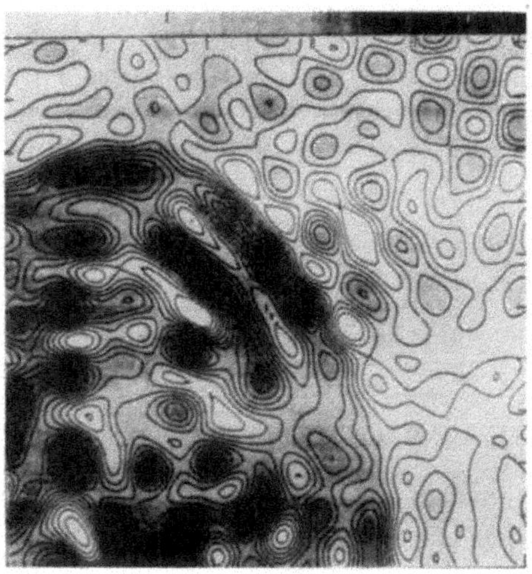

(g) Case C, $t_* = 2.0$

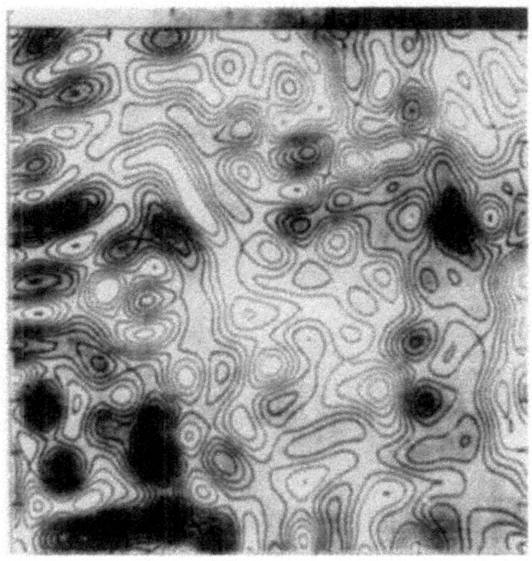

(h) Case C, $t_* = 3.0$

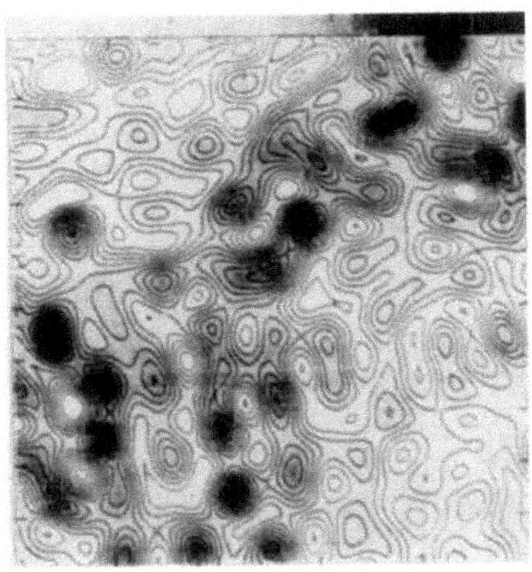

(i) Case C, $t_* = 4.0$

Fig. 1 (*Continued*)
(e) Case B, $t_* = 3.0$, (f) Case B, $t_* = 4.0$, (g) Case C, $t_* = 2.0$, (h) Case C, $t_* = 3.0$, (i) Case C, $t_* = 4.0$.

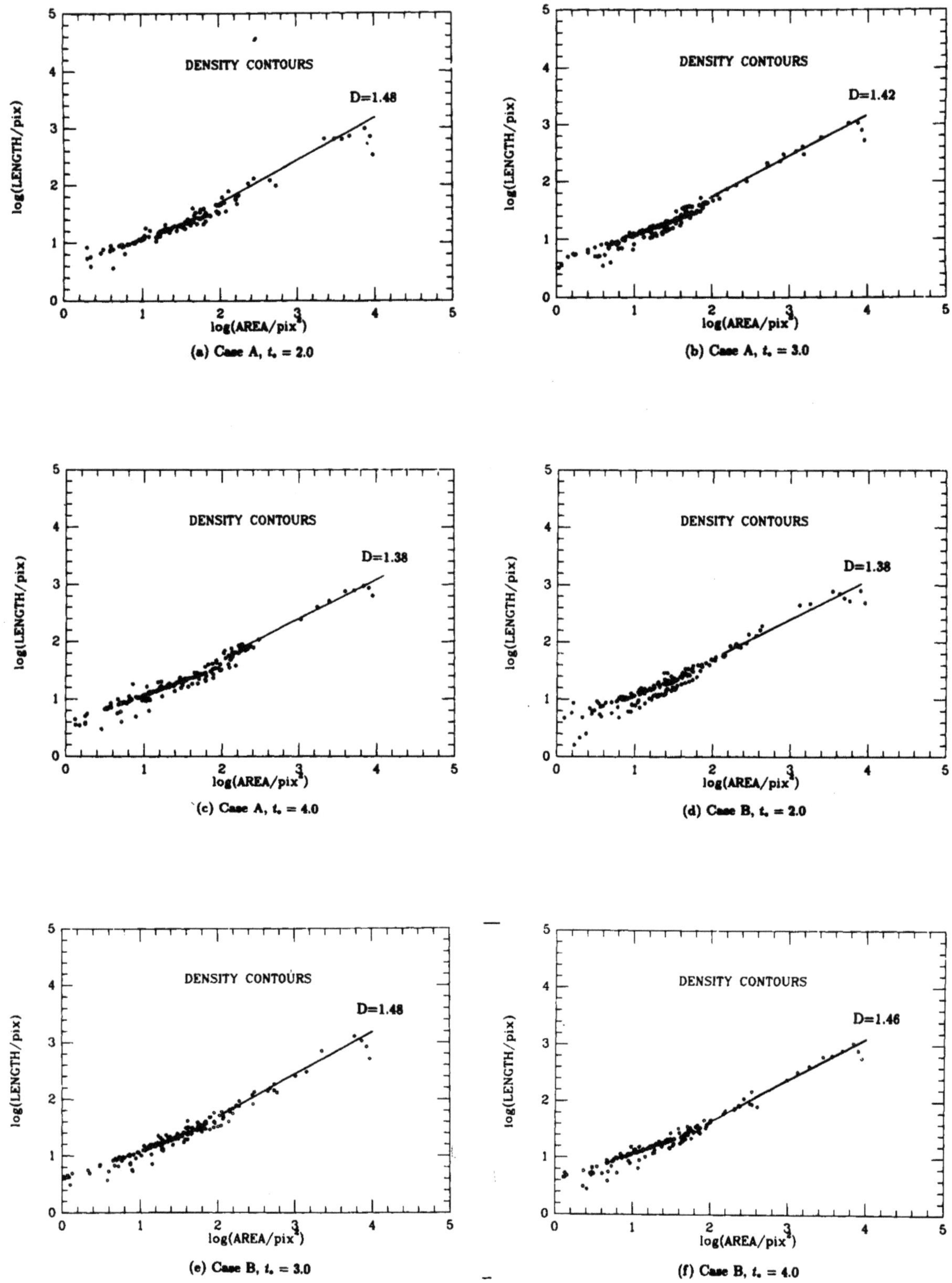

(a) Case A, $t_* = 2.0$ (b) Case A, $t_* = 3.0$ (c) Case A, $t_* = 4.0$ (d) Case B, $t_* = 2.0$ (e) Case B, $t_* = 3.0$ (f) Case B, $t_* = 4.0$

Figs. 2(a)–2(i) The area-perimeter relations and the fractal dimensions for **(a)** Case A, $t_* = 2.0$, **(b)** Case A, $t_* = 3.0$, **(c)** Case A, $t_* = 4.0$, **(d)** Case B, $t_* = 2.0$, **(e)** Case B, $t_* = 3.0$, **(f)** Case B, $t_* = 4.0$.

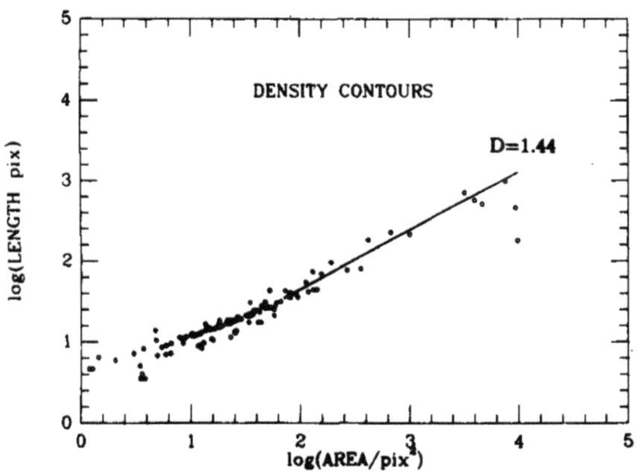

(g) Case C, $t_* = 2.0$

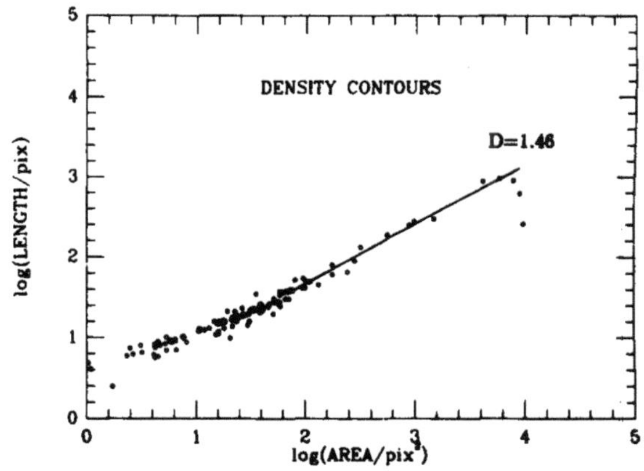

(h) Case C, $t_* = 3.0$

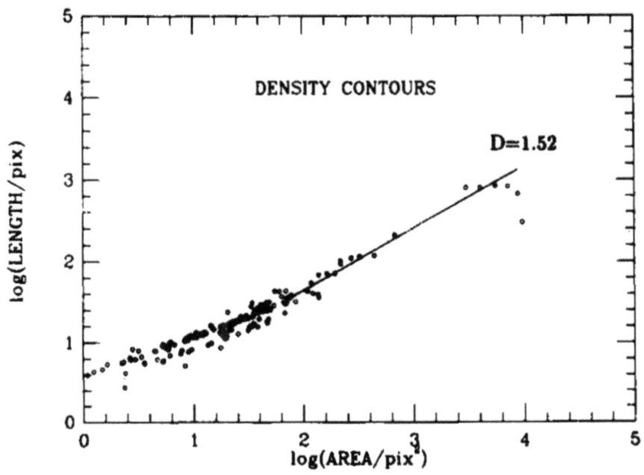

(i) Case C, $t_* = 4.0$

Fig. 2 (*Continued*)
(g) Case C, $t_* = 2.0$, (h) Case C, $t_* = 3.0$, (i) Case C, $t_* = 4.0$. Note that only the points above the "bending point," most of which represent filaments with turns and twists along their boundaries, make contribution to the fractal dimension. The few circles with largest areas in each case represent open contours and should also be excluded in the calculation of the fractal dimension.

ACKNOWLEDGMENTS

The authors are grateful to Mr. Th. Zimmermann for the use of his program for making the density contour maps and the fractal dimension analysis. Z. Y. Y. gratefully acknowledges the support he received from the Deutsche Forschungsgemeinschaft through research grant SFB 301.

REFERENCES

1. B. G. Elmegreen, in *Physics of Stra Formation and Early Stellar Evolution*, eds. N. Kylafis and C. J. Lada (Kluwer, Dordrecht, 1991), p. 35.
2. J. M. Scalo, in *Protostars and Planets*, eds. D. C. Black and M. S. Mathews (University of Arizona Press, 1985), p. 201.
3. J. M. Scalo, in *Physical Processes in Fragmentation and Star Formation*, eds. R. Capuzzo-Dolcetta, C. Chiosi and A. Di Fazio (Kluwer Academic Publisher, 1990), p. 151.

4. E. Falgarone and M. Pérault, in *Physical Processes in Interstellar Clouds*, eds. G. E. Morfill and M. Scholer (Reidél, Dordrecht, 1987), p. 59.
5. E. Falgarone and T. G. Phillips, in *Fragmentation of Molecular Clouds and Star Formation*, eds. E. Falgarone, F. Boulanger and G. Duvert (IAU Symp. No. 147, 1990), p. 119.
6. E. A. Lada, Ph.D. Thesis, The University of Texas at Austin, 1990.
7. Th. Zimmermann, in preparation (1992).
8. Z. Y. Yue, B. Zhang, G. Winnewisser and J. Stutzki, *Ann. Physik* **2**, 9 (1993).
9. B. B. Mandelbrot, *The Fractal Geometry of Nature* (W. H. Freeman and Company, 1983), p. 110.
10. M. Beech, *Ap. Sp. Sci.* **133**, 193 (1987).

APPENDIX A
AN OUTLINE OF THE DERIVATION OF EQS. (1)–(6)

As in Ref. 8, we consider the following equilibrium state for a compressible self-gravitating medium: The medium has a uniform density ρ_0 and is rotating about the z axis with a uniform angular velocity:

$$\Omega = (2\pi G\rho_0)^{1/2} \tag{A1}$$

where G is the universal gravitational constant, and a uniform magnetic field $\mathbf{B}_0$ along the z direction is assumed to exist. The two-dimensional and thus "horizontal" motions of the medium with respect to a frame of reference which corotates with the equilibrium state are governed by the following equations:

$$\frac{\partial\rho}{\partial t} + \frac{\partial(\rho u)}{\partial x} + \frac{\partial(\rho v)}{\partial y} = 0 \tag{A2}$$

$$\frac{\partial u}{\partial t} + u\frac{\partial u}{\partial x} + v\frac{\partial u}{\partial y} = -\frac{1}{\rho}\frac{\partial p_{\text{total}}}{\partial x} - \frac{\partial\psi}{\partial x} + 2\Omega v \tag{A3}$$

$$\frac{\partial v}{\partial t} + u\frac{\partial v}{\partial x} + v\frac{\partial v}{\partial y} = -\frac{1}{\rho}\frac{\partial p_{\text{total}}}{\partial y} - \frac{\partial\psi}{\partial y} - 2\Omega u \tag{A4}$$

$$\left(\frac{\partial^2}{\partial x^2} + \frac{\partial^2}{\partial y^2}\right)\psi = 4\pi G(\rho - \rho_0) \tag{A5}$$

where Eq. (A2) is the continuity equation, Eqs. (A3) and (A4) are the components of the equation of motion in the x and y directions respectively, and Eq. (A5) is the Poisson equation. ψ in Eqs. (A3)–(A5) is the gravitational potential only due to the density perturbation $\rho - \rho_0$. The total pressure p_{total} in these equations stands for the sum of the magnetic and the thermal pressures. As in Ref. 8, we make the following further simplifications:

(i) We assume that the initial conditions are periodic in both x and y with the same basic wavelength λ_0, or the same basic wave number:

$$k_0 = \frac{2\pi}{\lambda_0} \tag{A6}$$

(ii) We assume that the initial conditions have a four-fold symmetry [$\rho(x, y) = \rho(-y, x)$, $u(x, y) = v(-y, x)$, $v(x, y) = -u(-y, x)$]. Thus, the solutions satisfy these properties all the time;

(iii) We assume that the magnetic pressure is dominant and the thermal pressure is negligible. Thus, the characteristic speed is the Alfven speed:

$$v_{A0} = \frac{B_0}{(4\pi\rho_0)^{1/2}} \tag{A7}$$

If we define the following dimensionless quantities:

$$\rho_* = \frac{\rho}{\rho_0}, \qquad (x_*, y_*) = k_0(x, y), \qquad \psi_* = \frac{k_0^2}{4\pi G\rho_0}\psi,$$
$$(u_*, v_*) = \frac{(u, v)}{v_{A0}}, \qquad t_* = k_0 v_{A0} t \tag{A8}$$

and make the following Fourier expansions:

$$\rho_* = \sum_{n,m} \rho_{nm}(t_*) \cos(nx_* + my_*) \tag{A9}$$

$$\psi_* = \sum_{n,m} \psi_{nm}(t_*) \cos(nx_* + my_*) \tag{A10}$$

$$u_* = \sum_{n,m} u_{nm}(t_*) \sin(nx_* + my_*) \tag{A11}$$

$$v_* = \sum_{n,m} v_{nm}(t_*) \sin(nx_* + my_*) \tag{A12}$$

we obtain Eqs. (1)–(6) in the text immediately. The Jeans wavelength λ_J is now based on the Alfven speed v_{A0} and is defined by:

$$\lambda_J = \left(\frac{\pi}{G\rho_0}\right)^{1/2} v_{A0} \tag{A13}$$

The Poisson equation takes a simple form in the Fourier space:

$$\psi_{nm} = -\frac{\rho_{nm}}{n^2 + m^2} \qquad (\text{for} \quad n^2 + m^2 \neq 0) \tag{A14}$$

In the specific computation, we also have to truncate the summation of the Fourier components at a large wave number, i.e., we have to make the following approximation:

$$\sum_{n,m=-\infty}^{\infty} \approx \sum_{(n^2+m^2)^{1/2} \leq N} \tag{A15}$$

where N is a large number. A fourth-order Runge-Kutta method is used for the truncated Eqs. (1)–(3). N is taken to be 20, and the time step is taken to be 0.002.

Fractals, Vol. 1, No. 4 (1993) 917–921

DYNAMICAL STRUCTURE FACTOR OF HEISENBERG ANTIFERROMAGNETS WITH FRACTAL STRUCTURES

TAKAMICHI TERAO, KOUSUKE YAKUBO and TSUNEYOSHI NAKAYAMA
Department of Applied Physics, Hokkaido University, Sapporo 060, Japan

Abstract

The dynamical structure factor $S(q, \omega)$ of Heisenberg antiferromagnets with random fractal structures is investigated in terms of a dynamical scaling argument. It is found that the asymptotic behavior of $S(q, \omega)$ can be characterized by the dynamic exponent z_a and a new exponent y. We have performed numerical calculations of $S(q, \omega)$ in order to confirm its predicted scaling behaviors for $d = 2$ percolating antiferromagnets. The calculated values for the exponents y and z_a are consistent with our theoretical prediction.

1. INTRODUCTION

The spin-dynamics of Heisenberg antiferromagnets with fractal structures has attracted much attention in recent years because of the peculiar dynamical properties due to their specific geometrical features. For ferromagnetic or vibrational excitations on percolating random-fractal networks, Alexander and Orbach[1] conjectured that the spectral (or fracton) dimension $\tilde{d}$, the key dimension for the dynamics of fractal structures, is equal to 4/3 for any Euclidean dimension d. It has been confirmed by large-scale simulations that the value of $\tilde{d}$ definitely takes a value very close to 4/3 for any Euclidean dimension $d(\geq 2)$.[2] These excitations are called *fractons*, which play a crucial role with regard to the dynamical properties of elastic percolating networks. Spin-waves excited on Heisenberg antiferromagnets are governed by a different type of equations of motion from those for ferromagnetic or vibrational system. This can be mapped onto the master equation for diffusing particles. Hence, the universality class of antiferromagnetic fractons is different from that of ferromagnetic or vibrational fractons.[3]

In this article, we investigate the dynamical structure factor $S(\vec{q}, \omega)$ of percolating Heisenberg antiferromagnets in order to clarify their scaling properties. We predict, at first, the frequency and the wavenumber dependence of $S(q, \omega)$ for percolating Heisenberg antiferromagnets in terms of a dynamical scaling argument. Secondly, we confirm the scaling behavior of $S(q, \omega)$ by numerical calculations for $d = 2$ bond-percolating antiferromagnets. The calculated results are in accord with our theoretical predictions.

2. SCALING ARGUMENT

The Hamiltonian of a percolating Heisenberg antiferromagnet is given by:

$$\mathcal{H} = \sum_{\langle ij \rangle} J_{ij} \boldsymbol{S}_i \cdot \boldsymbol{S}_j \,, \tag{1}$$

where $\boldsymbol{S}_i$ denotes the spin at the site i, and J_{ij} is the exchange coupling constant between nearest neighbor spins. We choose J_{ij} as $J_{ij} = 1$ if a bond between sites i and j is present on the percolating network and $J_{ij} = 0$ otherwise. The linearized equations of motion for spin deviation $S_i^+ \equiv S_i^x + iS_i^y$ from the Néel state is expressed, in units of $S/\hbar = 1$ where S is the magnitude of single spin, by:

$$i\frac{\partial S_i^+}{\partial t} = \sum_j D_{ij} S_j^+ \,. \tag{2}$$

Here, D_{ij} is defined by:

$$D_{ij} = \sigma_i \Big(J_{ij} + \delta_{ij} \sum_k J_{ik} \Big) \,. \tag{3}$$

where σ_i is a variable taking $+1$ for the site i belonging to the up-spin sublattice, and -1 to the down-spin sublattice. Equation (3) shows that the matrix $[D_{ij}]$ is non-Hermitian and is different from the dynamical matrix for lattice vibrations or ferromagnetic spin-waves due to the term σ_i and the sign of the second term in the parenthesis.

The dynamical structure factor $S(q, \omega)$ averaged over all possible realizations of percolating networks is characterized by a function of $q(= |\vec{q}|)$ due to the spherical symmetry of percolating structures. We take the single-length-scale postulate (SLSP) for $S(q, \omega)$, as in the case of vibrational fractons.[4,5] The SLSP implies that there is only one-length-scale to characterize fractons. Consequently, $S(q, \omega)$ for antiferromagnetic fractons is given by the following scaling form:

$$S(q, \omega) = q^{-y} F[q\Lambda(\omega)] \,, \tag{4}$$

where $F(x)$ is a scaling function, and y is the exponent characterizing $S(q, \omega)$. The lower and upper bound of the new exponent y are found in the following. From Eq. (4), the equal-time correlation function $C(q)$ defined by the integration of $S(q, \omega)$ over ω is given by $C(q) \sim q^{z_a - y}$ taking into account the form of the argument $q\Lambda(\omega) \to \omega q^{-z_a}$, where the dispersion relation $\omega \sim [\Lambda(\omega)]^{-z_a}$ is used. The value of $z_a - y$ should be negative from a static scaling argument with respect to the equal-time correlation function.[6] This leads to the lower bound of the exponent y as $y > z_a$. The spatial correlation function $C(r)$ obtained from the Fourier transform of $C(q)$ is proportional to r^{y-z_a-d}. Since the spatial correlation function $C(r)$ should decrease with increasing distance r, one has $y < z_a + d$, namely the upper bound of y.

It is reasonable to assume that $S(q, \omega)$ takes a Lorentzian form with respect to frequency ω. An analytic expression of $S(q, \omega)$ of percolation antiferromagnets, which satisfies the dynamical scaling hypothesis, was obtained for $q\xi \ll 1$ in terms of Green's function technique by Christou and Stinchcombe.[7] They have suggested that $S(q, \omega)$ takes a Lorentzian form with respect to frequency ω. From matching the long-wavelength modes ($q\xi \ll 1$) with the fractons ($q\xi \gg 1$), the frequency dependence of $S(q, \omega)$ takes a Lorentzian form even in the fracton regime $q\xi \gg 1$. Thus, $S(q, \omega)$ should be written as:

$$S(q, \omega) = G(q)\frac{\Gamma(q)}{[\omega - \omega_c(q)]^2 + \Gamma^2(q)}, \tag{5}$$

where $\Gamma(q)$ and $G(q)$ denote the width of lineshape and q-dependent function, respectively. The symbol $\omega_c(q)$ represents the peak position of $S(q, \omega)$ in frequency space. The SLSP requires that both $\omega_c(q)$ and $\Gamma(q)$ obey the same power law as $\omega_c(q) \propto q^{z_a}$ and $\Gamma(q) \propto q^{z_a}$, respectively, where z_a is the dynamic exponent for percolating antiferromagnets. This exponent is related to the fracton dimension $\tilde{d}_a$ as $z_a = D_f/\tilde{d}_a$ (D_f is the fractal dimension). Comparing Eq. (5) with Eq. (4), one obtains $G(q) \propto q^{z_a-y}$. The dynamic structure factor $S(q, \omega)$ takes the power law as $\omega^{-2}q^{-y+2z_a}$ for $q\Lambda(\omega) \ll 1$, and q^{-y} for $q\Lambda(\omega) \gg 1$.

3. NUMERICAL RESULTS

We have performed numerical calculations to confirm the above predictions, and determined the values of y and z_a. The dynamical structure factor $S(\vec{q}, \omega)$ is related to the spatial Fourier transform of two-point susceptibility $\chi_{ij}(\omega)$. The two-point susceptibility is defined by $\chi_{ij}(\omega) \equiv S_i^+(\omega)/h_j^+(\omega)$, where $S_i^+(\omega)$ and $h_j^+(\omega)$ are the temporal-Fourier transform of $S_i^+(t)$ and the transverse field $h_j^+(t)$, respectively. The matrix $[N_{ij}(\omega)]$ satisfies the relation $\sum_j N_{ij}(\omega)S_j^+(\omega) = -h_i^+(\omega)$, where $N_{ij}(\omega) \equiv \sigma_i(\omega\delta_{ij} - D_{ij})$.[8] Thus, $\chi_{ij}(\omega)$ can be obtained from the inverse of the matrix $N_{ij}(\omega)$ by $\chi_{ij}(\omega) = -[N^{-1}(\omega)]_{ij}$ and, consequently, $S(\vec{q}, \omega)$ is written as:

$$S(\vec{q}, \omega) = \lim_{\delta \to +0} \mathrm{Im}[\chi(\vec{q}, \omega + i\delta)], \tag{6}$$

where $\chi(\vec{q}, \omega)$ is the spatial Fourier transform of $\chi_{ij}(\omega)$. We have numerically diagonalized the matrix $[D_{ij}]$ to obtain eigenvectors, and calculated Eq. (6) from the eigenvectors directly.

We have prepared 54 samples of $d = 2$ bond-percolating networks formed on 62×62 square lattices at the percolation threshold ($p_c = 0.50$) under the periodic boundary conditions. To calculate the exponent y, we have computed the value of z_a from the frequency dependence of the localization length $\Lambda(\omega) \propto \omega^{-1/z_a}$, using the same networks as for $S(q, \omega)$. We have obtained $z_a = 1.83\pm0.08$, which indicates $\tilde{d}_{AF} = 1.03\pm0.04$ from the relation $\tilde{d}_{AF} = D_f/z_a$, where $D_f = 91/48$ for $d = 2$ is used. The value of $\tilde{d}_{AF}$ agrees well with our recent work.[3] The value of the exponent y in Eq. (4) is obtained by least squares fitting from the q-dependence of $S(q, \omega)$ with fixed values of $q\Lambda(\omega)$, and we have found $y = 3.0 \pm 0.3$. This value is consistent with the theoretical bound for the exponent y.

Figure 1 shows calculated values for the scaling function $F[q\Lambda(\omega)] = q^y S(q, \omega)$ as a function of $q\Lambda(\omega)$ using the above value of y. Filled circles represent the average values of

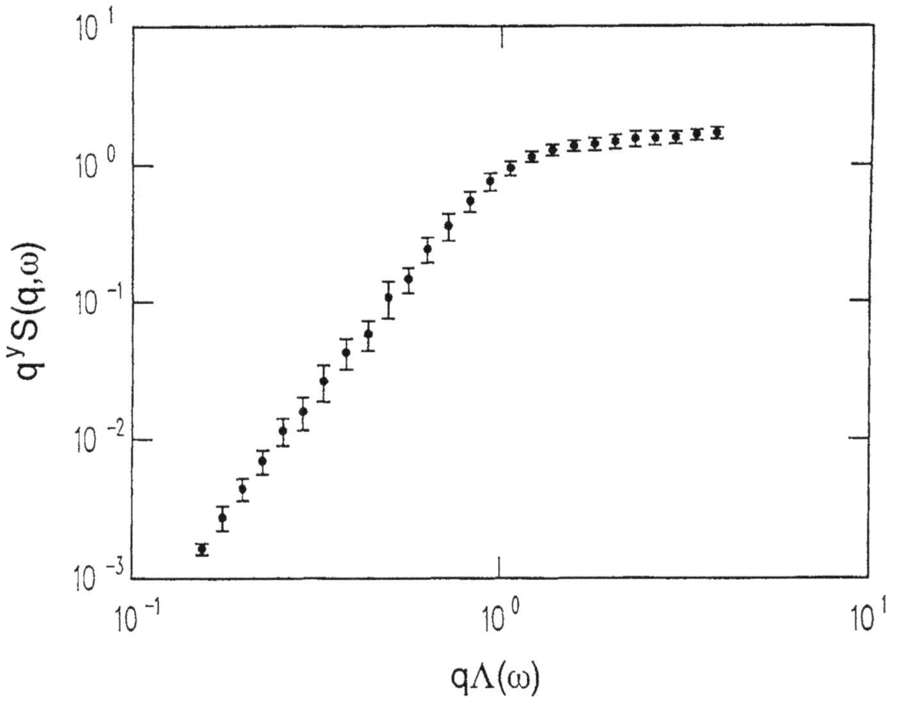

Fig. 1 The scaling function $F[q\Lambda(\omega)] \equiv q^{y}S(q, \omega)$ as a function of $q\Lambda(\omega)$. Filled circles and vertical bars indicate averaged values and statistical errors of $F[q\Lambda(\omega)]$, respectively.

$F[q\Lambda(\omega)]$ within a narrow range of the scaling variable $q\Lambda(\omega)$. The vertical bars indicate the statistical errors at each point. From the universal curve of $F[q\Lambda(\omega)]$, one sees that antiferromagnetic fractons surely satisfy the SLSP. The results for the scaling function $F(q\Lambda)$ in Fig. 1 indicate that the frequency and the wavenumber dependence of $S(q, \omega)$ behave asymptotically as $S(q, \omega) \propto \omega^{-1.9\pm0.1}q^{0.5\pm0.1}$ for $q\Lambda(\omega) \ll 1$, and $S(q, \omega) \propto q^{-2.8\pm0.1}$ for $q\Lambda(\omega) \gg 1$. If $z_a = 1.83 \pm 0.08$, obtained from the calculated frequency dependence of the localization length $\Lambda(\omega)$, and $y = 3.0 \pm 0.3$ are taken into account, one sees that the above asymptotic relations agree well with our theoretical predictions.

4. CONCLUSIONS

The dynamical structure factor $S(q, \omega)$ of percolating Heisenberg antiferromagnets at p_c has been investigated in terms of scaling arguments and direct numerical calculations. The asymptotic forms of $S(q, \omega)$ are discussed by a dynamic scaling argument. We have confirmed by direct numerical calculations the scaling behavior of $S(q, \omega)$ on $d = 2$ bond-percolating antiferromagnets. The value of the exponent y has also been determined from the numerical calculations, and is consistent with our theoretical prediction.

ACKNOWLEDGMENTS

This work was supported in part by a Grand-in-Aid for Scientific Research on the Special Projects, "Computational Physics as a New Frontier in Condensed Matter Research" from the Japan Ministry of Education, Science, and Culture. The Hokkaido University

Computing Center is acknowledged for the use of supercomputer facilities. One of the authors (T. T.) thanks JSPS Fellowships for Japanese Junior Scientists.

REFERENCES

1. S. Alexander and R. Orbach, *J. Phys. (Paris)* **43**, L625 (1982).
2. T. Nakayama, *Physica* **A191**, 386 (1992).
3. See an article by K. Yakubo, T. Terao and T. Nakayama, in this volume.
4. T. Nakayama and K. Yakubo, *J. Phys. Soc. Jpn.* **61**, 2601 (1992).
5. E. Stoll, M. Kolb and E. Courtens, *Phys. Rev. Lett.* **68**, 2472 (1992).
6. W. Gebhardt and U. Krey, *Phasenübergänge und kritische Phänomene* (Friedr. Vieweg & Sohn, Wiesbaden, 1980).
7. A. Christou and R. B. Stinchcombe, *J. Phys.* **C19**, 5917 (1986).
8. S. Kirkpatrick and A. B. Harris, *Phys. Rev.* **B12**, 4980 (1975).

Fractals, Vol. 1, No. 4 (1993) 922–929

EFFECTS OF FREE-END BOUNDARY CONDITION ON THE FREQUENCY SPECTRUM OF A VICSEK FRACTAL

SHI-YU WU and C. S. JAYANTHI
Department of Physics, University of Louisville, Louisville, KY 40292, USA

Abstract

In this paper, we investigate the effects of free-end boundary condition on the transverse vibrations of a Vicsek fractal. The unique feature of the eigenvalue spectrum corresponding to this case is that both non-degenerate and degenerate modes of a given generation are present in all subsequent generations, whereas in the case of fixed-end boundary condition only the degenerate modes persist. In the first case, the nature of the persistent degenerate modes is found to be either edge-localized or localized in the region between two adjacent clusters or adjacent sub-clusters, whereas in the second case the persistent modes are found to be edge-confined superlocalized modes. In both cases one finds that: (i) the frequency spectrum consists of atomic-like levels superimposed on a point-dense, cantor-set-like background; and (ii) there is side-by-side coexistence of the extended non-degenerate modes and superlocalized degenerate persistent modes.

1. INTRODUCTION

The possibility of obtaining an exact solution to the eigenvalue spectrum of standard fractals has prompted some studies on deterministic fractals.[1,2] It turns out that these fractals possess some interesting and exotic features. Recently, in a series of paper,[3–5] we studied the transverse vibrations of a general Vicsek fractal with nearest-neighbor interactions and fixed-end boundary condition. Our results indicate that the Vicsek fractal indeed has some unusual properties. These include: (i) the existence of degenerate persistent modes leading to a spectrum of atomic-like levels superimposed on point-dense, cantor-set-like background spectrum; and (ii) the side-by-side coexistence of the extended non-degenerate modes and superlocalized degenerate persistent modes. The Vicsek fractal has a tree-like structure. Any branch of the fractal can be severed from its main body by removing one connecting bond. The question is then, What is the effect of the boundary condition on the unusual outcomes of our investigation? In this paper, we report our study on the dynamics of the same Vicsek fractal but with free-end boundary condition. We found that there are substantive changes in both the frequency spectrum and the nature of the eigenmodes.

2. THE SYSTEM

One unique feature of the Vicsek fractal is that the dynamical matrix of the nth stage Vicsek can be constructed from that of the $(n-1)$th-stage following a simple pattern. In the case of the free-end boundary condition, the dynamical matrix of the transverse vibrations of the nth-stage can be written as:

$$\mathbf{H}_n = \begin{pmatrix} \mathbf{H}^c_{n-1} & \mathbf{V} & \mathbf{V} & \mathbf{V} & \mathbf{V} \\ \mathbf{V}^T & \mathbf{H}^s_{n-1} & \mathbf{0} & \mathbf{0} & \mathbf{0} \\ \mathbf{V}^T & 0 & \mathbf{H}^s_{n-1} & \mathbf{0} & \mathbf{0} \\ \mathbf{V}^T & \mathbf{0} & \mathbf{0} & \mathbf{H}^s_{n-1} & \mathbf{0} \\ \mathbf{V}^T & \mathbf{0} & \mathbf{0} & \mathbf{0} & \mathbf{H}^s_{n-1} \end{pmatrix} \tag{1}$$

where $\mathbf{V}$'s are the matrices describing the interactions between the central $(n-1)$th-stage component cluster and the four equivalent $(n-1)$th-stage side clusters in the nth-stage fractal. $\mathbf{H}^c_{n-1}$, the matrix of the central cluster, and $\mathbf{H}^s_{n-1}$, the matrix of one of the side clusters, differ from $\mathbf{H}_{n-1}$, the dynamical matrix of an isolated $(n-1)$th stage fractal, only slightly. The differences arise as a result of the change in the bonding structure of the sites connecting the component clusters to form the nth-stage fractal. In general, one can write:

$$\mathbf{H}^c_{n-1} = \mathbf{H}_{n-1} + \mathbf{H}'_{n-1} \quad \text{and} \quad \mathbf{H}^s_{n-1} = \mathbf{H}_{n-1} + \mathbf{H}''_{n-1} \tag{2}$$

where $\mathbf{H}'_{n-1}$ is a matrix of the same dimension as $\mathbf{H}_{n-1}$ but with only four non-vanishing elements along the diagonal which are associated with the four sites in the central cluster connected directly to the side clusters, and $\mathbf{H}''_{n-1}$ is a matrix with only one non-vanishing element along the diagonal for the site in the cluster which connects directly to the central cluster. The magnitudes of all these elements are equal to 1 in the reduced unit of $\gamma/m = 1$, with γ being the force constant and m the mass of the particle. In this unit, the magnitude of the only non-vanishing element of $\mathbf{V}$, corresponding to the interaction between the two adjacent bordering sites, one in the central cluster and the other one in the side cluster is -1.

3. THE EFFECTS OF THE FREE-END BOUNDARY CONDITION

Let ω_i be an eigenfrequency of $\mathbf{H}_n$ with the eigenvector $\mathbf{u}_n$ given by

$$\mathbf{u}^T_n = (\mathbf{u}^T_{n-1}(0),\, \mathbf{u}^T_{n-1}(1),\, \mathbf{u}^T_{n-1}(2),\, \mathbf{u}^T_{n-1}(3),\, \mathbf{u}^T_{n-1}(4)) \tag{3}$$

where $\mathbf{u}_{n-1}(k)$ is the displacement field of the kth component cluster ($k = 0, 1, 2, 3, 4$), with 0 denoting the central $(n-1)$th stage cluster. Using Eqs. (1)–(3), the eigenvalue equation: $(\mathbf{H}_n - \omega_i^2\mathbf{I}_n)\mathbf{u}_n = 0$ can be written as:

$$(\mathbf{H}_{n-1} - \omega_i^2\mathbf{I}_{n-1})\mathbf{u}_{n-1}(0) + \mathbf{H}'_{n-1}\mathbf{u}_{n-1}(0) + \sum_{k=1}^{4}\mathbf{V}\mathbf{u}_{n-1}(k) = 0 \tag{4}$$

$$\mathbf{V}^T\mathbf{u}_{n-1}(0) + \mathbf{H}''_{n-1}\mathbf{u}_{n-1}(k) + (\mathbf{H}_{n-1} - \omega_i^2\mathbf{I}_{n-1})\mathbf{u}_{n-1}(k) = 0 \quad (k = 1, 2, 3, 4) \tag{5}$$

Because of the structures of the matrices $\mathbf{H}'_{n-1}$, $\mathbf{H}''_{n-1}$ and $\mathbf{V}$'s, the column vector: $\mathbf{H}'_{n-1}\mathbf{u}_{n-1}(0) + \sum_{k=1}^{4}\mathbf{V}\mathbf{u}_{n-1}(k) = 0$ has only four non-vanishing elements $\mathbf{u}_{n-1}(0, k) - \mathbf{u}_{n-1}(k, 1)$ ($k = 1, 2, 3, 4$) while the column vector $\mathbf{V}^T\mathbf{u}_{n-1}(0) + \mathbf{H}''_{n-1}\mathbf{u}_{n-1}(k)$ has only one non-vanishing element $-\mathbf{u}_{n-1}(0, k) + \mathbf{u}_{n-1}(k, 1)$. Here $\mathbf{u}_{n-1}(0, k)$, ($k = 1, 2, 3, 4$), is the displacement of the site in the central cluster adjacent to the kth side-cluster, and $\mathbf{u}_{n-1}(k, 1)$ the displacement of the site in the kth side-cluster adjacent to the central cluster. From Eqs. (4) and (5), it can be seen that, if:

$$\mathbf{u}_{n-1}(0, k) = \mathbf{u}_{n-1}(k, 1) \quad (k = 1, \ldots, 4) \tag{6}$$

then

$$(\mathbf{H}_{n-1} - \omega_i^2\mathbf{I}_{n-1})\mathbf{u}_{n-1}(k) = 0 \quad (k = 0, 1, \ldots, 4). \tag{6a}$$

This then suggests that an eigenfrequency ω_i of H_{n-1}, be it a degenerate mode (as in the case of fixed-end boundary condition) or a non-degenerate mode, will persist and be an eigenfrequency of $\mathbf{H}_n$ with its eigenvector $\mathbf{u}_n$ given by Eq. (3). The components of $\mathbf{u}_n$, $\mathbf{u}_{n-1}(k)$ ($k = 0, \ldots, 4$), are eigenvectors of $\mathbf{H}_{n-1}$ corresponding to the same eigenfrequency ω_i, subject to the condition given by Eq. (6).

If ω_i is a non-degenerate mode of H_{n-1}, Eq. (6) requires that:

$$\mathbf{u}_{n-1}(0) = \mathbf{u}_{n-1}(1) = \mathbf{u}_{n-1}(2) = \mathbf{u}_{n-1}(3) = \mathbf{u}_{n-1}(4) \tag{7}$$

because they are the eigenvectors corresponding to a certain non-degenerate mode of identical eigenvalue equations. Hence, when this eigenfrequency persists and becomes a non-degenerate mode of H_n, the eigenvector $\mathbf{u}_n$ is a perfect patch-up of five identical eigenvectors $\mathbf{u}_{n-1}(k)$ in the five component clusters. The mode in question is therefore definitely an extended mode. The persistence of the non-degenerate mode from one generation to the next is a new feature that is not found in the case of the Vicsek fractal with the fixed-end boundary condition.

If ω_i is a degenerate mode of H_{n-1}, Eqs. (3)–(5) indicate that it will persist and be a degenerate mode of H_n. The pattern of evolution, however, will depend on the condition [Eq. (6)] which is not present in the case of the fixed-end boundary condition. In Sec. 4, we shall examine the effect of Eq. (6) on the evolution of the degeneracy for the persistent degenerate mode.

4. EVOLUTION OF THE DEGENERATE MODES

Following Ref. 4, we partition the matrix $\mathbf{H}_n$ given in Eq. (1) as:

$$\mathbf{H}_n = \begin{pmatrix} h_n^c & \tilde{\mathbf{V}} & \tilde{\mathbf{V}} & \tilde{\mathbf{V}} & \tilde{\mathbf{V}} \\ \tilde{\mathbf{V}}^T & \tilde{\mathbf{H}}_n & \mathbf{0} & \mathbf{0} & \mathbf{0} \\ \tilde{\mathbf{V}}^T & \mathbf{0} & \tilde{\mathbf{H}}_n & \mathbf{0} & \mathbf{0} \\ \tilde{\mathbf{V}}^T & \mathbf{0} & \mathbf{0} & \tilde{\mathbf{H}}_n & \mathbf{0} \\ \tilde{\mathbf{V}}^T & \mathbf{0} & \mathbf{0} & \mathbf{0} & \tilde{\mathbf{H}}_n \end{pmatrix} \tag{8}$$

Here $h_n^c = 4$ is a scalar quantity describing the particle at the central site of the nth-stage fractal, $\tilde{\mathbf{H}}_n$ the matrix describing one of the four equivalent branches surrounding the central site, and $\tilde{\mathbf{V}}$ the matrix describing the interaction of the central site with one of the branches. Let the eigenvector $\mathbf{u}_n$ of $\mathbf{H}_n$ corresponding to the eigenfrequency ω_i be represented by:

$$\mathbf{u}_n^T = (u_n^c, \tilde{\mathbf{u}}_n^T(a), \tilde{\mathbf{u}}_n^T(b), \tilde{\mathbf{u}}_n^T(c), \tilde{\mathbf{u}}_n^T(d)) \tag{9}$$

where u_n^c is the displacement at the central site, and $\tilde{\mathbf{u}}_n(\alpha)$ the displacement field of the α branch ($\alpha = a, b, c, d$). According to Eq. (8), the eigenvalue equation of $\mathbf{H}_n$ can then be expressed as:

$$(h_n^c - \omega_i^2)u_n^c + \sum_\alpha \tilde{V}\tilde{u}_n(\alpha) = 0 \tag{10}$$

$$(\tilde{\mathbf{H}}_n - \omega_i^2\tilde{\mathbf{I}}_n)\tilde{\mathbf{u}}_n(\alpha) + \tilde{\mathbf{V}}^T u_n^c = 0\,, \quad \alpha = a,\, b,\, c,\, d\,. \tag{11}$$

4.1 The Degenerate Mode of Lowest Degree of Degeneracy

If $u_n^c = 0$, Eq. (11) becomes:

$$(\tilde{\mathbf{H}}_n - \omega_i^2\tilde{\mathbf{I}}_n)\tilde{\mathbf{u}}_n(\alpha) = \mathbf{0}\,, \quad \alpha = a,\, b,\, c,\, d \tag{12}$$

and Eq. (10) reduces to:

$$\tilde{u}_n(a,\, 1) + \tilde{u}_n(b,\, 1) + \tilde{u}_n(c,\, 1) + \tilde{u}_n(d,\, 1) = 0 \tag{13}$$

where $\tilde{u}_n(\alpha, 1)$ is the displacement at the site in the branch α, which is adjacent to the central site. From Eqs. (9) and (12), a non-trivial solution of $\mathbf{u}_n$ requires that ω_i must also be an eigenfrequency of $\tilde{\mathbf{H}}_n$. Furthermore, since $\tilde{\mathbf{u}}_n(\alpha)$ for $\alpha = a, b, c, d$ individually satisfies the same eigenvalue equation [Eq. (12)] with only one condition [Eq. (13)] among them, there are many possible ways to construct the eigenvector $\mathbf{u}_n$ of $\mathbf{H}_n$ corresponding to ω_i using Eq. (9). Hence ω_i must be a degenerate mode of $\mathbf{H}_n$ if $u_n^c = 0$. At the same time, ω_i must also be an eigenfrequency of $\tilde{\mathbf{H}}_n$.

The simplest degenerate mode of $\mathbf{H}_n$ (i.e., the one with the lowest degree of degeneracy) occurs when ω_i is a non-degenerate mode of $\tilde{\mathbf{H}}_n$. In this case, the four displacement fields $\tilde{\mathbf{u}}_n(\alpha)$ must be proportional to each other because they are eigenvectors corresponding to a non-degenerate eigenvalue. When they are combined to form the eigenvector $\mathbf{u}_n$ of $\mathbf{H}_n$ according to Eq. (9), there can be only three independent eigenvectors $\mathbf{u}_n$ for the eigenfrequency ω_i of $\mathbf{H}_n$ because $\tilde{\mathbf{u}}_n(\alpha)$'s are related by Eq. (13). Hence ω_i is a three-fold degenerate mode of $\mathbf{H}_n$ when it is a non-degenerate mode of $\tilde{\mathbf{H}}_n$.

4.2 The Evolution of the Degree of Degeneracy of Persistent Modes

Let ω_i be a degenerate mode of $\mathbf{H}_{n-1}$. When five copies of the $(n-1)$th-stage clusters are assembled to form the nth-stage fractal described by $\mathbf{H}_n$, the discussion leading to [Eqs. (6a) and (6)] indicates that ω_i will be an eigenfrequency of $\mathbf{H}_n$ subject to the four conditions imposed by Eq. (6) on the component fields of the five clusters $\mathbf{u}_{n-1}(k)$, $(k = 0, 4)$. Each of these five displacement fields individually is an eigenvector of $\mathbf{H}_{n-1}$ corresponding to the same eigenfrequency ω_i. Hence if D_{n-1} is the degree of degeneracy of a degenerate mode ω_i of $\mathbf{H}_{n-1}$, then this mode will persist and become a mode with a degree of degeneracy D_n of H_n where:

$$D_n = 5D_{n-1} - 4 \tag{14}$$

Since the lowest degree of degeneracy of a degenerate mode in any stage is 3, the path of evolution of the degree of degeneracy of a persistent mode will follow the pattern: 3, 11, 51, 251, ...

5. THE FREQUENCY SPECTRUM

We can now predict the nature of the frequency spectrum of an nth-stage Vicsek fractal with a free-end boundary condition. The frequency spectrum is composed of both non-degenerate and degenerate components. Since all the non-degenerate modes of the previous stage will persist into the next stage, the non-degenerate component of the spectrum must consist of both persistent non-degenerate modes from previous stages and "new" non-degenerate modes. The degenerate component will consist of "new" 3-fold degenerate modes and persistent modes from previous stages. The persistent modes will include D_2-fold degenerate modes, D_3-fold degenerate modes, ..., and D_n-fold degenerate modes. Specifically, the D_n-fold degenerate mode is the descendant of the 3-fold degenerate mode of the first stage, a D_{n-1}-fold degenerate mode is the descendant of a 3-fold degenerate mode of the second stage and so on.

6. THE NATURE OF THE EIGENVECTORS OF THE PERSISTENT MODES

Let ω_i be a degenerate mode of $\mathbf{H}_{n-1}$ so that it will become a persistent mode of $\mathbf{H}_n$. Then we must have $u_n^c = 0$, $u_{n-1}^c = 0$ and ω_i is also an eigenfrequency for both $\tilde{\mathbf{H}}_{n-1}$ and $\tilde{\mathbf{H}}_n$. To determine the nature of the persistent mode in the nth-stage, we must understand the effect of Eq. (6) at the boundary between the central cluster and the side cluster. For that purpose, we examine the eigenvalue equation of $\tilde{\mathbf{H}}_n$ corresponding to ω_i.

Since the nth-stage fractal is constructed by assembling five $(n-1)$th-stage clusters, any one of the branches of the nth-stage (say, branch a) must be composed of the branch a of the central cluster and one of the side clusters (say cluster 1). Therefore, the matrix $\tilde{\mathbf{H}}_n(a)$ can be expressed as:

$$\tilde{\mathbf{H}}_n(a) = \begin{pmatrix} \tilde{\mathbf{H}}_{n-1}^c(0,a) & \mathbf{V}_{ac} & \mathbf{0} & \mathbf{0} & \mathbf{0} & \mathbf{0} \\ \mathbf{V}_{ca} & \tilde{\mathbf{H}}_{n-1}^s(1,c) & \tilde{\mathbf{V}}^T & \mathbf{0} & \mathbf{0} & \mathbf{0} \\ \mathbf{0} & \tilde{\mathbf{V}} & h_{n-1}^c(1) & \tilde{\mathbf{V}} & \tilde{\mathbf{V}} & \tilde{\mathbf{V}} \\ \mathbf{0} & \mathbf{0} & \tilde{\mathbf{V}}^T & \tilde{\mathbf{H}}_{n-1}(1,a) & \mathbf{0} & \mathbf{0} \\ \mathbf{0} & \mathbf{0} & \tilde{\mathbf{V}}^T & \mathbf{0} & \tilde{\mathbf{H}}_{n-1}(1,b) & \mathbf{0} \\ \mathbf{0} & \mathbf{0} & \tilde{\mathbf{V}}^T & \mathbf{0} & \mathbf{0} & \tilde{\mathbf{H}}_{n-1}(1,d) \end{pmatrix} \tag{15}$$

In Eq. (15), $\tilde{\mathbf{H}}^c_{n-1}$ and $\tilde{\mathbf{H}}^s_{n-1}$ are explicitly used to distinguish them from $\tilde{\mathbf{H}}_{n-1}$ because of the modification of the number of bonds at the adjacent bordering sites in branch a of the central cluster and branch c of the side cluster 1. Specifically,

$$\tilde{\mathbf{H}}^{c,s}_{n-1} = \tilde{\mathbf{H}}_{n-1} + \tilde{\mathbf{H}}'_{n-1} \tag{16}$$

where $\tilde{\mathbf{H}}'_{n-1}$ is a matrix with only one non-vanishing element along the diagonal which is associated with the bordering site in question. The magnitude of the element is equal to 1. The matrix $\mathbf{V}_{ac}$ has only one non-vanishing element with magnitude -1 corresponding to the interaction between the bordering sites of the central cluster and side cluster 1.

Using the representation:

$$\tilde{\mathbf{u}}^T_n(a) = (\tilde{\mathbf{u}}_{n-1}(0,a), \tilde{\mathbf{u}}_{n-1}(1,c), u^c_{n-1}(1), \tilde{\mathbf{u}}_{n-1}(1,a), \tilde{\mathbf{u}}_{n-1}(1,b), \tilde{\mathbf{u}}_{n-1}(1,d)) \tag{17}$$

where $u^c_{n-1}(1)$ is the displacement at the central site of the side cluster 1, and $\tilde{\mathbf{u}}_{n-1}(k,\alpha)$, the displacement field of the α branch in the $(n-1)$th-stage cluster k, the eigenvalue equation of $\tilde{\mathbf{H}}_n$ can be broken down into:

$$(\tilde{\mathbf{H}}^c_n(0,a) - \omega_i^2 \tilde{\mathbf{I}}_{n-1})\tilde{\mathbf{u}}_{n-1}(0,a) + \mathbf{V}_{ac}\mathbf{u}_{n-1}(1,c) = \mathbf{0} \tag{18}$$

$$\mathbf{V}_{ca}\tilde{\mathbf{u}}_{n-1}(0,a) + (\tilde{\mathbf{H}}^s_{n-1}(1,c) - \omega_i^2 \tilde{\mathbf{I}}_{n-1})\tilde{\mathbf{u}}_{n-1}(1,c) + \tilde{\mathbf{V}}^T \tilde{\mathbf{u}}^c_{n-1}(1) = \mathbf{0} \tag{19}$$

$$(h^c_{n-1}(1) - \omega_i^2)u^c_{n-1}(1) + \sum_\alpha \tilde{V}\tilde{u}_{n-1}(1,\alpha) = \mathbf{0} \tag{20}$$

$$\mathbf{V}^T u^c_{n-1}(1) + (\tilde{\mathbf{H}}_{n-1}(1,\alpha) - \omega_i^2 \tilde{\mathbf{I}}_{n-1})\tilde{\mathbf{u}}_{n-1}(1,\alpha) = \mathbf{0}\,, \quad \alpha = a, b, d \tag{21}$$

with $u^c_{n-1}(1) = 0$ and using Eq. (16), Eqs. (18) and (19) can be written as:

$$(\tilde{\mathbf{H}}_{n-1} - \omega_i^2 \tilde{\mathbf{I}}_{n-1})\tilde{\mathbf{u}}_{n-1}(0,a) + (\tilde{\mathbf{H}}'_{n-1}\tilde{\mathbf{u}}_{n-1}(0,a) + \mathbf{V}_{ac}\mathbf{u}_{n-1}(1,c)) = \mathbf{0} \tag{22}$$

$$\tilde{\mathbf{H}}'_{n-1}\tilde{\mathbf{u}}_{n-1}(1,c) + \mathbf{V}_{ca}\tilde{\mathbf{u}}_{n-1}(0,a) + (\tilde{\mathbf{H}}_{n-1} - \omega_i^2 \tilde{\mathbf{I}}_{n-1})\tilde{\mathbf{u}}_{n-1}(1,c) = \mathbf{0} \tag{23}$$

Because of the structure of the matrices $\tilde{\mathbf{H}}'_{n-1}$, $\mathbf{V}_{ac}$ and $\mathbf{V}_{ca}$, the column vector $(\tilde{\mathbf{H}}'_{n-1}\tilde{\mathbf{u}}_{n-1}(0,a) + \mathbf{V}_{ac}\mathbf{u}_{n-1}(1,c))$ has only one non-vanishing element $u_{n-1}(0,1) - u_{n-1}(1,1)$ and the column vector $(\tilde{\mathbf{H}}'_{n-1}\tilde{\mathbf{u}}_{n-1}(1,c) + \mathbf{V}_{ca}\tilde{\mathbf{u}}_{n-1}(0,a))$ has only one non-vanishing element $u_{n-1}(1,1) - u_{n-1}(0,1)$. Hence Eqs. (22) and (23) can be satisfied if $\tilde{\mathbf{u}}_{n-1}(0,a)$ and $\tilde{\mathbf{u}}_{n-1}(1,\mathbf{c})$ are eigenvectors of $\tilde{\mathbf{H}}_{n-1}(0,a)$ and $\tilde{\mathbf{H}}_{n-1}(1,c)$ respectively subject to the boundary condition:

$$u_{n-1}(0,1) = u_{n-1}(1,1) \tag{24}$$

which is simply one of the conditions imposed by Eq. (6). Equation (20) can be put in its scalar form:

$$\tilde{u}_{n-1}(1,a,1) + \tilde{u}_{n-1}(1,b,1) + \tilde{u}_{n-1}(1,c,1) + \tilde{u}_{n-1}(1,d,1) = 0 \tag{25}$$

where $\tilde{u}_{n-1}(1, \alpha, 1)$ ($\alpha = a, b, c, d$) is the displacement at the site in the branch $(1, \alpha)$ adjacent to the central site of the cluster 1. Finally, Eq. (21) reduces to:

$$(\tilde{\mathbf{H}}_{n-1}(1, \alpha) - \omega_i^2 \tilde{\mathbf{I}}_{n-1})\tilde{\mathbf{u}}_{n-1}(1, \alpha) = 0 \quad \alpha = a, b, d \tag{26}$$

The picture presented by Eqs. (22)–(26) is that the component fields $\tilde{\mathbf{u}}_{n-1}(0, a)$, $\tilde{\mathbf{u}}_{n-1}(1, c)$, and $\tilde{\mathbf{u}}_{n-1}(1, \alpha)$ ($\alpha = a, b, d$) of the eigenvector $\tilde{\mathbf{u}}_n(a)$ of $\tilde{\mathbf{H}}_n(a)$ are eigenvectors of $\tilde{\mathbf{H}}_{n-1}$ subject to the conditions given by Eqs. (24)–(25). Since the persistent modes are the descendants of the 3-fold degenerate modes, we shall analyze the situation when ω_i is a 3-fold degenerate mode of $\tilde{\mathbf{H}}_{n-1}$ and a non-degenerate mode of $\tilde{\mathbf{H}}_{n-1}$. In this case, because $\tilde{\mathbf{u}}_{n-1}(0, a)$ and $\tilde{\mathbf{u}}_{n-1}(1, c)$ are the eigenvectors to the same eigenvalue equation corresponding to a non-degenerate mode, they can differ only by a constant multiplication factor. With the condition given by Eq. (24), the constant of proportionality must equal to 1. Hence:

$$\tilde{\mathbf{u}}_{n-1}(0, a) \equiv \tilde{\mathbf{u}}_{n-1}(1, c) \tag{27}$$

Equation (27) is the relation of the linkage between the central cluster and the side cluster. There exist two possibilities: (i) If the prominent hump of the 3-fold mode in one of the clusters is in the region of linkage, Eq. (27) then requires that the prominent hump of the cluster which is linked to the cluster in question must also be in the region of linkage. This alignment produces a bridge-resonant mode in the region of linkage, a new feature which was not observed in the case of the fixed-end boundary condition; and (ii) If none of the prominent peaks of the bordering clusters is in the region of linkage, Eq. (27) simply provides the link for the mode to extend from the prominent hump to the rest of the system. The resulting mode is similar to the one observed in the case of fixed-end boundary condition. In both situations, the signature of the original 3-fold degenerate mode is carried by all the component fields $\tilde{\mathbf{u}}_{n-1}(k, \alpha)$. As the mode persists into the $(n+1)$th-stage and beyond, the magnitude of the displacements will be reduced after each linkage (i.e., no resonance) according to Eq. (25). Hence the persistent modes will become more confined as it evolves into higher degenerate modes. The persistent modes is therefore still superlocalized in a specific region of the fractal.

7. DISCUSSION

In this work, we found that the free-end boundary condition produces substantive changes in both the frequency spectrum and the nature of the eigenmodes. Although the general features of the frequency spectrum such as the existence of superlocalized and highly degenerate persistent modes and the side-by-side coexistence of the superlocalized persistent modes and extended non-degenerate modes seem to be only dependent on the tree-like structure of the Vicsek fractal, the detailed nature of the frequency spectrum and the eigenmodes, however, are dependent on the boundary condition. Many important properties of a system are closely related to the fine structure of the frequency spectrum and the nature of the eigenmodes. For example, the spectral dimension is important for the understanding of the low temperature behavior of a system. The spectral dimension is considered as dependent only on the fractal dimension and the force law.[6] Our conclusion seems to suggest that the spectral dimension for the Vicsek fractal may also depend on the boundary condition. This and some other aspects of the Vicsek fractal are actively being investigated.

ACKNOWLEDGMENTS

This work was supported both by the Research Corporation and the National Science Foundation (Grant no: EHR-9108764). One of the authors (SYW) would also like to acknowledge Professor F. Flores and other faculty members at the Departamento de Fisica Materia Condensada, Universidad Autonoma de Madrid for their kind hospitality and the support received from DGIYCT.

REFERENCES

1. R. Rammal, *J. Physique* **45**, 191 (1984).
2. E. Domany, S. Alexander, D. Bensimon and L. Kadanoff, *Phys. Rev.* **B28**, 3110 (1983).
3. C. S. Jayanthi, S. Y. Wu and J. Cocks, *Phys. Rev. Lett.* **69**, 1955 (1992).
4. C. S. Jayanthi and S. Y. Wu, *Phys. Rev.* **B48**, 10188 (1993).
5. C. S. Jayanthi and S. Y. Wu, *Phys. Rev.* **B48**, 10199 (1993).
6. S. H. Liu, in *Solid State Physics*, eds. H. Ehrenreich, F. Seitz and D. Turnbull (Academic, New York, 1986), vol. 39, p. 207.

Fractals, Vol. 1, No. 4 (1993) 930–938

FRACTAL ASPECTS OF INTERSTELLAR CLOUDS

THOMAS ZIMMERMANN and JÜRGEN STUTZKI
I. Physikalisches Institut, Universität zu Köln, Zülpicher Str. 77, 50937 Köln, Germany

Abstract

Interstellar matter consisting of gas and dust shows structures which appear to have self-similar projected shapes, similar to the properties of terrestrial clouds. Moreover, the completely different physics in these interstellar clouds reveals several additional self-similar/fractal aspects which are investigated by extensive astronomical observations, aiming especially at a better understanding of the processes of star formation which take place in interstellar molecular clouds.

1. INTRODUCTION

Only a short time after the fractal concept had been established by Mandelbrot,[1] it was applied to the shape of terrestrial clouds.[2–4] Based on the apparent similarity between terrestrial and interstellar clouds (ICs) of gas and dust (the naming of "clouds" not being accidental), subsequent work applied the concept to ICs.[5–13] Understanding the structure of interstellar molecular clouds (IMCs) and its possible links to turbulence (e.g. Ref. 14) is crucial for understanding IMC evolution and star formation. As stars (range of masses: 0.1–30 $M_\odot$) form out of IMCs (10–10^6 $M_\odot$), the structure of clouds and the cloud mass spectrum are likely to be closely linked to such fundamental quantities as the star formation rate and efficiency and the stellar initial mass distribution.[15–17]

Our investigations are based on an extended data set covering many size scales, as is necessary to extract fractal and hierarchical properties of ICs. This includes data from the most extended carbon monoxide (CO) maps of IMCs, covering 100 pc size structures, down to observations of the closest molecular cloud L1457 at 65 pc distance, which provides the smallest scales at 0.005 pc resolution (Sec. 2). The self-similar 2-dimensional projected shapes of IMCs suggest a possible self-similarity in the 3-dimensional structure. This is stressed by the internal velocity distribution of IMCs which can be measured along the line of sight (Sec. 3). In addition, IMCs show scaling and power-law relations in their mass distribution and excitation conditions, which can be regarded as another aspect of the fractal structure of IMCs (Sec. 4). Since the limitation to only 2-dimensional projections is a crucial point in all astronomical observations, we model 3-dimensional distributions of IMCs by fractal/hierarchical subdivision and compare their projections to observations by a box-counting method, which in addition yields an independent characterization of IMC structure (Sec. 5). Summarizing our results, we find the fractal structure of IMCs being reflected in several more aspects than just in their projected shapes, which, regarded as a whole, might be crucial to understanding the inner structure of IMCs and the physical processes leading to coagulation and star formation (Sec. 6).

2. OBSERVATIONS OF INTERSTELLAR MOLECULAR CLOUDS ON DIFFERENT SIZE SCALES

Star formation occurs in IMCs which constitute the densest regions of interstellar matter (composition: 1% dust, 99% gas; the latter: 50% molecular, 50% atomic/ionized gas) (e.g. see Refs. 18 and 19). Although the general processes governing star formation are understood, the details still remain unclear. Observations of molecular gas spectral line emission not only trace the overall projected spatial distribution of IMCs in our and in other galaxies, but also yield information about the velocity distribution (along the line of sight) by the Doppler-relation, and the excitation conditions (temperature, density, chemical abundancies) by observing different molecular species and isotopomers in their lowest rotational transitions (e.g. see Ref. 20). During the last decade, the second most abundant interstellar molecule, carbon monoxide (CO), and its isotopomers have proven to be the most important trace molecules.

Observations of molecular clouds are carried out with millimeter- and submillimeter-radiotelescopes which limit the highest possible angular resolution ε by diffraction to $\varepsilon \propto \lambda/D$, where D is the main reflector diameter and λ the wavelength. Although in principle one can mimic the low angular resolution of a small telescope with a large one, this is not possible and infinitum in practice, due to the limited stability of the integration and the data transfer time needed. Thus the resolution, i.e., telescope and wavelength used for observations, has to be appropriate with respect to the extent of the region to be mapped. We started our investigations with the Taurus-Perseus-Auriga region [Fig. 1(a)], containing some of the nearest IMCs, which has been mapped by Ungerechts and Thaddeus[21] as part of the whole Milky Way CO survey.[22] For more detailed observations, we chose at the edge of the Taurus region the closest known IMC, L1457, (distance: 65 pc) [Figs. 1(b) and 1(c)] which is particularly well suited for studying IMC structures and dynamics down to very small linear scales. Between 1987 and 1992, we mapped L1457 with successive higher resolutions in different isotopomers and transitions of CO[23–25] to get a more complete picture of molecular gas over a wide range of size scales (spatial resolution: Taurus region: 4–0.6 pc, L1457: 0.08–0.02 pc, L1457 core: 0.005 pc).

3. SELF-SIMILARITY OF SPATIAL AND VELOCITY DISTRIBUTION

A first indication of ICs' self-similarity was found in the projected shapes of ICs, i.e., by log(P)–log(A) plots (P: perimeter, A: area) of iso-column density contours from different data sets (visible extinction maps,[5,13] IRAS[26] survey,[6–8,10] HI surveys,[9,10] CO surveys,[11,12]) similar to the work of Lovejoy[2] and Rys and Waldvogel[4] for terrestrial clouds. From the relation $P^{1/D_H} \propto A^{1/2}$, these plots yield fractal dimensions D_H in a range between 1.2–1.7. The same fractal dimensions are derived independently when rescaling data sets with different resolutions ε to the same scale by the relation $P(\varepsilon) \propto \varepsilon^{1-D_H}$.[11,12] However, the significance of these numbers is still being discussed since they may be influenced by effects like noise[7] or finite resolution[12]; moreover, it is unknown in which way certain physical conditions lead to certain fractal dimensions of ICs' shapes.

The fractal shapes of iso-column density contours suggest fractal iso-density surfaces and thus a fractal spatial distribution, although due to the projection along the line of sight, it is impossible to observe the 3-dimensional structure of ICs directly. Fortunately, the velocity distribution of cloud substructures can be observed by their Doppler-shifts in the spectra; this yields information about substructures along the line of sight assuming different structures having different velocities. Looking at CO spectra at different spatial resolutions (Fig. 2), we find that the lines, showing the dynamical distribution of the molecular gas, split up with increasing resolution into several narrow lines, similar to the spatial structures. Although no general relation between velocity and spatial distribution of molecular gas along

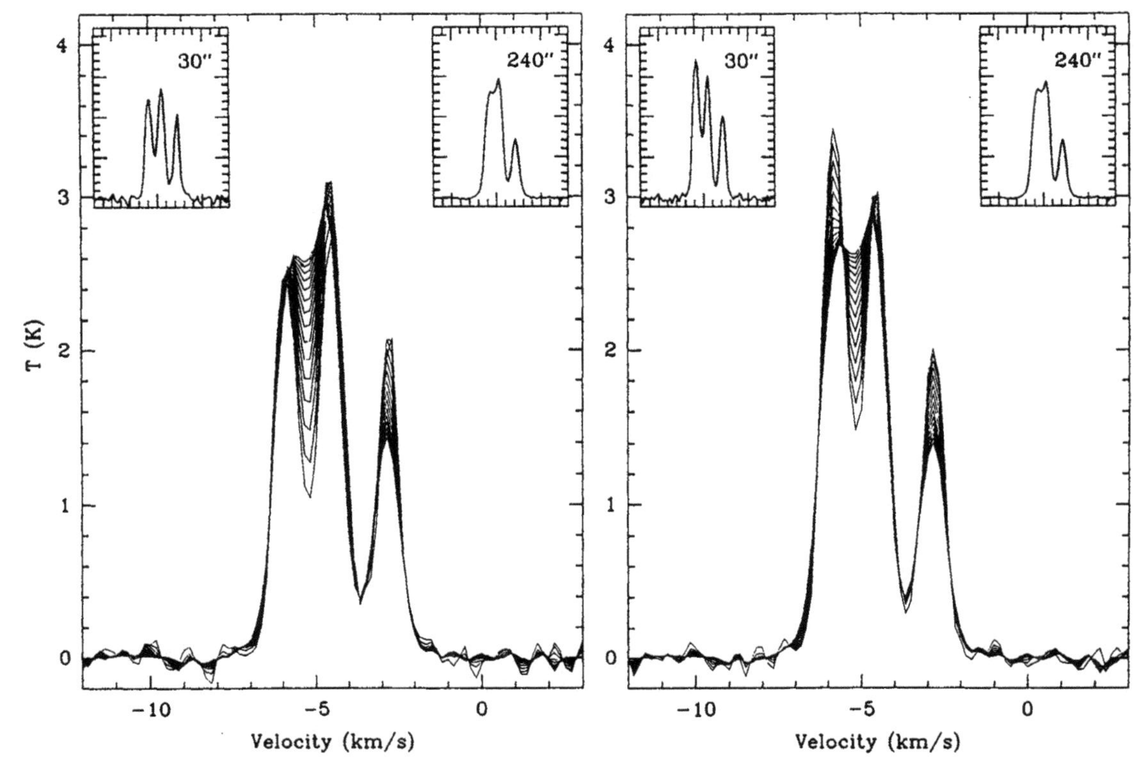

Fig. 2 Variation of typical CO spectra in the direction of L1457 at different spatial resolutions (30″(15″)240″) (intensity in units of Kelvin vs. velocity), showing a split up of the left line component with increasing resolution.

the line of sight can be assumed, the overall pressure of this phenomenon strongly points to a self-similar distribution of molecular gas in the direction perpendicular to the projection plane, and thus to a self-similar 3-dimensional distribution.

4. MASS DISTRIBUTION AND SCALING RELATIONS

The variation of physical excitation conditions (like density and temperature) and physical parameters (like mass and velocity dispersion) with size scale can be perceived by analyzing separable substructures, or even by averaging over arbitrary subregions of observed IMCs. The latter is not surprising, keeping in mind that the definition of a "substructure of its own" is appropriate only within the respective spatial resolution, because structures dissolve and split up into smaller substructures when changing to higher resolutions.

Nevertheless, authors have attempted to determine definite entities called "fragments" and "clumps"[21,23,27,28] to derive substructure mass spectra[27,29] which could be related to star formation characteristics like the stellar initial mass distribution.[15–17] We followed this idea and used an algorithm of Stutzki and Güsten[27] to decompose the CO map of the L1457 core [see Fig. 1(c)] into Gaussian-shaped substructures. The mass spectrum of this decomposition (Fig. 3) shows the power-law relation $dN/dM \propto M^{-1.73\pm0.06}$ (marked by solid line), where the largest structures have masses close to 0.5 solar masses ($M_{\odot}$).

For physical parameters of the Gaussian-shaped substructures power-law scaling relations with size R according to the Larson-relations[30] are expected. The dispersion of the velocity distribution σ_v follows for structures larger than the resolution limit (dashed line) a power-law $\sigma_v(\mathrm{kms}^{-1}) \propto R(\mathrm{pc})^{\alpha}$ (Fig. 4) with α between 0.38 (thin line,[30]) and 0.50 (fat line,[31,32]) which were found for samples of other ICs. This is expected if the turbulent energy scales linear with the size: $\sigma_v^2 \propto R$. If this scaling relation still holds down to thermal velocity dispersion, a lower limit for substructure sizes will be given.

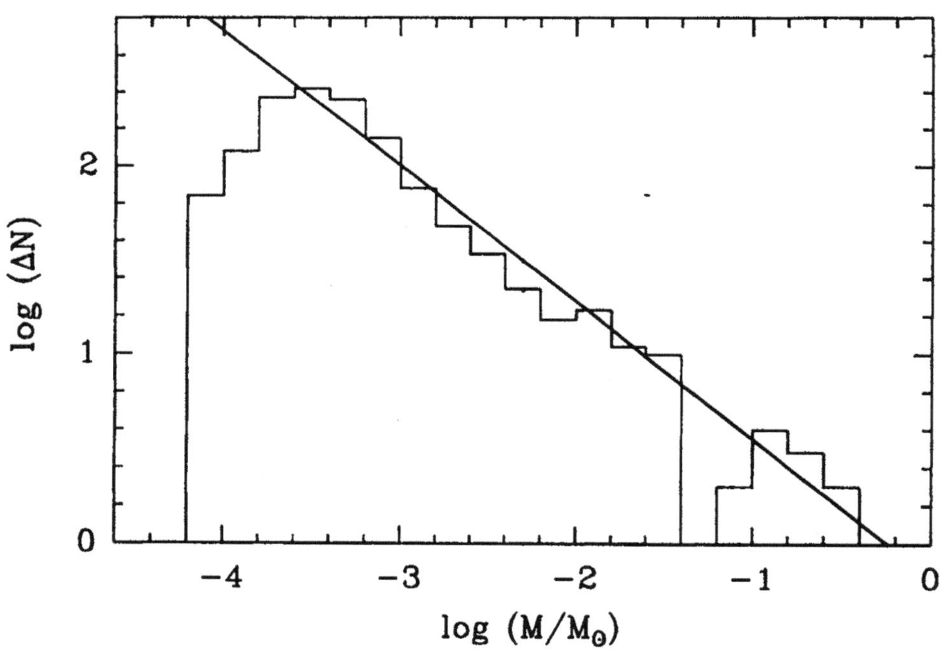

Fig. 3 Mass spectrum of the decomposition of the CO map of the L1457 core [see Fig. 1(c)] into Gaussian-shaped substructures, showing the power-law relation $dN/dM \propto M^{-1.73\pm0.06}$ (solid line).

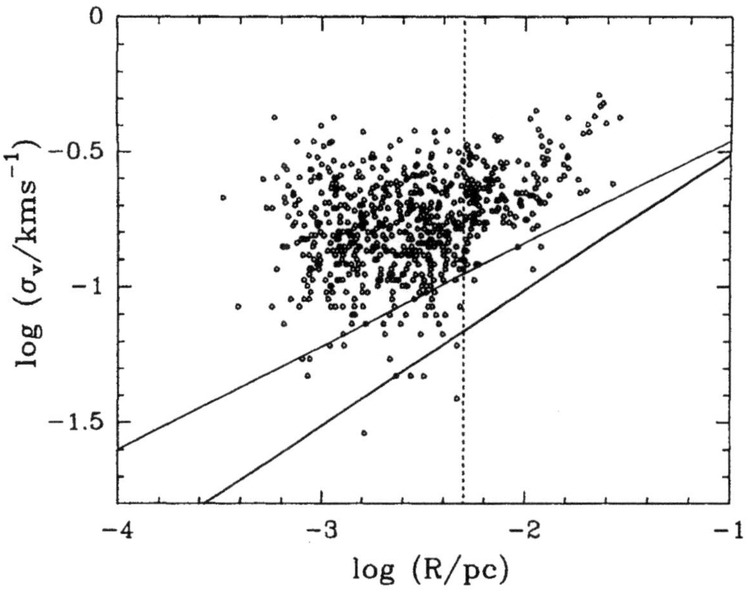

Fig. 4 Velocity dispersion σ_v vs. size R for Gaussian-shaped substructures (thick line: $\alpha = 0.50$,[31,32]; thin line: $\alpha = 0.38$,[30]); dashed line: resolution limit.

5. MODELLING THE 3-DIMENSIONAL SPATIAL STRUCTURE OF INTERSTELLAR CLOUDS

Being led by indications of 3-dimensional self-similarity of IMCs as discussed in Sec. 3, we have modeled 3-dimensional distributions of molecular gas and compared them with the observations. Our model-space is a cube of 64×64×64 cells, which is divided into 8 subcubes, each of which again into 8 subcubes, and so on, until the smallest subcubes consists of only 1 cell. The model-space is assumed to be filled with matter, which is distributed randomly onto N_1 of the 8 subcubes at division level L_1, i.e., $8 - N_1$ subcubes remain empty. Thus on each level L_n ($n = 1(1)6$) only the matter in filled cubes can be further distributed onto subcubes, until the lowest level L_6 is reached. Since the filled volume decreases from level to level, the density increases with smaller scales like those observed in real IMCs. This process produces a 3-dimensional Cantor-dust and is called "curdling" by Mandelbrot.[1] Only if the number of filled subcubes is the same on each level will this result in a (random) fractal distribution; in general, it is just hierarchical and not self-similar. Already Hoyle[33] tried to explain the distribution of galaxies with a similar 2-dimensional model, and Hetem and Lépine[13] used a method like this to model ICs structures as well.

To compare models with observations we apply a box-counting method to the observed map and the 2-dimensional projection of the model, i.e., we count the number of boxes needed to cover map areas with intensities higher than definite values [Fig. 5(a): 10%(10%)50% of peak intensity], and doing this for successive smaller box sizes. Figure 5(a) shows log-log plots of the number of boxes needed vs. inverse boxsize for the Taurus map and its best fit model which we obtained by varying the numbers N_n ($n = 1(1)6$) of matter filled boxes on each level. In Fig. 5(b), a part of the Taurus map is compared with the projection of its model. Whereas for Taurus and the whole map of L1457 [Figs. 1(a) and 1(b)] we find relatively high volume filling at largest and smallest scales, and relatively small filling

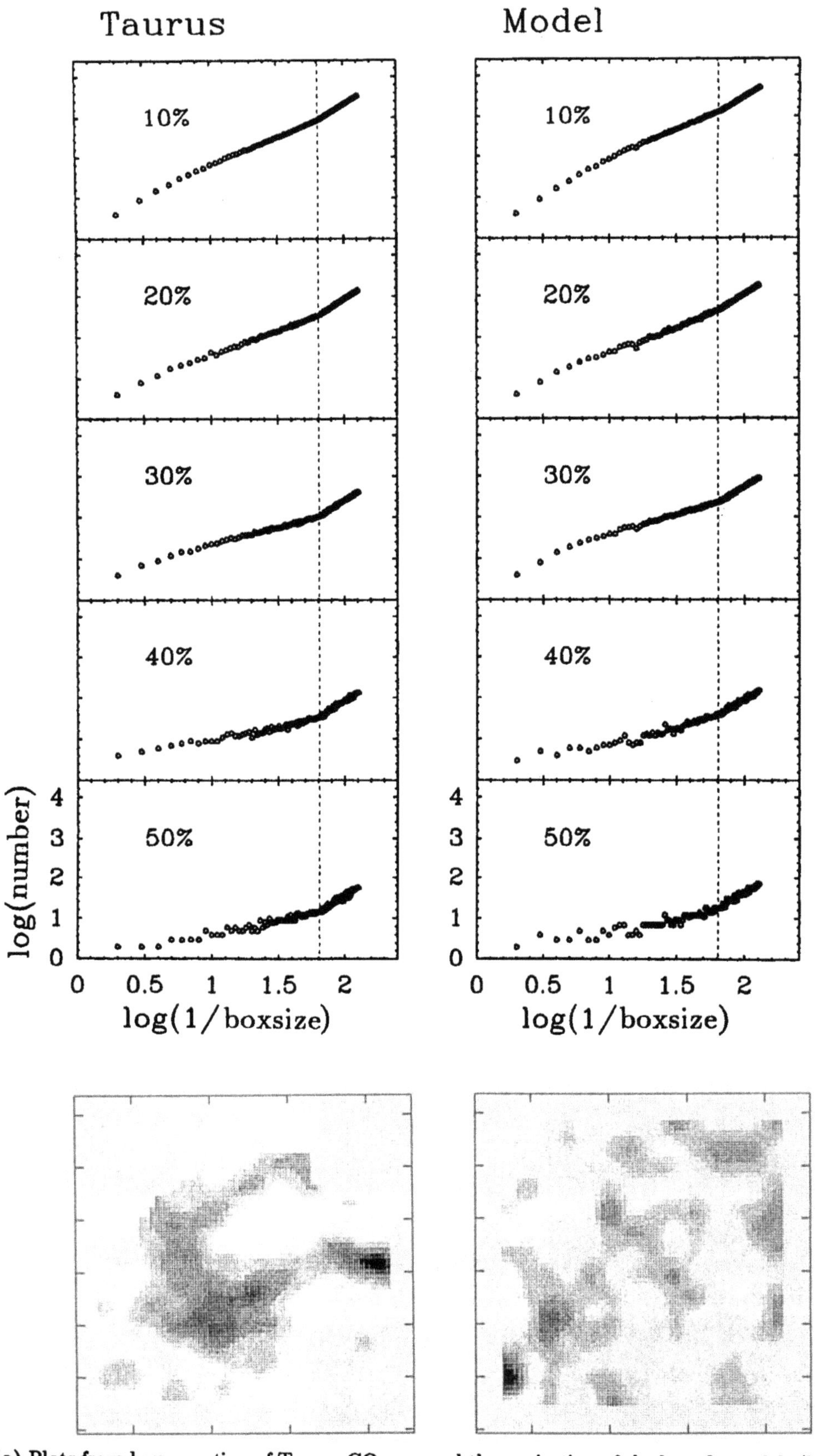

Fig. 5 (a) Plots from box-counting of Taurus CO map and the projection of the best fit model; (b) Taurus CO map compared to the projection of the best fit model.

in between, the best fit model of L1457 core [Fig. 1(c)] has the same number of 7 filled subcubes on all levels except one and therefore can be characterized by a fractal dimension $D_H = \log 7/\log 2 = 2.81$.

The projections of our models are not thought to have the exact morphology of the observations, but to show the same fractal/hierarchical properties in their area filling which is measured over all size scales, inherent in an observed map, by means of the box-counting. From the models we can extract typical density profiles along the line of sight. Assuming a certain velocity field along the line of sight we can calculate line profiles and compare them with the observed ones; e.g. for a constant velocity gradient the line profiles are identical to the density profiles.

6. CONCLUSIONS

From investigating the observations and modelling its 3-dimensional structures we draw the following conclusions:

1. Our observations of IMCs cover a wide range of size scales between 100 pc and 0.005 pc, which is necessary to investigate their hierarchical and fractal properties.
2. ICs appear to have self-similar projected shapes with fractal dimensions $D_H = 1.2$–1.7 derived from $\log(P)$–$\log(A)$ plots.
3. We find that molecular spectral line profiles split up with increasing spatial resolution into several narrow lines, pointing to a self-similar distribution of molecular gas along the line of sight as well.
4. Conclusion 2 and the overall presence of 3, point to a self-similar or at least hierarchical 3-dimensional fractal distribution of molecular gas.
5. IMCs substructures show a mass spectrum $dN/dM \propto M^{-1.73\pm0.06}$ and a power-law scaling relation of the velocity dispersion σ_v with size R of $\sigma_v \propto R^\alpha$, where $\alpha = 0.38 - 0.50$, which can be regarded as another aspect of the fractal structure.
6. Comparing observations and models by box-counting we succeed in modeling the observations by a method of hierarchical subdivision; from the models, density profiles along the line of sight and realistic spectral line profiles can be derived.

REFERENCES

1. B. B. Mandelbrot, *The Fractal Geometry of Nature* (1978, 1982).
2. S. Lovejoy, *Science* **216**, 185 (1982).
3. H. G. E. Hentschel and I. Procaccia, *Phys. Rev.* **A29**, 1461 (1984).
4. F. S. Rys and A. Waldvogel, *Phys. Rev. Lett.* **56**, 784 (1986).
5. M. Beech, *Astrophys. Space. Sci.* **133**, 193 (1987).
6. D. Bazell and F. X. Désert, *Astrophys. J.* **333**, 353 (1988).
7. R. L. Dickman et al., *Astrophys. J.* **365**, 586 (1990).
8. J. M. Scalo, in *Physical Processes in Fragmentation and Star Formation* (1990).
9. B. P. Wakker, Ph.D. thesis (University Groningen) (1990).
10. M. G. R. Vogelaar et al., in *Fragmentation of Molecular Clouds and Star Formation*, IAU Symp. No. 147 (1991).
11. E. Falgarone et al., *Astrophys. J.* **378**, 186 (1991).
12. T. Zimmermann and J. Stutzki, *Physica* **A191**, 79 (1992).
13. A. Jr. Hetem and J. R. D. Lépine, *Astron. Astrophys.* **270**, 451 (1993).

14. E. Falgarone and T. G. Phillips, *ApJ* **359**, 344 (1990).
15. R. B. Larson, *MNRAS* **256**, 641 (1992).
16. J. Stutzki, *Rev. Mod. Astron.* **6**, (1992).
17. E. A. Lada et al., in *Protostars & Planets III* (1993).
18. F. H. Shu et al., *Ann. Rev. Astron. Astrophys.* **25**, 23 (1987).
19. N. J. Evans, in *Frontiers of Stellar Evolution* (1990).
20. G. Winnewisser et al., in *Spectroscopy of the Earth's Atmosphere and Interstellar Medium* (1992).
21. H. Ungerechts and P. Thaddeus, *Astrophys. J. Suppl.* **63**, 645 (1987).
22. T. M. Dame et al., *Astrophys. J.* **322**, 706 (1987).
23. T. Zimmermann and H. Ungerechts, *Astron. Astrophys.* **238**, 337 (1990).
24. T. Zimmermann et al., in *Evolution of Interstellar Matter and Dynamics of Galaxies* (1992).
25. T. Zimmermann, Ph.D. thesis (University Köln) (1993).
26. C. A. Beichmann et al., eds., *IRAS (Infrared Astronomical Satellite) Catalogs and Atlases Explanatory Supplement* (1985).
27. J. Stutzki and R. Güsten, *Astrophys. J.* **356**, 513 (1990).
28. R. Herbertz et al., *Astron. Astrophys.* **249**, 483 (1991).
29. E. A. Lada et al., *Astrophys. J.* **368**, 432 (1991).
30. R. B. Larson, *MNRAS* **194**, 809 (1981).
31. P. C. Myers, *Astrophys. J.* **270**, 105 (1983).
32. T. M. Dame et al., *Astrophys. J.* **305**, 892 (1986).
33. F. Hoyle, *Astrophys. J.* **118**, 513 (1953).

Fractals, Vol. 1, No. 4 (1993) 939–946

ON THE FRACTAL STRUCTURE OF ELECTRON AVALANCHES

Z. DONKÓ and I. PÓCSIK
Research Institute for Solid State Physics of the Hungarian Academy of Sciences, H-1525 Budapest P. O. Box 49, Hungary

Abstract

The motion of electrons in helium gas in the presence of a homogeneous external electric field was studied. Moving between the two electrodes, the electrons participate in elastic and inelastic collision processes with gas atoms. In ionizing collisions, secondary electrons are also created and in this way self-similar electron avalanches build up. The statistical distribution of the fractal dimension and electron multiplication of electron avalanches was obtained based on the simulation of a large number of electron avalanches. The fractal dimension shows a power-law dependence on electron multiplication with an exponent of ≈ 0.33.

1. INTRODUCTION

The spectacular electric breakdown phenomena is a famous example displaying the occurrence of fractal structures in natural processes.[1,2] However, this breakdown is just a final event in a long chain of events; much less attention has yet been paid to the previous processes from the fractal point of view. These previous processes are very important, even from the practical point of view, as most of our light sources are using the effects of electric discharges in gases for light production. On the other hand, these effects have a much broader range of applications, and in general have interesting fractal features which we would like to demonstrate here, by relating the parameter of dimension to physical parameters.

To illustrate the most simple case, let us consider two parallel plane electrodes with the space between them filled with a buffer gas, as shown in Fig. 1. If the current density through the gas is in the range of 10^{-12}–10^{-9} A/cm^2, the so-called *Townsend discharge* is formed.[3,4] The electric field created by the external voltage source is homogeneous between

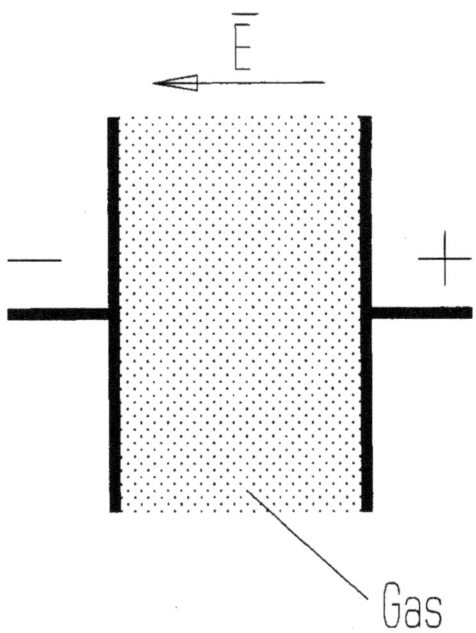

Fig. 1 Basic model arrangement of the gas discharge.

the electrodes, because the current density is so low that the space charges are negligible. In Townsend discharges, positive ions are believed to play the major role in the maintenance of the discharge by liberating electrons from the cathode surface.

When a single electron is emitted from the cathode (with a few electronvolts of energy), it is accelerated in the external electric field and it may excite and ionize the atoms of the background gas. In ionization processes, secondary electrons are created and thus the number of electrons multiplies with the result that an *electron avalanche* builds up. The number of electrons in an electron avalanche (including the primary one) is called *multiplication* and is denoted by M. If we neglect recombination within and on the walls of the discharge tube we can assume that all the $M-1$ ions created in the avalanche move towards the cathode under the influence of the electric field and finally reaching the cathode surface. The discharge is called *self-sustained* if the average number of ions (per emitted electron) falling onto the cathode surface is just enough to liberate a new electron. The *self-sustainment criteria* of the Townsend discharge may be formulated as[4]:

$$(M-1)\cdot\gamma = 1\,, \tag{1}$$

where γ is the electron emission coefficient of the cathode which depends on the given cathode material-buffer gas combination.

Hydrodynamic electron transport is defined as when the energy distribution of electrons depends only on the local value of the reduced electric field E/n (electric field to gas density ratio). In this case, the motion of electrons can be adequately described by transport coefficients (mobility, diffusion coefficients, etc.) depending only on the local value of E/n.[5] However, the presence of physical boundaries (e.g. electrodes) which act as a source or a sink of electrons, as well as high values of E/n may induce non-equilibrium effects.[6] When the motion of electrons lacks hydrodynamic equilibrium, it cannot be characterised by transport coefficients. One of the possible approaches to the study of electron motion in non-hydrodynamic regions is the Monte Carlo type simulation.[6,7]

2. THE METHOD OF SIMULATION

The Monte Carlo simulation is well suited to describe the motion of particles (electrons) under non-hydrodynamic conditions. Our simulation was carried out in $d = 2$ dimensions (one parallel and one perpendicular to the electric field). The following collision processes were taken into account (Q denotes the various cross sections):

$$\mathrm{He} + \mathrm{e}^- \rightarrow \mathrm{He} + \mathrm{e}^- \cdots \text{elastic electron scattering } (Q_{\mathrm{ela}}),$$

$$\mathrm{He} + \mathrm{e}^- \rightarrow \mathrm{He}^* + \mathrm{e}^- \cdots \text{electron impact excitation } (Q_{\mathrm{exc}}),$$

$$\mathrm{He} + \mathrm{e}^- \rightarrow \mathrm{He}^+ + \mathrm{e}^- + \mathrm{e}^- \cdots \text{electron impact ionization } (Q_{\mathrm{ion}}).$$

The cross sections of the collision processes were taken from Refs. 8 and 9. The scattering of electrons in elastic collisions and electronic excitation is assumed to be isotropic. The energy loss of electrons in elastic collisions is neglected. In the case of ionization, the directions of the scattered and the ejected (secondary) electrons are assumed to be perpendicular. The energy dependence of the cross sections of different collision processes (see Fig. 2) was taken into account in the simulation.

The electrons are traced one by one in the simulation. The first electron emitted from the cathode is traced as it participates in collision processes and finally being absorbed by the anode. The secondary electrons created in ionizing collisions are all "memorized" and they are traced independently as well. In the Monte Carlo simulation, many parameters are determined by random numbers:

- the free path of electrons in the collisionless motion,
- the type of collision that actually occurs after collisionless periods,
- the scattering angle of electron(s) after the collision process.

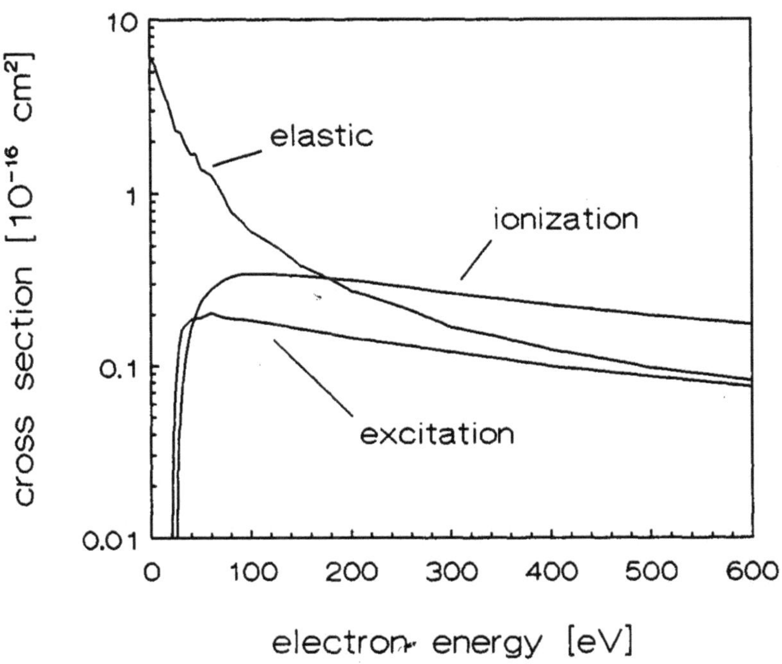

Fig. 2 Energy dependent cross sections for elastic scattering, electron impact excitation and ionization (helium).

The free path between collisions is determined by a random number in the following way. Let us consider first the case of an electric field free region. The concentration of the background gas is denoted by n; the *sum* of the cross sections of all different collision processes is $Q(\varepsilon) = Q_{\mathrm{ela}}(\varepsilon) + Q_{\mathrm{exc}}(\varepsilon) + Q_{\mathrm{ion}}(\varepsilon)$. In the field-free case, the ε energy of the electron does not change during the collisionless motion. The probability of that the free path of an electron is between s and $s + ds$ is given by:

$$p(s)ds = nQ(\varepsilon) \cdot \exp\{-nQ(\varepsilon)s\}ds\,. \tag{2}$$

The series of free path values s can be generated by the following expression:

$$s = \frac{1}{nQ(\varepsilon)} \ln(1 - R) \tag{3}$$

where R denotes a random variable having a uniform distribution in the $[0, 1)$ interval. In the presence of an electric field, the energy of the electron changes along its free path and so change the cross sections of collision processes. It can be shown that in this case, the following equation can be used to generate the sequence of free paths[7]:

$$\int_{s_o}^{s_c} nQ[\varepsilon(s)]ds = -\ln(1 - R)\,, \tag{4}$$

where s_o and s_c are the positions of the last and next collisions, respectively, measured along a curvilinear co-ordinate. Since Q depends on the energy of the electron which itself depends on the position, the determination of free path values is more complicated than in the field free case [Eq. (3)].

The type of collision which occurs is determined by a random number taking into account the cross sections of different processes at the value of the actual electron energy at s_c.

3. RESULTS AND DISCUSSION

Using the simulation described in Sec. 2, the properties of electron avalanches have been investigated for given discharge conditions. A representative example of an electron avalanche generated by one electron is displayed on Fig. 3(a), while the result of the box counting calculation is shown in Fig. 3(b) — this shows a nice linear dependence over a wide scale, with $D \cong 1.37$ Hausdorff dimension.

As all electron avalanches are different, we may expect to obtain different fractal dimensions for different avalanches. The simulation and box counting were carried out for 1000 individual avalanches and the resulting histogram of the fractal dimension is shown in Fig. 4. Avalanches with $D \cong 1$ correspond to the case where a small number of collisions has occurred or the primary electron was backscattered to the cathode. The asymmetric distribution has a peak around $D \cong 1.45$, and the mean and the standard deviation of the distribution were found to be $\langle D \rangle = 1.27$ and $\sigma_D = 0.81$, respectively by excluding the first peak in the distribution, somewhat different values were obtained: $\langle D \rangle = 1.34$ and $\sigma_D = 0.78$).

The statistical distribution of the multiplication (M) — displayed in Fig. 5 — was calculated from the simulation of 20 000 electrons. It can be seen in Fig. 5 that there is again a substantial number of electron avalanches characterised by non-, or few ionizing collisions. This feature, however, falls off soon, and a similarly asymmetric histogram at higher values

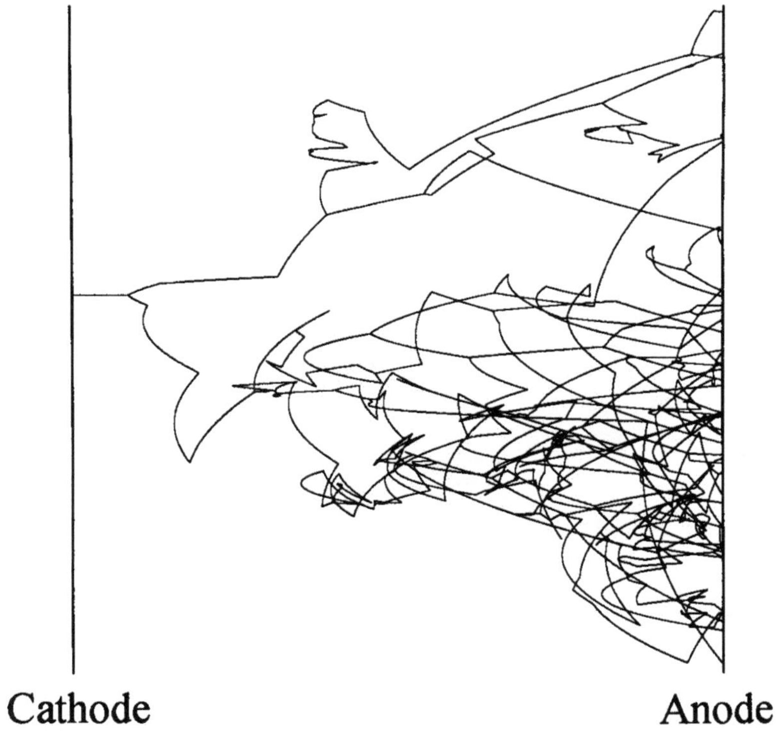

(a)

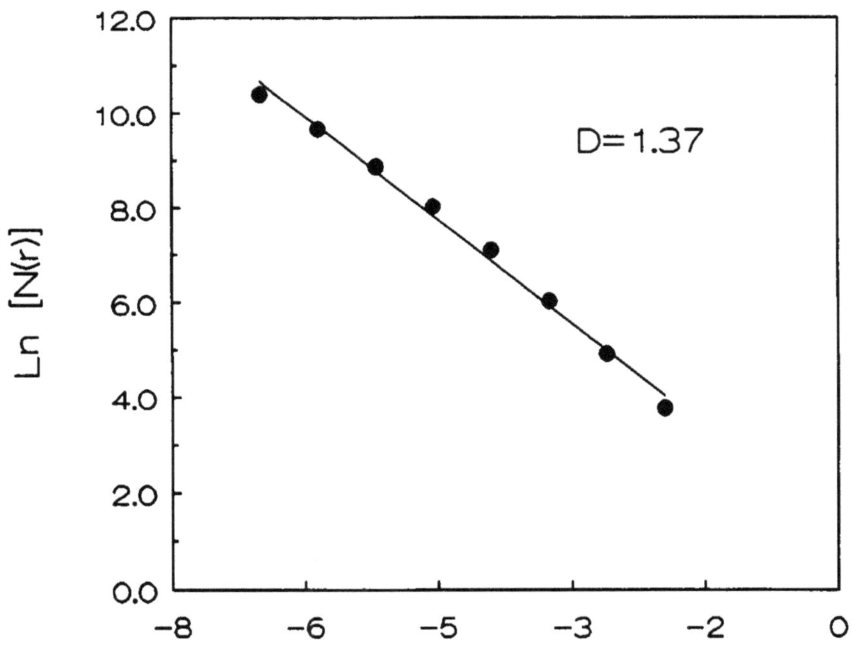

(b)

Fig. 3 **(a)** A plot of a typical electron avalanche obtained at the following values of parameters: 1 cm electrode gap, 500 V gap voltage, 2.5 mbar helium pressure, 10 eV initial electron energy. **(b)** The result of the box counting applied to the electron avalanche plotted in Fig. 3(a) resulting in a Hausdorff dimension of $D \cong 1.37$. (r denotes the linear size of the boxes and $N(r)$ is the number of boxes of size r occupied by the electron trajectories.)

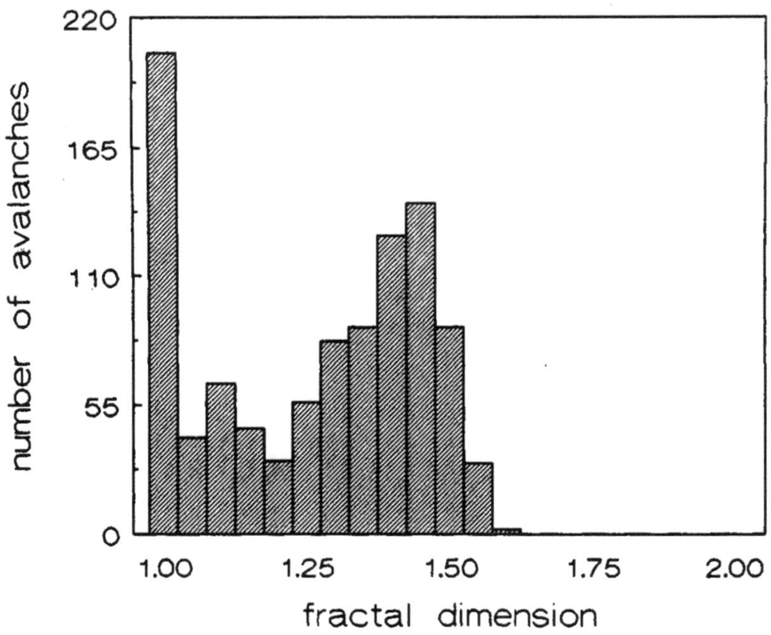

Fig. 4 The statistical distribution of the Hausdorff dimension (D) obtained from the simulation of 1000 electrons.

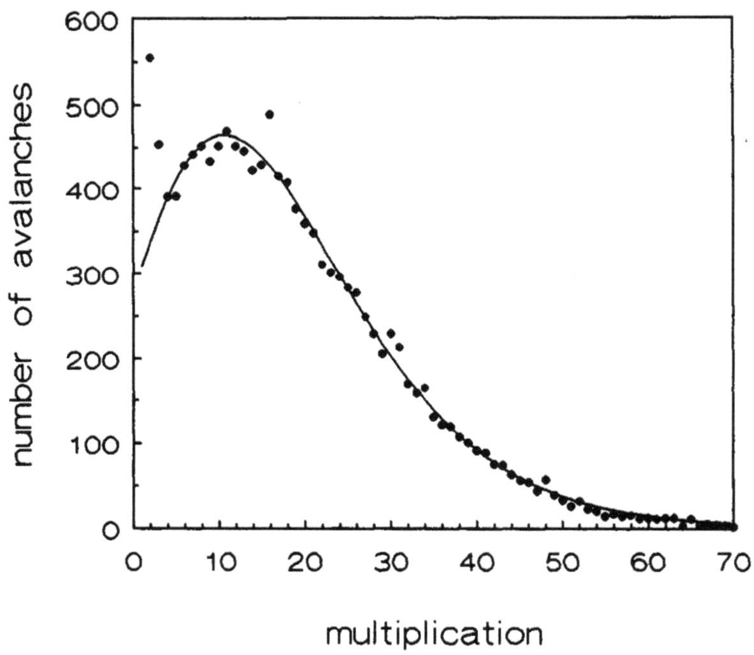

Fig. 5 The statistical distribution of the multiplication (M) calculated from the simulation of 20 000 electrons. The dots represent simulation results; the solid line is the best-fitting lognormal function.

of M shows the statistical homogeneity of the processes. The mean and the standard deviation of the distribution were found to be $\langle M \rangle = 12.67$ and $\sigma_M = 9.66$, respectively (when the $M = 1 - 4$ interval is excluded: $\langle M \rangle = 20.54$ and $\sigma_M = 12.48$).

The strongly asymmetric shape of the distribution of multiplication suggested that it

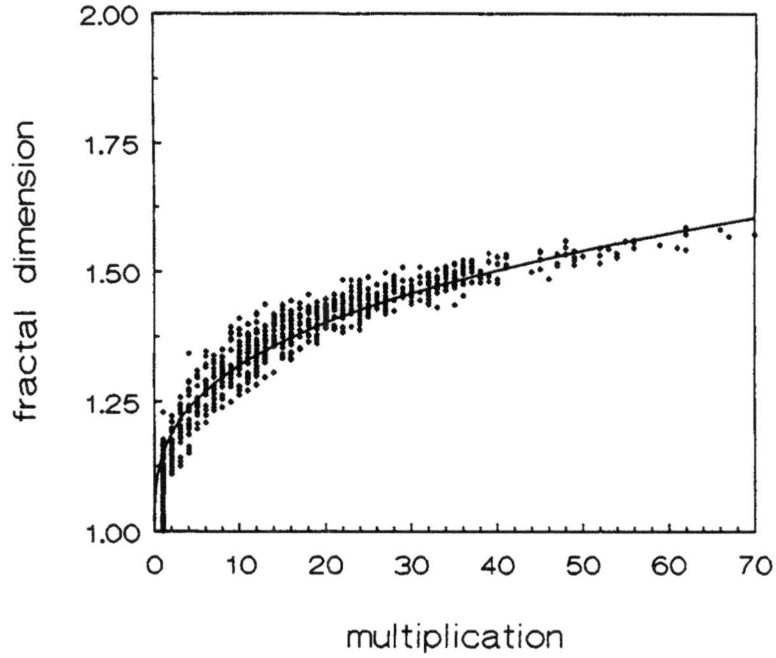

Fig. 6 The correlation between the Hausdorff dimension (D) of electron avalanches and the multiplication (M) of electrons. The dots represent individual avalanches; the solid line is the best-fitting power-low function.

may have a relation to the lognormal distribution (the lognormal statistics is based on the multiplicational connections between elementary probability processes,[10] in this case the number of collisions in which the electron participates). A lognormal curve of the following general form was fitted to the distribution of multiplications:

$$\frac{d\Lambda(x)}{dx} = \frac{1}{(x-\tau)\sigma\sqrt{2\pi}} \exp\left\{-\frac{[\ln(x-\tau)-\mu]^2}{2\sigma^2}\right\},$$

where Λ is the probability density, x is a continuous variable corresponding to the multiplication (M), τ, μ and σ are the shift, logarithm of the mean and geometrical standard deviation of the model distribution, respectively. The best fitting parameters were found to be: $\mu = 3.79$, $\tau = -29.7$ and $\sigma = 0.303$. The result of the curve fitting is also plotted in Fig. 5.

The number of splittings in an electron avalanche should have a strong impact on its fractal dimension. On the other hand, the number of splittings on the path is equal to the number of ionizations, (i.e., $M - 1$) meaning that some correlation is expected between the dimension of the electron path and the multiplication. This correlation can be clearly seen in Fig. 6. A power-law curve was fitted for this dataset and the best fitting exponent was found to be 0.33. The used mathematical expression with the best fitting parameters was:

$$D = 0.15 \cdot M^{0.33} + 1.$$

4. CONCLUSIONS

Monte Carlo simulation of the electron motion in helium gas was carried out, with convincing results, that the electron paths show fractal feature. A large number of individual electron avalanches was traced by the simulation and the parameters of the statistical distribution of the multiplication (M) and the fractal dimension (D) were calculated. Both the dimension of the electron avalanches and the electron-multiplication of the process show asymmetric distribution, which have successfully fitted by lognormal statistics. A surprisingly strong power-law dependence was found between the multiplication and the fractal dimension of electron avalanches.

ACKNOWLEDGMENTS

This work was supported by the Hungarian Science Foundation under Contract Nos.: OTKA-F-7475 and OTKA-T-1975.

REFERENCES

1. E. I. Knizhnik, A. D. Onisko and A. V. Gaydamaka, *Radiat. Phys. Chem.* **19**, 473 (1982).
2. L. Niemeyer, L. Pietronero and H. J. Wiesmann, *Phys. Rev. Lett.* **52**, 1033 (1984).
3. A. von Engel, in *Encyclodepia of Physics*, ed. S. Flügge (Springer, 1956), Vol. XXI, p. 504.
4. P. F. Little, in *Encyclodepia of Physics*, ed. S. Flügge (Springer, 1956), Vol. XXI, p. 573.
5. L. C. Pitchford, J. P. Boeuf, P. Segur and E. Marode, in *Nonequilibrium Effects in Ion and Electron Transport*, eds. J. W. Gallagher et al. (Plenum Press, New York, 1990), p. 1.
6. E. Marode and J. P. Boeuf, in *Proc. XVI. ICPIG Int. Conf. on Phenomena in Ionized Gases*, eds. W. Bötticher, H. Wenk and E. Schulz-Gulde. (Düsseldorf, 1983), p. 206.
7. J. P. Boeuf and E. Marode, *J. Phys. D: Appl. Phys.* **15**, 2169 (1982).
8. F. J. de Heer and R. H. J. Jansen, *J. Phys. B: At. Mol. Phys.* **10**, 3741 (1977).
9. F. J. de Heer, R. H. J. Jansen and W. van der Kaay, *J. Phys. B: At. Mol. Phys.* **12**, 979 (1979).
10. I. Pócsik, *Z. Phys.* **D20**, 395 (1991).

Fractals, Vol. 1, No. 4 (1993) 959–962

A NEW TYPE OF DYNAMICAL EXCITATIONS FOR PERCOLATION CLUSTERS

M. KOLB
Laboratoire Chimie Théorique, Ecole Normale Supérieure, 69364 Lyon, France
Institut de Recherches sur la Catalyse CNRS 2, Ave. A. Einstein, 69626 Villeurbanne Cedex, France

Abstract

By increasing the local connectivity of percolation clusters, structures with different scaling properties are generated. Both the topology, characterized by the chemical distance, and the dynamics, measured by the spectral dimension, change. For bond percolation in two (three) dimensions, new exponents have been determined by means of numerical simulations: $d_{\text{min}} = 1.02(4)$ $(1.09(6))$ and, for the scalar model, $d_s = 1.56(7)$ $(1.76(8))$.

1. INTRODUCTION

Because of their fractal nature, the vibrational modes of percolation clusters are localized.[1,2] Two important, and related, properties of these excitations are the density of states, $N(\omega) \sim (\omega)^{d_s-1}$, and the dispersion relation, $\lambda \sim \omega^{d_s/D}$, where λ is a characteristic size of the fracton, d_s the spectral dimension, and D the fractal dimension. For percolation clusters, the dynamical exponent d_s is well known: $d_s \cong 4/3$ for the scalar model, both in two- and three-dimensions, and $d_s \cong 0.8$ ($d_s \cong 0.9$) for the vector model in two (three) dimensions.[1,3]

Experimental investigations, notably scattering experiments on aerogels,[4] have measured the spectral dimension. While they are consistent with theoretical predictions, they tend to favor somewhat larger values for d_s. In the present calculation, we aim at finding percolation-like structures with such larger values for the spectral dimension. Qualitatively speaking, this can be achieved by increasing the connectivity of existing percolation clusters. We show that a slight modification of the local connectivity is sufficient to induce a change in the global connectivity, and as a consequence, the dynamics.

2. THE MODEL

Percolation clusters were generated by standard methods on square and cubic lattices at the percolation threshold.[5] Bond percolation was used instead of site percolation because of its extend scaling range.[6] This may be crucial for dynamical calculations. In two (three) dimensions, linear sizes up to $L = 2896$ ($L = 128$) on a hypercubic lattice were used for statics and $L = 90$ ($L = 32$) for dynamics. Both fixed and periodic boundaries were used, depending on the quantity measured. The modification of the connectivity with respect to standard percolation is that additional bonds are placed on the incipient infinite cluster (after removing the finite clusters). Two rules were applied: (1) bonds were placed between all nearest neighbor sites (*NN*), and (2) additional bonds were placed between all next nearest neighbor sites (*NNN*). Obviously, these modifications only change the connectivity, but not the mass distribution.

3. RESULTS

The radius of gyration as a function of the chemical distance, $R(l)$, was determined for bond percolation clusters with the original bond structure and for the *NN* and *NNN* bond modifications introduced above. The results show that the local *NN* and *NNN* modifications have a global effect. Figure 1 shows $R(l)$ for the unmodified, the *NN* modified

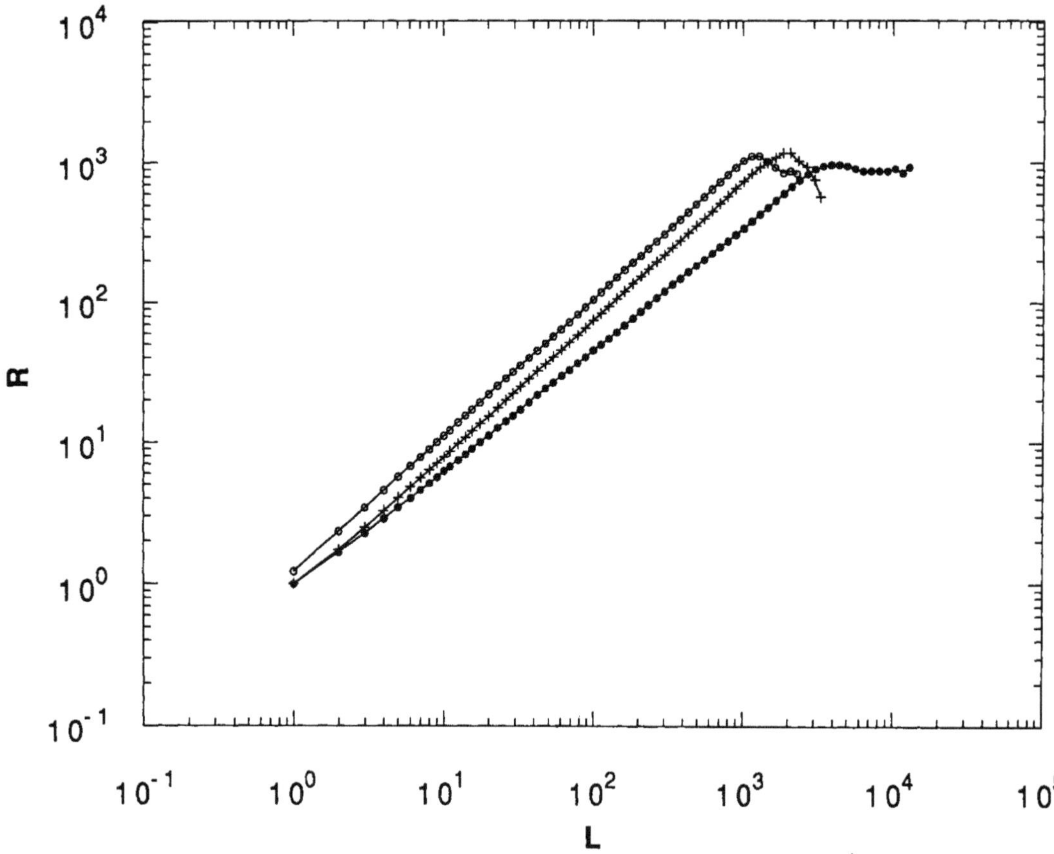

Fig. 1 Radius of gyration R as a function of the chemical distance L, for bond percolation. The unmodified incipient infinite cluster (dot) is compared with the *NN* modified (plus) and the *NNN* modified (open dot) cluster, for $L = 2048$.

and the *NNN* modified clusters in two dimensions. There is a change of the exponent d_{min} from $d_{min} = 1.16(4)$, consistent with the site percolation exponent,[7] to $d_{min} = 1.02(4)$ for *NN* and *NNN* bonded clusters. No significant change is observed between *NN* and *NNN* bonding. A similar result holds in three dimensions: d_{min} drops from $d_{min} = 1.41(6)$ to $d_{min} = 1.09(6)$, both for *NN* and *NNN* modifications.

The most interesting consequence of the changing connectivity is an expected change in the spectral properties. They were calculated for unmodified and modified bondings for the scalar percolation model. The results are entirely consistent with the static ones: the spectral dimension increases from $d_s = 1.34(3)$ to $d_s = 1.56(6)$ in two dimensions, and from $d_s = 1.36(4)$ to $d_s = 1.76(8)$ in three dimensions (both for *NN* and *NNN* modifications). Figure 2 shows the integrated density of states $N(E)$ for the three cases, with $E = \omega^2$. Again, *NN* and *NNN* modifications clearly scale in the same way, but not like the $N(E)$ of the original cluster. The relation $\lambda \sim \omega^{d_s/D}$ is also satisfied for the modified bonding rules.

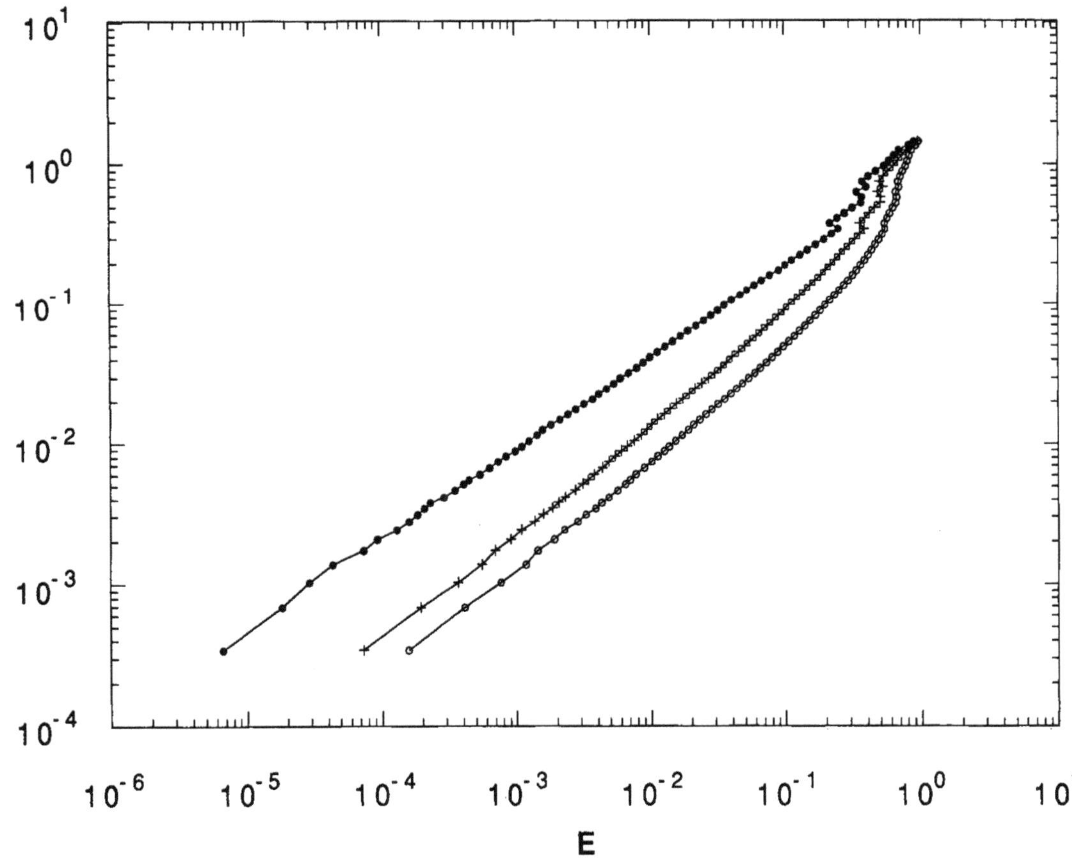

Fig. 2 Integrated density of states N as a function of $E = \omega^2$, for bond percolation. The unmodified incipient infinite cluster (dot) is compared with the *NN* modified (plus) and the *NNN* modified (open dot) cluster, for $L = 90$.

DISCUSSION

Adding additional bonds has been observed to change the fractal properties of percolation hulls for site percolation in two dimensions.[8] It has been shown that this result does not extend to three dimensions.[9] By considering properties of the cluster itself rather than the surrounding void space, the present calculation shows that changing the local connectivity does affect the scaling properties, both in two and in three dimensions.

Previous calculations for site percolation have shown that d_s varies with the degree of connectivity.[10] Our results show that a minimal change of the connectivity leads to an abrupt change of dynamical scaling.

REFERENCES

1. S. Alexander and R. Orbach, *J. Phys. Lett.* **43**, L625 (1982).
2. R. Rammal and G. Toulouse, *J. Phys. Lett.* **44**, L13 (1983).
3. I. Webman and G. S. Grest, *Phys. Rev.* **B31**, 1689 (1985).
4. E. Courtens, R. Vacher and E. Stoll, *Physica* **D38**, 41 (1989), and references therein.
5. D. Stauffer, *Introduction to Percolation Theory* Taylor and Francis, 1985.
6. E. Stoll, M. Kolb and E. Courtens, *Phys. Rev. Lett.* **68**, 2472 (1992).
7. H. J. Herrmann and H. E. Stanley, *J. Phys.* **A21**, L829 (1988).
8. T. Grossman and A. Aharony, *J. Phys.* **A19**, L746 (1986).
9. J.-F. Gouyet, M. Rosso and B. Sapoval, *Phys. Rev.* **B37**, 1832 (1988).
10. E. Stoll and E. Courtens, *Z. Phys.* **B81**, 1 (1990).

Fractals, Vol. 1, No. 4 (1993) 963–967

NONLINEAR STOCHASTIC DIFFERENTIAL EQUATIONS AND SELF-ORGANIZED CRITICALITY

ALBERT DIAZ-GUILERA*
Departament de Física Fonamental, Universitat de Barcelona,
Diagonal 647, 08028 Barcelona, Spain

Abstract

Several nonlinear stochastic differential equations have been proposed in connection with self-organized critical phenomena. Due to the threshold condition involved in its dynamic evolution, an infinite number of nonlinearities arise in a hydrodynamic description. We study two models with different noise correlations which make all nonlinear contributions to be equally relevant below the upper critical dimension. The asymptotic values of the critical exponents are estimated from a systematic expansion in the number of coupling constants by means of the dynamic renormalization group.

1. INTRODUCTION

Recently a lot of attention has been paid to the phenomenon known as self-organized criticality (SOC). Bak, Tang, and Wiesenfeld (BTW)[1] studied a cellular automaton model as a paradigm for the explanation of two widely occurring phenomena in nature: $1/f$ noise and fractal structures. Both have in common a lack of characteristic scales. This scale invariance suggests that the system is critical in analogy with classical equilibrium critical phenomena, but in self-organized criticality one deals with dynamical nonequilibrium statistical properties. On the other hand, the system evolves naturally to the critical state without any tuning of external parameters.

*E-mail: albert@ulyses.ub.es

Several cellular automata models exhibiting self-organized criticality have been reported in the literature. In the original sandpile model of Bak. et al.,[1] the system is perturbed externally by a random addition of sand grains; once the slope between two contiguous cells has reached a threshold condition, sand is transferred to its neighbors in a fixed amount. Taking this model as a reference, different dynamical rules have been investigated leading to a wide variety of universality classes: directed flows,[2] threshold condition imposed on the height, on the gradient or even on the Laplacian,[3] continuous variables with a full transfer from a cell instead of a fixed discrete amount,[4-7] deterministic perturbations in a nonconservative system,[8-10] anisotropy,[11] as some examples.

On the other hand, some authors have attempted to connect these cellular automata models showing SOC to nonlinear stochastic differential equations.[12-14] These continuous descriptions are built according to the symmetry rules obeyed by the discrete models. Anomalous diffusion equations with singularities in the diffusion coefficient have been considered in order to study the dynamics of the avalanches generated in the critical state.[15,16]

In this paper, we study two nonlinear stochastic differential equations whose deterministic parts keep the symmetries of the usual sandpile BTW models.[7,17] However, they have different noise correlations from which one gets different critical behavior. Both models are treated independently and the noise correlations are obtained by different limits of a general Ornstein-Uhlenbeck process and in some way they represent different aspects of SOC. For both models, an infinite number of nonlinearities are relevant below their upper critical dimension $d_c = 4, 2$; by taking a finite number of nonlinearities, we estimate the values of the critical exponents from an extrapolation. For one of the models we get the dynamical exponent obtained in the numerical simulations of systems exhibiting SOC.

2. STOCHASTIC DIFFERENTIAL EQUATIONS

The nonlinear stochastic differential equation we are going to study is:

$$\frac{\partial E(\mathbf{r}, t)}{\partial t} = D\nabla^2 E(\mathbf{r}, t) + \sum_{n=1}^{\infty} \lambda_{2n+1}\nabla^2 E^{2n+1}(\mathbf{r}, t) + \eta(\mathbf{r}, t) \tag{1}$$

where D is the bare diffusion coefficient and λ_i is the coupling constant accounting for the nonlinearities of the model. The variable E which we call energy can have different physical interpretations.[5] The difference between the models lies in the noise correlation:

$$\langle \eta(\mathbf{r}, t)\eta(\mathbf{r}', t')\rangle = 2\frac{\Gamma}{T}\delta^d(\mathbf{r} - \mathbf{r}')\,. \tag{2a}$$

$$\langle \eta(\mathbf{r}, t)\eta(\mathbf{r}', t')\rangle = 2\Gamma\delta(t - t')\delta^d(\mathbf{r} - \mathbf{r}')\,. \tag{2b}$$

Γ being the intensity of the noise and T is a macroscopic time to be discussed in the last section.

From Eq. (1) we can see that both models are translational and rotational invariant and that the energy is conserved in its deterministic part. An additional symmetry ($E \to -E$ invariance) has been introduced, since in Ref. 17 it has been shown that it is generated by a renormalization group procedure when Eq. (1) is derived from an equation involving a threshold condition.

3. DYNAMIC RENORMALIZATION GROUP

The scaling behavior of the system described by the nonlinear stochastic differential equation is obtained from the correlation function:

$$\langle (E(\mathbf{r}_0, t_0) - E(\mathbf{r}_0 + \mathbf{r}, t_0 + t))^2 \rangle^{\frac{1}{2}} \sim r^{\chi} f\left(\frac{t}{r^z}\right) \tag{3}$$

where χ and z are the "roughening exponent" and the dynamical exponent respectively. The relevance of the different coupling constants can be checked by naive dimensional analysis: a change of scale $\mathbf{r} \to e^l \mathbf{r}$ is followed by $t \to e^{zl} t$ and $E \to e^{\chi l} E$ in order for Eq. (3) be satisfied. Under this scaling transformation, z and χ are chosen to keep the linear equation invariant. With these values one can see that all coupling constants are of the same order in $\varepsilon = d_c - d$ and hence all nonlinear terms are equally relevant for $d < d_c$. The upper critical dimension d_c are 4 and 2 for the models, i.e., Eqs. (2a) and (2b), respectively.

The dynamic-renormalization-group procedure consists of two steps: an elimination of the faster short wavelength modes followed by a rescaling of the remaining modes.[18,19] The recursion relations after an infinitesimal transformation for the coefficients are[17]:

$$\frac{dD}{dl} = D\left[z - 2 + 3\frac{\Gamma \Lambda^{d-4} \lambda_3}{D^3}\right] \tag{4a}$$

$$\frac{d\left(\frac{\Gamma}{T}\right)}{dl} = \frac{\Gamma}{T}[2z - 2\chi - d] \tag{4b}$$

$$\frac{d\lambda_n(l)}{dl} = \lambda_n[(n-1)\chi + z - 2)] + \sum_j \gamma_j \Gamma \Lambda^{d-4} D^{a_j} \prod_{i=2}^{n+2} \lambda_i^{b_{ij}} \tag{4c}$$

and

$$\frac{dD}{dl} = D\left[z - 2 + 3\frac{\Gamma \Lambda^{d-2} \lambda_3}{D^2}\right] \tag{5a}$$

$$\frac{d\Gamma}{dl} = \Gamma[z - 2\chi - d] \tag{5b}$$

$$\frac{d\lambda_n(l)}{dl} = \lambda_n[(n-1)\chi + z - 2)] + \sum_j \gamma_j \Gamma \Lambda^{d-2} D^{a_j} \prod_{i=2}^{n+2} \lambda_i^{b_{ij}} \tag{5c}$$

for the models, i.e., Eqs. (2a) and (2b), respectively. In the previous equations, γ_j is numerical factor and b_{ij} a natural number accounting for the vertex λ_i entering the renormalization of the vertex λ_n in the j-th term of the sum.

Since all terms in this expansion are equally relevant below the respective upper critical dimension, a calculation of the critical exponents would involve all the contributions. Nevertheless the exponents can be obtained to first order in $\varepsilon = d_c - d$ as a series in the number of coupling constants which are taken into account. For details of the calculations for model (2a), see Ref. 17 where one obtains:

$$\begin{aligned} &z = 2 - 0.32(4-d); \quad \chi = 0.18(4-d) \quad \text{for } d \le d_c = 4 \\ &z = 2; \qquad\qquad\qquad \chi = \frac{4-d}{2} \qquad\quad \text{for } d \ge d_c = 4 \end{aligned} \tag{6a}$$

On the other hand, the results for model (2b) can be obtained in an equivalent way to be:

$$
\begin{aligned}
&z = 2 - 0.33(2-d); \quad &\chi = 0.34(2-d) \quad &\text{for } d \le d_c = 2 \\
&z = 2; \quad &\chi = \frac{2-d}{2} \quad &\text{for } d \ge d_c = 2
\end{aligned}
\tag{6b}
$$

4. RELATION WITH SELF-ORGANIZED CRITICALITY

Up to now it is not clear yet the relation between the cellular automata models exhibiting self-organized criticality and the nonlinear stochastic differential equations analyzed in the present work. In the original sand-pile model, and in others derived from it, the system is perturbed externally until a site reaches a threshold condition and then an avalanche is triggered. During the evolution of the avalanche, there is no external perturbation onto the system. This introduces two time scales in the model: a fast time scale (the diffusion scale) and a slow one (the noise scale).

For the purpose of an analytical treatment, one can assume the external noise to be described by a Gaussian process with zero mean and correlation function given by:

$$
\langle \eta(\mathbf{r}, t)\eta(\mathbf{r}', t')\rangle = \frac{2\Gamma}{\tau} e^{-|t-t'|/\tau}\delta^d(\mathbf{r}-\mathbf{r}') . \tag{7}
$$

This is an Ornstein-Uhlenbeck process, with τ the correlation time. This general case can describe two different limits depending on whether τ is a microscopic or a macroscopic time. In the former case, we get Eq. (2b) whereas in the latter, one obtains Eq. (2a) with an intensity of the noise that scales with $1/\tau$.

When analyzing the SOC models within a mesoscopic time scale, we deal with the dynamics of single avalanches. In this case model, i.e., Eq. (2a) is more appropriate since the noise scale is larger than the time scale we are interested in. A quenched noise of low intensity does not affect the evolution of the avalanche since it only depends on whether a site is above its threshold. This is the reason for which the estimation of the dynamical exponent [Eq. (6a)] agrees with the values obtained in the numerical simulations and from scaling arguments[4] concerning the evolution of single avalanches. Nevertheless, it can modify the roughness of the interface since it can make two distant sites to be correlated in an artificial way; however, the effect of the nonlinearities is to lower the value of the roughening exponent with respect to the linear model.

SOC models can be analyzed from a different point of view by considering the avalanches to be instantaneous. In this case one is mainly interested in the form of the interface after the relaxation events (avalanches) have taken place and not in its dynamical evolution. This situation is properly described by Eqs. (1) and (2b), but the parameters entering the deterministic evolution diverge since they are related to the diffusion coefficient. Only finite values of the parameters make the problem suitable of an analytical treatment; in doing so, we make the two time scales discussed above to be of the same order.[14] Although it is not the situation one deals with in the simulations, it is interesting to note that the roughening exponent χ is lowered with respect to the previously discussed model. Actually, one should expect a roughening exponent $\chi \le 0$ due to the fact that the energy at each site is bounded after the avalanches are finished. But as the two dynamics (diffusion and noise) overlap, we can still get a rough interface, at least for low dimensionality.

To summarize, we have analyzed two nonlinear stochastic differential equations by means of the dynamic renormalization group. Both equations share the deterministic evolution but differ in the noise correlation. Due to the symmetries of the cellular automata models they are derived from, an infinite number of relevant nonlinearities must be taken into account. By taking a finite number of these nonlinear couplings, we extrapolate to infinity and get an estimation of the critical exponents. These two models allow the description of different, as well as complementary, characteristics of cellular automata models exhibiting SOC. The quenched noise model reproduces the dynamical exponent obtained in the evolution of single avalanches but introduces additional correlations in the interface profile. On the other hand, the delta-correlated noise model gives a roughening exponent which is closer to what one would expect in the profile of the interface after the avalanches have taken place.

ACKNOWLEDGMENTS

This work has been supported by CICyT of the Spanish Government, grants #PB89-0233 and #PB92-0863.

REFERENCES

1. P. Bak, C. Tang and K. Wiesenfeld, *Phys. Rev.* **A38**, 364 (1988).
2. L. P. Kadanoff, S. R. Nagel, L. Wu and S. Zhou, *Phys. Rev.* **A39**, 6524 (1989).
3. S. S. Manna, *Physica* **179A**, 249 (1991).
4. Y.-C. Zhang, *Phys. Rev. Lett.* **63**, 470 (1989).
5. L. Pietronero, P. Tartaglia and Y.-C. Zhang, *Physica* **A173**, 22 (1991).
6. Z. Fodor and I. M. Janosi, *Phys. Rev.* **A44**, 1386 (1991).
7. A. Díaz-Guilera, *Phys. Rev.* **A45**, 8551 (1992).
8. H. J. S. Feder and J. Feder, *Phys. Rev. Lett.* **66**, 2669 (1991).
9. K. Christensen, Z. Olami and P. Bak, *Phys. Rev. Lett.* **68**, 2417 (1992).
10. Z. Olami and K. Christensen, *Phys. Rev.* **A46**, R172 (1992).
11. I. M. Janosi, *Phys. Rev.* **A42**, 769 (1990).
12. T. Hwa and M. Kardar, *Phys. Rev. Lett.* **62**, 1813 (1989).
13. G. Grinstein, D.-H. Lee and S. Sachdev, *Phys. Rev. Lett.* **64**, 1927 (1990).
14. T. Hwa and M. Kardar, *Phys. Rev.* **A45**, 7002 (1992).
15. J. M. Carlson, J. T. Chayes, E. R. Grannan and G. H. Swindle, *Phys. Rev. Lett.* **65**, 2547 (1990).
16. P. Bantay and I. M. Janosi, *Phys. Rev. Lett.* **68**, 2058 (1992).
17. A. Diaz-Guilera, (preprint).
18. D. Forster, D. R. Nelson and M. J. Stephen, *Phys. Rev.* **A16**, 732 (1977).
19. E. Medina, T. Hwa, M. Kardar, Y.-C. Zhang, *Phys. Rev.* **A39**, 3053 (1990).

Fractals, Vol. 1, No. 4 (1993) 968–976

PATTERN FORMATION IN NONCOHESIVE AND COHESIVE GRANULAR MEDIA

E. LEMAIRE
IRSID, 34 rue de la Croix de Fer, F-78105 Saint-Germain en Laye, France
Y. OULD MOHAMED ABDELHAYE, J. LARUE, R. BENOIT
P. LEVITZ and H. VAN DAMME*
Centre de Recherche sur la Matière Divisée, CNRS-Université d'Orléans, F-45071 ORLEANS cedex 02, France

Abstract

We studied pattern formation in granular media in two distinct states: in the dry and non-cohesive (powder) state, on the one hand, and in the wet and cohesive (paste) state, on the other. In the first case, we have shown that gas injection in a thin layer of powder within a Hele Shaw cell leads to fractal patterns remarkably similar to viscous fingering patterns obtained with complex fluids. In the second case, we have shown that the tensile cohesive viscoplastic fracture of a layer of paste leads to self-affine rough surfaces with a Hurst exponent close to 0.88, very close to the value obtained for fragile fracture by other authors.[18] Our observations reinforce the universality of two fractal growth processes and add a new element to the ambivalent nature of the granular state of matter.

1. INTRODUCTION

Granular media may exhibit two distinct types of behavior. In the dry state (powders), they are cohesionless media the complex behavior of which is dominated by friction forces.[1] Their most characteristic property, described by Reynolds more than a hundred years ago,

*to whom correspondence should be addressed.

is dilatancy. Flow or even simple deformation of a randomly packed granular medium is impossible without allowing the material to expand. On the other hand, in the wet state (pastes and slurries), they may exhibit cohesion due to capillary forces if the slurry or paste is unsaturated (intergranular gas-liquid interfaces), or to interparticular short-range attractive forces (as in cement pastes). Concomitantly, the role of friction may be diminished by introducing even shorter-range repulsive forces (by electrostatic effects or polymer adsorption for instance), leading to a crossover from dilatant to truly plastic behavior.[2]

To what extent dry powders may behave as non-Newtonian fluids and to what extent pastes may behave as ductile solids is still a matter of investigation. We report here a preliminary experimental study of two types of interfacial instabilities which may be considered as typical of the fluid and solid states of matter, respectively: viscous fingering on one hand, and fracture on the other. However, unconventionally, we performed the viscous fingering experiments in a layer of dry powder, whereas the fracture experiments were performed in pastes. As will be shown below, both processes lead to fractal patterns or surfaces which are remarkably similar to those obtained in non-granular fluids or solids, although some specific features were observed.

2. VISCOUS FINGERING IN POWDERS

Viscous fingering (VF) is an interfacial instability which occurs when a low-viscosity fluid is forced under pressure into a high-viscosity fluid within a medium where flow is dominated by friction with the walls, such as a porous medium or a Hele-Shaw cell (two plates separated by a narrow gap).[3] Viscous forces tend to destabilize the interface, whereas interface tension, T (or other stresses stemming from gradients such as concentration gradients, for instance[3,4]) tends to stabilize it. A suitable control parameter with immiscible fluids and a very large viscosity contrast is the capillary number, $C_a = \mu U/T$, where U is the interface velocity and μ the viscosity. C_a expresses the ratio of viscous over capillary forces. In Newtonian fluids within axial or radial Hele-Shaw cells, the VF instability leads to the growth of smooth fingers,[5,6] whose width and splitting depend on C_a. The use of a shear-thinning non-Newtonian fluid introduces an additional instability and allows for increasing considerably the effective capillary number.[7] This results in a drastic decrease of the wavelength of maximal growth and an increase of branching,[8] leading to fractal arborescences[9–11] with a mass fractal exponent of 1.65 in radial geometry. As will be shown below, gas injection in a quasi-2D powder bed in a Hele-Shaw leads to similar patterns.

The experiments were performed by injecting air at an upstream pressure of 4×10^5 Pa into a layer of zirconia powder (grain diameter: 140 ± 20 micrometers) in a 40 cm × 40 cm Hele-Shaw cell made of two 1 cm thick plexiglass plates. Filling a Hele-Shaw cell homogeneously with a powder is much more difficult than with ordinary fluids. Our cell was filled by pouring the powder on the bottom plate in order to make a central heap and gently sliding and rotating the upper plate until a spacing of 0.5 ± 0.1 mm is reached. After filling, the cell was clamped on its four edges.

Two regimes are observed. At low inlet valve opening, the powder layer behaves as a rigid porous medium allowing only for gas permeation in its pores, all granular movements being blocked. On the other hand, at full valve opening, the particles were repelled by the gas flux, leading to a branched pattern of powder-free channels remarkably similar to the fractal viscous fingering patterns obtained with non-Newtonian fluids in radial cells.[9–11] The front velocity estimated from the growing sequences is always high, of the order of 0.3 m s^{-1}. A typical growing sequence is shown in Fig. 1. The interface between the gas channels and the powder is in general very sharp (Fig. 2). This permits the easy determination of the mass fractal dimension of the patterns from the area vs. radius relationship (Fig. 3). It is 1.65 ± 0.05. The interface itself may be either straight or very wavy (Fig. 2), suggesting that several instability modes are acting.

Fig. 1 A growth sequence of a fingering pattern. The time between exposures is 0.05 s. Note the circular compacted domain around the pattern.

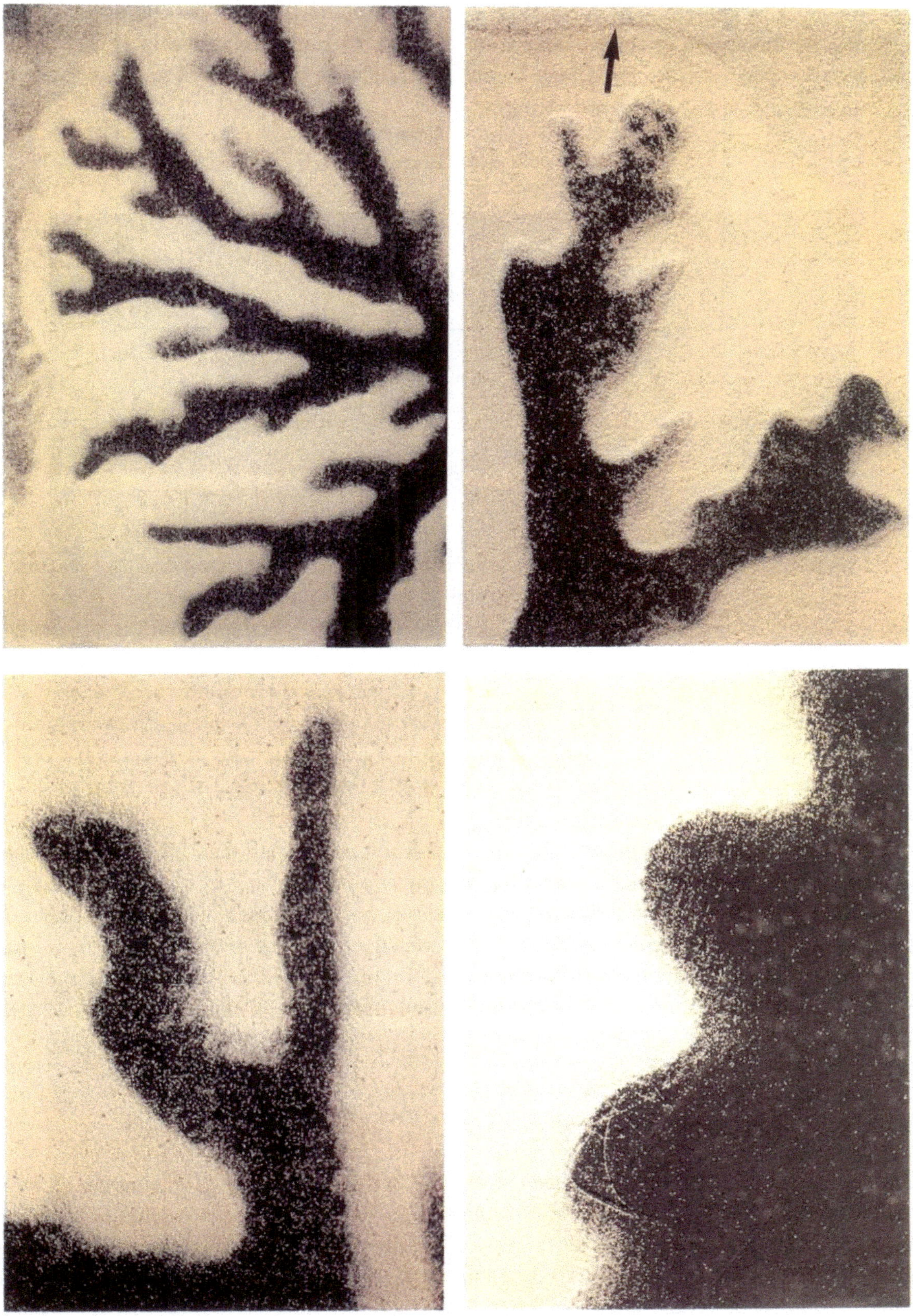

Fig. 2 Enlarged view of a pattern and of fingers. The arrow indicates the boundary between the dense and unperturbed granular regions. Note the sharpness of the interface and its shape which may be very straight or, on the contrary, quasi-periodically wavy.

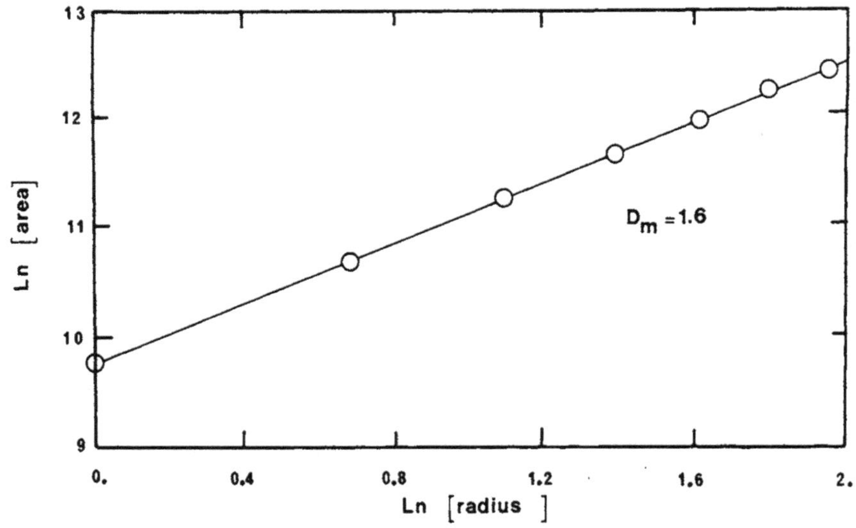

Fig. 3 Determination of the mass fractal dimension of the pattern of Fig. 1 from the area-radius relationship.

Undoubtedly, this granular flow is made possible by the (not yet quantified) separation of the cell plates under gas pressure. Close examination of the patterns shows that, in fact, two regions may be detected in the invaded granular layer. The fingering pattern is surrounded by a compacted zone which is itself surrounded by the unperturbed region (Fig. 1 and, more clearly, Fig. 2). The interface between the compacted and the unperturbed regions is as sharp as that between the channels and the compacted region. The shape of the compacted zone is roughly circular. The distance between the branch tip and the boundary of the dense region is of the same order as the finger width, which is itself of the order of $20b$, b being the cell spacing.

In viscous fingering with miscible fluids, the mechanism of wavelength selection is related to energy dissipation or, more generally, entropy production.[3,4,12,13] It is likely that the same type of mechanism is acting in the present case. A reasonable suggestion based on Bagnold's model for shear flow of powders[14] is that a wavelength of maximum growth may be obtained by minimizing the rate of frictional energy dissipation, but more data on the relationships between finger width, cell deformation, particle roughness and particle velocity are needed to put this on a quantitative basis.

3. THE FRACTURE OF PASTES

The second instability that we studied is fracture. In fracture, the destabilizing factor is the release of elastic stress whereas the stabilizing factors are, like in viscous fingering, interfacial energy or energy dissipation. The fracture of fragile or ductile solids has been shown by several authors to produce self-affine rough surfaces.[15,19] In ductile fracture with aluminium alloys,[18] the fractal dimension of the fracture surface determined by the slit-island method was found to be in all cases close to 2.2, corresponding to a roughness exponent $H = 3 - D$ of 0.8. On the other hand, for brittle fracture on a variety of materials including steel, graphite, porcelain, bakelite and plaster, K. J. Maloy et al.[19] found significantly higher roughness exponents (lower local fractal dimension), with an average value of 0.87.

Fig. 4 Oblique and grazing views of a layer of paste after rupture and hardening.

Our experiments on viscoplastic fracture were performed by separating along a vertical axis two horizontal 30 cm × 30 cm glass plates separated by a 1 cm thick layer of fresh paste (tensile mode l fracture). The paste was a mortar prepared by mixing sand (average particle size: 200 μm) with an acrylic resin (Produits Chimiques du Nord) at a sand/resin weight ratio of 2. In order to avoid slipping of the paste on the glass (adhesive fracture), the plates were first made rough by coating them with a thin (less than 1 mm) layer of paste which was left to harden. This was enough to ensure cohesive fracture of the paste layer. The two parts left after separation were not strictly equal. Due to gravitation, the layer of paste left on the bottom plate was always somewhat thicker than the layer on the top plate. The separation velocity was controlled in a range going from 1 to 10 mm s^{-1}. After hardening of the paste, the roughness of the fracture surfaces was analyzed by determining the roughness exponents or fractal dimensions of the one-dimensional profiles obtained by sawing the surface into vertical slices (Fig. 5). We used both the power spectrum or the box counting method. No significant difference was observed between slow and fast fracture. The average value for the roughness exponents for 5 experiments at different rupture velocities (three profiles

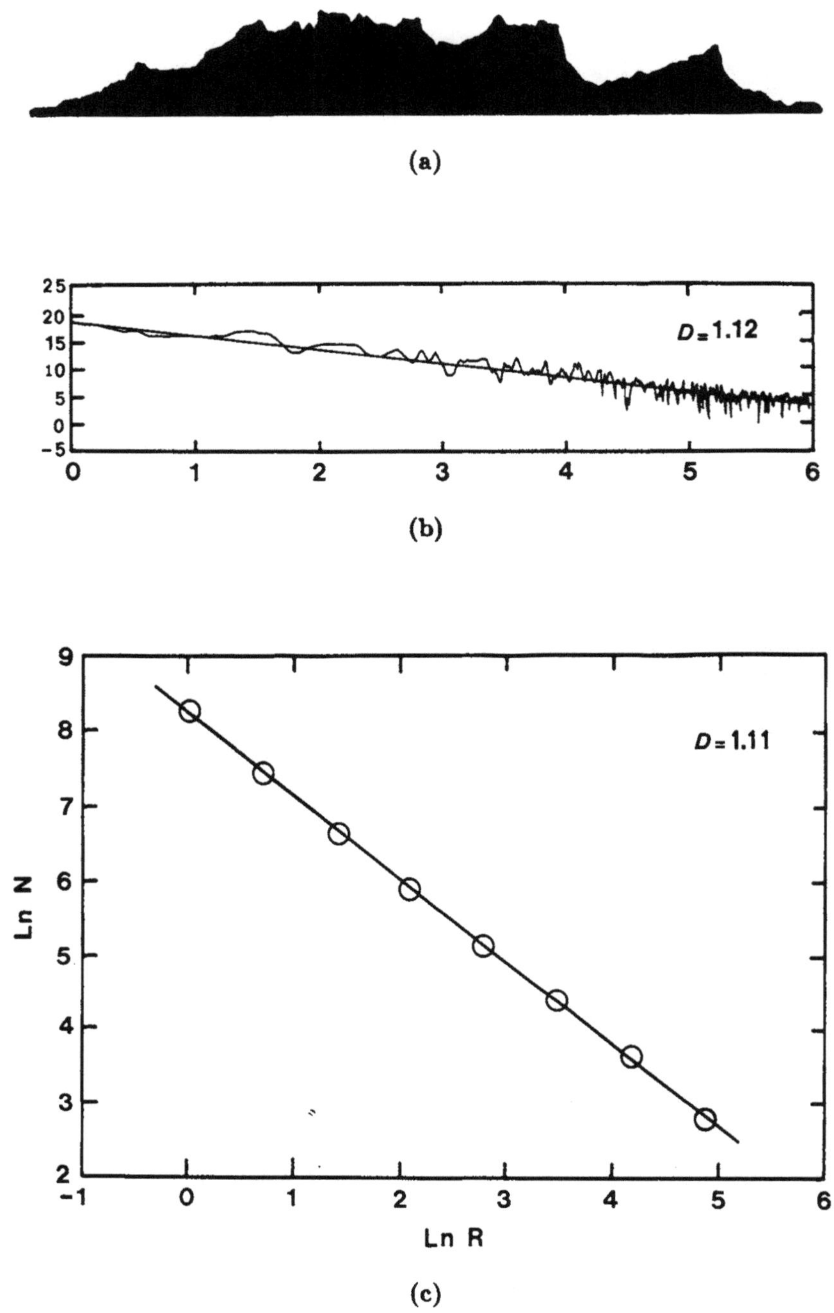

Fig. 5 (a) Profile of a slice sawed from the rough layer of Fig. 4; (b) determination of the roughness exponent from the Fourier power spectrum; (c) determination of the fractal dimension by the box counting method.

were analyzed in each experiment) was found to be 0.88 ± 0.05, which corresponds to an average local fractal dimension of 1.12 for the profiles and 2.12 for the surfaces. This value is remarkably close to the value obtained for brittle fracture by Maloy et al.[19]

The mechanism by which the layer of paste breaks is not yet clear. When this experiment is done with a colloidal fluid such as a clay suspension of much lower viscosity than the granular paste, it leads to a pattern which is, in fact, the result of viscous fingering due to air penetration from the boundary of the cell into the broadening gap. This leads to a pattern of smooth and rounded crests of uniform height where the valleys are all accessible

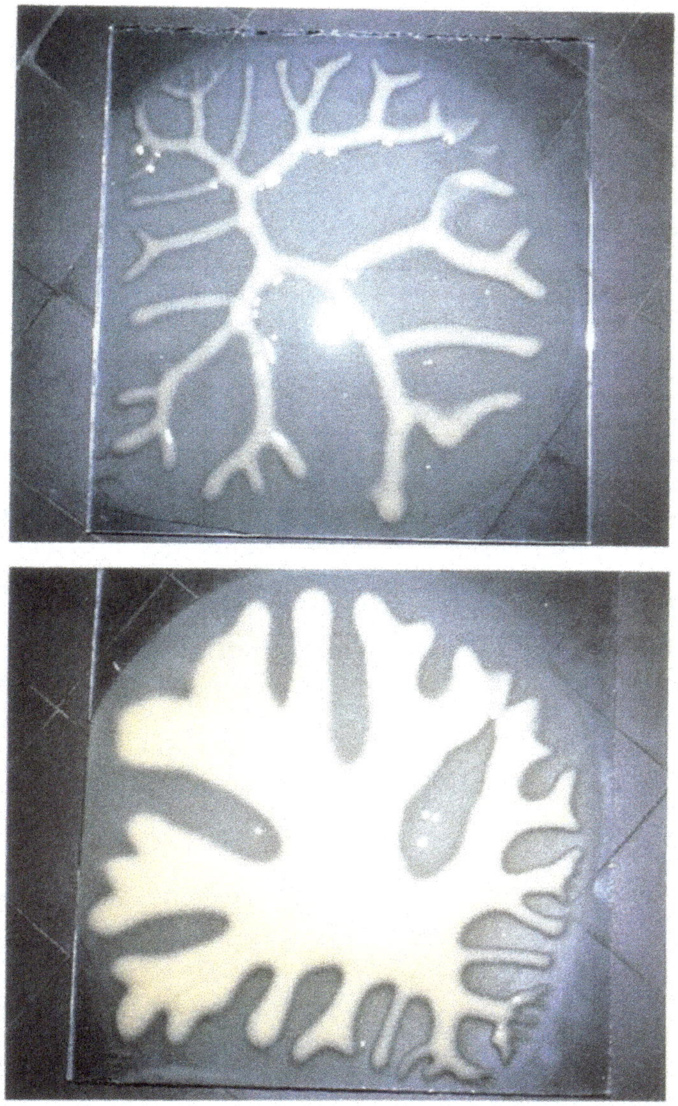

Fig. 6 Patterns obtained by tearing away two plates separated by a layer of colloidal clay suspension (5% bentonite in water, w/w).

from the boundary of the cell (Fig. 6). This is not the case in the paste fracture surfaces, suggesting that fracture is initiated by cavitation or bubble growth at many places.

4. CONCLUSION

What we have shown in this preliminary paper is that two fractal pattern growth phenomenona which are typical of the liquid and solid state of matter may also be observed in dry and wet granular media, respectively. Furthermore, the fractal dimensions obtained are, within experimental error, the same as those obtained in the reference experiments (viscous fingering in viscous liquids and fracture in brittle solids). The main questions that this result raises are the mechanism of wavelength selection at the gas-powder interface on the one hand (in viscous fingering), and the origin of the universality of height correlations in brittle as well as viscoplastic fracture on the other.

REFERENCES

1. For a recent review, see H. M. Jaeger and S. R. Nagel, *Science* **255**, 1523 (1992).
2. K. Kendall, in *Tribology in Particulate Technology*, eds. B. J. Briscoe and M. J. Adams (Adam Hilger, Bristol, UK, 1987), pp. 91–102.
3. G. M. Homsy, *Ann. Rev. Fluid Mech.* **19**, 271 (1987).
4. D. D. Joseph, *Eur. J. Mech. B/Fluids* **9**, 565 (1990).
5. P. G. Saffman and G. I. Taylor, *Proc. Roy. Soc. London, Ser.* **A245**, 312 (1958).
6. L. Paterson, *J. Fluid Mech.* **113**, 513 (1981).
7. H. Van Damme, E. Alsac, C. Laroche and L. Gatineau, *Europhys. Lett.* **5**, 25 (1988).
8. H. Van Damme, E. Alsac and C. Laroche, *C.R. Acad. Sc.* **309**, serie II, 11 (1989).
9. G. Daccord, J. Nittmann and H. E. Stanley, *Phys. Rev. Lett.* **56**, 336 (1986).
10. H. Van Damme, F. Obrecht, P. Levitz, L. Gatineau and C. Laroche, *Nature* **320**, 731 (1986).
11. H. Van Damme, C. Laroche and L. Gatineau, *Rev. Phys. Appl. (Paris)* **22**, 241 (1987).
12. L. Paterson, *Phys. Fluids* **28**, 26 (1985).
13. H. Van Damme and E. Lemaire, in *Disorder and Fracture*, eds. J. C. Charmet, S. Roux and E. Guyon (NATO ASI Series, B235, Plenum Press, New York, 1990).
14. R. A. Bagnold, *Proc. Roy. Soc. London, Ser.* **A225**, 49 (1954) and **295**, 219 (1966).
15. B. B. Mandelbrot, D. E. Passoja and A. J. Paullay, *Nature* **308**, 721 (1984).
16. C. S. Pande, L. R. Richards and S. Smith, *J. Mater. Sc. Lett.* **6**, 295 (1987).
17. D. L. Davidson, *J. Mater. Sc.* **24**, 681 (1989).
18. E. Bouchaud, G. Laplasset and J. Planes, *Europhys. Lett.* **13**, 73 (1990).
19. K. J. Maloy, A. Hansen, E. L. Hinrichssen and S. Roux, *Phys. Rev. Lett.* **68**, 213 (1992).

Fractals, Vol. 1, No. 4 (1993) 985–991

QUANTIFYING MORPHOLOGY OF SCALE-INVARIANT STRUCTURES BEYOND THE FRACTAL DIMENSION

RAPHAEL BLUMENFELD[†] and ROBIN C. BALL[‡]
[†] *Princeton Materials Institute, Princeton University, Bowen Hall, 70 Prospect Avenue, Princeton 08540-5211, NJ, USA.*
[‡] *Cavendish Laboratory, Cambridge University, Madingley Road, Cambridge CB3 0HE, UK.*

Abstract

We present a correlation scheme to quantify the morphology beyond the standard fractal dimension which corresponds to information from the pair correlation function. The method consists of analyzing hierarchical correlations in log-space thus summing contributions from higher order correlations in the usual space coordinates. The scheme gives information on the characteristics of structure which can be used as a fingerprint to distinguish between structures with the same fractal dimension. The method is also sensitive to oscillations in logarithmic scales, which are admissible solutions for renormalisation equations. Such oscillations appear as resonances, thus making this scheme suitable to analyze such phenomena experimentally. Illustrative examples are given for all those applications by analyzing numerically grown structures. The case of the unrestricted Brownian walk is exactly calculated. We discuss an application of this scheme to check recent analytic results obtained for scale-invariant branching mechanisms in slow cracking patterns and in noise-reduced diffusion-limited-aggregation. We propose that this method is a suitable candidate to quantify the presently qualitative concept of morphology.

1. INTRODUCTION

Following the systematic observations of Mandelbrot,[1] an overwhelming abundance of fractal structures in nature has been firmly established in the last decade. The symmetry underlying these structures is their scale-invariance, namely, under a transformation $r \to r' = Rr$ the structure can be mapped back onto (a subset of) itself. In doing so, the "measure" must be rescaled by a factor of R^D, where the index D is the fractal dimension. Such structures have been very useful in modeling disordered systems, phase transitions and critical phenomena, and stochastic growth processes. D is a key characteristic of the structure and is used to both distinguish between different fractal structures and to compare measurements made on different scales. Due to the scale-invariance various properties of the structure are governed by power laws. This is the case for many connective, topological, transport and mechanical properties. The long list of exponents has been taken as a way to characterize structures better than just by the index D. This unsystematic specification clearly calls for some organizing principles to exactly characterize the structure of fractals. Such characterization is essential when there is a need to distinguish between different morphologies, which requires a quantitative definition of morphology. The few attempts that have been made focused on specifying the variance of the spatial distribution of voids in systems of mass points.[1-3] Here we develop and apply a systematic method to characterize fractal structures and different morphologies in terms of the correlations between subsets of the structure at different scales. We show that this approach enables distinction to be made between structures with similar fractal dimension at already small system sizes. For the exactly solvable case of Brownian random walks and randomly branched polymers, the method identifies corrections to scaling from the *large scaling regime* and we propose that this feature is general to other fractals. We show that this correlation scheme is sensitive to logarithmic periodicities and can be used to analyze branching processes and complex fractal dimensions.

Let us consider the implications of various symmetries on measurements of n-point normalized correlation functions. The one-point function is just the local density $\phi_1(\mathbf{r})$, whose volume averaged in a translational invariant system is constant ϕ_1. The two-point correlation function is $\phi_2(\mathbf{r}, \mathbf{r}') = \phi_1 C_2(\mathbf{r} - \mathbf{r}')$ and if the system is translational and rotational invariant then $C_2 = C_2(|\mathbf{r} - \mathbf{r}'|)$. If in addition, the system is scale-invariant then $C_2(r) = Ar^{D-d}$ and only the constants A and D can be determined. For the three-point function, translational and rotational invariance give:

$$\phi_3(\mathbf{r}, \mathbf{r}', \mathbf{r}'') = \phi_1 C_3(|\mathbf{r} - \mathbf{r}'|, |\mathbf{r} - \mathbf{r}''|, \theta), \tag{1.1}$$

where θ is the angle between the two vectors. Dilational invariance then reduces C_3 to depend only on $|\mathbf{r} - \mathbf{r}'|/|\mathbf{r} - \mathbf{r}''|$ and θ:

$$\phi_3(\mathbf{r}, \mathbf{r}', \mathbf{r}'') = \phi_1 C_2(|\mathbf{r} - \mathbf{r}'|) C_2(|\mathbf{r} - \mathbf{r}''|) \mathcal{L}\left[\ln \frac{|\mathbf{r} - \mathbf{r}'|}{|\mathbf{r} - \mathbf{r}''|}, \theta\right]. \tag{1.2}$$

So in a statistically isotropic structure with a continuous dilational symmetry, the three-point correlations reduce to a new non-trivial scaling function $\mathcal{L}$ of just two variables. Because of the additional symmetry, we expect three-point correlations to show the signature of fractal structures. This is analogous to the use of the pair distribution in the study of isotropic and translational invariant fluids. In that context, the internal structure shows up in the two-point correlation (Fourier spectrum, structure factor). In our case the scale-invariance reduces the pair correlation function to describe only the scaling of the average, and therefore to detect internal details requires "climbing" one point up the ladder. The scale invariance also suggests that it is useful to analyze the correlations in *logarithmic* coordinates, effectively summing contribution from *all* the n-point correlations in the usual coordinates.

In the following we reformulate the reduced three-point correlation function in operatively measurable terms. We analyze $\mathcal{L}$ for several structures, showing that: (i) it detects features characteristic to the structure; (ii) it distinguishes between structures of similar fractal dimension; and (iii) it is sensitive to logarithmic periodicities. This latter feature can be used to probe scale-invariant branched structures such as diffusion limited aggregates complex fractal dimensions.

2. THE LACUNARITY FUNCTION

Consider a set of points in d dimensions, each of which is of a unit mass. Choosing an arbitrary point i we construct concentric shells around it that are of uniform width in *logarithm* of the distance $r = |\mathbf{r}|$ from i. We measure $\delta S_i(r)$ = number of other points in a spherical shell at radius r from it. The average of this measurement over all choices of i, normalized by the volume of the shell, is the usual pair correlation function C_2:

$$C_2(r) \equiv \langle \delta S_i(r) \rangle / \delta V(r) \tag{2.1}$$

where $\langle\,\rangle$ stands for averages over i. The relative fluctuations in the mass distribution about site i in terms of the relative shell masses are $\delta\sigma_i(r) = \delta S_i(r)/\langle \delta S(r) \rangle$. For a fractal the fluctuations in $\delta\sigma_i(r)$ are of order unity and are correlated over multiplicative ranges of radius, namely, over a simple range in $\rho = \ln r$. We now define a correlation matrix:

$$M(\rho, \rho') = \langle \delta\sigma_i(r) \delta\sigma_i(r') \rangle \, . \tag{2.2}$$

We expect that due to the scale-invariance $M(\rho, \rho') = \mathcal{L}(\rho - \rho')$. Thus $\mathcal{L}$ is the average of the autocorrelation function of the relative shell masses, and we term it a (radial) Lacunarity Function (LF). Note that an earlier definition of lacunarity[2] is simply the point $\mathcal{L}(0)$. In Fig. 1, we show contour plots of M for several two-dimensional systems: A 4000-monomer-long self-avoiding walk (SAW), a 4096-particle cluster-cluster aggregate (CCA) and a sample of 10^4 randomly chosen particles from a 10^5-particle diffusion-limited aggregate (DLA). Disregarding end effects at the largest and smallest separations, one can easily detect contours parallel to the diagonal that are the signature of scale-invariance and show that the correlations depend only on the relative logarithmic separation $|\rho - \rho'|$. The corresponding LFs are shown in Fig. 2 and show nontrivial structures. In all three cases $\mathcal{L}$ starts from a value appreciably larger than the one at the origin, and saturates to unity for $|\rho - \rho'| > \lambda_s$. λ_s resembles a correlation length in traditional systems in that when $r'/r > \exp(\lambda_s)$ subsets of the structure are uncorrelated.

Gefen et al.[2] and Mandelbrot[4] introduced the concept of Lacunarity to distinguish between fractal patterns with similar fractal dimension, a task that proved to be surprisingly difficult to achieve by a simple analytic tool. Figure 2 shows that $\mathcal{L}$ indeed distinguishes between CCA and SAW that have very similar fractal dimensions (1.42 and 1.34 in 2d, respectively), which has been achieved for 4000 particles already! Note that to separate the structures by traditional measurements of the fractal dimensions sufficiently accurately, one needs to analyze systems of tens of thousands of particles. The difference between the structures is observed by the different characteristics of the decays of $\mathcal{L}$ to unity. To check that indeed on this scale the fractal dimensions cannot be conclusively separated, we have

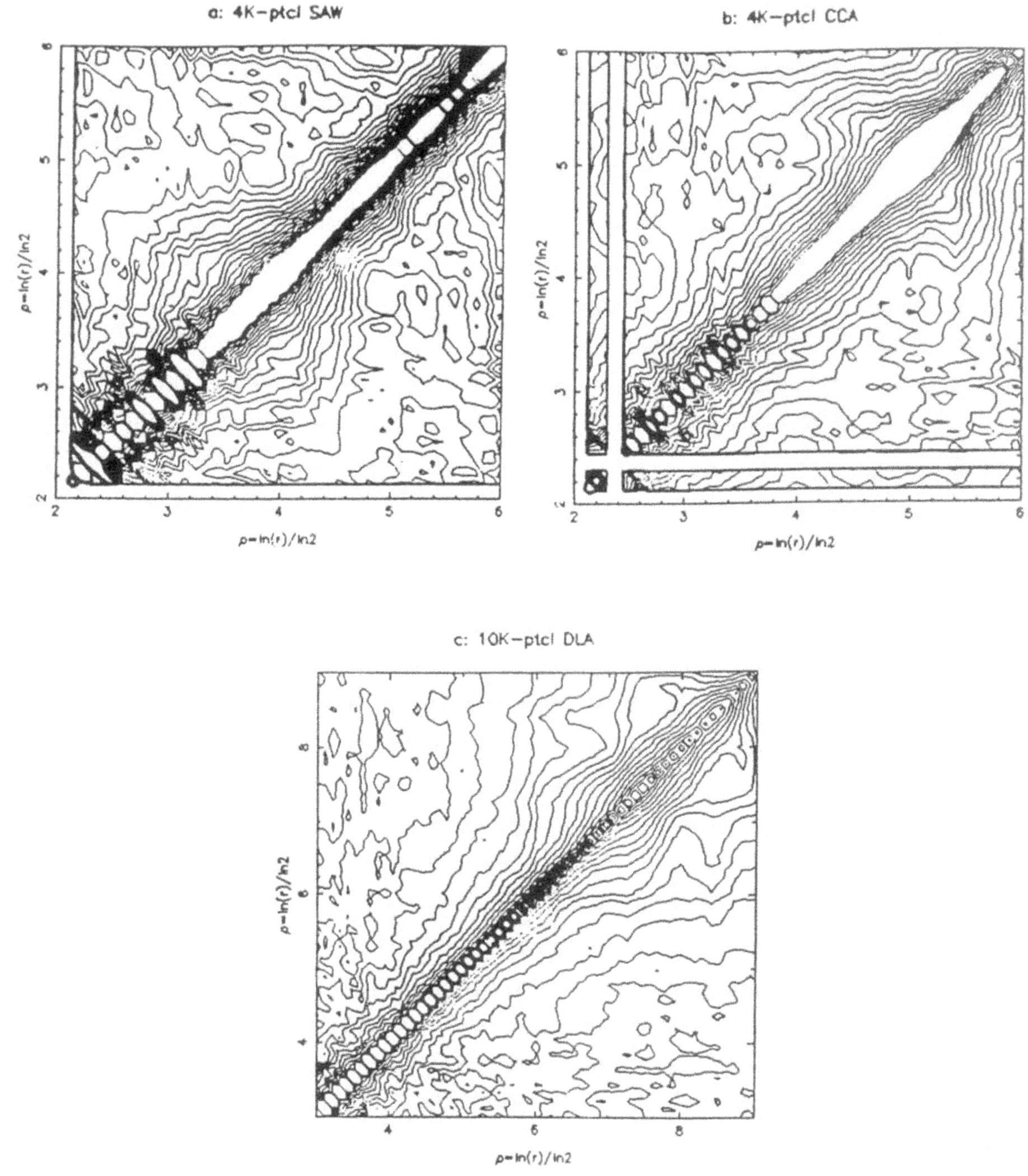

Fig. 1 A contour plot of the correlation matrix $M(\rho, \rho')$ for: **(a)** 4000-monomer-long SAW; **(b)** a 4096-particle CCA, and **(c)** a sample of 10^4-particles chosen randomly from a 10^5-particle DLA.

measured these and found $D_{\mathrm{CCA}} = 1.45 \pm 0.15$ and $D_{\mathrm{SAW}} = 1.3 \pm 0.2$, which establishes our claim.

Previously[5] we have shown that such a formulation can be generalized to take account of directions by correlating $\delta S_i(\mathbf{r})$ also at fixed vector separation rather than just fixed radius. The resulting angle-dependent LF contains the full three-point correlation function for a scale invariant and isotropic structure. From it, the radial $\mathcal{L}$ function can be recovered as its (solid angle weighted) angular average. It is interesting to calculate $\mathcal{L}$ for the usual

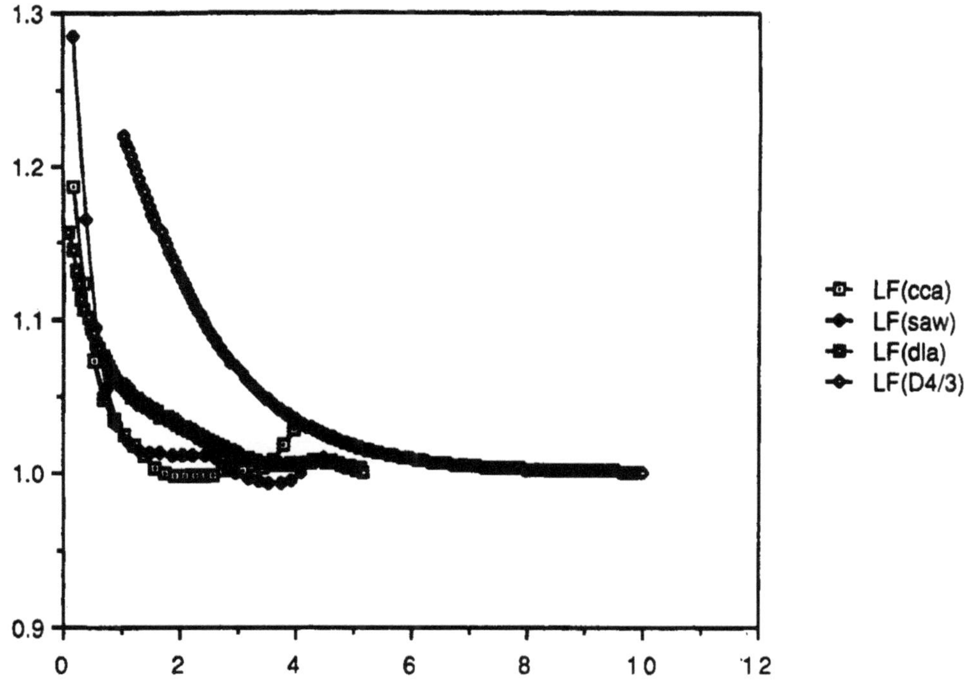

Fig. 2 The Lacunarity Function $\mathcal{L}$ for: SAW, CCA, DLA and from Eq. (2.5).

Kirkwood approximation:

$$C_3(\mathbf{r}, \mathbf{r}') = \frac{1}{2}[C_2(\mathbf{r})C_2(\mathbf{r}') + C_2(\mathbf{r}' - \mathbf{r})C_2(\mathbf{r}) + C_2(\mathbf{r} - \mathbf{r}')C_2(\mathbf{r}')], \tag{2.3}$$

where an explicitly exact result can be obtained. Equation (2.3) holds for structures produced by a Markovian process, of which infinite Brownian walks,[6] linear and randomly branched polymers, and galactic correlations[7] are but a few examples. Substituting the power-law form of C_2 gives for the angle-dependent LF:

$$\mathcal{L}_a(\ln R, \theta) = \frac{[1 + (1 + R^{d-D})(1 + R^2 - 2R\cos\theta)^{(D-d)/2}]}{2}, \tag{2.4}$$

where $R \equiv r/r' = \exp(\rho - \rho')$. Integrating over the angular variable gives:

$$\mathcal{L}_a(\lambda) = \frac{1}{2} + \cosh\frac{(d-D)\lambda}{2}[2\sinh\lambda]^{(D-d)/2}\frac{P^{\mu}_{\nu}(\coth\lambda)}{P^{\mu}_{-\nu}(\coth\lambda)}, \tag{2.5}$$

where $\lambda = |\ln R|$, $\mu = 1 - d/2$ and $\nu = D/2 - 1$. Shown in Fig. 2 for $D = 1.33$, $\mathcal{L}_a$ looks smooth and without any particular structure, unlike in real systems. This shows that $\mathcal{L}$ is generally non-trivial and cannot be reduced to $C_2(r)$ as in Markovian processes.

3. LOGARITHMIC PERIODICITIES AND CORRECTIONS TO SCALING

Some stochastic growths entertain logarithmic periodicities that show up as geometrically increasing spacing between major structural subsets.[8–10] The particular spacing patterns depend on the basic selection mechanisms. We claim that the function $\mathcal{L}$ is naturally suited

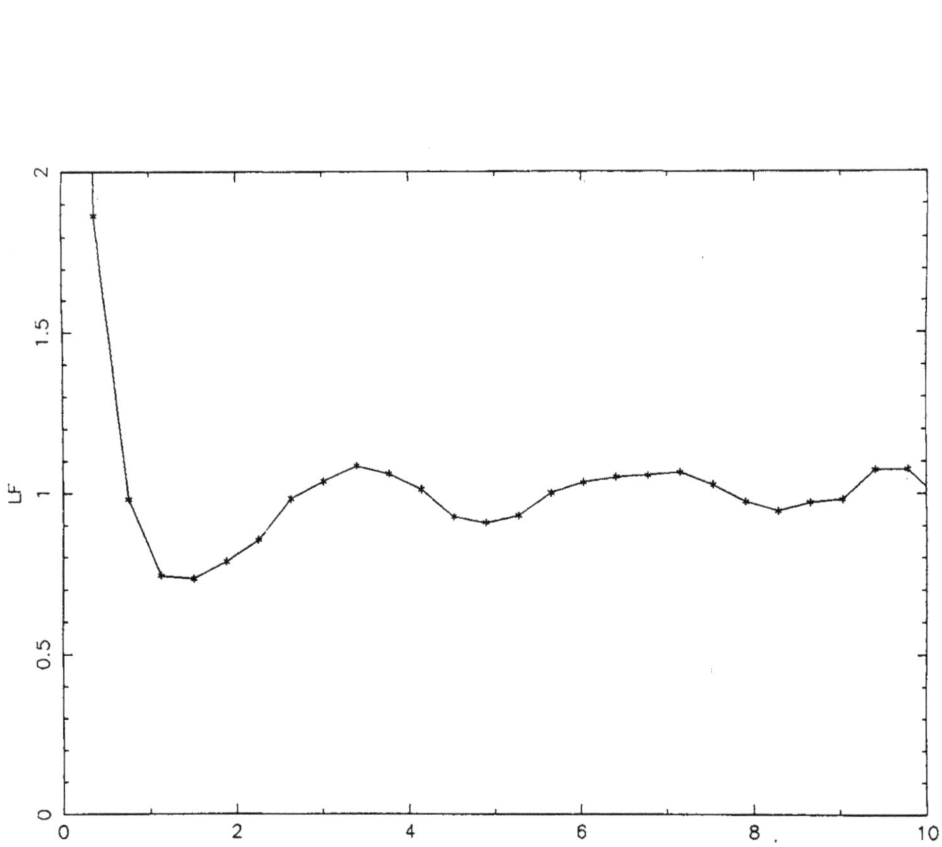

Fig. 3 The LF of a disordered Cantor set displaying the characteristic oscillations.

to detect such oscillations, which should appear as distinct peaks in the $\mathcal{L}(\lambda)$ plot, indicating that at some ratio of lengthscales, there is a repetition in the structure regardless of absolute scale. To test this, we have examined randomly disordered Cantor sets. Although to the eye, the resulting structures look very disordered, Fig. 3 shows that $\mathcal{L}$ can still detect the underlying logarithmic oscillation, though damped by the disorder.

We now note that straight line segments in $\mathcal{L}$ for $\lambda > 1$ indicate that it is of the form:

$$\mathcal{L} = 1 + A_p(r'/r)^{-p} + O((r'/r)^{-q}), \quad q > p > 0. \tag{3.1}$$

This we have tested for the exact Markovian case to indeed confirm that Eq. (3.1) holds with $p = c$ and $q = 2$, which correspond to corrections to scaling in $C_2(r)$ from finite length and discrete steps, respectively. We therefore conjecture that the power p, and other corrections, should match ordinary corrections to scaling for general fractal structures too. To the best of our knowledge, this is the first case where corrections to scaling can be obtained from the *asymptotic* structure in what has been traditionally considered the pure scaling regime.

4. CONCLUSION

This new correlation scheme has been shown to distinguish between structures with similar fractal dimensions but different morphologies, as demonstrated for the case of cluster-cluster aggregation and self-avoiding walks. This distinction can already be achieved for sizes as small as 4000 particles, where measurements of fractal dimensions are still too noisy for separation. The scheme is well suited to analyse logarithmic periodicities in stochastic patterns, as demonstrated for disordered Cantor sets. This technique can therefore assist in better understanding evolution of biological systems with complex dimensions such as lungs[10] and branching processes such as noise-reduced DLA[9] and slow cracking,[8] to which we hope to apply the scheme. We have shown that for linear and randomly branched polymers $\mathcal{L}$ can retrieve corrections to scaling from the asymptotic regime, and propose that this feature can be used for other fractal structures too. Generalizations of the formalism have been proposed for the angle-dependent LF.

ACKNOWLEDGMENTS

We thank P. B. Warren for providing the data on CCA and P. Meakin for providing the data on DLA. R. Blumenfeld acknowledges support by the US Department of Energy.

REFERENCES

1. B. B. Mandelbrot, *The Fractal Geometry of Nature* (Freeman, San Francisco, 1982).
2. Y. Gefen, B. B. Mandelbrot and A. Aharony, *Phys. Rev. Lett.* **45**, 855 (1980); Y. Gefen, A. Aharony and B. B. Mandelbrot, *J. Phys.* **A17**, 1277 (1984).
3. R. Blumenfeld and B. B. Mandelbrot, unpublished.
4. B. B. Mandelbrot, *C. R. Acad. Sci. Ser.* **A288**, 81 (1979).
5. R. Blumenfeld and R. C. Ball, *Phys. Rev.* **E47**, 2298 (1993).
6. B. B. Mandelbrot, *Comptes Rendus* (Paris) **A280**, 1551 (1975).
7. P. J. E. Peebles, *The Large-Scale Structure of the Universe* (Princeton, New Jersey 1980) pp. 157, 224, 248; See also in *Fractals in Physics* ed. L. Pietronero (North Holland, Amsterdam, 1990) pp. 278. Note however that the factor claimed in this study 1.3 rather than 1/2.
8. R. C. Ball and R. Blumenfeld, *Phys. Rev. Lett.* **65**, 1784 (1990); R. Blumenfeld and R. C. Ball, *Physica* **A177**, 407 (1991); R. Blumenfeld and R. C. Ball, *Mod. Phys. Lett.* **B5**, 1567 (1991).
9. J. Eckmann, P. Meakin, I. Procaccia and R. Zeitak, *Phys. Rev.* **A39**, 3185 (1989); R. C. Ball, P. W. H. Barker and R. Blumenfeld, *EuroPhys. Lett.* **16**, 47 (1991).
10. M. F. Shlesinger and B. J. West, *Phys. Rev. Lett.* **67**, 2106 (1991).

Fractals, Vol. 1, No. 4 (1993) 1002–1007

SCALE-INVARIANT DYNAMICS AND UNIVERSALITY CLASSES IN LAPLACIAN FRACTAL GROWTH

A. VESPIGNANI and L. PIETRONERO
Università di Roma "La Sapienza"-P.le A.Moro, 2-00185 Roma, Italy

Abstract

We consider the problem of the universality and general structure of the scale invariant dynamics in Laplacian fractal growth models. In particular, we show that growth having sticking probability not equal to one is renormalized asymptotically into effective growth rules with unit sticking probability (standard growth rule). This result shows why this modified model belongs to the same universality class as the standard DLA, and contributes to clarifying the asymptotic form of the scale invariant growth rule. These conclusions about the universality classes can also be extended to reaction-limited cluster-cluster aggregation.

1. INTRODUCTION

Many of the fractal growth processes observed in nature can be modeled by using growth rules which are non-local, i.e., the growth is influenced by distant points of the system. Prototype models are the Laplacian growth models such as the Diffusion Limited Aggregation (DLA) and the Dielectric Breakdown Model (DBM).[1–3] These systems are called Laplacian because they are based on field like quantities, e.g. the probability of finding diffusing particles or the electric potential, which satisfy the Laplace equation with moving boundary condition. The rules governing the growth of DLA and DBM are rather simple. However, the structures generated are very complex fractal patterns which result from a self-organizing stochastic process based on the Laplace equation. A large number of theoretical approaches extending standard theoretical methods turned out to be rather problematic.[1] For this reason, we introduced a new theoretical approach named Fixed Scale Transformation,[4] based on the dynamical evolution of the system on the same scale. The use of small scale growth rules allowed good quantitative results to be obtained. Nevertheless, the Fixed Scale Transformation requires in principle the use of the asymptotic scale invariant growth rules. This focuses attention on the essential properties of the scale invariant dynamics and their characterization. Moreover, the knowledge of the effective dynamics is one of the key points in understanding why growth models give rise to fractal structures. In particular, the problem of universality and self-criticality of DLA and DBM are strictly related to their effective asymptotic dynamics.

Laplacian growth models are intrinsically critical: i.e., under scale transformation the small scale dynamics flows into the asymptotic scale invariant one without tuning any parameter. The problem is then to define the effective dynamics (growth rules) that applies to coarse-grained variables. In order to address this question, we have introduced the space of growth rules where each point represents a given growth rule.[5] The space is identified by the relevant parameters of the dynamics and, in general, it is highly complex. By defining a renormalization group (RG) transformation that acts in this space, it is possible to find its fixed point, whose existence and nature (stable as well unstable) is responsible for the universality class of these models. In addition, the quantitative study of the fixed point allows the asymptotic growth rule to be characterized in detail.

In this way, we have obtained several results about the scale invariant structure of the dynamics, showing the universality of the DLA and DBM with respect to the introduction of new parameters in the minimal scale growth rule.[5,6] A key parameter for the renormalization of the growth rules is the effective noise reduction parameter.[6]

In this paper, we want to address the question of the universality of DLA with respect to the modification of the small scale growth rules by the introduction of a sticking probability different from one.[7] Numerical simulations[7] have shown that the introduction of the sticking probability does not change the universality class of DLA. This means that the growth rule evolves towards the same fixed point in the space of growth rules.

The problem of the asymptotic behavior of sticking probability P_s will be addressed by constructing a scheme of renormalization in real space. We show that the RG equation for P_s has an attractive fixed point at the trivial value $P_s = 1$ (i.e., the sticking probability influence has disappeared). This result indicates that in the scale invariant growth rule, the sticking probability must be considered as irrelevant, thus explaining why this parameter does not change the universality class of the model. In addition, this result contributes to clarifying the asymptotic form of the DLA and DBM growth rules and the general question of the universality class of these models. It is interesting to note that the previous results are valid also for different growth models such as the Reaction Limited Cluster-Cluster Aggregation Model.

This paper is organized as follows. In Sec. 2, we consider the problem of the universality class of DLA and DBM and we relate it to the scale invariant dynamics of the models. We define the space of growth rules and summarize the results obtained for the invariant form of the growth rules. In Sec. 3, we construct a renormalization scheme for the sticking probability and we derive the renormalization equation. This allows the study of the asymptotic behavior of the sticking probability. In Sec. 4, we summarize and discuss the results obtained.

2. UNIVERSALITY AND SCALE INVARIANT DYNAMICS

If DLA produced a simple fractal structure with universal properties, the fractal dimension should not be changed under the influence of parameters such as anisotropy, type of lattice, growth geometry, etc. While the situation is not quite clear with respect to different growth geometries, the question of the universality of fractal dimension can be clearly addressed by investigating modifications of the original model.

The standard DLA[2] is defined with a site rule in which the diffusing particle sticks to the surface when it reaches a site adjacent to the aggregate. The following modified versions can be considered:

a) DLA defined per bond. The particles stick when, from a surface site, they move into a site occupied by the aggregate. In DBM the growth probability is associated with each bond.[3]
b) DLA with *diagonal coupling*[1,7] (next-to-nearest neighbor interaction). In this version, the particles stick as soon as they reach a site which is next-nearest neighbor to the aggregate. In DBM this corresponds to considering a diagonal bond probability proportional to the electric field.
c) DLA with sticking probability less than one.[1,7] The particles stick to the surface with probability P_s and continue to diffuse with probability $1 - P_s$.

The above modifications are irrelevant from the point of view of the fractal dimension. Thus, we can say that all these modifications belong to the same universality class and evolve towards the same fixed point. This means that under scale transformation, the growth rules flow automatically in the subspace of the relevant parameters, where there is the attractive fixed point that identifies the value of the parameters relative to the asymptotic form of the growth rules.

In previous work[5,6] using real space renormalization schemes, it has been possible to show the following results:

— The scale invariant growth probabilities should be defined per *bond* and not per site. Growth probabilities defined per site evolve, under scale change, into bond probabilities.
— An eventual diagonal coupling is effectively eliminated in the scale invariant regime.
— The noise reduction parameter has been identified as the most relevant parameter automatically generated by a scale transformation. A divergence of this parameter would lead to a cross-over towards a compact fractal structure. Actually, the fixed point for this parameter is attractive and the fixed point value S^* is of the order of unity ($S^* \simeq 2,4$).

These results provide an understanding of the self-critical nature of the DLA (attractiveness of the fixed point dynamics) and they show that the usual DLA growth rules are rather close to the scale invariant ones.

Clearly, the previous statements do not characterize completely the growth rule. Indeed, in the highly complex space of growth rules, it is not possible to consider all the parameters. For example, the sticking probability parameter has been not included in previous studies. In the next section, we shall introduce a Real Space Renormalization Group (RSRG) scheme to study the behavior under scale transformation of this parameter.

3. RENORMALIZATION OF THE STICKING PROBABILITY

In DLA with sticking probability, the particles stick irreversibly to the surface with a probability P_s less than one and continue to diffuse with probability $1 - P_s$. This probability

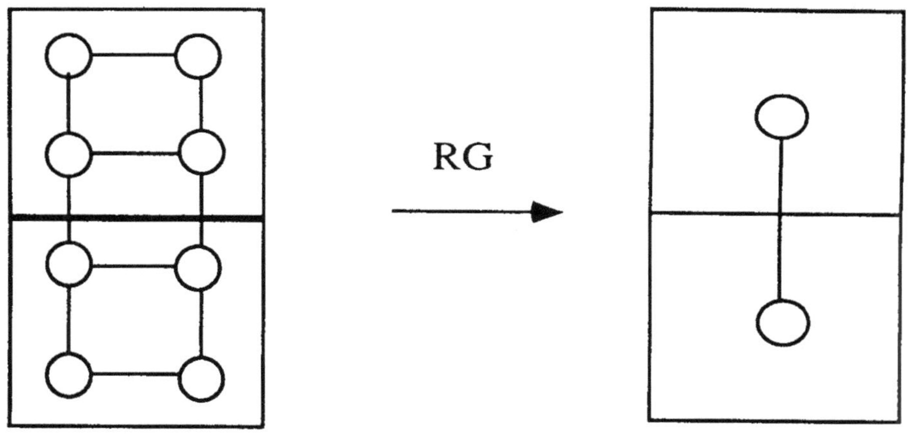

Fig. 1 Cell-to-bond renormalization scheme.

is introduced to mimic the chemical repulsive barrier that can prevent the particles from joining to the cluster even if they are very close. In a certain sense, it is the inverse of the average time needed for the formation of a bond. However, simulations show that the structures generated have the same fractal dimension for any value of P_s different from zero. In the language of the renormalization group, the sticking probability is an irrelevant parameter that flows automatically towards the fixed point.

Here we develop a simple real-space renormalization scheme that allows the scale change behavior of the sticking probability to be studied. Let us define the usual "cell-to-bond" renormalization scheme (Fig. 1). We consider that a coarse-grained bond is occupied (grown) if there is a path which spans the small scale cell. At a generic scale, the growth process of DLA with sticking probability is given by two steps. First, the diffusing particle came into contact with the aggregate; this will occur with a probability given by the scale invariant growth probability. Then, with a probability P_s the particle will stick irreversibly; otherwise, it continue to diffuse.

In order to find the rescaled sticking probability, we consider the growth of the coarse-grained bond. The arrival of the particle (first step) corresponds at the small scale to the growth process that generates a spanning path. What has occurred before in a cell is not relevant, because the rescaled sticking probability is evaluated conditionally to the fact that the rescaled bond can definitively grow. Thus, we consider only the cases where the arrival of the next particle will give rise to a spanning path in the small scale cell.

Let us consider the situation depicted in Fig. 2. The growth of the rescaled bond corresponds to the sticking of a particle on the site 1. This will occur with probability P_s, while with probability $1 - P_s$ the particle continues to diffuse. After this first process, we have $P_s' = P_s$. However, if the particle diffuses in site 2 of Fig. 2, it is still in the cell and at the coarse-grained level, it has not diffused out of the growing rescaled bond. The particle can now diffuse again in site 1 and try again to stick to the structure. We have in this way contribution by further diffusive processes with the only condition that the particles do not exit out of the cell.

Therefore, the rescaled sticking probability will be given by the following series:

$$P_s' = P_s + \alpha(1 - P_s)P_s + \cdots \qquad (1)$$

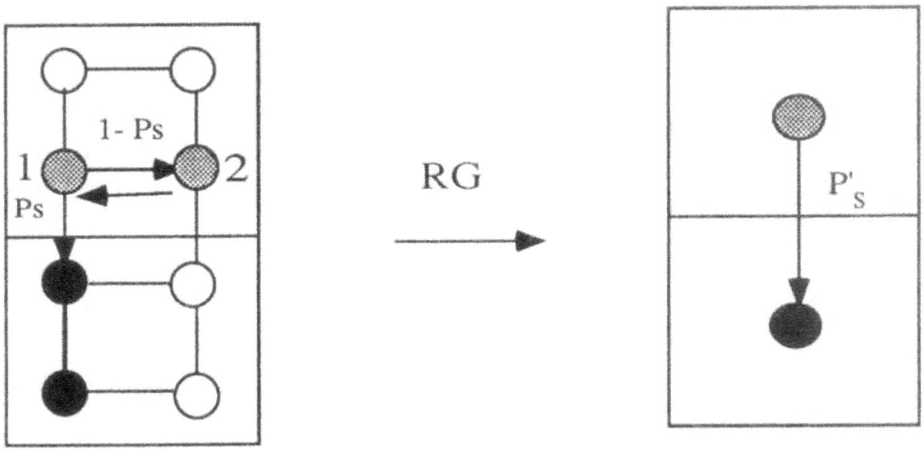

Fig. 2 Renormalization scheme for the sticking probability in the case of a particular configuration. The particle can stick to the structure at the first trial, or after successive diffusion processes inside the cell.

where the second term represents the probability that the particle diffuses twice in the cell before it sticks irreversibly. The series can be continued in order to take into account three diffusion processes and so on. The constant α appearing in Eq. (1) is the probability that the particle does not diffuse out of the cell and it depends on the scale invariant dynamics of the model. It should be noted that in the series in Eq. (1), we have to consider all the possible configurations of the cell of Fig. 2. However, for every starting cell the first diffusion process will give the term P_s. Second order contributions will depend on the cells. Therefore the constant α is the average over the various configurations. The Eq. (1) has fixed points in:

$$P_s = 0; \quad P_s = 1 \tag{2}$$

However, the only stable fixed point is $P_s = 1$, where:

$$\left.\frac{dP'_s}{dP_s}\right|_{P_s=1} = 1 - \alpha < 1 \tag{3}$$

The sticking probability at the microscopic scale will then disappear in the asymptotic scale invariant growth rules. In fact, for each value of the probability P_s, the scale transformation leads asymptotically towards the trivial fixed point $P_s = 1$. Therefore, all the modifications of the standard DLA with sticking probability belong to the same universality class. On the contrary, the unstable fixed point $P_s = 0$ defines a different universality class. As pointed out by some authors[1] in the limit $P_s = 0$ the model is analogous to the Eden Model where each surface site is filled with equal probability. In this case, the fractal dimension is equal to the Euclidean dimension.

The previous results can be extended also to different models such as the Reaction Limited Cluster-Cluster Aggregation (RLCA),[8] explaining why it belongs to the same universality class as the ordinary Cluster-Cluster Aggregation model.[1,9] In fact, the previous discussion does not depend on the nature of the diffusion process and can be rephrased for the RLCA. This is in agreement with the numerical simulations[8] which show that the RLCA fractal dimension is the same for every value of P_s.

4. SUMMARY AND CONCLUSIONS

In this paper, we have related the universality class of the Laplacian fractals Models to their asymptotic scale invariant dynamics. We address the problem of the effective dynamics that applies to coarse-grained variables defining a Renormalization Group approach in the space of the growth rules. The fixed point of this Renormalization Transformation is responsible for universality and self-critical properties of these models.

In particular, we study the behavior under scale transformation of the sticking probability P_s that is introduced into the model to mimic the velocity of the chemical aggregation process. By developing a Real space renormalization scheme, it is possible to show that the sticking probability has an attractive fixed point at the trivial value $P_s = 1$. This explains why this modification of the growth rule is irrelevant for the fractal dimension and why DLA with $P_s \neq 0$ belongs to the same universality class as the standard DLA. In addition, this contributes to clarifying the invariant form and symmetry of the scale invariant growth rules of DLA and DBM, allowing a clearer description of the properties of the dynamics giving rise to the fractal structure to be made. Finally, these results can also be applied to the study of the universality classes of different models such as the reaction limited cluster-cluster aggregation model, having a behavior similar to DLA with respect to sticking probability.

ACKNOWLEDGMENTS

We would like to thank P. Tartaglia and M. Vergassola for useful discussions.

REFERENCES

1. T. Vicsek, *Fractal Growth Phenomena* (World Scientific Pub., Singapore, 1992).
2. T. A. Witten and L. M. Sander, *Phys. Rev. Lett.* **47**, 1400 (1981).
3. L. Niemeyer, L. Pietronero and H. J. Wiesmann, *Phys. Rev. Lett.* **52**, 1033 (1984).
4. L. Pietronero, A. Erzan and C. Evertsz, *Phys. Rev. Lett.* **61**, 861 (1988).
5. R. De Angelis, M. Marsili, L. Pietronero, A. Vespignani and H. J. Wiesmann, *Europhys. Lett.* **16**, 417 (1991).
6. R. Cafiero, L. Pietronero and A. Vespignani, *Phys. Rev. Lett.*, in press (1993).
7. P. Meakin, *Phys. Rev.* **A26**, 1495 (1983).
8. R. Jullien and M. Kolb, *J. Phys.* **17**, L639 (1984).
9. P. Meakin, *Phys. Rev. Lett.* **51**, 1119 (1983).

Fractals, Vol. 1, No. 4 (1993) 1008–1021

SEDIMENTATION OF GLASS BEADS UNDER THE INFLUENCE OF GRAVITY

M. L. KURNAZ, K. V. McCLOUD and J. V. MAHER
Department of Physics and Astronomy, University of Pittsburgh, Pittsburgh, Pennsylvania 15260, USA

Abstract

The quasi-two-dimensional sedimentation of 0.06 cm diameter silica spheres in viscous oil (between glass plates separated by a 1 mm gap) results in quasi-one dimensional rough surfaces. These surfaces are rough on all length-scales between the particle size and the cell size, but different roughness exponents are observed in each of two well defined regimes ($\alpha_s = 0.81 \pm 0.03$ for $L \leq 2$ cm and $\alpha_l = 0.92 \pm 0.05$ for $L \geq 2$ cm). The dynamical exponent, β, is measured to be 0.48 ± 0.16. The role and range of hydrodynamic forces must be considered before this system can be related to available rough interface theories. A first attempt at assessing the effect of hydrodynamics yields a noise spectrum whose probability distribution is close to Gaussian (with significant excess of large fluctuations) and whose correlation function resembles that measured for density-density correlations in the oil far above the growing interface.

1. INTRODUCTION

In recent years, considerable interest has developed in the growth of rough surfaces.[1–45] This problem is not only potentially of practical importance, but it also involves fundamental nonequilibrium statistical physics. Theoretical efforts[1–34] have emphasized the simple Kardar-Parisi-Zhang (KPZ) surface growth equation[2] with various kinds of added noise[18,22,27,32–34] to accommodate the possible variety of forces which might complicate the dynamics of growth in any one physical system. Few experiments are available,[35–45] and these have tended to measure roughness alone with little information about the dynamics of growth. However, the available experiments have not in general produced results in good agreement with theoretical expectations, producing a challenge to understanding how much more physics must be put into the problem before physically attainable observations can be understood.

In this paper, we report on the study of a quasi-two-dimensional problem in which glass beads settle through a viscous fluid and form a quasi-one-dimensional interface at the bottom of the cell. The hydrodynamics of this problem is very complicated and should involve long range forces caused by back-flows from the motion of neighboring particles out to a distance set by viscosity and the density of particles in the settling suspension.[46,47] Even though we will make some remarks below about what is known about the hydrodynamics, for present purposes we regard the interfaces formed at the bottom of our cell by the settled particles to be interesting rough surfaces. That is, we assume these surfaces to be formed under dynamical rules where the complicated interparticle forces provide a noise which must be added to the basic sedimentation-in-vacuum rules of Edwards and Wilkinson[1] with the functional form and range of that noise important for the applicability of available theories to this problem.

In Sec. 2, we describe our experiment and present examples of the patterns we observe. In Sec. 3, we describe the analyses we have performed and present the results of those analyses. In Sec. 4, we discuss our results in comparison with other interfacial growth experiments and available theoretical results.

2. EXPERIMENT

All measurements have been performed with a closed cell 40 cm by 40 cm by 0.1 cm. The walls of the cell are of 1/4″ float glass, and these are held 1 mm apart by sealed side frames of precision machined plexiglass. The cell was filled with heavy paraffin oil whose viscosity, η, is approximately 200 cps. A very large number ($\sim 40\,000$) of 0.06 cm diameter monodisperse silica spheres[48] were also placed in the cell before it was closed; these particles fill approximately 3% of the volume of the cell. The problem is quasi-two-dimensional in that the layer of settled particles at the bottom of the cell is only one particle thick in its gap direction. The cell can be rotated about a horizontal axis perpendicular to the gap direction. When the cell is rotated, the particles which had been at rest at the bottom must fall through the oil. The initial breakup of the settled array involves the formation of a set of finger-like structures with a characteristic spacing of approximately 1 cm; this stage of the dynamics will not be discussed in this paper. Later stages of the flow involve very complicated patterns with backflow effects very obvious. Characteristic lengths which impress the eye are much coarser than the initial fingering patterns. While one correlation function will be presented below showing length scales for particle-density fluctuations in the center of the cell, for the most part the results of our study of this hydrodynamic regime will not be presented in this paper. It takes approximately 20 minutes for the first particle to reach the bottom of the cell, at which time we arbitrarily set our clock to zero. All times presented below are time differences from the arrival time of the first particle. After the arrival of the first particle, approximately 13 minutes elapse before enough particles have arrived to cover the bottom of the container with a layer at least one particle thick at each point. All data presented below are from measurements made after the bottom of the container was completely covered. The last particles typically joined the interface 60 minutes after the arrival of the first particle. We have measured three time series in which patterns were measured at regular intervals from the earliest completely-covering array until the last of the particles settled down. We have also measured 50 independent runs in which we measured only the final configuration of the settled particles. As particles join the interface, they tend to settle in gently on a verticle line with no obvious lateral movement as they approach the surface and no observable rearrangement of the surface as they hit it. Overhangs are never formed.

Our measurements were made with a 35 mm slr camera producing slides which were digitized to a resolution of 2031 × 1348 pixels by a Nikon LS 3500 35 mm film scanner. Individual particles were typically resolvable, and thus the position of the interface could be traced very accurately. Digitized interfaces were then analyzed with the various statistical tests described below. Particle density fluctuations in the fluid above the interface could also be determined from the intensity distribution in the digitized photographs, with intensity vs. position typically stored for several horizontal and vertical lines defined on the image of the cell.

An obvious possibility is that the final configuration of particles might depend in some very important way on the initial configuration of particles at the top of the cell. No such obvious connection was observed. We found that, if we tilted and shook the cell to force an initial configuration of particles which was very atypical, then the system required approximately four resettlings of the particles before the atypical configuration's features were completely obliterated. However, if the initial configuration was not prepared to be atypical, no obvious correlation between initial and final configurations could be discerned by eye. In an attempt to make this insight quantitative and to look for subtler correlations, we defined a mean absolute difference function[34] between initial and later configurations to be:

$$\Delta h_n = \sum_{x_i} |h_n(x_i) - h_1(x_i)| \tag{1}$$

where n is the number of the configuration in the sequence and $h_n(x_i)$ is the height of the interface at horizontal position x_i after the nth run. Figure 1(a) shows this function for 50 sedimentation runs following a very atypical configuration in which all particles formed a triangle on one side of the cell. Figure 1(b) shows data from most of the same runs, but the

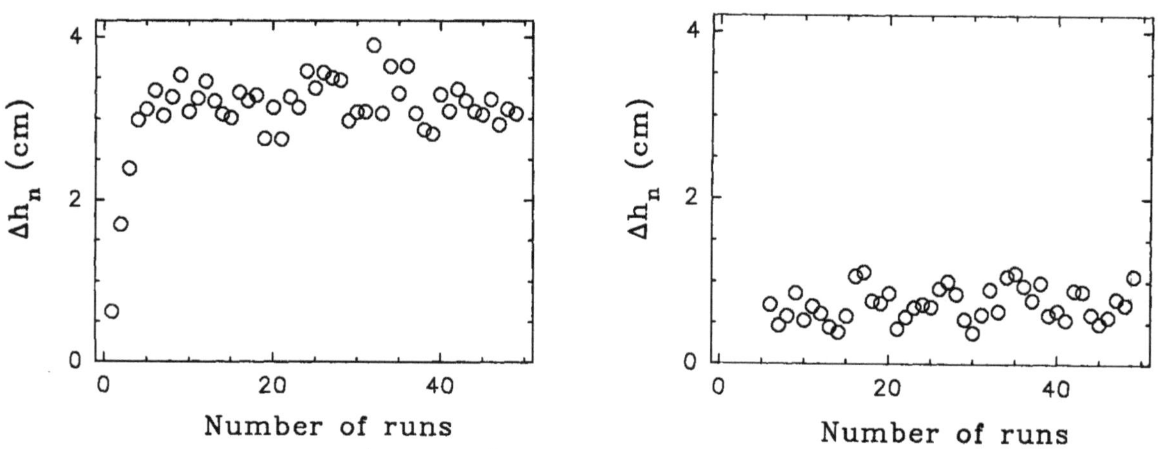

(a) (b)

Fig. 1 Mean absolute difference function defined in Eq. (1) using: (a) an atypical triangular initial configuration for the first run, and (b) a "typical" configuration for the first run.

sixth run from Fig. 1(a) is used as the initial configuration and compared with all subsequent runs. No obvious correlations appear once the triangular configuration is forgotten by the system. Figure 2 shows several typical configurations at the end of sedimentation runs. There was a strong tendency to have minima near the walls of the cell, so we have only used the middle 25 cm of interface for the analyses presented below. Within the middle 25 cm there is, on the average, only a gentle variation of surface height across the cell. Figure 3 shows the average position of the middle 25 cm of the interface at the end of a run; this average has been constructed using 46 runs.

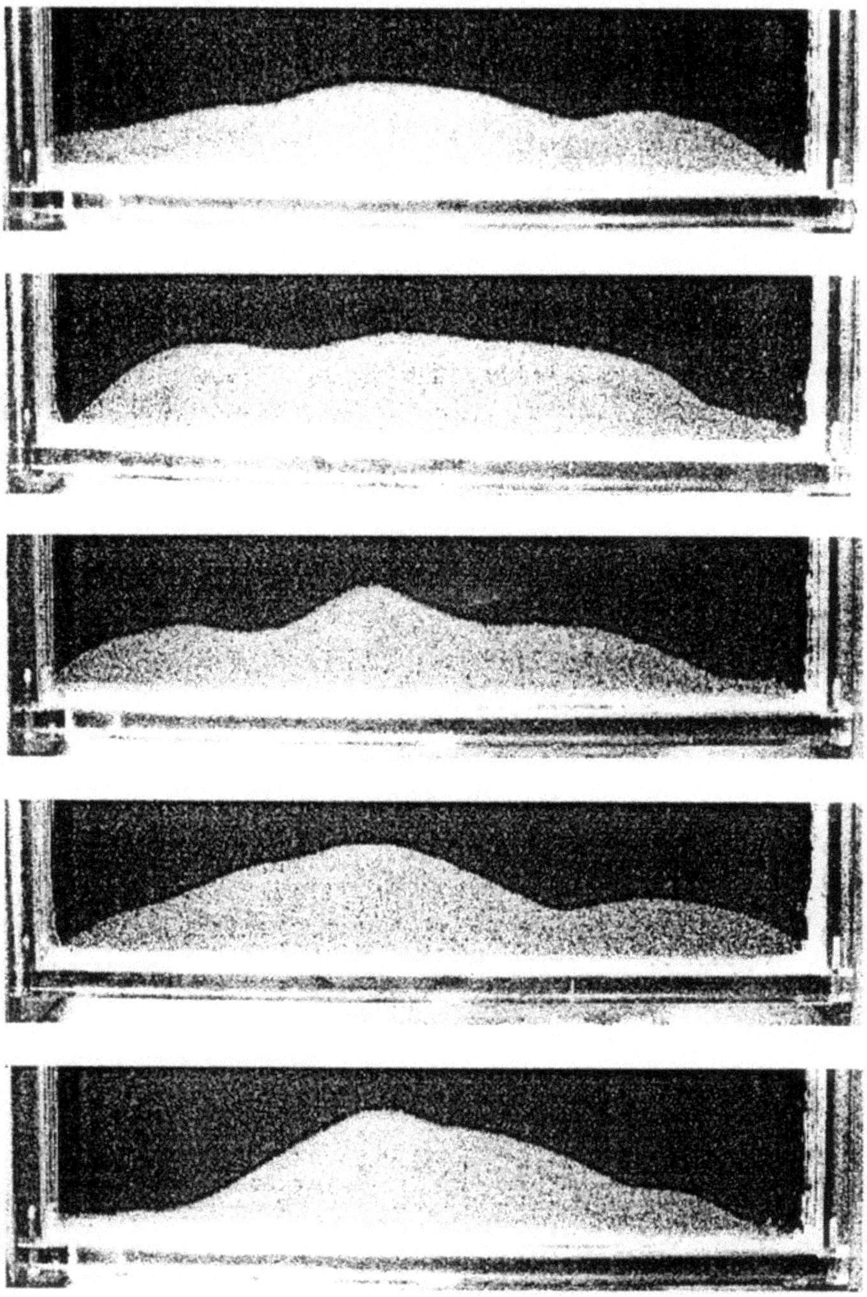

Fig. 2 Examples of typical final interfaces after all the particles have settled.

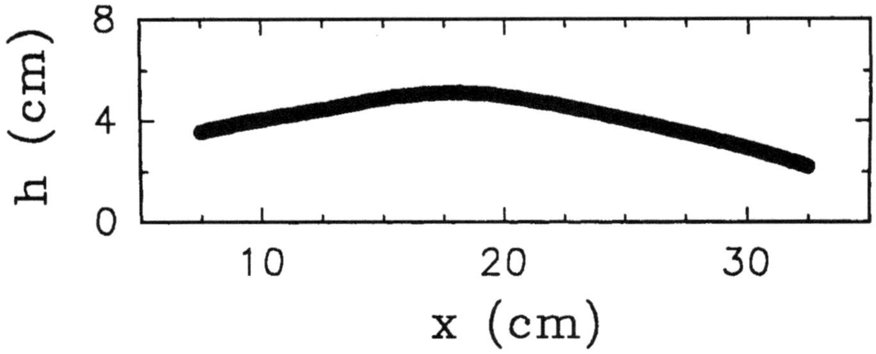

Fig. 3 Interface height as a function of position x averaged over 46 independent final configurations. As discussed in the text, 7.5 cm have been eliminated on each side of the cell.

3. ANALYSIS

The set of final configurations of all 46 runs can be used to define the roughness of the interface with great accuracy. Defining the rms thickness of the interface to be:

$$W(L,t) = \left[\frac{1}{N}\sum_{i=1}^{N} \tilde{h}(x_i,t)^2\right]^{\frac{1}{2}} \tag{2}$$

where

$$\tilde{h}(x_i,t) = h(x_i,t) - \bar{h}(t) \tag{3}$$

and

$$\bar{h}(t) = \frac{1}{N}\sum_{i=1}^{N} h(x_i,t) \tag{4}$$

where N is the total number of surface sites, x_i, in a segment of horizontal length L. We look for late-time roughness quantified by exponent α where $W \sim L^\alpha$. That is, following the scaling ansatz for rough interface growth,[3] we expect

$$W(L,t) = L^\alpha f\left(\frac{t}{L^{\alpha/\beta}}\right) \tag{5}$$

where the exponents α and β are the static and dynamic scaling exponents. The function $f(t/L^{\alpha/\beta})$ is not known in detail but is expected to have asymptotic forms such that:

$$W(L,t) \sim t^\beta \quad \text{for} \quad t \ll L^{\alpha/\beta} \tag{6}$$

and

$$W(L,t) \sim L^\alpha \quad \text{for} \quad t \gg L^{\alpha/\beta} \tag{7}$$

In the analysis presented below, we assume that our late time and final configuration measurements are suitable for use with Eq. (7) to determine α. (The final configuration has an average thickness $\bar{h}_\infty = 2.8$ cm, approximately 11% of the effective 25 cm width, L_{max}, of the cell.)

Figure 4 shows a typical example of $W(L)$ at late times (final configuration). There are clearly two domains of roughness. At small length scales ($L \leq 2$ cm), there is roughly

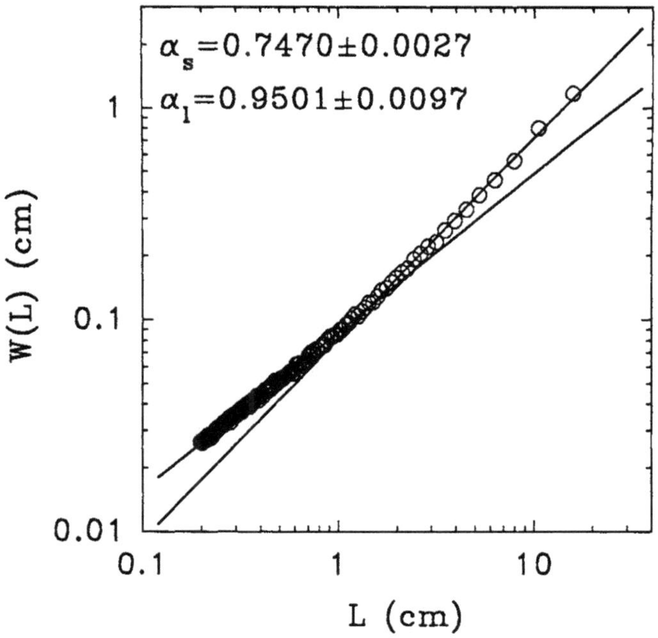

Fig. 4 Roughness function $W(L, t)$ at the end of a typical run.

a decade in L wherein the data appear to follow a power law whose exponent is approximately 0.8. We call this exponent α_s, and its value averaged over the entire set of runs is 0.81 ± 0.03. At longer length scales, ($L \geq 2$ cm), there is roughly a decade over which a different exponent can be identified. This exponent, $\alpha_l = 0.92 \pm 0.05$.

While the roughness of the final-state interface is robust (subject to only small fluctuations in the exponents with α_l always larger than α_s), the dynamics by which the growing interface develops appears to be subject to significant fluctuations. In an attempt to extend the analysis to observe early stages of the developing interface, time series of measurements were made for three flow realizations. At early times either of, or both, α_s and α_l can be as small as 0.5 and either or both can take on its final value at the earliest times. An example can be seen in Figs. 5 and 6 wherein α_l starts at a very low value and α_s has its final value from the beginning. In Fig. 5, we show the interfacial thickness $W(L, t)$ for three different times in the same run. At early times, the exponent is clearly lower for long scales than for short ones. This long scale exponent can be seen to grow through and eventually exceed the exponent for short length scales. In Fig. 6, we show the two roughness exponents for all times in the measurement. Table 1 lists the early values, α_s' and α_l', of the roughness exponents for each of the three observed time series. In Runs 1 and 3, $\alpha_s' \simeq \alpha_s$, which means that this exponent did not vary during the sedimentation process. Similarly $\alpha_l' \simeq \alpha_l$ in Run 3. However, in Run 1 α_l' starts much smaller than α_l and in Run 2, both exponents are much smaller at early times than at the end of the run. The non-reproducibility of this feature of the interfacial development is discouraging; it may require an enormous number of flow realizations to construct a meaningful "ensemble" of measurements.

In the hope of extracting more information from the growth dynamics, we have also constructed the height-height correlation function:

$$c(L, t) = \langle [\tilde{h}(x', t') - \tilde{h}(x' + L, t' + t)]^2 \rangle_{x', t'} \tag{8}$$

Table 1 Early-time Roughness Exponents α'_s and α'_l and Growth Exponent β

Run Number	α'_s	α'_l	β
1	0.70 ± 0.05	0.50 ± 0.04	0.52 ± 0.05
2	0.52 ± 0.04	0.51 ± 0.04	0.61 ± 0.05
3	0.68 ± 0.05	0.88 ± 0.05	0.30 ± 0.03

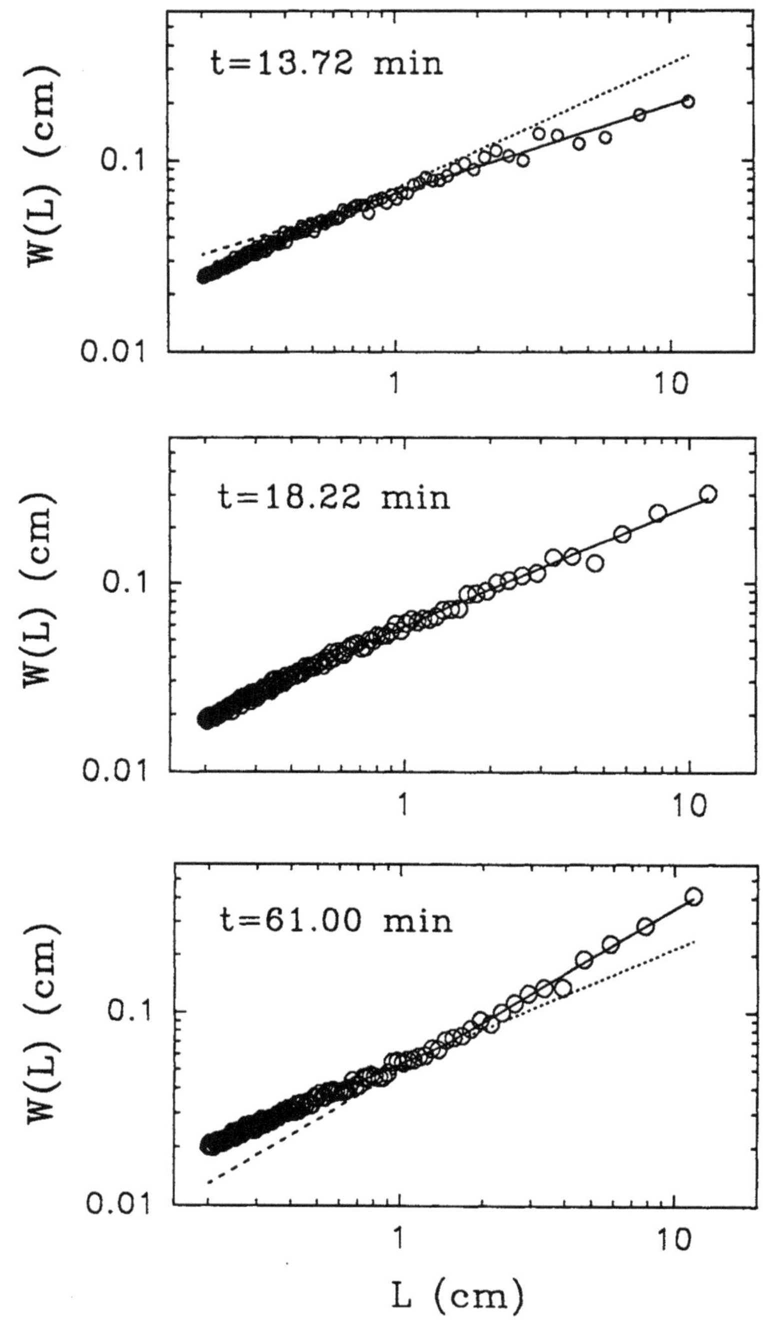

Fig. 5 $W(L, t)$ is shown for three values of the time in time series #1 of Table 1.

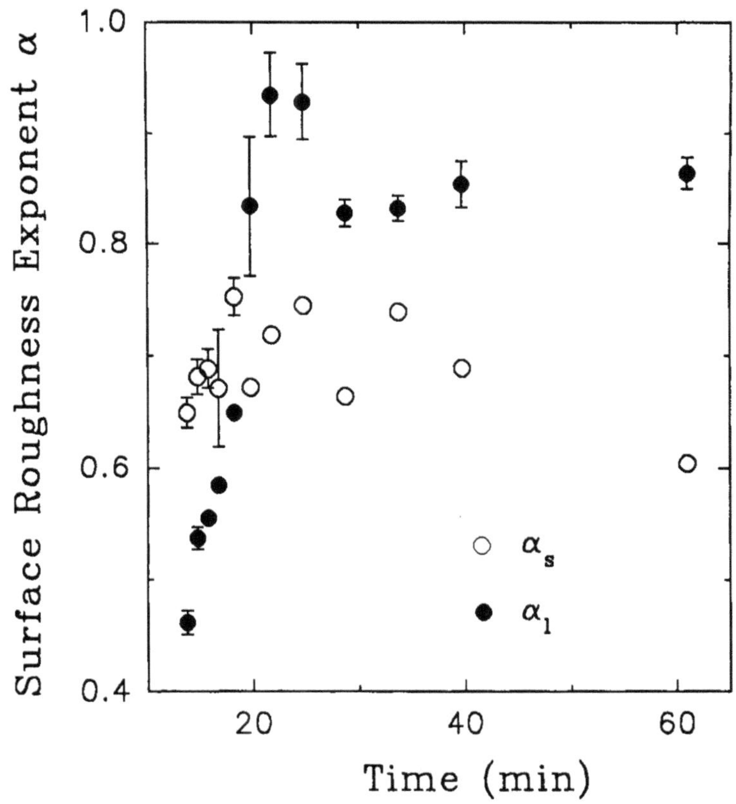

Fig. 6 The time evolution of the two exponents α_s and α_l is shown. The final value for α_s is uncharacteristically low for this run (run 1 of Table 1).

Using the scaling ansatz of Family and Vicsek,[3] we expect $c(L, 0) \sim L^{2\alpha}$ for L large but significantly less than the length of the cell and $c(0, t) \sim t^{2\beta}$ for short time differences. This approach provides an alternative weighting of the data to determine average exponents. Figures 7 and 8 show $c(L, 0)$ vs. L and $c(0, t)$ vs. t, respectively, for Run 2 of Table 1. $c(L, 0)$ yields $\alpha_s = 0.84 \pm 0.04$, in good agreement with the analysis above despite the fact that early time data contribute to the average. This agreement does not extend to large length-scales where the correlation function is expected to become somewhat less accurate and where the averaging of the data of different roughness indicates a poorly defined but small (~ 0.5) value of α_l. We have concluded that, in this case where the roughness exponents can change as the interface develops, much more accurate data are needed before $c(L, 0)$ can be expected to give as reliable a measure of α as can be extracted through use of Eq. (7).

The value of β determined from Fig. 8 is 0.61 ± 0.05. Table 1 lists β values for all three time series runs. As can be seen from the table, the value of β fluctuates significantly from run to run. The three runs combine to yield a highly uncertain value $\beta = 0.48 \pm 0.16$. It was not possible to use Eq. (6) to extract reliable values of β from the roughness functions, because the analysis is too sensitive to even small temporal variations in α. The averaging over the entire data set which is inherent in the correlation function appears to calm the fluctuations in the temporal dynamics enough to elucidate the underlying growth rate at least crudely, even though it washes out the long-scale roughness exponents in the same averaging.

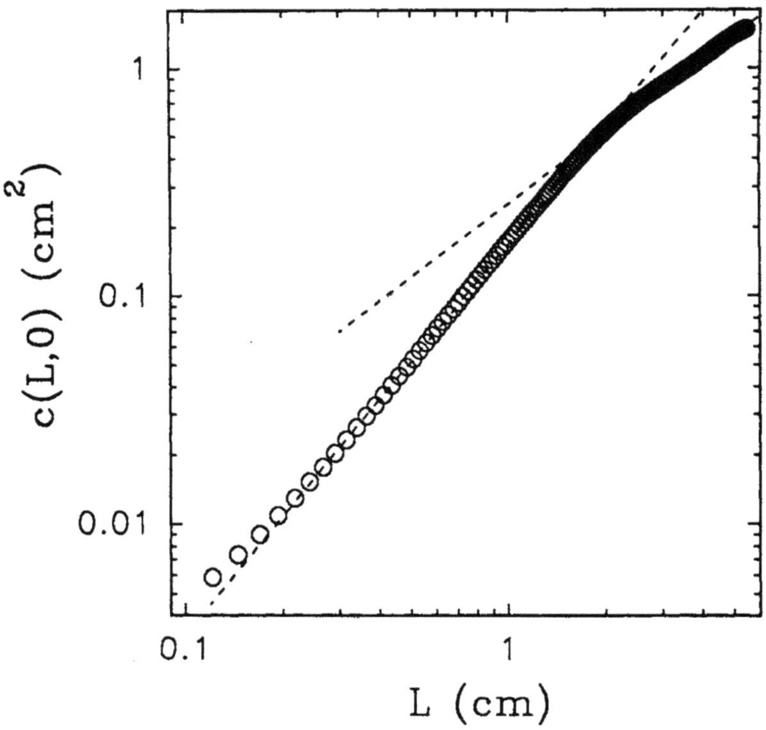

Fig. 7 Correlation function [defined in Eq. (8)] $c(L, 0)$ for run 2 of Table 1.

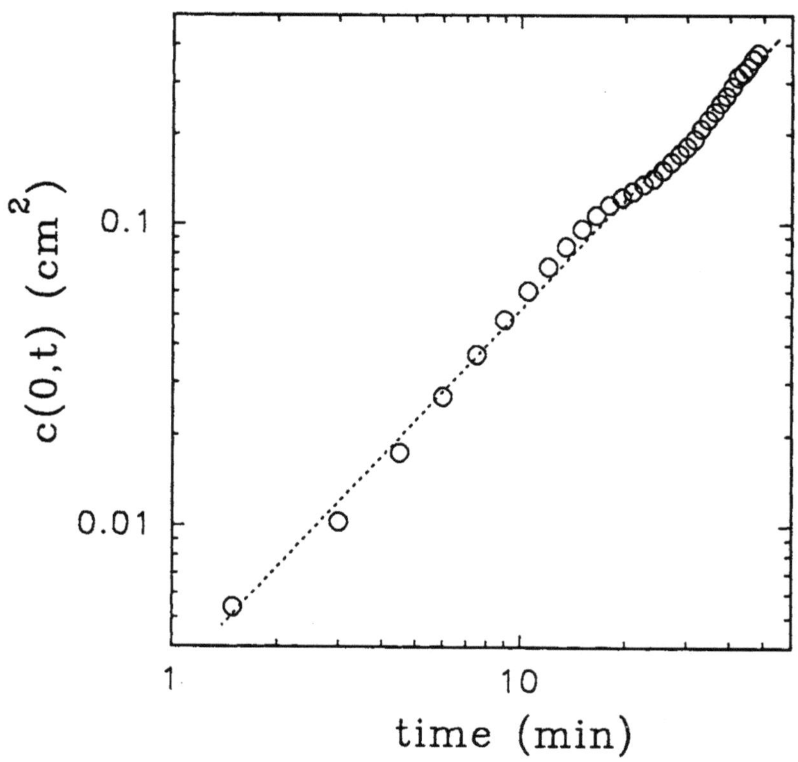

Fig. 8 Correlation function [defined in Eq. (8)] $c(0, t)$ for run 2 of Table 1.

Interestingly, the short-length-scale analysis gives $\alpha_s + \frac{\alpha_s}{\beta} = 2.50 \pm 0.57$, roughly consistent with the value of 2 which would be expected for a system with Galilean invariance. We do not know how to extract a separate β for the long-length-scale regime, in the event that the regime is to be regarded as an important scaling regime. If we assume β to be the same for both regimes, we obtain $\alpha_l + \frac{\alpha_l}{\beta} = 2.83 \pm 0.65$.

In an attempt to quantify the noise in the problem we have constructed a spatio-temporal fluctuation function:

$$\eta(x, t_1) = \frac{\tilde{h}(x, t_2) - \tilde{h}(x, t_1)}{\bar{h}(t_2) - \bar{h}(t_1)} \tag{9}$$

An example of the spatial correlation function of $\eta(x, t)$ is shown in Fig. 9. The noise falls off at short length scales with a characteristic distance of approximately 2 cm and is anticorrelated at large length scales (≥ 4 cm). The probability distribution of fluctuations is shown in Fig. 10 vs. η^2. If the noise were pure Gaussian, the data in Fig. 10 would fall on a straight line. While the data do not fall on a line, they definitely do not follow a power law (see inset to Fig. 10) and should be considered to be Gaussian with some superimposed long range distribution. The data of time series run 2 were used to produce Figs. 9 and 10; results from the other time series are similar. The interfacial-fluctuation correlation function shown in Fig. 9 resembles the correlations seen in the fluid far above the growing interface. Figure 11 shows a density-density correlation function for a typical case of sedimentation dynamics in the fluid at a height 6 cm above the interface.

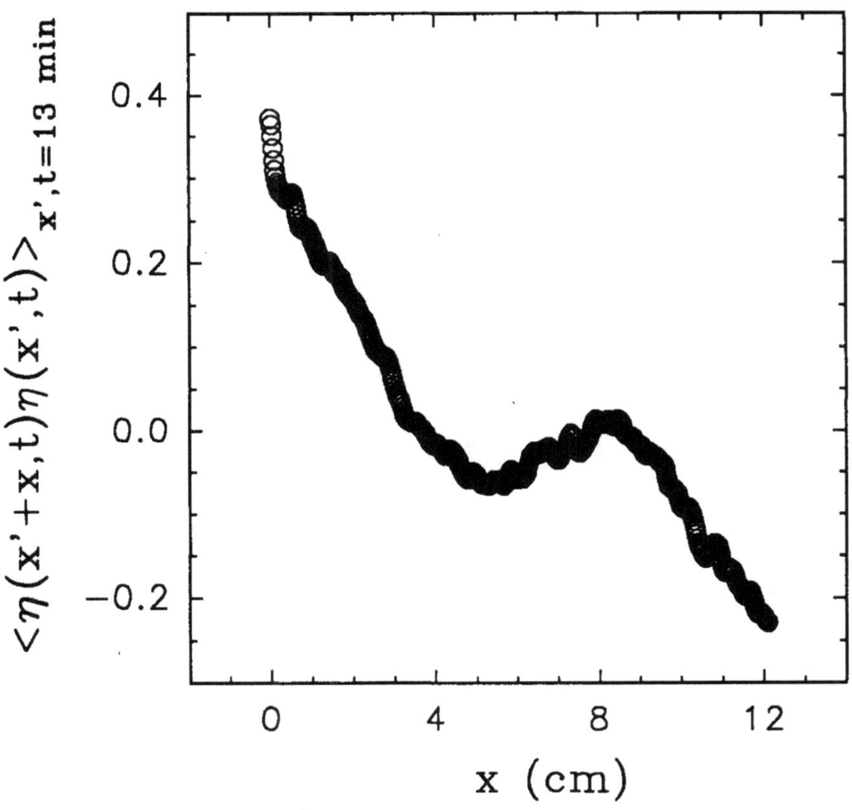

Fig. 9 Spatial correlation function of spatio-temporal fluctuation function $\eta(x, t)$ [defined in Eq. (9)] for arbitrarily chosen $t = 13$ min in run 2 of Table 1.

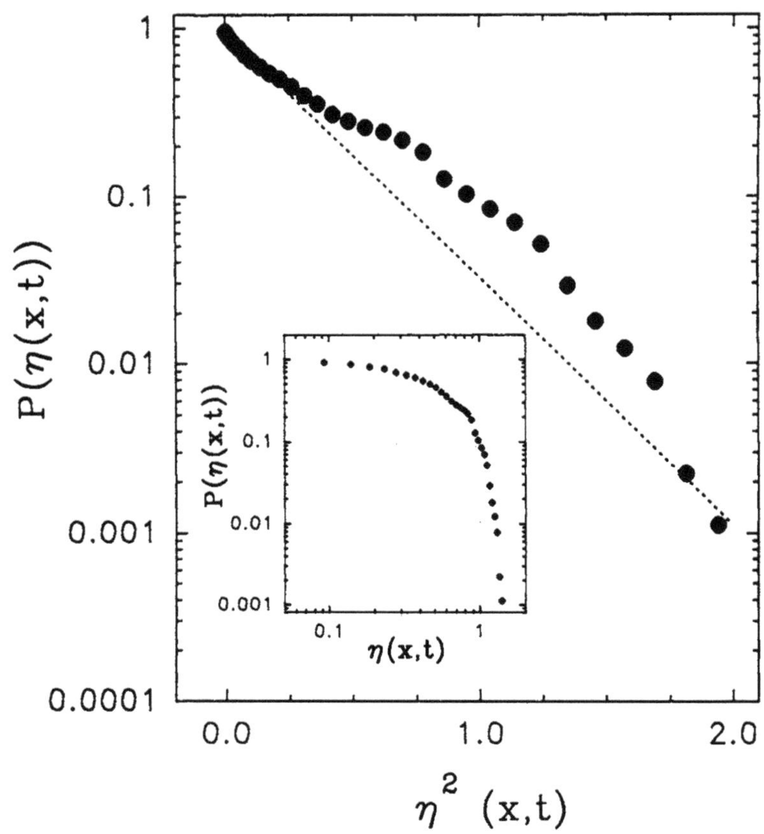

Fig. 10 Probability distribution function of spatio-temporal fluctuation function $\eta(x, t)$ [defined in Eq. (9)] for arbitrarily chosen $t = 13$ min in run 2 of Table 1.

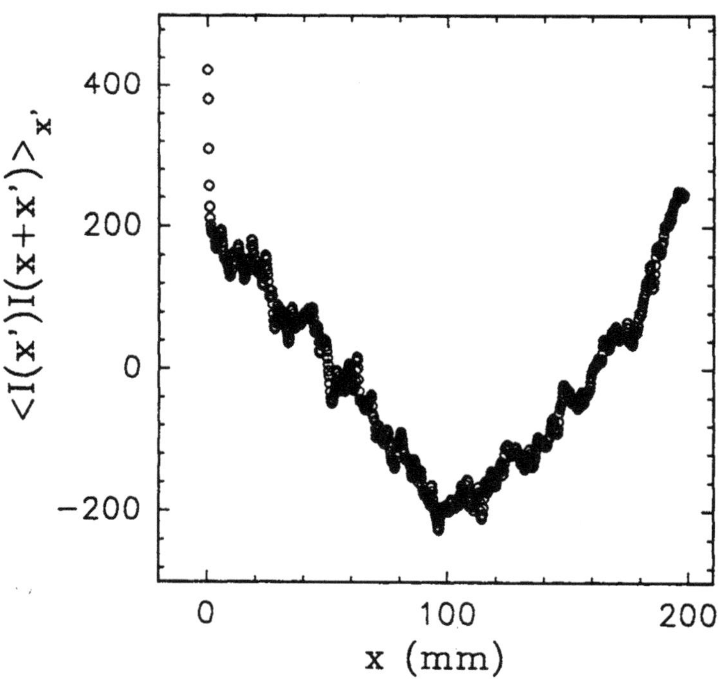

Fig. 11 Density-density correlation function of sedimenting silica spheres in oil 6 cm above a growing interface at $t = 13$ min in run 2 of Table 1.

4. DISCUSSION

From the results presented above, it is clear that we have two regimes of interfacial growth, one on length scales below approximately 2 cm and one on longer length scales. Either could be in a scaling regime, or both could be system specific, and on currently available information, we can only speculate as to possible universal features to this problem. One might argue, based on the similarity between the correlations in the fluid and the correlations in the interfacial noise, that the short length scale phenomena are dominated by the long-range hydrodynamic forces in the sedimentation problem. In this case, comparison with available interfacial growth equations based on the KPZ approach could only be defended for this regime if one argued that the long range hydrodynamic forces, complicated as they must be, can be represented by short range forces plus noise. If this argument were to hold, then the departures of our noise from pure-Gaussian form (Fig. 10) would have to be analyzed to see whether it can move the system's roughness exponent up from the KPZ value of 0.5 to 0.8. Our measured value of β has definitely not been extracted at early times, since our analysis does not begin until a continuous interface is defined over the entire bottom of the cell. However, one can speculate that the correlation function [Eq. (8)] can pull β out of local dynamical rules which follow the same growth law even at later times.

While structure at the longer length scales (> 2 cm) might also be affected by long range hydrodynamic forces, the hydrodynamic correlations are negative over much of this regime and the problem can more plausibly be argued to be one of surface growth with short range forces than in the case of the shorter length-scales. While one must always worry about finite-cell-size effects in such a system, our decision to analyze only the central 25 cm of the 40 cm cell is conservative in protecting us from wall effects, and the one decade length-scale range from 2 cm to 20 cm is sufficient to make the results in Figs. 5 and 6 reasonably firm. Since the interface-dynamic analog of critical phenomena should only be reached for an infinitely large system, one might expect the scaling regime associated with any growth law analysis to correspond to our larger-length scale regime unless that regime is dominated by cell-wall effects.

The results presented above show a rough interface problem which complements the previously available experimental results.[35–45] In the previously available works, Rubio et al.[35] and Horvath et al.[36] have performed experiments on immiscible displacement by a wetting fluid in a porous medium. Frette et al.[40] used a nonwetting fluid to investigate buoyancy-driven fluid migration in porous media. Pfeifer et al.[42] looked at multilayer absorption and Vicsek et al.[44] studied the development of bacterial colonies. Eklund et al.[37] and Krim et al.[39] used scanning tunneling microscopy to study rough surfaces formed by ion bombardment. Both You et al.[45] and Chiarello et al.[38] used x-ray reflectivity to study rough surfaces of sputter-deposited gold films and vapor-deposited films respectively. Kahanda et al.[41] performed electrochemical deposition of copper, and finally, Licoppe et al.[43] used solid state recrystallization of amorphous GaAs films to investigate surface roughness. All of these experiments have special experimental peculiarities which introduce extra forces not explicitly considered in simple growth laws like the KPZ equation. While it might be hoped that most of these extra effects might prove to be irrelevant to the rough interface growth exponents, the experiments have in fact almost always shown exponents which differ from the KPZ expectations. This has resulted in studies of the mathematical forms of noise,[18,22,27,32–34] which can change the KPZ exponents, allowing accommodation to the experimental results without adding yet more complicated terms to the growth equation. The present experiment presents results which resemble most of the other experimental results, with roughness exponents well above 0.5. It will be some time before the peculiarities of all the experiments are sorted out and the relations to theory well defined. In most cases, there is not enough control over the process to hope to elucidate the nature of the forces involved in the unique features of the experiments. In the meantime, the present experiment, even though it is almost certainly affected by the long-range hydrodynamic forces evident in the fluid above the interface, provides a case in which one can envision adding controlled modifications, influencing the range of the forces and measuring the changes in the noise spectrum. Experiments are underway to vary system parameters to try to resolve some of the issues just discussed. By varying the aspect ratio of the cell it should be possible to learn whether finite-cell-size effects can be blamed for the results seen in the large-length scale regime. If the crossover regime at 2 cm does not move with system size, then that regime is almost certainly set by the range of hydrodynamic forces. By changing the viscosity of the fluid, it is possible to change the characteristic length scale of the sedimentation hydrodynamics, and when this is done, it will be instructive to observe the behavior of the crossover length scale for the interfacial growth. (While in principle, changing viscosity and changing aspect ratio can be argued to be equivalent, cooperative screening effects in this complicated system are very likely to separate the two effects.) These follow-up experiments will take some time to perform, so for the present we have a fascinating and partially complete picture with two regimes of rough surface growth, each of which may be relatable to asymptotic scaling and both of which may turn out to be crucially system dependent.

5. SUMMARY AND CONCLUSIONS

We have studied quasi-one-dimensional interfaces formed by the sedimentation of glass beads in a viscous oil. The surfaces showed roughness at all length scales, but the roughness exponents were different in two well-defined regimes, one for length scales below $\sim$ 2 cm and one for longer length scales. Both roughness exponents ($\alpha_s = 0.81 \pm 0.03$ and $\alpha_l = 0.92 \pm 0.05$) are much larger than the value $\frac{1}{2}$ which would result from the Edwards and Wilkinson[1] model for sedimentation without hydrodynamics, or from the KPZ growth law which should hold for growth dominated by short range forces with Gaussian noise. The dynamic exponent β is found to be 0.48 ± 0.16. The noise has also been measured and found to be close to Gaussian with some possibly important deviations at larger values of the fluctuations. While it may be some time before the relation between theory and experiment is clarified for interface growth problems, this problem yields results which are similar to those found in other experiments and has the attractive feature that the system parameters can be controllably varied in ways not accessible to many other experiments.

ACKNOWLEDGMENTS

We appreciate helpful discussions with F. Family and D. Jasnow. We are grateful to D. Bideau for giving us samples of monodisperse silica particles. This work is supported by the US D.O.E on grant # DE-FG02-84ER45131.

REFERENCES

1. S. F. Edwards and D. R. Wilkinson, *Proc. Royal Soc. London* **A381**, 17 (1982).
2. M. Kardar, G. Parisi and Y.-C. Zhang, *Phys. Rev. Lett.* **56**, 889 (1986).
3. F. Family and T. Vicsek, *J. Phys.* **A18**, L75 (1985).
4. P. Meakin, P. Ramanlal, L. M. Sander and R. C. Ball, *Phys. Rev.* **A34**, 5091 (1986).
5. F. Family, *J. Phys.* **A19**, L441 (1986).
6. M. Plischke, Z. Rácz and D. Liu, *Phys. Rev.* **B35**, 3485 (1987).
7. P. Meakin and R. Jullien, *J. Physique* **48**, 1651 (1987).
8. R. Jullien and P. Meakin, *Europhys. Lett.* **4**, 1385 (1987).
9. R. Baiod, D. Kessler, P. Ramanlal, L. M. Sander and R. Savit, *Phys. Rev.* **A38**, 3672 (1988).
10. R. Jullien and P. Meakin, *J. Phys.* **A22**, L1115 (1989).
11. J. M. Kim and J. M. Kosterlitz, *Phys. Rev. Lett.* **62**, 2289 (1989).
12. J. Kertész and D. E. Wolf, *Phys. Rev. Lett.* **62**, 2571 (1989).
13. J. G. Amar and F. Family, *Phys. Rev. Lett.* **64**, 543 (1990).
14. D. E. Wolf and J. Villain, *Europhys. Lett.* **13**, 389 (1990).
15. L. Golubović and R. Bruinsma, *Phys. Rev. Lett.* **66**, 321 (1991).
16. S. Das Sarma and P. Tamborenea, *Phys. Rev. Lett.* **66**, 325 (1991).
17. J. Krug and H. Spohn, *Phys. Rev.* **A38**, 4271 (1988).
18. E. Medina, T. Hwa, M. Kardar and Y.-C. Zhang, *Phys. Rev.* **A39**, 3053 (1989).
19. T. Sun, H. Guo and M. Grant, *Phys. Rev.* **A40**, 6763 (1989).
20. J. G. Amar and F. Family, *Phys. Rev.* **A41**, 3399 (1990).
21. T. Hwa, M. Kardar and M. Paczuski, *Phys. Rev. Lett.* **66**, 441 (1991).
22. Y.-C. Zhang, *J. Physique* **51**, 2129 (1990).
23. N. Martys, M. Cieplak and M. O. Robbins, *Phys. Rev. Lett.* **66**, 1058 (1991).
24. H. Yan, D. Kessler and L. M. Sander, *Phys. Rev. Lett.* **64**, 926 (1990).
25. P. Devillard and H. E. Stanley, *Physica* **A160**, 298 (1989).
26. C.-H. Lam, L. M. Sander and D. W. Wolf, *Phys. Rev.* **A46**, R6128 (1992).
27. S. V. Buldyrev, S. Havlin, J. Kertész, H. E. Stanley and T. Vicsek, *Phys. Rev.* **A43**, 7113 (1991).
28. D. A. Huse, J. G. Amar and F. Family, *Phys. Rev.* **A41**, 7075 (1990).
29. T. Halpin-Healy and A. Assdah, *Phys. Rev.* **A46**, 3527 (1992).
30. P. Meakin and J. Krug, *Phys. Rev.* **A46**, 3390 (1992).
31. D. A. Kessler, H. Levine and L. M. Sander, *Phys. Rev. Lett.* **69**, 100 (1992).
32. V. K. Horváth, F. Family and T. Vicsek, *Phys. Rev. Lett.* **67**, 3207 (1991).
33. J. G. Amar, P.-M. Lam and F. Family, *Phys. Rev.* **A43**, 4548 (1991).
34. J. G. Amar and F. Family (unpublished).
35. M. A. Rubio, C. A. Edwards, A. Dougherty and J. P. Gollub, *Phys. Rev. Lett.* **63**, 1685 (1989).
36. V. K. Horváth, F. Family and T. Vicsek, *J. Phys. A: Math. Gen.* **24**, L25 (1991).
37. E. A. Eklund, R. Bruinsma, J. Rudnick and R. S. Williams, *Phys. Rev. Lett.* **67**, 1759 (1991).
38. R. Chiarello, V. Panella, J. Krim and C. Thompson, *Phys. Rev. Lett.* **67**, 3408 (1991).
39. J. Krim, I. Heyvaert, C. van Haesendonck and Y. Bruynseraede, *Phys. Rev. Lett.* **70**, 57 (1993).
40. V. Frette, J. Feder, T. Jøssang and P. Meakin, *Phys. Rev. Lett.* **68**, 3164 (1992).
41. G. L. Kahanda, X.-Q. Zou, R. Farrell and P. Z. Wong, *Phys. Rev. Lett.* **68**, 3741 (1992).
42. P. Pfeifer, Y. J. Wu, M. W. Cole and J. Krim, *Phys. Rev. Lett.* **62**, 1997 (1989).
43. C. Licoppe, Y. I. Nissim and C. d'Anterroches, *Phys. Rev.* **B37**, 1287 (1988).
44. T. Vicsek, M. Cserzö and V. K. Horváth, *Physica* **A167**, 315 (1990).
45. H. You, R. P. Chiarello, H. K. Kim and K. G. Vandervoort, *Phys. Rev. Lett.* **70**, 2900 (1993).
46. R. H. Davis and A. Acrivos, *Ann. Rev. Fluid Mech.* **17**, 91 (1985) and references therein.
47. F. M. Auzerais, R. Jackson and W. B. Russel, *J. Fluid Mech.* **195**, 437 (1988) and references therein.
48. D. Bideau, private communication.

Fractals, Vol. 1, No. 4 (1993) 1022–1029

SELF ORGANIZATION IN A FOREST-FIRE MODEL

B. DROSSEL and F. SCHWABL
Physik-Department der Technischen Universität München,
D-85747 Garching, Germany

Abstract

We generalize the forest-fire model of P. Bak et al., which contains a tree nearest growth probability p and fire spreading to the neighbors, by including a lightning probability f and an immunity g which is the probability that a tree catches no fire although one of its neighbors is burning. The model becomes self-organized critical in the limit $f/p \to 0$, provided the time scales of tree growth and burning down of forest clusters are separated. The size distribution of forest clusters obeys a power law. We calculate the critical exponents in one dimension. A continuous phase transition is observed in the general forest-fire model when g reaches its critical value. We determine the critical line $g_C(p)$ and show that the critical fire propagation represents a new type of percolation. Finally, we point out similarities between the forest-fire model and excitable media, which comprise such different systems as chemical reactions, spreading of diseases and populations, and propagation of electrical activity in neurons.

1. INTRODUCTION AND DEFINITION OF THE MODEL

In 1990, P. Bak et al.[1] introduced a forest-fire model which contains a tree growth probability p and fire spreading to nearest neighbors. In the limit of very slow tree growth, this model becomes more and more deterministic[2,3] and shows spiral-shaped fire fronts (see Fig. 1). We investigate a generalized version of this model, including a lightning parameter f and an immunity g. Depending on the range of parameters, the generalized model shows a variety of different phenomena. It is defined as follows: each site of a d-dimensional hypercubic lattice of size L^d is occupied by a tree, a burning tree, or it is empty. The rules for one timestep are:

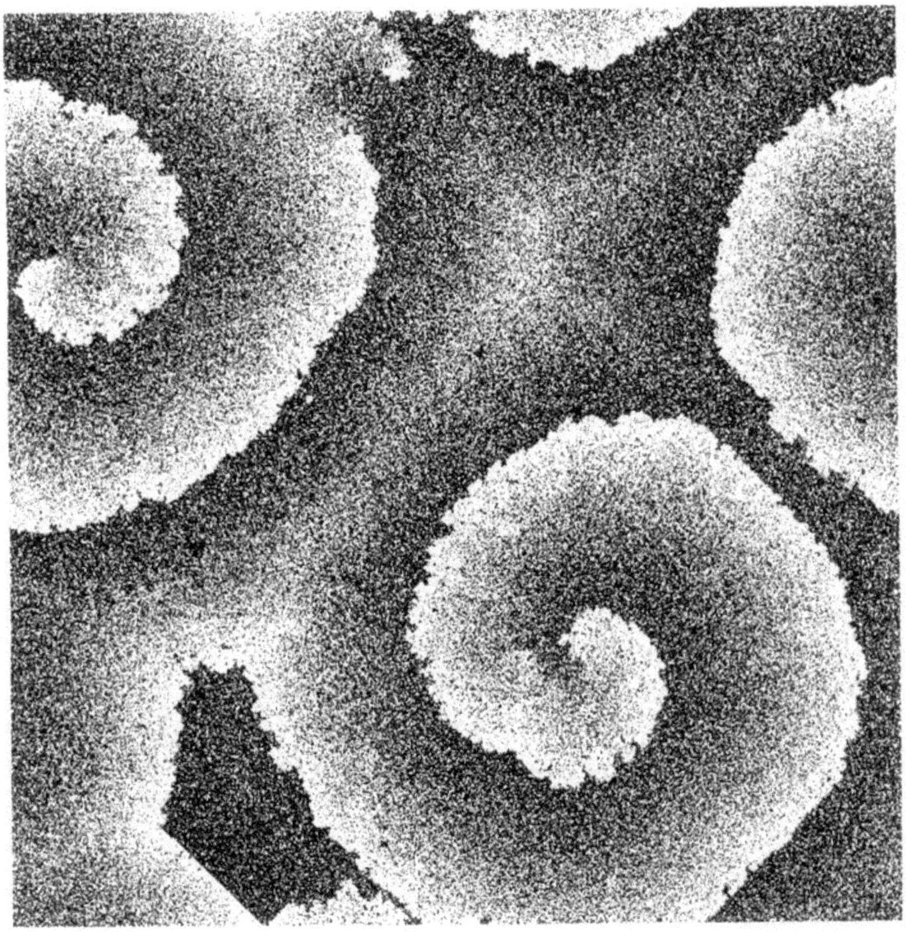

Fig. 1 Snapshot of the Bak et al. forest-fire model in the steady state for $p = 0.005$ and $L = 800$. Trees are black, empty sites are white (burning trees are situated at the sharp black-white transition but are too small to be seen).

(i) burning tree → empty site;
(ii) tree → burning tree with probability $1 - g$ if at least one nearest neighbor is burning;
(iii) tree → burning tree with probability $f(l - g) \ll 1$ if no neighbor is burning;
(iv) empty site → tree with probability p.

Starting with arbitrary initial conditions, this system approaches, after a short transition period, a steady state the properties of which depend on the parameter values. Let W_{-1}, W_0, and W_1 be the mean density of empty sites, of trees, and of burning trees in the steady state. These densities are related by the equations:

$$W_{-1} + W_0 + W_1 = 1 \tag{1}$$

and

$$W_1 = pW_{-1} . \tag{2}$$

The second equation says that the mean number of growing trees equals the mean number of burning trees in the steady state.

The most interesting behavior in this model occurs for those regions in the parameter space where the fire density decreases to zero. Here phase transitions and critical behavior can be expected. Equation (2) indicates that the fire density decreases to zero when p approaches zero. In the case $f = 0$, the fire density also decreases to zero when the immunity g approaches a critical value $g_C(p)$. In this paper, we focus on two parameter regions: in Sec. 2, we describe self-organized critical behavior which arises for $(g = 0; p \to 0; f/p \to 0)$ under the condition that the time scales of tree growth and burning down of forest clusters are separated.[4] In Sec. 3, we investigate the percolation-like transition which takes place for $g \to g_C(p)$.[5] We have already mentioned the spiral-shaped fire fronts which show up for $(f, g = 0; p \to 0)$. Since this state is not critical but quasideterministic, it is not described further in this paper. In the last section, we discuss possible applications of the forest-fire model to excitable systems.

2. SELF-ORGANIZED CRITICAL BEHAVIOR IN THE FOREST-FIRE MODEL

In this section, we set $g = 0$ and consider the case $p \ll 1$, $f \ll p$. In order to understand how self-organized critical behavior arises, let us first simplify dynamics and assume that a whole forest cluster is burned down instantaneously, i.e., during one time step if one of its trees is struck by lightning. Then the essential dynamics of the system depend only on the ratio f/p. Let $\bar{s}$ be the mean number of trees destroyed by a lightning stroke. It equals in the steady state the mean number of trees growing between two lightning strokes and is given by:

$$\bar{s} = p(1 - W_0)/fW_0 \,. \tag{3}$$

In the limit $f/p \to 0$, this number diverges proportionally to $(f/p)^{-1}$, since W_0 must approach a constant value for small f/p. Equation (3) represents a power law indicating a critical point in the limit $f/p \to 0$. Let $n(s)$ be the mean number of forest clusters per unit volume consisting of s trees. Then the mean forest density is:

$$W_0 = \sum_1^{\infty} sn(s) \,, \tag{4}$$

and the mean number of trees destroyed by a lightning stroke is:

$$\bar{s} = \sum_1^{\infty} s^2 n(s)/W_0 \,. \tag{5}$$

Since $W_0(f/p \to 0)$ is finite and $\bar{s}$ diverges $\propto (f/p)^{-1}$, these equations imply that $n(s)$ decreases at least like s^{-2} but not faster than s^{-3}. We conclude that:

$$n(s) \propto s^{-\tau} C(s/s_{\mathrm{max}}) \tag{6}$$

with $2 \le \tau \le 3$ and

$$s_{\mathrm{max}}(f/p) \propto (f/p)^{-\lambda} \propto \bar{s}^{\lambda} \,. \tag{7}$$

The cutoff function $C(x)$ is more or less constant for $x \le 1$ and decreases to zero for $x > 1$. Equations (4)–(6) yield $\bar{s} \propto s_{\mathrm{max}}^{3-\tau}$, which leads to the scaling relation:

$$\lambda = 1/(3 - \tau) \,. \tag{8}$$

In the case $\tau = 2$, the right-hand side of Eq. (6) acquires a factor $1/\ln(s_{\max})$, since the forest density given by Eq. (4) cannot diverge in the limit $f/p \to 0$. The mean number of forest clusters per unit volume $\sum_1^\infty n(s)$ therefore decreases to zero for $f/p \to 0$ and $\tau = 2$.

Let us now return to the real forest-fire model, where a forest cluster is not burned down instantaneously but during a finite time interval which depends on the size of the forest cluster. For a given ratio f/p, it is possible to choose a value of p which is so small that forest clusters burn down before trees grow at their edge; in this case the forest clusters still burn down rapidly, and the above Eqs. (3), (6), (7) and (8) are still valid. We conclude that the forest-fire model assumes a critical state with a power-law distribution of forest cluster sizes for $f \ll p$ and sufficiently small p. Since this critical state is reached independently of the initial conditions and for a wide range of parameter values, it is called a self-organized critical state.[6] The conditions $f \ll p$ and p sufficiently small describe a double separation of time scales: trees grow fast compared to the occurrence of lightning in the system, and forest clusters burn down much faster than trees grow.

Figure 2 shows a snapshot of the system for $f/p = 1/200$. Figure 3 shows the distribution $sn(s)$ obtained in a $2d$ computer simulation.[7] The exponent τ is $\tau \simeq 2.14$.[7,8] In one dimension, the model can be solved analytically. The probability that n neighboring sites (with $n \ll s_{\max}$) are all occupied by trees is:

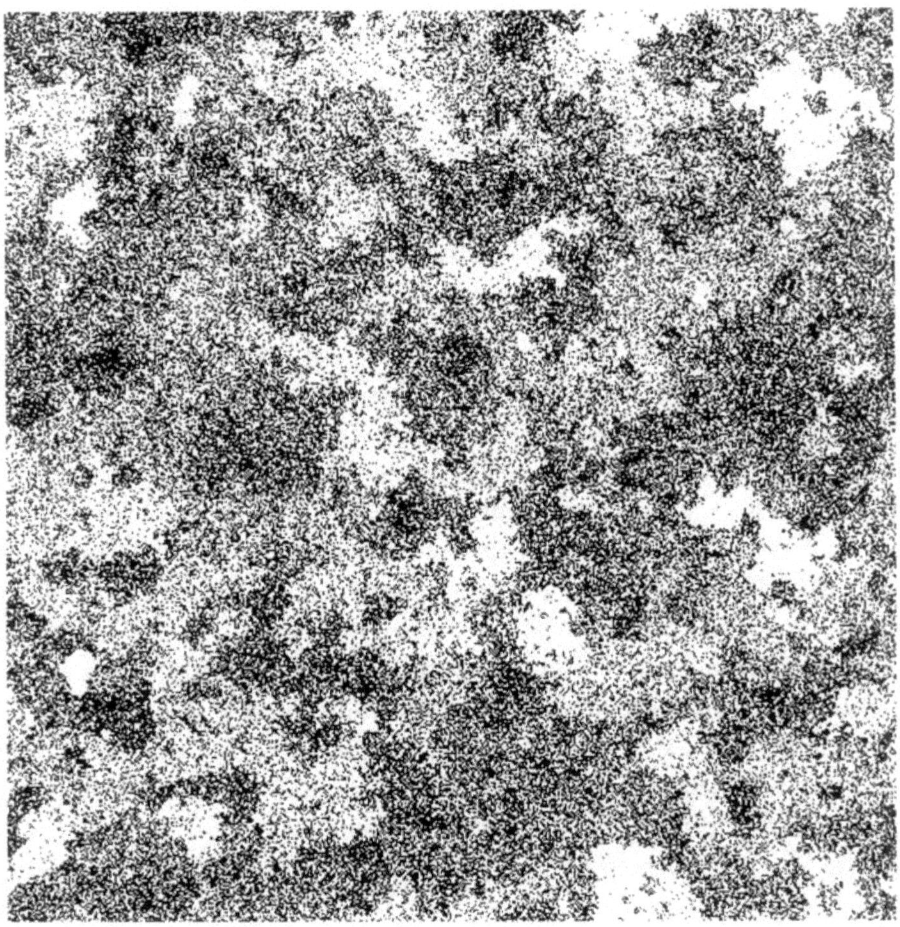

Fig. 2 Snapshot of the self-organized critical forest-fire model for $f/p = 1/200$ and $L = 500$. Trees are black, empty sites are white.

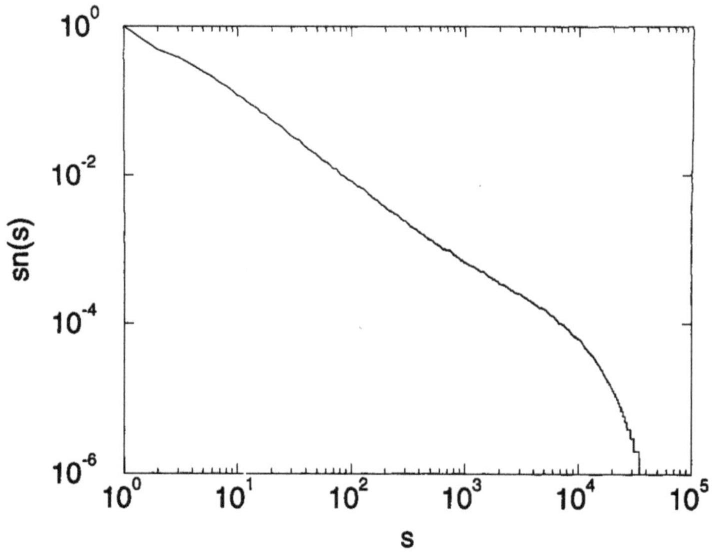

Fig. 3 Size distribution $sn(s)$ of the forest clusters that are struck by lightning for $f/p = 1/1000$.

$$P_n = 1 - (1 - W_0)\sum_{m=1}^{n} m^{-1}$$
$$\simeq 1 - (1 - W_0)\ln(n) \quad \text{for large } n\,.$$

In the limit $f/p \to 0$, $s_{\max}$ diverges and n may become large. Nevertheless P_n must be ≥ 0. It follows $(1 - W_0) \propto [\ln(s_{\max}]^{-1}$, i.e., the forest density approaches logarithmically the value one in the limit $f/p \to 0$, and the mean number of forest clusters per unit length decreases to zero. We conclude that $\tau = 2$ and [(via Eq. (8)] $\lambda = 1$.

3. PERCOLATION IN THE FOREST-FIRE MODEL

We now consider the case $f = 0$ and $g > 0$. Watching a 2-dimensional system on a video screen, one realizes that the fire fronts present for $g = 0$ become more and more fuzzy with increasing immunity g and that the forest grows denser. Figure 4 shows a snapshot of the system for a value of g just below g_C. At the critical value $g = g_C(p)$, the fire density goes to zero, and the critical fire spreading resembles percolation. For $g > g_C$, the steady state of the system is a completely green forest. Figure 5 shows the density of empty sites $W_{-1} = W_1/p$ as a function of g for $p = 0.1$ as determined from computer simulations. The fire density depends linearly on $(g_C - g)$ near the critical line. This linear dependence can already be obtained from mean-field theory.[5] Figure 6 shows the critical line $g_C(p)$ which was also obtained from computer simulations. The critical immunity g_C increases with increasing p. This is plausible since with increasing p the fire can sooner return to sites where it already has been. The main features of $g_C(p)$ can be derived by a simple calculation. If the immunity is at its critical value and if the initial state is an insolated burning tree in an infinitely large forest, the mean number of burning trees after a long time interval is again a single burning tree, i.e., the fire is just able to survive. We therefore started with an isolated burning tree with one empty neighbor (the site where the fire has been one timestep ago) and calculated the mean number $F(t, p, g)$ of burning trees after t timesteps. An approximate value of g_C is given by the condition $F(t, p, g_C) = 1$. The

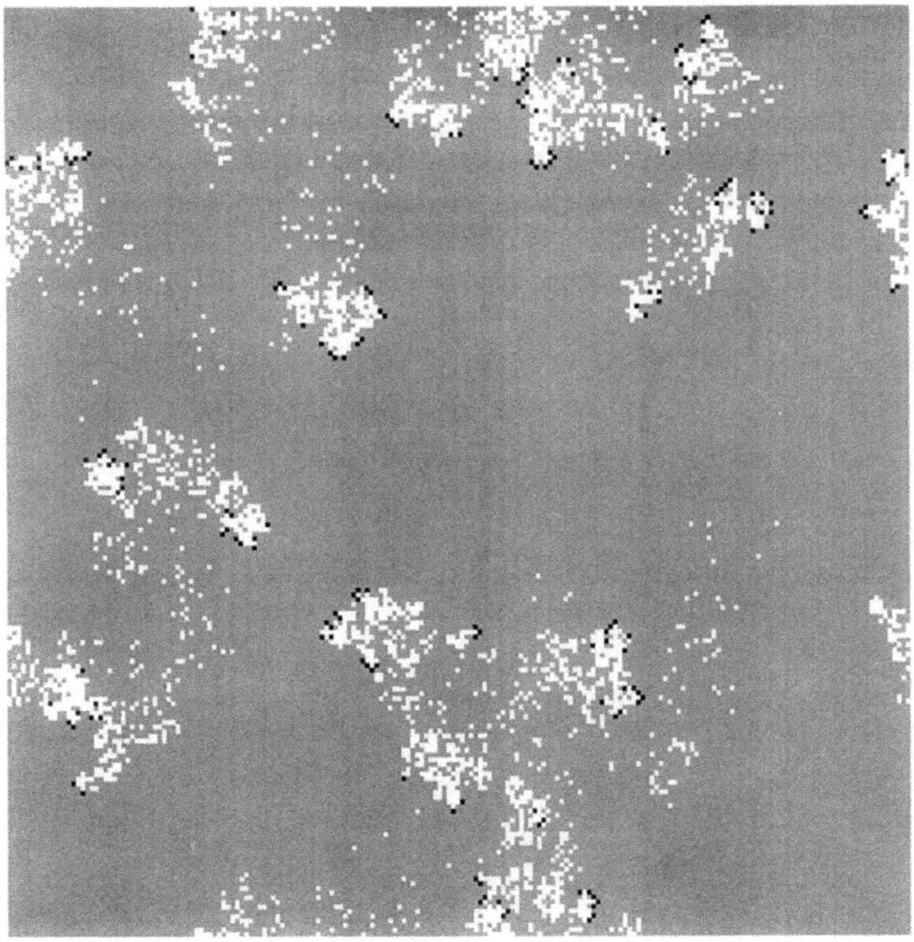

Fig. 4 Snapshot of the general forest-fire model near the critical immunity for $p = 0.1$, $g = 0.48$, $L = 200$. Trees are grey, empty sites are white, and burning trees are black.

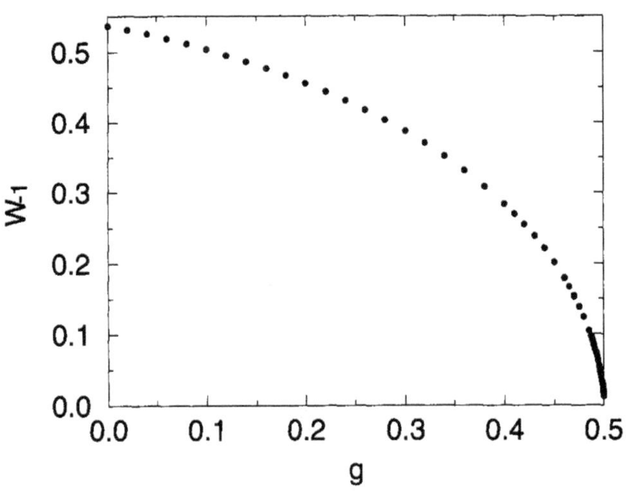

Fig. 5 Mean density of empty sites as function of immunity for $p = 0.1$.

larger t, the more accurate is the result. Figure 7 shows $g_C(p)$ for $t = 4$. This curve is linear for small p and flat for $p \to 0$, as obtained from the simulations.

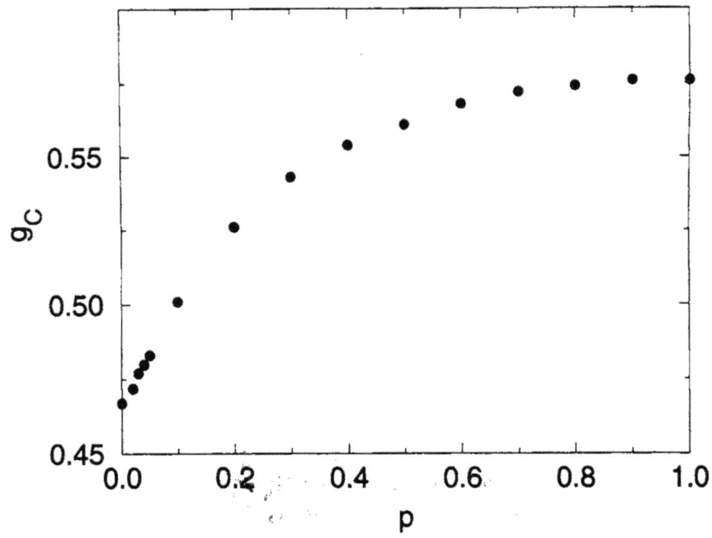

Fig. 6 The critical line $g_C(p)$.

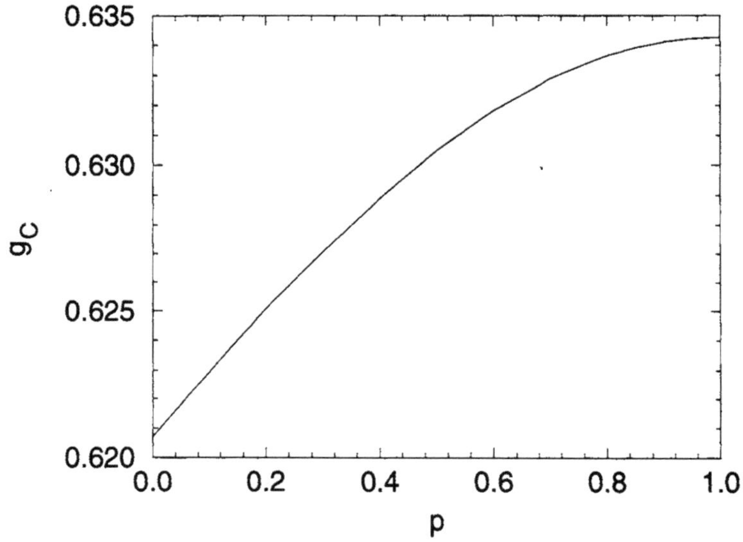

Fig. 7 Critical immunity $g_C(p)$ as calculated for $t = 4$.

Finally, we comment on the nature of the percolation phenomenon occurring at g_C. From the simulations we obtained $\lim_{p\to 0} g_C(p) \simeq 0.467$. Had we introduced the immunity by making sites or bond between sites permanently immune, we would have obtained a value $g_C(p = 0) = 0.41$ or 0.5, which is just one minus the percolation threshold for site or bond percolation. The critical immunity in this case were independent of p since there could not exist an infinite path for the fire through the system for $p > 0$ if it did not for $p = 0$. In our model, sites are not permanently immune and the fire can take paths that it cannot take in site percolation. The percolation threshold $1 - g_C(p = 0)$, therefore, is lower than in site percolation. We call the new kind of percolation observed in this model "fluctuating site percolation".

4. THE FOREST-FIRE MODEL AND EXCITABLE SYSTEMS

The general forest-fire model presented in this paper can be considered as a cellular automation for excitable systems (for a review on excitable systems, see e.g. Refs. 9 and 10). Excitable systems have three possible states: quiescent, excited, and refractory. Excitation spreads to neighboring sites, and after excitation, the medium cannot be excited for a certain period. Examples of excitable systems are the spreading of diseases, chemical reactions, propagation of electrical activity in neurons and even spiral galaxies. So far mainly "regular" behavior of excitable systems like spreading waves, target patterns, or spiral waves has been investigated, phenomena which occur also in the forest-fire model. Some years ago, after chaos theory has been founded, people also looked for chaotic behavior[11] in excitable systems and found it by solving numerically the partial differential equations which describe those systems. The results of this paper suggest that self-organized critical and percolation-like behavior can also be observed in excitable systems when the appropriate parameter regions are investigated.

ACKNOWLEDGMENTS

We thank S. Clar for producing Figs. 1, 2, 3 and 4, and S. Clar, W. Moßner, and N. Knoblauch for writing the simulation codes.

REFERENCES

1. P. Bak, K. Chen and C. Tang, *Phys. Lett.* **A147**, 297 (1990).
2. P. Grassberger and H. Kantz, *J. Stat. Phys.* **63**, 685 (1991).
3. W. Moßner, B. Drossel and F. Schwabl, *Physica* **A190**, 205 (1992).
4. B. Drossel and F. Schwabl, *Phys. Rev. Lett.* **69**, 1629 (1992).
5. B. Drossel and F. Schwabl, *Physica* **A199**, 183 (1993).
6. P. Bak, C. Tang and K. Wiesenfeld, *Phys. Rev.* **A38**, 364 (1988).
7. S. Clar, B. Drossel and F. Schwabl, to be published.
8. P. Grassberger, *J. Phys.* **A26**, 2081 (1993).
9. J. J. Tyson and J. P. Keener, *Physica* **D32**, 327 (1988).
10. E. Meron, *Phys. Rep.* **218**, 1 (1992).
11. C. F. Olsen and H. Degn, *Quart. Rev. Biophys.* **18**, 165 (1985).

Fractals, Vol. 1, No. 4 (1993) 1030–1036

SOC IN A CLASS OF SANDPILE MODELS WITH STOCHASTIC DYNAMICS

S. LÜBECK and K. D. USADEL
Theoretische Physik and SFB 166,
Universität Duisburg, Lotharstrasse 1, 47048 Duisburg, FRG

Abstract

We have studied one-dimensional cellular automata with updating rules depending stochastically on the difference of the heights of neighboring cells. The probability for toppling depends on a parameter λ which goes to one with increasing slope, i.e., the dynamics can be varied continuously. We have investigated the scaling properties of the model using finite-size scaling analysis. A robust power-law behavior is observed for the probability density of the size of avalanches in a certain range of λ values. The two exponents which determine the dependence of the probability density on time and system size both depend continuously on λ, i.e., the model exhibits nonuniversal behavior. We also measured the roughness of the surface of the sandpile and here we obtained a universal behavior, i.e., a roughness exponent of about 1.75 for all values of λ. For the temporal behavior of the mass, a $f^{-\phi}$ spectrum is obtained with an exponent ϕ close to 2, again for all λ-values.

1. INTRODUCTION

Bak, Tang and Wiesenfeld[1] recently introduced the notion of self-organized criticality (SOC) as a framework to understand the dynamics of extended dissipative dynamical systems. The dynamics drive the system into a state where relaxation processes occur over all time and length scales. It was suggested[1] that sandpiles were a particularly clear example of a self-organized system.

Theoretical sandpile models have been investigated by several authors.[1–6] They usually involve two kinds of steps. First a particle is added at randomly chosen sites, then the particles are eventually distributed to neighboring sites according to dynamical rules. Usually, these rules are deterministic: topping of particles may or may not occur, depending on the local occupation sites. However, real sandpiles behave in a different fashion. In real sandpiles, the particles have different shapes, various sizes and they can stick together. As a matter of fact, large and very unstable slopes can develop with slides of particles in an apparently stochastic fashion: a certain slope may or may not lead to a slide depending on a great number of internal degrees of freedom like the abovementioned different shapes and sizes of the particles, the friction between them, the local variations in packing, etc. To incorporate some of these features into a theoretical model, we propose as a first step a model with dynamical rules which are not deterministic: topping of particles occurs with a probability function which is small if the slope to neighboring sites is small and which increases to one with increasing slope. In the present paper, we report results of extensive simulations of this model. We restrict ourselves to the one-dimensional case which already shows an interesting and nontrivial behavior.

2. THE MODEL AND SIMULATIONS

The model we consider is a one-dimensional "nonlocal unlimited" sandpile model[2] with the following rules. We assume integer heights h_i at lattice sites $i = 1, 2, 3, \ldots, L$, where L is the size of the system. The local slope σ_i at the site i is defined as the height difference between two neighboring cells, $\sigma_i = h_i - h_{i+1}$. The dynamics obey the following rules: If the slope σ_i exceeds a threshold value σ_c an avalanche can be generated. With a certain probability $p(\sigma_i)$ grains drop from site i to the neighboring cell with:

$$\begin{aligned} h_i &\to h_i - n_i \\ h_{i+j} &\to h_{i+j} + 1 \qquad \text{for} \qquad j = 1, 2, \ldots, n_i \end{aligned} \tag{1}$$

The number n_i of grains which topple grows with increasing σ_i, $n_i = \sigma_i - N$ (unlimited model). Since there is toppling to n_i right neighbors, the model is called nonlocal. Note that from Eq. (1), it follows that all slopes in the system are non-negative.

The simulation starts by adding particles at the top until the stability condition $\sigma_1 < \sigma_c$ is violated. Then, with probability $p(\sigma_1)$, n_i grains flow to the right neighbor according to Eq. (1) and an avalanche develops. If no toppling takes place, another grain is added to site 1. If topping takes place, the next site will be visited and so on. An avalanche stops at site i if this site does not topple or if its slope to site $i + 1$ is too small. Thus an avalanche only starts at the top. The length of an avalanche is defined as the number of sliding sites. At the right edge of the system, grains can leave the pile, i.e., h_i is set equal to zero for $i > L$.

The most important new ingredients in our model is the probability function $p(\sigma_i)$ which determines whether a certain site i topples or not. This function should obey the following conditions:

$$\begin{aligned} p(\sigma < \sigma_c) &\equiv 0 \\ p(\sigma') > p(\sigma) \qquad &\text{for} \qquad \sigma' > \sigma \\ p(\sigma) \to 1 \qquad &\text{for} \qquad \sigma \to \infty \end{aligned} \tag{2}$$

The first condition states that there is only toppling if a threshold value is reached; the second condition means that there is increasing probability for toppling if the slopes increase; and the third equation means that toppling always takes place for very large slopes, i.e., the heights h_i are finite for any system of finite length. In our simulation we use:

$$p(\sigma_i) = 1 - e^{-\lambda(\sigma_i - 1)} \tag{3}$$

λ is called the toppling parameter. Thus there are three parameters in our model: σ_c which determines the stability condition, N which determines the number of sliding grains in one event, and most importantly the toppling parameter λ with which we can tune the dynamics continuously. In our simulations we choose $\sigma_c = 2$, $N = 1$ and study in particular the dependence of the model on λ. Note that for $\lambda \to \infty$, one gets a trivial model: $\sigma_i = 1$ for all i and any avalanche reaches the edge of the pile.

The most important quantity to study is the probability density for the lengths of the avalanches for which a finite-size scaling analysis is performed. As usual we assume for this probability density $P(t, L)$ a scaling form:

$$P(t, L) = L^{-\beta} g(L^{-\nu} t) \tag{4}$$

where t is the number of toppling sites. For $t \ll L$, we expect a power-law behavior, $P(t, L) \sim t^{-\kappa}$ independent of L. Thus β, ν and κ must obey the following scaling relation[2]:

$$\beta = \nu\kappa \tag{5}$$

Another important quantity is the roughness exponent ζ which is a measure of the fluctuations about the average height profile and which reflects the critical nature of the steady state. It is defined by[3]:

$$\xi = \sqrt{L^{-1} \sum_{i=1}^{L} \langle (h_i - \langle h_i \rangle)^2 \rangle} \sim L^{\zeta} \tag{6}$$

Finally we measure the total mass of the pile:

$$M(t) = \sum_{i=1}^{L} h_i(t) \tag{7}$$

and subsequently calculated its Fourier spectrum $|M(f)|^2$ to determine the exponent ϕ.

3. RESULTS

The simulations were performed for lattice sizes of $L = 50, 100, 200, 500, 1000$. Before performing a measurement, a large number of particles (around 10^8) were added, starting from an empty lattice to reach the critical state. Note that all measured values are independent of the initial conditions, i.e., the critical state is an attractor of the dynamics. First, we measured the probability density of the length of the avalanches $P(t, L)$ for a lattice size $L = 100$ and examined how $P(t, L)$ depends on the toppling parameter λ. A power-law behavior is observed in the range $1 < \lambda < 3$. Outside this range, we obtained on a log-log plot curves which deviate visibly from a power-law behavior. We have not investigated this

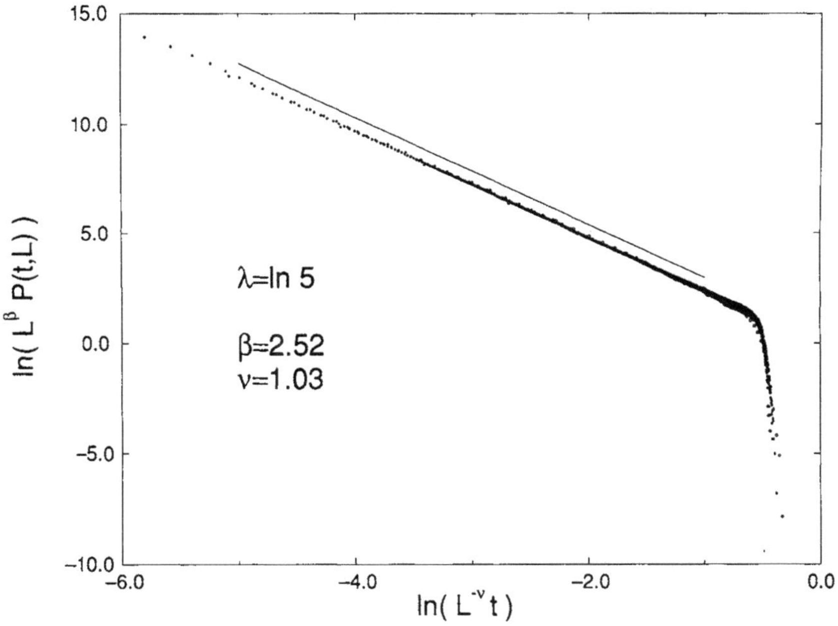

Fig. 1 Finite-size scaling fit for $\lambda = \ln 5$ and five different lattice sizes $L = 50, 100, 200, 500, 1000$. The solid line corresponds to a power-law with an exponent $\kappa = 2.453$.

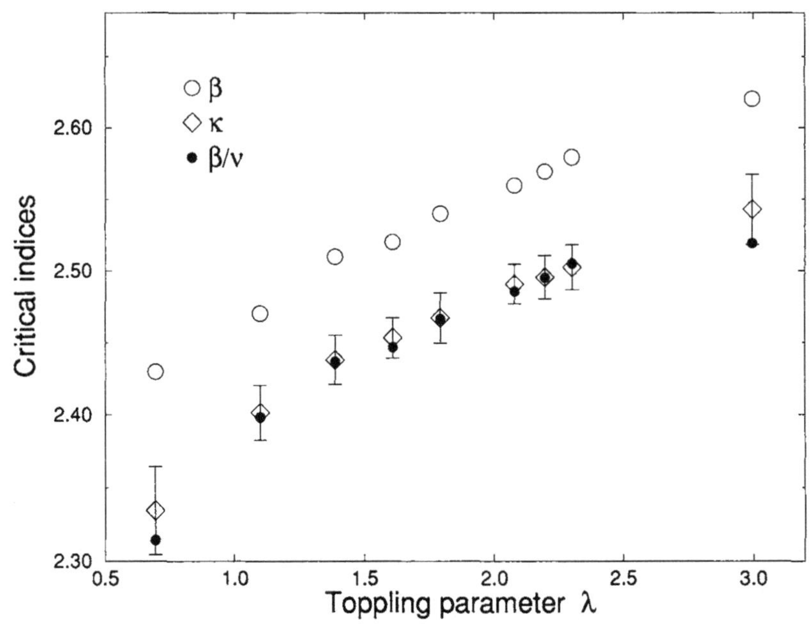

Fig. 2 Scaling exponent β and power-law exponent κ as a function of the toppling parameter λ. Solid circles correspond to $\nu^{-1}\beta$. Equation (5) is fulfilled within numerical accuracy.

problem any further but restricted ourselves to the λ-values where power-law behavior is clearly observed. Then the probability density $P(t, L)$ was measured for various values of λ and lattice sizes. As can be seen from Fig. 1, finite-size scaling works extremely well. For different values of λ, we get $\nu \approx 1$ but the scaling exponent β changes strongly [Fig. 2]. Thus we expect a varying exponent κ. κ is calculated using regression analysis for lattice

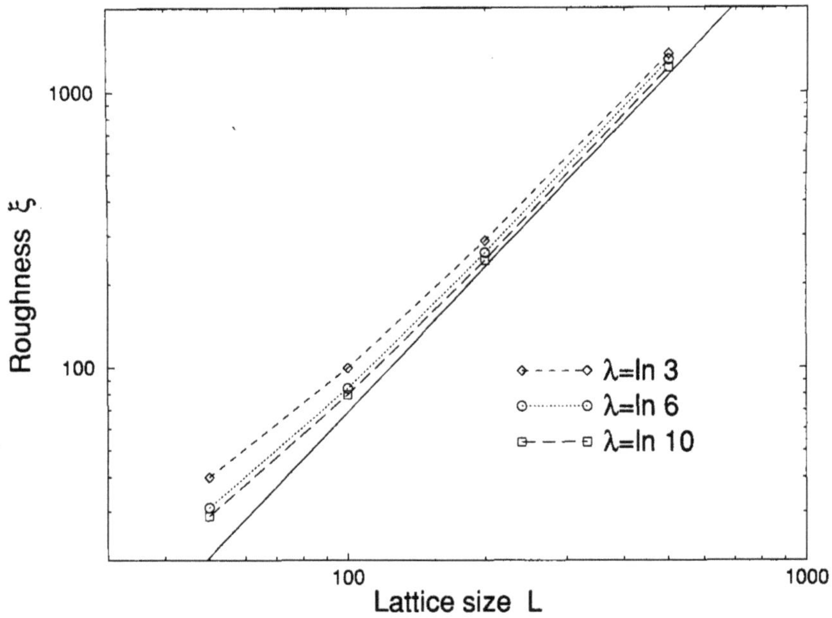

Fig. 3 Roughness ξ as a function of lattice size for different values of the toppling parameter λ. The solid line corresponds to a power-law $\xi \sim L^{\zeta}$ with $\zeta = 1.75$.

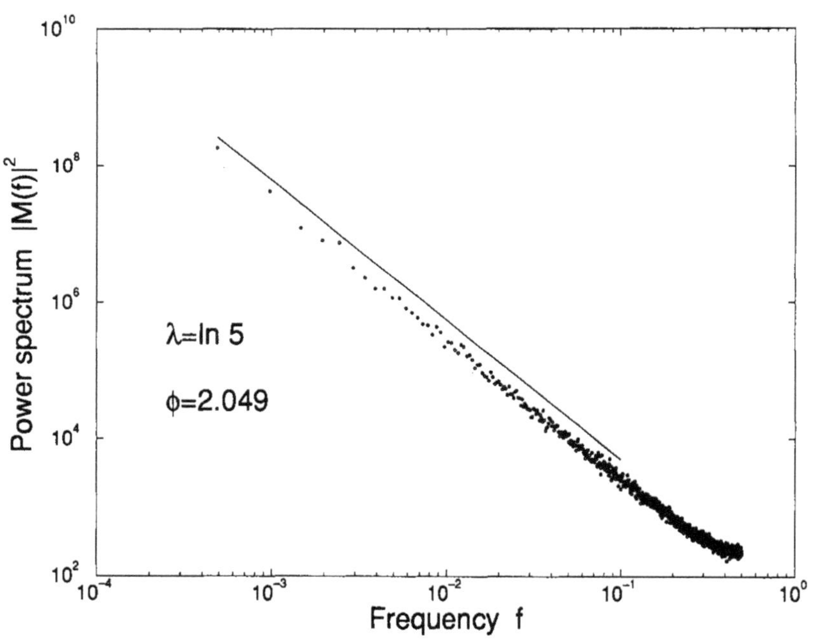

Fig. 4 Power spectrum of the time evolution of the total mass of the pile for $\lambda = \ln 5$. The spectrum is an average over 40 measurements with 2048 events. Solid line corresponds to $|M(f)|^2 \sim f^{-2.049}$.

sizes $L = 100, 200, 500, 1000$. The averaged value of κ is shown in Fig. 2. Note that the change of the exponent cannot be explained by statistical errors since they are far too small. The relation between the exponents, Eq. (5), is always fulfilled.

Table 1 The Critical Exponents for Different Values of the Toppling Parameter λ

λ	β	ν	κ	ζ	ϕ
ln 2	2.43	1.03	2.334	1.59	2.048
ln 3	2.47	1.03	2.401	1.70	2.061
ln 4	2.51	1.03	2.438	1.75	2.051
ln 5	2.52	1.03	2.453	1.80	2.049
ln 6	2.54	1.03	2.467	1.76	2.053
ln 8	2.56	1.03	2.491	1.75	2.064
ln 9	2.57	1.03	2.495	1.77	2.058
ln 10	2.58	1.04	2.503	1.76	2.112
ln 20	2.62	1.05	2.543	1.75	2.030

To determine the roughness exponents ζ, we limit our simulation to lattice sizes $L \leq 500$. After reaching the critical scale we add around 10^8 particles to calculate ξ for different values of λ. Figure 3 shows the results. One can see that for $L \geq 100$, the roughness ξ scales with the system size as L^{ζ} and ζ is independent of the toppling parameter λ. However, one should note that for a very accurate determination of ζ, one needs simulations with system sizes larger than 500, since the data points shown in Fig. 3 are still not on a straight line. But those measurements would demand more than 10^9 events and have not yet been done. With increasing lattice size, the roughness grows very fast, because $\zeta \geq 1$. The average height profile scales as $\langle h_i \rangle \sim L^{\alpha_1}$ with $\alpha_1 \approx 1.8$ for all i and λ.[a] The exponents α_1 and ζ should obey the relation $\alpha_1 \geq \zeta$, which is fulfilled in our model. It is very interesting to see whether the roughness exponent ζ is equal to the scaling exponent of the averaged height α_1. Finally, we have calculated the time dependence of the total mass of the pile [Eq. (7)]. We used a lattice size $L = 50$, and measured the mass $M(t)$ for 2048 events. Figure 4 shows the average over 40 of the corresponding fourier-spectra. The power spectrum $|M(f)|^2$ scales as $f^{-\phi}$ with $\phi \approx 2$ for all λ. The measured exponents are summarized in Table 1.

4. CONCLUSIONS

In conclusion, we have simulated a one-dimensional cellular automaton with nonlocal, unlimited and stochastic dynamics which exhibits SOC behavior in a certain range of λ-values. All trademarks of SOC are observed: finite-size scaling and a power-law behavior of the avalanche probability as well as $f^{-\phi}$ behavior for the power spectrum. However, our model displays nonuniversality: the exponents β and κ change continuously with λ. On the other hand, the exponents which describe the fluctuations, ϕ and ζ, are universal, i.e., independent of λ. Kertész et al.[8] have shown that a great class of sandpile automata yield

[a]Note that the average height profile is not a linear function of i, but obeys the scaling relation $\langle h_i \rangle = L^{\alpha_1} f(L^{-\alpha_2} i)$ with $f = a_0 + a_1 e^{-a_2 i^2}$.[7]

f^{-2} behavior. Thus it is understandable why ϕ is constant. In contrast, the reasons why the roughness exponent is independent of the toppling parameter are not known to us. Generalization of this model to higher dimensions and an investigation of the behavior for very small and very large values of λ, where power-law behavior apparently breaks down, is underway.

REFERENCES

1. P. Bak, C. Tang and K. Wiesenfeld, *Phys. Rev. Lett.* **59**, 381 (1987); P. Bak, C. Tang and K. Wiesenfeld, *Phys. Rev.* **A38**, 364 (1988).
2. L. P. Kadanoff, S. R. Nagel, L. Wu and S. M. Zhou, *Phys. Rev.* **A39**, 6524 (1988).
3. J. Krug, J. E. S. Socolar and G. Grinstein, *Phys. Rev.* **A46**, 4479 (1992).
4. T. Hwa and M. Kardar, *Phys. Rev.* **A45**, 7002 (1992).
5. D. Dhar and R. Ramaswamy, *Phys. Rev. Lett.* **63**, 1659 (1989).
6. K. Christensen and Z. Olami, *Phys. Rev.* **A46**, 1829 (1992).
7. S. Lübeck and K. D. Usadel, to be published.
8. J. Kertész and L. B. Kiss, *J. Phys.* **A23**, 433 (1990).

Fractals, Vol. 1, No. 4 (1993) 1051–1058

THE STATISTICS OF CRACK BRANCHING DURING FAST CRACK PROPAGATION

E. BOUCHAUD[†,*], J.-P. BOUCHAUD[**], J. PLANÈS[†], and G. LAPASSET[†]
[†]*ONERA (OM), 29, Avenue de la Division Leclerc, B.P. 72,*
92322 Châtillon Cedex, France
[*] *University of Cambridge, Dept. of Mat. Sci. and Metall.,*
Pembroke Street, CB2 3QZ Cambridge, England
[**]*SPEC, CEA, CEN-Saclay, F-91191 Gif-sur-Yvette Cedex, France*
and Cavendish Laboratory, Madingley Road,
CB3 0HE Cambridge, England

Abstract

We present simple theoretical ideas which allow an understanding of part of the scaling laws recently observed on branched fracture surfaces. Some of these are argued to be common to all critical branched structures, which barely survive when propagating. This is exemplified by numerical results on the directed percolation cluster, which serves as a good toy-model. Finally, we propose a Langevin equation for unbranched cracks in three dimensions, which naturally leads to self-affine structures.

1. INTRODUCTION

There has been a renewed interest in the statistical description of fracture surfaces, mainly motivated by the pioneering work of Mandelbrot et al.[1], who first concluded that these surfaces are *self-affine* reliefs, i.e., that the local height z scales with the in-plane displacement r as $z \propto r^{\zeta}$, where $\zeta < 1$ is the *roughness exponent*. Recent experiments spurred further interest: not only the "fractal" region is surprisingly large (over three decades in some cases:

0.5 μm to 0.5 mm^2), but also the roughness exponent ζ seems to be quite independent of the material and of the fracture mode.[2,3] Experiments using very different techniques, on various materials, prone to ductile or to brittle failure, all yield $\zeta \simeq 0.8$.[1,4,2,5,6,7]. This *universal* behavior is predicted by numerical simulations of random fuse models,[8,9] leading to roughness exponents independent of the microscopic disorder. These results strongly suggest that the model describing crack formation and propagation must exhibit scaling properties at large length scales,[10] much as is found in nonlinear diffusion equations describing, for example, growing surfaces,[11] or flame fronts.[12] We shall indeed propose, in Sec. 3 a *Langevin equation* for the propagation of unbranched cracks in three dimensions, which leads to this scaling behavior.

On the other hand, it is well known, although not yet fully understood, that fast running cracks produce secondary branches.[13] New experiments on polycrystalline Ni_3Al have analyzed the statistics of fracture surfaces containing secondary cracks.[14,15] A roughness index can be defined for such surfaces also, and is found to be equal to: $\zeta = 0.83 \pm 0.04$.[14,15] Furthermore it is found[14] that the conditional probability distribution $P(r, z)$ that the surface passes through the point of coordinates (r, z) knowing that the origin belongs to it exhibits a scaling behavior of the type:

$$P(r, z) = r^{-\zeta} f\left(\frac{z}{r^\zeta}\right) \tag{1}$$

with $f(z/r^\zeta) \sim r^\zeta/z$ as soon as $z > r^\zeta$. It is argued in Sec. 2 that this behavior is fairly general, which is confirmed by the numerical study of the branched structure of the bidimensional directed percolation (DP) cluster, presented in Sec. 1.

The typical number of secondary cracks in the case of Ni_3Al is shown (Sec. 1) to exhibit a power-law increase with an exponent $b \simeq 0.13$. b is shown to be the ratio of two microscopic lengths linked respectively to the probability of branching along the main crack path and to the typical length of a secondary crack (Sec. 2). Such a crack is shown to be a "*critical*" structure, for which secondary branches barely survive.

2. EXPERIMENTAL AND NUMERICAL RESULTS

We briefly present in this Section experimental results obtained on polycrystalline Ni_3Al[14,15] and a numerical analysis of the bidimensional (DP) cluster.

2.1 Experimental Results on Ni_3Al

The material used, as well as the practical procedure are described in Ref. 14. The "number of secondary branches" $n(r)$ at a given distance r in the "horizontal" plane is characterized through its fluctuations $N(r) = \langle (n(r+x) - n(x))^2 \rangle_x^{1/2}$.

For the three experiments, $N(r)$ is determined for each image, and shown to exhibit a power-law increase for distances r smaller than the correlation length ξ. Figure 1 shows $N^2(r)/N^2(\xi)$ as a function of r/ξ in a log-log plot for the three experiments under consideration. One has: $N(r) \sim r^b$ with $b = 0.13 \pm 0.03$.

It will be emphasized in Sec. 2 that such a behavior should be expected on general grounds.

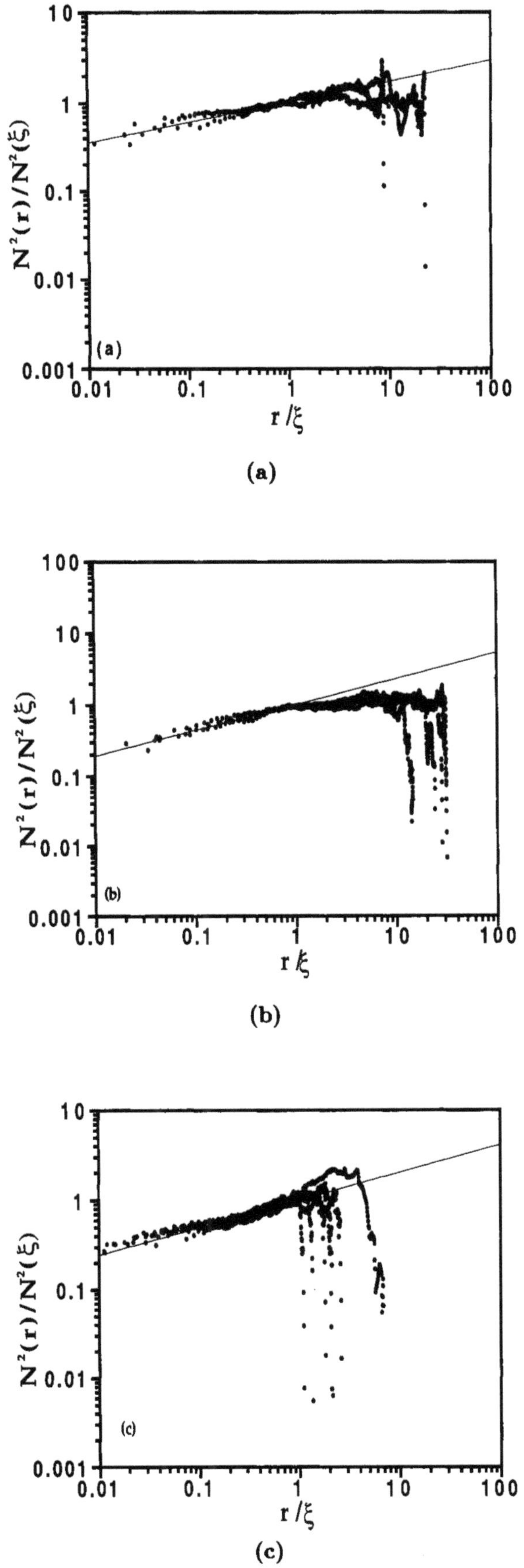

Fig. 1 $N^2(r)/N^2(\xi)$ as a function of r/ξ. Curves (a), (b) and (c) correspond to experiments 1, 2 and 3 described in Ref. 14.

2.2 Numerical Analysis of the Bidimensional Directed Percolation "Cracks"

It was first suggested by Kertész and Vicsek[16] that (DP) was a good toy-model for unbranched fracture in two dimensions. We show here that, although the values of the various exponents are very different from the ones found experimentally on the 3-d system, (DP) contains the important ingredients to reproduce the main properties reported above and in Ref. 14.

Let us thus start by analyzing a minimal model of crack branching in two dimensions. It is assumed that an existing crack may either propagate as a random walk, die, or give birth to a secondary crack. Furthermore, when two cracks come close to each other, one of them dies. The coarse-grained equation governing the "time" evolution of the crack density $\rho(r, z)$ may thus be written as:

$$\frac{\partial \rho(r, z)}{\partial r} = D \frac{\partial^2 \rho(r, z)}{\partial z^2} + A\rho(r, z) - B\rho^2(r, z) \tag{2}$$

where D is the "diffusion coefficient", A measures the difference between the branching probability and the death probability, and B models the competition between cracks. It is reasonable to assume that as the external stress is increased, A increases while B and D are roughly constant. Equation (2) has been thoroughly discussed in the literature.[17–19] For A less than a certain value A_c, the density ρ vanishes almost surely for large times: the crack does not propagate. For $A > A_c$, on the other hand, the density reaches a finite value. $A = A_c$ is thus a second order phase transition point, which is in the universality class of directed percolation. A typical "crack" obtained for $A = A_c$ is shown in Fig. 2. The critical exponents for this case are well known[17–20]: one finds in particular that the number of branches $N(r)$ grows as $r^{b_{DP}}$ with $b_{DP} = 0.31$, and that the "width" of the structure

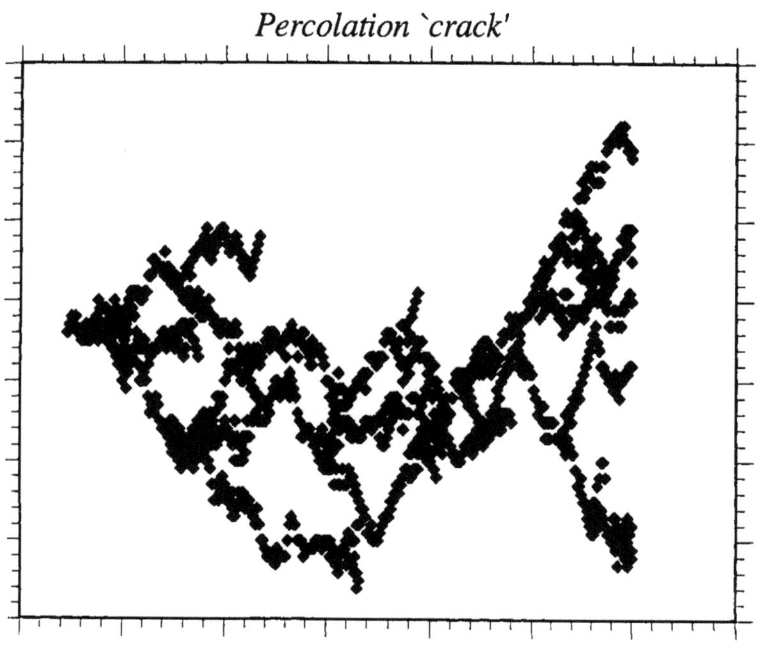

Fig. 2 "Crack" generated by numerically evolving a population of walkers obeying Eq. (2), with $A = A_c$.

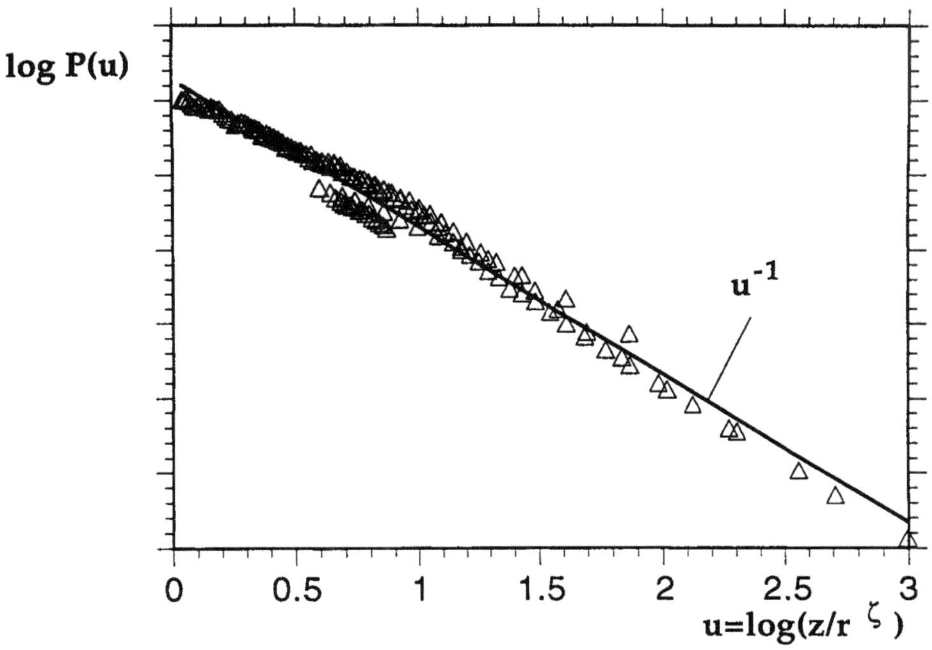

Fig. 3 Plot of $f(u)$ (defined by Eq. (1)), for different values of r, for the directed percolation cluster problem (to be compared with Fig. 7 of Ref. 14.) The full line is the best power-law fit, which gives $\mu = -0.01$.

grows as $r^{\zeta_{DP}}$ with $\zeta_{DP} = 0.63$. These values were confirmed by the analysis used in the experimental case.

On the other hand, the full conditional probability distribution $P(r, z)$ is determined with the algorithm used to analyse the experimental binary profiles images.[14] It is found that it is precisely given by Eq. (1), with $\zeta = \zeta_{DP}$, and $f(u) \simeq u^{-1}$ for large u (see Fig. 3).

Before turning to more general statements, let us add one remark. Equation (2) with a *random* $A(r, z)$ and $B = 0$ is the equation for the partition function of directed polymers in a random environment (i.e., an elastic string interacting with random impurities[21]), which was already proposed in the context of *unbranched* cracks.[22,23,9] The idea expressed[23,9] is that a crack might *minimize globally* the fracture energy along its path, and hence should be identical to the ground state of a directed polymer in a random environment. This leads to $\zeta_{2d} = 2/3$, which is rather close to the (DP) value and to 2-d experiments.[24] In three dimensions, however, the "*minimum surface*" exponent is close to $\zeta_{3d} \simeq 0.4$–0.5,[25] far below the experimental value $\zeta = 0.8$ (see however Ref. 3). This whole argument assumes that the crack can sample all the possible paths before propagating, which can be realized if there is branching — branching is a well known method to sample phase space thermodynamically (and is commonly used in Monte-Carlo simulations[26]).

3. CRITICAL BRANCHED STRUCTURES

We now show how both the experimental results on Ni_3Al and the numerical simulations on (DP) can be understood within the same framework.

Because the experimental cuts studied are perpendicular to the direction of propagation of the main crack, it is natural to assume a uniform crack velocity within these sections, and thus a uniform probability of branching $\frac{1}{l_1}$ per unit length along the main crack path. The probability that a new off-spring born at r' survives until r is denoted $K(r-r')$. Assumption of self-similarity of the whole structure leads to the following self-consistent equation:

$$N(r) = \int_0^r dr' \frac{1}{l_1} K(r-r')N(r-r') \tag{3}$$

which leads to the following differential equation:

$$\frac{dN(r)}{dr} = \frac{1}{l_1} K(r)N(r) \tag{4}$$

If $K(r)$ decreases faster than $1/r$, then, from Eq. (4), $N(r)$ tends to a constant when r grows to infinity. On the contrary, if $K(r)$ decreases more slowly than $1/r$, $N(r)$ undergoes an exponential growth when r goes to infinity. $K(r) = 1/r$ thus appears as the critical point for which the secondary cracks barely survive. This behavior was found as well in the problem of branching on a Bethe lattice (for which $b = 2$),[27] and on various tree structures studied numerically.[28,29]

Writing $K(r) = l_2/r$ then leads to: $N(r) = r^b$ with $b = \frac{l_2}{l_1}$. Note that l_2 is related to the average length $\bar{l}$ of a secondary branch, which diverges logarithmically: $\bar{l} = \int_0^\infty dr\ r dK(r)/dr \equiv \int_0^\infty dr K(r)$. A small exponent b corresponding to a typical size of secondary branches l_2 smaller than the typical distance l_1 between branching points fits the visual aspect of the cracks (see Fig. 2 in Ref. 4).

Now, defining $N(r, z)$ the number of points belonging to the structure with coordinates (z, r), one has, from the definition of $P(r, z)$, $N(r, z) \equiv N(r)P(r, z)$. A general equation obeyed by $N(r, z)$ can be written:

$$N(r, z) = \int_0^r \frac{dr'}{l_1} K(r-r')\left[\int_{-\infty}^{+\infty} dz' P(z', r')N(z-z', r-r')\right] \tag{5}$$

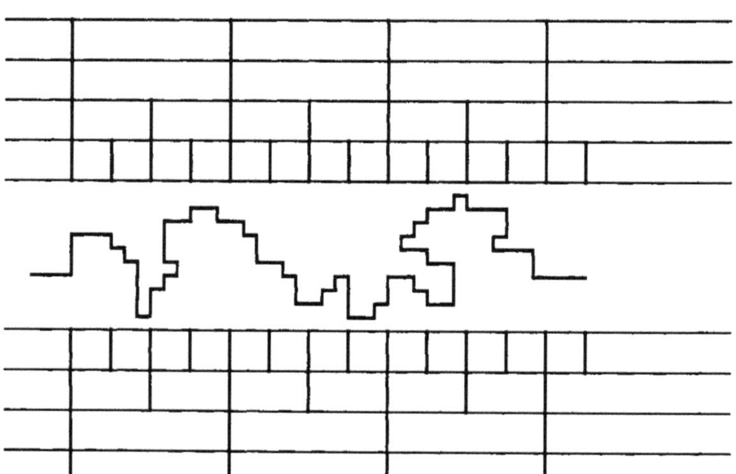

Fig. 4 Sketch of the *fluffy fractal.* $P(r, z)$ assuming the scaling form given in Eq. (1), is simply equal to $1/z$ as soon as point (r, z) is no more on the backbone, but still in its vicinity (i.e. $z \geq r^\zeta$). This decreasing density is depicted by thick segments.

Assuming the scaling form given in Eq. (1) and $K(r) = l_2/r$, it appears that the function f has to fulfill a nonlinear equation,[30] the solution of which reads: $f(u) \simeq u^{-1}$ for large u, i.e., precisely the form of f suggested by the experimental data[14] and the numerical results on (DP) clusters.

In both cases, the branched structure appears to be a *fluffy fractal* (see Fig. 4), with branches attached to a "skeleton" of roughness $\zeta = 0.8$, leading to a density independent of r as soon as one is no longer on the "skeleton", but still in its vicinity (i.e., for $z \geq r^{\zeta}$).

4. DISCUSSION

We thus have shown how branched cracks observed experimentally behave as "*critical*" systems. In that respect, although it is obviously too simple to describe 3-fracture, 2-d DP is an excellent toy-model, which retains this essential feature. Further experiments studying the cross-over between non-branching and branching cracks could lead to quantitative relations between the measured quantities and the macroscopic parameters (crack velocity, ΔK in fatigue ...) which control the transition.

From the theoretical point of view, on the other hand, the next step would be to build a minimal model for the three dimensional problem. Although we have not found how to generalize Eq. (2) for three dimensional *branched* cracks, it is interesting to discuss the case of unbranched cracks. The simplest non-trivial equation describing the propagation of a line (the crack front) in a disordered medium is the following[30,31]:

$$\frac{\partial x}{\partial t} = V + \gamma \frac{\partial^2 x}{\partial y^2} + \frac{\lambda_{xx}}{2}\left(\frac{\partial x}{\partial y}\right)^2 + \frac{\lambda_{xz}}{2}\left(\frac{\partial z}{\partial y}\right)^2 + \eta_x(y, t) \tag{6a}$$

$$\frac{\partial z}{\partial t} = \gamma \frac{\partial^2 z}{\partial y^2} + \lambda_z \left(\frac{\partial x}{\partial y}\right)\left(\frac{\partial z}{\partial y}\right) + \eta_z(y, t) \tag{6b}$$

where x is the direction of crack propagation, y is along the crack front and z is the tensile axis. V is the nominal crack velocity, γ is the "line tension" and $\eta_{x,z}$ are noises describing the local disorder in the material, due to microstructure. Finally, the nonlinear terms $\{\lambda\}$ describe the dependence of the local crack velocity on the local curvature of the crack, and are the only ones allowed by the symmetries of the problem: $(y \to -y)$, $(z \to -z)$. The trace of the crack front is then the fracture surface, the equation of which is $z(x, y)$. Interestingly enough, Eq. (6) was recently discussed in the context of driven vortex lines in superconductors.[32] One finds that $z(x, y)$ is an *anisotropic* self-affine surface: $\langle [z(x, y) - z(0, 0)]^2 \rangle = y^{2\zeta} G(\frac{x}{y^{\phi}})$ with $G(0) = \text{cst}$ and $G(u) \simeq u^{2\zeta/\phi}$ for large u. Quite surprisingly, the exponents ζ and ϕ are found to depend on the value of the nonlinear couplings $\{\lambda\}$: for Gaussian noises $\eta_{x,z}$, Ertas and Kardar[32] find $\zeta \simeq 0.5$ and $\phi \simeq 1.5$–2.0 in most regions of the parameter space $\{\lambda\}$, but $\zeta \simeq 0.75$, $\phi \simeq 1.5$ when $\lambda_{xz} \ll \lambda_{xx}, \lambda_z$. The latter result is quite close to the experimental values[1,4,2,5,6] $\zeta \simeq 0.8$ (although no significant anisotropy was detected in the experiments of Ref. 2, suggesting rather $\phi \simeq 1$). Although this coincidence is intriguing, we feel that more work on the very rich structure of Eq. (6) is needed to argue convincingly that these equations may offer a key to the fractal nature of unbranched cracks. We believe, however, that Eq. (6) is at least of pedagogical value, since it shows how a reasonable equation for crack propagation naturally leads to self-affine fracture surfaces.

ACKNOWLEDGMENTS

E. B. was supported by DRET contract No. 92.1437A000 and by CNRS (Laboratoire d'Eticoles des Microstructures). She wishes to express warm gratitude to Patrick Veyssiere.

REFERENCES

1. B. B. Mandelbrot, D. E. Passoja and A. J. Paullay, *Nature (London)* **308**, 721 (1984).
2. E. Bouchaud, G. Lapasset and J. Planès, *Europhys. Lett.* **13**, 73 (1990).
3. Although weaker values $\zeta \simeq 0.4$–0.6 have recently been reported, either for low-cycle fatigue: P. Mc Anulty, L. V. Meisel and P. J. Cote, *Phy. Rev.* **A46**, 3523 (1992); or for much smaller length scales (nanometers): V. Y. Milman, N. A. Stelmashenko, R. Blumenfeld and R. C. Ball, Cavendish Laboratory (Cambridge, UK) preprint.
4. J. J. Mecholsky, D. E. Passoja and K. S. Feinberg-Ringel, *J. Am. Ceram. Soc.* **72**, 60 (1989); Z. Q. Mu and C. W. Lung, *J. Phys. D.: Appl. Phys.* **21**, 848 (1988); R. H. Dauskardt, F. Haubensak and R. O. Ritchie, *Acta Metall. Mater.* **38**, 143 (1990).
5. K. J. Måløy, A. Hansen, E. L. Hinrichsen and S. Roux, *Phys. Rev. Lett.* **68**, 213 (1992).
6. A. Imre, T. Pajkossy and L. Nyikos, *Acta Metall. Mater.* **40**, 1819 (1992).
7. E. Bouchaud, "Complex Microstructures Analysis: Fracture Surfaces as an Example", in Proceedings of the International Congress *Dislocations 93* (Aussois, France, 1993).
8. L. de Arcangelis and H. J. Herrmann, *Phys. Rev.* **B39**, 2678 (1989); *Statistical Models for the Fracture of Disordered Media*, eds. by H. J. Herrmann and S. Roux (Elsevier, Amsterdam, 1990). See, however, C. Vanneste and D. Sornette, *J. Physique (France)* **I2**, 1621 (1992), where a model of cracks with non-universal fractal dimensions is studied.
9. A. Hansen, E. L. Hinrichsen and S. Roux, *Phys. Rev. Lett.* **66**, 2476 (1991), S. Roux and D. François, *Scripta Metall.* **25**, 1092 (1991).
10. R. C. Ball and R. Blumenfeld, *Phys. Rev. Lett.* **65**, 1784 (1990).
11. M. Kardar, G. Parisi and Y.-C. Zhang, *Phys. Rev. Lett.* **56**, 889 (1986).
12. See e.g. S. Zaleski, *Physica* **D34**, 427 (1989).
13. B. R. Lawn and T. R. Wilshaw, *Fracture of Brittle Solids* (Cambridge University Press, Cambridge, 1975) Chap. 5, p. 100; E. H. Yoffe, *Phil. Mag.* **42**, 673 (1951); K. Ravi-Chandar and W. G. Knauss, *Int. J. of Fracture* **26**, 65 (1984); ibid. **26**, 141 (1984); J. Fineberg, S. P. Gross, M. Marder and H. L. Swinney, *Phys. Rev. Lett.* **67**, 457, 1991.
14. J. Planès, E. Bouchaud and G. Lapasset, *Statistics of Fracture Surfaces*, this conference.
15. E. Bouchaud, G. Lapasset, J. Planès and S. Naveos, *Phys. Rev.* **B48**, 3451, (1993).
16. J. Kertész and T. Vicsek, *J. Phys.* **C13**, L343 (1980).
17. P. Grassberger and A. de la Torre, *Ann. Phys. (NY)* **122**, 373 (1979).
18. J. L. Cardy and R. L. Sugar, *J. Phys.* **A13**, L423 (1980).
19. I. Jensen, *Phys. Rev.* **E47**, R1 (1993).
20. J. W. Essam, A. J. Guttmann and K. de Bell, *J. Phys.* **A21**, 3815, 1988.
21. D. A. Huse and C. L. Henley, *Phys. Rev. Lett.* **54**, 2708 (1985); D. S. Fisher, D. A. Huse and C. L. Henley, *Phys. Rev. Lett.* **55**, 2924 (1985), M. Kardar, *Nucl. Phys.* **B290 [FS20]**, 582 (1987).
22. A. Chudnovsky and B. Kunin, *J. Appl. Phys.* **62**, 4124 (1987).
23. M. Kardar, in *Disorder and Fracture*, eds. J. C. Charmet, S. Roux and E. Guyon (Plenum, New York, 1990).
24. C. Poirier, M. Ammi, D. Bideau and J.-P. Troadec, *Phys. Rev. Lett.* **68**, 216 (1992); J. Kertész, V. K. Horvath and F. Weber, *Fractals* **1**, 67 (1993).
25. See e.g. M. Mézard and G. Parisi, *J. Phys. (France)* **I1**, 809 (1991); T. Halpin-Healy, *Phys. Rev.* **A42**, 711 (1990).
26. D. Ceperley and M. Kalos, in *Monte Carlo Methods in Statistical Physics*, ed. K. Binder (Springer, Berlin, 1979). See also T. Garel and H. Orland, *J. Phys.* **A23**, L621 (1990).
27. P. Alstrom, *Phys. Rev.* **A38**, 4905 (1988).
28. E. Perlsman and M. Schwartz, *Europhysics Lett.* **17**, 11 (1992).
29. N. N. Pang and T. Halpin-Healy, *Phys. Rev.* **E47**, R784 (1993).
30. J.-P. Bouchaud, E. Bouchaud, G. Lapasset and J. Planès, *Phys. Rev. Lett.* **71**, 2240 (1993).
31. For another proposal, see J. A. Hodgdon and J. P. Sethua, *Phys. Rev.* **B47**, 4831 (1993).
32. D. Ertas and M. Kardar, *Phys. Rev. Lett.* **69**, 889 (1992).

Fractals, Vol. 1, No. 4 (1993) 1059–1067

STATISTICS OF FRACTURE SURFACES

J. PLANÈS, E. BOUCHAUD and G. LAPASSET
ONERA (OM), 29, Avenue de la Division Leclerc,
B.P. 72, F-92322 Châtillon Cedex, France

Abstract

Experimental measurements of the roughness exponent ζ of some fracture surfaces are reported. Whatever the fracture mode, for unbranched (aluminium alloy) or branched (Ni_3Al) surfaces, ζ is found quite close to 0.8, in accordance with other experimental determinations, but not with theoretical predictions of standard 3-d models. The complete height distribution for the complex branched structures is shown to slowly decrease with increasing altitude. It implies a power-law behavior for the low order moments which is experimentally recovered. Exponents reveal different from that of gaussian distributions.

1. INTRODUCTION

Numerous experimental measurements of the fractal dimension of fracture surfaces have been performed for a few years on various materials using various experimental methods,[1–4] renewing the pioneering work of Mandelbrot et al.[5] They surprisingly all lead to the conclusion that fracture surfaces are reasonably well described by self-affine surfaces with a roughness exponent ζ equal to 0.8 ± 0.1, which yields a new great interest in the theoretical understanding of fracture mechanisms (see discussions in Refs. 6–8).

This paper briefly reviews our experimental work on two metallic alloys, the commercial 7475 Aluminium and the intermetallic Ni_3Al. We will focus on the latter which presents new features compared to the former: secondary branched cracks and overhangs. It has been shown that it is necessary to go beyond the simple determination of the fractal dimension in order to give a more accurate description of these complex surfaces. The basic object to be analyzed is the quantity $P(r, z)$, namely, the probability of the surface going through the point of abscissa r and height z — z being the anisotropy axis — conditionally to its passing at (0,0). The self-affinity is expressed through the scaling relation:

$$P(r, z) = \frac{1}{r^\zeta} f\left(\frac{z}{r^\zeta}\right) \tag{1}$$

which is verified for $\zeta = 0.8$. Conversely to other experiments,[9] the f function is not found to be gaussian-like but is slowly decaying for high values of the argument: $f(x) \sim x^{-(1+\mu)}$ with $\mu \sim 0$. A second quantity of interest is the "thickness of the surface" which is found to behave like:

$$z_{\text{max}}(r) \simeq z_0 + z_1 r^\zeta \tag{2}$$

(here r is the width of the analyzed zone). The relations obeyed by $P(r, z)$ and z_{max} are shown to yield a scaling behavior for the finite-order moments of the P distribution: $M_q(r) \sim r^{\zeta(q)}$. Diverging from Gaussian distribution,[9] the ratios $[M_q(r)]^{\frac{1}{q}}/[M_2(r)]^{\frac{1}{2}}$ are r-dependent. These moments have been computed for $2 \leq q \leq 6$ and the scaling relations verified with $\zeta(q)/\zeta(2) \neq q/2$.

This anomalous statistics is thought to be related to dynamical phenomena involved in the fracture process, which are known to be responsible for instabilities and growing of crack branching. This is discussed in a second contribution.[8]

2. EXPERIMENTAL

2.1 Specimen Preparation

The aluminium alloy was submitted to four different solution heat treatments in order to vary the fracture toughness (K_Q ranging from 28.5 to 45.0 $MPa\sqrt{m}$) as well as the predominant fracture mode (from inter- to trans-granular). The Ni_3Al alloy was only homogenized at high temperature. Its microstructure reveals much more anisotropic than that of the 7475 alloy, which could be at the origin of the "mode II" cracks that appear during mechanical testing. More details and micrographs can be found in Refs. 1 and 6.

Fracture surfaces were electrochemically coated, making sure of good wetting and high contrast for backscattered electrons. Aluminium surfaces were then polished perpendicularly to the anisotropy axis (longitudinal cuts), thus exhibiting islands and lakes in the coating background (Fig. 1). Ni_3Al polishing planes, on the contrary, contain the anisotropy axis (transverse cuts), thus showing the metal-deposit interface along the main crack propagation (Fig. 2). In both cases, multiple cuts ensure statistical reliability.

2.2 Methods of Analysis

Thanks to the high contrast of the SEM micrographs, separation lines are easily extracted by image processing, giving a numerical description of cuts of the fracture surfaces. The sizes of the pictures are 768 × 512 and 703 × 512 pixels. The auto-correlation of longitudinal cuts is computed and rotationally averaged to reach the so-called *return probability* $P(r)$ at a given height. Let us recall that it measures the probability that the altitude of the surface is the same at distance r as at origin, and that obviously $P(r) \equiv P(r, 0)$ as defined in Eq. (1). In the case of transverse cuts, averages concern heights instead of angles. This $P(r)$ is known to behave as $r^{-\zeta}$ for self-affine fractals.[10]

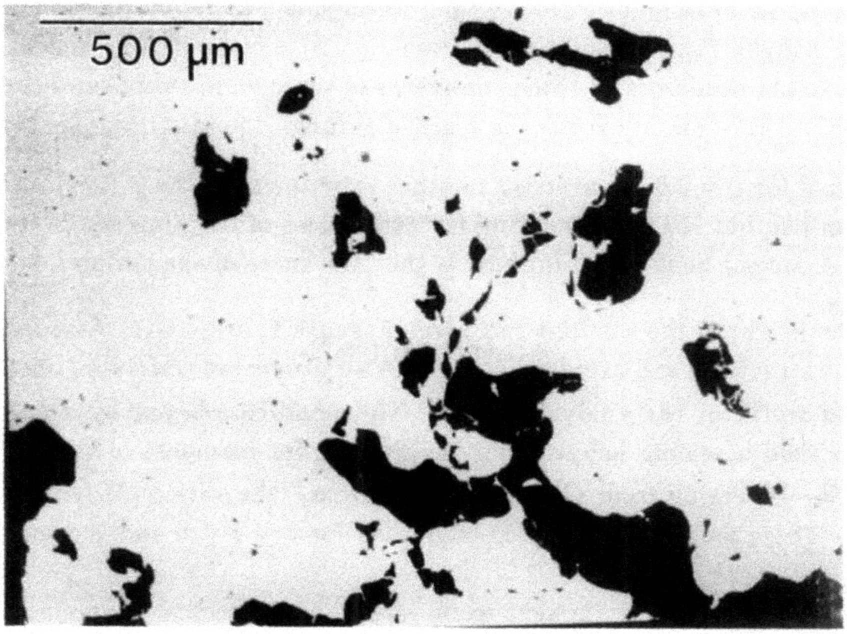

Fig. 1 Scanning electron micrograph of a longitudinal cut of the aluminium alloy fracture surface, as observed with backscattered electrons. The tensile axis is perpendicular to the image plane. The alloy appears in black, the nickel coating in white.

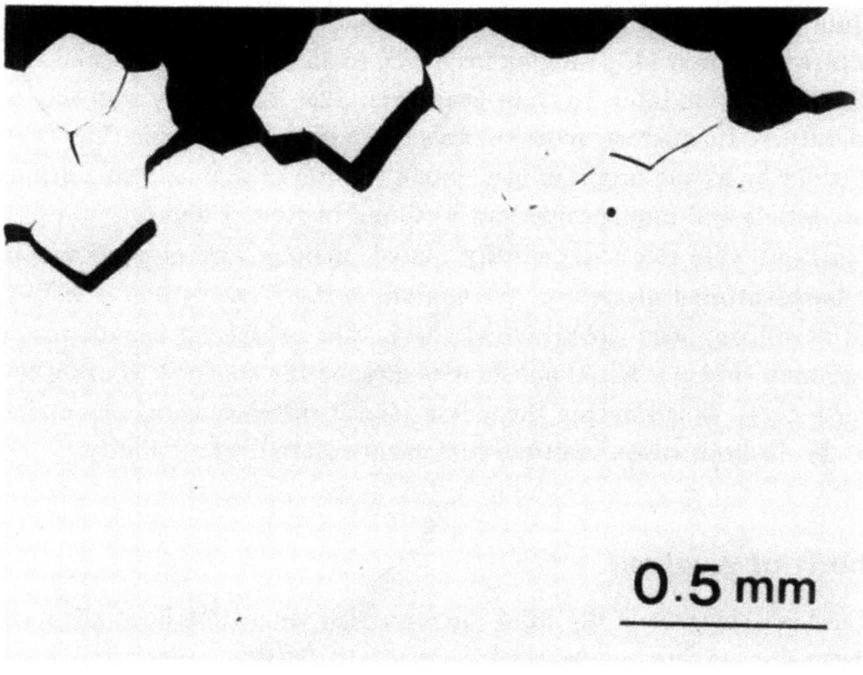

Fig. 2 Optical micrograph of a transverse cut of Ni_3Al. The tensile axis lies vertical in the image plane. Coating is opaque resin. Secondary cracks and overhangs lying beside the polishing plane clearly appear. Results quoted in the text concern another sample coated with Ni-Pd and observed in SEM. This sample gives similar results with a lower precision (lower accuracy and more finite size effects).

For Ni_3Al, where $z(r)$ cannot be defined (multiple z values at given r because of secondary cracks and overhangs), the complete distribution $P(r, z)$ is computed as well as $P_m(r, \bar{z})$. $\bar{z}$ is the average height at abscissa r. Some moments of these distributions are computed too. The "thickness" of the fracture surface is reached through the infinite order moment of P:

$$z_{\max}(r) = \left\langle \max_{[x;x+r]}(z) - \min_{[x;x+r]}(z) \right\rangle_x \tag{3}$$

which gives the height of the smallest rectangle of length r, originating at x, containing all branches of the structure, and averaged over x. Other orders are computed on P_m following the definition:

$$M_q(r) = \langle (|\bar{z}(x+r) - \bar{z}(x)|)^q \rangle_x \tag{4}$$

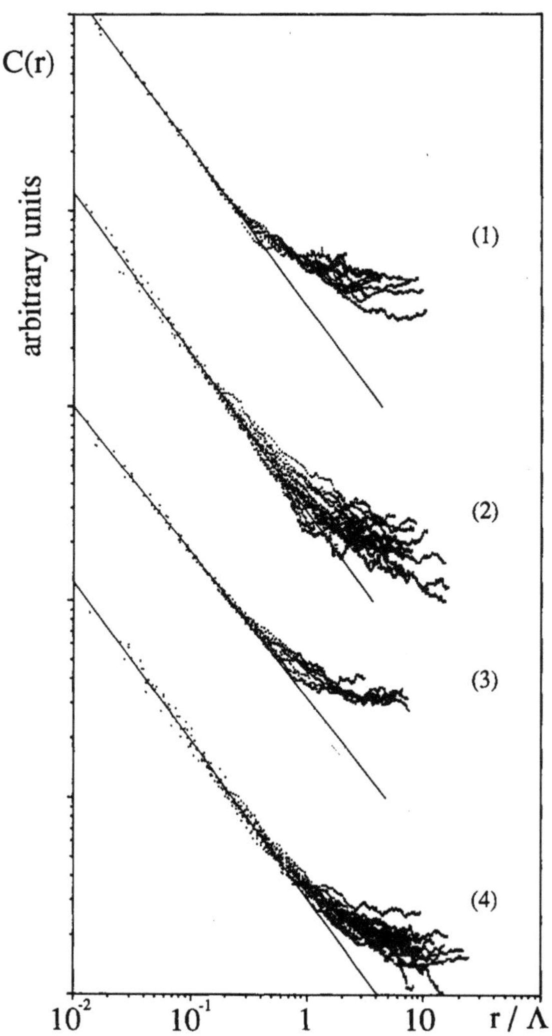

Fig. 3 Aluminium alloy 7475. $C(r)$ (= $P(r)$) vs. r/Λ on a log-log plot, for the four studied samples. Ten to 20 pictures were analyzed for each sample. The linear regimes extend over two decades in r/Λ, which correspond to three decades in real scale: $r = 0.5$ μm to 0.5 mm. The four slopes are equal to -0.8, within experimental accuracy.

3. RESULTS

3.1 Aluminium Alloy

The computed return probability corresponding to various magnifications of the same sample were rescaled: $P(r)/P(\Lambda)$ was plotted vs. r/Λ, where Λ is approximately equal to the size of the micrograph. Figure 3 shows the spectra corresponding to the four samples on a log-log plot. On each plot, a linear regime extends over two decades, which corresponds to *three decades in real scale*: 0.5 μm to 0.5 mm. The four slopes were determined and were found to be equal, within experimental accuracy (error bars were estimated from the dispersion of the experimental points in the power-law regime). Thus the averaged exponent over the four samples is equal to: $\zeta = 0.80 \pm 0.05$.

3.2 Ni_3Al

Return probability. Because of anisotropy, longitudinal cuts are not analyzed here. For each micrograph, the probability $P(r)$ is plotted vs. r/ξ, where the "correlation length" ξ is indicative of the size of the scaling domain, the value of which lying within the range of grain widths (200–300 μm). It can be seen in Fig. 4 that the scaling zone spreads over two decades with $\zeta = 0.83 \pm 0.04$.

Height probability distribution. The complete abovementioned distributions P and P_m are plotted in Figs. 5 and 6, which gather the results of five images at different magnifications. Both exhibit a long-tail behavior, although with different exponents: $\mu(P) \sim 0.1$ and $\mu(P_m) \sim 0.67$. The curves $r^\zeta P(r, z)$ as a function of z/r^ζ can be shown to reasonably superimpose on each other for separate values of r (see Fig. 8 in Ref. 6). Because we are

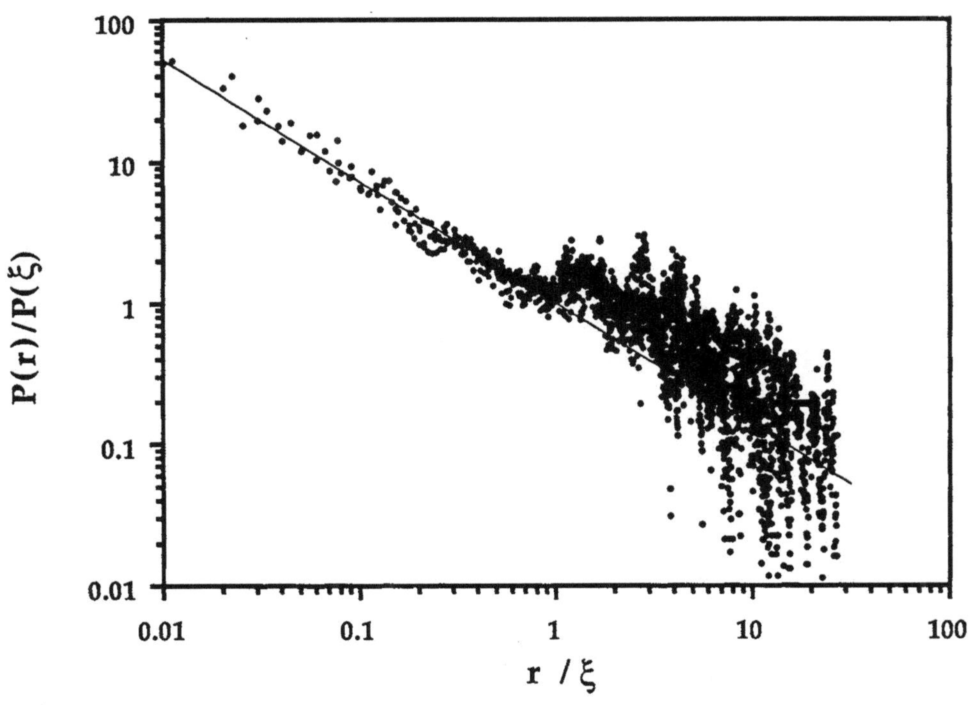

Fig. 4 Ni_3Al: log-log plot of the return probability $P(r)/P(\xi)$ as a function of r/ξ, with 5 superimposed analyses. The linear regime extends approximately over two decades in r/ξ. The measured roughness index ξ is 0.82 ± 0.3, and the estimated correlation lengths ξ fluctuate from 150 to 250 μm.

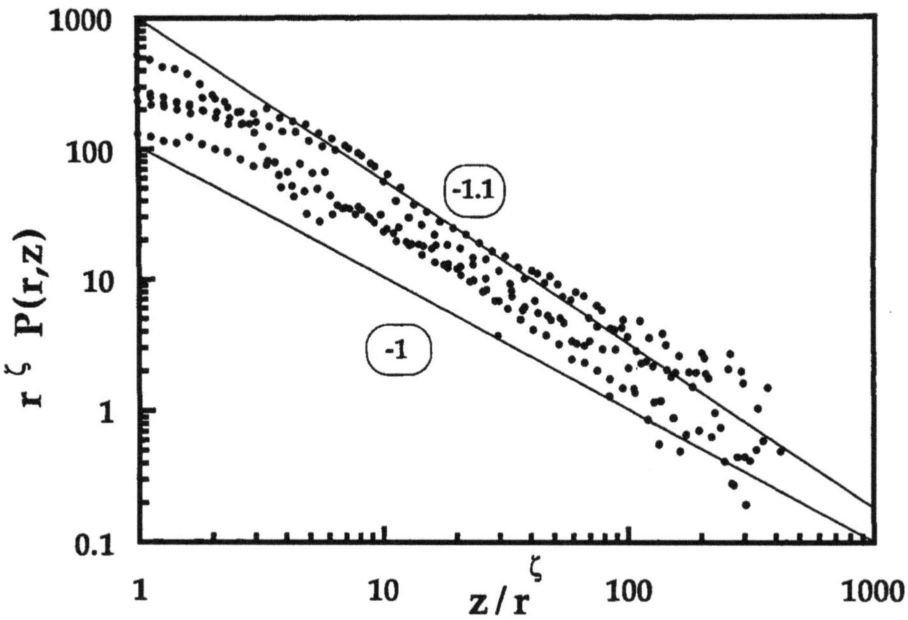

Fig. 5 Ni_3Al: $r^\zeta P(r, z)$ as a function of z/r^ζ on a log-log plot, for values of r in the range 1–100 pixels. $r^\zeta P(r, z)$ was averaged over all the values of r which lead to close values of z/r^ζ. Linear regimes on the log-log plot extend approximately over two decades for each micrograph. On average, the exponent is close to -1.1. A line of slope -1 is also drawn to show that the actual exponent is slightly smaller than -1, but this might be due to logarithmic terms for which we could not have any evidence.

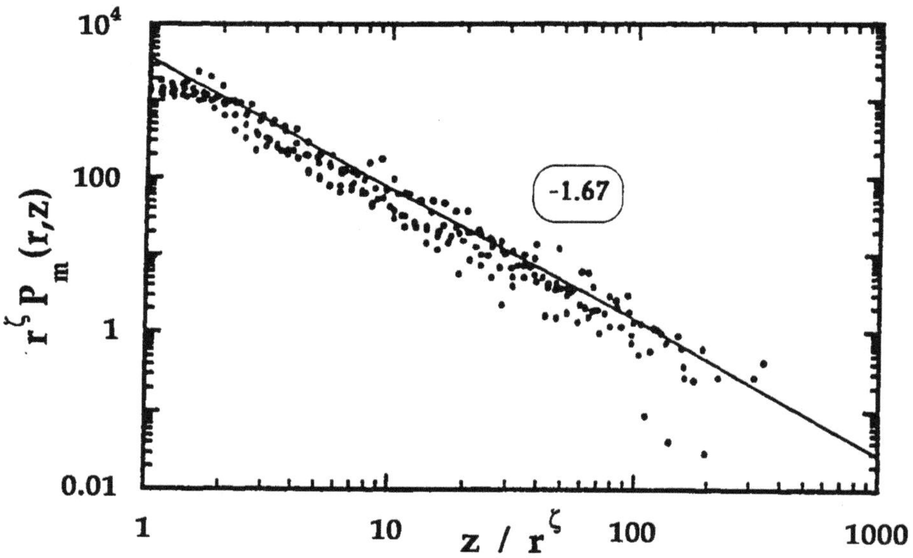

Fig. 6 Same as Fig. 5 for $r^\zeta P_m(r, \bar{z})$. The linear regime extending over two decades has a slope equal to: -1.67 ± 0.07.

interested in the asymptotic behavior of $f(z/r^\zeta)$ only small values ($r < 100$ pixels) are relevant. In Fig. 5, ordinates have been averaged over closed values of z/r^ζ. Two lines with slopes -1.1 (the fitted one) and -1 are drawn to show that the actual exponent should lie close to -1, with possible (logarithmic?) corrections we are not able to measure.

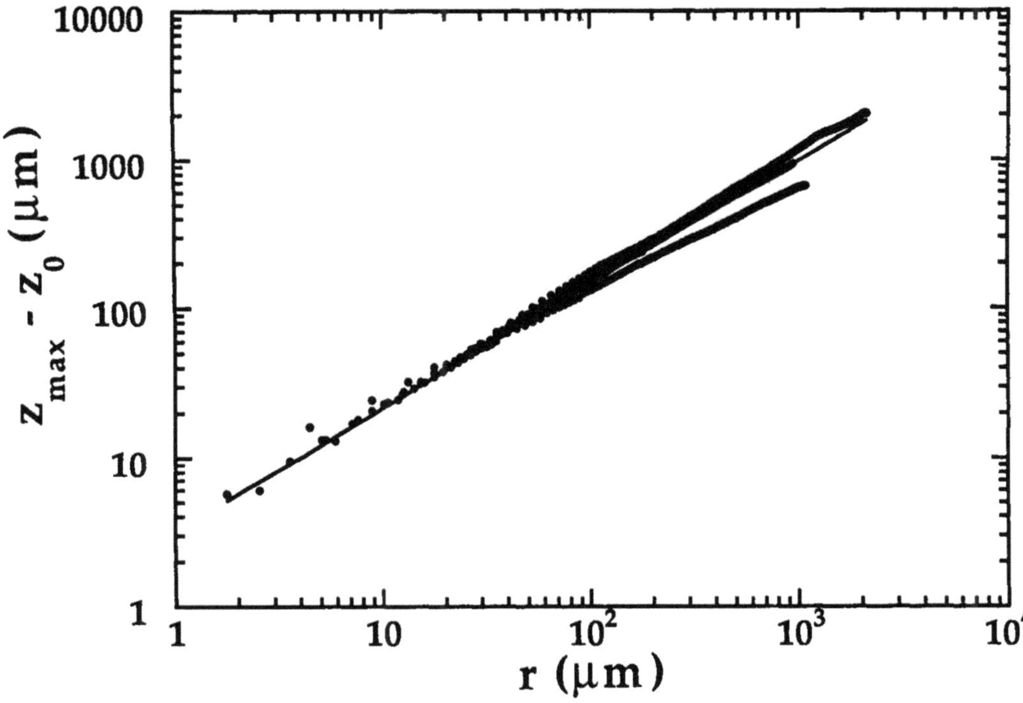

Fig. 7 Ni_3Al: infinite order moment $z_{max}(r) - z_0$ as a function of r, on a log-log plot. Length z_0 is of the order of the whole width occupied by the branches, and, as a consequence, its value varies slightly from image to image. The linear regime extends over three decades in real scale. The slope is 0.83, i.e., equal to ζ with very good accuracy.

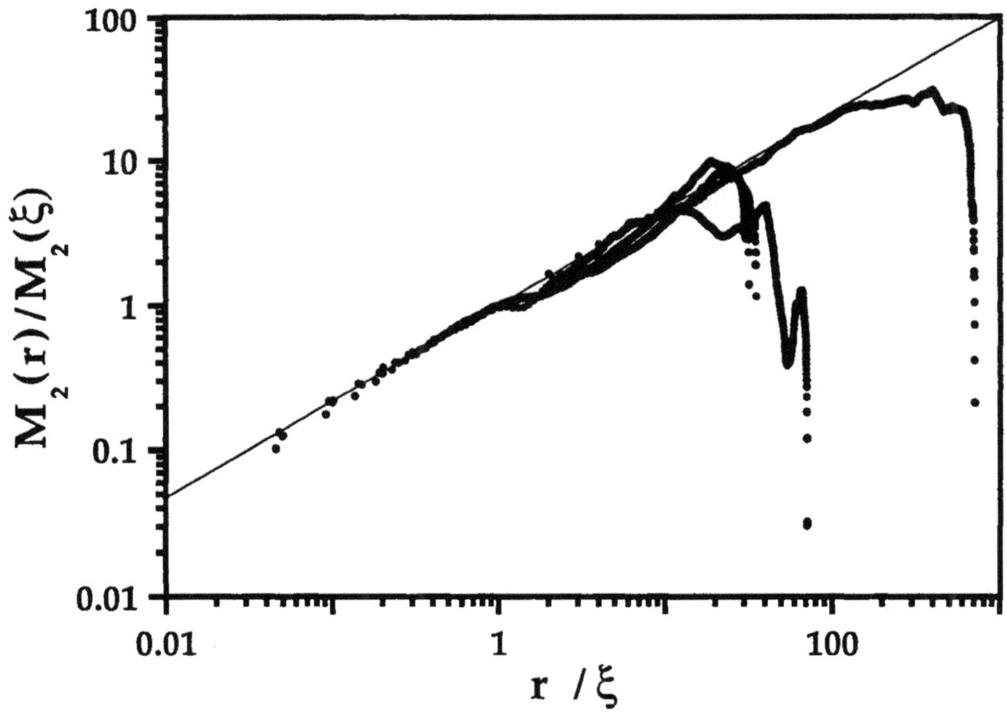

Fig. 8 Ni_3Al: second order moment $(M_2^2(r)/M_2^2(\xi))$ of the distribution $P_m(r, \bar{z}(r))$ of averaged heights as a function of r/ξ in a log-log plot. The linear regime extending over approximately three decades allows for the determination of exponent $\zeta(2) = 0.71 \pm 0.02$. Other orders present similar behaviors and the exponents definitely do not follow the "gaussian" identities $\zeta(q)/q = \zeta(2)/2$ (see values in the text).

Figure 6 corresponds to Fig. 5 for the P_m distribution. We will see in the following that the exponent is quite compatible with the computed moments.

Thickness $z_{\max}$. The quantity $(z_{\max} - z_0)$, as defined in Eq. (3), is plotted vs. r/ξ. A power law is found over *three decades.* z_0 represents an "intrinsic" width and can thus slightly vary from image to image. Its value is of a few hundreds of microns, and never becomes negligible compared to the r-depending term; this is in contrast to the results of Ref. 3.

Finite order moments of P_m. Moments M_q [Eq. (4)] were determined for finite values of q ($q = 2$ to 6) for the distribution of averaged heights described above by the probability $P_m(r, \bar{z}(r))$. Figure 8 shows that the second order ($q = 2$) moments of P_m behaves like a power-law with exponent: $\zeta(2) = 0.71 \pm 0.02$. Ratios of the type:

$$\frac{(M_q(r))^{\frac{1}{q}}}{(M_2(r))^{\frac{1}{2}}} \tag{5}$$

with $q = 3$ to 6, are computed and exhibit power-law behaviors with exponents:

$$\frac{\zeta(3)}{3} - \frac{\zeta(2)}{2} = -0.11 \pm 0.01 \quad \frac{\zeta(4)}{4} - \frac{\zeta(2)}{2} = -0.18 \pm 0.02$$

$$\frac{\zeta(5)}{5} - \frac{\zeta(2)}{2} = -0.19 \pm 0.02 \quad \frac{\zeta(6)}{6} - \frac{\zeta(2)}{2} = -0.24 \pm 0.02$$

Although these exponents are rather small, the ratios defined by Eq. (5) do depend on r. Other experiments[9] do find constant ratios, characteristic of gaussian distributions. The method of analysis is however different, consisting of a mechanical profilometer, which only reaches the hull of the fracture surface.

4. DISCUSSION

The results detailed above first confirm the tendency for universality of the roughness exponent ζ which is found to be very close to 0.8 for the aluminium alloys and Ni_3Al, giving rise respectively to ductile and brittle fracture, as well as for a large class of brittle materials.[2-4] Traditional *minimum surface* models, which suppose that the surface *globally* minimizes energy, lead to $\zeta = 0.4 - 0.5$.[11] Experimental results invalidate this hypothesis. Moreover, it is a well-known result that crack branching appears when a critical propagation velocity is overpassed.[12-14] It thus seems necessary to take dynamical effects into account (see Ref. 8).

We tried to get an insight into the complexity of Ni_3Al fracture surface by studying the averaged height distribution probability which was found to verify:

$$P_m(r, \bar{z}(r)) = \frac{1}{r^\zeta}\left(\frac{\bar{z}}{r^\zeta}\right)^{-1-\mu'} \tag{6}$$

for large values of $\bar{z}/r^\zeta$. Computing the moments with the value $\mu' = 0.67$ in Eq. (6) gives:

$$\langle z^q \rangle = \int_a^{z_{\max}} \frac{dz}{r^\zeta} z^q \left(\frac{z}{r^\zeta}\right)^{-1-\mu'} = r^{\mu'\zeta}(z_{\max}(r))^{q-\mu'}) \tag{7}$$

As the leading term in $z_{\max}$ is constant, apparent exponents $\zeta(q)$ are equal to:

$$\zeta(q) \simeq \mu'\zeta + \zeta_{\max}(q - \mu') \tag{8}$$

where $\zeta_{\max} \simeq 0.1$ is an "effective" exponent characterizing $z_{\max}$:

$$z_{\max}(r) \simeq z_0 + z_1 \left(\frac{r}{\xi}\right)^{\zeta} \sim r^{0.1} \tag{9}$$

This is in rather good agreement with the results expressed in the relations found for $\zeta(q)/q$. This non-gaussian behavior, although not yet confirmed by other experiments, however, appear in the resolution of one of the most fruitful model of dynamic roughening, the Kardar-Parisi-Zhang model, when the noise has a broad probability distribution (a power law with exponent $-(1+\mu)$). In particular, $\zeta \simeq 0.8$ and q-dependent $\zeta(q)/q$ ratios are found for the moments of the height correlation functions for small μ.[15] Those results echo the analysis done in the second paper.[8]

Another quantity of interest is closely related to the P histogram: the number of branches at abscissa r. It is studied in Ref. 8 and is shown to be perfectly compatible with the behavior of $P(z, r) \sim r^{-\zeta}(z/r^{\zeta})^{-1} \sim z^{-1}$.

From an experimental point of view, it would be interesting to be able to compare the statistics of branched and non-branched fracture structures, for instance on aluminium alloys subjected to fatigue which are likely to present a "branching transition". In-situ observation of the propagation front in a transparent material could also help in the study of the dynamics.

ACKNOWLEDGMENTS

The referee is acknowledged for pointing out Ref. 15. Part of this work has been achieved while one of us (E. B.) was on leave at the University of Cambridge, under a contract of the French Ministry of Defence, DRET $n°$ 92/1437/A000. The authors are indebted to CNRS and P. Veyssière for support.

REFERENCES

1. E. Bouchaud, G. Lapasset and J. Planès, *Europhys. Lett.* **13**, 73 (1990).
2. R. H. Dauskardt, F. Haubensak and R. O. Ritchie, *Acta Metall. Mater.* **38**, 143 (1990).
3. K. J. Måløy, A. Hansen, E. L. Hinrichsen and S. Roux, *Phys. Rev. Lett.* **68**, 213 (1992).
4. A. Imre, T. Pajkossy and L. Nyikos, *Acta Metall. Mater.* **40**, 1819 (1992).
5. B. B. Mandelbrot, D. E. Passoja and A. J. Paullay, *Nature* (London) **308**, 721 (1984).
6. E. Bouchaud, G. Lapasset, J. Planès and S. Navéos, *Phys. Rev.* **B48**, 2917 (1993).
7. J. P. Bouchaud, E. Bouchaud, G. Lapasset and J. Planès, to appear in *Phys. Rev. Lett.*
8. E. Bouchaud, J. P. Bouchaud, G. Lapasset and J. Planès, this conference.
9. A. Hansen, C. Poirier, J.-P. Troadec, K. J. Måløy, S. Roux and E. L. Hinrichsen, submitted to *Nature* (London).
10. J. Feder, *Fractals* (Plenum Press, New York and London, 1988).
11. S. Roux and D. Francois, *Scripta Metall.* **25**, 1092 (1991).
12. B. R. Lawn and T. R. Wilshaw, *Fracture of Brittle Solids* (Cambridge University Press, Cambridge, 1975), Chap. 5, p. 100.
13. J. Fineberg, S. P. Gross, M. Marder and H. L. Swinney, *Phys. Rev. Lett.* **67**, 457 (1991) and *Phys. Rev.* **B45**, 5146 (1992).
14. J. S. Langer, *Phys. Rev.* **A46**, 3123 (1992).
15. Y. C. Zhang, *J. Phys.* (France) **51**, 2129 (1990); A. L. Barabási, R. Bourbonnais, M. Jensen, J. Kertész, T. Vicsek and Y. C. Zhang, *Phys. Rev.* **A45**, R6951 (1992).

Fractals, Vol. 1, No. 4 (1993) 1075–1079

TRACER TIME DISTRIBUTION OF PARTICLES IN POROUS MEDIA

MARIELA ARAUJO
Departamento de Yacimientos, Intevep S.A.,
Apartado 76343, Caracas 1070-A, Venezuela

Abstract

We study the transit time distributions of tracer particles in a porous medium through which a constant flow rate is established. Our model assumes that non-Gaussian dispersion is due to the presence of low velocity zones or channels in parallel with a faster flow path. Each channel is represented as a trap and simulates the existence of variable permeability blocks inside the porous medium. The time the tracer particle spends inside each channel is related to the heterogeneity of the sample, and is assumed here to have a power-law distribution. We compare the transit time distribution of these particles for the case in which the traps are Poisson distributed with the one in which the trap distribution is a power-law function.

1. INTRODUCTION

Many important physical processes involve the transport of diffusive particles in the presence of a conduction field. A number of technological applications rely on an accurate understanding of dispersion in porous media, including secondary oil recovery, underground water management, waste disposal and chromatography.[1–5]

Here, we consider a simple model of a stratified medium with broadly varying permeabilities. The interest is focused in the distribution of transit times (t_L), the latter being the times spent by a dispersing particle to travel across a heterogeneous sample of length L. Commonly, dispersion has been regarded as a Gaussian process, i.e., a transport mechanism whereby the distribution of transit times is a normal distribution. There are many interesting situations, however, where a non-Gaussian behavior can be expected. Anomalous diffusion and/or dispersion, for instance, takes place in poorly connected artificial porous media and consolidated rocks. This non-Gaussian behavior has sometimes been associated with finite-size effects. Truly anomalous dispersion, however, also emerges from strong sample heterogeneities.[6–8]

Anomalous behavior is seen here to be reflected in the transit times distributions ($P(t_L)$). We find that if the media are sufficiently disordered $P(t_L)'s$ have long tails related to the presence of both slow and fast individual events, where a tracer particle enters a region and is released after a time much larger than the typical convection time. In the limit of a stationary fluid, dispersion is due to ordinary molecular diffusion whereas for finite fluid velocities, tracer particles are carried along the flow, and the dispersion depends on the magnitude of the velocity. The relative importance of convective to diffusive transport is characterized by the Péclet number.

We consider a simple picture of a heterogeneous porous medium consisting of a number of parallel layers with variable permeability. An external pressure difference imposes a flow across the sample parallel to the layers. The tracer particles are initially located at one extreme of the sample, and spread into it. The time required to reach the other extreme is measured in numerical simulations and the $P(t_L)$ is estimated from a large number of single events.

It is assumed that dispersion in the direction perpendicular to the layers is too small for the tracer particles to sample evenly the whole stack of layers while they travel. Therefore, particles located initially inside a relatively high permeance or high velocity channel or strata will stay in it during the entire path through the sample (in a layered porous media in which an external pressure difference is applied, the permeance is proportional to the convective fluid velocity).

As a further simplification, it is assumed that the particles can either enter a channel with probability p, or move along a main path by convection in the external flow with probability $1-p$. Thus, basically, the description of the tracer motion is reduced to the motion within two layers: the main path and the channel (Fig. 1).

The channels may represent spatial poorly connected regions, in which the particles are temporarily trapped. If the only way out a channel is by its entrance, the tracer particle can leave only by diffusion, and therefore it can spend large times inside it.

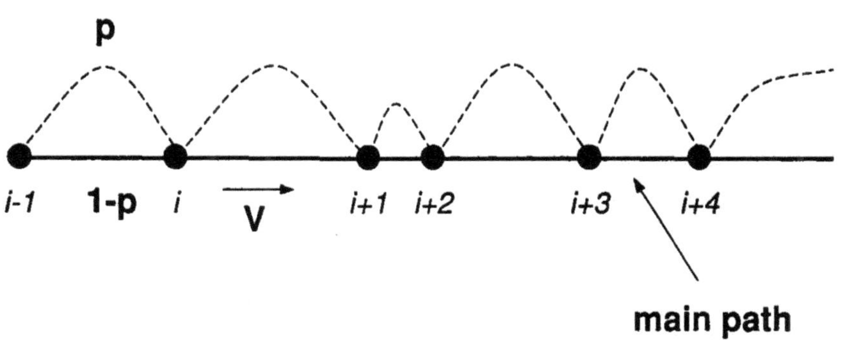

Fig. 1 Schematic view of the flow paths. The dots represent the traps and the solid line the main path. Dotted lines represent slow channels in parallel with the main path. p is the probability that the particle enters a channel.

The entrance of a channel is marked by the location of a "trap". In this model, we have a sequence of traps located along the main path as indicated schematically in Fig. 1.

In the simulations, the tracer particles at $t = 0$ are located at one extreme of the sample that we label $x = 0$. At the position of each trap, they can either enter a channel with probability p or move along the main path with probability $1 - p$. These probabilities are taken from a uniform distribution. We consider cases in which the average distance between traps is a constant, specifically the case where the traps are Poisson distributed along the main path, and the case in which the distance between traps is taken from a power-law distribution of the form $p(l) \sim \nu l^{-(1+\nu)}$ with $0 < \nu < 1$.

If the particle "decides" to move along the main path, it is convected by the external flow with velocity V. On the other hand, inside a channel the particle is temporarily trapped. We study the cases where once a tracer enters a channel, it is convected along it with a random velocity v, uniformly distributed between $0 < v < V$ (V is the convective velocity due to the external flow), and the case where the times that the tracer spends inside it is taken from a power-law distribution of the form $p(\tau) \sim \mu\tau^{-(1+\mu)}$ with $0 < \mu < 1$.

The transit time t_L, is the sum of the waiting times that the tracer particles spend between traps. The experiment is repeated a large number of times ($\sim 10^5$), and from these values of $t_L's$, the transit-time distribution is estimated.

Three different cases are considered: (a) traps Poisson distributed along the main path and inside the channels, particles are convected with a random velocity v, uniformly distributed between $0 < v < V$; (b) traps Poisson distributed along main path, and the particles wait inside each channel a waiting time taken from a power-law distribution of the form $p(\tau) \sim \mu\tau^{-(1+\mu)}$, with $0 < \mu < 1$; and (c) the distance between traps is taken from a power-law distribution $p(l) \sim l^{-(1+\nu)}$, with $0 < \nu < 1$, and the waiting times for particles traveling inside channels are also taken from a power-law distribution $p(\tau)$ as in case (b). We consider that the Péclet number is of order one; thus, diffusion and convection are both relevant.

We start with the case in which the traps are Poisson distributed along the main path. The average separation between traps l, is given by the inverse of the concentration of traps. The time the particles spend inside a channel is thus $\tau_i = l/v_i$, v_i being a random variable uniformly distributed between 0 and V. The transit time t_L is the sum of random variables with a minimum value, which corresponds to the convection time, $\tau_{\min} = l/V$, and a maximum value that corresponds to the situation in which the particles move inside the channel only by diffusion $\tau_{\max} = l^2/D$, D being the diffusion constant. For this case, the distribution of the sum of these random variables is Gaussian as can easily be seen from the Central Limit Theorem (CLT).[9] This result is also valid for all possible trap distributions with some characteristic length scale.

For case (b), the same distribution of traps (Poisson) is considered, and the time a particle spends inside the channel is taken from a power-law distribution $p(\tau) \sim \mu\tau^{-(1+\mu)}$. Here, the transit time $t_L = \sum_{n=1}^{N_{traps}} \tau_n$ (with τ_n taken from $p(\tau)$ for $0 < \mu < 1$) behaves typically as $L^{1/\mu}$. We find that the transit time distribution is of a power-law form with the same exponent μ as the distribution of waiting times inside the channels. Similar results in a different context have been previously reported.[10]

For the third case (c), both the spatial distribution of traps and the waiting times distribution inside the channels are of power-law form $p(l) \sim \nu l^{-(1+\nu)}$ and $p(\tau) \sim \mu\tau^{-(1+\mu)}$, respectively, with ν and μ between 0 and 1. In this case, the transit time is the sum of random variables taken from two different power-law distributions (note that when the particle moves along the main path, the time it spends in this spatial region before it encounters the next trap is just the ratio of the distance between two consecutive traps (taken from $p(l)$) and the convective velocity V). We find that the transit time t_L, has also a power-law distribution:

$$P(t_L) \sim t_L^{-(1+\sigma)}$$

with $\sigma = \min\{\nu, \mu\}$. That is, the more singular distribution dominates. Figure 2 shows a plot of the logarithm of $P(t_L)$ as a function of the logarithm of t_L for values of $\mu = 0.25$ and $\nu = 0.5$. The straight line behavior in this double logarithmic scale clearly supports the above statement.

The non-Gaussian behavior of the transit time distribution obtained in the second and the third cases is obviously a consequence of the distribution of waiting times of the particles inside the channels chosen here. Previous works assumed that the particles were temporarily trapped and then released from them exponentially.[6] There, the distribution of transit times was found to be non-Gaussian for small systems, and it turned into a Gaussian distribution as the size of the system was increased. Here, we considered systems of size L up to 10–50 $\times$ 10^6 sites and measure the values of transit times for different values of L. The non-Gaussian behavior reported here was definitely not related to finite size effects.

It has been found experimentally that for some systems the transit times distribution is truly non-Gaussian[6,8]; however, no simple model was presented to account for the non-Gaussian behavior, which is precisely what we do here. Physically, we can justify the use of a power-law distribution for the waiting times inside the channels by the fact that porous media are highly irregular, with poorly connected regions where the tracer particles can spend large times.

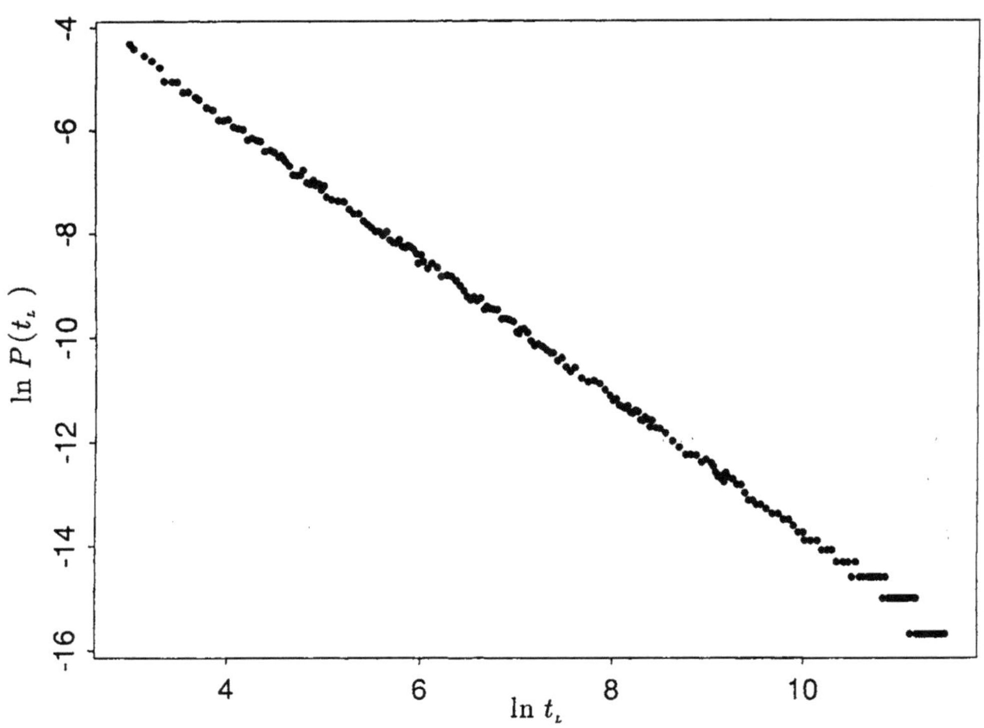

Fig. 2 Double logarithmic plot of the transit times distribution $P(t_L)$ as a function of t_L for a system of $L = 50 \times 10^6$ sites. Here, $\mu = 0.25$ and $\nu = 0.5$.

ACKNOWLEDGMENTS

The author gratefully acknowledged Intevep S.A. for permission to publish this paper and thanks Rafael Angulo for a critical reading of the manuscript.

REFERENCES

1. E. Guyon, J. P. Nadal and Y. Pomeau, eds., *Disorder and Mixing*, (Kluwer Acad. Pub., Amsterdam, 1988).
2. D. L. Freyburg, *Water Resour. Res.* **22**, 2031 (1986); G. Sposito, W. A. Jury, and V. K. Gupta, *Water Resour. Res.* **22**, 77 (1986).
3. G. Dagan, *Annu. Rev. Fluid Mech.* **19**, 183 (1987).
4. J. Bear, *Dynamics of Fluids in Porous Media* (Elsevier Pub., New York, 1972).
5. J. C. Bacri, N. Rakatomalala and D. Salin, *Phys. Rev. Lett.* **58**, 2035 (1986).
6. J. C. Bacri, J. P. Bouchaud, A. Georges, E. Guyon, J. P. Hulin, N. Rakotomalala and D. Salin, in *Hydrodynamics of Disperse Media*, eds. J. P. Hulin, A. M. Cazabat, E. Guyon and F. Carmona, (Elsevier, North Holland, 1990).
7. D. L. Koch and J. F. Brady, *Phys. Fluids* **31**, 965 (1988); *J. Fluid Mech.* **180**, 387 (1987).
8. J. P. Hulin and D. Salin: Experimental Study of Tracer Dispersion in Model and Natural Porous Media, in *Proceedings of the Cargese NATO School on Mixing and Disorder*, eds. E. Guyon, J. P. Nadal and Y. Pomeau (Martinus Nijhoff, Dordrecht, 1987), p. 399.
9. W. Feller, *An Introduction to Probability Theory*, Vols. 1 and 2, (Wiley, New York, 1971).
10. J. P. Bouchaud and A. Georges, *Phys. Rep.* **195**, 127 (1990).

Fractals, Vol. 1, No. 4 (1993) 1080–1085

TURBULENT FLOW IN VIBRATED BED OF POWDER: NEW TARGET TO INVESTIGATE TURBULENT FLOW

Y.-H. TAGUCHI
Department of Physics, Tokyo Institute of Technology,
Oh-okayama, Meguro-ku, Tokyo 152, Japan

Abstract

The flow patterns in the vibrated bed of powder are investigated numerically in two dimensions. They turn out to exhibit turbulent flow like that observed in fully developed turbulence. The power spectrum of their flow lines has $k^{-5/3}$ power, where k is the wave number. Since the numerical investigation of powder requires very few computational resources, it provides us with a new tool to study turbulence easily.

1. INTRODUCTION

Recently, nonlinear systems with many degrees of freedom are the central issues for statistical physicists. This is because they provide us with many new ideas on statistical physics. In particular, fluid turbulence has provided such ideas. For example, chaos has been found in the model equation of fluid,[1] period doubling is confirmed in the thermal convection of fluid,[2] and fractal has been proposed to describe intermittent structure of fluid turbulence.[3]

On the other hand, fluid turbulence is itself hard to investigate in detail. The fundamental equation of fluid turbulence, the Navier-Stokes equation, is a typical nonlinear partial differential equation which we do not have any general methods to solve. Therefore, where the high nonlinearity, like turbulence, is dominant, the solution cannot be dealt with well. It is difficult, even using computer simulations, to reproduce the fully developed turbulent state, as it requires too much computational resources (large memory and long cpu time).

Because of the above reason, physicists introduced some phenomenological modelings to study turbulence. One of typical examples is amplitude equation or time dependent Gintzburg-Landau equations (TDGL). One can even deal with them analytically sometimes. TDGL can solve many nonlinear phenomena with many degrees of freedom, including phase ordering[4] and thermal[5] or electro hydrodynamic[6] convections. It can even reproduce defect turbulence.[7]

However, TDGL can only treat "weak" turbulence, since it is essentially perturbation expansion near the instability point. It cannot deal with highly nonlinear state that is far from the onset of instability. When one wishes to study real turbulent state, one has to find another phenomenology.

A typical one among these phenomenologies is coupled map lattice[8] (CML) or cell dynamical systems (CDS). CML and CDS are the sets of nonlinear maps like logistic maps. The maps are distributed on the lattice and coupled with each other, typically with diffusive interaction.

Recently, CML has been shown to be a good numerical simulator, especially in phase ordering; it has already become a standard tool to simulate phase ordering.[9,10] Moreover, this method has reproduced many interesting features in fluid dynamics. Yanagita and Kaneko[11] have succeeded in reproducing weak to hard turbulence transition in thermal convection. CML also exhibits traveling waves,[12] which is frequently observed in the experiment.

CML reproduces many other phenomena that are hard to reproduce with the partial differential equations. In addition to this, CML can be applied to weak nonlinear region other than hard turbulence. CML is a really surprising tool to study nonlinear systems with many degrees of freedom. But it has a weak point: It cannot achieve a strong straight relationship with the real phenomena, i.e., with experiment.

In this paper, I propose molecular dynamic simulation of powder as a good example with which to study turbulence with many degrees of freedom. The standard molecular dynamics simulation is not an efficient tool of simulation, since one has to simulate typically million particles to study the real situation that includes 10^{23} particles. But the powder system includes "only" a few million particles. Thousand or even hundred particles in the simulation are enough to reproduce experiment. Therefore, it can be simulated very easily, even with a cheap IBM-PC.

This paper is organized as follows. In Sec. 2, I define the model by which I study powder dynamics. Section 3 gives us the results obtained, while the conclusion appears in Sec. 4.

2. MODEL

The powder[13,14] I deal with in this paper is a set of monodisperse spheres. They collide with each other viscoelastically, and their surfaces have frictions. In the experiments, they use glass spheres with a diameter of 0.1 mm to 1 mm.

Cundal and Strack[15] have proposed a new method to simulate powder dynamics: Distinct Elementary Method, DEM. In the DEM, a powder particle is modeled as an

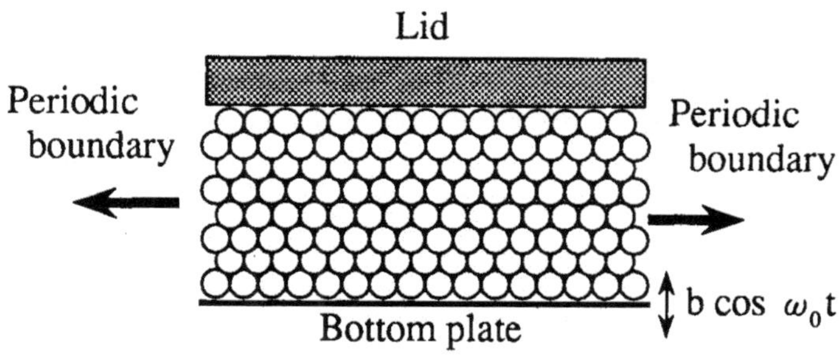

Fig. 1 A schematic figure for numerical setups.

elastic particle with viscosity. The model I employed is:

$$\ddot{\mathbf{x}}_i = -\sum_{j=1}^{N} \theta(d - |\mathbf{x}_i - \mathbf{x}_j|)\left[k_0\left\{\mathbf{x}_i - \mathbf{x}_j - d\frac{\mathbf{x}_i - \mathbf{x}_j}{|\mathbf{x}_i - \mathbf{x}_j|}\right\} + \eta(\mathbf{v}_i - \mathbf{v}_j)\right] - \mathbf{g}, \quad (1)$$

where N is the total number of spheres, $\mathbf{x}_i$ is the position vector of the ith sphere, $\mathbf{g}$ is the gravity and $\mathbf{v}$ is velocity. Each particle has a diameter of d. Because of a step function $\theta(x)$, particles interact only when they contact with each other. k_0 and η are the elastic constant and the viscosity coefficient, respectively. These two give coefficient of restitution, e, and collision time t_{col} when two particles collide head-on; $e = \exp(-\eta\pi/\omega)$ and $t_{col} = \pi/\omega$ where $\omega = (2k_0 - \eta^2)^{1/2}$.

Using this modeling, I succeeded in reproducing convective motion in granular materials of vibrated bed. This phenomenon was first discovered by Faraday[16] who founded electromagnetism. When one shakes the shallow vessel filled with powder particles, the powder spontaneously exhibits convective motions.[17 a]

Although it has remained unsolved over one hundred and fifty years, others[19] and I[20] independently succeeded in reproducing this convective motion with the model obeying Eq. (1). In the formation of convection, it turned out to be important to consider both friction among particles and that between particles and wall. When there is no friction between wall and particles, the direction of convection is reversed. Without side wall, we can never see the formation of convection. It exhibits random motions.[21]

In this paper, I concentrate on the turbulent nature of this powder flow. Figure 1 shows the schematic of the numerical set up. All calculations are done in the two dimensional space. This two dimensional space has lower bound that has elastic interaction with particles. The particles can never penetrate this lower bound. The lower bound oscillates as a function of time t, $b\cos\omega_0 t$.

Besides this, the two dimensional space has another boundary conditions. The lid with mass M exists at the top of the layer of powder particles. This lid has elastic interaction with the powder particles. Its motion has to obey the Newton equation of the material with mass M under gravity. It does not tilt.

Theoretically, the lid can be regarded as a constant pressure condition for the uppermost layer. Since the powder does not have any surface tensions, the particles

[a]The *roll* patterns like Bénard-structures are also observed in the experiment.[18]

near the surface region diffuse freely without the lid. This free diffusion prevents us from studying bulk motion among particles, which I am interested in. The lid is necessary to suppress this surface diffusion.

3. RESULTS

Figure 2 shows the typical snapshot of flow lines obtained in the simulation. In contrast to the flow lines with side walls,[20,19] the flow lines here are not clear convection patterns. Instead, the flow lines exhibit complicated patterns. Its appearance is similar to the fluid turbulence. The flow lines are continuous and eddy like structures can be seen.

In spite of the discrete nature of the powder, its flow lines are strongly connected. Since there is no free space for the particles to move to, all particle motions move cooperatively. Otherwise, no motion can occur.

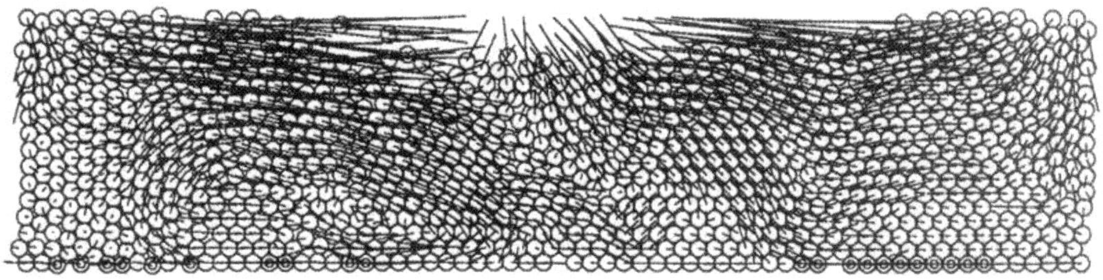

Fig. 2 Typical snapshot and displacement vectors.

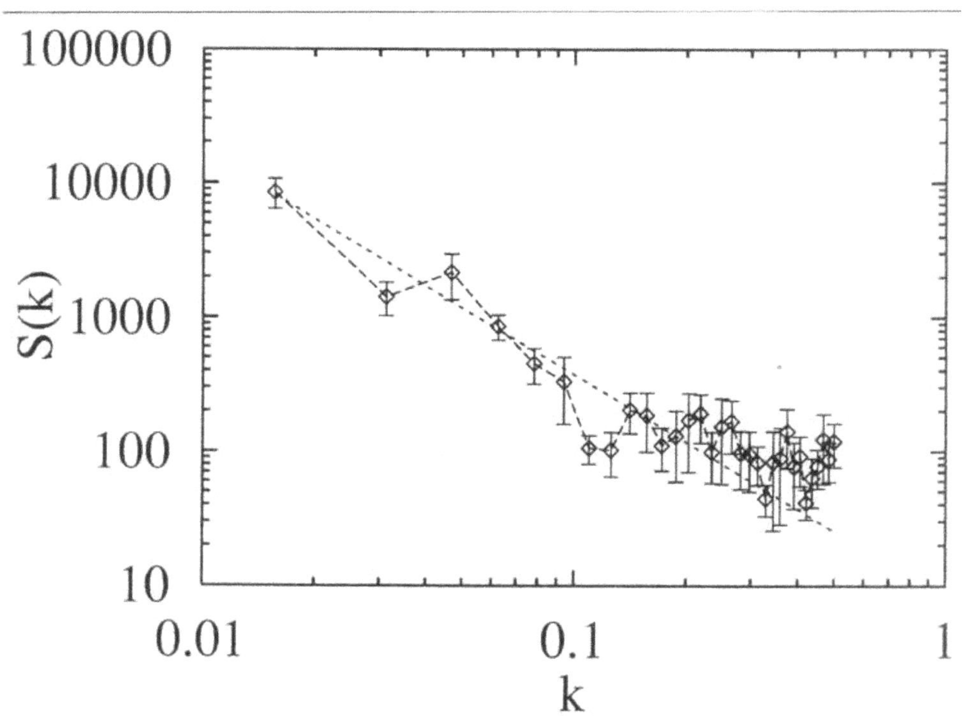

Fig. 3 Power spectrum of flow lines. The straight line shows $k^{-5/3}$ where k is the wave number.

Next I would like to know whether the flow lines are similar to those in the fluid turbulence. Figure 3 shows the power spectrum of flow lines belonging to the lowermost layer. It clearly reveals power behavior. Its power is close to -5/3 which is the same as that which fluid turbulence exhibits. Therefore the flow patterns are clearly similar to the fluid turbulence. Details will be published elsewhere.

4. CONCLUSION

In this paper, I have shown that the vibrated bed of powder exhibits turbulent structure like fluid. As shown in the above, the model has definitely finite number of degrees of freedom. We do not have to take infinite limit that causes difficulties in the numerical point for the Navier-Sokes equation.

The treatment of this model is much easier than the partial differential equations. The model has experimental direct correspondence that CML lacks. Therefore, powder dynamics is the new and good tools to study turbulent state.

ACKNOWLEDGMENTS

I acknowledge Prof. H. J. Herrmann who has suggested to investigate power spectrum of powder flow. I would like also to thank Profs. J. A. C. Gallas, D. C. Hong, A. Mehta, T. Pöschel, H. M. Jeager, S. R. Nagel and Drs. G. H. Ristow and J. Lee for sending their works before publication. NEC Software is also acknowledged for allowing me to use EWS-4800/220, in which all calculations were performed. This work is partly supported by Grant-in-Aid for Encouragement of Young Scientists (04750135, 05740252), from the Ministry of Education, Science and Culture, Japan. Multi-disciplinary Science Foundation (Japan) supports the trip to attend this workshop.

REFERENCES

1. E. N. Lorenz, *J. Atmos. Sci.* **20**, 130 (1963).
2. J. Maurer and A. Libchaber, *J. de Phys. Lett.* **40**, 419 (1979); A. Libchaber et al., *J. de Phys. Lett.* **43**, 221 (1982).
3. B. B. Mandelbrot, *The Fractal Geometry of Nature* (Freeman, San Francisco, 1982).
4. A. J. Bray, in *Statistical Physics — Invited Papers from STATPHYS 18*, ed. S. Hess (North-Holland, Amsterdam, 1993), p. 28 (*Physica* **A194**, Nos. 1–4).
5. P. Manneville, *Dissipative Structures and Weak Turbulence* (Academic Press, Boston, 1990).
6. S. Kai, ed., *Pattern Formation in Complex Dissipative Systems* (World Scientific, Singapore, 1992).
7. S. Sasa, *Prog. Theor. Phys.* **83**, 824 (1990); **84**, 1008 (1990).
8. K. Kaneko, *Prog. Theor. Phys.* **72**, 480 (1984); **74**, 1033 (1985); *Physica* **D34**, 1 (1989); **36**, 60 (1989).
9. Y. Oono and S. Puri, *Phys. Rev.* **A38**, 434 (1988).
10. H. Nishimori and T. Nukii, *J. Phys. Soc. Jpn.* **58**, 563 (1988).
11. T. Yanagita and K. Kaneko, *Phys. Lett.* **A175**, 415 (1993).
12. K. Kaneko, *Phys. Rev. Lett.* **69**, 905 (1992).
13. For reviews, H. M. Jeager and S. R. Nagel, *Science* **255**, 1523 (1992); *Dynamics of Powder Systems*, eds. Y.-H. Taguchi, H. Hayakawa, S. Sasa and H. Nishimori; *Int. J. Mod. Phys. B* Vol. 7, Nos. 9 and 10 (1993); C. S. Campbell, *Annu. Rev. Fluid Mech.* **22**, 57 (1990).
14. S. Sasa and H. Hayakawa, *Europhys. Lett.* **17**, 685 (1992).
15. P. A. Cundall and O. D. L. Strack, *Geotechnique* **29**, 47 (1979).

16. M. Faraday, *Phil. Trans. R. Soc. Lond.* **52**, 299 (1989).
17. P. Evesque and J. Rajchenbach, *Phys. Rev. Lett.* **62**, 44 (1989); C. Laroche, S. Douady and S. Fauve, *J. Phys. (Paris)* **50**, 699 (1989).
18. F. Dinkelacker, A. Hübler and E. Lüscher, *Biological Cybernetics* **56**, 51 (1987).
19. J. A. C. Gallas, H. J. Herrmann and S. Sokolowski, *Phys. Rev. Lett.* **69**, 1371 (1992); *Physica* **A189**, 437 (1992).
20. Y.-H. Taguchi, *Phys. Rev. Lett.* **69**, 1367 (1992).
21. Y.-H. Taguchi, *J. Phys.* **II 2**, 2103 (1992).

A. APPENDIX

A.1 The Method of Integration

I employ Euler scheme to integrate Eq. (1). The time step Δt changes at each time step. Suppose the ith particle has the acceleration $\mathbf{a}_i$ at time t. I approximate the velocity and position of the ith particle after the period Δt,

$$\mathbf{x}_i(t+\Delta t) = \mathbf{x}_i(t) + \mathbf{v}_i(t)\Delta t + \frac{1}{2}\mathbf{a}_i(t)\Delta t^2 \tag{2}$$

$$\mathbf{v}_i(t+\Delta t) = \mathbf{v}_i(t) + \mathbf{a}_i(t)\Delta t\,. \tag{3}$$

From Eq. (2), I calculate Δt as a function of $\Delta x_i(t) \equiv |x_i(t+\Delta t) - x_i(t)|$ and $\Delta y_i(t) \equiv |y_i(t+\Delta t) - y_i(t)|$, where x_i and y_i are the x and y component of $\mathbf{x}_i$,

$$\Delta t_{xi} = \frac{\sqrt{v_{xi}^2 + 2a_{xi}\Delta x_i} - v_{xi}}{a_{xi}} \tag{4}$$

$$\Delta t_{yi} = \frac{\sqrt{v_{yi}^2 + 2a_{yi}\Delta y_i} - v_{yi}}{a_{yi}}\,, \tag{5}$$

where v_{xi} and v_{yi} are the x and y components of $\mathbf{v}_i$ respectively and a_{xi} and a_{yi} are the x and y components of $\mathbf{a}_i$. Then Δt is equal to $\min_{\alpha=\{xi,yi\}} \Delta t_\alpha$. In the actual simulation, I use $\Delta x_i = \Delta y_i = 0.01d$. Of course, one should check whether this selection of Δx_i and Δy_i is suitable or not.

Or if one is interested in $\mathbf{v}_i$ itself, one should require the same restriction to $\mathbf{v}_i$ other than $\mathbf{x}_i$. In this paper, I am not interested in $\mathbf{v}_i$ itself.

Author Index

Milton Keynes UK
Ingram Content Group UK Ltd.
UKHW051625120924
448026UK00007B/48